# RNA Technologies

**Series Editors**

Jan Barciszewski, Nanobiomedical Center, Adam Mickiewicz University, Poznań, Poland

Institute of Bioorganic Chemistry of the Polish, Academy of Sciences, Poznań, Poland

Nikolaus Rajewsky, Max Delbrück Center for Molecular Medicine, Berlin Institute for Medical Systems Biology, Berlin-Buch, Berlin, Germany

**Founding Editor**

Volker A. Erdmann, Institute of Chemistry and Biochemistry, Free University of Berlin, Berlin, Germany

More information about this series at http://www.springer.com/series/8619

Stefan Jurga • Jan Barciszewski
Editors

# The DNA, RNA, and Histone Methylomes

Springer

*Editors*
Stefan Jurga
Nanobiomedical Center
Adam Mickiewicz University
Poznań, Poland

Jan Barciszewski
Nanobiomedical Center
Adam Mickiewicz University
Poznań, Poland

Institute of Bioorganic Chemistry of the Polish Academy of Sciences
Poznań, Poland

ISSN 2197-9731 ISSN 2197-9758 (electronic)
RNA Technologies
ISBN 978-3-030-14791-4 ISBN 978-3-030-14792-1 (eBook)
https://doi.org/10.1007/978-3-030-14792-1



This Springer imprint is published by the registered company Springer Nature Switzerland AG.
The registered company address is: Gewerbestrasse 11, 6330 Cham, Switzerland

# Preface

## Chromatin Methylome

In eukaryotes, the genome size is highly variable between organisms. It is organized into chromatin, a nuclear complex encompassing DNA, RNA, and associated proteins. Furthermore, chromatin is organized into two distinct domains, namely, euchromatin and heterochromatin, depending on the compaction state. Euchromatin exhibits relatively loose compaction and is typically transcriptionally permissive, whereas heterochromatin is more condensed, rich in repetitive sequences, and typically transcriptionally repressive. Modulation at each level enables chromatin-based information to vary, in order to respond to different signals for numerous gene regulatory functions. This determines chromatin plasticity as a means of generating a variety of properties for each cell type, during the process of development, and also when cells face different environmental and metabolic signals, senescence, disease, and death. The degree to which chromatin is organized and packaged is highly influenced by various reasons and factors, including chemical modifications to histones and DNA, particularly methylation.

There is a remarkable difference between the length of DNA and the size of the nucleus, and thus, the entire DNA molecule has to be efficiently compacted in order to fit inside the physically small space. Two meters of human DNA has to be compacted into the confines of a 2–10 micron nucleus and accessible to the protein machineries that utilize it for critical biological functions. So, there is a question of how these diverse genomic functions, such as transcription, repair, replication, and recombination of DNA, occur at the right place and time to promote cellular growth, differentiation, and development.

Methylation of DNA is a critical part of epigenetic regulation in eukaryotes. 5-methylcytosine ($m^5C$) is conserved in species ranging from vertebrates to fungi and protists. In mammals, $m^5C$ has well-described roles in regulating gene expression, and altered methylation patterns are hallmarks of normal embryonic development, as well as tumorigenesis. Current understanding of cytosine DNA methylation has benefited from chemical biology approaches that developed the mechanisms governing proteins which introduce or recognize methyl marks with DNA methyltransferases. The most accurate model for active demethylation in mammals involves sequential oxidation of $m^5C$ by ten-eleven translocation (TET) family enzymes, followed by base excision repair (BER) to regenerate unmodified cytosine. All three oxidized $m^5C$ bases have been detected in diverse cell types, mostly in the context of cytosine-guanine dinucleotides (CpG islands). In general, the genomic levels of 5-hydroxymethylcytosine ($hm^5C$) are 10- to 100-fold lower than $m^5C$, while levels of 5-formylcytosine ($f^5C$) and 5-carboxycytosine ($ca^5C$) are at least tenfold lower than $hm^5C$—approximately one in $10^5$–$10^6$ nucleotides.

The second group of cellular methylomes consists of RNA modifications. Compared with DNA modifications, they are largely neglected and have yet not drawn extensive attention until very recently. Chemical modifications to RNA were already established in the 1970s. Until now, over 170 posttranscriptional RNA modifications have been identified. They can control the turnover and/or translation of transcripts during cell-state transitions and therefore play important roles during tissue development and homeostasis. Although RNA modifications are highly diverse and can be found in all RNA species, the recent discoveries underpin an emerging common theme, namely how methylation of RNA coordinates translation of transcripts that encode functionally related proteins, when cells respond to differentiation or other cellular and environmental cues. For example, the $N^6$-methyladenosine ($m^6A$) modification of mRNA is an essential regulator of mammalian gene expression. Transcripts that maintain the cell state are cotranscriptionally decorated with $m^6A$ which promotes and coordinates the timely decay of these transcripts, which allows cells to differentiate. mRNA modifications also contribute to the survival and growth of tumor cells, further highlighting their importance in the regulation of cell fate decisions.

Posttranslational modifications of histones affect chromatin state and gene expression. An important breakthrough in the understanding of histone modification function was achieved through identification of the protein machineries that incorporate (write), remove (erase), and bind (read) histone substituents. Recently, the histone code concept emerged as a hypothesis to stimulate new thinking about how histone modifications might function. On the basis of the analysis of this variation and other histone modifications, where the associated functions were known, it was possible to infer that histone modifications might work solely, as well as in combination (on one or more histone tails) to mediate the distinct functions associated with them. Histone modifications can directly alter the biophysical properties of the target protein, provide a docking site for specific interaction partners, interfere with binding events of other factors, or act through a combination of these mechanisms.

A very important posttranslational modification of histones is the methylation of lysine and arginine residues. Protein methylation in living organisms is catalyzed by methyltransferases and involves the transfer of a $CH_3$ group from S-adenosylmethionine. Lysine can form mono-, di-, and trimethyllysines in methyltransferase-catalyzed reactions, while arginine can form mono- and dimethylarginine. These modified amino acids differ by size and hydrophobicity from the original residue. Histone methyltransferases are highly specific toward the nature of the amino acid residue (histone-lysine methyltransferases and histone-arginine methyltransferases) and the position of this residue in the polypeptide chain. Lysine and arginine methylation residues in histones constitute a very important element of the already mentioned histone code. One should also notice histone lysine acetylation on the ε-amine group. It not only neutralizes the positive charge of the amine group, enhances the hydrophobicity, and increases the size of the lysine side chain but also provides platforms for binding by proteins involved in chromatin and gene regulations.

Recent technological advances allow for genome-wide analysis of DNA and histone methylations, which affect their structures and have the potential to reveal the regulation mechanisms on a level beyond the primary structure. Chemical changes effected by methyl group induce various phenotypes encoded in chromatin structure, and that is just the perspective of epigenetics. A mechanistic understanding of chromatin and epigenetics plasticity in response to various cellular stress conditions may help to reveal the epigenetics contributions for genome and phenotype regulation.

To deeply discuss the key issues of the methylome, we brought together a diverse group of experts, who work on different aspects of chromatin methylations from mechanism to its biological consequences. The book includes 24 chapters.

Poznań, Poland — Stefan Jurga
Poznań, Poland — Jan Barciszewski

# Contents

# Establishment, Erasure and Synthetic Reprogramming of DNA Methylation in Mammalian Cells

**Renata Z. Jurkowska and Tomasz P. Jurkowski**

## Contents




**Abstract** DNA methylation is a crucial epigenetic modification involved in the control of cellular function and the balance between generation of DNA methylation and its removal is important for human health. This chapter focuses on the enzymatic machinery responsible for the processes of establishment, maintenance and removal of DNA methylation patterns in mammals. We describe the biochemical, structural and enzymatic properties of DNA methyltransferases and TET DNA hydroxylases,



R. Z. Jurkowska (✉)
BioMed X Innovation Center, Heidelberg, Germany
e-mail: jurkowska@bio.mx

T. P. Jurkowski (✉)
Abteilung Biochemie, Institut für Biochemie und Technische Biochemie, Universität Stuttgart, Stuttgart, Germany
e-mail: tomasz.jurkowski@ibc.uni-stuttgart.de





S. Jurga, J. Barciszewski (eds.), *The DNA, RNA, and Histone Methylomes*, RNA Technologies, https://doi.org/10.1007/978-3-030-14792-1_1

as well as their regulation in cells. We discuss how these enzymes are recruited to specific genomic loci, and how their chromatin interactions, as well as their intrinsic sequence specificities and molecular mechanisms contribute to the methylation pattern of the cell. Finally, we introduce the concept of epigenetic (re)programming, in which designer epigenetic editing tools consisting of a DNA targeting domain fused to an epigenetic editor domain can be used to edit the epigenetic state of a given locus in the genome in order to dissect the functional role of DNA methylation and demethylation. We discuss the promises of this emerging technology for studying epigenetic processes in cells and for engineering of cellular states.

**Keywords** DNA methylation · DNA demethylation · Synthetic epigenetics · TET · DNMT · dCas9 · Epigenetic editing

## 1 Introduction

Methylation of CpG sites in the DNA plays a vital role in mammalian development and has been studied extensively in the past decades. However, despite advances in understanding of the targeting and regulation of DNA methyltransferases in cells, the specific processes contributing to the generation, maintenance and erasure of DNA methylation patterns are not yet fully elucidated, and the exact molecular mechanisms leading to the aberrant methylation observed in human disease (like cancer) are only partially understood. The discovery of TET enzymes has changed the view on DNA methylation as a very stable modification, as it showed that active DNA demethylation can occur through stepwise oxidation of 5-methylcytosine (5mC) to 5-hydroxymethylcytosine (5hmC), 5-formylcytosine (5fC) and finally to 5-carboxylcytosine (5caC), followed by the removal of the oxidised bases by Thymine DNA glycosylase (TDG) and base excision repair mechanism. However, while the biological function of TET enzymes has been studied quite extensively, very little is known about their biochemical properties and their specificity, catalytic mechanism, as well as the contribution of different domains to enzymes targeting and regulation. In this chapter, we summarise the most important properties of both DNA methyltransferases and TET enzymes and describe some of the molecular pathways leading to their recruitment to the target sites. Finally, we introduce the concept of epigenetic editing as an elegant approach to dissect the function of DNA methylation and demethylation in a locus specific manner.

## 2 Setting of DNA Methylation

Since the discovery of methylated DNA bases in 1948, major advances have been made in our understanding of the biological role of DNA methylation, as well as the mechanisms regulating the function of DNA methyltransferases in cells. Through these discoveries, DNA methyltransferases emerged as key epigenetic enzymes

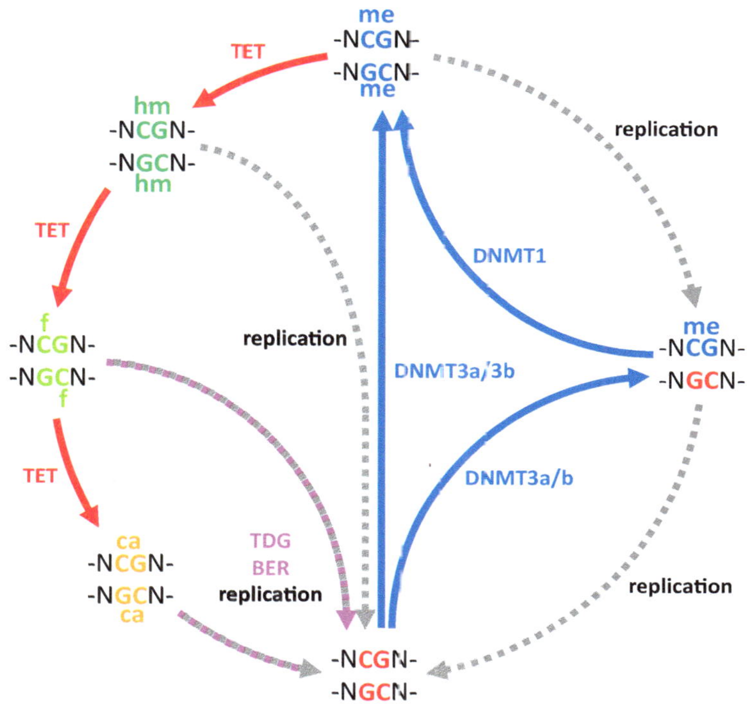


**Fig. 1** The methylation cycle in mammals. DNA methylation is set on unmethlyted cytosines by the combined action of DNMT3A and DNMT3B (blue) and maintained during replication by DNMT1, which has high preference for hemimethylated CpG sites. DNA methylation can be lost by a passive mechanism, getting diluted after consecutive cycles of DNA replication or through an active mechanism, involving oxidation of 5mC to 5hmC, 5fC and 5caC by TET enzymes (red). 5fC and 5caC can be recognised and excised by TDG and base excision repair enzymes, leading to the restoration of an unmethylated state (pink)

regulating mammalian development and cellular specialisation, which is clearly emphasised by the lethal phenotypes of the genetic knockouts of any of the DNA methyltransferase enzymes in mice and by the ever-growing number of diseases showing disturbed DNA methylation signatures.

DNA methylation in mammals occurs at the C5 position of the cytosine residues, primairly in the CpG sites, although non-CpG methylation is also present (albeit at lower levels). About 60–80% of the CpG sites in the human genome (corresponding to around 3–4% of all cytosines) are methylated in a tissue and cell type-specific pattern [reviewed in Schubeler (2015), Jurkowska et al. (2011a), Gowher and Jeltsch (2018), Ravichandran et al. (2018)]. Additionally, 5-hydroxymethylcytosine (5hmC), 5-formylcytosine (5fC) and 5-carboxylcytosine (5caC), which arise from the step-wise oxidation of the methyl group of 5-methylcytosine, have recently been discovered in mammalian genome DNA (Fig. 1) (Tahiliani et al. 2009; He et al. 2011; Ito et al. 2011). Three methyltranferase enzymes are responsible for the generation and maintenance of the global DNA methylation patterns in humans: DNMT3A, DNMT3B and DNMT1. DNMT3A and DNMT3B introduce DNA methylation during mammalian development and maturation of germ cells, with the assistance of a regulatory factor DNMT3L. After their establishment, DNA methylation patterns are largely preserved, with only small tissue-specific changes, but can get significantly altered in disease. During DNA replication, unmethylated DNA strands are synthetised, leading to the conversion of fully methylated CpG sites into hemimethylated sites that are then re-methylated by a maintenance methyltransferase DNMT1, which has high preference towards hemimethylated DNA (Fig. 1), and is ubiquitously and highly expressed in proliferating cells. This elegant inheritance mechanism enables DNA methylation function as a key epigenetic mark mediating long-term transcriptional silencing. In this respect, DNA methylation is involved in silencing of repetitive elements, genomic imprinting, X-chromosome inactivation and regulation of gene expression during development and cellular specialisation [reviewed in Smith and Meissner (2013), Bogdanovic and Lister (2017)]. DNA methylation can be lost by either passive mechanism, when maintenance MTase activity is absent, or via an active demethylation process (see below). Considering its important biological roles, it is not surprising that aberrant DNA methylation changes play a prominent role in the development of human diseases, including for example haematological cancer (Bergman and Cedar 2013; Yang et al. 2015).

## *2.1 Architecture of DNA Methyltransferases*

In the structure of the mammalian DNA methyltransferases, two functional parts can be identified, a large N-terminal part and a smaller C-terminal catalytic part (Fig. 2). The N-terminal part of DNMTs contains several distinct domains with targeting and regulatory functions. The C-terminal domain contains ten catalytic amino acid motifs conserved among prokaryotic and eukaryotic C5-DNA methyltransferases and folds into a conserved structure called AdoMet-dependent MTase fold, which consists of a mixed seven-stranded beta sheet, formed by six parallel beta strands and a seventh strand in an antiparallel orientation, inserted into the sheet between strands 5 and 6. Six helices are folded around the central beta sheet (Cheng and Blumenthal 2008). This domain is involved in binding of the cofactor *S*-adenosyl-*L*-methionine (AdoMet), recognition of the DNA substrate and in catalysis. Interestingly, the spatial arrangement of the various domains in DNMTs plays a prominent role in the regulation of enzymes' activity and specificity through allosteric control of the catalytic domain [reviewed in Jeltsch and Jurkowska (2016)].

### 2.1.1 Domain Composition of DNMT1

DNMT1 was the first mammalian DNA methyltransferase enzyme to be cloned and biochemically characterised. It is a large protein (1616 aa in humans), containing

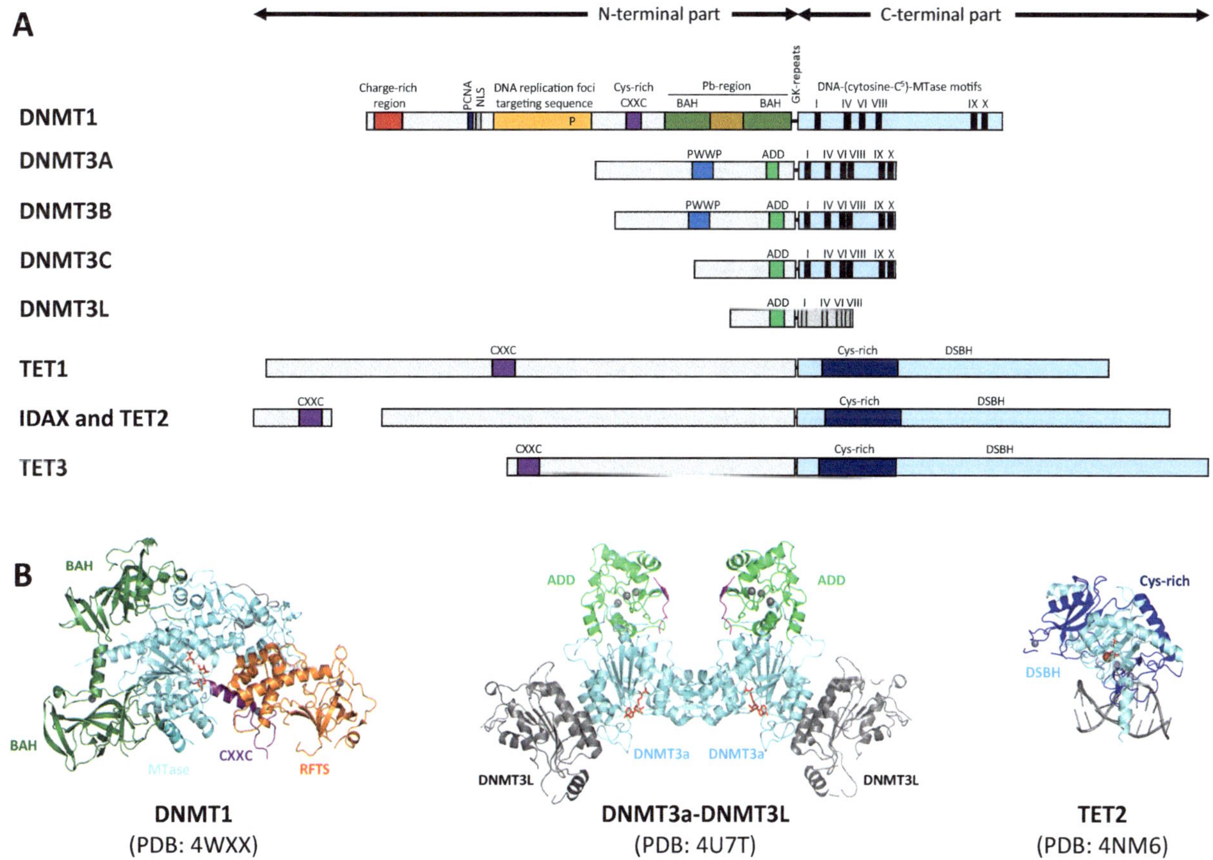


**Fig. 2** Domain composition of mammalian DNA methyltransferases and TET enzymes [adapted from Ravichandran et al. (2018)]. (**a**) The N-terminal regulatory part of the enzymes contains various chromatin and protein interacting domains, in the C-terminal domain of DNA methyltranferases, the catalytic

several distinct domains, which are listed below from the N-terminus to the C-terminus of the protein (Jurkowska and Jeltsch 2016) (Fig. 2):

- DMAPD (DNA methyltransferase-associated protein 1 interaction domain) is involved in the targeting of Dnmt1 to replication foci.
- PBD (PCNA—proliferating cell nuclear antigen—binding domain) recruits DNMT1 to the replication fork during S phase via interaction with PCNA.
- RFTD (replication foci-targeting domain) is involved in the targeting of DNMT1 to replication foci and to centromeric chromatin.
- CXXC domain binds unmethylated DNA and might be involved in the specificity of DNMT1
- BAH1 and BAH2 (bromo-adjacent homology 1 and 2) domains are necessary for the folding of the enzyme, but their exact biological function is unknown

The catalytic domain of DNMT1 is inactive in an isolated form despite presence of all conserved methyltransferase motifs required for catalysis, demonstrating that it is controlled by the N-terminal domain of the enzyme. Indeed, structural and biochemical studies confirmed that several domains in the N-terminal part of DNMT1 directly contact the catalytic domain, providing examples of the sophisticated allosteric regulation of DNMT1 [reviewed in Jeltsch and Jurkowska (2016)].

### 2.1.2 Domain Composition of DNMT3 Family

Human DNMT3 family comprises three members: DNMT3A (912 aa), DNMT3B (853 aa) and DNMT3L (387 aa). DNMT3A and DNMT3B are enzymatically active, whereas DNMT3L does not possess methyltransferase activity, but it stimulates the activity of DNMT3A and DNMT3B (Gowher et al. 2005; Chedin et al. 2002). Of note, a novel member of the rodent Dnmt3 family, Dnmt3c that arose from duplication of the DNMT3B gene, has been identified recently (Barau et al. 2016). This male germline-specific variant is required for methylation of retrotransposons during mouse spermatogenesis (Barau et al. 2016). However, its orthologue has not been identified in humans.

In the N-terminal part of the DNMT3 proteins, which differ significantly from the N-terminal part of DNMT1, three separate regions can be distinguished (Jurkowska et al. 2011a) (Fig. 2):

- very N-terminal segment of DNMT3A and DNMT3B, which is the most variable region between both proteins, binds DNA and is important for anchoring of the enzymes to nucleosomes. It seems to also play a role in the targeting of DNMT3A to the shores of bivalent CpG promoters.

---

**Fig. 2** (continued) motifs are indicated. For details, refer to text. (**b**) Crystal structures of the human DNMT1 (351–1600 fragment), the human DNMT3A and DNMT3L complex bound to the histone H3 tail and the human TET2 catalytic domain in complex with DNA. Distinct protein domains are color coded according to the color scheme used in (**a**)

- ADD (ATRX-DNMT3-DNMT3L) domain of DNMT3 proteins mediates the interaction with the N-terminal tails of histone H3, as well as other chromatin proteins; and is involved in the allosteric regulation of the enzymes' activity.
- PWWP domain of DNMT3A and DNMT3B interacts with histone H3 tails trimethylated at lysine 36 and is essential for targeting of the enzymes to pericentromeric chromatin and gene bodies. This domain is missing in DNMT3L.

The catalytic domains of DNMT3A, and DNMT3B share ~80% sequence identity and are active in isolated form (Gowher and Jeltsch 2002). In contrast, despite clear homology with the other family members, the C-terminal domain of DNMT3L is catalytically inactive due to amino acid exchanges and deletions within the conserved methyltransferase motifs. Structural studies of the DNMT3A and DNMT3L revealed that both proteins form a tetrameric complex, consisting of two molecules of DNMT3A in the centre and two molecules of DNMT3L at the edges of the tetramer (Jia et al. 2007; Guo et al. 2015; Zhang et al. 2018). These interfaces also support self-interaction of DNMT3A and contribute an interesting regulatory mechanism for the activity and localisation of DNMT3A [reviewed in Jeltsch and Jurkowska (2013)]. The arrangement of the two DNMT3A catalytic sites allows methylation of two adjacent CpG sites in one binding event (Jia et al. 2007; Jurkowska et al. 2008; Zhang et al. 2018). The long-awaited structure of the DNA-bound form of the complex revealed that the DNA binding interface of DNMT3A is formed by a specific loop from the target recognition domain, the catalytic loop and the homodimeric interface of DNMT3A (Zhang et al. 2018).

## 2.2 *Catalytic Properties of DNMTs*

All DNA cytosine-C5-methyltransferases share a similar catalytic mechanism, involving conserved amino acid motifs, for the transfer of the methyl group from the cofactor AdoMed to the target cytosine base [reviewed in Jurkowska et al. (2011a)]. Interestingly, they use base flipping to rotate the target cytosine out of the DNA duplex and insert it in the catalytic pocket. As in the case of other DNMTs, base flipping was observed in the structures of DNMT1 and DNMT3A with substrate DNA as well (Song et al. 2012; Zhang et al. 2018).

DNMT1 is a very processive enzyme, capable of methylating multiple CpG sites along the DNA without dissociating from the substrate (Vilkaitis et al. 2005; Hermann et al. 2004). This property fits well to the maintenance role of DNMT1 at the replication fork, as it allows very efficient methylation of the newly synthetized daughter strand before the chromatin is reassembled. The structure of DNMT1 revealed that the enzyme enwraps the DNA, enabling sliding of the protein along the substrate and catalysis of successive methylation reactions (Song et al. 2012).

In contrast to DNMT1, DNMT3A methylates DNA in a distributive manner (Norvil et al. 2018; Gowher and Jeltsch 2002), requiring enzyme dissociation after each round of methylation. In addition, it cooperatively binds to DNA, forming multimeric protein-DNA filaments reviewed in Jeltsch and Jurkowska (2013). Cooperative binding enables methylation of multiple sites on the same DNA molecule and increases DNMT3A activity (Emperle et al. 2014), leading to an efficient spreading of DNA methylation over a larger region (Stepper et al. 2017). Interestingly, binding of several DNMT3A along the DNA leads to an 8-10 bps periodicity in the methylation pattern (Jia et al. 2007; Jurkowska et al. 2008). In contrast to DNMT3A, DNMT3B is able to processively methylate multiple CpG sites and binds to the DNA in a non-cooperative manner (Norvil et al. 2018; Gowher and Jeltsch 2002), indicating that small sequence differences in the catalytic domains of DNMT3A and DNMT3B have a profound impact on the catalytic properties of these related enzymes.

## 2.3 *Intrinsic DNA Sequence Specificity of DNMTs*

For a long time, DNA methylation in mammals was thought to be largely restricted to CpG sites, however, recent studies revealed the presence of non-CpG methylation in several cell types and tissues, both in mouse and in humans [reviewed in He and Ecker (2015)]. The original DNA methylation pattern is set by DNMT3A and DNMT3B, which are classically designated as de novo MTases, as they do not display preference between unmethylated and hemi-methylated DNA. Although both enzymes preferentially methylate CpG dinucleotides, they can also modify cytosines in a non-CpG context, with a preference for CA $>>$ CT $>$ CC (Gowher and Jeltsch 2001; Zhang et al. 2018). Experiments using DNMT3s knockout in embryonic stem cells or ectopic expression of DNMT3A in cells lacking DNA methylation provided direct evidence that DNMT3 enzymes introduce methylation in non-CpG context also in vivo (Ramsahoye et al. 2000). Conversely, DNMT1 cannot efficiently methylate non-CpG sites, leading to the loss of the non-CpG methylation through cellular division in the absence of DNMT3 enzymes. Therefore, presence of the non-CpG methylation directly reflects DNMT3 enzyme activity in cells. Consistently, methylated non-CpG sites are widespread in cells and tissues, where DNMT3A and DNMT3B are highly expressed (like embryonic stem cells, induced pluripotent cells, oocytes and brain), but absent or present at only marginal levels in most somatic tissues and cells with low expression of these enzymes (Ziller et al. 2011; Varley et al. 2013).

Currently, it is unclear what is the biological role of the non-CpG methylation. It has been viewed as a by-product of the hyperactivity and low specificity of DNMT3 enzymes. However, depending on the experimental system, there is also evidence of its potential role in gene repression or expression. Most insights about the potential biological role of non-CpG methylation came from studies on brain [reviewed in Kinde et al. (2015), Jang et al. (2017)], where non-CpG methylation occurs at high levels and contributes to neuronal maturation and specification of brain cells.

However, further studies are required to elucidate the exact biological function of the non-CpG methylation.

Although DNMT3A and DNMT3B do not seem to have strong sequence specificity beyond CpG dinucleotides, both enzymes are sensitive to the sequences flanking their target cytosines. For example, DNMT3A prefers purine bases at the 5′ end of the CpG, whereas pyrimidines are favoured at their 3′ end (Handa and Jeltsch 2005; Jurkowska et al. 2011b). Interestingly, experimental flanking sequence preferences of DNMT3s correlate with the methylation level of CpG sites found in the human genome (Handa and Jeltsch 2005), suggesting that the inherent sequence preferences of de novo methyltransferases contribute to the selection of their target regions in the genome.

The mechanistic understanding of the flanking sequence preferences and specificity of DNMT3 enzymes towards CpG sites has long awaited the availability of the structure with bound substrate DNA, which has been obtained recently (Zhang et al. 2018). The structure of the DNMT3A-DNMT3L complex bound to DNA revealed that the guanine base of the target CpG site is accurately recognised by the R836 residue of DNMT3A, mutation of which results in a reduced preference of the enzyme for CpG methylation (Gowher et al. 2006; Zhang et al. 2018). Additionally, several residues directly contact the bases flanking the CpG dinucleotide, explaining the strong flanking sequence preferences of DNMT3A. Notably, in the structure, no protein contact with the cytosine base of the opposite strand was observed, explaining the lack of discrimination of DNMT3 enzymes between unmethylated and hemi-methylated DNA (Zhang et al. 2018).

In contrast to DNMT3 enzymes, DNMT1 shows strong preference towards hemi-methylated DNA over unmethylated substrate (Song et al. 2012; Goyal et al. 2006; Bashtrykov et al. 2012), which enables its function as the methylation copy machine at the replication fork. The structure of DNMT1 bound to hemi-methylated DNA provided molecular explanation for this preference and revealed that the methyl group of the cytosine is recognised by a hydrophobic pocket in the catalytic domain of DNMT1 and that both the 5mC and the corresponding G in the target DNA strand are recognized accurately (Song et al. 2012). This observation also explains the high specificity of DNMT1 towards CpG sites over non-CpG sites mentioned above.

## 2.4 *Recruitment of DNMT Enzymes to Chromatin and Replicating DNA*

Correct establishment and maintenance of DNA methylation patterns is crucial for human development and health, therefore mechanisms contributing to these processes have been extensively studied over the past decades. Several synergistic models, including both the inherent specificity of the methyltransferases, as well as the role of other proteins and chromatin modifications, have been proposed to explain how specific DNA methylation patterns are established [reviewed in Jurkowska et al. (2011a)].

### 2.4.1 Interaction of DNMT3s with Chromatin Marks

As DNA methylation is embedded in multifaced epigenetic network, direct interaction with specific chromatin marks has been proposed as a general mechanism involved in the recruitment of DNA methyltransferases to specific genomic regions.

The ADD domain, which is present in all DNMT3 proteins, interacts specifically with H3 tails unmethylated at lysine 4, modification (e.g. acetylation or di/trimethylation) of which prevents ADD binding (Noh et al. 2015; Otani et al. 2009; Zhang et al. 2010b). Importantly, binding of H3 tails to the ADD domain also allosterically activates DNMT3A (Guo et al. 2015; Li et al. 2011), thereby stimulating methylation of chromatin-bound DNA by DNMT3A. Since (tri)methylation of H3K4 is associated with active genes, its presence would repel DNA methyltransferases and prevent DNA methylation of active regions. Indeed, a strong inverse genome-wide correlation of DNA methylation and H3K4me3 modification was observed (Hodges et al. 2009; Meissner et al. 2008) and demethylation of K4 of H3 at enhancers of pluripotency genes was required for localization of DNMT3 enzymes in embryonic stem cells (Petell et al. 2016). The crucial role of the ADD domain in the targeting of DNMT3A to chromatin in vivo was further confirmed by an elegant study, which showed that engineering of the ADD domain of DNMT3A led to aberrant DNA methylation patterns in cells and disturbed differentiation programs of embryonic cells (Noh et al. 2015). Besides interacting with histone H3 tails, the ADD domain is a platform involved in DNMT3A interaction with other proteins, including transcription factors, histone methyltransferases and other chromatin proteins [reviewed in Ravichandran et al. (2018)]. Importantly, as ADD domain is involved in the allosteric regulation of DNMT3A, interaction with this domain may directly influence DNMT3A activity, as shown for H3 (Guo et al. 2015) and MeCP2 (Rajavelu et al. 2018).

The PWWP domain of DNMT3A and DNMT3B, which specifically recognizes H3 tails tri-methylated at K36 (H3K36me3) (Dhayalan et al. 2010), is the second DNMT3 domain directly contributing to the recruitment of methyltransferases to specific genomic regions, including pericentromeric chromatin and gene bodies. Strong correlation of both H3K36me3 and DNA methylation was observed in the body of active genes and at exon-intron boundaries (Vakoc et al. 2006; Kolasinska-Zwierz et al. 2009; Baubec et al. 2015). The central role of H3K36me3 recognition in targeting of DNA methylation was experimentally confirmed in a variety of cellular systems (Neri et al. 2017; Morselli et al. 2015; Baubec et al. 2015). For example, H3K36me3-dependent intragenic DNA methylation by DNMT3B is crucial to protect gene bodies from cryptic transcription initiation (Neri et al. 2017). Furthermore, a subset of heterochromatic repeats shows strong enrichment in H3K36me3, explaining the role of the DNMT3A PWWP domain in the heterochromatic localization of the enzyme (Ernst et al. 2011).

Besides interacting with histone tails, the PWWP domain of DNMT3 enzymes can also bind DNA (Qiu et al. 2002). A recent model for the methylation of nucleosomal DNA by DNMT3A suggested that the targeting occurs through a specific binding of H3K36me3 by the PWWP domain of DNMT3A, which is followed by an activation of the catalytic domain mediated by the binding of H3 tails to the ADD domain, resulting in the methylation of nearby cytosines by the catalytic domain (Rondelet et al. 2016).

### 2.4.2 Recruitment of DNMT1 to Replicating Chromatin

Several targeting mechanisms ensure the correct localization of DNMT1 to replicating DNA. First one involves PCNA, a component of the replication machinery that interacts and co-localizes with DNMT1 in vivo (Iida et al. 2002), indicating that it might directly recruit the methyltransferase to the replication fork and load it onto DNA. The PCNA-DNMT1 interaction contributes to the efficiency of DNA re-methylation in cells, but it is not essential for this process (Egger et al. 2006). Second factor essential for the recruitment of DNMT1 and the maintenance of DNA methylation patterns in mammals is UHRF1 (Sharif et al. 2007; Bostick et al. 2007). UHRF1 specifically binds to hemi-methylated DNA via its SET and RING-associated (SRA) domain (Hashimoto et al. 2008; Bostick et al. 2007; Arita et al. 2008) and recognizes histone H3 tails methylated at lysine 9 (H3K9me2/me3) via cooperative binding of its tandem Tudor domain (TTD) and its plant homeodomain (PHD) (Rothbart et al. 2012; Nady et al. 2011). The chromatin interactions of UHRF1 are necessary for the recruitment of DNMT1 to replicating chromatin, since UHRF1 mutations preventing histone binding abolished DNA methylation by DNMT1 in cells (Rothbart et al. 2012; Nady et al. 2011). Similarly, UHRF1 knockout in mice results in a genome-wide loss of DNA methylation (Bostick et al. 2007; Sharif et al. 2007). In addition to its role in targeting of DNMT1, UHRF1 was also shown to stimulate the catalytic activity of DNMT1 through a direct interaction (Bashtrykov et al. 2014).

A model of a direct recruitment of DNMT1 by histone marks is also plausible, as the methyltransferase preferentially associates with H3 tails ubiquitinated at K18 and K23 (Qin et al. 2015; Nishiyama et al. 2013). This interaction is mediated by the replication foci-targeting (RFTS) domain of DNMT1 and leads to the recruitment of the enzyme to newly replicated DNA and its simultaneous activation, providing another beautiful example of allosteric regulation of DNMTs. The ubiquitination of the H3 tail is introduced by UHRF1 and is stimulated UHFR1 binding to hemi-methylated DNA (Harrison et al. 2016). Ubiquitinated H3 accumulates during S-phase, leading to the recruitment of DNMT1 to newly replicated DNA (Qin et al. 2015; Nishiyama et al. 2013; Harrison et al. 2016). These data indicate an important additional connection between DNMT1 and UHRF1 chromatin interactions, which is essential for an efficient maintenance of DNA methylation.

# 3 Erasure of DNA Methylation

Since its discovery, 5mC was considered a very stable modification due to the chemical strength of the C-C bond. Therefore, DNA demethylation was expected only to occur passively, through a replication-dependent dilution in the absence or inhibition of the maintenance methylation machinery (Fig. 1). However, global genome-wide loss of DNA methylation occurring in a DNA replication-independent manner was observed in mouse zygotes (Mayer et al. 2000; Oswald et al. 2000) and during specification of primordial germ cells (Hajkova et al. 2002; Yamazaki et al. 2003), pointing towards existence of an active demethylation machinery. Furthermore, active DNA demethylation has also been observed at specific loci in T cells, neurons and other cells (Bruniquel and Schwartz 2003; Martinowich et al. 2003).

Despite the discovery of biological processes where active DNA demethylation occurs in the absence of DNA replication, the enzymatic machinery responsible for this process in mammals remained enigmatic until 2009, when a group of enzymes called Ten-Eleven Translocation (TET) was shown to oxidize 5mC to 5-hydroxymethylcytosine (5hmC) both *in vitro* and in mouse embryonic stem cells (Tahiliani et al. 2009). Moreover, TET enzymes are able to further oxidize 5hmC to 5-formylcytosine (5fC) and 5-carboxylcytosine (5caC) (He et al. 2011; Ito et al. 2011). These oxidized bases are recognized and excised by the thymine DNA glycosylase (TDG) triggering base excision repair pathway (BER) to replace the abasic site by an unmodified cytosine (He et al. 2011; Maiti and Drohat 2011), thereby completing the DNA demethylation cycle (Fig. 1). This finding suggested a plausible pathway for active DNA demethylation and opened a new dynamic field of research (Tahiliani et al. 2009).

## 3.1 *Architecture of TET Enzymes*

The mammalian TET family comprises three paralogous members (TET1, 2136 aa in human, TET2, 2002 aa and TET3, 1776 aa), which share similar domain architecture (Fig. 2). They all are big proteins harboring a large, mostly unstructured N-terminal part and the C-terminal catalytic domain. The core catalytic domain is composed of a cysteine-rich region and a following double-stranded β helix domain (DSBH) characteristic for $Fe^{2+}$/αKG dioxygenases and (Hu et al. 2013, 2015). In metazoan TETs, the DSBH domain is interrupted by a large unstructured region, which is believed to engage in protein-protein interactions. In their N-termini, TET1 and TET3 contain a CXXC domain, which interacts with DNA (Xu et al. 2012; Jin et al. 2016). The CXXC domain of TET2 was lost during evolution after gene duplication and inversion, and is now encoded as a separate protein IDAX (inhibition of the dvl and axin complex) (Iyer et al. 2009).

The recently solved crystal structure of the human TET2 catalytic domain in complex with 5mC containing DNA substrate (PDB ID: 4NM6) revealed that the

DSBH core domain forms a globular structure, which is stabilized by the Cys-rich region enwrapping the DSBH core (Hu et al. 2013). The Cys-rich region is crucial for the stability of the DSBH domain and consequently for catalysis (Hu et al. 2013). DNA is bound above the DSBH core in a groove that is enriched in basic and hydrophobic amino acids. Similar to DNMTs and DNA repair enzymes (Klimasauskas et al. 1994), TET enzymes utilize a base flipping mechanism to position the target base in the catalytic pocket for the oxidation reaction. Once the base is located in the catalytic pocket, the methyl-group is oriented towards the catalytic iron and α-KG, which facilitate the catalytic turnover (Hu et al. 2013). TET enzymes follow a conserved catalytic mechanism that is characteristic for other known $Fe^{2+}$/αKG-dependent dioxygenases, like histone lysine demethylases (JMJC-family) [reviewed in Hausinger and Schofield (2015)].

## 3.2 *Intrinsic DNA Sequence Specificity of TET Enzymes*

Most of the studies on TET enzymes were focused on elucidating their biological role and their reaction products; however, the intrinsic biochemical properties of TET enzymes that govern their function remain not well investigated. Recent reports showed that TET-dependent demethylation in zygotes represents only a small fraction of all demethylation events and that TET-associated demethylation seems to be locus specific (Guo et al. 2014; von Meyenn et al. 2016). Additionally, fine mapping of the genomic location of 5hmC using SCL-exo protocol showed that 5hmC is highly enriched within defined sequence context (Serandour et al. 2016), suggesting that TET enzymes could display some sequence preferences. It is however still unknown what is the molecular reason granting this sequence selectivity. Both DNMTs and TET enzymes modify CpG dinucleotides, yet DNMT3 enzymes can also efficiently methylate non-CpG sites (as discussed above). Unfortunately, little work was contributed to investigate the intrinsic preference of TET enzymes towards non-CpG sites. In the initial report that identified TET enzymes as 5mC hydroxylases, the authors showed that these enzymes are capable of oxidation of 5mC embedded in a CpG site, yet non-CpG substrates were not tested (Tahiliani et al. 2009). Later on, it has been showed that 5mCpA and 5mCpC sites were poor substrates for TET2, with conversion efficiencies of $<2\%$ and $<5\%$, respectively, as opposed to $>85\%$ for 5mCG sites in the same sequence context (Hu et al. 2013). Like TET2, TET1 preferentially oxidizes 5mCpG with some incidence of oxidation of 5mCpC sites. Structural studies provided molecular explanation for the observed preference of TET enzymes towards the CpG sites (Hu et al. 2013). The TET2-DNA crystal structure showed that the target 5mC is specifically recognized by two hydrogen bonds formed by the side chains of H1904 and N1387 and base endocyclic nitrogen atoms N3 and N4, respectively. The base-stacking interaction between TET2 Y1902 residue and the pyrimidine base of the 5mC additionally supports this recognition. Furthermore, base-stacking interaction between Y1294 residue and the G:5mC base pair in the DNA provides specific recognition of the following G:C base pair within the CpG dinucleotide. Intriguingly, TET2 enzyme does not make any contact with the methyl group of the target cytosine, suggesting that it could generate oxidation of 5hmC to 5fC and 5caC (Hu et al. 2013).

In the TET2:DNA co-crystal structure no protein-base specific contacts outside of the CpG site were observed, suggesting that the enzyme has weak or no flanking sequence specificity (Hu et al. 2013, 2015). Nevertheless, the bound DNA is strongly bent and distorted, giving the possibility of indirect readout of DNA sequence as observed with numerous other DNA binding proteins, restriction enzymes and bacterial MTases (Jurkowski et al. 2007; Little et al. 2008). Whether TET enzymes use indirect readout for sequence recognition remains to be addressed.

In addition, TET enzymes are also able to oxidize the methyl group of thymine (T) to 5-hydroxymethyl uracil (5hmU) (Pfaffeneder et al. 2014), however the efficiency of this reaction is rather low, and its physiological relevance still needs to be uncovered.

## 3.3 On Site and Lateral Processivity of TET Enzymes

Processivity of TET enzymes can be regarded in two different ways. First, as a serial oxidation of 5mC to 5hmC, 5fC and 5caC on a single CpG site without the enzyme dissociating from that site, which could be regarded as "on-site processivity". Second, which could be called "lateral processivity", is the consecutive oxidation of numerous CpGs on a single DNA molecule.

The isolated catalytic domain of human TET2 efficiently oxidizes 5mC to 5hmC, yet, further oxidation steps are inefficient, leading to the reaction stalling at the 5hmC state (Hu et al. 2015). Conversely, numerous reports showed that TET enzymes are capable of efficient conversion of 5mC to 5caC without being blocked at the 5hmC state (Tamanaha et al. 2016; Liu et al. 2017; Crawford et al. 2016). Moreover, the same group which reported stalling of the oxidation reaction at the 5hmC state, also showed that TET2 could convert 5mC all the way to 5caC (Hu et al. 2013). It is likely that the contradictory conclusions of the studies that investigated mouse TET2 "on-site" processivity could be potentially explained by differences in the reaction conditions and experimental setup (Tamanaha et al. 2016; Crawford et al. 2016).

The catalytic domains of TET1 and TET2 show no preference to modify neighboring CpG sites on the same DNA molecule, suggesting that TET enzymes are not laterally processive (Tamanaha et al. 2016). However, both TET1 and TET3 full-length enzymes contain an additional DNA binding domain, namely the CXXC domain, which can modify the enzymes behavior on DNA. Strikingly, TET3 CXXC domain preferentially binds 5caCpG sites, which represent the final TET reaction product. This observation led to proposal that the TET3 CXXC—5caCpG interaction could stimulate processive activity of the enzyme and consequently lead to spreading of the 5caC from the first oxidized CpG site. In the proposed model, the first 5mCpG site that is oxidized to 5caCpG gets bound by the CXXC domain of the enzyme, therefore keeping the catalytic domain in close proximity and promoting oxidation of nearby 5mCpGs (Jin et al. 2016). It is a very interesting hypothesis, which still requires further experimental validation.

### 3.4 *Oxidation of RNA Bases*

Mammalian TET enzymes were initially found to oxidize 5mC in genomic DNA, but have since been also shown to oxidize 5mC in RNA (Basanta-Sanchez et al. 2017; Fu et al. 2014). Moreover, the presence of TET homologues in organisms that do not possess any active DNA methyltransferase, like *D. melanogaster*, suggested that other substrate than 5mC in dsDNA could be processed by the enzyme (Dunwell et al. 2013). Indeed, *Drosophila* TET is responsible (at least in part) for the formation of 5hmC in the fly mRNAs, particularly in mRNAs involved in neuronal development. Consequently, blocking of the TET enzyme causes brain defects and is lethal. In vivo, RNA hydroxymethylation promotes mRNA translation (Delatte et al. 2016).

DNA and RNA represent different structural configurations, which impact the way TET enzymes can interact with them. DeNizio and colleagues performed a systematic survey aimed to compare the activity of TET2 CD on ds- and ss- DNA and RNA, as well as DNA/RNA hybrids. They discovered that 5mC in dsDNA is the most proficient substrate, ssRNA and ssDNA are well tolerated, whereas dsRNA is a very poor substrate for TET2 (DeNizio et al. 2018).

### 3.5 *Recruitment of TET Enzymes*

The mechanisms of locus specific recruitment and regulation of TET enzymes is much less understood than the genomic distribution and physiological relevance of the oxidized-5mC derivatives. The CXXC domains located in the N-termini of TET1 and TET3 are thought to be at least in part responsible for the targeting of the enzymes to the CpG-rich regions (CpG islands), as the CXXC domain has been shown to recruit DNMT1, MLL1, CFP1 to unmethylated CpG sites (Stroynowska-Czerwinska et al. 2018; Xu et al. 2018). Consistently, DNA binding studies showed that the CXXC domain of TET1 is able to bind to CpG-rich DNA irrespective of its modification state (C, 5mC or 5hmC) (Zhang et al. 2010a), whereas the CXXC domain of TET3 from *Xenopus* binds unmodified cytosines in both CpG and non-CpG context, with a slightly higher preference for CpG (Xu et al. 2012; Jin et al. 2016). Another interesting study demonstrated that the CXXC domain of TET3 can bind 5caCpG and that full-length TET3 preferentially binds to the transcriptional start sites (TSS) of genes involved in base excision repair (Jin et al. 2016). This suggests that TET3 may be specifically targeted to these loci through the CXXC domain or by other interacting proteins (Jin et al. 2016).

TET2, which lacks the CXXC domain may be more depend on other proteins, for example transcription factors (TFs), for locus specific recruitment. Supporting this idea, TET2 interacts with the transcription factor Wilms tumor (WT) and Early B cell factor 1 (EBF1), which modulate TET2 activity and target gene expression [reviewed in (Ravichandran et al.)]. Recently, several TFs important for cellular differentiation were reported to induce DNA demethylation by interacting with TET proteins. For example, RUNX1, an essential master transcription factor in hematopoietic development and an important regulator of immune functions, was shown to recruit TET2 and induce local DNA demethylation at its binding regions (Suzuki et al. 2017). Likewise, NANOG-dependent recruitment of TET1 and TET2 promotes expression of genes involved in reprogramming and lineage commitment (Costa et al. 2013). Furthermore, a study by Perera and colleagues in mouse retinal cells demonstrated that RE1-silencing transcription factor (REST) recruits an isoform of TET3 lacking the CXXC domain along with the histone methyltransferase NSD3 to activate its target genes (Perera et al. 2015). TET enzymes were shown to interact with proteins involved in base excision repair pathway such as TDG, PARP1, MBD4, NEIL (Muller et al. 2014). Furthermore, all three TET enzymes associate with O-linked ß-D-N-acetylglucosamine (O-GlcNAc) transferase (OGT). It has been suggested that TETs recruit OGT to the chromatin and that TET-OGT interaction promotes the OGT activity (Vella et al. 2013; Chen et al. 2013). In summary, it is increasingly clear that TET enzymes do not function alone but interact with multiple other proteins in a contextual manner and through this cooperation modulate gene expression.

## 4 Synthetic Programming of DNA Methylation

Rapid development of next-generation based sequencing technologies enabled genome-wide interrogation of cytosine methylation at single-base resolution, providing invaluable insights into the frequency and genomic distribution of 5mCs, as well as into the interplay between DNA methylation and other epigenetic mechanisms. Yet, the lack of tools for locus-specific manipulation of cytosine status has hampered the functional understanding of the role of DNA methylation and demethylation. Recent progress in programmable DNA binding domains has open new synthetic ways to study epigenetic regulation (Jurkowski et al. 2015), and in particular DNA methylation and demethylation. Fusing an active DNA methyltransferase or demethylase (or any other epigenetic enzyme) to a customizable DNA binding domain enables targeting of the methylation or demethylation functionality to selected places in the genome. From a mechanistic point of view this powerful technology permits not only to study the principles of how the enzymes set up or remove the methylation mark, but also to directly probe and dissect the epigenetic mechanisms and transcriptional consequences of DNA methylation or demethylation at a given genomic locus. On the application side, it allows verifying consequentiality of disease associated epigenetic changes or even their repair as a potential therapeutic strategy.

### 4.1 Programmable Genome Targeting Modules

Three different classes of programmable DNA binding domains have been employed so far in epigenetic editing. The C2H2 zinc fingers were the first example of predictable DNA interaction domains amenable to rational protein design [reviewed in Wolfe et al. (2000), Pabo et al. (2001)] and were first used for programmable sequence specific genome targeting of fused epigenetic enzymes (Xu and Bestor 1997). More recently, two additional programmable genome binders were discovered: the TAL effector arrays (TALE) (Boch et al. 2009) and CRISPR/Cas9 systems (Jinek et al. 2012). The transcription activating-like effectors (TALEs) are important virulence factors initially isolated from the bacterial plant pathogen *Xanthomonas* (Boch and Bonas 2010) and are composed of tandemly arranged 34 amino acid long highly similar repeats (Scholze and Boch 2010).

The newest and most exciting addition to the genome targeting toolbox repository is the CRISPR/Cas9 system (Hsu et al. 2014). CRISPR (*c*lustered *r*egularly *i*nterspaced *s*hort *p*alindromic *r*epeats) functions as a prokaryotic adaptive immune system that confers resistance to exogenous genetic elements such as plasmids and phages (Mojica et al. 2005). CRISPR/Cas9 proteins recognize their targets based on Watson/Crick base pairing and rely on complementarity of the recognized DNA and the guide RNA sequences which are used for targeting. Therefore, retargeting of the Cas9 protein to specific genomic location requires only a gRNA component specific for the desired target. However, because Cas9 is an active nuclease, for targeting of the epigenetic enzymes, a catalytically inactive Cas9 variant is used. It still recognizes and binds the target sequence, yet does not cleave it (Qi et al. 2013). Whereas each of the available programmable genome targeting domains offers unique advantages and disadvantages, due to the simplicity of target design and the possibility for multiplexing CRISPR/Cas9 system seems the most attractive.

### 4.2 Epigenetic Effector Domains

The epigenetic editing activity is provided by fusing active DNA methylating or demethylating enzymes to the targeting domain. Until now different DNA methyltransferases have been used [reviewed in Lau and Suh (2018), Lei et al. (2018)], which include bacterial CpG specific methyltransferases M.SssI (Xiong et al. 2017) or MQ1 (a modified CpG methyltransferase derived from *Mollicutes spiroplasma*) (Lei et al. 2017), the catalytic domains or full-length mammalian Dnmt3a (Vojta et al. 2016; Liu et al. 2016) or Dnmt3b (Lin et al. 2018) proteins, as well as an engineered Dnmt3a-Dnmt3L fusion protein, which in addition to the Dnmt3a methyltransferase contains the co-activator protein Dnmt3L (Stepper et al. 2017; Saunderson et al. 2017). For targeted DNA demethylation, all three mammalian TET enzymes have been used (Liu et al. 2016), yet TET1 CD is the most commonly used version. Interestingly, a direct removal of methylated cytosine has also been achieved recently by employing plant ROS1 DNA glycosylase, leading to the transcriptional increase of the target locus (Parrilla-Doblas et al. 2017).

## *4.3 Applications of Targeted DNA Methylation/ Demethylation*

Targeted DNA modification (both methylation and demethylation) has the potential to answer so far unapproachable questions in basic and translational research. It allows mechanistic dissection of the epigenetic signaling cascades and validation of the causality of epigenetic changes in diseases. It can widen our understanding of epigenetic dynamics and the basis of stability of DNA methylation signal, but also address the contribution of epigenetic changes to etiology of complex and simple diseases, through discovery and validation of disease-promoting epimutations and provide means for reverting them (Fig. 3).

### 4.3.1 How Do Epigenetic Changes Contribute to Disease Etiology?

As discussed above, widespread changes in DNA methylation patterns are commonly observed in diseases (Egger et al. 2004; Koch et al. 2018), including cancer, chronic or acute diseases. However, it is hard to evaluate whether these epigenetic changes are causal for the disease progression or are merely by-standers, reflecting the overall epigenetic dysregulation caused by the disease.

Epigenome-wide association studies (EWAS) are commonly used to derive associations between epigenetic variation and a particular identifiable phenotype (Birney et al. 2016). When epigenetic patterns, such as DNA methylation, change at specific loci, discriminating the phenotypically affected cases from the control individuals, this is considered an indication that epigenetic perturbation has taken place that is associated either causally or consequentially with the studied phenotype. However, EWAS results do not allow discriminating causal from consequential epigenetic changes, just merely their correlation with the screened phenotype. In such cases, targeted DNA methylation/demethylation could be used to study the causality of the observed changes towards the phenotype.

Aberrant promoter methylation is a well-recognized hallmark of cancer; however, it is unclear whether epigenetic changes are enough to drive cellular transformation. Sanderson (Saunderson et al. 2017) used CRISPR-based targeted DNA methylation to stably methylate and repress the CDKN2A, HIC1, PTEN and RASSF1 tumor suppressor genes in healthy primary breast cells. Furthermore, they show that targeted *de novo* methylation of the CDKN2A p16 transcript promoter prevented cells from entering senescence arrest, thus possibly facilitating tumor initiation.

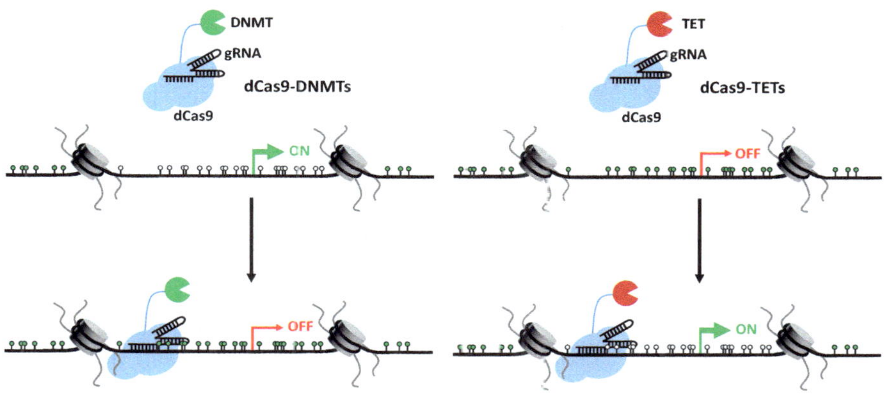


**Fig. 3** Targeted DNA methylation and demethylation controls gene expression. Locus-specific targeting of DNA methyltransferases (DNMTs) represses the target gene, whereas demethylation of a methylated promoter after targeting of TETs leads to gene activation. Green lollipops indicate methylated CpG sites, white lollipops—unmethylated CpG sites

### 4.3.2 Repair of Aberrant, Disease Causing Epigenetic States

Fragile X syndrome (FXS) is the most frequent form of inherited mental retardation (Sutcliffe et al. 1992). FMR-1 gene found in fragile X patients shows an increase in the number of CGG repeats and an abnormal methylation of a CpG island 250 bp proximal to this repeat. Liu et al., used the dCas9-TET1 CD construct to demethylate CGG repeats in FXS induced pluripotent stem cells (iPS) and reactivate the silenced FMR1 gene by demethylating and activating its promoter (Liu et al. 2018).

## *4.4 Limitations of Targeted DNA Methylation and Demethylation Tools*

Despite being such a very powerful technology, epigenetic editing has also its pitfalls and limitations. The promise of epigenetic regulation is that once DNA methylation is established or removed, cellular epigenetic mechanisms will maintain the new state of the locus, such that it can be inherited after semiconservative DNA replication (Jeltsch and Jurkowska 2014). Therefore, targeted DNA methylation or demethylation could provide a unique opportunity to heritably switch off gene expression (loss-of-function) (Siddique et al. 2013; Nunna et al. 2014; Stolzenburg et al. 2012). However, recent reports indicate that DNA methylation deposited at active gene promoters is not necessarily stably maintained and consequently gets diluted with DNA replication and cell division (Vojta et al. 2016; Kungulovski et al. 2015). Nevertheless, stable epigenetic reprogramming has also been achieved (Saunderson et al. 2017; Amabile et al. 2016), yet the epigenetic mechanisms which grant this stability are not well understood. It is plausible that the stability of the introduced DNA methylation will depend on the local environment of the targeted locus. As DNA methylation is just one layer of epigenetic regulation, targeting of multiple epigenetic marks simultaneously might improve the stability of introduced DNA methylation.

Unintended, off-target epigenetic modification can lead to misinterpretation of the epigenetic editing experiments in regard to the observed biological effects, therefore specificity of introduced epigenetic modification is of principal importance. The precision of targeting is even more important for potential therapeutic epigenetic interventions, as mistargeted modifications can disregulate other genes and cause additional diseases. Because of this, numerous studies have addressed this issue [reviewed in Lei et al. (2018)]. Two types of off-targeting can be distinguished: first one stems from the misrecognition of the targeting module (i.e. binding promiscuity of the dCas9 part to other near-cognate sequences in the genome) and the second one is the unintended modification by the epigenetic domains used.

In targeting experiments, dCas9 ChIP-seq coupled with bisulfite sequencing has been used to investigate the off-target methylation, and showed that even at the top ranking dCas9 binding sites dCas9-DNMT3a only marginally increased DNA methylation relative to the methylation observed at the intentionally targeted loci (Liu et al. 2016), suggesting that mis-targeting of dCas9 is not contributing strongly to off-target methylation. Other studies applied genome-wide sequencing technologies, including reduced-representation bisulfite sequencing (RRBS) and whole genome bisulfite sequencing to assess potential side effects of various methylation tools and reported no detectable off-target hypermethylation (Huang et al. 2017; Lei et al. 2017). Similarly, few off-target effects have been reported with demethylation tools.

In contrast, numerous studies also reported significant off-target methylation when targeting dCas9-DNMT3a CD. The extent of the off-targets effects varied vastly (Huang et al. 2017; Lei et al. 2017; McDonald et al. 2016). A recent study showed presence of extensive off-target genome-wide methylation in mouse ES cells (mESC) and somatic cells (Galonska et al. 2018) regardless of whether or not sgRNA was used for targeting. Expression level of the constructs might greatly influence the extent of off-targets, as once the "true" binding sites are occupied, the rest of the produced targeting constructs will be available to modify unspecific sites.

**Acknowledgements** We apologize to the authors whose work could not be cited due to the space limitations. Work in the authors laboratory has been supported by Boehringer Ingelheim (R.Z.J) and DFG SPP1784 (T.P.J).

## References

Amabile A, Migliara A, Capasso P et al (2016) Inheritable silencing of endogenous genes by hit-and-run targeted epigenetic editing. Cell 167:219–232, e214

Arita K, Ariyoshi M, Tochio H et al (2008) Recognition of hemi-methylated DNA by the SRA protein UHRF1 by a base-flipping mechanism. Nature 455:818–821

Barau J, Teissandier A, Zamudio N et al (2016) The DNA methyltransferase DNMT3C protects male germ cells from transposon activity. Science 354:909–912
Basanta-Sanchez M, Wang R, Liu Z et al (2017) TET1-mediated oxidation of 5-formylcytosine (5fC) to 5-carboxycytosine (5caC) in RNA. Chembiochem 18:72–76
Bashtrykov P, Jankevicius G, Smarandache A et al (2012) Specificity of Dnmt1 for methylation of hemimethylated CpG sites resides in its catalytic domain. Chem Biol 19:572–578
Bashtrykov P, Jankevicius G, Jurkowska RZ et al (2014) The UHRF1 protein stimulates the activity and specificity of the maintenance DNA methyltransferase DNMT1 by an allosteric mechanism. J Biol Chem 289:4106–4115
Baubec T, Colombo DF, Wirbelauer C et al (2015) Genomic profiling of DNA methyltransferases reveals a role for DNMT3B in genic methylation. Nature 520:243–247
Bergman Y, Cedar H (2013) DNA methylation dynamics in health and disease. Nat Struct Mol Biol 20:274–281
Birney E, Smith GD, Greally JM (2016) Epigenome-wide association studies and the interpretation of disease -omics. PLoS Genet 12:e1006105
Boch J, Bonas U (2010) Xanthomonas AvrBs3 family-type III effectors: discovery and function. Annu Rev Phytopathol 48:419–436
Boch J, Scholze H, Schornack S et al (2009) Breaking the code of DNA binding specificity of TAL-type III effectors. Science 326:1509–1512
Bogdanovic O, Lister R (2017) DNA methylation and the preservation of cell identity. Curr Opin Genet Dev 46:9–14
Bostick M, Kim JK, Esteve PO et al (2007) UHRF1 plays a role in maintaining DNA methylation in mammalian cells. Science 317:1760–1764
Bruniquel D, Schwartz RH (2003) Selective, stable demethylation of the interleukin-2 gene enhances transcription by an active process. Nat Immunol 4:235–240
Chedin F, Lieber MR, Hsieh CL (2002) The DNA methyltransferase-like protein DNMT3L stimulates de novo methylation by Dnmt3a. Proc Natl Acad Sci U S A 99:16916–16921
Chen Q, Chen Y, Bian C et al (2013) TET2 promotes histone O-GlcNAcylation during gene transcription. Nature 493:561–564
Cheng X, Blumenthal RM (2008) Mammalian DNA methyltransferases: a structural perspective. Structure 16:341–350
Costa Y, Ding J, Theunissen TW et al (2013) NANOG-dependent function of TET1 and TET2 in establishment of pluripotency. Nature 495:370–374
Crawford DJ, Liu MY, Nabel CS et al (2016) Tet2 catalyzes stepwise 5-methylcytosine oxidation by an iterative and de novo mechanism. J Am Chem Soc 138:730–733
Delatte B, Wang F, Ngoc LV et al (2016) RNA biochemistry. Transcriptome-wide distribution and function of RNA hydroxymethylcytosine. Science 351:282–285
DeNizio JE, Liu MY, Leddin E et al (2018) Selectivity and promiscuity in TET-mediated oxidation of 5-methylcytosine in DNA and RNA. Biochemistry. https://doi.org/10.1021/acs.biochem.8b00912
Dhayalan A, Rajavelu A, Rathert P et al (2010) The Dnmt3a PWWP domain reads histone 3 lysine 36 trimethylation and guides DNA methylation. J Biol Chem 285:26114–26120
Dunwell TL, McGuffin LJ, Dunwell JM et al (2013) The mysterious presence of a 5-methylcytosine oxidase in the Drosophila genome: possible explanations. Cell Cycle 12:3357–3365
Egger G, Liang G, Aparicio A et al (2004) Epigenetics in human disease and prospects for epigenetic therapy. Nature 429:457–463
Egger G, Jeong S, Escobar SG et al (2006) Identification of DNMT1 (DNA methyltransferase 1) hypomorphs in somatic knockouts suggests an essential role for DNMT1 in cell survival. Proc Natl Acad Sci U S A 103:14080–14085
Emperle M, Rajavelu A, Reinhardt R et al (2014) Cooperative DNA binding and protein/DNA fiber formation increases the activity of the Dnmt3a DNA methyltransferase. J Biol Chem 289:29602–29613

Ernst J, Kheradpour P, Mikkelsen TS et al (2011) Mapping and analysis of chromatin state dynamics in nine human cell types. Nature 473:43–49
Fu L, Guerrero CR, Zhong N et al (2014) Tet-mediated formation of 5-hydroxymethylcytosine in RNA. J Am Chem Soc 136:11582–11585
Galonska C, Charlton J, Mattei AL et al (2018) Genome-wide tracking of dCas9-methyltransferase footprints. Nat Commun 9:597
Gowher H, Jeltsch A (2001) Enzymatic properties of recombinant Dnmt3a DNA methyltransferase from mouse: the enzyme modifies DNA in a non-processive manner and also methylates non-CpG [correction of non-CpA] sites. J Mol Biol 309:1201–1208
Gowher H, Jeltsch A (2002) Molecular enzymology of the catalytic domains of the Dnmt3a and Dnmt3b DNA methyltransferases. J Biol Chem 277:20409–20414
Gowher H, Jeltsch A (2018) Mammalian DNA methyltransferases: new discoveries and open questions. Biochem Soc Trans 46:1191–1202
Gowher H, Liebert K, Hermann A et al (2005) Mechanism of stimulation of catalytic activity of Dnmt3A and Dnmt3B DNA-(cytosine-C5)-methyltransferases by Dnmt3L. J Biol Chem 280:13341–13348
Gowher H, Loutchanwoot P, Vorobjeva O et al (2006) Mutational analysis of the catalytic domain of the murine Dnmt3a DNA-(cytosine C5)-methyltransferase. J Mol Biol 357:928–941
Goyal R, Reinhardt R, Jeltsch A (2006) Accuracy of DNA methylation pattern preservation by the Dnmt1 methyltransferase. Nucleic Acids Res 34:1182–1188
Guo F, Li X, Liang D et al (2014) Active and passive demethylation of male and female pronuclear DNA in the mammalian zygote. Cell Stem Cell 15:447–459
Guo X, Wang L, Li J et al (2015) Structural insight into autoinhibition and histone H3-induced activation of DNMT3A. Nature 517:640–644
Hajkova P, Erhardt S, Lane N et al (2002) Epigenetic reprogramming in mouse primordial germ cells. Mech Dev 117:15–23
Handa V, Jeltsch A (2005) Profound flanking sequence preference of Dnmt3a and Dnmt3b mammalian DNA methyltransferases shape the human epigenome. J Mol Biol 348:1103–1112
Harrison JS, Cornett EM, Goldfarb D et al (2016) Hemi-methylated DNA regulates DNA methylation inheritance through allosteric activation of H3 ubiquitylation by UHRF1. elife 5:e17101
Hashimoto H, Horton JR, Zhang X et al (2008) The SRA domain of UHRF1 flips 5-methylcytosine out of the DNA helix. Nature 455:826–829
Hausinger RP, Schofield CJ (2015) 2-oxoglutarate-dependent oxygenases. RSC metallobiology, vol 3. Royal Society of Chemistry, Cambridge
He Y, Ecker JR (2015) Non-CG methylation in the human genome. Annu Rev Genomics Hum Genet 16:55–77
He YF, Li BZ, Li Z et al (2011) Tet-mediated formation of 5-carboxylcytosine and its excision by TDG in mammalian DNA. Science 333:1303–1307
Hermann A, Goyal R, Jeltsch A (2004) The Dnmt1 DNA-(cytosine-C5)-methyltransferase methylates DNA processively with high preference for hemimethylated target sites. J Biol Chem 279:48350–48359
Hodges E, Smith AD, Kendall J et al (2009) High definition profiling of mammalian DNA methylation by array capture and single molecule bisulfite sequencing. Genome Res 19:1593–1605
Hsu PD, Lander ES, Zhang F (2014) Development and applications of CRISPR-Cas9 for genome engineering. Cell 157:1262–1278
Hu L, Li Z, Cheng J et al (2013) Crystal structure of TET2-DNA complex: insight into TET-mediated 5mC oxidation. Cell 155:1545–1555
Hu L, Lu J, Cheng J et al (2015) Structural insight into substrate preference for TET-mediated oxidation. Nature 527:118–122
Huang YH, Su J, Lei Y et al (2017) DNA epigenome editing using CRISPR-Cas SunTag-directed DNMT3A. Genome Biol 18:176

Iida T, Suetake I, Tajima S et al (2002) PCNA clamp facilitates action of DNA cytosine methyltransferase 1 on hemimethylated DNA. Genes Cells 7:997–1007

Ito S, Shen L, Dai Q et al (2011) Tet proteins can convert 5-methylcytosine to 5-formylcytosine and 5-carboxylcytosine. Science 333:1300–1303

Iyer LM, Tahiliani M, Rao A et al (2009) Prediction of novel families of enzymes involved in oxidative and other complex modifications of bases in nucleic acids. Cell Cycle 8:1698–1710

Jang HS, Shin WJ, Lee JE et al (2017) CpG and non-CpG methylation in epigenetic gene regulation and brain function. Genes (Basel) 8(6). pii: E148

Jeltsch A, Jurkowska RZ (2013) Multimerization of the dnmt3a DNA methyltransferase and its functional implications. Prog Mol Biol Transl Sci 117:445–464

Jeltsch A, Jurkowska RZ (2014) New concepts in DNA methylation. Trends Biochem Sci 39:310–318

Jeltsch A, Jurkowska RZ (2016) Allosteric control of mammalian DNA methyltransferases – a new regulatory paradigm. Nucleic Acids Res 44:8556–8575

Jia D, Jurkowska RZ, Zhang X et al (2007) Structure of Dnmt3a bound to Dnmt3L suggests a model for de novo DNA methylation. Nature 449:248–251

Jin SG, Zhang ZM, Dunwell TL et al (2016) Tet3 reads 5-carboxylcytosine through its CXXC domain and is a potential guardian against neurodegeneration. Cell Rep 14:493–505

Jinek M, Chylinski K, Fonfara I et al (2012) A programmable dual-RNA-guided DNA endonuclease in adaptive bacterial immunity. Science 337:816–821

Jurkowska RZ, Jeltsch A (2016) Enzymology of mammalian DNA methyltransferases. Adv Exp Med Biol 945:87–122

Jurkowska RZ, Anspach N, Urbanke C et al (2008) Formation of nucleoprotein filaments by mammalian DNA methyltransferase Dnmt3a in complex with regulator Dnmt3L. Nucleic Acids Res 36:6656–6663

Jurkowska RZ, Jurkowski TP, Jeltsch A (2011a) Structure and function of mammalian DNA methyltransferases. Chembiochem 12:206–222

Jurkowska RZ, Siddique AN, Jurkowski TP et al (2011b) Approaches to enzyme and substrate design of the murine Dnmt3a DNA methyltransferase. Chembiochem 12:1589–1594

Jurkowski TP, Anspach N, Kulishova L et al (2007) The M.EcoRV DNA-(adenine N6)-methyltransferase uses DNA bending for recognition of an expanded EcoDam recognition site. J Biol Chem 282:36942–36952

Jurkowski TP, Ravichandran M, Stepper P (2015) Synthetic epigenetics-towards intelligent control of epigenetic states and cell identity. Clin Epigenetics 7:18

Kinde B, Gabel HW, Gilbert CS et al (2015) Reading the unique DNA methylation landscape of the brain: non-CpG methylation, hydroxymethylation, and MeCP2. Proc Natl Acad Sci U S A 112:6800–6806

Klimasauskas S, Kumar S, Roberts RJ et al (1994) HhaI methyltransferase flips its target base out of the DNA helix. Cell 76:357–369

Koch A, Joosten SC, Feng Z et al (2018) Analysis of DNA methylation in cancer: location revisited. Nat Rev Clin Oncol 15:459–466

Kolasinska-Zwierz P, Down T, Latorre I et al (2009) Differential chromatin marking of introns and expressed exons by H3K36me3. Nat Genet 41:376–381

Kungulovski G, Nunna S, Thomas M et al (2015) Targeted epigenome editing of an endogenous locus with chromatin modifiers is not stably maintained. Epigenetics Chromatin 8:12

Lau CH, Suh Y (2018) In vivo epigenome editing and transcriptional modulation using CRISPR technology. Transgenic Res 27:489–509

Lei Y, Zhang X, Su J et al (2017) Targeted DNA methylation in vivo using an engineered dCas9-MQ1 fusion protein. Nat Commun 8:16026

Lei Y, Huang YH, Goodell MA (2018) DNA methylation and de-methylation using hybrid site-targeting proteins. Genome Biol 19:187

Li BZ, Huang Z, Cui QY et al (2011) Histone tails regulate DNA methylation by allosterically activating de novo methyltransferase. Cell Res 21:1172–1181

Lin L, Liu Y, Xu F et al (2018) Genome-wide determination of on-target and off-target characteristics for RNA-guided DNA methylation by dCas9 methyltransferases. GigaScience 7:1–19
Little EJ, Babic AC, Horton NC (2008) Early interrogation and recognition of DNA sequence by indirect readout. Structure 16:1828–1837
Liu XS, Wu H, Ji X et al (2016) Editing DNA methylation in the mammalian genome. Cell 167:233–247 e217
Liu MY, Torabifard H, Crawford DJ et al (2017) Mutations along a TET2 active site scaffold stall oxidation at 5-hydroxymethylcytosine. Nat Chem Biol 13:181–187
Liu XS, Wu H, Krzisch M et al (2018) Rescue of fragile X syndrome neurons by DNA methylation editing of the FMR1 gene. Cell 172:979–992 e976
Maiti A, Drohat AC (2011) Thymine DNA glycosylase can rapidly excise 5-formylcytosine and 5-carboxylcytosine: potential implications for active demethylation of CpG sites. J Biol Chem 286:35334–35338
Martinowich K, Hattori D, Wu H et al (2003) DNA methylation-related chromatin remodeling in activity-dependent BDNF gene regulation. Science 302:890–893
Mayer W, Niveleau A, Walter J et al (2000) Demethylation of the zygotic paternal genome. Nature 403:501–502
McDonald JI, Celik H, Rois LE et al (2016) Reprogrammable CRISPR/Cas9-based system for inducing site-specific DNA methylation. Biol Open 5:866–874
Meissner A, Mikkelsen TS, Gu H et al (2008) Genome-scale DNA methylation maps of pluripotent and differentiated cells. Nature 454:766–770
Mojica FJ, Diez-Villasenor C, Garcia-Martinez J et al (2005) Intervening sequences of regularly spaced prokaryotic repeats derive from foreign genetic elements. J Mol Evol 60:174–182
Morselli M, Pastor WA, Montanini B et al (2015) In vivo targeting of de novo DNA methylation by histone modifications in yeast and mouse. elife 4:e06205
Muller U, Bauer C, Siegl M et al (2014) TET-mediated oxidation of methylcytosine causes TDG or NEIL glycosylase dependent gene reactivation. Nucleic Acids Res 42:8592–8604
Nady N, Lemak A, Walker JR et al (2011) Recognition of multivalent histone states associated with heterochromatin by UHRF1 protein. J Biol Chem 286:24300–24311
Neri F, Rapelli S, Krepelova A et al (2017) Intragenic DNA methylation prevents spurious transcription initiation. Nature 543:72–77
Nishiyama A, Yamaguchi L, Sharif J et al (2013) Uhrf1-dependent H3K23 ubiquitylation couples maintenance DNA methylation and replication. Nature 502:249–253
Noh KM, Wang H, Kim HR et al (2015) Engineering of a histone-recognition domain in Dnmt3a alters the epigenetic landscape and phenotypic features of mouse ESCs. Mol Cell 59:89–103
Norvil AB, Petell CJ, Alabdi L et al (2018) Dnmt3b methylates DNA by a noncooperative mechanism, and its activity is unaffected by manipulations at the predicted dimer interface. Biochemistry 57:4312–4324
Nunna S, Reinhardt R, Ragozin S et al (2014) Targeted methylation of the epithelial cell adhesion molecule (EpCAM) promoter to silence its expression in ovarian cancer cells. PLoS One 9: e87703
Oswald J, Engemann S, Lane N et al (2000) Active demethylation of the paternal genome in the mouse zygote. Curr Biol 10:475–478
Otani J, Nankumo T, Arita K et al (2009) Structural basis for recognition of H3K4 methylation status by the DNA methyltransferase 3A ATRX-DNMT3-DNMT3L domain. EMBO Rep 10:1235–1241
Pabo CO, Peisach E, Grant RA (2001) Design and selection of novel Cys2His2 zinc finger proteins. Annu Rev Biochem 70:313–340
Parrilla-Doblas JT, Ariza RR, Roldan-Arjona T (2017) Targeted DNA demethylation in human cells by fusion of a plant 5-methylcytosine DNA glycosylase to a sequence-specific DNA binding domain. Epigenetics 12:296–303
Perera A, Eisen D, Wagner M et al (2015) TET3 is recruited by REST for context-specific hydroxymethylation and induction of gene expression. Cell Rep 11:283–294

Petell CJ, Alabdi L, He M et al (2016) An epigenetic switch regulates de novo DNA methylation at a subset of pluripotency gene enhancers during embryonic stem cell differentiation. Nucleic Acids Res 44:7605–7617

Pfaffeneder T, Spada F, Wagner M et al (2014) Tet oxidizes thymine to 5-hydroxymethyluracil in mouse embryonic stem cell DNA. Nat Chem Biol 10:574–581

Qi LS, Larson MH, Gilbert LA et al (2013) Repurposing CRISPR as an RNA-guided platform for sequence-specific control of gene expression. Cell 152:1173–1183

Qin W, Wolf P, Liu N et al (2015) DNA methylation requires a DNMT1 ubiquitin interacting motif (UIM) and histone ubiquitination. Cell Res 25:911–929

Qiu C, Sawada K, Zhang X et al (2002) The PWWP domain of mammalian DNA methyltransferase Dnmt3b defines a new family of DNA-binding folds. Nat Struct Biol 9:217–224

Rajavelu A, Lungu C, Emperle M et al (2018) Chromatin-dependent allosteric regulation of DNMT3A activity by MeCP2. Nucleic Acids Res 46:9044–9056

Ramsahoye BH, Biniszkiewicz D, Lyko F et al (2000) Non-CpG methylation is prevalent in embryonic stem cells and may be mediated by DNA methyltransferase 3a. Proc Natl Acad Sci U S A 97:5237–5242

Ravichandran M, Jurkowska RZ, Jurkowski TP (2018) Target specificity of mammalian DNA methylation and demethylation machinery. Org Biomol Chem 16:1419–1435

Rondelet G, Dal Maso T, Willems L et al (2016) Structural basis for recognition of histone H3K36me3 nucleosome by human de novo DNA methyltransferases 3A and 3B. J Struct Biol 194:357–367

Rothbart SB, Krajewski K, Nady N et al (2012) Association of UHRF1 with methylated H3K9 directs the maintenance of DNA methylation. Nat Struct Mol Biol 19:1155–1160

Saunderson EA, Stepper P, Gomm JJ et al (2017) Hit-and-run epigenetic editing prevents senescence entry in primary breast cells from healthy donors. Nat Commun 8:1450

Scholze H, Boch J (2010) TAL effector-DNA specificity. Virulence 1:428–432

Schubeler D (2015) Function and information content of DNA methylation. Nature 517:321–326

Serandour AA, Avner S, Mahe EA et al (2016) Single-CpG resolution mapping of 5-hydroxymethylcytosine by chemical labeling and exonuclease digestion identifies evolutionarily unconserved CpGs as TET targets. Genome Biol 17:56

Sharif J, Muto M, Takebayashi S et al (2007) The SRA protein Np95 mediates epigenetic inheritance by recruiting Dnmt1 to methylated DNA. Nature 450:908–912

Siddique AN, Nunna S, Rajavelu A et al (2013) Targeted methylation and gene silencing of VEGF-A in human cells by using a designed Dnmt3a-Dnmt3L single-chain fusion protein with increased DNA methylation activity. J Mol Biol 425:479–491

Smith ZD, Meissner A (2013) DNA methylation: roles in mammalian development. Nat Rev Genet 14:204–220

Song J, Teplova M, Ishibe-Murakami S et al (2012) Structure-based mechanistic insights into DNMT1-mediated maintenance DNA methylation. Science 335:709–712

Stepper P, Kungulovski G, Jurkowska RZ et al (2017) Efficient targeted DNA methylation with chimeric dCas9-Dnmt3a-Dnmt3L methyltransferase. Nucleic Acids Res 45:1703–1713

Stolzenburg S, Rots MG, Beltran AS et al (2012) Targeted silencing of the oncogenic transcription factor SOX2 in breast cancer. Nucleic Acids Res 40:6725–6740

Stroynowska-Czerwinska A, Piasecka A, Bochtler M (2018) Specificity of MLL1 and TET3 CXXC domains towards naturally occurring cytosine modifications. Biochim Biophys Acta Gene Regul Mech 1861:1093–1101

Sutcliffe JS, Nelson DL, Zhang F et al (1992) DNA methylation represses FMR-1 transcription in fragile X syndrome. Hum Mol Genet 1:397–400

Suzuki T, Shimizu Y, Furuhata E et al (2017) RUNX1 regulates site specificity of DNA demethylation by recruitment of DNA demethylation machineries in hematopoietic cells. Blood Adv 1:1699–1711

Tahiliani M, Koh KP, Shen Y et al (2009) Conversion of 5-methylcytosine to 5-hydroxymethylcytosine in mammalian DNA by MLL partner TET1. Science 324:930–935

Tamanaha E, Guan S, Marks K et al (2016) Distributive processing by the iron(II)/alpha-ketoglutarate-dependent catalytic domains of the TET enzymes is consistent with epigenetic roles for oxidized 5-methylcytosine bases. J Am Chem Soc 138:9345–9348

Vakoc CR, Sachdeva MM, Wang H et al (2006) Profile of histone lysine methylation across transcribed mammalian chromatin. Mol Cell Biol 26:9185–9195

Varley KE, Gertz J, Bowling KM et al (2013) Dynamic DNA methylation across diverse human cell lines and tissues. Genome Res 23:555–567

Vella P, Scelfo A, Jammula S et al (2013) Tet proteins connect the O-linked N-acetylglucosamine transferase Ogt to chromatin in embryonic stem cells. Mol Cell 49:645–656

Vilkaitis G, Suetake I, Klimasauskas S et al (2005) Processive methylation of hemimethylated CpG sites by mouse Dnmt1 DNA methyltransferase. J Biol Chem 280:64–72

Vojta A, Dobrinic P, Tadic V et al (2016) Repurposing the CRISPR-Cas9 system for targeted DNA methylation. Nucleic Acids Res 44:5615–5628

von Meyenn F, Iurlaro M, Habibi E et al (2016) Impairment of DNA methylation maintenance is the main cause of global demethylation in naive embryonic stem cells. Mol Cell 62:983

Wolfe SA, Nekludova L, Pabo CO (2000) DNA recognition by Cys2His2 zinc finger proteins. Annu Rev Biophys Biomol Struct 29:183–212

Xiong T, Meister GE, Workman RE et al (2017) Targeted DNA methylation in human cells using engineered dCas9-methyltransferases. Sci Rep 7:6732

Xu GL, Bestor TH (1997) Cytosine methylation targetted to pre-determined sequences. Nat Genet 17:376–378

Xu Y, Xu C, Kato A et al (2012) Tet3 CXXC domain and dioxygenase activity cooperatively regulate key genes for Xenopus eye and neural development. Cell 151:1200–1213

Xu C, Liu K, Lei M et al (2018) DNA sequence recognition of human CXXC domains and their structural determinants. Structure 26:85–95 e83

Yamazaki Y, Mann MR, Lee SS et al (2003) Reprogramming of primordial germ cells begins before migration into the genital ridge, making these cells inadequate donors for reproductive cloning. Proc Natl Acad Sci U S A 100:12207–12212

Yang L, Rau R, Goodell MA (2015) DNMT3A in haematological malignancies. Nat Rev Cancer 15:152–165

Zhang H, Zhang X, Clark E et al (2010a) TET1 is a DNA-binding protein that modulates DNA methylation and gene transcription via hydroxylation of 5-methylcytosine. Cell Res 20:1390–1393

Zhang Y, Jurkowska R, Soeroes S et al (2010b) Chromatin methylation activity of Dnmt3a and Dnmt3a/3L is guided by interaction of the ADD domain with the histone H3 tail. Nucleic Acids Res 38:4246–4253

Zhang ZM, Lu R, Wang P et al (2018) Structural basis for DNMT3A-mediated de novo DNA methylation. Nature 554:387–391

Ziller MJ, Muller F, Liao J et al (2011) Genomic distribution and inter-sample variation of non-CpG methylation across human cell types. PLoS Genet 7:e1002389

# Origin and Mechanisms of DNA Methylation Dynamics in Cancers

**Hariharan Easwaran and Stephen B. Baylin**

## Contents



**Abstract** DNA methylation is a key epigenetic mark in the heritable regulation of gene expression, with important roles in normal development and disease. Genomewide alterations in DNA methylation patterns are universal feature across cancers. Studies in the last few years have shown that similar alterations occur during various normal physiological processes, such as aging. Understanding mechanisms involved in DNA methylation alterations is critical for understanding cancer etiology. In this chapter, we discuss recent work on the nature of the genomic region-specific DNA methylation alterations, its functional implications, and the mechanisms underlying these alterations.

**Keywords** DNA methylation · CpG islands · Hypermethylation · Hypomethylation · Cancer · Epigenetics · Chromatin · Tumor suppressor · Aging · Inflammation

H. Easwaran (✉) · S. B. Baylin
Department of Oncology, The Sidney Kimmel Comprehensive Cancer Center at Johns Hopkins, The Johns Hopkins University School of Medicine, Baltimore, MD, USA
e-mail: heaswar1@jhmi.edu; sbaylin1@jhmi.edu



S. Jurga, J. Barciszewski (eds.), *The DNA, RNA, and Histone Methylomes*, RNA Technologies, https://doi.org/10.1007/978-3-030-14792-1_2

# 1 Introduction

Cancer is an outcome of deviant gene expression programs, which helps hijacking every normal cellular process to the advantage of cancer cells. This relies on tilting the balance of oncogenic and tumor suppressive mechanisms by the altered, but heritable, gene expression programs. Epigenetic processes are central mechanisms involved in the maintenance of regulated gene expression programs between cell generations. Aberrant alterations to the epigenetic machinery are universally observed across cancers, and these alterations have prominent roles in tumorigenesis. Recent studies have revealed how epigenetic mechanisms may undergo perturbations as a result of genetic alterations, exposure to environmental agents and during the normal course of aging. These perturbations lead to changes in gene expression program that create the permissive state for functioning of cancer driver mutations. The sum effect of perturbations to epigenetic machinery and the occurrence of key cancer driver mutations lead to the uncontrolled growth and dissemination of cancer cells.

A vast range of epigenetic mechanisms are involved that can cause oncogenic and tumor suppressor inactivation. These range from a plethora of histone modifications and DNA modifications, the end result of which is the fine control of gene expression by modulating accessibility of protein factors to the chromatin, the higher order chromatin arrangement and nuclear architecture. Among these, DNA methylation is one of the foremost important functional epigenetic mark in normal development and cancers. Recent studies have highlighted other key derivative modifications of cytosine methylation (example 5-hydroxymethylcytosine), which may play important roles. This review will focus on DNA methylation, and chiefly discuss the generation and functional impact of aberrant DNA modifications in cancers. After a brief overview on epigenetics, this review will highlight the long-known importance of DNA methylation abnormalities in cancers. With this as the pivot, and special emphasis on DNA hypermethylation events, we will discuss its possible origins and functional implications, and the molecular machinery involved in methylation.

# 2 Epigenetics Overview

The current definition of epigenetics, i.e. a stable and heritable change in gene expression without any changes in DNA sequence (Bird 2007), implies that the molecular determinants of the epigenetic mark be faithfully replicated along with DNA replication at every round of replicative cell division. Chromatin modifications, such as histone post-translational modifications (PTMs) methylation, acetylation (to name a few), and DNA methylation are major constituents of epigenetic information subject to heritable silencing (Almouzni and Cedar 2016). Historically, DNA methylation is the first proposed epigenetic mark, which due to its biochemical properties—namely covalent linkage to DNA and a repertoire of enzymes that maintain the methylation at every round of DNA replication (discussed later)—is the most agreed upon modification that fits the definition of epigenetic information. In mammals, DNA methylation mainly occurs as a covalent modification of cytosines in the context of the palindromic 5′-CpG-3′ dinucleotide ($^{m}$CpG), and is propagated on both parent and nascent strands after DNA replication. Further a complex relationship of DNA methylation with chromatin modifications, involving mutually exclusive and inclusive interactions (Kouzarides 2007), is proposed to enable methylation as a stable carrier of epigenetic information. For example, during development, differentiation and disease, the transcription factor (TF)-network and associated chromatin modifiers establish the epigenome defined by chromatin modifications. Once chromatin modifications and ensuing gene expression patterns are established, DNA methylation could provide a basis for an efficient way to re-establish this information during subsequent cell division cycles and mediate heritable transcriptional silencing.

## 3 DNA Methylation Abnormalities in Cancer

Due to the widespread gene expression changes in cancers, it is not surprising that the epigenome is highly perturbed compared to normal cells, which help maintain the altered expression state over the course of cancer initiation and progression. Thus, all forms of epigenetic information, viz. DNA methylation, histone PTMs, nucleosomal positioning and higher order chromosomal structure, are altered in cancers. Aberrant DNA methylation in cancers is an early change showing genomic region-specific patterns of gains and losses in DNA methylation. Recent advances in mapping methylation pattern across the whole genome ("methylome") have provided deep insights into its normal composition and large-scale alteration in cancers. In this section we highlight the key regions in the genome that harbor altered DNA methylation and how these alterations play important roles during tumorigenesis.

### *3.1 Global Hypomethylation*

The earliest reported DNA methylation change in cancers is that of methylation losses at structural elements of genes. Normal human cells have about 70–80% of CpG residues in the genome methylated, while in cancers this reduces drastically (Ehrlich and Lacey 2013). It is now well established that gene bodies, inter-genic regions and various repetitive elements (like LINE-1) undergo hypomethylation in cancers (Ehrlich and Lacey 2013). In the earlier stages of tumorigenesis, hypomethylation can cause LOH by inducing genomic instability, but in later stages it suppresses tumor formation possibly by preventing epigenetic silencing of tumor suppressor genes (discussed later) (Yamada et al. 2005). Global DNA hypomethylation has been shown to disrupt various aspects of the normal regulation of the genome—activation of oncogenes, altered transcription start sites, loss of imprinting, genomic instability by heterochromatin loss and reactivation of transposable elements (Chen et al. 1998; Eden et al. 2003; Ehrlich and Lacey 2013; Gaudet et al. 2003; Holm et al. 2005; Hur et al. 2014). Each of these abnormalities plays roles in tumorigenesis by altering gene expression, destabilizing the genome and increasing mutational rates.

Early studies analyzing total methylated CpG content in normal and cancer tissues have shown that DNA hypomethylation in cancers is prevalent equally across the repetitive and unique DNA sequence fractions of the genome. In accordance with the nature of genomic elements that are hypomethylated, genomewide DNA methylation analyses have confirmed these early observations by showing that hypomethylation occurs in about 50% of the genome in blocks of contiguous genomic regions termed partially-methylated domains (PMDs) that are greater than >100 kb in size (Berman et al. 2011; Hansen et al. 2011). The majority (70–80%) of the genomic CpG sites are otherwise methylated, thus resulting in genomic regions containing highly methylated domains (HMDs) interspersed by the PMDs. Key features of PMDs are that they have low gene density, they are embedded in the late replicating domains, and their boundaries associate with the nuclear lamina domain (LAD) and insulator proteins like CTCF (Bergman and Cedar 2013; Berman et al. 2011; Hansen et al. 2011) (Fig. 1a). PMDs are present in differentiated primary cells and immortalized cell lines, but not in embryonic stem cells (Lister et al. 2011). Recent work has determined a DNA sequence signature, individual units of the sequence "WCGW" (W is adenine or thymine), that is most prone to hypomethylation. Mapping CpG methylation changes in this signature sequence throughout the genome has revealed that PMDs are conserved and universal feature of all normal lineage committed proliferating primary tissues and cells in culture (Zhou et al. 2018). The degree of the PMD-HMD contrast, driven by the depth of hypomethylation of the signature sequence, is very pronounced in cancer cells compared to their normal cell counterparts. The contrasting retention of methylation in HMD and loss of methylation in PMD indicates that the regulators and enzyme machinery responsible for epigenetic maintenance have differential activity in different regions of the genome. Especially it indicates that the methylation aberrations in cancers are not due to a general global loss or gain in activity of the DNA methyltransferase (DNMT) enzymes responsible for catalyzing the methylation of cytosine, but that their local recruitment and activity in defined regions in the genome is altered.

As mentioned above, detailed analyses of the nature of the PMD domains with respect to other genomic features have revealed that they correlate best with late replicating domains in the genome (Fig. 1a). During S-phase when DNA is replicated, the newly replicated, unmethylated daughter strands are re-methylated to copy the methylation pattern of the parent strand by the DNMT1 enzyme (Jones and Liang 2009). Given the high rate at which DNA is replicated in the replication fork (about 0.03 s per nucleotide) (Jackson and Pombo 1998) and the slow rate of DNA methylation by recombinant DNMT1 (~73–433 s per methyl group transfer) (Pradhan et al. 1999), maintenance methylation by DNMT1 at each round of cell division needs to keep up with the quick rate of DNA synthesis. DNMT1 acts in

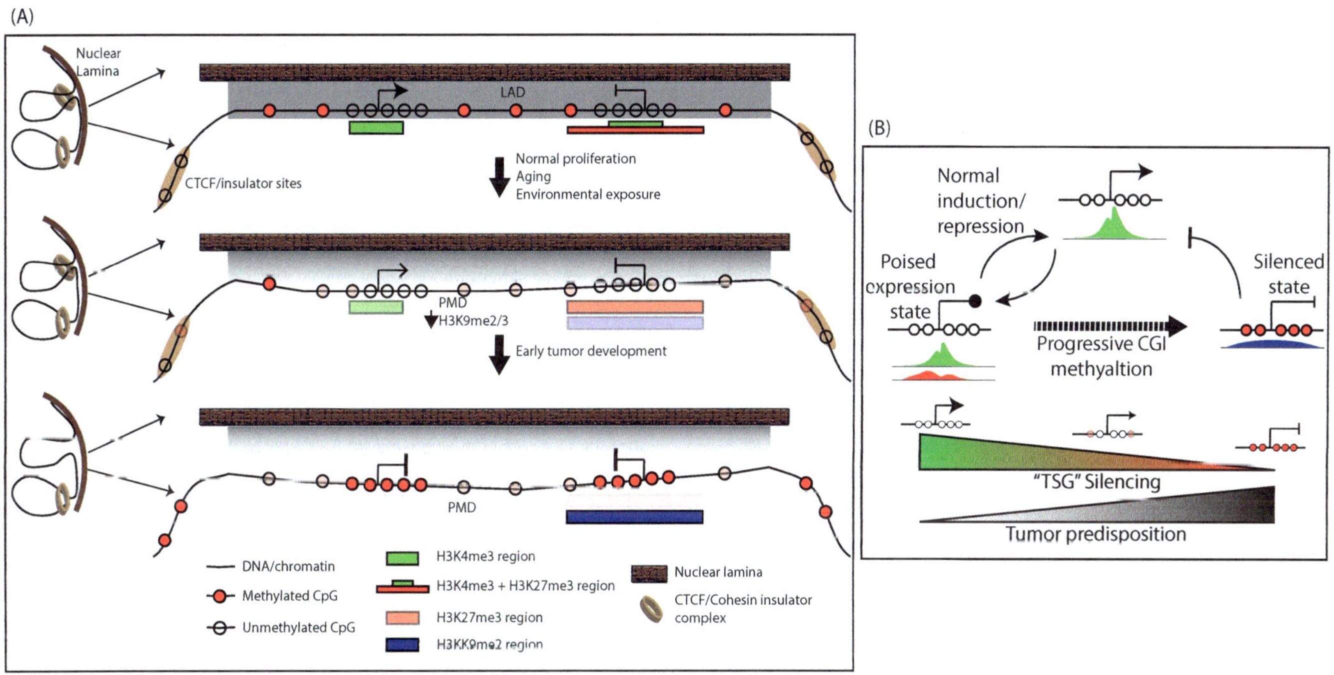


**Fig. 1** (**a**) Regions of DNA showing aberrant methylation changes. DNA/chromatin is normally organized in domains associated with the nuclear lamina (lamin associated domains, LAD), and other regions forming higher ordered looped domains in the nuclear interior (left panel). During progressive cell divisions (in association with aging, environmental exposure, inflammation, etc.), DNA methylation decreases across the LAD forming the partially methylated domains (PMD). Simultaneously local promoter CGI (increased density of red circles) gain methylation, which is associated with silencing. Increasing DNA methylation and decreasing H3K4me3 and H3K27me3 at CGI promoters may be involved in a feedback regulatory mechanism causing progressive gain in DNA methylation and loss of these histone marks, causing increased gene silencing. Progressive loss of CpG methylation across the PMD, along with changes in other key heterochromatic histone modifications (like loss of H3K9me2/me3) is associated with loss of lamin association, and global gene expression changes.

concert with the replication machinery via being recruited to these sites by combined interaction with hemi-methylated DNA, PCNA and UHRF1 (Chuang et al. 1997; Easwaran et al. 2004; Leonhardt et al. 1992; Sharif et al. 2007). At these sites DNMT1 continues to remain associated with the late-replicated DNA post S-phase in the G2/M phases (Easwaran et al. 2004, 2005), and its methylation maintenance function continues even during G2 phase (Schermelleh et al. 2007). The fact that the hypomethylated PMDs are embedded in late replicating domains most likely indicates that the maintenance function is not efficient, and as a result the methylation mark is eroded at every cell cycle. PMDs, which encompasses both unique (containing genes and regulatory elements) and repetitive regions of the genome (retroviral elements), could be the central players subject to altered gene expression and genomic instability during progressive mitotic cycles in normal aging and cancer cells.

Beyond the promiscuous nature of methylation maintenance at the PMDs, whether or not other mechanisms involving the epigenetic machinery may actively regulate these regions remains to be studied. In somatic cells, maintenance of DNA methylation relies on the combined activity of the three major DNA methyltransferases, DNMT1, DNMT3A and B (Jones and Liang 2009), and the demethylation activities of TET enzymes (Wu and Zhang 2014). Whether or not activity of these enzymes, or the multitude of epigenetic regulators mutated in cancers (Dawson 2017), may accentuate hypomethylation in PMDs needs to be determined. Interestingly the PMDs are enriched for genes that undergo focal CpG island (CGI) promoter hypermethylation (discussed in next section). In the normal scenario, PMDs form during differentiation (Lister et al. 2011), but without focal CGI-hypermethylation (Berman et al. 2011). Mechanistically it is pertinent to explore if the global hypomethylation and focal hypermethylation are coupled processes occurring during aging and cancer. Further, since current studies suggest that the hypomethylation at PMDs are generated due to methylation erosion during successive cell division cycles, it is suggested that the degree of hypomethylation in PMD could indicate the mitotic history of a cell (Zhou et al. 2018). Highly proliferative tumors will thus have deeper PMD-HMD domains. The utility of using PMD as a "mitotic clock" is an exciting possibility to track history and rate of cell division in tumors at primary tumor and metastatic sites.

---

**Fig. 1** (continued) Methylation of boundary insulator elements cause inhibition of CTCF binding resulting in further large-scale structural changes. (**b**) Promoter CGI methylation. Majority CGI promoters methylated in cancers are usually marked with H3K4me3 (green) and H3K27me3 (red) (bivalent state) in normal stem progenitor cells, representing a poised expression state. During normal cellular homeostasis (normal differentiation, exposure to stress, etc.) these genes are subject to regulated induction and repression. Progressive promoter CGI methylation accumulation causes promoter silencing and non-responsiveness to induction. Promoter silencing by CGI methylation is also associated with heterochromatic H3K9me2/3 marks (blue). A "continuum" model of "TSG" silencing wherein progressive silencing of important developmental and differentiation regulators (shown in the bottom) will lead to a gradient of expression states that will accordingly be associated with increasing predisposition to tumorigenesis

## 3.2 Gene Promoter CpG-Hypermethylation/Hypomethylation

About 60% of mammalian gene promoters have more than the expected occurrence of the palindromic CpG dinucleotide sequence in narrow contiguous regions around the transcription start site (TSS), termed CpG islands (CGI) (Suzuki and Bird 2008). The concept of CGI was derived from early observations that the methylation-sensitive restriction enzyme HpaII recognizing and cleaving CCGG sequence generates unexpectedly small fragments in the mouse genome (HpaII-tiny fragments, HTF) indicating that these sites are concentrated in certain regions of the genome (Illingworth and Bird 2009). These regions, termed "islands" are distinct from the rest of the genome in that they: (1) are unique sequences ranging in size ~1kb; (2) contain tenfold higher HpaII sites; (3) are GC-rich without depletion of CpG-dinucleotide sequence (which otherwise is depleted in the vertebrate genome); (4) are unmethylated in all normal tissues (except inactive-X chromosome, imprinted and germline genes); (5) are generally present in the 5′-promoter region of genes. A more formal and practical definition of CGI based on genomic sequencing is that these are regions around gene promoters devoid of Alu-repetitive elements, which are greater than 500 bp, with a GC content greater than 55% and observed CpG/expected CpG ratio above 0.65 (Takai and Jones 2002). CGI in their unmethylated state are subject to regulated gene expression, while in their methylated state are subject to permanent silencing (Baylin and Jones 2016). Thus whereas most protein-coding genes are in a transcriptionally permissive chromatin state harboring active chromatin marks (like H3K4me3 and H3K9, 14Ac) with productive transcriptional initiation by RNA Pol II, methylated genes are distinctly devoid of the active marks (Sen et al. 2016) and RNA Pol II indicating a non-permissive chromatin state (Deaton and Bird 2011).

Absence/loss of methylation of CpGs in promoters is associated with gene activation, while presence/gain of methylation is associated with gene silencing (Fig. 1b). This inverse relation is better correlated for genes with CGI (Baylin and Jones 2016). Although also observed for non-CGI genes (Han et al. 2011; Hartung et al. 2012), in somatic cells methylation of non-CGI promoters does not rule out gene expression (Weber et al. 2007). Thus it appears that the role of methylation in non-CGI promoters, vis-à-vis the role of histone modifications, is less relevant, and needs further studies. Methylation of the CpGs in the CGI and in the sequences around TSS promoter region alters chromatin structure, which inhibits binding of the transcription machinery (Baubec et al. 2013; Baylin and Jones 2016; Deaton and Bird 2011). The most well studied methylation changes in cancers are CGI-promoter alterations because ~98% of the CGI promoters are unmethylated in normal somatic cells, and because of the established inverse correlation between expression and CGI methylation. Both aberrant hypomethylation and hypermethylation of the CGI promoter genes are observed in cancers. Hypomethylation of certain CGI promoters in cancers is associated with activation of growth-promoting cell cycle regulators (Mazor et al. 2015) and genes otherwise expressed specifically in the germline, whose expression in tumors may be associated with proto-oncogenic functions (Van

Tongelen et al. 2017). On the other hand, larger numbers of genes in cancers harbor promoter CGI hypermethylation causing de novo gene silencing, or for many genes that are already silenced, CGI methylation will result in blocking of induction of these genes in response to normal differentiation cues (Fig. 1b).

The effects of CGI hypomethylation and hypermethylation at promoters in cancers parallel those of genetic alterations, namely the "gain-of-function" oncogenic activation and "loss-of-function" tumor suppressor gene (TSG) inactivation, respectively. Since CGI hypermethylation is a predominant epigenetic change in cancers, affecting promoters of about 5–10% of CGI containing genes, the mechanisms underlying their methylation and their roles in tumorigenesis are of great interest and studied extensively. In this regard, a key question is whether a given hypermethylated gene is a "driver" of the cancer phenotype. Importantly, in this regard, DNA methylation has been observed in almost every genetically identified TSG (Herman and Baylin 2003). In the context of classic TSG's, these harbor truncating genetic alterations, such as mutations, insertions and deletions, completely inactivating the genes. Since deposition of DNA methylation marks at regulatory elements mainly causes gene silencing, this constitutes a key alternative mechanism to inactivate tumor suppressor genes, wherein one copy of the TSG is inactivated by genetic alteration and the second hit is an epigenetic alteration. A strict definition of TSG requires that these genes oppose mechanisms involved in promoting tumorigenesis and that both copies of the genes should be inactivated in a "two-hit" model of tumor initiation (Knudson 2001). Since deposition of DNA methylation mark at regulatory elements mainly causes gene silencing, in many cancer cases it constitutes a key alternative mechanism to inactivate tumor suppressor genes, wherein one copy of the TSG is inactivated by genetic alteration and the second hit is an epigenetic alteration (Herman and Baylin 2003). Importantly, DNA methylation has been observed in almost every genetically identified TSG. Often important driver genes, such as *CDKN2A* and *MLH1*, are inactivated by methylation than mutation, and far more number of genes than that are genetically altered are epigenetically altered (Gao and Teschendorff 2017; Schuebel et al. 2007).

In addition to the above classic TSG's, many more promoter hypermethylated genes are seldom mutated in cancer raising the driver versus passenger question more poignantly than for the above discussed genes. Many of the affected genes identified belong to important cancer processes. Important insight for the importance of the bulk of hypermethylated genes comes from examining their potential relevance in tumor signaling pathways. In the case of genetic mutations, extensive characterization has indicated that about 2–8 mutations typically represent driver genes in any given tumor, which play key roles in tumor initiation and stepwise progression (Vogelstein et al. 2013). Importantly, more genes in key cancer related pathways are inactivated by epigenetic silencing than by mutations (Schuebel et al. 2007). How many of the methylated genes and what roles they exactly play in the context of the cancer driver events is largely unknown. A compelling case for epigenetic changes playing important roles in tumor development are pediatric cancers which have lower mutational load (Lawrence et al. 2013; McKenna and Roberts 2009), particularly the ependymomas, which are childhood brain tumors that lack genetic alterations but show extensive DNA methylation changes of differentiation genes (Mack et al. 2014). Although these studies show the indisputable role for epigenetic changes in cancers, a lot needs to be learnt about the roles of the affected genes, especially during the course of tumor development.

To understand the function of CGI-promoter hypermethylated genes, it is important to consider that many of these genes do not fit the classic definition of TSG proposed in "Knudson's two-hit" model, in that its complete inactivation might not be the central driver of tumor development. A refined definition of TSG by Knudson and colleagues may help define the extent of the role of epigenetic silencing of such genes in tumorigenesis (Berger et al. 2011). In the revised definition, TSG is viewed as a concept wherein full inactivation of involved genes is not required, but rather a "continuum" of partial silenced states may play critical roles in tumorigenesis. This view of the role of epigenetic modulation of gene silencing is especially relevant when considering the multitude of promoter CGI hypermethylated genes without direct roles in cell cycle and DNA repair checkpoints (classic tumor suppression), but those that are involved in developmental pathways (Easwaran et al. 2012; Ohm et al. 2007; Schlesinger et al. 2007; Widschwendter et al. 2007) and other key processes such as immune checkpoint mechanisms (Wrangle et al. 2013). In this regard, it is interesting to note that 60–70% of the de novo CGI-promoter methylated genes in cancer cells are differentiation and developmental regulators marked by PcG-associated silencing in normal embryonic stem cells and adult progenitor cells (Easwaran et al. 2012). By the combined actions of trithorax (responsible for the activating H3K4me3 modification) and polycomb (PcG, responsible for the activating H3K27me3 modification) mediated bivalent chromatin marks (simultaneous presence of both H3K4me3 and H3K27me3 in the same nucleosome) in embryonic and adult stem cells, these genes are maintained at poised expression state amenable to repression or activation upon normal differentiation cues (Bernstein et al. 2006) (Fig. 1b). Thus many of the genes that are methylated in cancers are already in silent/low-expression state in the corresponding normal cells (Easwaran et al. 2012). This observation has called into question the actual benefit of promoter CGI methylation in cancer development (O'Donnell et al. 2014). However, as mentioned above, it is very important to recognize that many of the cancer-specific promoter CGI methylated genes are regulators of development and differentiation, and expressed in response to normal differentiation cues (Mikkelsen et al. 2007; Squazzo et al. 2006). The preponderance of promoter DNA methylation in such genes for which low but inducible expression is important for the balance between stem cell maintenance and differentiation is an important class of TSGs which does not fit the classic TSG definition but fits the above mentioned refined "continuum" model of TSGs. When silenced by DNA methylation, these genes may not be appropriately reactivated and thus hamper proper differentiation (Fig. 1b). Their silencing by promoter methylation-mediated non-inducible, low expression state may help in the stem-like state of cancers. Key examples of such genes are the developmental and differentiation regulators, such as *SFRPs* and *SOX17*, that are almost never mutated but are hypermethylated in many cancers and may act as TSGs by antagonizing Wnt-signaling. We recently showed that silencing these genes by themselves are not the driver events, but their inactivation causes the necessary and sufficient defects in differentiation that creates a permissive state for tumorigenesis by cancer driver mutations, such as oncogenic BRAF (Tao et al. 2019). Increased stem cell maintenance and decreased or abnormal differentiation potential of cancers is thus an important outcome of such abnormal epigenetic silencing. Detailed characterization of the function of CGI methylated genes during tumorigenesis will provide insights into early tumorigenesis and potential of using these as early biomarkers of cancer promoting defects in normal differentiation. A key challenge in understanding the various roles of these genes in tumor development will be to use an array of experimental approaches, involving in vitro, ex vivo and mouse models to characterize the roles of these genes. Further, DNA methylation alterations track with the type of driver mutation in a cancer, indicating that roles of these genes should be analyzed in the context of the key driver mutations and the tumor signaling pathways involved.

## 3.3 Structurally Ordered Genomic Domains and Distal Regulatory Elements

The human genome is organized into modular domains called "topologically associating domains" (TADs) delineated by discrete boundaries of insulator elements. TADs, which range in the megabase-scale, are observed with specialized techniques like HiC that map proximity of stretches of distant genomic elements, thus providing a spatial organization of chromatin in the nucleus (Dixon et al. 2016). Through structured organization of chromatin, distal regulatory elements (enhancers) in the genome interact in a specific and regulated manner to orchestrate regulation of gene expression. Specialized proteins bind to the boundaries (insulator elements) of the TADs and help maintain the discreteness of these domains by acting as borders to limit spreading of chromatin factors and histone modifications. The organization of chromosomes into TADs facilitates physical proximity of chromatin elements that otherwise are localized over considerable distances on the linear DNA. Chromosome looping within the TADs brings promoters and distal regulatory elements in close proximity and their interaction is mediated by transcription factors such as the mediator complex, cohesins and CTCF. Looping itself is governed by insulator binding proteins such as CTCF (Rao et al. 2014). Such 3D-dimensional organization of the chromatin plays important roles in regulating gene expression during differentiation and development as indicated by the fact that tissue-specific genes are frequently in association with distal enhancers in a cell-type specific manner. CRISPR-mediated re-engineering in mice of a chromosomal rearrangement, which in human patient families disrupts a TAD locus associated with limb malformations, severely phenocopies the human condition in mice (Lupianez et al. 2015). Thus the mere disruption of TADs results in profound alteration of gene expression. The underlying DNA sequence in insulator elements at the boundaries of TADs are subject to DNA methylation, which has profound implications on their activity via altering the binding of proteins, such as CTCF, which are essential for maintaining TADs. In cancers, given the genome wide alterations of DNA methylation that can be linked to genome instability and rearrangements, there are profound implications resulting from disruption of TADs, and thereby the regulatory elements, as is being discovered and appreciated in recent years and discussed below.

An important class of the distal regulatory elements is the enhancers making up ~10% of the human genome, which are cell type specific genomic sequences that regulate transcriptional activity of genes that are thousands to million bases away (Jin et al. 2013; Li et al. 2012; Rao et al. 2014). Although enhancers are mainly regulated at the level of TF binding, chromatin composition and histone modifications, and some of these features distinguishes them from gene promoters, (Heintzman et al. 2007; Kim and Shiekhattar 2015; Shlyueva et al. 2014), there is increasing evidence that activity of enhancers is linked to, and may also be regulated by, DNA methylation (Jones 2012). Activity of enhancers is tissue type specific, and importantly CpG methylation in enhancers is also observed to be tissue type specific. In genomewide analyses of tissue type specific methylation patterns, about 26% of the cell type specific DNA methylation sites overlap with putative enhancers, and another 40% of such sites overlap with DNAse I hypersensitive sites which are features of regulatory elements (Ziller et al. 2013). About 90% of hypomethylated regions in colon cancer compared to normal colon contain enhancers (Berman et al. 2011). Importantly, DNA methylation changes at enhancers are better correlated to expression changes of target genes than promoter-CGI methylation (Aran et al. 2013; Leadem et al. 2018). This latter observation is due to the fact that CGI methylation is rare in normal non-neural somatic cells, wherein expression of genes is regulated by chromatin modifications at the CGI promoters and rarely by methylation alteration of promoter CGI (Baylin and Jones 2016; Suzuki and Bird 2008). Majority of enhancers have low density of CpG dinucelotides, which when methylated are associated with absence of the active enhancer histone mark (H3K4me1). Thus it is suggested that DNA methylation state of enhancer can also direct histone modifications (Fig. 2), and play deterministic roles in the activity of enhancers. Unmethylated or partially methylated enhancers can recruit a class of histone methyltransferases that catalyze the H3K4me1 mark, thus marking enhancer regions for poised or active state (Sharifi-Zarchi et al. 2017). Alternately, methylation of CTCF sites in insulator elements can alter chromatin looping, allowing interaction of enhancer elements with target gene promoters (Fig. 2). These studies highlight the importance of understanding DNA methylation alterations in enhancers, in addition to the promoter methylation changes, to understand epigenetic deregulation of gene expression in tumor development.

Since enhancer regions are enriched for transcription factor (TF) binding sites, and are regulated by the TFs, DNA methylation alterations will interfere with TF binding and alter regulation of target genes. In concordance with the global hypomethylation in cancers, large-scale analysis of enhancers across multiple tumor types showed that majority of enhancers undergo hypomethylation in cancers. Analyses of array based DNA methylation data, which limits analysis to only a

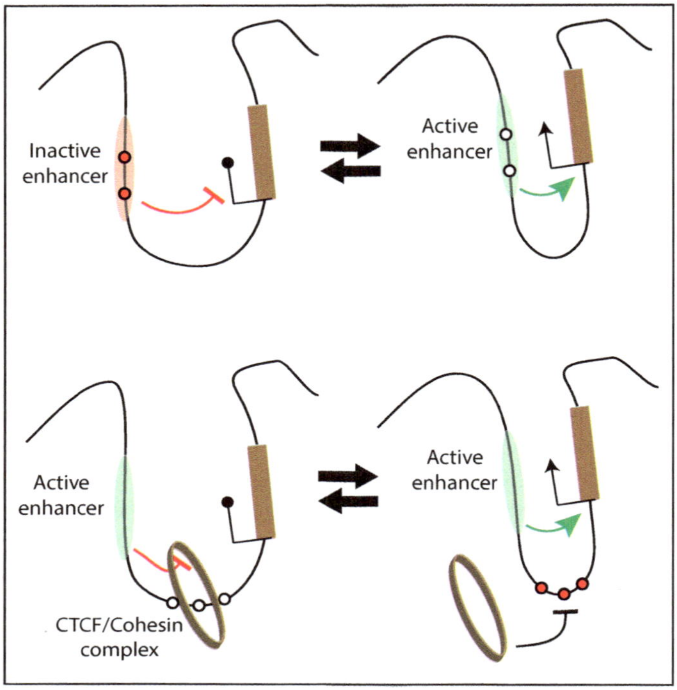


**Fig. 2** Distal regulatory elements are subject to regulation by DNA methylation. Methylation alterations in non-promoter regions, aberrant hypermethylation or hypomethylation of enhancers, can cause deregulated activation or inactivation of genes. *Top panel:* Methylation of CpG dinucleotides in enhancer elements causes inactivation of enhancers and prevents activation of target genes, which may be in poised expression state. Hypomethylation of enhancer causes recruitment of DNA binding factors and activating histone modifications, which allows target gene activation. *Bottom panel:* Methylation of CpG sites in insulator regions bound by CTCF/cohesin complexes prevents trans-activation of target gene by enhancers. Altered methylation of the CpG sites at the insulator elements causes interaction of distal enhancers and genes, causing activation of genes

subset of all putative enhancers for which probes are available, showed ~6000 and 1200 enhancers undergo hypo- or hypermethylation, respectively, impacting thousands of genes across various tumors (Yao et al. 2015). As would be expected, hypomethylation of an enhancer is correlated with upregulation of the potential target genes while hypermethylation is associated with downregulation. Interestingly, known tumor suppressor genes (like *CDKN1A*, *SPRY2*) were downregulated in association with the corresponding enhancer methylation while various oncogenes (like *MYC*, *TERT)* were upregulated in association with hypomethylated enhancers. In the context of the PMDs described earlier, enhancer deregulation due to DNA hypomethylation might be pervasive during successive rounds of cell divisions in cancer cells, as well as normal aging cells.

As with enhancers, CTCF sites that mark insulator elements controlling chromosomal looping are under tight regulation by DNA methylation. CTCF binding sites contain CpG-sites that when methylated abrogate binding of CTCF. Thus, considering the genomewide changes in DNA methylation in cancers, it then only remains a matter of exploration of these regions to understand the extent of their disruption, the effect on the TADs and the ensuing gene expression changes. A key study showed how in IDH mutant cases of gliomas, which are associated with increased CpG-methylation, a CTCF-binding site gets aberrantly methylated disrupting binding of CTCF resulting in abrogation of the tight insulator function. As a result, the oncogene *PDGFRA* is able to interact with an enhancer located ~900 kb away in a neighboring TAD resulting in increased *PDGFRA* expression. Importantly, this enhancer otherwise does not interact with *PDGFRA* in normal and gliomas without the IDH mutation (Flavahan et al. 2016). Such abnormal activation of oncogenes, and silencing of tumor suppressor genes, will probably prove to be a theme across various cancer types because of the large-scale DNA methylation deregulation.

## 4 Origin of Cancer Methylation Changes from Normal Physiological Processes

The global methylation changes described above in cancers are observed in multiple normal physiological processes involving continued mitotic cycling of somatic cells, such as during tissue regeneration associated with aging, and other processes like inflammation, immortalization and senescence. Continuous cycling of cells during aging involves both global hypomethylation and promoter hypermethylation (Issa 2014). One of the earliest observations in the field is that de novo promoter methylation of ER and IGF2 genes occurs in the colonic mucosa of aged individuals, and methylation of these progressively increased in cancers (Issa 2014). Thereon, multiple gene promoters methylated in cancers have been shown to be also methylated in aging. Global analyses in multiple tissues in mice and human have demonstrated profound genome-wide changes in the DNA methylation levels during aging (Christensen et al. 2009; Hannum et al. 2013; Heyn et al. 2012; Maegawa et al. 2010, 2014, 2017; Rakyan et al. 2010; Teschendorff et al. 2010). Since promoter methylation events in cancers are very early and frequent events, similar methylation patterns observed during aging have suggested a potential for transition of aged cells with altered methylation patterns to tumor initiating cells (Fig. 3). Multiple aspects of age-related methylation, beyond the genes that are specifically affected, compels understanding age-related methylation important for understanding the etiology of cancer. Comparison of methylation alterations occurring during in vitro immortalization and transformation of cells by serial expression of hTERT, SV40 large T and HRAS in relation to cells undergoing senescence showed that the transformation-associated methylation arise stochastically, while senescence-associated methylation arise in a defined and programmatic manner. Importantly, genes that get stochastically methylated during transformation, compared to those specifically methylated during senescence, are more likely to be also methylated in

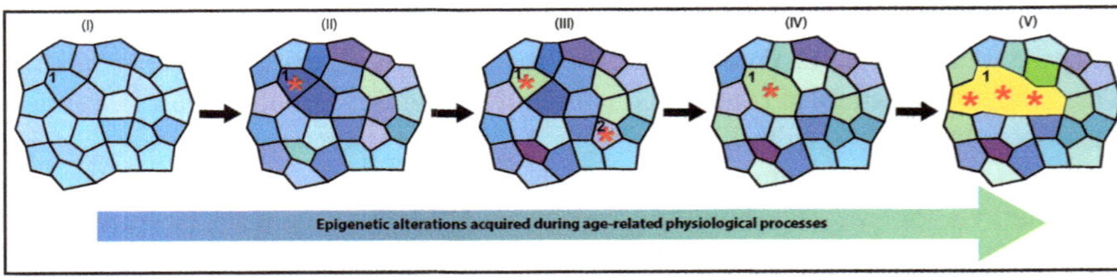


**Fig. 3** Progressive epigenetic changes during aging provide permissive state for oncogenic driver effect of pre-existing or acquired mutations. The figure shows schematic of regions in tissues/cell subpopulations harbor related DNA methylation patterns (colored polygons), which diverge with aging. Most likely this methylation heterogeneity is maintained in different long-living stem cells that will give rise to subclonal populations carrying similar epigenetic marks. Oncogenic mutations (region 1) is initially not tumorigenic (I, II) unless a permissive epigenetic state is achieved (III). Mutations occurring in cells with non-permissive epigenetic background (region 2, III) are not transformative and lost. Epigenetic states keep diverging further, and continue to evolve in the tumor cells (IV, V)

aging tissues, suggesting that the methylation changes observed in cancers most likely derive from normally dividing, aging cells (Xie et al. 2018).

Since stem-cells are the long-lived components of aging tissues, alterations in maintenance of DNA methylation are expected to be fixed, and further evolve in the stem cells during continuous prolonged divisions during tissue replenishment and aging (Fig. 3). Compelling evidence for the origin of age-related methylation alterations arising in long-lived continuously dividing stem cells is the contrasting age-related methylation changes in the continuously dividing colon stem cell population vs. the rarely dividing stem cells in the hair follicles (Shibata 2009). Whereas age-related progressive methylation changes are observed in the colonic stem cells, hair follicle stem cells show no such changes during the lifetime of an individual. The observation that many of the genes methylated during aging belong to the PcG targets also methylated in cancer (Rakyan et al. 2010; Teschendorff et al. 2010) may innocuously suggest that such methylation changes in both cancer and aging is just a mere consequence of an inherent bias for these genes to be methylated during multiple mitotic divisions. Although this could be true for many genes by nature of them being PcG marked, the same sets of genes epigenetically altered in aging and cancers, potentially impacting tumor suppressors and stem cell pathways (Easwaran et al. 2012; West et al. 2013), additionally suggest that age-related methylation of these genes may increase cell fitness allowing for clonal expansion and neoplasms to develop (Fig. 3). In regard to this, methylation of CpG residues is highly polymorphic in multiple primary tissues, which increase in tumors and in primary cells in culture (Landan et al. 2012). The diversity in methylation patterns arising from polymorphic methylation is suggested to provide the necessary variation for Darwinian selection of fitter clones (Hansen et al. 2011) (Fig. 3). For example, DNA methylation occurs at key colorectal cancer (CRC) and adenoma related control genes, such as *APC*, *AXIN2*, *DKK1*, *HPP1*, *N33*, *CDKN2A/p16*, *SFRP1*, *SFRP2* and *SFRP4*, during ageing (Belshaw et al. 2008). Some of these genes are otherwise mutated, a key example being *CDKN2A/p16*, which is a critical gene that prevents oncogene-induced transformation. Preexisting epigenetic silencing of multiple such genes in the same cell, which may occur due to stochastically arising polymorphic methylation patterns during aging, will sufficiently block cell cycle checkpoint and activate stem-cell pathways allowing oncogenic-driver effects. Recently we showed that simultaneous inactivation of genes subject to epigenetic silencing in colon cancers, namely *CDX2*, *SFRP4*, *SOX17* and *CDKN2A*, sufficiently creates the permissive state for oncogenic-BRAF induced transformation of colon derived organoid cultures (Tao et al. 2019).

How cancer-related methylation changes come about to be is tightly linked to understanding tumor initiation. Various age-related physiological processes, like chronic inflammation and carcinogen induced genomic stress, acutely trigger epigenetic changes observed in cancers (Asada et al. 2015; Blanco et al. 2007; Hahn et al. 2008; Niwa and Ushijima 2010; O'Hagan et al. 2011; Vaz et al. 2017). As discussed above, age-related methylation of CGI promoters and other genomic elements (both hypo- and hypermethylation) will help in the initial stages of tumorigenesis, and these methylation patterns may get selected and further expanded during continuous cell divisions (Fig. 3). In the context of the current framework of oncogenesis, sequential occurrence of mutations allow expansion of fitter cells by both neutral evolution and selection causing genetically heterogeneous tumors (McGranahan and Swanton 2017). Undoubtedly random somatic mutations that accumulate during aging are central drivers in this framework of tumorigenesis. However, substantial numbers of somatic mutations in mice occur during the growth phases early in life, and the rate of mutation accumulation slows down once stem cell divisions decrease and as organs enter maintenance mode (Vijg et al. 2005) [discussed in Rozhok and DeGregori (2016)]. Importantly cancer mutations precede tumor incidence by years to decades (Brucher and Jamall 2016; Desai et al. 2018; Forsberg et al. 2013; Mori et al. 2002; Vogelstein et al. 2013), whereas the latter increases exponentially with age. This is contrasted for methylation alterations that continue to deviate and accumulate during aging (Fraga et al. 2005; Hannum et al. 2013). Thus in the current framework of oncogenesis, a key question regarding the age-related cancer risk is what non-genetic factors allow the impact of driver mutations to unfold as a function of age. Progressive age associated epigenetic modifications may be one of the key factors in this (Fig 3), and for most age-related cancer incidences, methylation patterns observed in cancers may thus originate from subpopulations of aging cells that carry epigenetic alterations that creates a permissive state for tumor initiation (Fig. 3).

## 5 Molecular Mechanisms of Methylation Patterning

Establishment and maintenance of DNA methylation relies on combined action of the three major DNA methyltransferase, DNMT1, DNMT3A and B. The latter two enzymes have been proposed to mainly play roles in de novo DNA methylation. During embryonic development in mice, Dnmt3a is implicated in establishing methylation at imprinted genes discriminating genes by parent of origin (Kaneda et al. 2004; Okano et al. 1999) while Dnmt3b is involved in methylation of pericentromric repeats (Okano et al. 1999; Xu et al. 1999). The de novo methylation activity of Dnmt3 enzymes is directed to defined chromatin regions by specialized protein sequences in its N-terminal domain. DNMT3A, and possibly DNMT3B too, exist in an auto-inhibitory inactive state that specifically is activated by direct interaction with H3 histone tail unmethylated at lysine-4 (H3K4me0) (Guo et al. 2015). Various active chromatin modifications, including H3K4me3, inhibit interaction of DNMT3 with the H3-tail. In CGI promoters, the H3K4me3 mark is enriched by targeting the H3K4 methyltransferase SET1 complex (MLL1/2) via specific binding of the CXXC-domain containing CFP1 to unmethylated CpGs (Baubec and Schubeler 2014; Clouaire et al. 2014; Thomson et al. 2010). Hence the underlying CpG density of CGI promoters itself attracts H3K4me3 marks, which in turn antagonizes methylation by the DNMT3 enzymes, thus maintaining CGI promoters in an unmethylated state. In contrast to the DNMT3-chromatin interactions ensuring methylation free zone at CGI, the PWWP domain of DNMT3 directs the enzyme to H3K36me2/3 regions of the chromatin, which corresponds to gene bodies and heterochromatin (Dhayalan et al. 2010; Zhang et al. 2010). Further, DNMT3A/B interaction with HP1 and G9A recruits it to H3K9me3 residues in pericentric heterochromatin (Lehnertz et al. 2003) and gene promoters (Epsztejn-Litman et al. 2008) respectively. Additional control of the DNMT3 activity towards unmethylated DNA is imposed by secluding DNMT3 to the methylated DNA/chromatin in heterochromatic fraction, and degradation of free-floating enzymes (Jeong et al. 2009; Sharma et al. 2011). In somatic cells, where the expression of DNMT3 is markedly reduced, via the above mechanisms of recruitment to non-H3K4me3-containing, CpG methylated chromatin, the activity of DNMT3 enzymes is restrained to already methylated regions where they cooperate with DNMT1 in maintaining the methylation patterns (Jeltsch and Jurkowska 2014; Jones and Liang 2009; Sharma et al. 2011). The inherent ability of DNMT3 enzymes to read the 'histone code' thus seems to be a prominent mechanism in establishing methylation patterns during development. The established methylation patterns are then maintained during successive rounds of replication mainly by DNMT1, but also by the DNMT3 enzymes by virtue of their affinity for methylated CpG containing nucleosomes (Jones and Liang 2009).

In line with the above paradigm for establishment of CpG methylation patterns by the histone code, the underlying DNA sequence has an important role in determining if regions of the genome will or will not be methylated (Lienert et al. 2011; Stadler et al. 2011). This role of the DNA sequence may precede or work in parallel with the histone code. For example, cis-acting sequences (~700–1000 bp) that have affinity to DNA binding transcription factors protect DNA from methylation independent of transcriptional activity or CpG density (Lienert et al. 2011). Importantly, the cis-acting sequences containing TF binding sites can protect exogenous DNA from CpG methylation, as well as can cause demethylation of exogenously introduced methylated DNA. In this model, TF binding per se may sterically hinder access of DNA methyltransferases to DNA, or more likely the TF could mediate recruitment of the MLL class of histone methyltransferases leading to H3K4me3, which as described above will prevent de novo CpG methylation (Demers et al. 2007; Rao and Dou 2015). Further, as described earlier, silencing of CGI promoters during normal development and differentiation is mediated by H3K27me3 mark, which is regulated by the polycomb repressive complex (PRC) 1 and 2. Presence of the H3K27me3 mark is normally anti-correlated with DNA methylation in normal and cancer cells (Easwaran et al. 2012; Kouzarides 2007; Sen et al. 2016). Accordingly the polycomb mark has been linked to preventing DNA from methylation. A component of the PRC1 complex, FBXL10/KDM2B, containing the CXXC domain that can bind to unmethylated CpG-rich sequence, plays the primary role in this anti-correlation between DNA methylation and H3K27me3. Specifically, in ES cells FBXL10/KDM2B occupies ~90% of all promoter-associated CGI, and an equal proportion of CGI promoters that are silenced by PRC1/2 complexes (Farcas et al. 2012; He et al. 2013; Wu et al. 2013). FBXL10 prevents CpG methylation by binding to the PRC regulated CGI promoters, as loss of FBXL10 results in rapid methylation of only those promoters that are polycomb regulated (Boulard et al. 2015). It is important to note that, as introduced earlier, polycomb-regulated genes are more often methylated in cancers. Deregulation of FBXL10 mechanics in cancers, for example it is mutated in diffuse B-cell lymphomas (Pasqualucci et al. 2011), may cause aberrant promoter CGI methylation for some of the genes during aging and tumorigenesis.

In summary, the cellular transcription program established during development and differentiation, as defined by the expressed repertoire of transcription and DNA binding factors, sufficiently is able to establish and maintain the DNA methylation epigenetic program (Burger et al. 2013; Stadler et al. 2011). In this model, absence of DNA binding factors to cis-elements may trigger DNA methylation, and presence will protect from DNA methylation. The specific molecular details of the dynamics of TF binding, histone modifications and recruitment of DNMTs in mediating methylation patterns still needs elucidation. A noteworthy aspect from the above discussion is that modes of both gene activation and silencing of promoter CGIs, viz. by H3K4me3 and H3K27me3 respectively have central roles in keeping DNA methylation at bay. And the evidence points that there is a concerted role of underlying DNA sequence (TF binding sites and unmethylated CpG dense regions) directed histone code in establishing the methylation patterns. Although mechanics of the CpG methylation patterning is better worked out for the CGI promoters, the remainder of genome may permit CpG methylation due to lack of such DNA sequences and the specific activating histone marks.

## 6 Road to Cancer Methylome

A key question is how the normal mechanics of DNA methylation establishment and maintenance are perturbed in cancers, and its precursor aging cells. Above discussions indicate that there are layers of molecular deregulation in response to external

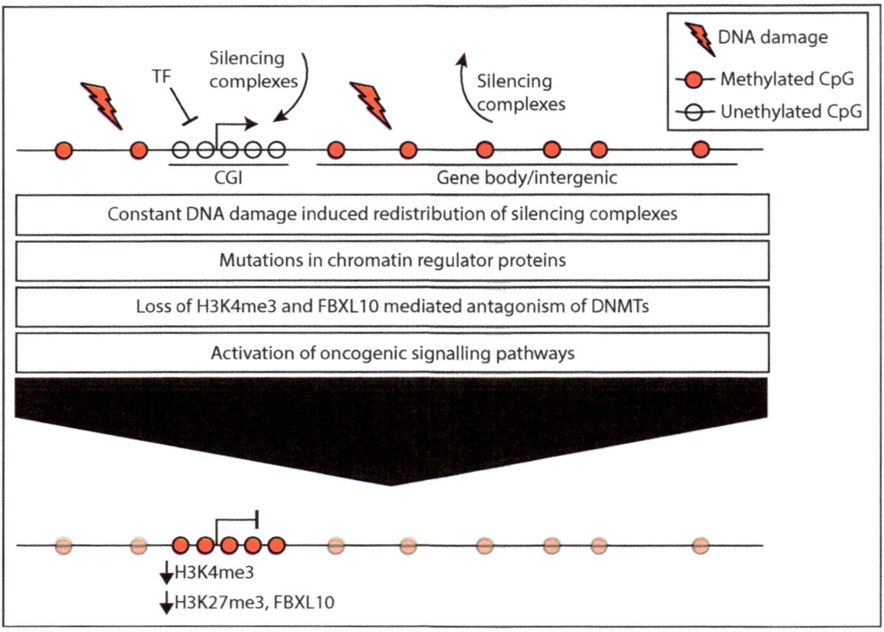


**Fig. 4** Molecular basis for altered methylation in cancers. Multiple layers of deregulation of chromatin proteins mediate methylation alterations during aging and tumorigenesis. These alterations in general result in changes in recruitment of silencing complexes to CGI promoters, and their simultaneous loss from gene body and intergenic regions

stimuli (exposure to carcinogenic genotoxic stressors), microenvironmental chemokine changes (inflammation, aging), cell intrinsic pathway alterations (oncogenic and epigenetic modifier mutations), that in conjunction with selection of methylation profiles lead to the ultimate methylation landscape in cancers. Individual layers in this regulation are as follows (Fig. 4):

(a) DNA damage causes genomewide chromatin changes involving altered recruitment of DNMTs, SIRT1, PRC components, CHD4 (NuRD silencing complex component) to damaged CGI promoters, which helps maintain repressed chromatin at CGI promoters. At the same time, DNMTs and SIRT1 is released from the remainder of the genome, potentially causing hypomethylation. Thus continuous DNA damage caused due to cell intrinsic ROS levels, environmental chemical exposure, inflammatory microenvironment and oncogenic stress will lead to gradual genomewide aberrations in DNA methylation (O'Hagan et al. 2011; Vaz et al. 2017). Particularly, these processes can be key drivers of stochastic methylation changes during aging and inflammation.
(b) Direct loss- and gain-of-function mutations in chromatin regulator proteins will impact the methylome. In this regard, mutations of chromatin proteins are the most common class of cancer mutations, and are observed across various cancers

(Dawson 2017; Shen and Laird 2013). For majority of these mutations, their roles in modulating the methylome need to be investigated. Leading examples of mutations that directly impact methylation are the IDH and TET enzymes. IDH1 and 2 are metabolic enzymes that are not directly involved in chromatin modulation, but their mutation in cancers causes neomorphic enzyme activity that converts α-ketoglutarate (α-KG) to D-2-hydroxyglutarate (D-2-HG). The latter is an oncometabolite that inhibits various Fe(II)/2-oxoglutarate-dependent dioxygenases (Dang et al. 2009; Xu et al. 2011). These include various histone demethylases that protect against DNA methylation by diminishing chromatin marks that attract DNA methylation, and the TET family of enzymes that catalyze DNA demethylation by oxidation of 5-methylcytosine (5mC) to 5-hydroxymethylcytosine (5hmC). This has directly been implicated in increased DNA methylation phenotype, called the CpG-island methylator phenotype, in acute myeloid leukemia (AML) (Figueroa et al. 2010) and gliomas (Noushmehr et al. 2010; Turcan et al. 2012). On the other hand, enzymes involved in DNA demethylation discussed earlier, like *TET2*, are also mutated in various cancers, especially prominent in the hematological malignancies (Scourzic et al. 2015). *TET2* mutations majorly cause methylation of distal regulatory enhancer elements (Rasmussen et al. 2015).

(c) Cancers involve significant expression changes in transcription factor repertoire, a striking example being the almost universally overexpressed MYC factor in various cancers (Dang 2012). This will alter occupancy of TFs at CGI promoters and distal regulatory elements. Altered presence or absence of TFs will cause changes to the histone code thereby causing changes to DNA methylation patterns (Gebhard et al. 2010). Redistribution of DNA binding factors can be directly linked to oncogenic mutations, such as the frequent MEK-ERK pathway activating mutations in RAS/RAF. In ES cells, inhibition of MEK, has been shown to reduce global DNA methylation by reducing DNMT3 enzymes and activating TET1 (Sim et al. 2017). Similarly, growth-factor signalling pathways, such as FGF and Wnt that are activated in many epithelial cancers, have been proposed to induce PRC2-dependent CGI methylation. These latter implications have been made from investigations on the methylation dynamics occurring in extraembryonic tissue of the trophoectoderm lineage, which has revealed de novo promoter CGI methylation of the same developmental regulator genes methylated in cancers (Smith et al. 2017). In these studies, signalling by FGF and WNT increases the promoter CGI methylation. This conservation of CGI methylation patterns in the extraembryonic ectoderm and human cancers indicate strong parallels in activation of oncogenic signalling pathways and PRC-dependent CGI methylation. In concordance with these latter studies, we showed recently that continued culturing, over several months, of colon derived organoids in Wnt-enriched medium used for organoid growth results in an aging-like accumulation of DNA methylation at key developmental regulator promoters, which is necessary for oncogenic-BRAF induced transformation (Tao et al. 2019).

(d) The observations that PcG-regulated promoters are most prone to get methylated indicate a role for the PcG components in mediating recruitment of DNMTs. A key finding supporting this is that loss of FBXL10/KDM2B causes methylation of only those promoters associated with PRC1/2 (Boulard et al. 2015). Whether, and how, loss of FBXL10 from some PcG occupied promoters occurs during tumor development remains to be studied. In this regard, how PRC-mediated recruitment of DNMTs occurs in the context of H3K4me3 and H3K27me3 marked bivalent chromatin promoters is important to understand. Presence of active or stalled RNA Pol II, which occur with H3K4me3, is associated with protection from CGI methylation while presence of H3K27me3 predisposes to CGI methylation (Takeshima et al. 2009). Thus, a sequential step may involve removal of transcriptionally poised state to a PcG regulated promoter, which subsequently may acquire methylation by altered activity of FBXL10.

## 7 Conclusions and Future Directions

Accumulation of DNA methylation alterations occurs during various normal processes, importantly during aging. Increasing evidence suggests that these alterations have a role in predisposing to tumorigenesis. How all the mechanisms suggested above interlace to produce the epigenetic drift during aging, and in cancer, is important to understand the specific means by which various cancer predisposition factors work through modulating the epigenome. Equally important is to understand the role of the epigenetically modified genes during the early steps of tumorigenesis. We predict that development of appropriate markers that can differentiate epigenetically altered cell populations, in response to intrinsic (such as inflammation) and extrinsic (such as environmental exposures) factors, carrying aberrant methylation of functionally relevant genes holds promise in identifying cancer risk states.

## References

Almouzni G, Cedar H (2016) Maintenance of epigenetic information. Cold Spring Harb Perspect Biol 8. pii: a019372

Aran D, Sabato S, Hellman A (2013) DNA methylation of distal regulatory sites characterizes dysregulation of cancer genes. Genome Biol 14:R21

Asada K, Nakajima T, Shimazu T et al (2015) Demonstration of the usefulness of epigenetic cancer risk prediction by a multicentre prospective cohort study. Gut 64:388–396

Baubec T, Schubeler D (2014) Genomic patterns and context specific interpretation of DNA methylation. Curr Opin Genet Dev 25:85–92

Baubec T, Ivanek R, Lienert F et al (2013) Methylation-dependent and -independent genomic targeting principles of the MBD protein family. Cell 153:480–492

Baylin SB, Jones PA (2016) Epigenetic determinants of cancer. Cold Spring Harb Perspect Biol 8. pii: a019505

Belshaw NJ, Elliott GO, Foxall RJ et al (2008) Profiling CpG island field methylation in both morphologically normal and neoplastic human colonic mucosa. Br J Cancer 99:136–142

Berger AH, Knudson AG, Pandolfi PP (2011) A continuum model for tumour suppression. Nature 476:163–169

Bergman Y, Cedar H (2013) DNA methylation dynamics in health and disease. Nat Struct Mol Biol 20:274–281

Berman BP, Weisenberger DJ, Aman JF et al (2011) Regions of focal DNA hypermethylation and long-range hypomethylation in colorectal cancer coincide with nuclear lamina-associated domains. Nat Genet 44:40–46

Bernstein BE, Mikkelsen TS, Xie X et al (2006) A bivalent chromatin structure marks key developmental genes in embryonic stem cells. Cell 125:315–326

Bird A (2007) Perceptions of epigenetics. Nature 447:396–398

Blanco D, Vicent S, Fraga MF et al (2007) Molecular analysis of a multistep lung cancer model induced by chronic inflammation reveals epigenetic regulation of p16 and activation of the DNA damage response pathway. Neoplasia 9:840–852

Boulard M, Edwards JR, Bestor TH (2015) FBXL10 protects Polycomb-bound genes from hypermethylation. Nat Genet 47:479–485

Brucher BL, Jamall IS (2016) Somatic mutation theory – why it's wrong for most cancers. Cell Physiol Biochem 38:1663–1680

Burger L, Gaidatzis D, Schubeler D et al (2013) Identification of active regulatory regions from DNA methylation data. Nucleic Acids Res 41:e155

Chen RZ, Pettersson U, Beard C et al (1998) DNA hypomethylation leads to elevated mutation rates. Nature 395:89–93

Christensen BC, Houseman EA, Marsit CJ et al (2009) Aging and environmental exposures alter tissue-specific DNA methylation dependent upon CpG island context. PLoS Genet 5:e1000602

Chuang LS, Ian HI, Koh TW et al (1997) Human DNA-(cytosine-5) methyltransferase-PCNA complex as a target for p21WAF1. Science 277:1996–2000

Clouaire T, Webb S, Bird A (2014) Cfp1 is required for gene expression-dependent H3K4 trimethylation and H3K9 acetylation in embryonic stem cells. Genome Biol 15:451

Dang CV (2012) MYC on the path to cancer. Cell 149:22–35

Dang L, White DW, Gross S et al (2009) Cancer-associated IDH1 mutations produce 2-hydroxyglutarate. Nature 462:739–744

Dawson MA (2017) The cancer epigenome: concepts, challenges, and therapeutic opportunities. Science 355:1147–1152

Deaton AM, Bird A (2011) CpG islands and the regulation of transcription. Genes Dev 25:1010–1022

Demers C, Chaturvedi CP, Ranish JA et al (2007) Activator-mediated recruitment of the MLL2 methyltransferase complex to the beta-globin locus. Mol Cell 27:573–584

Desai P, Mencia-Trinchant N, Savenkov O et al (2018) Somatic mutations precede acute myeloid leukemia years before diagnosis. Nat Med 24:1015–1023

Dhayalan A, Rajavelu A, Rathert P et al (2010) The Dnmt3a PWWP domain reads histone 3 lysine 36 trimethylation and guides DNA methylation. J Biol Chem 285:26114–26120

Dixon JR, Gorkin DU, Ren B (2016) Chromatin domains: the unit of chromosome organization. Mol Cell 62:668–680

Easwaran H, Schermelleh L, Leonhardt H et al (2004) Replication-independent chromatin loading of Dnmt1 during G2 and M phases. EMBO Rep 5:1181–1186

Easwaran H, Leonhardt H, Cardoso M (2005) Cell cycle markers for live cell analyses. Cell Cycle 4:453–455

Easwaran H, Johnstone SE, Van Neste L et al (2012) A DNA hypermethylation module for the stem/progenitor cell signature of cancer. Genome Res 22:837–849

Eden A, Gaudet F, Waghmare A et al (2003) Chromosomal instability and tumors promoted by DNA hypomethylation. Science 300:455

Ehrlich M, Lacey M (2013) DNA hypomethylation and hemimethylation in cancer. Adv Exp Med Biol 754:31–56

Epsztejn-Litman S, Feldman N, Abu-Remaileh M et al (2008) De novo DNA methylation promoted by G9a prevents reprogramming of embryonically silenced genes. Nat Struct Mol Biol 15:1176–1183

Farcas AM, Blackledge NP, Sudbery I et al (2012) KDM2B links the Polycomb Repressive Complex 1 (PRC1) to recognition of CpG islands. elife 1:e00205

Figueroa ME, Abdel-Wahab O, Lu C et al (2010) Leukemic IDH1 and IDH2 mutations result in a hypermethylation phenotype, disrupt TET2 function, and impair hematopoietic differentiation. Cancer Cell 18:553–567

Flavahan WA, Drier Y, Liau BB et al (2016) Insulator dysfunction and oncogene activation in IDH mutant gliomas. Nature 529:110–114

Forsberg LA, Absher D, Dumanski JP (2013) Republished: Non-heritable genetics of human disease: spotlight on post-zygotic genetic variation acquired during lifetime. Postgrad Med J 89:417–426

Fraga MF, Ballestar E, Paz MF et al (2005) Epigenetic differences arise during the lifetime of monozygotic twins. Proc Natl Acad Sci U S A 102:10604–10609

Gao Y, Teschendorff AE (2017) Epigenetic and genetic deregulation in cancer target distinct signaling pathway domains. Nucleic Acids Res 45:583–596

Gaudet F, Hodgson JG, Eden A et al (2003) Induction of tumors in mice by genomic hypomethylation. Science 300:489–492

Gebhard C, Benner C, Ehrich M et al (2010) General transcription factor binding at CpG islands in normal cells correlates with resistance to de novo DNA methylation in cancer cells. Cancer Res 70:1398–1407

Guo X, Wang L, Li J et al (2015) Structural insight into autoinhibition and histone H3-induced activation of DNMT3A. Nature 517:640–644

Hahn MA, Hahn T, Lee DH et al (2008) Methylation of polycomb target genes in intestinal cancer is mediated by inflammation. Cancer Res 68:10280–10289

Han H, Cortez CC, Yang X et al (2011) DNA methylation directly silences genes with non-CpG island promoters and establishes a nucleosome occupied promoter. Hum Mol Genet 20:4299–4310

Hannum G, Guinney J, Zhao L et al (2013) Genome-wide methylation profiles reveal quantitative views of human aging rates. Mol Cell 49:359–367

Hansen KD, Timp W, Bravo HC et al (2011) Increased methylation variation in epigenetic domains across cancer types. Nat Genet 43:768–775

Hartung T, Zhang L, Kanwar R et al (2012) Diametrically opposite methylome-transcriptome relationships in high- and low-CpG promoter genes in postmitotic neural rat tissue. Epigenetics 7:421–428

He J, Shen L, Wan M et al (2013) Kdm2b maintains murine embryonic stem cell status by recruiting PRC1 complex to CpG islands of developmental genes. Nat Cell Biol 15:373–384

Heintzman ND, Stuart RK, Hon G et al (2007) Distinct and predictive chromatin signatures of transcriptional promoters and enhancers in the human genome. Nat Genet 39:311–318

Herman JG, Baylin SB (2003) Gene silencing in cancer in association with promoter hypermethylation. N Engl J Med 349:2042–2054

Heyn H, Li N, Ferreira HJ et al (2012) Distinct DNA methylomes of newborns and centenarians. Proc Natl Acad Sci U S A 109:10522–10527

Holm TM, Jackson-Grusby L, Brambrink T et al (2005) Global loss of imprinting leads to widespread tumorigenesis in adult mice. Cancer Cell 8:275–285

Hur K, Cejas P, Feliu J et al (2014) Hypomethylation of long interspersed nuclear element-1 (LINE-1) leads to activation of proto-oncogenes in human colorectal cancer metastasis. Gut 63:635–646

Illingworth RS, Bird AP (2009) CpG islands – a rough guide. FEBS Lett 583:1713–1720

Issa JP (2014) Aging and epigenetic drift: a vicious cycle. J Clin Invest 124:24–29

Jackson DA, Pombo A (1998) Replicon clusters are stable units of chromosome structure: evidence that nuclear organization contributes to the efficient activation and propagation of S phase in human cells. J Cell Biol 140:1285–1295
Jeltsch A, Jurkowska RZ (2014) New concepts in DNA methylation. Trends Biochem Sci 39:310–318
Jeong S, Liang G, Sharma S et al (2009) Selective anchoring of DNA methyltransferases 3A and 3B to nucleosomes containing methylated DNA. Mol Cell Biol 29:5366–5376
Jin F, Li Y, Dixon JR et al (2013) A high-resolution map of the three-dimensional chromatin interactome in human cells. Nature 503:290–294
Jones PA (2012) Functions of DNA methylation: islands, start sites, gene bodies and beyond. Nat Rev Genet 13:484–492
Jones PA, Liang G (2009) Rethinking how DNA methylation patterns are maintained. Nat Rev Genet 10:805–811
Kaneda M, Okano M, Hata K et al (2004) Essential role for de novo DNA methyltransferase Dnmt3a in paternal and maternal imprinting. Nature 429:900–903
Kim TK, Shiekhattar R (2015) Architectural and functional commonalities between enhancers and promoters. Cell 162:948–959
Knudson AG (2001) Two genetic hits (more or less) to cancer. Nat Rev Cancer 1:157–162
Kouzarides T (2007) Chromatin modifications and their function. Cell 128:693–705
Landan G, Cohen NM, Mukamel Z et al (2012) Epigenetic polymorphism and the stochastic formation of differentially methylated regions in normal and cancerous tissues. Nat Genet 44:1207–1214
Lawrence MS, Stojanov P, Polak P et al (2013) Mutational heterogeneity in cancer and the search for new cancer-associated genes. Nature 499:214–218
Leadem BR, Kagiampakis I, Wilson C et al (2018) A KDM5 inhibitor increases global H3K4 trimethylation occupancy and enhances the biological efficacy of 5-Aza-2'-deoxycytidine. Cancer Res 78:1127–1139
Lehnertz B, Ueda Y, Derijck AA et al (2003) Suv39h-mediated histone h3 lysine 9 methylation directs DNA methylation to major satellite repeats at pericentric heterochromatin. Curr Biol 13:1192–1200
Leonhardt H, Page AW, Weier HU et al (1992) A targeting sequence directs DNA methyltransferase to sites of DNA replication in mammalian nuclei. Cell 71:865–873
Li G, Ruan X, Auerbach RK et al (2012) Extensive promoter-centered chromatin interactions provide a topological basis for transcription regulation. Cell 148:84–98
Lienert F, Wirbelauer C, Som I et al (2011) Identification of genetic elements that autonomously determine DNA methylation states. Nat Genet 43:1091–1097
Lister R, Pelizzola M, Kida YS et al (2011) Hotspots of aberrant epigenomic reprogramming in human induced pluripotent stem cells. Nature 471:68–73
Lupianez DG, Kraft K, Heinrich V et al (2015) Disruptions of topological chromatin domains cause pathogenic rewiring of gene-enhancer interactions. Cell 161:1012–1025
Mack SC, Witt H, Piro RM et al (2014) Epigenomic alterations define lethal CIMP-positive ependymomas of infancy. Nature 506:445–450
Maegawa S, Hinkal G, Kim HS et al (2010) Widespread and tissue specific age-related DNA methylation changes in mice. Genome Res 20:332–340
Maegawa S, Gough SM, Watanabe-Okochi N et al (2014) Age-related epigenetic drift in the pathogenesis of MDS and AML. Genome Res 24:580–591
Maegawa S, Lu Y, Tahara T et al (2017) Caloric restriction delays age-related methylation drift. Nat Commun 8:539
Mazor T, Pankov A, Johnson BE et al (2015) DNA methylation and somatic mutations converge on the cell cycle and define similar evolutionary histories in brain tumors. Cancer Cell 28:307–317
McGranahan N, Swanton C (2017) Clonal heterogeneity and tumor evolution: past, present, and the future. Cell 168:613–628

McKenna ES, Roberts CW (2009) Epigenetics and cancer without genomic instability. Cell Cycle 8:23–26
Mikkelsen TS, Ku M, Jaffe DB et al (2007) Genome-wide maps of chromatin state in pluripotent and lineage-committed cells. Nature 448:553–560
Mori H, Colman SM, Xiao Z et al (2002) Chromosome translocations and covert leukemic clones are generated during normal fetal development. Proc Natl Acad Sci U S A 99:8242–8247
Niwa T, Ushijima T (2010) Induction of epigenetic alterations by chronic inflammation and its significance on carcinogenesis. Adv Genet 71:41–56
Noushmehr H, Weisenberger DJ, Diefes K et al (2010) Identification of a CpG island methylator phenotype that defines a distinct subgroup of glioma. Cancer Cell 17:510–522
O'Donnell AH, Edwards JR, Rollins RA et al (2014) Methylation abnormalities in mammary carcinoma: the methylation suicide hypothesis. J Cancer Ther 5:1311–1324
O'Hagan HM, Wang W, Sen S et al (2011) Oxidative damage targets complexes containing DNA methyltransferases, SIRT1, and polycomb members to promoter CpG Islands. Cancer Cell 20:606–619
Ohm JE, McGarvey KM, Yu X et al (2007) A stem cell-like chromatin pattern may predispose tumor suppressor genes to DNA hypermethylation and heritable silencing. Nat Genet 39:237–242
Okano M, Bell DW, Haber DA et al (1999) DNA methyltransferases Dnmt3a and Dnmt3b are essential for de novo methylation and mammalian development. Cell 99:247–257
Pasqualucci L, Trifonov V, Fabbri G et al (2011) Analysis of the coding genome of diffuse large B-cell lymphoma. Nat Genet 43:830–837
Pradhan S, Bacolla A, Wells RD et al (1999) Recombinant human DNA (cytosine-5) methyltransferase. I. Expression, purification, and comparison of de novo and maintenance methylation. J Biol Chem 274:33002–33010
Rakyan VK, Down TA, Maslau S et al (2010) Human aging-associated DNA hypermethylation occurs preferentially at bivalent chromatin domains. Genome Res 20:434–439
Rao RC, Dou Y (2015) Hijacked in cancer: the KMT2 (MLL) family of methyltransferases. Nat Rev Cancer 15:334–346
Rao SS, Huntley MH, Durand NC et al (2014) A 3D map of the human genome at kilobase resolution reveals principles of chromatin looping. Cell 159:1665–1680
Rasmussen KD, Jia G, Johansen JV et al (2015) Loss of TET2 in hematopoietic cells leads to DNA hypermethylation of active enhancers and induction of leukemogenesis. Genes Dev 29:910–922
Rozhok AI, DeGregori J (2016) The evolution of lifespan and age-dependent cancer risk. Trends Cancer 2:552–560
Schermelleh L, Haemmer A, Spada F et al (2007) Dynamics of Dnmt1 interaction with the replication machinery and its role in postreplicative maintenance of DNA methylation. Nucleic Acids Res 35:4301–4312
Schlesinger Y, Straussman R, Keshet I et al (2007) Polycomb-mediated methylation on Lys27 of histone H3 pre-marks genes for de novo methylation in cancer. Nat Genet 39:232–236
Schuebel KE, Chen W, Cope L et al (2007) Comparing the DNA hypermethylome with gene mutations in human colorectal cancer. PLoS Genet 3:1709–1723
Scourzic L, Mouly E, Bernard OA (2015) TET proteins and the control of cytosine demethylation in cancer. Genome Med 7:9
Sen S, Block KF, Pasini A et al (2016) Genome-wide positioning of bivalent mononucleosomes. BMC Med Genet 9:60
Sharif J, Muto M, Takebayashi S et al (2007) The SRA protein Np95 mediates epigenetic inheritance by recruiting Dnmt1 to methylated DNA. Nature 450:908–912
Sharifi-Zarchi A, Gerovska D, Adachi K et al (2017) DNA methylation regulates discrimination of enhancers from promoters through a H3K4me1-H3K4me3 seesaw mechanism. BMC Genomics 18:964

Sharma S, De Carvalho DD, Jeong S et al (2011) Nucleosomes containing methylated DNA stabilize DNA methyltransferases 3A/3B and ensure faithful epigenetic inheritance. PLoS Genet 7:e1001286
Shen H, Laird PW (2013) Interplay between the cancer genome and epigenome. Cell 153:38–55
Shibata D (2009) Inferring human stem cell behaviour from epigenetic drift. J Pathol 217:199–205
Shlyueva D, Stampfel G, Stark A (2014) Transcriptional enhancers: from properties to genome-wide predictions. Nat Rev Genet 15:272–286
Sim YJ, Kim MS, Nayfeh A et al (2017) 2i maintains a naive ground state in ESCs through two distinct epigenetic mechanisms. Stem Cell Reports 8:1312–1328
Smith ZD, Shi J, Gu H et al (2017) Epigenetic restriction of extraembryonic lineages mirrors the somatic transition to cancer. Nature 549:543–547
Squazzo SL, O'Geen H, Komashko VM et al (2006) Suz12 binds to silenced regions of the genome in a cell-type-specific manner. Genome Res 16:890–900
Stadler MB, Murr R, Burger L et al (2011) DNA-binding factors shape the mouse methylome at distal regulatory regions. Nature 480:490–495
Suzuki MM, Bird A (2008) DNA methylation landscapes: provocative insights from epigenomics. Nat Rev Genet 9:465–476
Takai D, Jones PA (2002) Comprehensive analysis of CpG islands in human chromosomes 21 and 22. Proc Natl Acad Sci U S A 99:3740–3745
Takeshima H, Yamashita S, Shimazu T et al (2009) The presence of RNA polymerase II, active or stalled, predicts epigenetic fate of promoter CpG islands. Genome Res 19:1974–1982
Tao Y, Kang B, Petkovich DA et al (2019) Aging-like spontaneous epigenetic silencing facilitates Wnt activation, stemness and BrafV600E-induced tumorigenesis. Cancer Cell 35(2):315–328
Teschendorff AE, Menon U, Gentry-Maharaj A et al (2010) Age-dependent DNA methylation of genes that are suppressed in stem cells is a hallmark of cancer. Genome Res 20:440–446
Thomson JP, Skene PJ, Selfridge J et al (2010) CpG islands influence chromatin structure via the CpG-binding protein Cfp1. Nature 464:1082–1086
Turcan S, Rohle D, Goenka A et al (2012) IDH1 mutation is sufficient to establish the glioma hypermethylator phenotype. Nature 483:479–483
Van Tongelen A, Loriot A, De Smet C (2017) Oncogenic roles of DNA hypomethylation through the activation of cancer-germline genes. Cancer Lett 396:130–137
Vaz M, Hwang SY, Kagiampakis I et al (2017) Chronic cigarette smoke-induced epigenomic changes precede sensitization of bronchial epithelial cells to single-step transformation by KRAS mutations. Cancer Cell 32:360–376 e366
Vijg J, Busuttil RA, Bahar R et al (2005) Aging and genome maintenance. Ann N Y Acad Sci 1055:35–47
Vogelstein B, Papadopoulos N, Velculescu VE et al (2013) Cancer genome landscapes. Science 339:1546–1558
Weber M, Hellmann I, Stadler MB et al (2007) Distribution, silencing potential and evolutionary impact of promoter DNA methylation in the human genome. Nat Genet 39:457–466
West J, Beck S, Wang X et al (2013) An integrative network algorithm identifies age-associated differential methylation interactome hotspots targeting stem-cell differentiation pathways. Sci Rep 3:1630
Widschwendter M, Fiegl H, Egle D et al (2007) Epigenetic stem cell signature in cancer. Nat Genet 39:157–158
Wrangle J, Wang W, Koch A et al (2013) Alterations of immune response of non-small cell lung cancer with azacytidine. Oncotarget 4:2067–2079
Wu H, Zhang Y (2014) Reversing DNA methylation: mechanisms, genomics, and biological functions. Cell 156:45–68
Wu X, Johansen JV, Helin K (2013) Fbxl10/Kdm2b recruits polycomb repressive complex 1 to CpG islands and regulates H2A ubiquitylation. Mol Cell 49:1134–1146
Xie W, Kagiampakis I, Pan L et al (2018) DNA methylation patterns separate senescence from transformation potential and indicate cancer risk. Cancer Cell 33:309–321 e305

Xu GL, Bestor TH, Bourc'his D et al (1999) Chromosome instability and immunodeficiency syndrome caused by mutations in a DNA methyltransferase gene. Nature 402:187–191

Xu W, Yang H, Liu Y et al (2011) Oncometabolite 2-hydroxyglutarate is a competitive inhibitor of alpha-ketoglutarate-dependent dioxygenases. Cancer Cell 19:17–30

Yamada Y, Jackson-Grusby L, Linhart H et al (2005) Opposing effects of DNA hypomethylation on intestinal and liver carcinogenesis. Proc Natl Acad Sci U S A 102:13580–13585

Yao L, Shen H, Laird PW et al (2015) Inferring regulatory element landscapes and transcription factor networks from cancer methylomes. Genome Biol 16:105

Zhang Y, Jurkowska R, Soeroes S et al (2010) Chromatin methylation activity of Dnmt3a and Dnmt3a/3L is guided by interaction of the ADD domain with the histone H3 tail. Nucleic Acids Res 38:4246–4253

Zhou W, Dinh HQ, Ramjan Z et al (2018) DNA methylation loss in late-replicating domains is linked to mitotic cell division. Nat Genet 50:591–602

Ziller MJ, Gu H, Muller F et al (2013) Charting a dynamic DNA methylation landscape of the human genome. Nature 500:477–481

# CpG Islands Methylation Alterations in Cancer: Functionally Intriguing Security Locks, Useful Early Tumor Biomarkers

**Eleonora Loi and Patrizia Zavattari**

## Contents



**Abstract** DNA methylation is an epigenetic modification consisting in the addition of a methyl group to the position 5 of a cytosine in a CpG context. In normal mammalian cells, while CpG islands, mostly concentrated at promoter regions, are protected from DNA methylation, intergenic and repetitive regions are normally hypermethylated. In cancer cells, a massive change in the global methylation pattern occurs. Intergenic and repetitive regions of the genome become hypomethylated leading to the reactivation of transposable elements and genomic instability. In contrast, a focal hypermethylation of CpG islands at promoter regions occurs and it is normally associated to gene expression downregulation. Thus, aberrant DNA methylation is one of the most striking features of cancer cells and several studies have demonstrated that cancer-specific methylation patterns exist. For this reason, DNA methylation represents an extremely useful biomarker for several applications, including cancer risk definition, prediction of clinical outcomes, treatment response and cancer relapse. Finally, the association between DNA methylation and gene expression, although notoriously recognized, is not yet fully known, and the study of DNA methylation alterations in cancer and their consequences can help elucidate the mechanisms underlying this relationship.

E. Loi · P. Zavattari (✉)
Department of Biomedical Sciences, Unit of Biology and Genetics, University of Cagliari, Cittadella Universitaria di Monserrato, Monserrato, Cagliari, Italy
e-mail: pzavattari@unica.it



S. Jurga, J. Barciszewski (eds.), *The DNA, RNA, and Histone Methylomes*, RNA Technologies, https://doi.org/10.1007/978-3-030-14792-1_3

**Keywords** DNA methylation alterations in cancer · CpG islands · Repressed genes · Epigenetic switching · Predictive biomarkers · Diagnostic biomarkers · Prognostic biomarkers

# 1 Introduction

As known, the tumorigenic process is characterized by the onset and accumulation of genetic, cytogenetic and epigenetic alterations (Kinzler and Vogelstein 1996). The knowledge of the latter, protagonists in the regulation of gene expression, has been made possible above all by recent genomic approaches, thus allowing the development of "epigenomics". It has been so elucidated that, in addition to genetic mutations, cancer is characterized by alterations in chromatin compaction, both at histone level and in DNA (Jones and Baylin 2007). A cancer epigenome is characterized by extended genome hypomethylation, hypermethylation of CpG islands (CGIs) located in specific gene promoters, and loss of imprinting (Jones and Baylin 2002). Although we still investigate the mechanisms that induce these changes, it is now quite certain that these are early events during tumorigenesis. Extended hypomethylation mainly affects regions comprising repetitive elements, retrotransposons, poor CpG promoters, introns and gene deserts (Rodriguez et al. 2006). Hypomethylation promotes chromosomal rearrangements, leading to increased genomic instability (Eden et al. 2003), which can be caused for example by the activation, by hypomethylation, of retrotransposons, and their subsequent translocation into other genomic regions (Howard et al. 2008).

# 2 Promoter-CpG Islands Aberrant Methylation and Gene Expression in Cancer

The linkage between DNA methylation and gene expression has been investigated since the 1980s when several promoter-centric studies have found a strong association between hypermethylation and decreased expression levels of the downstream genes leading to the birth of the well-known dogma stating that aberrant DNA methylation is associated to gene repression.

## *2.1 DNA Methylation as a Mechanism to Inactivate Tumour Suppressor Genes*

The first studies about DNA methylation in cancer were candidate gene studies describing promoter-CGIs hypermethylation in tumour suppressor genes (TSGs) and linking TSGs inactivation to aberrant DNA methylation. In fact, genes involved in processes important for the development and progression of cancer such as DNA repair, cell cycle, cell adhesion, apoptosis, angiogenesis have been found hypermethylated both in hereditary cancer syndromes, where aberrant DNA methylation can act as a "second hit" to completely inactivate a TSG, and in sporadic cancers. *RB1* (Greger et al. 1989), *CDKN2A*, *MLH1* and *BRCA1* (Baylin 2005) genes, are all examples of TSGs hypermethylated and downregulated in cancer.

## 2.2 DNA Methylation as a Mechanism to Regulate the Epigenome

In other cases, hypermethylation is associated to downregulation of transcription factors, and the consequent silencing of their targets. This is the case of *RUNX3* in cancer of the esophagus (Long et al. 2007), *GATA-4* and *GATA-5* in colorectal and gastric tumors (Akiyama et al. 2003). Another example is represented by homeobox genes. These genes contain a conserved DNA sequence encoding for a DNA binding domain called homeodomain (HD). HD proteins play a fundamental role during development acting as transcriptional factors regulating the expression of genes involved in cellular adhesion, proliferation and differentiation and their expression is finely regulated by Polycomb repressive complex. Homeobox genes have been frequently found hypermethylated in several types of cancer (Rauch et al. 2006, 2007; Ohm et al. 2007; Schlesinger et al. 2007; Widschwendter et al. 2007; Gal-Yam et al. 2008) and their inhibition by promoter CGI hypermethylation contributes to the inactivation of regulatory or DNA repair genes, concurring to tumorigenesis. In fact, in agreement with the model proposed by Timp et al., epigenetic alterations contribute to carcinogenesis targeting genes regulating the epigenome itself (Timp and Feinberg 2013).

## 2.3 DNA Methylation Targets Are Normally Repressed Genes

Until recently, DNA methylation was believed to be the cause of TSGs silencing in cancer. Several experiments demonstrating that the removal of DNA methylation, through 5′-aza-2′-deoxycytidine treatment, cause gene re-activation, seemed to support this hypothesis. However, the advent of genome-wide methylation and expression studies has led to the finding that most hypermethylated genes in cancer are already repressed or lowly expressed in the normal tissues that give rise to tumors questioning the causative role of DNA methylation as a mechanism to induce gene silencing (Keshet et al. 2006).

### 2.3.1 DNA Methylation as a Mechanism to Maintain and Stabilize Transcriptional Inactivation: No Effect on Gene Expression or Further Downregulation of the Target Genes?

The fact that DNA methylation targets silenced genes has been already known. In fact, in the X-chromosome inactivation, genes targeted by DNA methylation are already silenced and DNA methylation acts maintaining their silenced state. Several studies have found that microenvironmental changes in tumor progression, such as hypoxia, induce gene downregulation of tumor suppressor genes, such as *E-CAD* (Krishnamachary et al. 2006), *BRCA1* (Bindra et al. 2005), *MLH1* (Bindra and Glazer 2007), and *RUNX* (Lee et al. 2009), frequently hypermethylated in cancer. The analysis of the normal expression of *RUNX3* gene, frequently hypermethylated in gastric cancer, revealed that it is never expressed in the cells that give rise to tumors sustaining the hypothesis that genes targeted by DNA methylation in cancer are fully repressed prior hypermethylation (Levanon et al. 2011). Furthermore, it has been demonstrated that a transient reduction in gene expression triggers a pathway for gene silencing involving first histone modifications and only later DNA methylation (Oyer et al. 2009). Thus, DNA methylation seems to be a later event, respect to histone modifications, although still an early event in the tumorigenesis, unnecessary for gene repression induction but fundamental for the maintenance and stabilization of a transcriptional inactive state. This hypothesis is supported by the fact that only genes repressed in a lineage-specific fashion and not tissue-specific or housekeeping genes, are targeted by DNA methylation (Sproul et al. 2011, 2012). Moreover, a meta-analysis of a large dataset of 672 matched cancerous and healthy methylomes and gene expression data across 3 types of tissues, including colon, breast and lung, from The Cancer Genome Atlas (TCGA) found that genes partially methylated in cancer showed a significant lower expression than other genes and that these genes were hypomethylated in normal tissues. However, a comparison between gene expression profiles in normal and breast cancer tissues revealed that the genes were already lowly expressed in the normal tissues (Moarii et al. 2015). It might be also possible that genes with promoter-associated CGI hypermethylation are expressed at low levels in normal tissues but become further downregulated in cancer. A study of colon cancer found that only about 7% (112 genes) of genes hypermethylated in CIMP-high tumors were also transcriptionally downregulated compared to normal adjacent tissues and 48/112 genes were not methylated but downregulated in non-CIMP subgroups (Hinoue et al. 2012). Noteworthy, the gene expression analysis conducted in this study was performed using gene expression microarrays where genes displaying low levels of expression are close to the background intensity level of hybridization to probe, making them unsuitable for differential gene expression analysis of lowly expressed genes. In fact, in a study conducted by our research group, we found that 72 loci, out of 74 regions altered by methylation in colorectal cancer, showing CGI hypermethylation, did not show gene expression dysregulation in the tumor, when analyzed by microarrays, since all those genes displayed an extremely low level of expression, close to the background

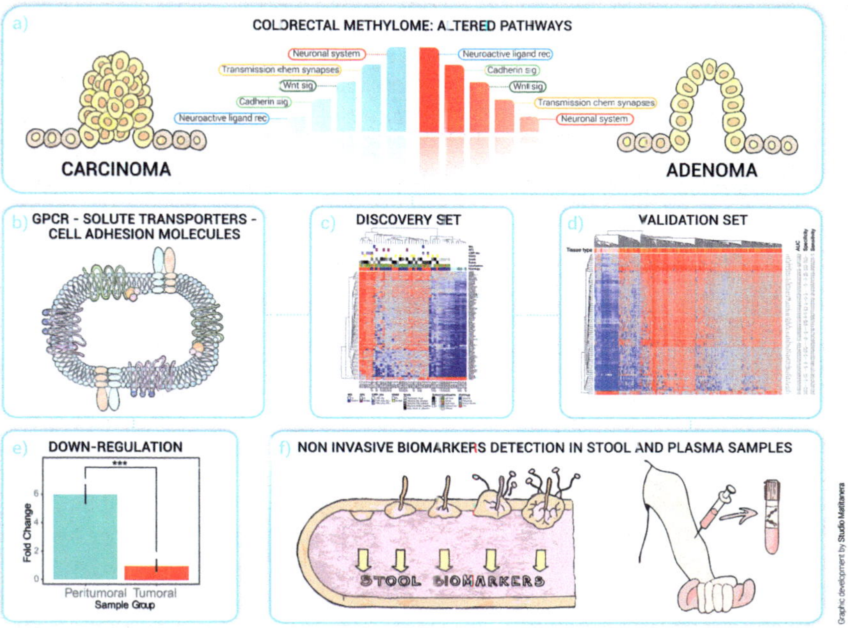


**Fig. 1** (**a**) Methylation alterations in colorectal cancer (CRC) characterizing both early and carcinoma stages, belong to the same pathways. (**b**) The most affected genes encode for proteins involved in the crosstalk between cell and surrounding environment. (**c**) In this study, a panel of 74 altered CpG islands was identified, (**d**) able to discriminate CRCs and adenomas from peritumoral and normal mucosa, with very high specificity (100%) and sensitivity (99.9%). (**e**) Over 70% of the hypermethylated islands resulted in downregulation of gene expression. (**f**) The alterations were also tested and detected through non-invasive techniques, both in stool DNA and in cfDNA from plasma

intensity level (Fadda et al. 2018). On the other hand, in silico RNA seq data revealed that 70% of genes showing CGI hypermethylation were downregulated in colorectal cancer and a differential gene expression analysis conducted by qRT-PCR for some selected genes revealed that they are further downregulated in colorectal cancer samples compared to normal colon samples (Fig. 1e). Using western blot, conducted for some selected loci, we found that protein levels also decrease in the tumor tissue (data unpublished). A recent colon cancer work supports the view that on a subset of genes enriched in common cancer pathways, methylation is significantly associated to gene expression. In particular, the authors showed that majority of hypermethylated regions in both promoter and gene body were related to downregulation (Klett et al. 2018).

In summary, the association between promoter aberrant methylation and gene expression in cancer is extremely complex: although it is possible that rare TSGs are silenced by DNA methylation, an increasing number of evidences has shown that

DNA methylation targets promoter-associated CGIs of genes that are in some cases fully repressed and in other cases lowly expressed in normal tissues where the tumors arise. Moreover, several studies have shown that many of these genes are not differentially expressed between tumor and normal samples concluding that aberrant methylation does not precede transcriptional inactivation. Nevertheless, other studies detected a significant downregulation of genes hypermethylated in tumor samples compared to normal samples.

## 2.4 How Do DNA Methylation Alterations Choose Their Targets?

It remains to be clarified how these genes are chosen as targets for the methylation alterations. A hypothesis predicts that silencing by hypermethylation of specific genes gives to the cells an advantage in terms of growth, resulting in their clonal selection and proliferation. As a confirm, several types of cancer present methylation specific patterns to specific genes (Berdasco and Esteller 2010). Another hypothesis foresees that during tumorigenesis, DNA methyl transferases (DNMTs) are guided towards specific target sequences through their association with histone marks. DNA methylation and histone modifications act independently and in concert in regulating gene expression, for example giving rise to a rigid repressive chromatin structure that induces reduced cellular plasticity. In embryonic stem cells, Polycomb proteins are responsible for silencing genes critical for development in a reversible manner, by tri-methylation of lysine 27 on histone H3 (H3K27me3) (Schlesinger et al. 2007; Widschwendter et al. 2007; Ohm et al. 2007); this silencing is maintained even after differentiation by EZH2, a histone-lysine N-methyltransferase, the functional enzymatic component of the Polycomb Repressive Complex 2 (PRC2). In the tumor cells the genes marked by PRC2-deposited H3K27me3 undergo a specific de novo DNA methylation by DNMTs (Viré et al. 2006). This so-called "epigenetic switching" guarantees a more stable and permanent silencing of genes, important for cell proliferation and tumorigenesis (Gal-Yam et al. 2008). This link between stem cells and tumorigenesis supports the "cancer stem cell" hypothesis, according to which epigenetic modifications, characteristic of stem cells, occur very early in tumor cells, undergo an undifferentiation towards progenitor cells (Sharma et al. 2010). Reassuming stem cell capacity, preneoplastic cells would likely give rise to a high-risk aberrant progenitor cell population that can undergo transformation with accumulation of genetic mutations (Baylin and Ohm 2006).

It is therefore not surprising that, by studying the methylation alterations in the genome of different types of cancer (colorectal, gastric, biliary tract cancers, gliomas and chronic lymphocytic leukemia), also our group has found prevalently hypermethylation at the CGIs associated with the promoter of normally repressed genes in the tissues where the tumor occurs (Fadda et al. 2018; Antonelli et al. 2018). In particular, in colorectal neoplasms our investigations have shown that the aforementioned alterations, associated with genes already normally repressed, are shared by carcinomas and adenomas, therefore likely early events in tumorigenesis. These results support the aforementioned hypothesis according to which the epigenetic events that characterize tumorigenesis, would re-establish a gene-expression regulation pattern typical of that tissue's stem cells.

## 3 Which Pathways Are Most Affected by Methylation Changes in Cancer?

In the study conducted by our research group on colorectal cancer, the list of genes whose associated CGIs were significantly altered, was subjected to a gene enrichment and candidate gene prioritization analysis by Toppgene, allowing the identification of the pathways most affected by aberrant methylation (Fadda et al. 2018). The crosstalk between tumor cells and surrounding environment resulted particularly involved, in terms of membrane receptors, solute transporters and cell adhesion molecules. Functional annotation analysis has highlighted the enrichment of protocadherins, integrins, members of the solute carrier family, and, extensively, G-protein coupled receptors (GPCRs) involved in the transduction of neuroactive signals (Fig. 1a, b). Although the large family of GPCRs mediates many biological processes, and although they have been demonstrated to act either as oncogenes or as tumor suppressors, their importance in tumorigenesis is undervalued. The GPCRs actually control many of the functions of tumorigenesis, including cell-mediated immunity, proliferation, invasion and survival at the secondary site (Feigin 2013). The analysis of functional annotations carried out by Toppgene on other types of cancer (gastric, BTC, CLL) has allowed us to highlight that the pathways most affected by alterations of CGI methylation are mainly the same described for colorectal neoplasms (manuscript in preparation). This result would suggest that the tumorigenic mechanisms involved in the epigenetic reprogramming process, as said characteristic of tumor cells, would be extremely common among the different types of cancer.

## 4 DNA Methylation Alterations as Predictive, Diagnostic and Prognostic Tumor Biomarkers

Besides the remarkable functional relevance that methylation alterations play in the tumorigenic process, they also arouse great interest for clinical use. In fact, a better knowledge of these epigenetic alterations could not only help clarify which cellular processes are affected by such early changes during tumorigenesis, but it might also provide potential tumor biomarkers. Almost any biological tissue sample or bodily fluid can be used for DNA methylation analysis. DNA methylation is the most robust epigenetic mark and will survive most sample storage conditions. The robustness of

DNA methylation marks makes DNA methylation analysis very attractive in a clinical environment for the early detection of cancer and easy-to-access tissues or bodily fluids can be collected. Such samples include venous peripheral blood, buccal epithelium or saliva, urine, stools, bronchial aspirates, and, even in some cases, muscle or adipose tissue. Moreover, genes altered by methylation not only might represent potential biomarkers for early detection of CRC (Church et al. 2014), but could also be important prognostic and predictive markers to improve therapeutic interventions (Phipps et al. 2015; Lam et al. 2016; Barault et al. 2018). The panel of 74 altered CGIs above mentioned (Fig. 1c), identified by our research group in colorectal adenomas and carcinomas, discriminates CRCs and adenomas from peritumoral and normal mucosa with very high specificity and sensitivity (Fig. 1d). The performance ability of the panel was cross-validated *in silico* by analysing it in over five hundred samples, including colorectal carcinomas, adenomas, normal counterparts and other tumor types (Fig. 1d). The identified panel appears very robust and informative (sensitivity 99.99%), specific for CRC (specificity 100%) from early to metastatic stages. To establish the possible usefulness of these non-invasive markers for detection of colorectal cancer, we selected three biomarkers and identified the presence of altered methylation in stool DNA and plasma cell-free circulating DNA from patients (Fig. 1f). Overall, the panel showed a good diagnostic and prognostic value even in the non-invasive assessment, strengthening its potential value in screening and follow-up of colorectal cancer patients.

Recently, a predictive role of these kind of epigenetic markers has been suggested, observing that typical methylation alterations of certain tumors (such as mature B-cell neoplasms) would be detectable even in blood samples collected years before diagnosis (Wong Doo et al. 2016). Data submitted for publication, obtained by our research group, confirm the existence of very early methylome alterations in chronic lymphocytic leukemia, which allow an extremely early diagnosis, show a correlation between methylation rate and tumor aggressiveness, and may even predict the pathology many years before the onset. These observations add to the diagnostic and prognostic role of methylome alterations in cancer, also the ability to predict the disease risk, thus further increasing their value as important tumor biomarkers.

## References

Akiyama Y, Watkins N, Suzuki H et al (2003) GATA-4 and GATA-5 transcription factor genes and potential downstream antitumor target genes are epigenetically silenced in colorectal and gastric cancer. Mol Cell Biol 23:8429–8439

Antonelli M, Fadda A, Loi E et al (2018) Integrated DNA methylation analysis identifies topographical and tumoral biomarkers in pilocytic astrocytomas. Oncotarget 9:13807–13821

Barault L, Amatu A, Siravegna G et al (2018) Discovery of methylated circulating DNA biomarkers for comprehensive non-invasive monitoring of treatment response in metastatic colorectal cancer. Gut 67:1995–2005

Baylin SB (2005) DNA methylation and gene silencing in cancer. Nat Clin Pract Oncol 2:S4–S11

Baylin SB, Ohm JE (2006) Epigenetic gene silencing in cancer – a mechanism for early oncogenic pathway addiction? Nat Rev Cancer 6:107–116
Berdasco M, Esteller M (2010) Aberrant epigenetic landscape in cancer: how cellular identity goes awry. Dev Cell 19:698–711
Bindra RS, Glazer PM (2007) Co-repression of mismatch repair gene expression by hypoxia in cancer cells: role of the Myc/Max network. Cancer Lett 252:93–103
Bindra RS, Gibson SL, Meng A et al (2005) Hypoxia-induced down-regulation of *BRCA1* expression by E2Fs. Cancer Res 65:11597–11604
Church TR, Wandell M, Lofton-Day C et al (2014) Prospective evaluation of methylated SEPT9 in plasma for detection of asymptomatic colorectal cancer. Gut 63:317–325
Eden A, Gaudet F, Waghmare A et al (2003) Chromosomal instability and tumors promoted by DNA hypomethylation. Science 300:455–455
Fadda A, Gentilini D, Moi L et al (2018) Colorectal cancer early methylation alterations affect the crosstalk between cell and surrounding environment, tracing a biomarker signature specific for this tumor. Int J Cancer 143:907–920
Feigin ME (2013) Harnessing the genome for characterization of G-protein coupled receptors in cancer pathogenesis. FEBS J 280:4729–4738
Gal-Yam EN, Egger G, Iniguez L et al (2008) Frequent switching of Polycomb repressive marks and DNA hypermethylation in the PC3 prostate cancer cell line. Proc Natl Acad Sci U S A 105:12979–12984
Greger V, Passarge E, Höpping W et al (1989) Epigenetic changes may contribute to the formation and spontaneous regression of retinoblastoma. Hum Genet 83:155–158
Hinoue T, Weisenberger DJ, Lange CPE et al (2012) Genome-scale analysis of aberrant DNA methylation in colorectal cancer. Genome Res 22:271–282
Howard G, Eiges R, Gaudet F et al (2008) Activation and transposition of endogenous retroviral elements in hypomethylation induced tumors in mice. Oncogene 27:404–408
Jones PA, Baylin SB (2002) The fundamental role of epigenetic events in cancer. Nat Rev Genet 3:415–428
Jones PA, Baylin SB (2007) The epigenomics of cancer. Cell 128:683–692
Keshet I, Schlesinger Y, Farkash S et al (2006) Evidence for an instructive mechanism of de novo methylation in cancer cells. Nat Genet 38:149–153
Kinzler KW, Vogelstein B (1996) Lessons from hereditary colorectal cancer. Cell 87:159–170
Klett H, Balavarca Y, Toth R et al (2018) Robust prediction of gene regulation in colorectal cancer tissues from DNA methylation profiles. Epigenetics 13 386–397
Krishnamachary B, Zagzag D, Nagasawa H et al (2006) Hypoxia-inducible factor-1-dependent repression of *E-cadherin* in von Hippel-Lindau tumor suppressor–null renal cell carcinoma mediated by TCF3, ZFHX1A, and ZFHX1B. Cancer Res 66:2725–2731
Lam K, Pan K, Linnekamp JF et al (2016) DNA methylation based biomarkers in colorectal cancer: a systematic review. Biochim Biophys Acta Rev Cancer 1866:106–120
Lee SH, Kim J, Kim W-H et al (2009) Hypoxic silencing of tumor suppressor RUNX3 by histone modification in gastric cancer cells. Oncogene 28:184–194
Levanon D, Bernstein Y, Negreanu V et al (2011) Absence of Runx3 expression in normal gastrointestinal epithelium calls into question its tumour suppressor function. EMBO Mol Med 3:593–604
Long C, Yin B, Lu Q et al (2007) Promoter hypermethylation of the RUNX3 gene in esophageal squamous cell carcinoma. Cancer Investig 25:685–690
Moarii M, Boeva V, Vert J-P et al (2015) Changes in correlation between promoter methylation and gene expression in cancer. BMC Genomics 16:873
Ohm JE, McGarvey KM, Yu X et al (2007) A stem cell–like chromatin pattern may predispose tumor suppressor genes to DNA hypermethylation and heritable silencing. Nat Genet 39:237–242
Oyer JA, Chu A, Brar S et al (2009) Aberrant epigenetic silencing is triggered by a transient reduction in gene expression. PLoS One 4:e4832

Phipps AI, Limburg PJ, Baron JA et al (2015) Association between molecular subtypes of colorectal cancer and patient survival. Gastroenterology 148:77–87.e2
Rauch T, Li H, Wu X et al (2006) MIRA-assisted microarray analysis, a new technology for the determination of DNA methylation patterns, identifies frequent methylation of homeodomain-containing genes in lung cancer cells. Cancer Res 66:7939–7947
Rauch T, Wang Z, Zhang X et al (2007) Homeobox gene methylation in lung cancer studied by genome-wide analysis with a microarray-based methylated CpG island recovery assay. Proc Natl Acad Sci U S A 104:5527–5532
Rodriguez J, Frigola J, Vendrell E et al (2006) Chromosomal instability correlates with genome-wide DNA demethylation in human primary colorectal cancers. Cancer Res 66:8462–9468
Schlesinger Y, Straussman R, Keshet I et al (2007) Polycomb-mediated methylation on Lys27 of histone H3 pre-marks genes for de novo methylation in cancer. Nat Genet 39:232–236
Sharma S, Kelly TK, Jones PA (2010) Epigenetics in cancer. Carcinogenesis 31:27–36
Sproul D, Nestor C, Culley J et al (2011) Transcriptionally repressed genes become aberrantly methylated and distinguish tumors of different lineages in breast cancer. Proc Natl Acad Sci U S A 108:4364–4369
Sproul D, Kitchen RR, Nestor CE et al (2012) Tissue of origin determines cancer-associated CpG island promoter hypermethylation patterns. Genome Biol 13:R84
Timp W, Feinberg AP (2013) Cancer as a dysregulated epigenome allowing cellular growth advantage at the expense of the host. Nat Rev Cancer 13:497–510
Viré E, Brenner C, Deplus R et al (2006) The Polycomb group protein EZH2 directly controls DNA methylation. Nature 439:871–874
Widschwendter M, Fiegl H, Egle D et al (2007) Epigenetic stem cell signature in cancer. Nat Genet 39:157–158
Wong Doo N, Makalic E, Joo JE et al (2016) Global measures of peripheral blood-derived DNA methylation as a risk factor in the development of mature B-cell neoplasms. Epigenomics 8:55–66

# Histone and DNA Methylome in Neurodegenerative, Neuropsychiatric and Neurodevelopmental Disorders

Harsha Rani and Vijayalakshmi Mahadevan

## Contents



**Abstract** Genome-environment interaction and epigenome plasticity significantly influence the pathogenesis of neurodegenerative and neuropsychiatric disorders. Recent advancements in the field to study genome wide chromatin modifications provide a comprehensive view of the epigenome. Dysregulation of epigenetic machinery has emerged as a major genetic driver of neuro developmental and neuro degenerative disorders, intellectual disabilities and autism spectrum disorders.

H. Rani · V. Mahadevan (✉)
Chromatin and Epigenetics Group, Institute of Bioinformatics and Applied Biotechnology (IBAB), Bangalore, India
e-mail: mviji@ibab.ac.in



S. Jurga, J. Barciszewski (eds.), *The DNA, RNA, and Histone Methylomes*, RNA Technologies, https://doi.org/10.1007/978-3-030-14792-1_4

Emerging evidences point to the involvement of the epigenome in the onset and progression of Alzheimer's disease, Parkinson's disease and Huntington's disease. This review focusses on the changes in epigenetic machinery, specifically on the histone methylation and DNA methylation patterns during the onset and progression of neurodegenerative diseases and neuropsychiatric disorders. The power of epigenetic inhibitors to function as potential diagnostic and therapeutic markers is also discussed.

**Keywords** Epigenome · Neuropsychiatry · Brain plasticity · Epigenetic inhibitors · Schizophrenia · Huntington's disease · Alzheimer's disease · Parkinson's disease · Obsessive compulsive disorder · Bipolar disorder

# 1 Introduction

## *1.1 Dynamics of Histone Modification*

Chromatin is an active and dynamic substrate for transcriptional and developmental processes and resides inside the cell nucleus in eukaryotes. The strategic and hierarchical wrapping of the 146bp of DNA around the octameric scaffold of core histones H2A, H2B, H3 and H4 forms a single unit of nucleosome (Luger et al. 1997; Kornberg 1974), which forms a higher order structure with repeating nucleosomal subunits and linker histone H1 (Luger et al. 1997). The higher order structure of chromatin is an impediment for transcription factors to gain access to DNA. Increasing evidences in the field have established chromatin as a dynamic entity that regulates gene programs and cellular functions through alteration of its structure and architecture via enzymatic modifications of histone tails or through nucleosome remodelling. Post translational modifications of histones include acetylation, methylation, phosphorylation, ubiquitination, citrullination and ADP-ribosylation that take place on the tail domains of core histones (Hottiger 2011; Cao and Yan 2012).

The sites of modifications are predominantly clustered in the first few amino acids of the core histones H3, H4, H2A and H2B though a few residues inside the core of the nucleosomes have been identified. These combinatorial patterns of histone modifications create a 'histone code' (Strahl and David Allis 2000) and control a variety of biological functions, including recruitment of DNA replication, transcription and repair (Abmayr and Workman 2012). Individual histone modifications have been shown to crosstalk with histone-enzyme interaction where nearby or distant PTMs interdependently recruit or release enzymes required for modifications (Daujat et al. 2002).

Histone methylation is a highly dynamic event which regulates diverse biological processes including cell-cycle regulation, DNA damage, stress response, development and differentiation (Pedersen and Helin 2010; Greer and Shi 2012). Methylation of histone generally occurs on arginines, lysines and histidines on the N-terminal tail of histones. Lysines can be monomethylated (me1), di-methylated(me2) or trimethylated (me3) on their ε- amino group (Ng et al. 2009). However, arginines can either be monomethylated (me1) or symmetrically (me2s)/asymmetrically di-methylated (me2a) on their guanidinyl group. Symmetrical di-methylation of arginine refers to the addition of one methyl group to each nitrogen of the guanidinium group, whereas asymmetrical di-methylation refers to the addition of both methyl groups to one nitrogen of the guanidinium group (Borun et al. 1972). On the other hand, histidines have been found to be mono methylated although it seems to be a rare event. The effect of methylation of histones is dependent on the location of methylation residues on the histone tail and degree of methylation (Heintzman et al. 2007). SAMe (S-adenosyl methionine) is a major methyl donor which functions through cellular transmethylation pathways and methylates many substrates including those for DNA methylation and histone methylation. Generation of SAMe involves a bicyclic cellular pathway consisting of folate and methionine (1 carbon cycle). Thus, it takes part in critical epigenetic mechanism and connects nurture based metabolism with brain development (Mentch et al. 2015; Gao et al. 2018).

Genomewide location analysis (GWLA) of histone H3 methylation (me) patterns at different lysines (Ks) using Chromatin Immuno Precipitation (ChIP) revealed that these methylation patterns (unlike the acetylation marks) are predominantly enriched over broad genomic regions rather than being restricted to the promoter regions (except H3K4me). This study also provided new information on the distribution patterns of lysine methylation across the coding regions of human genes (Miao and Natarajan 2005). A study on the dynamics of distinct methylation marks (both lysine and arginine residues) in HeLa cells using heavy methyl stable isotope labelling by amino acids in cell culture (SILAC) revealed that different methylation states within the same peptide have different rates of formation and is found to be enriched mainly over broad genomic regions (Zee et al. 2010)

## *1.2 Readers, Writers and Erasers of Methylation Machinery*

### 1.2.1 Epigenetic Writers

Addition of methyl groups donated from S-adenosylmethionine to histones is catalysed by histone methyl transferases (HMTs). Three families of histone methyl transferases have been classified so far which includes the SET-domain containing proteins, DOT1-l like proteins and arginine N-methyltransferase (PRMT) family proteins (Table 1). The SET domain is a sequence motif (named after Su(var) 3-9, Enhancer of Zeste, Trithorax) regulating lysine methylation and is found in several chromatin associated proteins, including members of both the Trithorax group and Polycomb group (Rea et al. 2000). The non-SET domain DOT-1 (disruptor of telomeric silencing: also called Kmt4) and its mammalian homolog, DOT1L (DOT1-like) possess histone methyltransferase activity towards histone H3Lys79. PRMT family in turn is specific for arginine methyl transferase activity (Feng et al. 2002).

**Table 1** Members of the SET-domain containing family with common and unique domains of each member of the family

| SET family | Members associated | Domains common to the family in addition to the SET domain | Domains unique to particular members | References |
|---|---|---|---|---|
| SUV39 family | | Pre-SET (9 Cys, 3 Zn), post-SET ($CXCX_4C$) | | Rea et al. (2000) |
| | SUV39H1 | | 4 Cys, chromo | |
| | SUV39H2 | | 4 Cys, chromo | |
| | G9a | | E/KR-rich, NRSF-binding, ankyrin repeats | |
| | GLP1 (EuHMT1) | | Same as G9a | |
| | ESET (SETDB1) | | Tudor, MBD | |
| | CLLL8 (SETDB2) | | MBD | |
| SET1 family | | Post-SET ($CXCX_4C$) | | Lee and Skalnik (2005) |
| | MLL1 (HRX, ALL1) | | AT hook, Bromo PHD, CXXC | |
| | HRX2 (MLL4) | | Same as above | |
| | ALR (MLL2) | | PHD, ring finger | |
| | MLL3 | | PHD, ring finger | |
| | SET1 (ASH2) | | RRM, poly-S/E/P | |
| | SET1L | | RRM, poly-S/E/P | |
| SET2 family | | Pre-SET (7-9 Cys); post-SET ($CXCX_4C$) | | Kizer et al. (2005) |
| | WHSC1 (NSD2) | | PWWP, PHD, HMG, ring finger | |
| | WHSCL1 (NSD3) | | PWWP, PHD, ring finger | |
| | NSD1 | | PWWP, PHD, ring finger | |
| | HIF1 (HYPB) | | WW | |
| | ASH1 | | AT hook, bromo, BAH, PHD | |

(continued)

**Table 1** (continued)

| SET family | Members associated | Domains common to the family in addition to the SET domain | Domains unique to particular members | References |
|---|---|---|---|---|
| RIZ family | RIZ (PRDM2) | | C2H2 zinc finger | Jiang and Huang (2000) |
| | BLIMP1 (PRDM1) | | C2H2 zinc finger | |
| SMYD family | | Post-SET ($CXCX_2C$) | | Hamamoto et al. (2004) |
| | SMYD3 | | Zf-MYND | |
| | SMYD1 | | Zf-MYND | |
| EZ family | | Pre-SET (~15 Cys) | | Margueron et al. (2008) |
| | EZH1 | | 2 SANT | |
| | EZH2 | | 2 SANT | |
| SUV4-20 family | | Post-SET ($CXCX_2C$) | | Wu et al. (2013) |
| | SUV4-20H1 | | | |
| | SUV4-20H2 | | | |

DOT-1 and its homologs share a conserved region with four sequence motifs-I, post I, II and III of the SAM methyl transferase. Although the catalytic domain of DOT1 proteins is structurally similar to arginine methyltransferases, these family of proteins catalyse methylation preferentially at H3K79 in the core of the nucleosome. Since H3K79 methylation plays an important role during embryonic development, over expression of Dot1 was found to disrupt telomeric silencing in yeast screens. Knock out of mDOT1L results in lethality during the time frame of organogenesis in cardiovascular development.

### 1.2.2 Epigenetic Erasers

Two families of demethylases including the amine oxidases and jumonji C (JmjC)-domain have been documented so far. The Jumonji C(JmjC)-domain contains iron ($Fe^{2+}$) and alpha-ketoglutarate-dependent dioxygenase which can reverse lysine methylation and has various functional roles in biological processes including DNA/RNA repair pathways. Demethylation of monomethyl arginines to citrulline has been shown to be catalysed by protein arginine deiminase type 4 (PADI4). However, this enzyme is not an arginine demethylase as it works on both the methylated and unmethylated arginines (Cuthbert et al. 2004). These enzymes are highly conserved from yeast to humans and demethylate histone and non-histone substrates. The histone modifications catalysed by the Jumonji Domain are documented in Table 2.

### *1.3 Histone Demethylation by LSD1*

The activity of LSDI enzyme is limited to di-methylated and mono-methylated lysine residues. Each demethylation cycle requires electrons to be shuttled to molecular oxygen via an FAD/FADH moiety and through production of hydrogen peroxide. Isolation of LSD1 demethylase complexes from mammalian cells revealed that it requires Co-REST, a chromatin associated transcriptional repressor, to demethylate nucleosomal substrates (Lee et al. 2005). LSD1 functions both as an activator and repressor. Association of LSD1 with Co-REST leads to transcriptional repression of neuronal genes in non-neuronal cell lineages. Association of LSD1 with androgen receptor (AR) converts LSD1 to an H3K9 demethylase, allowing it to function as transcriptional activator of androgen receptor in response to hormonal stimulus (Metzger et al. 2005).

Epigenetic mechanisms regulate the function and homeostasis of the central nervous system. Dysregulation of epigenetic machinery has emerged as a major genetic driver of neurodevelopmental and neurodegenerative disorders, intellectual disabilities and autism spectrum disorders. Such epigenomic changes cause perennial alterations in cells of the central nervous system and influence neuronal function and physiology. Brain Derived Neurotropic Factor plays a crucial role in the development, maintenance and plasticity of the CNS and has been associated with several neuropsychiatric disorders like Schizophrenia, Bipolar Disorder and depression (Cohen-Cory et al. 2010; Zagrebelsky and Korte 2014). Methylation patterns are dynamically regulated in neurons by experiential stimuli which in turn regulate memory related genes (Lattal and Wood 2013).

Emerging evidences point to the involvement of the epigenome in the onset and progression of Alzheimer's disease, Parkinson's disease and Huntington's disease. This review focusses on the changes in epigenetic machinery, specifically on the histone methylation and DNA methylation patterns during the onset and progression of neurodegenerative diseases and neuropsychiatric disorders. The power of epigenetic inhibitors as potential diagnostic and therapeutic markers is also discussed.

## 2 Epigenetic Alterations in Neurodegenerative Disorders

**Neurodegenerative disorders** manifest as neuronal disabilities accompanied by massive neuronal loss and accumulation of toxic proteins (such as β amyloid in AD and Huntingtin in HD) with progression of the disease (Forman et al. 2004).

**Table 2** Epigenetic machinery of Jumonji family of proteins

| Subfamily | Name | Synonym | Lysine Demethylase | Arginine Demethylase | References |
|---|---|---|---|---|---|
| JHDM1 | JHDM1A | KDM2A | H3 (K36me1/2) | | Frescas et al. (2008) |
| | JHDM1B | KDM2B | H3(K4me3/K36me2) | | He et al. (2008) |
| PHF2/PHF8 | JHDM1D | KDM7A | H3(K9me1/2) | | Tsukada et al. (2010) |
| | PHF8 | KDM7B | H3(K9me1/2) H4(K20me1) | | Qi et al. (2010) |
| | PHF2 | JHDM1E | H3(K9me1) H4(K20me3) | | Baba et al. (2011) |
| JHDM2 | HR | | H3(K9me1/2) | | Liu et al. (2014) |
| | JMJD1A | KDM3A | H3(K9me1/2) | | Yamane et al. (2006) |
| | JMJD1B | KDM3B | H3(K9me1/2) | | Yamane et al. (2006) |
| | JMJD1C | KDM3C | H3(K9me1/2) | | Chen et al. (2015) |
| JMJD2/JHDM3 | JMJD2A | KDM4A | H3(K4me3/K27me3/K36me3) | H3(R2me2a) | Whetstine et al. (2006) |
| | JMJD2B | KDM4B | H3(K9me3/K36me3) | | Katoh and Katoh (2007) |
| | JMJD2C | KDM4C | H3(K9me3) | | Pedersen et al. (2014) |
| | JMJD2D | KDM4D | H3(K9me2/3) | | Krishnan and Trievel (2013) |
| JARID | JARID1A | KDM5A | H3(K4me2/3) | | Horton et al. (2016) |
| | JARID1B | KDM5B | H3(K4me1/2/3) | | Zhang et al. (2014) |
| | JARID1C | KDM5C | H3(K4me2/3) | H3(R2me1/2a/2s/R8me2a/2s) H4(R3me2a/2s) | Iwase et al. (2007) |
| | JARID1D | KDM5D | H3(K4me2/3) | | Li et al. (2016) |
| | JARID2 | | | | Pasini et al. (2010) |

(continued)

**Table 2** (continued)

| Subfamily | Name | Synonym | Lysine Demethylase | Arginine Demethylase | References |
|---|---|---|---|---|---|
| JmjC domain only | JMJD5 | KMD8 | H3(K36me2) | | Hsia et al. (2010) |
| | JMJD7 | | | | |
| | TYW5 | | | | |
| | HSPBAP1 | | | | |
| | HIF1AN | FIH | | | |
| | JMJD4 | | | | |
| | JMJD6 | | | H3(R2me2a/2s) H4(R3me1/2a/2s) | Chang et al. (2007) |
| | JMJD8 | | | | |
| | NO66 | RIOX1 | H3(K4me1/3/K36me2/3) | | Eilbracht et al. (2004) |
| | MINA | RIOX2 | | | |
| UTX/UTY | JMJD3 | KDM6B | H3(K27me2/3) | | Xiang et al. (2007) |
| | UTX | KDM6A | H3(K27me1/2/3) | | Agger et al. (2007) |
| | UTY | KDM6C | H3(K27me3) | | Walport et al. (2014) |

Diseases such as Alzheimer's accumulate intracellular beta-amyloid plaques and inter cellular neurofibrillary tangles that regulate neuronal death and loss of cognitive abilities leading to dementia (Vila and Przedborski 2003). Mitochondrial alterations, defects in axonal transport and alterations in dendrite pathology are observed in neurons undergoing such transition (Schon and Przedborski 2011; Kweon et al. 2017).

The complex neuronal pathophysiology during ageing and neuronal loss strongly implies distinct roles of chromatin states in regulating neuronal function and identity. It is well established that epigenetic factors are vital regulators of ageing, lifespan and health span in yeast and *C. elegans*. It is known that histone acetylation regulates learning and decline in memory with age in mouse models of Alzheimer's disease (Kawahara et al. 2009: Gräff and Tsai 2013). Chromatin Immunoprecipitation followed by Sequencing (ChIP seq) and single cell sequencing studies have helped unravel changes in chromatin during neurodegenerative processes. Mutations in chromatin related factors, transcriptional regulators like FMR1, alterations in histone acetylome and methylome profiles are distinctly associated with neurological disorders, intellectual disabilities and autism (Bourgeron 2015; Sun et al. 2016). The ability of epigenetic mechanisms to integrate diverse environmental and physiological inputs to generate adaptive long-lasting brain functions regulates multifactorial diseases such as Parkinson's, Alzheimer's, Amylotrophic lateral sclerosis (ALS), Multiple sclerosis and even epilepsy. Ever since it has been proposed that DNA methylation age measures the cumulative effect of an epigenetic maintenance system (EMS) and hence genomic stability, the epigenetic clock and the ratio of S-adenosyl methionine (SAM) /S-adenosylhomocysteine (SAH) have been used to measure the age of tissues based on methylation markers (Horvath 2013; Levine et al. 2015).

## 2.1 *Alzheimer's Disease*

Alzheimer's disease (AD) is a complex neurodegenerative disorder that involves multiple pathological processes characterized clinically by progressive loss of memory and neuronal loss. The presence of amyloid beta (Aβ) plaques and neurofibrillary tangles (NFTs) composed of hyperphosphorylated Tau protein are characteristic hallmarks of Alzheimer's disease. Alzheimer's disease can be classified into late-onset AD (LOAD) and early onset AD (EOAD) depending on the age of onset of the disease. LOAD, the more common form of AD affects people above 65 years of age. Mutations in APP (Amyloid precursor protein), PS1 (Presenilin1), PS2 (Presenilin2) and APOε4 are involved in the early onset of familial AD (fAD) which occurs in less than 2% of the cases reported with AD (Wijsman et al. 2005).

Genetic linkage and association studies in the more common form of sporadic AD (sAD) have identified several genetic variants that shows mild or moderate increase in the risk of sAD (Bertram et al. 2007). A recent meta-analysis of four genome wide association studies (GWAS) totalling 17,008 cases and 37,154 controls for probing additional genetic risk factors responsible for AD identified 11 susceptibility risk loci for LOAD. These newly associated loci predicted newer pathways involving hippocampal synaptic function and axonal transport in AD patients (Lambert et al. 2013).

## 2.2 *Epigenetics of Alzheimer's Disease*

### 2.2.1 Alteration in Histone Methylation Profiles in Alzheimer's Disease

Human neurodegenerative disorders such as Alzheimer's disease, Parkinson's disease, Huntington's disease and ALS (Amyotropic Lateral Sclerosis) show significant changes in the transcription profile pointing to the involvement of dysregulated chromatin in such cases. Since ageing is a predominant risk factor in neurodegenerative diseases, chromatin alterations and epigenetic changes in ageing brains are vital targets of neurodegeneration. Aberrant changes in histone acetylation and methylation have been implicated in age related neurodegeneration (Akbarian et al. 2013; Nativio et al. 2018). Ageing models of animals show increase in the levels of H3K9me2, H3K9me3 and H3K27me3 in the cerebral cortex and hippocampal regions of these models. Active chromatin marks such as H3K27ac and H3K36me3 show decreased levels in the same regions of the brain in these animal models. Early and late pathological investigations in the hippocampus of mouse models showed coordinated downregulation of synaptic plasticity genes and upregulation of immune response genes targeted by regulators of the electron transport system. ChIP sequencing attempted on hippocampus CK-p25 mouse models of AD and CK wild type littermates for seven histone markers (active marks, enhancers, repressive marks and marks associated with gene bodies) observed interesting behaviour of the active mark H3K4me3. In these models. 3667 of the upregulated genes corresponding to H3K4me3 peaks were enriched for immune and stimulus response while the down regulated genes corresponding to H3K4me3 peaks showed synaptic and learning functions (Gjoneska et al. 2015).

The active chromatin regulator H3K4me3 is of interest in studies of neurodegenerative disease including Alzheimer's disease because of its influence on synaptic transmission and learning and memory. Presence of H3K4me3 in the nucleus has been shown to directly influence the efficiency of post initiation processes of active transcription and is found globally ($\sim$90%) at RNA polymerase II binding sites.

Progression of Alzheimer's disease from early to final stage has been defined on the basis of the progressive presence of NFT in successive brain regions, with stage I affecting limbic or brain stem regions, and widespread neocortical stage VI and during the end stage of the disease (Braak and Braak 1991). H3K4me3 generally localizes within the nuclear compartment of the cell along with other epigenetic regulators and co-regulates chromatin structure. However, reduced nuclear and increased cytoplasmic localization of H3K4me3 has been observed in the autopsy tissues from hippocampal regions of AD patients (Mastroeni et al. 2015). The observed cytoplasmic localization of H3K4me3 is associated with and even precedes tau markers examined in early Braak stages. The function of the ectopically present molecules in the cytoplasm is not yet known, but existing data suggest that these molecules and H3K4me3 might be involved in Tau hyperphosphorylation and axonal transport. The loss of H3K4me3 from the nucleus might be responsible for the overall decrease in the expression of synaptic genes. Brain derived neurotrophic factor (BDNF) plays an important role in memory formation. Several studies show that BDNF expression is downregulated in AD brains in humans. H3K9 methylation, a determining factor in lower *Bdnf* expression showed age dependent elevation in levels in non-transgenic neurons and further increase in cortical neurons cultured from the hippocampal regions of 3xTg-AD mouse models (Walker et al. 2012).

At initial asymptomatic stages, neurofibrillary tangles (NFT) in cerebrum are restricted to the trans-entorhinal, entorhinal cortices and CA1 region of the hippocampus. Nucleolin (NCL), nucleophosmin (NPM) are major nucleolar proteins acting as histone-binding chaperones and are required for chromatin compaction and regulation of rRNA transcription through H3K9me2. The levels of H3K9me2 and H4K12ac showed decline in CA1 and DG regions of hippocampus neurons of post mortem tissues of AD patients (Hernández-Ortega et al. 2016).

Post translational histone modifications are triggered in response to Aβ, a signalling molecule derived from dysregulated amyloid precursor protein (APP) processing. Aβ oligomers are potent signalling molecules that indirectly modulate transcription by acetylating and methylating H3 lysine residues (AcH3 and H3me2) (Lithner et al. 2013). Aβ also induces genome wide hypomethylation in cerebral endothelial cell cultures, causing specific hypermethylation and repression of the gene for neprilysin which triggers Aβ deposition.

### 2.2.2 Histone Demethylases in Alzheimer's Disease

Recent work in the field has suggested a critical role for lysine demethylases in neurodegenerative diseases. Pharmacological inhibition of LSDI in a variety of neuroblastoma cells has been shown to block the mTORC1 pathway in a dose dependent manner. Inhibition of LSD1 was shown to trigger mTOR dependent activation of autophagy in neuroblastoma cells by transcriptionally activating the expression of SESN2. Chromatin Immuno Precipitation (ChIP) experiments showed direct binding of LSD1 to the Transcription Start Site (TSS) of the SESN2 promoter followed by activation of H3 acetylation and decrease of H3K27me3 in neuroblastoma cells. This establishes a novel neuroepigenetic mechanism that may offer new therapeutic routes targetting the autophagy-lysosomal pathway in neurodegeneration (Ambrosio and Majello 2018).

LSD1/KDM1A is an amine histone demethylase which in conjunction with the Co-REST complex, specifically demethylates mono-methylation and di-methylation of K4 on H3K4me1/2 but not on H3K4me3. LSD1 has many roles throughout development and can also be found in terminally differentiated cells throughout the brain. Recent studies showed that the loss of LSD1 in $LSD1^{CAGG}$ mice results in widespread hippocampus and cortex neuronal cell death. LSD1 is continuously required to prevent neuro degeneration that leads to learning and memory defects.

Gene Ontology and Gene Set Enrichment Analysis on transcriptome sequencing datasets with loss of LSD1 have implicated common pathways leading to neuronal cell death which includes activation of genes in the microglia and immune pathways, defect in oxidative phosphorylation, loss of synaptic transmission and failure to maintain cell cycle arrest. Thus loss of LSD1 affects multiple neurodegenerative pathways simultaneously with one or more of these pathways leading to neuronal cell death (Christopher et al. 2017).

### 2.2.3 DNA Methylation in Alzheimer's Disease

DNA methylation occurs due to the covalent addition of a methyl group from S-adenosyl methionine to the 5′ position of cytosines (5mC) linked to guanines (CpG dinucleotides). About 70% of the promoters in human genome are frequently enriched in CpGs forming the CpG islands. Recent observation suggests that the methylation status of CpGs is associated with transcription repression while methylation of CpGs in gene bodies promoted transcription. DNA methylation is dynamically regulated in the human cerebral cortex throughout the lifespan and involves differentiated neurons.

Diverse cell lines have shown lower levels of DNA methylation associated with AD. While global hypomethylation was detected in the entorhinal cortex region of post-mortem tissues from AD patients (Liang et al. 2008) widespread hyper methylation patterns correlating with higher levels of 5hmC and 5mC were observed in the middle frontal gyrus (MFG) and the middle temporal gyrus (MTG) regions of AD patients observed in a different study (Coppieters et al. 2014). Studies on single monozygotic twins discordant for AD showed a significant loss of DNA methylation in the temporal neo-cortex neuronal nuclei of the AD twin (Mastroeni et al. 2009).

Decrease in the levels of 5hmC was observed in the post-mortem tissues of hippocampal regions of AD patients when compared to their normal controls and also in the AD twin considered for the study (Chouliaras et al. 2013). However, recent study involving genome wide profiling of 5hmC using post-mortem brain samples of AD patients identified 517 differentially hydroxylated methylated regions (DhMRs) annotated to 321 distinct genes involved in formation of neuritic plaques (NPs) and 60 DhMRs annotated to 49 distinct genes associated with the formation of neurofibrillary tangles (NFTs). This suggests a new dimension of epigenetic regulation by 5hmC that might play an important role in brain aging and neurodegenerative disorders (Zhao et al. 2017).

Various studies on neuronal cells, patient tissues and animal models have recorded aberrant alterations in DNA methylation patterns associated with multiple genes in Alzheimer's disease (Table 3). Such changes in methylation were found to differ among transcription factor binding sites of tau promoter. Folate/methionine/homocysteine metabolism plays an important role in DNA methylation mechanisms. B2 dependent MTHFR (methylenetetrahydrofolate) catalyses the conversion of 5,10 methylenetetrahydrofolate to 5-MTHF which is the methyl donor for the re-methylation of homocysteine (Hcy). Studies on post-mortem prefrontal cortex tissue and peripheral lymphocytes of AD patients show hypermethylation in the promoter region of the MTHFR gene. It is hence well established that the methylation of DNA is critical to epigenetic processes associated not only with the normal brain function and aging but also with AD. Changes in expression of individual genes aids understanding of the pathways and mechanisms involved in AD. The APOὲ gene represents a bimodal structure with a hypomethylated CpG- poor promoter and a fully methylated 3′ CpG- island, containing the sequence for the ε4- haplotype (genetic risk factor for LOAD) (Wang et al. 2008).

Epigenome wide association studies in prefrontal cortex and superior temporal lobe from 147 AD patient sets identified an extended region of elevated DNA methylation in the HoxA gene clusters across a 48 kb region spanning 208 differentially methylated positions (DMPs) in CpG sites adding to the growing evidence of the involvement of Hox gene in Alzheimer's disease (Smith et al. 2018).

Recent genome wide association studies (GWAS) on DNA methylation in the supratemporal gyrus of 34 patients with AD and 34 controls identified 479 autosomal differential methylated regions (DMRs), the majority of which were hypermethylated in AD cases. These identified DMRs colocalise with other functional epigenetic signatures in brain tissues, most notably hypermethylated DMRs were enriched in poised promoters, characterized by the presence of both H3K4me3 and H3K27me3 (Watson et al. 2016).

**Table 3** Gene specific aberrant DNA methylation across different regions of the brain

| Sl. No | Gene | Regions of the brain | Methylation status | References |
|---|---|---|---|---|
| 1. | ANK1<br>Ankyrins | Human (entorhinal, temporal and prefrontal cortex | Increase in methylation in the gene body | De Jager et al. (2014), Lunnon et al. (2014) |
| 2. | APOε4<br>Genetic risk factor for LOAD | Human prefrontal cortex and lymphocytes | Increase in methylation in the promoter regions | Wang et al. (2008) |
| 3. | APP<br>Amyloid precursor protein | Human prefrontal cortex | Decrease in methylation in promoter region | West et al. (1995), Barrachina and Ferrer (2009) |
| 4. | TREM2<br>Triggering receptor expressed in myeloid cells 2 | Human hippocampus | Increase in methylation in the promoter region | Celarain et al. (2016) |
| 5. | NF-κB, COX2<br>Pro-inflammatory cytokines | Human frontal cortex | Decrease in methylation pattern in promoter region | Rao et al. (2012) |
| 6. | BDNF<br>A member of the nerve growth factor family of proteins | Human frontal cortex | Decrease in methylation pattern in the promoter region | Rao et al. (2012) |
| 7. | CDH3<br>Cadherin protein | Human (entorhinal, temporal and prefrontal cortex) | Increase in methylation in the gene body | De Jager et al. (2014), Lunnon et al. (2014) |
| 8. | CREB<br>Transcription factor involved in synaptic plasticity and cognition | Human frontal cortex | Increase in methylation in the promoter region | Rao et al. (2012) |
| 9. | CLU (APOJ)<br>Clusterin (Third most associated LOAD risk gene) | Human neural cells | Unknown | Nuutinen et al. (2005) |
| 10. | DUSP22<br>Dual specificity Phosphatase-22 | Human hippocampus | Increase in methylation pattern in the promoter region | Sanchez-Mut et al. (2013) |
| 11. | KDM2B<br>Lysine demethylase of H2B | Human entorhinal, temporal and prefrontal cortex | Increase in methylation pattern in gene body | De Jager et al. (2014), Lunnon et al. (2014) |
| 12. | MTHFR<br>Convert 5,10—MTHF to 5-MTHF | Human pre-frontal cortex and lymphocytes | Increase in methylation in promoter region | Wang et al. (2008) |

(continued)

**Table 3** (continued)

| Sl. No | Gene | Regions of the brain | Methylation status | References |
|---|---|---|---|---|
| 13. | PCNT (DIP2)<br>Pericentrin localizes to the centrosome and recruit proteins to the pericentriolar matrix | Human (entorhinal, temporal and prefrontal cortex) | Decrease in methylation in gene body | De Jager et al. (2014) |
| 14. | PP2A<br>Dephosphorylation of Tau | Neuroblastoma cells | Decrease in methylation in the promoter region | Vafai and Stock (2002), Zhou et al. (2008) |
| 15. | RHBDF2, HLA-DRB5<br>Involved in inflammatory responses in AD | Human entorhinal, temporal and prefrontal cortex | Increase in methylation in gene body | De Jager et al. (2014) |
| 16. | S100A2<br>S100 family of calcium binding proteins | Human cerebral cortex | Decrease in methylation in the promoter region | Siegmund et al. (2007) |
| 17. | SLC2A4<br>Involved in neural development | Human prefrontal cortex | Increase in methylation in the promoter regions | Yu et al. (2015) |
| 18. | SORBS3<br>Encoding a cell adhesion expressed in neurons and glia | Human (entorhinal, frontal cortex, temporal), APP/PS1 and 3Xtg-AD | Increase in methylation pattern in promoter region | Siegmund et al. (2007), Sanchez-Mut et al. (2013) |
| 19. | SPTBN4<br>Spectrin beta 4 | APP/PS1,3Xtg-AD and human frontal cortex | Increase in methylation in the promoter | Sanchez-Mut et al. (2013) |
| 20. | TBXA2R<br>Thromboxane A2 receptor | 3Xtg-AD, APP/PS1 and human | Increase methylation in promoter region | Sanchez-Mut et al. (2013) |
| 21. | IGFBP7<br>Insulin- like growth factor binding protein 7 | APPPS1-21 and human frontal cortex | Increase in methylation in promoter region | Agbemenyah et al. (2014) |
| 22. | BACE<br>β-Site APP-cleaving enzyme | TgCRND8 Mice models | Decrease in methylation in promoter region | Fuso et al. (2008) |
| 23. | PSEN1<br>Component of γ-secretase complex | TgCRN8 Mice models | Decrease in methylation in promoter region | Fuso et al. (2008) |
| 24. | Neprilysin<br>An Aβ degrading enzyme | Murine cerebral endothelial cells | Increase in methylation in promoter region | Chen et al. (2009) |

## 3 Huntington's Disease (HD)

Huntington's disease is a rare and progressive neurodegenerative disorder often identified by polyglutamine (Poly Q) repeats on the genome. This disease is inherited as a fully penetrant autosomal dominant trait caused by expansion of CAG repeats within exon1 of the Huntingtin gene. A toxic gain of function (GoF) mutant of Huntingtin protein disrupts multiple intracellular pathways, leading to cognitive impairments and motor disorders, involving the hallmark feature of chorea (involuntary jerky movements of the face and limbs) and gait abnormalities accompanying progressive neurodegeneration. This mutation either leads to depletion of normal Htt (which plays an important role in endocytosis and vesicle trafficking) disrupting synaptic functions or leads to the formation of a misfolded mutant protein (mHtt) which impedes vesicular trafficking and diverse intracellular processes (Zuccato et al. 2010). This repeat instability is regulated by various epigenetic mechanisms which include changes in histone modifications, alterations in DNA methylation patterns and chromatin remodelling factors which in turn influence the degree of striatal degeneration and the age of onset of Huntington's disease (Bedford and Brindle 2012).

### *3.1 Histone Methylome in Huntington's Disease*

**Huntington's disease** is characterised by transcriptional repression of key neuronal transcripts like neurotransmitters, growth factors and their receptors. Repression of dopamine receptor 2(Drd2), pre-enkephalin (Penk1), cannabinoid receptor (Cb2) and brain derived neurotrophic factor (Bdnf) are implicated in the pathogenesis of Huntington's disease. A critical event underlying transcriptional dysregulation of these key genes is the alteration in the chromatin structure in regulatory regions of these genes. Hence it is important to understand HD pathogenesis through the dimension of regulation of chromatin structure and epigenetic modifications (Zuccato and Cattaneo 2007).

Alterations in H3 methylation are implicated in cognition impairment and intellectual disabilities in Huntington's disease. Early studies of aberrant methylation of histones in HD on mice models R6/2 and N171-82Q demonstrate elevated levels of H3K9me2 and H3K9me3 in the striatum and cerebellum. The levels of histone methylation were shown to decrease on treatment with mithramycin, which prevents H3 hypermethylation in the R6/2 mouse cell line (Ferrante 2004). Mithramycin was shown to prevent brain atrophy, ventricular atrophy and striatal neuronal atrophy seen in R6/2 mice.

Chromatin immunoprecipitation on the HD locus of R6/1 and R6/2 HD transgenic mouse lines has shown correlation in the levels of H3K9me2 (heterochromatin), H3K9ac (euchromatin) and H3K4me3 (Transcription initiation) with the expression levels in the striatum and cerebellum. Also, the levels of H3K36me3 (mark associated with active transcription) and phosphorylated serine of RNA PolII corelated strongly with CAG instability in R6/1 and R6/2 mice. Furthermore, RNA Pol II at the promoter-proximal region of the HD locus was increased in the striatum when compared to cerebellum contributing to the tissue specific instability of CAG repeats as found in HD (Goula et al. 2012).

Studies on ERG-associated with SET domain (ESET), a histone H3K9 methyl transferase has yielded interesting insights on ESET regulation of neuronal survival in HD models. The levels of ESET/SETDB1 and H3K9me3 are altered in the striatal neurons of HD patients and are insignificantly increased in caudate nucleus in HD brains as compared to control HD striatal tissue brain samples. Combined administration of mithramycin and cystamine were found to significantly reduce the expression level of ESET in R6/2 mice and H3K9me3.This combinatorial treatment also conferred extended survival (by 40%,) enhanced body weight and improved motor activity ameliorating neuropathological conditions in R6/2 mice models (Ryu et al. 2006).

ChIP sequencing experiments on NeuN-selected neuronal cell nuclei from postmortem prefrontal cortical samples for six HD cases and six non-neurologic controls showed an average of 63% of total H3K4me3 reads mapping to transcriptional start site- proximal peaks and 36% of the distal peaks colocalizing to known enhancer sites. Distal peaks showed differential enrichment of six transcription factors and chromatin remodellers including EZH2 and SUZ12 of the PRC2 (Polycomb Repressive) complex. In HD, PRC2 inhibition is associated with upregulated H3K4me3 (Dong et al. 2015).

Knockdown of histone demethylase (JARID1C) in R6/2 mice models and human HD brains showed that mutant HTT acts to activate cell signalling pathways that impact H3K4me3, which spreads broadly downstream of the transcription start site (TSS). Reduction in the levels of JARID1C or SMCX in primary neurons was found to reverse the downregulation of key neuronal genes triggered by the expression of mutant HTT. This implies the consideration of JARID1C as a potential target for epigenetic therapy for Huntington's disease (Vashishtha et al. 2013). Genome wide mapping approach identified a large number of epigenetically altered loci in the neuronal HD genome, including loss of H3K4me3 and excessive DNA methylation on the hairy and enhancer of split 4 (HES4) promoter as well as altered expression of HES4 and its target genes MASH1 and p21 involved in striatal development. The epigenetic changes at the HES4 gene may hence be used as a novel biomarker for clinical and histopathological outcomes in Huntington's disease (Bai et al. 2014).

The involvement of chromatin remodeling complexes in the pathogenesis of Huntington's disease has been well established. Mutant form of huntingtin (mHtt) has been shown to induce the transcription of α- thalassemia/mental retardation X linked (ATRX), a DNA dependent ATPase/helicase belonging to the Rad54-like subfamily of SWI/SNF chromatin remodelling proteins. Knock down of ATRX was shown to decrease the levels of promyelocytic leukemia nuclear body (PMLNB) and H3K9me3 suggesting that ATRX mediated organization of pericentromeric heterochromatin through increase in H3K9me3 in striatal cells plays a vital role in HD pathogenesis. Elevation in the expression levels of chromatin remodeller ATRX protein in white blood cells of pre-symptomatic and symptomatic HD is a distinct epigenetic signature of Huntington's disease (Lee et al. 2012).

The modulatory polyglutamine region of the huntingtin protein facilitates the activity of epigenetic silencing complex PRC2 and its methyltransferase activity which are important for normal murine embryonic development. Full length endogenous huntingtin was found to be associated with PRC2 subunits in wildtype murine embryoid bodies and with H3K27me3 at HoxB9 while embryos lacking huntingtin showed distinct impairment of PRC2 regulation of Hox gene expression and chromatin silencing function in embryos. Lack of huntingtin protein led to impaired PRC2 epigenetic gene and chromatin silencing function in murine embryos and impaired reestablishment of global histone H3K27me3 in developing embryos, whereas full length recombinant human huntingtin specifically stimulated tri-methyl transferase activity of polycomb repressive complex 2 (PRC2) a multi protein complex with histone methyltransferase activity both in Hdh(Q111) embryoid bodies of mouse and in vitro (Seong et al. 2010).

Recent studies on HD models of *Drosophila melanogaster* showed that loss of function mutation of EZH2, the catalytic subunit of PRC2 responsible for H3K27me3 (mark for facultative heterochromatin), enhances neurodegeneration. Furthermore, epigenetic marks such as H3K27me3 (facultative heterochromatin) showed specific effects on HD pathology with reduction of demethylases Utx1 which rescues HTT induced pathology. Reduction in the levels of key methylase components of PRC2 complex led to aggressive pathology. Manipulation of enzymes which regulates histone marks representative of constitutive heterochromatin like PR-SET7 and HMT420 showed no effects on HD pathology (Song et al. 2018).

Microarray data analysis of HD brain revealed that the RE1 silencing Transcription factor (REST) bound genes are preferentially repressed in HD patients. REST, a master regulator of neuronal genes is highly expressed in immature central nervous system cells and in mature neurons, and is linked to HTT. Wild type huntingtin protein sequesters REST protein in the cytoplasm denying access of REST to its cis-regulatory elements on its target genes such as BDNF. Wild type Htt affects BDNF gene transcription by stimulating the activity of specific promoter of the complex BDNF gene. H3K4me3 enrichment was reduced at the REST/NRSF promoter II, thus suggesting that reduced transcription could be the consequence of changes in chromatin structure at REST binding site and BDNF locus (Buckley et al. 2010).

Recent studies on nuclear lamins showed increased levels of Lamin B in the putamen of Huntington's disease patients as well as in the striatum of R6/1 mouse models of HD. R6/1 mouse model showed increase in the levels of lamin B1 and B2 in the striatum and cortex from the early stages of Huntington's disease while showing elevated levels in hippocampus only at late stages. Lamin A and C were also found to be enhanced in the striatum and hippocampus at late stages but were not altered in the cortex. However, protein levels of the lamin B receptor remained unchanged (Alcala et al. 2014).

**Table 4** Region wise gene expression changes in HD/HD models

| Gene | Source: | HD model: | Expression status | References |
|---|---|---|---|---|
| *Dnmt3a* | Striatum | R6/2 mice | Decrease | Ng et al. (2013) |
| *Dnmt1* | STHdhQ111 cells | Cells | Decrease | Ng et al. (2013) |
| *DNMT1* | Cortex | Human | Disrupted coexpression | Narayanan et al. (2014) |
| *DNMT3A* | Cortex | Human | Disrupted coexpression | Narayanan et al. (2014) |
| *Gadd45a* | Striatum | R6/2 mice | Decrease | Tang et al. (2011) |
| *Gadd45b* | Muscle | N171-82Q mice | Decrease | Ng et al. (2013) |
| *Gadd45g* | STHdhQ111 cells | Cells | Increase | Ng et al. (2013) |
| *Rnf4* | Striatum | R6/2 mice | Decrease | Tang et al. (2011) |
| *Rnf4* | Muscle | N171-82Q mice | Increase | Jia et al. (2015) |

## 3.2 DNA Methylation in Huntington's Disease

Several studies have observed aberrant DNA methylation patterns in Huntington's patients and HD model systems. Adenosine $A_{2A}$ receptor ($A_{2A}R$), a GPCR which stimulates adenylyl cyclase is highly expressed in the striatum especially in the GABAergic medium sized neurons (MSNs) that express enkephalin. The receptor is severely affected in Huntington's disease (HD). Reduced levels of $A_{2A}R$ were observed in the putamen of HD patients and striatum of R6/1 and R6/2 mice model investigated at later stages of Huntington's disease. Furthermore, an increase in 5mC levels and reduction in 5hmC levels in the 5'UTR of ADORA2AR in the putamen of HD patients was observed. This has led to a gene therapy designed with DNA methyl transferase inhibitors targeted to increase the levels of A receptor in animal models. Expression levels of different genes involved in Huntington's disease specific to brain region are shown in Table 4.

5hmC plays an important role in neurodevelopment. Genome wide reduction of 5-hmC signal in the striatum and cortex of YAC128 (Yeast chromosome transgene with 128 CAG repeats) HD mice has been reported and disease specific differentially hydroxymethylated (DhMRs) in gene body have been identified. These DhMRs associated genes are involved in a number of canonical pathways including neuronal development/differentiation and neuronal function and survival. Alterations of these pathways could play role in HD, thus suggesting that reduction of 5-hmC marker is a novel epigenetic signature in HD featuring impairment of neurogenesis, neural function and survival (Wang et al. 2013).

Recent genome wide DNA methylation profiling of human cortex tissues with a subset of matched liver tissues, from a cohort of HD and control individuals identified novel site-specific differential DNA methylation patterns spanning the promoter and intragenic regions of the HTT, including a differentially methylated

CTCF-binding site in the HTT promoter. This CTCF site displayed increased occupancy in cortex tissue, with higher HTT expression than the one in the liver (De Souza et al. 2016). DNA methylation profiles investigated from whole blood of Huntington's patients however showed no recognizable changes in methylation patterns in HD implying that blood compartments are not strong enough to prove as a viable biomarker to predict age-of-onset of HD (Zadel et al. 2018).

## 4 Parkinson's Disease

Parkinson's Disease (PD) is the second most common neurodegenerative disorder after Alzheimer's disease and affects more than 6 million people across the world. The pathological hallmark of PD involves motor dysfunctions due to loss of dopamine producing neurons in nigro-striatal pathways (Braak et al. 2002), lack of control of voluntary movements, tremor, instability in postures and muscular rigidity. The loss of dopaminergic neurons in substantia nigra pars compacta (SNpc) results in the impairment of the execution of co-ordinated movements. The disease stage is also accompanied by formation of fibrillary cytoplasmic inclusions, known as Lewy bodies which contain ubiquitin and α-synuclein.

### *4.1 DNA Methylation Profiles in Parkinson's Disease*

Genome wide association studies (GWAS) have revealed variations in two of the familial PD genes SNCA and LRRK2 as an important risk factors for sporadic PD (Satake et al. 2009). Mutations in Parkin (Park2) and Pink1 (PTEN induced kinase protein 1) are predominant risk factors for PD (Urdinguio et al. 2009). More than 90% of clinically reported cases of PD are sporadic implying a strong interplay of epigenetic factors in the pathogenesis of Parkinson's disease (Gapp et al. 2014). SNP's in the promoter region or in the 3'UTR of the SNCA gene have been identified in PD patients. SNCA encodes presynaptic protein α-synuclein. Point mutations and multiplications of SNCA causes familial Parkinsonian syndromes with high penetrance. Investigations on the influence of epigenetic changes of SNCA expression on SNpc, putamen and cortex of sporadic PD samples of patients revealed hypomethylation of CpG islands at the promoter and intron 1 of the SNCA gene. Hypomethylation of SNCA intron1 was observed on peripheral blood samples of 490 patients with sporadic PD (de Boni et al. 2015).

High resolution methylation study on alpha synuclein gene (SNCA) reveals no significant difference in the methylation pattern of promoter and intron 1. This inconsistency might be due to the difference in sequencing techniques involved or due to different CpG sites investigated. This cannot be considered as a specific biomarker for PD as similar patterns have also been found in dementia with Lewy bodies and Alzheimer's Disease (Funahashi et al. 2017). Mislocalisation of the DNA methyl transferase DNMT1 has been observed in post mortem brain samples of PD and LBD patient brain samples. DNMT1 was found to be sequestered in the cytoplasm leading to reduced nuclear fraction in the cells. A global hypomethylation of genes was consistent with the decreased levels of DNMT1 (Desplats et al. 2011).

DNA methylation in sporadic PD is mostly based on homocysteine cycle dysregulation. Comparison of genome wide methylation profile of sporadic PD cases with aged and sexed matched healthy controls revealed a single hypomethylated gene CYP2E1, in the putamen and cortex region of the brain during the later stages of the disease (Kaut et al. 2012). CYP2E1 is predominantly expressed in neurons and colocalized to tyrosine hydroxylase in rat substantia nigra. Enhanced CYP2E1 activity facilitates the formation of potentially toxic metabolites like isoquinolines which are structurally related to dopaminergic neuron 1—methyl 4 phenyl 1,2,3,6 tetra hydro pyridine (MPTP). Thus, altered methylation of genes such as CYP2E1 may contribute to individual susceptibility to PD.

Parkinson's patients undergo a circadian fluctuation with symptoms that involve worsening of motor symptoms during afternoon and evening. Body temperature, blood pressure and cortisol synthesis are also affected in PD. Several clock genes including period (PER1, PER2 and PER3) cryptochrome (CRY1 and CRY2), CLOCK, aryl hydrocarbon receptor nuclear translocator like (ARNT L1, also called BMAL1) and NPAS 2 have been identified in PD pathology. The promoters of seven clock genes examined through Methylation Specific PCR showed CpG islands associated with some of these genes. While most of the clock gene promoters were devoid of methylation, the methylation levels were detectable only in the CRY1 and NPAS2 promoters. The methylation frequency of the NPAS2 was significantly decreased in patients (Lin et al. 2012).

TNF-α, an important inflammatory factor has been also implicated in the pathogenesis of PD. Widespread hypomethylation of TNF-α promoter in the SNpc compared to cortex both in PD patients and in neurologically healthy controls, indicating increased susceptibility of neurons located in SNpc to TNF-α mediated inflammation. Increased concentration of plasma total homocysteine (tHcy) in patients with Parkinson's disease is responsible for cognitive impairment, neuropathy and depression in these patients.

Markers of neurodegeneration (APP, α synuclein) are also associated with cognitive impairment. Blood samples of 87 patients with PD analysed for tHcy, methylmalonic acid (MMA), vitamin B, folate, S-adenosyl methionine (SAM), S-adenosyl homocysteine (SAH), and amyloid-β showed that PD patients with no cognitive impairment had a higher plasma SAM/SAH ratio than with patients with mild or severe cognitive impairment. This relates the cognitive function in patients with Parkinson disease to a higher methylation potential (SAM/SAH ratio) and higher plasma vitamin B6. Vitamin B6 on cognitive function shows an indirect relation to enhanced methylation status and reduction in amyloid β production. The concentrations of tHcy, MMA, and SAH in plasma were higher in patients receiving single treatment with L-dopa compared to the other treatment groups. cognitive function in patients with Parkinson disease was related to a higher methylation potential (SAM/SAH ratio) and higher plasma vitamin B6 (Obeid et al. 2009).

More recent analysis of blood samples from PD patients and controls suggest an increased age acceleration preceding the onset of motor and non-motor symptoms which can be used as a biomarker for PD (Horvath and Ritz 2015).

Epigenome wide association studies (EWAS) from blood samples of PD patients and PD patients with anxiety led to the identification of more than 12,000 genes with differential methylation patterns. These genes are involved in brain centric pathways such as neuroactive ligand-receptor interaction, neutrotrophin signalling, in neurodevelopment and in neuronal apoptosis (FANCC and TNKS2) (Moore et al. 2014). Recent studies on mitochondrial DNA (mtDNA) showed a significant loss of 5-methyl cytosine levels in the D-loop region of mitochondria found in substantia nigra in Parkinson's disease suggesting that mtDNA epigenetic modulation plays an important role in various neurodegenerative disorders including Parkinson's (Blanch et al. 2016).

## 4.2 *Histone Methylation in Parkinson's Disease*

Transcription factor Nurr1, which plays a key role in the development and maintenance of the midbrain dopamine cells, plays a part in the pathogenesis of PD and provide and provide an important link to chromatin modifying complexes. Nurr1 is significantly reduced in patients affected with PD. The Co-REST repressor complex which plays a critical role in Nurr1-mediated transcriptional repression, recruits a group of proteins consisting of HDACs, the histone methyl transferase G9a and LSD1, which target promoters leading to transcriptional repression (Saijo et al. 2009).

Microglial activation states can produce either detrimental or beneficial effects in the Central Nervous System. Dysregulated microglial activation state amplifies neuronal damage and contributes to the pathogenesis of Parkinson's disease. Microglial activation states have been classified into two major phenotypes M1 (classical activation) and M2 (alternative activation). Activated microglia are present in the vicinity of degenerating neurons in the substantia nigra regions of PD patients. These activation states may change throughout the pathological process of PD. H3K27me3 demethylase Jumonji domain containing 3 (Jmjd3) plays an important role in M2 polarization. Suppression of Jmjd3 in murine N9 microglial cells and in the substantia nigra region of C57BL/6 mice model for PD inhibited M2 polarisation. The inhibition of M2 polarisation and exaggerated M1 microglial inflammatory responses led to extensive neuronal death. This suggests that Jmjd3 is able to enhance the polarization of M2 microglia by modifying histone H3K27me3, which plays a pivotal role in the switch of microglia phenotypes contributing to the pathogenesis of PD (Tang et al. 2011). PINK 1, which functions as a regulator of mitochondrial homeostasis and apoptosis, encodes PINK1 protein which interacts and phosphorylates ectoderm development Polycomb histone methylated modulator (EED/WAIT 1), inducing relocalisation of EED/WAIT1 to mitochondria. This interaction may regulate H3K27 tri methylation through positive and negative effects on EED/WAIT1 (Berthier et al. 2013).

Recent investigation using αS (alpha- synuclein) transgenic *Drosophila melanogaster* and human neuroblastoma SH-SY5Y cells showed that αS selectively enhances H3K9 mono- and di-methylation. Epigenetic silencing affects the neural cell adhesion molecule L1 and the synaptosomal-associated protein SNAP25. Eukaryotic Histone Methyl Transferase 2 (EHMT2) might be a key regulator of this modification. Further investigation on REST target genes harbouring RE1 sites, revealed that the promoter region of *SNAP25* occupied with H3K9me2 upon overexpression of αS results in reduced gene expression and ultimately lower protein levels. Thus overexpression of alpha synuclein alters the distribution of histone marks on genes associated with the REST complex resulting in disturbed synaptic activities (Sugeno et al. 2016).

# 5 Regulation of Histone Methylation in Neuro psychiatric Disorders

## *5.1 Epigenetics of the Neuropsychiatric Disorders*

Neuropsychiatric disorders represent a complex and heterogenous group of disorders involving a variety of factors that regulate pathophysiology of such diseases. Hence, it is difficult to correlate the pathophysiology of these disorders to a single gene. There is significantly increasing evidence that epigenetic mechanisms mediate gene-environment interactions during critical periods of the lifespan and manifest as mental illness (Kendler 2001; McEwen 2000). Though the influence of epigenetic mechanisms on neuropsychiatric disorders are primarily been understood through alteration in acetylation patterns and DNA methylation, the role of histone methylation in regulating Schizophrenia, ADHD, OCD and Bipolar Disorders are emerging in the field.

Alterations in brain transcriptomes in mood and psychosis spectrum disorders are associated with alterations in histone lysine methylation and other epigenetic regulators of gene expression. Cognition decline is often associated with age-dependent decline of synaptic function in brain regions such as hippocampus and prefrontal cortex which are crucial for memory formation and consolidation. Human prefrontal cortex (PFC) plays an important role in complex cognitive behaviour, personality, decision making and orchestration of thoughts and actions. Both the hippocampus and prefrontal cortex regions of the brain are frequently impacted in the neural circuitry of mood and psychosis spectrum disorders. Histone methyl transferase MLL1 is predominantly expressed in the anterior subventricular zone (SVZ) and olfactory bulb of the hippocampus and facilitates proliferation and neurogenesis of Neural Stem Cells (Lim et al. 2009).

## 5.2 Schizophrenia

Schizophrenia is one of the major psychiatric disorders whose onset begins at adolescence, though cognitive disturbances are evident at much earlier phases. Schizophrenia is highly heterogeneous and manifests through major symptoms like psychosis with delusions, hallucinations and disorganised thoughts, cognitive dysfunction and depressed mood and negative symptoms including anhedonia, social withdrawal and poor thought and speech output (Ibrahim and Tamminga 2011). The dopamine hypothesis of Schizophrenia states that the hyperactivity of dopamine D2 receptor neurotransmission in subcortical and the limbic brain regions contribute to positive symptoms of Schizophrenia while the negative and cognitive symptoms are caused due to hypo-functionality of dopamine D1 receptor neurotransmission in the prefrontal cortex (Carlsson and Lindqvist 2009). Further, alterations in GABAergic mRNA expression play a key role for prefrontal dysfunction in Schizophrenia and other neuro developmental disorders.

Anti-psychotics prescribed for Schizophrenia target the dopaminergic or serotonergic receptor system and show therapeutic value in nearly 75% of the patients. The disabling and significant feature of Schizophrenia is cognitive impairment for which no pharmacological intervention has shown therapeutic benefits so far.

### 5.2.1 Histone Methylation in Schizophrenia

In human brain, glutamatergic neurotransmission is mediated through ligand-gated ion channels, NMDA (*N*-methyl-D-aspartate), AMDA (α-amino-3-hydroxy-5-methyl-4-isoxazolepropionic acid), kainate (KA) and G-protein coupled metabotropic receptors (m-GluR). Gene promoter specific histone lysine methylation is involved in developmental regulation and maintenance of expression of ionotropic and metabotropic glutamate receptors. Native ChIP assays on cerebral cortex region of samples from diverse age groups have identified histone methylation marks at proximal promoters of 16 ionotropic and metabotropic glutamate receptors (GRIN1, 2A-D, GRM1, 3, 4, 6, 7) genes. H3K4me2 and H3K4me3 (active chromatin marks) showed significant correlation with mRNA levels in immature and mature cerebral cortex while H3K27me3 and H4K20me3 (Silencing Chromatin marks) were upregulated in adult cerebellum and do not correlate with transcription. Differential histone H3-K4 methylation at gene promoters of glutamate receptor gene could thus be considered as a chromatin marker for transcriptional dysregulation in various neuropsychiatric disorders (Stadler et al. 2005).

Hypometabolism and altered gene expression in the prefrontal cortex are associated with negative symptoms and cognitive deficits of Schizophrenia. Cellular metabolism regulates chromatin structure including covalent histone modifications, myelination and other functions. Studies involving histone and gene transcript profiling in the post-mortem prefrontal cortex of 41 subjects with Schizophrenia and 41 matched controls identified high levels of H3R17me associated with downregulated metabolic gene expression in the prefrontal cortex of a subset of subjects with Schizophrenia (Akbarian et al. 2005).

Study involving peripheral blood lymphocytes from 19 healthy controls and 25 patients with Schizophrenia has provided evidence that this disease is associated with a restrictive chromatin state. Elevated levels of H3K9me2 were observed in Schizophrenia patients as compared to controls. Pharmacological treatment with Trichostatin A (inhibitor of class I, II and IV HDACs effectively preventing the deacetylation of H3K9) decreases the levels of H3K9me2 in lymphocyte cultures from patients with healthy controls (Gavin and Sharma 2009).

In contrast to the methylation patterns observed in specific gene promoters, an interesting study on Schizophrenia patients showed an overall increase in the levels of H3K9me2 in both the lymphocyte and post mortem parietal cortex of patients with Schizophrenia as compared to their non-psychiatric controls. The mRNA expression profiles of G9a and GLP responsible for bulk of genomic H3K9me2 modification and SETDB1, (the only euchromatic HMT to specifically di or tri-methylate H3K9) was also significantly increased thus positively correlating with H3K9me2 levels in the brain regions investigated. Sex-dependent restrictive epigenome studies on 74 participants, (40 patients with Schizophrenia (19 women, 21 men) and 34 healthy individuals (19 women, 15 men) indicate that men with Schizophrenia expressed the highest levels of G9a, SETDB1 mRNA and H3K9me2 protein levels as compared to women (Chase et al. 2013). These studies establish histone methyl transferases as potential therapeutic targets for Schizophrenia and for diagnosis of the disorder.

Maturation of human PFC and rodent cerebral cortex are accompanied by progressive increase in GABAergic mRNA levels, including *GAD1*, which encodes a key enzyme for GABA synthesis. These developmentally regulated changes in mRNA levels were associated with chromatin remodeling at *GAD1/Gad1* and other GABAergic gene loci, which correlate with increase in trimethylation of H3K4 in both systems. Post mortem samples from dorso-rostral pole of the frontal lobe of Schizophrenia patients showed significant deficits for *GAD1* mRNA and elevation in H3K4me3 levels in females, but not in males. Decrease in GAD1 mRNA levels corresponded to a decrease in H3K4me3 levels (open chromatin mark) and increase in the levels of the repressive mark H3K27me3. MLL1 is expressed in cortical interneurons and regulates H3K4 methylation at GABAergic gene promoters. Study on male C57BL/6 mice using clozapine (an atypical anti-psychotic drug) showed threefold increase in GAD1 associated H3K4me3 in comparison to the controls. MLL1 occupancy at the Gad1 promoter showed a significant twofold increase after a single dose of clozapine. Thus, clozapine induced histone methylation at the GAD1 locus increases MLL1 expression and its recruitment to the GABAergic promoter (Huang et al. 2007).

A recent study that interrogated histone modifications associated with open chromatin in neurons versus the dorsolateral prefrontal cortex and anterior cingulate cortex of Schizophrenia patients has provided interesting insights on the epigenetic implications of cell-type specific genome organisation and function in the human brain and other tissues. 157 reference maps were generated from dorsolateral prefrontal cortex (PFC), anterior cingulate cortex (ACC) from neuronal (NeuN+), neuron depleted (NeuN-) and bulk tissue chromatin for epigenetic marks H3K4me3 and H3K27ac. Non-neuronal chromatin was concordant with epigenomic signatures of cortical homogenates from multiple sources investigated in the study while a significant epigenomic distance was observed in histone methylation and acetylation profiles obtained from ACC and PFC neurons. H3K4me3 and H3K27ac signatures in neuronal chromatin were significantly overrepresented by risk variants for Schizophrenia and other neuropsychiatric disorders (Girdhar et al. 2018).

Exome sequencing data from 231 Schizophrenic patients and 34 control trios, identified two de novo loss of function (LoF) variants in the SETD1A gene (which encodes a subunit of histone methyl transferase) which provide evidence for a more general role of chromatin regulators in Schizophrenia (Takata et al. 2014; Girdhar et al. 2018). Recent study involving whole-exome sequences of 4264 Schizophrenia patients identified a strong genome wide association between loss of function (LoF) variants and Schizophrenia risk in SETD1A implying epigenetic dysregulation in the H3K4 methylation pathway in Schizophrenia (Singh et al. 2016).

Whole genome association studies from peripheral venous blood of a homogeneous population of China (119 Schizophrenia patients, 119 recruited from homogenous population in China) identified JARID2 (Jumonji AT rich interactive domain 2) within the Schizophrenia susceptibility locus on chromosome 6p22 to confer genetic risk in multiple populations (Liu et al. 2009). JARID2 plays an essential role for binding of PcG proteins (PRC 2) to target genes leading to transcriptional repression through catalysing the di and tri methylation of H3K27. GABAergic neuronal markers including GAD67 and REELIN (RELN) have been shown to be markedly downregulated in Schizophrenia. REELIN (RELN) an extracellular matrix glycoprotein that controls neuronal cell migration and the lamination of the cortico-limbic structures during embryonic development plays a major role in brain development and maturation. Post mortem brain cohorts have demonstrated reduced RELN expression (by 50%) in the prefrontal cortex (PFC), temporal cortex, hippocampus and caudate nuclei of patients with Schizophrenia (Guidotti et al. 2000).

An interesting study on pregnant Swiss Albino ND4 mice models exposed to prenatal restraint stress (PRS) showed that offspring born from stressed mothers display Schizophrenia like behavioural endophenotypes as compared to their controls (offspring born from mothers without stress). Decrease in levels of RELN, GAD67 and BDNF expression and increased levels of DNMT1 and DNMT3a were observed in the GABAergic neurons of the frontal cortex and hippocampus of the offspring born from PRS exposed mice (Matrisciano et al. 2013). Heterozygous reeler mice haplo-insufficient in RELN when treated with HDAC inhibitors Trichostatin-A and Valproic Acid for 15 days showed increase in DNA demethylase activity and restored RELN expression.

GAD67 is an enzyme that catalyses the decarboxylation of glutamate to form GABA in chandelier type GABA interneurons and is associated with working memory deficits in Schizophrenia (Lewis et al. 2005). Neurons in the dorsolateral prefrontal cortex (DLPFC) of post mortem samples of Schizophrenics showed a pronounced decrease in GAD mRNA levels in the neurons of layer I (40%) and layer

II (48%) and an overall 30% decrease in the layer III to VI (Akbarian 1995). The decrease in the level of GABAergic transmission in Schizophrenia is associated with the increased expression of DNMT1. mRNA levels of DNMT1, DNMT3a and DNMT3b were measured in Broadman's area 10 (BA 10), Caudate nucleus (CN) and putamen from post mortem Schizophrenia patient samples. A two fold increase in the mRNA levels of DNMT1 was observed in GABAergic neurons of BA10 layers and the neurons of CN and PT in SCZ while increased expression of DNMT3a was restricted to cortical layer I and II GABAergic neuron in SCZ (Zhubi et al. 2009).

## 5.3 *Bipolar Disorders*

Bipolar disorder is a chronic depressive condition characterized by manic-depressive illness, unusual shifts in mood and energy, activity levels and hypomanic episodes reflecting in inability to carry out day to day tasks. Four basic types of bipolar disorders have been documented, with majority of patients being diagnosed either with Bipolar Disorder I (BPI) manic or mixed episodes or Bipolar Disorder II (BPII) with depressed episodes.

### 5.3.1 Histone Methylation in Bipolar Disorders

The Synapsin family of neuronal phosphoproteins composed of three genes (SYN1, SYN2 and SYN3) are involved in synaptogenesis, synaptic transmission and synaptic plasticity and play significant roles in several disorders such as Schizophrenia, Bipolar disorder and epilepsy. Recent investigations using Chromatin Immuno Precipitation assays on the Broadmann Area 10 (BA10) of the prefrontal cortex of post mortem brains of 13 BD patients showed significant increase in the expression profiles of synapsin variants (SYN1a and SYN2a). The upregulation in the synapsin genes corresponded to a significant enrichment of H3K4me3 (open chromatin mark) levels at the synapsin promoters (Cruceanu et al. 2013).

Depressive illness is correlated with dysregulation of epigenetic regulatory mechanisms, particularly the transcriptionally repressive di-and tri -methylation of histone 3 lysine 9 (H3K9me2/me3) in nucleus acumens (NAc), region involved in the development of anhedonia, the hallmark for depression. Study on C57B1/6 male mice models showed that repeated cocaine abuse potentiated depressive behaviour through reduction in H3K9me2 and G9a/GLP levels in NAc which enhances susceptibility to subsequent social stress.

Lysine demethylases, specifically the Jumonji domain containing demethylases 2 (JmjD2) family that act on H3K9 and H3K36 methylation machinery are critical epigenetic regulators of etiopathology of depression and related disorders. Study on C7B1/6 mice model showed that except JmjD2, the expression of all other known members of JmjD2a, b and c were downregulated in depressed mice in the NAc region of the mice. Systemic administration of JMJD inhibitor (DMOC) induces depression like symptoms in mice resulting in significant increase in the levels of H3K9me2 and H3K9me3 in NAc region (Pathak et al. 2017). However, direct correlation of these epigenetic dysregulation with bipolar disorder has not been studied yet.

Genome wide association studies across over 60,000 participants from the Psychiatric Genomics Consortium investigating common pathways across Schizophrenia, Bipolar disorder and major depression revealed the strongest association with histone methylation. Histone H3-K4 methylation featured among the top-hits in the Bipolar disorder along with association of multiple immune and neuronal signalling pathways (The Network and Pathway Analysis Subgroup of the Psychiatric Genomics Consortium 2015).

Bipolar disorders and other major psychosis disorders involve dysfunction in GABAergic neurotransmission. Downregulation of glutamic acid decarboxylase regulatory network (GAD1) causes a decrease in the expression of the glutamic acid decarboxylase and impaired gamma aminobutyric acid neurotransmission in brain. The decreased GABAergic neurotransmission is related to the cognitive dysfunction.

DNA methylation changes play a role in the pathophysiology of psychotic disorders. Epigenetic association study targeting GAD1 regulatory network genes from post mortem hippocampal human brain tissue of 8 patients with Bipolar disorder identified DNA methylation patterns to be distinct across circuit locations within the tri-synaptic pathway. 11% of CpG sites within GAD1 regulatory network were identified as DMPs suggesting that DNA methylation is an active process in the dysregulation of GABAergic inter neuronal function. Genes MSX1, CCND2 and DAXX with differential methylation profiles within the GAD1 regulatory network were identified in disease association. *MSX1* encodes Msh homeobox 1 and is a regulator of early central nervous system and craniofacial development and is unique to the hippocampus. *MSX1* is expressed at higher levels in the adult hippocampus than in the foetal hippocampus. It interacts with SUZ12, a component of the Polycomb Repressive Complex 2, to direct H3K27me3 to targeted genomic locations (Ruzicka et al. 2015).

Genome wide methylome analysis from peripheral blood samples of three patients with bipolar disorders using Methyl-DNA immunoprecipitation in association with high-throughput sequencing (MeDIP Seq), identified thousands of differentially methylated regions preferentially located in promoter 3'UTRs and 5'UTR of the genes. Distinct patterns of aberrant DNA methylation around Transcription Start site (TSS) were observed frequently upto 2kb from CGI (CpG island shores) as well as in promoters that lack CGIs. Furthermore, changes in 56 genes obtained from peripheral blood showed consistency with post-mortem brain samples including DNMT1, CACNA1S, PRAME, MYT1L and STAB1. Among these genes CACNA1S on 1q32 and PRAME on 22q11.22 are considered as hotspots for Bipolar Disorders (Li et al. 2015).

Post mortem frontal cortex (Broadmann area 9) shows upregulation of mRNA and protein levels of neuroinflammatory and Arachidonic acid (AA) cascade markers such as AA selective calcium-independent cytosolic phospholipase

$A_2$($cPLA_2$), secretory $PLA_2$ ($sPLA_2$-IIA and cyclooxygenase-2 (COX2) along with the loss of synaptic proteins synaptophysin and debrin in patients with Bipolar disorders. Epigenetic modifications are associated with upregulated mRNA and protein levels of AA cascade, neurotrophic and synaptic protein markers. Increased Cox-2 expression in the BD corelated with hypomethylated state of the Cox-2 CpG promoter region. However, other AA cascade markers did not have DNA promoter methylation changes. Furthermore, global hypermethylated DNA in BD brains was observed suggesting decreased transcriptional activity in these disorders. These changes were also associated with significant increase in H3 phosphorylation suggesting an onset of apoptosis (Rao et al. 2012).

A recent study which analysed the methylation status of peripheral venous blood from 150 patients with bipolar disorders identified low levels of methylation at the promoter region of COMT(Catechol- O methyltransferase) and PPIEL (Peptidyl -prolyl isomerase E like). Lower levels of methylation of COMT and PPIEL can hence be closely related to Bipolar disorder and could regulate the level of dopamine (Zhang et al. 2018).

## 5.4 *Obsessive Compulsive Disorder*

Obsessive Compulsive Disorder (OCD) is a chronic neuro developmental and psychiatric disorder characterized by uncontrollable recurring thoughts and behaviours and affects 3% of the general population. The glutamatergic system which includes glutamate ionotropic receptor NMDA types (GRINs) are the most central nodes with highest degree of connections. 57 such genes are involved in 29 pathways with greatest number of genes involved in hetero trimer G protein signalling pathways and others including the dopaminergic, serotonergic, GABAergic, opioidergic, adrenergic, cholinergic and glutamatergic systems involved in the pathogenesis of the disease (Bozorgmehr et al. 2017).

Whole exome sequencing studies involving 20 simplex OCD parent-child trios have estimated the rate of de-novo (DN) single nucleotide variation in OCD was $2.51 \times 10^{-8}$ per base per generation. This study also identified 19 DN SNVs (11 missense mutations and one nonsense mutation). Most of the genes harbouring DN SNVs in OCD were located in the human brain and revealed enrichment of immunological and CNS functioning and development pathways (Cappi et al. 2016a).

### 5.4.1 DNA and Histone Methylation in Obsessive Compulsive Disorder

A Genome wide DNA methylation study of OCD of 65 patients from Chinese Han population with OCD resulted in identification of 2190 unique genes differentially methylated between OCD and healthy control subjects. 4013 of these loci were located in CpG islands and 2478 were in promoter regions.

Pathway enrichment analysis revealed the involvement of actin cytoskeleton, cell adhesion molecules, actin binding, transcription regulator activity to be associated with the risk of OCD (Yue et al. 2016). Gamma aminobutyric acid (GABA) B receptor1 in blood samples at birth, estrogen receptor 1(ESR1), the myelin oligodendrocyte glycoprotein (MOG) and the brain derived neurotrophic factor (BDNF) in blood samples at the time of diagnosis showed significant association with OCD (Nissen et al. 2016).

Recent study investigating the common and unique architecture of ASD, SCZ, BD and OCD identified 10 genes (BDNF, CACNA1C, CHRNA7, DRD2, HTR2A, MAOA, MTHFR, NOS1AP, SLO6A3 and TPH2) to be commonly associated with the aetiology of the disease. These genes are predominantly involved in the dopaminergic and serotonergic pathways, the voltage gated calcium ion channel gene network, folate metabolism, regulation of hippo signalling pathway and the regulation of gene silencing and expression. Hippo signalling pathway was found to be commonly associated with these neuropsychiatric disorders, implicating neural development and neuronal maintenance as key factors in disorder psychopathology (O'Connell et al. 2018; Zhang et al. 2018).

Oxytocin, the most abundant neuropeptide in the brain which acts as a neuromodulator and hormone to its G-protein coupled receptor (OXTR) is linked to neuro-behaviour functions. Oxytocin has been found to be associated with the pathophysiology of OCD. DNA methylation studies from peripheral blood leucocytes on 43 OCD patients and 34 healthy controls investigating methylation pattern of OXTR revealed hypermethylation in CpG Sites of two sequences targets located in the exon III, suggesting that at some critical point of development, environmental factors led to hypermethylation of the OXTR in OCD patients (Cappi et al. 2016b).

Histone methylation patterns in OCD are yet to be investigated in detail. However increase in anxiety and deficits in cognition and memory could be linked to SETDB1 expression in brain. Increased expression and activity of SETDB1 histone methyltransferase in forebrain neurons is associated with an antidepressant-like phenotype in behavioural paradigms related to anhedonia, despair and helplessness. Chromatin conformation capture (3C) and SETDB1 ChIP revealed a loop formation tethering the *NR2B/Grin2b* promoter to the SETDB1 target site positioned 30kb downstream of the transcription start site. SETDB1 -mediated repressive histone methylation at *NR2B/Grin2b* was associated with decreased NR2B expression in hippocampus and ventral striatum, suggesting the role for neuronal SETDB1 in the regulation of affective and motivational behaviours through repressive chromatin remodelling at a select set of target genes (Jiang et al. 2010).

Behavioural problems, including OCD have been shown to be associated with LoF mutations in SETD5. SETD5 encodes a histone methyltransferase that lies within the critical interval for 3p25.Analysis of blood samples of children and young adults recruited to the genetics of learning disability study with moderate to severe intellectual disability provide evidence that the loss of function SETD5 is a relatively frequent cause of intellectual diability and the affected individuals showed phenotypic similarity to those previously reported with a deletion in the critical region of 3p25 (Grozeva et al. 2014).

## 6 Genome Wide Changes in Autism Spectrum Disorders

Autism spectrum disorders (ASD) are a group of early-onset neurodevelopmental syndromes characterized by symptoms of two categories—defective behavioural impairments including social communication problems and restrictive repetitive behaviours. Neuropathological alterations in autism include megalencephaly (whole brain enlargement) and increased head circumference including increased cortical thickness and abnormalities in cortical morphology. Genetic contribution to autism comes from the studies of mono-zygotic (identical) and di-zygotic (fraternal) twins which showed that the monozygotic twins have a 50% or higher concordance for autism and dizygotic twins have 3% probability for autism.

Autism is a complex disorder that involves large number of genes associated with disease risk and involves interplay of common and rare variants. Whole-Exome Sequencing studies involving 3871 autism cases and 9937 ancestor matched or parental controls using Transmission and De novo association (TADA) identified 33 autosomal genes with false discovery ratio (FDR)$<$ 0.1 and 107 genes with FDR $<$0.3. Out of the total 33 genes, 15 are known ASD risk genes, 11 have been reported previously with mutations but were not classified as true risk genes while 7 are novel genes. The newly discovered genes include ASH1L and MLL3 which play an important role in chromatin remodelling. However, the 107 gene sets with FDR $<$0.3 show evolutionary constraint and incur de novo loss of function mutations in 5% of autistic patients. Furthermore, many genes identified in this study encode proteins for synaptic, transcriptional and chromatin remodeling pathways (De Rubeis et al. 2014).

Primary causes of ASD are highly heterogeneous, however it appears to converge on shared downstream epigenomic changes associated with specific functions. These shared chromatin alterations could in turn be responsible for some of the shared symptoms of ASD.

A genome wide study that involved the redistribution process of H3K4me3 during the transition from early infancy ($<$1) to older ages. This study identified two fold change (503 increased loci and 208 decreased loci) in the genome transcription start site. Furthermore, overlap of 711 differentially obtained H3K4me3 peaks in autistic patients with previously annotated autism risk loci showed significant correlation thus confirming significant overlap between genetic and epigenetic risk architecture in autism (Shulha et al. 2012).

Recent evidence for shared pathways and functional themes among differentially acetylated loci in the autism spectrum disorders comes from the histone acetylome wide association study (HAWAS) that involves chromatin immunoprecipitation sequencing (ChIP-seq) of H3K27ac mark on post-mortem samples from ASD patients. This study revealed aberrations (over 5000 enhancer/promoter loci) in histone acetylation patterns which are widespread in ASD cerebral cortex. Function enrichment analysis of differentially acetylated (DA) peaks in prefrontal cortex and temporal cortex showed similar functional profiles. The increased levels of H3K27ac, showed strong enrichment for genes related to ion channels, synaptic function and epilepsy/neuronal excitability all of which was previously dysregulated in this disorder and the decreased acetylation pattern was found to be associated with digestive tract morphogenesis, chemokine signaling, HDAC activity and immune responses to microglia. Furthermore, correlating histone acetylation with genotype, greater than 2000 histone acetylation quantitative trait loci (haQTLs) were discovered including casual variants for psychiatric diseases (Sun et al. 2016). Recent study involving genome wide integrative analysis of miRNA expression in postmortem brain from ASD patients and controls, identified miRNAs like has-miR-21-3p and co-regulated modules that are disrupted in ASD. This include hsa-miR-21-3p miRNA the second most abundant miRNA in ASD (Wu et al. 2016).

Despite remarkable advances in genetics and genomics the etiology of around 70% of ASD cases remains unknown. The epigenome-wide association study integrated with the transcriptome study in blood samples from the cohort of idiopathic ASD patients showed significant hypomethylation pattern caused by rare meSNVs at six loci as well as a few clustered epimutations in single-ASD patient. Furthermore, this study also revealed a significant load of deleterious mutations affecting ERMN in ASD as compared with controls thus suggesting ERMN as a novel gene involved in ASD (Homs et al. 2016).

Parallel study involving multiple gene expression profile comparisons with human Alu-inserted genes in ASD samples identified four studies that showed association between Alu-inserted genes and differentially expressed genes (DEGs) in ASD. It was further identified that intronic Alu insertion corresponded DEGs in ASD. Biological functions associated with 320 DEGs with Alu insertion significantly associated with neurodevelopmental disorders and neurological functions involved in ASD. Alu methylation analysis using combined restriction analysis (COBRA) of lympho blastoid cell lines and Alu expression analysis using qRT-PCR also showed that the dysregulation of Alu methylation and expression was not observed in all cases but only in ASD subgroups. This suggests that the classification of ASD individuals into subgroups will help reduce heterogeneity and may lead to the discovery of novel mechanisms associated with Alu element in ASD subgroups (Saeliw et al. 2018).

# 7 Conclusion

## *7.1 Therapeutic Interventions Based on the Histone and DNA Methylome and the Challenges*

Neurodegenerative and neuropsychiatric disorders present complex aetiology and are regulated by combinations of genetic and environmental risks. Since epigenetic events are the regulators of environmental impact on the genome, investigating epigenetic regulation in the brain is a key to understand the onset and progression of neurodegenerative and neuropsychiatric disorders. The histone and DNA methylome have provided novel and newer insights into mechanisms of neural development, disease and ageing. Emerging evidences on mutations and functional alterations in the epigenetic machinery have increased our understanding of Alzheimer's disease, Parkinson's disease and Huntington's disease, besides neuro-psychiatric disorders. Genome wide alterations in epigenetic modifications therefore show potential to be developed as biomarkers for brain disorders.

DNA methylation patterns on specific genes have been explored as epigenetic markers for Parkinson's disease. However, DNA methylation levels at the α-synuclein intron 1 promoter from substantia nigra of PD patients and blood from PD patients had observed conflicting results. Hence minor differences in DNA methylation levels in patient tissues and selection of genomic loci and CpG sites pose significant challenges in the development of epigenetic biomarkers. The specificity and robustness of α-synuclein based epigenetic biomarkers can be enhanced by reducing variability across patient groups, by analyzing subtypes of PD and by employing this biomarker for early diagnosis.

It is well established that changes in DNA methylation influence the expression of APP, PS1 and Aβ which are intermediates in Alzhiemer's Disease and the hypomethylation of promoters of PS1 leads to over expression of Aβ (Mulder et al. 2005). Hence DNA methylation is a viable target for therapy in Alzhiemer's Disease. Administration of S-adenosyl methionine adjunct to regular antidepressants has been shown to improve cognitive symptoms and memory in patients with depression (Levkovitz et al. 2012).

In neuropsychiatric disorders like Schizophrenia, the clinical manifestations initiate at the prodromal stage followed by a first episode in adolescence and deteriorate further after this episode. Since Schizophrenia shows fluctuating changes with increasing episodes, the identification of epigenetic biomarkers like DNA methylation or demethylation in peripheral blood cells could potentially help in prophylactic treatment leading to the prevention of prodromal phase or the onset of the first epidome or a relapse.

A major challenge in the management of neurodegenerative and neuropsychiatric disorders is early diagnosis. There is no established criterion for early and accurate detection of these disorders through reference value of biomarkers from patient blood or cerebral spinal fluid or through imaging approaches. Epigenetic alterations on the histone and the DNA methylome offer potential diagnostic tools for these diseases and would aid screening of such modifications at early stages and further reversal through epigenetic therapy.

## References

Abmayr SM, Workman JL (2012) Holding on through DNA replication: histone modification or modifier? Cell 150:875–877

Agbemenyah HY, Agis-Balboa RC, Burkhardt S et al (2014) Insulin growth factor binding protein 7 is a novel target to treat dementia. Neurobiol Dis 62:135–143

Agger K, Cloos PAC, Christensen J et al (2007) UTX and JMJD3 are histone H3K27 demethylases involved in HOX gene regulation and development. Nature 449:731–734
Akbarian S (1995) Gene expression for glutamic acid decarboxylase is reduced without loss of neurons in prefrontal cortex of schizophrenics. Arch Gen Psychiatry 52:258
Akbarian S, Ruehl MG, Bliven E et al (2005) Chromatin alterations associated with down-regulated metabolic gene expression in the prefrontal cortex of subjects with schizophrenia. Arch Gen Psychiatry 62:829
Akbarian S, Beeri MS, Haroutunian V (2013) Epigenetic determinants of healthy and diseased brain aging and cognition. JAMA Neurol 70:711–718
Alcala R, Creus-Muncunill J, Azkona G et al (2014) B10 nuclear lamina is differentially altered in huntington's disease brain regions. J Neurol Neurosurg Psychiatry 85:A12–A12
Ambrosio S, Majello B (2018) Targeting histone demethylase LSD1/KDM1a in neurodegenerative diseases. J Exp Neurosci 12:1179069518765743
Baba A, Ohtake F, Okuno Y et al (2011) PKA-dependent regulation of the histone lysine demethylase complex PHF2-ARID5B. Nat Cell Biol 13:668–675
Bai G, Cheung I, Shulha HP et al (2014) Epigenetic dysregulation of hairy and enhancer of split 4 (HES4) is associated with striatal degeneration in postmortem Huntington brains. Hum Mol Genet 24:1441–1456
Barrachina M, Ferrer I (2009) DNA methylation of Alzheimer disease and tauopathy-related genes in postmortem brain. J Neuropathol Exp Neurol 68:880–891
Bedford DC, Brindle PK (2012) Is histone acetylation the most important physiological function for CBP and p300? Aging 4:247–255
Berthier A, Jiménez-Sáinz J, Pulido R (2013) PINK1 regulates histone H3 trimethylation and gene expression by interaction with the polycomb protein EED/WAIT1. Proc Natl Acad Sci U S A 110:14729–14734
Bertram L, McQueen MB, Mullin K et al (2007) Systematic meta-analyses of Alzheimer disease genetic association studies: the AlzGene database. Nat Genet 39:17–23
Blanch M, Mosquera JL, Ansoleaga B et al (2016) Altered mitochondrial DNA methylation pattern in Alzheimer disease-related pathology and in Parkinson disease. Am J Pathol 186:385–397
Borun TW, Pearson D, Paik WK (1972) Studies of histone methylation during the HeLa S-3 cell cycle. J Biol Chem 247:4288–4298
Bourgeron T (2015) From the genetic architecture to synaptic plasticity in autism spectrum disorder. Nat Rev Neurosci 16:551–563
Bozorgmehr A, Ghadirivasfi M, Shahsavand Ananloo E (2017) Obsessive-compulsive disorder, which genes? Which functions? Which pathways? An integrated holistic view regarding OCD and its complex genetic etiology. J Neurogenet 31:153–160
Braak H, Braak E (1991) Alzheimer's disease affects limbic nuclei of the thalamus. Acta Neuropathol 81:261–268
Braak H, Del Tredici K, Bratzke H et al (2002) Staging of the intracerebral inclusion body pathology associated with idiopathic Parkinson's disease (preclinical and clinical stages). J Neurol 249(Suppl 3):III/1–III/5
Buckley NJ, Johnson R, Zuccato C et al (2010) The role of REST in transcriptional and epigenetic dysregulation in Huntington's disease. Neurobiol Dis 39:28–39
Cappi C, Brentani H, Lima L et al (2016a) Whole-exome sequencing in obsessive-compulsive disorder identifies rare mutations in immunological and neurodevelopmental pathways. Transl Psychiatry 6:e764–e764
Cao J, Yan Q (2012) Histone ubiquitination and deubiquitination in transcription, DNA damage response, and cancer. Front Oncol 2:1–9
Cappi C, Diniz JB, Requena GL et al (2016b) Epigenetic evidence for involvement of the oxytocin receptor gene in obsessive-compulsive disorder. BMC Neurosci 17:79
Carlsson A, Lindqvist M (2009) Effect of chlorpromazine or haloperidol on formation of 3-methoxytyramine and normetanephrine in mouse brain. Acta Pharmacol Toxicol 20:140–144

Celarain N, Sánchez-Ruiz de Gordoa J, Zelaya MV et al (2016) TREM2 upregulation correlates with 5-hydroxymethycytosine enrichment in Alzheimer's disease hippocampus. Clin Epigenetics 8:37
Chang B, Chen Y, Zhao Y, Bruick RK (2007) JMJD6 is a histone arginine demethylase. Science 318:444–447
Chase KA, Gavin DP, Guidotti A, Sharma RP (2013) Histone methylation at H3K9: evidence for a restrictive epigenome in schizophrenia. Schizophr Res 149:15–20
Chen K-L, Wang SS-S, Yang Y-Y et al (2009) The epigenetic effects of amyloid-beta(1-40) on global DNA and neprilysin genes in murine cerebral endothelial cells. Biochem Biophys Res Commun 378:57–61
Chen M, Zhu N, Liu X et al (2015) JMJD1C is required for the survival of acute myeloid leukemia by functioning as a coactivator for key transcription factors. Genes Dev 29:2123–2139
Chouliaras L, Mastroeni D, Delvaux E et al (2013) Consistent decrease in global DNA methylation and hydroxymethylation in the hippocampus of Alzheimer's disease patients. Neurobiol Aging 34:2091–2099
Christopher MA, Myrick DA, Barwick BG et al (2017) LSD1 protects against hippocampal and cortical neurodegeneration. Nat Commun 8:805
Cohen-Cory S, Kidane AH, Shirkey NJ, Marshak S (2010) Brain-derived neurotrophic factor and the development of structural neuronal connectivity. Dev Neurobiol 70:271–288
Coppieters N, Dieriks BV, Lill C et al (2014) Global changes in DNA methylation and hydroxymethylation in Alzheimer's disease human brain. Neurobiol Aging 35:1334–1344
Cruceanu C, Alda M, Nagy C et al (2013) H3K4 tri-methylation in synapsin genes leads to different expression patterns in bipolar disorder and major depression. Int J Neuropsychopharmacol 16:289–299
Cuthbert GL, Daujat S, Snowden AW et al (2004) Histone deimination antagonizes arginine methylation. Cell 118:545–553
Daujat S, Bauer U-M, Shah V et al (2002) Crosstalk between CARM1 methylation and CBP acetylation on histone H3. Curr Biol 12:2090–2097
de Boni L, Riedel L, Schmitt I et al (2015) DNA methylation levels of α-synuclein intron 1 in the aging brain. Neurobiol Aging 36:3334.e7–3334.e11
De Jager PL, Srivastava G, Lunnon K et al (2014) Alzheimer's disease: early alterations in brain DNA methylation at ANK1, BIN1, RHBDF2 and other loci. Nat Neurosci 17:1156–1163
De Rubeis S, He X, Goldberg AP et al (2014) Synaptic, transcriptional and chromatin genes disrupted in autism. Nature 515:209–215
De Souza RAG, Islam SA, McEwen LM et al (2016) DNA methylation profiling in human Huntington's disease brain. Hum Mol Genet 25:2013–2030
Desplats P, Spencer B, Coffee E et al (2011) Alpha-synuclein sequesters Dnmt1 from the nucleus: a novel mechanism for epigenetic alterations in Lewy body diseases. J Biol Chem 286:9031–9037
Dong X, Tsuji J, Labadorf A et al (2015) The role of H3K4me3 in transcriptional regulation is altered in Huntington's disease. PLoS One 10:e0144398
Eilbracht J, Reichenzeller M, Hergt M et al (2004) NO66, a highly conserved dual location protein in the nucleolus and in a special type of synchronously replicating chromatin. Mol Biol Cell 15:1816–1832
Feng Q, Wang H, Ng HH et al (2002) Methylation of H3-lysine 79 is mediated by a new family of HMTases without a SET domain. Curr Biol 12:1052–1058
Ferrante RJ (2004) Chemotherapy for the brain: the antitumor antibiotic mithramycin prolongs survival in a mouse model of Huntington's disease. J Neurosci 24:10335–10342
Frescas D, Guardavaccaro D, Kuchay SM et al (2008) KDM2A represses transcription of centromeric satellite repeats and maintains the heterochromatic state. Cell Cycle 7:3539–3547
Forman MS, Trojanowski JQ, Lee VM-Y (2004) Neurodegenerative diseases: a decade of discoveries paves the way for therapeutic breakthroughs. Nat Med 10:1055–1063
Funahashi Y, Yoshino Y, Yamazaki K et al (2017) DNA methylation changes at SNCA intron 1 in patients with dementia with Lewy bodies. Psychiatry Clin Neurosci 71:28–35

Fuso A, Nicolia V, Cavallaro RA et al (2008) B-vitamin deprivation induces hyperhomocysteinemia and brain S-adenosylhomocysteine, depletes brain S-adenosylmethionine, and enhances PS1 and BACE expression and amyloid-beta deposition in mice. Mol Cell Neurosci 37:731–746

Gao J, Cahill CM, Huang X et al (2018) S-Adenosyl methionine and transmethylation pathways in neuropsychiatric diseases throughout life. Neurotherapeutics 15:156–175

Gapp K, Woldemichael BT, Bohacek J, Mansuy IM (2014) Epigenetic regulation in neurodevelopment and neurodegenerative diseases. Neuroscience 264:99–111

Gavin DP, Sharma RP (2009) Chromatin from peripheral blood mononuclear cells as biomarkers for epigenetic abnormalities in schizophrenia. Cardiovasc Psychiatry Neurol 2009:409562

Girdhar K, Hoffman GE, Jiang Y et al (2018) Cell-specific histone modification maps in the human frontal lobe link schizophrenia risk to the neuronal epigenome. Nat Neurosci 21:1126–1136

Gjoneska E, Pfenning AR, Mathys H et al (2015) Conserved epigenomic signals in mice and humans reveal immune basis of Alzheimer's disease. Nature 518:365–369

Goula A-V, Stys A, Chan JPK et al (2012) Transcription elongation and tissue-specific somatic CAG instability. PLoS Genet 8:e1003051

Gräff J, Tsai L-H (2013) Histone acetylation: molecular mnemonics on the chromatin. Nat Rev Neurosci 14:97–111

Greer EL, Shi Y (2012) Histone methylation: a dynamic mark in health, disease and inheritance. Nat Rev Genet 13:343–357

Grozeva D, Carss K, Spasic-Boskovic O et al (2014) De novo loss-of-function mutations in SETD5, encoding a methyltransferase in a 3p25 microdeletion syndrome critical region, cause intellectual disability. Am J Hum Genet 94:618–624

Guidotti A, Auta J, Davis JM et al (2000) Decrease in reelin and glutamic acid decarboxylase67 (GAD67) expression in schizophrenia and bipolar disorder: a postmortem brain study. Arch Gen Psychiatry 57:1061–1069

Hamamoto R, Furukawa Y, Morita M et al (2004) SMYD3 encodes a histone methyltransferase involved in the proliferation of cancer cells. Nat Cell Biol 6:731–740

He J, Kallin EM, Tsukada Y-I, Zhang Y (2008) The H3K36 demethylase Jhdm1b/Kdm2b regulates cell proliferation and senescence through p15(Ink4b). Nat Struct Mol Biol 15:1169–1175

Heintzman ND, Stuart RK, Hon G et al (2007) Distinct and predictive chromatin signatures of transcriptional promoters and enhancers in the human genome. Nat Genet 39:311–318

Hernández-Ortega K, Garcia-Esparcia P, Gil L et al (2016) Altered machinery of protein synthesis in Alzheimer's: from the nucleolus to the ribosome. Brain Pathol 26:593–605

Homs A, Codina-Solà M, Rodríguez-Santiago B et al (2016) Genetic and epigenetic methylation defects and implication of the ERMN gene in autism spectrum disorders. Transl Psychiatry 6: e855

Horton JR, Liu X, Gale M et al (2016) Structural basis for KDM5A histone lysine demethylase inhibition by diverse compounds. Cell Chem Biol 23:769–781

Horvath S (2013) DNA methylation age of human tissues and cell types. Genome Biol 14:R115

Horvath S, Ritz BR (2015) Increased epigenetic age and granulocyte counts in the blood of Parkinson's disease patients. Aging 7:1130–1142

Hottiger MO (2011) ADP-ribosylation of histones by ARTD1: an additional module of the histone code? FEBS Lett 585:1595–1599

Hsia DA, Tepper CG, Pochampalli MR et al (2010) KDM8, a H3K36me2 histone demethylase that acts in the cyclin A1 coding region to regulate cancer cell proliferation. Proc Natl Acad Sci U S A 107:9671–9676

Huang H-S, Matevossian A, Whittle C et al (2007) Prefrontal dysfunction in schizophrenia involves mixed-lineage leukemia 1-regulated histone methylation at GABAergic gene promoters. J Neurosci 27:11254–11262

Ibrahim HM, Tamminga CA (2011) Schizophrenia: treatment targets beyond monoamine systems. Annu Rev Pharmacol Toxicol 51:189–209

Iwase S, Lan F, Bayliss P et al (2007) The X-linked mental retardation gene SMCX/JARID1C defines a family of histone H3 lysine 4 demethylases. Cell 128:1077–1088

Jia H, Morris CD, Williams RM et al (2015) HDAC inhibition imparts beneficial transgenerational effects in Huntington's disease mice via altered DNA and histone methylation. Proc Natl Acad Sci U S A 112:E56–E64

Jiang GL, Huang S (2000) The yin-yang of PR-domain family genes in tumorigenesis. Histol Histopathol 15:109–117

Jiang Y, Jakovcevski M, Bharadwaj R et al (2010) Setdb1 histone methyltransferase regulates mood-related behaviors and expression of the NMDA receptor subunit NR2B. J Neurosci 30:7152–7167

Katoh Y, Katoh M (2007) Comparative integromics on JMJD2A, JMJD2B and JMJD2C: preferential expression of JMJD2C in undifferentiated ES cells. Int J Mol Med 20:269–273

Kaut O, Schmitt I, Wüllner U (2012) Genome-scale methylation analysis of Parkinson's disease patients' brains reveals DNA hypomethylation and increased mRNA expression of cytochrome P450 2E1. Neurogenetics 13:87–91

Kawahara TLA, Michishita E, Adler AS et al (2009) SIRT6 links histone H3 lysine 9 deacetylation to NF-kappaB-dependent gene expression and organismal life span. Cell 136:62–74

Kendler KS (2001) Twin studies of psychiatric illness: an update. Arch Gen Psychiatry 58:1005–1014

Kizer KO, Phatnani HP, Shibata Y et al (2005) A novel domain in Set2 mediates RNA polymerase II interaction and couples histone H3 K36 methylation with transcript elongation. Mol Cell Biol 25:3305–3316

Kornberg RD (1974) Chromatin structure: a repeating unit of histones and DNA. Science 184:868–871

Krishnan S, Trievel RC (2013) Structural and functional analysis of JMJD2D reveals molecular basis for sitespecific demethylation among JMJD2 demethylases. Structure 21:98–108

Kweon JH, Kim S, Lee SB (2017) The cellular basis of dendrite pathology in neurodegenerative diseases. BMB Rep 50:5–11

Lambert JC, Ibrahim-Verbaas CA, Harold D et al (2013) Meta-analysis of 74,046 individuals identifies 11 new susceptibility loci for Alzheimer's disease. Nat Genet 45:1452–1458

Lattal KM, Wood MA (2013) Epigenetics and persistent memory: implications for reconsolidation and silent extinction beyond the zero. Nat Neurosci 16:124–129

Lee J-H, Skalnik DG (2005) CpG-binding protein (CXXC finger protein 1) is a component of the mammalian Set1 histone H3-Lys4 methyltransferase complex, the analogue of the yeast Set1/COMPASS complex. J Biol Chem 280:41725–41731

Lee MG, Wynder C, Cooch N, Shiekhattar R (2005) An essential role for CoREST in nucleosomal histone 3 lysine 4 demethylation. Nature 437:432–435

Lee J, Hong YK, Jeon GS et al (2012) ATRX induction by mutant huntingtin via Cdx2 modulates heterochromatin condensation and pathology in Huntington's disease. Cell Death Differ 19:1109–1116

Levine ME, Lu AT, Bennett DA, Horvath S (2015) Epigenetic age of the pre-frontal cortex is associated with neuritic plaques, amyloid load, and Alzheimer's disease related cognitive functioning. Aging 7:1198–1211

Levkovitz Y, Alpert JE, Brintz CE et al (2012) Effects of S-adenosylmethionine augmentation of serotonin-reuptake inhibitor antidepressants on cognitive symptoms of major depressive disorder. J Affect Disord 136:1174–1178

Lewis DA, Hashimoto T, Volk DW (2005) Cortical inhibitory neurons and schizophrenia. Nat Rev Neurosci 6:312–324

Li Y, Zhang Y, Li S et al (2015) Genome-wide DNA methylome analysis reveals epigenetically dysregulated non-coding RNAs in human breast cancer. Sci Rep 5:8790

Li N, Dhar SS, Chen T-Y et al (2016) JARID1D is a suppressor and prognostic marker of prostate cancer invasion and metastasis. Cancer Res 76:831–843

Liang WS, Dunckley T, Beach TG et al (2008) Altered neuronal gene expression in brain regions differentially affected by Alzheimer's disease: a reference data set. Physiol Genomics 33:240–256
Lim DA, Huang Y-C, Swigut T et al (2009) Chromatin remodelling factor Mll1 is essential for neurogenesis from postnatal neural stem cells. Nature 458:529–533
Lin Q, Ding H, Zheng Z et al (2012) Promoter methylation analysis of seven clock genes in Parkinson's disease. Neurosci Lett 507:147–150
Lithner CU, Lacor PN, Zhao W-Q et al (2013) Disruption of neocortical histone H3 homeostasis by soluble Aβ: implications for Alzheimer's disease. Neurobiol Aging 34:2081–2090
Liu Y, Chen G, Norton N et al (2009) Whole genome association study in a homogenous population in Shandong peninsula of China reveals JARID2 as a susceptibility gene for schizophrenia. J Biomed Biotechnol 2009:1–7
Liu L, Kim H, Casta A et al (2014) Hairless is a histone H3K9 demethylase. FASEB J 28:1534–1542
Luger K, Mäder AW, Richmond RK et al (1997) Crystal structure of the nucleosome core particle at 2.8 Å resolution. Nature 389:251–260
Lunnon K, Smith R, Hannon E et al (2014) Methylomic profiling implicates cortical deregulation of ANK1 in Alzheimer's disease. Nat Neurosci 17:1164–1170
Margueron R, Li G, Sarma K et al (2008) Ezh1 and Ezh2 maintain repressive chromatin through different mechanisms. Mol Cell 32:503–518
Mastroeni D, McKee A, Grover A et al (2009) Epigenetic differences in cortical neurons from a pair of monozygotic twins discordant for Alzheimer's disease. PLoS One 4:e6617
Mastroeni D, Delvaux E, Nolz J et al (2015) Aberrant intracellular localization of H3k4me3 demonstrates an early epigenetic phenomenon in Alzheimer's disease. Neurobiol Aging 36:3121–3129
Matrisciano F, Tueting P, Dalal I et al (2013) Epigenetic modifications of GABAergic interneurons are associated with the schizophrenia-like phenotype induced by prenatal stress in mice. Neuropharmacology 68:184–194
McEwen BS (2000) Effects of adverse experiences for brain structure and function. Biol Psychiatry 48:721–731
Mentch SJ, Mehrmohamadi M, Huang L et al (2015) Histone methylation dynamics and gene regulation occur through the sensing of one-carbon metabolism. Cell Metab 22:861–873
Metzger E, Wissmann M, Yin N et al (2005) LSD1 demethylates repressive histone marks to promote androgen-receptor-dependent transcription. Nature 437:436–439
Miao F, Natarajan R (2005) Mapping global histone methylation patterns in the coding regions of human genes. Mol Cell Biol 25:4650–4661
Moore K, McKnight AJ, Craig D, O'Neill F (2014) Epigenome-wide association study for Parkinson's disease. NeuroMolecular Med 16:845–855
Mulder C, Schoonenboom NSM, Jansen EEW et al (2005) The transmethylation cycle in the brain of Alzheimer patients. Neurosci Lett 386:69–71
Narayanan M, Huynh JL, Wang K et al (2014) Common dysregulation network in the human prefrontal cortex underlies two neurodegenerative diseases. Mol Syst Biol 10:743
Nativio R, Donahue G, Berson A et al (2018) Publisher correction: dysregulation of the epigenetic landscape of normal aging in Alzheimer's disease. Nat Neurosci 21:1018–1018
Ng SS, Yue WW, Oppermann U, Klose RJ (2009) Dynamic protein methylation in chromatin biology. Cell Mol Life Sci 66:407–422
Ng CW, Yildirim F, Yap YS et al (2013) Extensive changes in DNA methylation are associated with expression of mutant huntingtin. Proc Natl Acad Sci U S A 110:2354–2359
Nissen JB, Hansen CS, Starnawska A et al (2016) DNA methylation at the neonatal state and at the time of diagnosis: preliminary support for an association with the estrogen receptor 1, gamma-aminobutyric acid B receptor 1, and myelin oligodendrocyte glycoprotein in female adolescent patients with OCD. Front Psychiatry 7:35

Nuutinen T, Suuronen T, Kyrylenko S et al (2005) Induction of clusterin/apoJ expression by histone deacetylase inhibitors in neural cells. Neurochem Int 47:528–538

O'Connell KS, McGregor NW, Lochner C et al (2018) The genetic architecture of schizophrenia, bipolar disorder, obsessive-compulsive disorder and autism spectrum disorder. Mol Cell Neurosci 88:300–307

Obeid R, Schadt A, Dillmann U et al (2009) Methylation status and neurodegenerative markers in Parkinson disease. Clin Chem 55:1852–1860

Pasini D, Cloos PAC, Walfridsson J et al (2010) JARID2 regulates binding of the Polycomb repressive complex 2 to target genes in ES cells. Nature 464:306–310

Pathak SS, Maitra S, Chakravarty S, Kumar A (2017) Histone lysine demethylases of JMJD2 or KDM4 family are important epigenetic regulators in reward circuitry in the etiopathology of depression. Neuropsychopharmacology 42:854–863

Pedersen MT, Helin K (2010) Histone demethylases in development and disease. Trends Cell Biol 20:662–671

Pedersen MT, Agger K, Laugesen A et al (2014) The demethylase JMJD2C localizes to H3K4me3-positive transcription start sites and is dispensable for embryonic development. Mol Cell Biol 34:1031–1045

Qi HH, Sarkissian M, Hu G-Q et al (2010) Histone H4K20/H3K9 demethylase PHF8 regulates zebrafish brain and craniofacial development. Nature 466:503–507

Rao JS, Keleshian VL, Klein S, Rapoport SI (2012) Epigenetic modifications in frontal cortex from Alzheimer's disease and bipolar disorder patients. Transl Psychiatry 2:e132–e132

Rea S, Eisenhaber F, O'Carroll D et al (2000) Regulation of chromatin structure by site-specific histone H3 methyltransferases. Nature 406:593–599

Ruzicka WB, Subburaju S, Benes FM (2015) Circuit- and diagnosis-specific DNA methylation changes at γ-Aminobutyric acid-related genes in postmortem human hippocampus in schizophrenia and bipolar disorder. JAMA Psychiatry 72:541–551

Ryu H, Lee J, Hagerty SW et al (2006) ESET/SETDB1 gene expression and histone H3 (K9) trimethylation in Huntington's disease. Proc Natl Acad Sci U S A 103:19176–19181

Saeliw T, Tangsuwansri C, Thongkorn S et al (2018) Integrated genome-wide Alu methylation and transcriptome profiling analyses reveal novel epigenetic regulatory networks associated with autism spectrum disorder. Mol Autism 9:27

Saijo K, Winner B, Carson CT et al (2009) A Nurr1/CoREST pathway in microglia and astrocytes protects dopaminergic neurons from inflammation-induced death. Cell 137:47–59

Sanchez-Mut JV, Aso E, Panayotis N et al (2013) DNA methylation map of mouse and human brain identifies target genes in Alzheimer's disease. Brain 136:3018–3027

Satake W, Nakabayashi Y, Mizuta I et al (2009) Genome-wide association study identifies common variants at four loci as genetic risk factors for Parkinson's disease. Nat Genet 41:1303–1307

Schon EA, Przedborski S (2011) Mitochondria: the next (neurode)generation. Neuron 70:1033–1053

Seong IS, Woda JM, Song J-J et al (2010) Huntingtin facilitates polycomb repressive complex 2. Hum Mol Genet 19:573–583

Shulha HP, Cheung I, Whittle C et al (2012) Epigenetic signatures of autism: trimethylated H3K4 landscapes in prefrontal neurons. Arch Gen Psychiatry 69:314–324

Siegmund KD, Connor CM, Campan M et al (2007) DNA methylation in the human cerebral cortex is dynamically regulated throughout the life span and involves differentiated neurons. PLoS One 2:e895

Singh T, Kurki MI, Curtis D et al (2016) Rare loss-of-function variants in SETD1A are associated with schizophrenia and developmental disorders. Nat Neurosci 19:571–577

Smith RG, Hannon E, De Jager PL et al (2018) Elevated DNA methylation across a 48-kb region spanning the HOXA gene cluster is associated with Alzheimer's disease neuropathology. Alzheimers Dement. https://doi.org/10.1016/j.jalz.2018.01.017

Song W, Zsindely N, Faragó A et al (2018) Systematic genetic interaction studies identify histone demethylase Utx as potential target for ameliorating Huntington's disease. Hum Mol Genet 27:759–759
Stadler F, Kolb G, Rubusch L et al (2005) Histone methylation at gene promoters is associated with developmental regulation and region-specific expression of ionotropic and metabotropic glutamate receptors in human brain. J Neurochem 94:324–336
Strahl BD, David Allis C (2000) The language of covalent histone modifications. Nature 403:41–45
Sugeno N, Jäckel S, Voigt A et al (2016) α-Synuclein enhances histone H3 lysine-9 dimethylation and H3K9me2-dependent transcriptional responses. Sci Rep 6:36328
Sun W, Poschmann J, Cruz-Herrera Del Rosario R et al (2016) Histone acetylome-wide association study of autism spectrum disorder. Cell 167:1385–1397.e11
Takata A, Xu B, Ionita-Laza I et al (2014) Loss-of-function variants in schizophrenia risk and SETD1A as a candidate susceptibility gene. Neuron 82:773–780
Tang B, Seredenina T, Coppola G et al (2011) Gene expression profiling of R6/2 transgenic mice with different CAG repeat lengths reveals genes associated with disease onset and progression in Huntington's disease. Neurobiol Dis 42:459–467
The Network and Pathway Analysis Subgroup of the Psychiatric Genomics Consortium (2015) Psychiatric genome-wide association study analyses implicate neuronal, immune and histone pathways. Nat Neurosci 18:199–209
Tsukada Y-I, Ishitani T, Nakayama KI (2010) KDM7 is a dual demethylase for histone H3 Lys 9 and Lys 27 and functions in brain development. Genes Dev 24:432–437
Urdinguio RG, Sanchez-Mut JV, Esteller M (2009) Epigenetic mechanisms in neurological diseases: genes, syndromes, and therapies. Lancet Neurol 8:1056–1072
Vafai SB, Stock JB (2002) Protein phosphatase 2A methylation: a link between elevated plasma homocysteine and Alzheimer's disease. FEBS Lett 518:1–4
Vashishtha M, Ng CW, Yildirim F et al (2013) Targeting H3K4 trimethylation in Huntington disease. Proc Natl Acad Sci U S A 110:E3027–E3036
Vila M, Przedborski S (2003) Targeting programmed cell death in neurodegenerative diseases. Nat Rev Neurosci 4:365–375
Walker MP, LaFerla FM, Oddo SS, Brewer GJ (2012) Reversible epigenetic histone modifications and Bdnf expression in neurons with aging and from a mouse model of Alzheimer's disease. Age 35:519–531
Walport LJ, Hopkinson RJ, Vollmar M et al (2014) Human UTY(KDM6C) is a male-specific Nε-methyl lysyl demethylase. J Biol Chem 289:18302–18313
Wang S-C, Oelze B, Schumacher A (2008) Age-specific epigenetic drift in late-onset Alzheimer's disease. PLoS One 3:e2698
Wang F, Yang Y, Lin X et al (2013) Genome-wide loss of 5-hmC is a novel epigenetic feature of Huntington's disease. Hum Mol Genet 22:3641–3653
Watson CT, Roussos P, Garg P et al (2016) Genome-wide DNA methylation profiling in the superior temporal gyrus reveals epigenetic signatures associated with Alzheimer's disease. Genome Med 8:5
West RL, Lee JM, Maroun LE (1995) Hypomethylation of the amyloid precursor protein gene in the brain of an Alzheimer's disease patient. J Mol Neurosci 6:141–146
Whetstine JR, Nottke A, Lan F et al (2006) Reversal of histone lysine trimethylation by the JMJD2 family of histone demethylases. Cell 125:467–481
Wijsman EM, Daw EW, Yu X et al (2005) APOE and other loci affect age-at-onset in Alzheimer's disease families with PS2 mutation. Am J Med Genet B Neuropsychiatr Genet 132B:14–20
Wu H, Siarheyeva A, Zeng H et al (2013) Crystal structures of the human histone H4K20 methyltransferases SUV420H1 and SUV420H2. FEBS Lett 587:3859–3868
Wu YE, Parikshak NN, Belgard TG, Geschwind DH (2016) Genome-wide, integrative analysis implicates microRNA dysregulation in autism spectrum disorder. Nat Neurosci 19:1463–1476
Xiang Y, Zhu Z, Han G et al (2007) JMJD3 is a histone H3K27 demethylase. Cell Res 17:850–857

Yamane K, Toumazou C, Tsukada Y-I et al (2006) JHDM2A, a JmjC-containing H3K9 demethylase, facilitates transcription activation by androgen receptor. Cell 125:483–495

Yu L, Chibnik LB, Srivastava GP et al (2015) Association of brain DNA methylation in SORL1, ABCA7, HLA-DRB5, SLC24A4, and BIN1 with pathological diagnosis of Alzheimer disease. JAMA Neurol 72:15–24

Yue W, Cheng W, Liu Z et al (2016) Genome-wide DNA methylation analysis in obsessive-compulsive disorder patients. Sci Rep 6:31333

Zadel M, Maver A, Kovanda A, Peterlin B (2018) DNA methylation profiles in whole blood of Huntington's disease patients. Front Neurol 9. https://doi.org/10.3389/fneur.2018.00655

Zagrebelsky M, Korte M (2014) Form follows function: BDNF and its involvement in sculpting the function and structure of synapses. Neuropharmacology 76:628–638

Zee BM, Levin RS, Xu B et al (2010) In vivo residue-specific histone methylation dynamics. J Biol Chem 285:3341–3350

Zhang Y, Yang H, Guo X et al (2014) The PHD1 finger of KDM5B recognizes unmodified H3K4 during the demethylation of histone H3K4me2/3 by KDM5B. Protein Cell 5:837–850

Zhang H-S, Ke X-Y, Hu L-L et al (2018) Study on the epigenetic methylation modification of bipolar disorder major genes. Eur Rev Med Pharmacol Sci 22:1421–1425

Zhao J, Zhu Y, Yang J et al (2017) A genome-wide profiling of brain DNA hydroxymethylation in Alzheimer's disease. Alzheimers Dement 13:674–688

Zhou X-W, Gustafsson J-A, Tanila H et al (2008) Tau hyperphosphorylation correlates with reduced methylation of protein phosphatase 2A. Neurobiol Dis 31:386–394

Zhubi A, Veldic M, Puri NV et al (2009) An upregulation of DNA-methyltransferase 1 and 3a expressed in telencephalic GABAergic neurons of schizophrenia patients is also detected in peripheral blood lymphocytes. Schizophr Res 111:115–122

Zuccato C, Cattaneo E (2007) Role of brain-derived neurotrophic factor in Huntington's disease. Prog Neurobiol 81:294–330

Zuccato C, Valenza M, Cattaneo E (2010) Molecular mechanisms and potential therapeutical targets in Huntington's disease. Physiol Rev 90:905–981

# DNA Methylation in Neuronal Development and Disease

**Emily C. Bruggeman and Bing Yao**

## Contents



**Abstract** DNA methylation is an epigenetic modification that spatially and temporally regulates gene expression and has essential roles in controlling neuronal development and function. DNA methylation is generally associated with heterochromatin and repression of gene transcription. Methylated cytosine residues can also undergo demethylation by ten-eleven translocation (TET) enzymes, resulting in 5-hydroxymethylation and its downstream derivatives. Once thought of as an intermediary in the demethylation process, 5-hmC has been found to be a unique and stable epigenetic mark. 5-hmC is more highly enriched in mammalian brains than in other somatic cells, indicating its critical roles in the central nervous system. Unlike methylation, hydroxymethylation is usually associated with euchromatin and gene activation. Much progress has been made in the past few decades in defining the roles of methylation and hydroxymethylation, and

E. C. Bruggeman · B. Yao (✉)
Department of Human Genetics, Emory University School of Medicine, Atlanta, GA, USA
e-mail: bing.yao@emory.edu

 
S. Jurga, J. Barciszewski (eds.), *The DNA, RNA, and Histone Methylomes*, RNA Technologies, https://doi.org/10.1007/978-3-030-14792-1_5

their molecular machinery in the brain and nervous system. In this chapter, we provide a comprehensive review of the roles of methylation and hydroxymethylation in brain development, functions and their dysregulation in brain disorders. First, we discuss the current understanding of these epigenetic marks in normal neuronal development as well as brain function. DNA methylation and hydroxymethylation have also been implicated in the development and progression of neurodegenerative diseases, such as Alzheimer's and Parkinson's diseases. We discuss consistencies and inconsistencies of the available data in human, mouse and in vitro studies that link methylation and hydroxymethylation to neurodegeneration. Finally, we explore the potential of these neuronal epigenetic marks and their molecular machinery to provide novel therapeutic targets in neurodegenerative diseases.

**Keywords** DNA methylation · DNA hydroxymethylation · Neurodevelopment · Neurodegenerative diseases · TET · DNMT

# 1 Introduction

5-methylcytosine (5-mC), the addition of a methyl group to the cytosine nucleotide, is an epigenetic modification that plays an intrinsic and essential role in biological development and processes. 5-mC was first reported as identified in cow thymus in 1948 and was referred to as "epi-cytosine" (Hotchkiss 1948). In early studies of DNA methylation, restriction enzymes were used to investigate DNA methylation patterns, since certain enzymes were found to be selective in cutting nucleotide sites only if they were unmodified (Singer et al. 1979; Waalwijk and Flavell 1978). More advanced techniques have since been developed and have rapidly expanded our understanding of epigenetic functions and mechanisms. In mammals, 5-mC appears most often at CpG dinucleotides [~70–80% are methylated (Ehrlich et al. 1982; Doskocil and Sorm 1962)], but they are also found at non-CpG sites (e.g., CpA) (Xie et al. 2012). 5-mC is enriched in repetitive regions, gene bodies and intergenic regions and its functions depend on the location (Wen et al. 2014; Szulwach et al. 2011). Some of the major functions include regulation of gene expression by inhibiting transcription, genomic stability and imprinting (Zhou and Robertson 2016; Brandeis et al. 1993). The frequent inverse correlations between methylation of CpG islands and gene expression led to the understanding that 5-mC plays mostly a repressive role in gene regulation (Keshet et al. 1985), especially when at the promoter region of a gene (Langner et al. 1984). DNA methylation is associated with heterochromatin (tightly packed genetic material that is not readily accessible to transcription factors) and thus inhibits gene activation (Choy et al. 2010). DNA methylation maintains genomic stability by silencing repeat elements (Zhou and Robertson 2016; Nichol and Pearson 2002) and transposable elements through methylation, which would otherwise disrupt genomic integrity if they were expressed (Yoder et al. 1997). DNA methylation controls imprinting, in which gene expression from one parental allele is repressed, allowing for expression of the allele from the other parent (Stewart et al. 2016). In addition, it is responsible for inactivation of the X chromosome, a critical genetic event that ensures that females have only one active copy of the sex chromosome instead of two (Brandeis et al. 1993). 5-mC is particularly enriched in brain tissue and its roles in the brain will be discussed in later sections (Ehrlich et al. 1982).

In contrast, DNA hydroxymethylation is primarily associated with active gene transcription and euchromatin (Ficz et al. 2011). Unlike 5-mC, 5-hmC is enriched at poised enhancers, exon-intron boundaries and gene bodies of actively transcribed genes (Wen et al. 2014; Stroud et al. 2011; Mellen et al. 2012). Elevations in 5-hmC do not always correlate with reductions in 5-mC, as would be expected since 5-hmC is a modification of methylated cytosine (Hahn et al. 2013). Thus, 5-hmC is considered to be a stable epigenetic mark in its own right and not just an intermediary step towards demethylation (Hahn et al. 2013). While 5-mC is critical to proliferating cells, 5-hmC appears to be essential to differentiating cells, although the role seems dependent on the cell type. For example, 5-hmC undergoes significant loss during differentiation of embryonic stem cells (ESCs) (Tahiliani et al. 2009; Ruzov et al. 2011; Szwagierczak et al. 2010; Kinney et al. 2011), but it increases during differentiation in adult neural stem cells (aNSCs) (Li et al. 2017b). 5-hmC is associated with cells that are in a pluripotent state and is highest in the embryo within the first 2 weeks following fertilization (Messerschmidt et al. 2014; Ruzov et al. 2011). Importantly, the brain has the highest levels of 5-hmC compared to any other tissue examined (Globisch et al. 2010; Li and Liu 2011; Kriaucionis and Heintz 2009) and it is dynamically regulated during both embryonic neurodevelopment and neurogenesis (Szulwach et al. 2011; Santiago et al. 2014).

Both 5-mC and 5-hmC regulate cellular proliferation and differentiation by temporal and spatial control of gene expression and thus are critical to organ development (Brandeis et al. 1993; Roost et al. 2017). The brain in particular has one of the highest methylation and hydroxymethylation levels of any tissue (Ehrlich et al. 1982; Li and Liu 2011), and both modifications are essential to proper neuronal and synaptic functions (Miller and Sweatt 2007; Sweatt 2016; Fasolino and Zhou 2017; Rudenko et al. 2013). It is key that we expand our knowledge of neuronal methylation in order to further understand neuronal development and disease.

# 2 Methylation Machinery

## *2.1 Methylation Writers*

DNA methylation is catalyzed by DNA methyltransferases (DNMTs), also known as DNA methylation writers, which transfer a methyl group from S-adenosylmethionine to the 5′ position of the cytosine residue (Kumar et al. 1994) (Fig. 1). Although five enzymes have been described, the major players are DNMT1, DNMT3A and DNMTB (Turek-Plewa and Jagodzinski 2005). These three writers regulate DNA methylation during critical embryonic and fetal developmental periods and their expression is dynamic throughout early development (Uysal et al. 2017). By

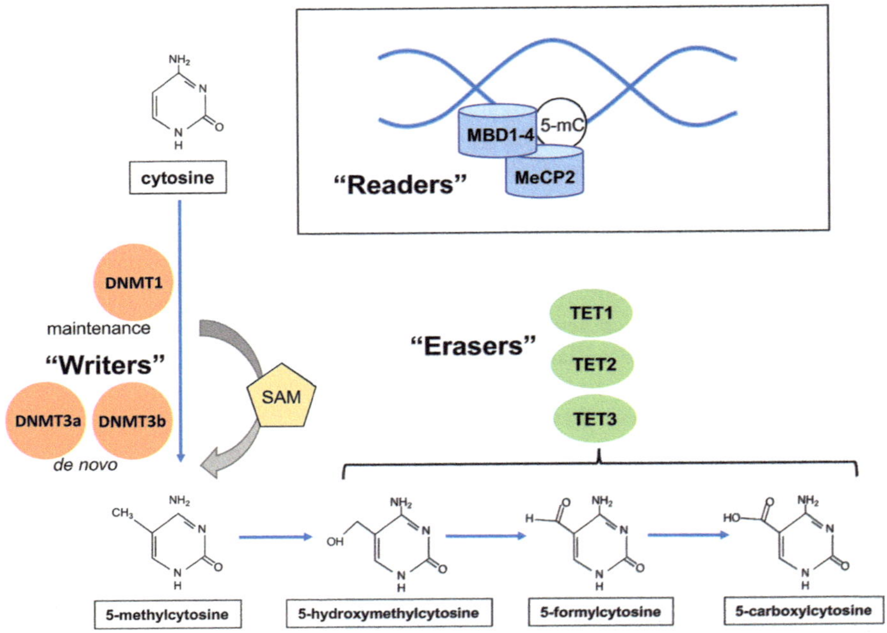


**Fig. 1** A schematic of the DNA methylation process and the machinery involved. DNA methyltransferases (DNMTs) catalyze the addition of a methyl group to the 5′ position of the cytosine, using S-adenosylmethionine (SAM) as the methyl donor. Ten-eleven translocation enzymes catalyze the oxidation of 5-methylcytosine into 5-hydroxymethylcytosine, and catalyze further reactions into 5-formylcytosine and 5-carboxylcytosine, in the demethylation process. Methylation readers methyl-CpG-binding protein 2 (MeCP2) and methyl-CpG binding domain proteins 1–4 (MBD1–4) facilitate the functional effects of methylation on gene expression and chromatin structure

controlling DNA methylation, they regulate the inhibition or activation of genes. DNA methyltransferase activity is high in preimplantation embryos during which time global loss and subsequent re-establishment of DNA methylation occurs (Carlson et al. 1992). DNMTs are required for imprinting and X chromosome inactivation in the embryo, as well (Howell et al. 2001; Biniszkiewicz et al. 2002). DNMT1, DNMT3A and DNMT3B are expressed ubiquitously in most somatic tissues, although the expression level varies (Robertson et al. 1999).

The writers differ in their cellular functions. Broadly speaking, DNMT1 is a maintenance enzyme and DNMT3A and DNMT3B are de novo enzymes. DNMT1 maintains methylation patterns in dividing cells, is involved in DNA repair, and has specific activity on hemi-methylated DNA (Bashtrykov et al. 2012). While the canonical function of DNMT1 is to maintain methylation, there is some evidence that it may have a secondary role in de novo methylation (e.g. DNMT1 shows in vitro activity on unmethylated DNA) (Jeltsch and Jurkowska 2014). Global deficiency of DNMT1 causes embryonic lethality in mice, underscoring its essential role for maintaining methylation patterns during embryonic development (Li et al.

1992). DNMT1 is also highly expressed in adult postmitotic neurons (Inano et al. 2000). On the other hand, DNMT3A and DNMT3B add on new methyl groups to DNA in response to environment or experience (Okano et al. 1999). Both DNMT3 writers are highly expressed in undifferentiated embryonic stem cells (ESCs), but their expression drops upon differentiation (Okano et al. 1998). In adulthood, both enzymes show reduced expression, with DNMT3A being ubiquitously low and DNMT3B barely detectable, in most tissues (Okano et al. 1998). There are some key differences between the de novo enzymes, however. During embryogenesis, DNMT3B is most heavily expressed in the brain, while DNMT3A is expressed throughout the entire embryo (Okano et al. 1999). Global DNMT3B knockout impairs neural tube development, resulting in embryonic death (Okano et al. 1999). Global DNMT3A knockout does not seem to impair gross development, but these mice have impaired postnatal neurogenesis and die within a month (Okano et al. 1999; Wu et al. 2010). Importantly, DNMT3A is required for methylation of non-CpG sites, in particular CpA (Guo et al 2014). Thus, overall, DNMT3A plays a larger role in developed brain function, whereas DNMT3B expression has more critical roles in early development.

## 2.2 *Methylation Readers*

Methylation readers are required to translate the methylation code into a functional action for certain genes. The methyl-CpG binding domain (MBD) family of proteins includes MBD1-4 and methyl-CpG-binding proteins 1 and 2 (MeCP1 and MeCP2) (Ballestar and Wolffe 2001). MeCP2 was the first and most-thoroughly studied protein in the MBD family and was found to repress gene transcription by interacting with histone deacetylases and subsequently modifying the chromatin structure to a heterochromatic state (Nan et al. 1998; Jones et al. 1998). MeCP2 and MBD1 both facilitate the methylation of histone 3 lysine 9 (H3K9), which promotes heterochromatin and transcriptional silencing (Sarraf and Stancheva 2004; Fuks et al. 2003). Interestingly, MeCP2 has been demonstrated to also associate with 5-hmC and is the primary 5-hmC binding protein in brain (Mellen et al. 2012). MeCP2 is crucial to neural development and function. Mutations in *MeCP2* result in a severe neurological disorder called Rett Syndrome (which occurs only in females because the *MeCP2* gene is X-linked, and the mutation is lethal in males) (Pohodich and Zoghbi 2015). Rett syndrome presents as significant intellectual, language and motor impairments that appear after the individual is at least 1 year old, and then continually progress (Ip et al. 2018). In fact, a mutation of one particular residue of MeCP2 that occurs in Rett Syndrome impairs this protein's binding specifically with 5-hmC (Mellen et al. 2012). Deletion of MBD1 in mice impairs neuronal differentiation, neurogenesis, synaptic plasticity and cognitive function (Zhao et al. 2003).

Other methyl-binding proteins have functions beyond gene repression; for example, MBD4 is a thymine glycosylase and mediates DNA repair (Bogdanovic and Veenstra 2009). While MBD3 does not directly bind to methylated DNA, it mediates methylation function by interacting with other factors, such as histone deacetylases or MBD2 (Ballestar and Wolffe 2001). MBD3 is an essential element of the methylation machinery, as deficiency of this protein is embryonically lethal (Hendrich et al. 2001). In particular, MBD3 is expressed in embryonic neuroepithelial cells (unlike its similar counterpart, MBD2) and continues to be expressed in specific forebrain structures postnatally (Jung et al. 2003). In summary, whether mediating gene repression or facilitating other mechanisms, methylation readers are essential biological components and apparently highly influential to neurological function.

### 2.3 *Methylation Erasers*

DNA methylation is both stable and reversible. Demethylation occurs through a series of steps driven by the ten-eleven translocation (TET) family of dioxygenases (TET1, TET2 and TET3; DNA methylation erasers). In the first step of this process, 5-methylcytosine is oxidized to produce 5-hydroxymethylcytosine (5-hmC) (Tahiliani et al. 2009). The demethylation process is completed through several more steps, in which TET proteins convert 5-hmC into 5-formylcytosine (5-fC) and 5-carboxylcytosine (5-caC). Both 5-fC and 5-caC can be excised by thymine DNA glycosylase (TDG) to result in an unmodified cytosine (Ito et al. 2011; Maiti and Drohat 2011). The TET enzymes vary in their substrate preference. TET1 and TET2 demonstrate greater enzymatic activity on 5-mC as compared to 5-hmC or 5-fC (Hu et al. 2015), whereas TET3 shows a stronger affinity for 5-caC (Jin et al. 2016). The three TET enzymes also show some key distinctions in the contexts in which they form 5-hmC. For example, in mouse ESCs, TET1 is responsible for 5-hmC marks near transcription start sites, while TET2 primarily maintains 5-hmC that is enriched in gene bodies (Huang et al. 2014). During the early wave of global demethylation within the zygote, TET3 is responsible for rapidly demethylating the paternal genome (Messerschmidt et al. 2014). Finally, imbalances in the TET enzymes disrupt neural function and cognitive processes (discussed in a later section), making them essential to normal brain function (Kaas et al. 2013; Rudenko et al. 2013; Zhang et al. 2013).

## 3 Methylation in the Brain

### 3.1 *5-Methylcytosine*

The brain is a phenomenally complex organ, with numerous regions and nuclei, each of which express a unique gene profile to carry out specific functions. In the cerebral cortex alone, there are six different layers, characterized by differences in cell morphology and in neuronal inputs and projections. Careful orchestration of gene expression at the right time, and in the right place, is paramount to the proper development of the brain.

In human embryonic stem cells, 5-mC is decreased during early differentiation into neural precursor cells (Kim et al. 2014). Although the overall abundance of 5-mC does not significantly change from neural precursor cells (NPCs) to neurons (Hahn et al. 2013), the methylation landscape is highly dynamic, controlling production and differentiation of various brain cells in a loci-specific manner. Following neurogenesis, NPCs switch to producing astrocytes (Martynoga et al. 2012) and the methylation profile of these NPCs is distinct from the profile at the time of neurogenesis (Sanosaka et al. 2017). Glial fibrillary acidic protein (GFAP; an astrocyte marker) has reduced promoter methylation at specific transcription factor binding sites following neurogenesis, allowing for increased expression of GFAP (Takizawa et al. 2001; Teter et al. 1996; Condorelli et al. 1997). 5-mC becomes enriched in the promoters of genes related to pluripotency, which are downregulated during differentiation (Kim et al. 2014). DNMT1 has been shown to be critical to the migration of interneurons to their specific destination in the cerebral cortex, in part by regulating expression of paired box 6 (*PAX6*) (Pensold and Zimmer 2018; Pensold et al. 2017). Knockdown of DNMT1 in embryonic neural progenitor cells disrupts the timing of astrogliogenesis, causing earlier production of these cells (Fan et al. 2005). DNMT3B seems to be critical to early neuronal development, as global knockout mouse embryos have neural tube defects and die before birth (Okano et al. 1999).

Methylation plays a critical role in the adult brain by regulating synaptic plasticity, and different neuron types show distinct methylation patterns (Mo et al. 2015). Neuronal activation induces acute changes in the methylation of genes involved in processes such as synaptic function, calcium signaling and protein phosphorylation (Guo et al. 2011). Inhibiting DNA methyltransferase activity alters the electrophysiological properties of cultured neurons (Meadows et al. 2016) and DNMT activity in the amygdala is necessary for the neural plasticity that occurs during fear learning and memory (Maddox et al. 2014). DNMT1 expression remains at a surprisingly detectable level for a maintenance enzyme, even though most neuronal cells are postmitotic and are no longer proliferating (Goto et al. 1994; Inano et al. 2000). Deficiency of DNMT1 in forebrain neurons has been shown to cause impairments in long-term potentiation of cortical neurons (Golshani et al. 2005) and progressive degeneration of neurons in the cortex and hippocampus, in one particular knockout mouse model (Hutnick et al. 2009). Deficiency of both DNMT1 and DNMT3A in forebrain neurons (using a different knockout strategy) impairs synaptic plasticity in hippocampal neurons and produces learning and memory deficits in mice (Feng et al. 2010). DNMT3A is required for maintaining methylation of CpA sites in the adult hippocampus (Guo et al. 2014). Mice lacking the methylation reader MBD1 show reduced neurogenesis, reduced synaptic plasticity and spatial learning deficits in adulthood (Zhao et al. 2003). MeCP2 is essential to adult brain function, and as mentioned previously, mutations in MeCP2 lead to severe neurological deficits (Rett Syndrome) in humans (Pohodich and Zoghbi 2015; McGraw et al. 2011). In addition, overexpression of MeCP2 enhances learning and synaptic plasticity in early adulthood, but causes seizures and premature death later on (Collins et al. 2004), underscoring the delicate balance required in DNA methylation dynamics for brain function. Thus, 5-methylcytosine and its machinery have complex roles in both neuronal development and adult brain function.

### 3.2 5-Hydroxymethylcytosine

5-hmC exhibits a very different pattern from 5-mC in both embryonic and adult neurons. In contrast to 5-mC, 5-hmC increases as NPCs differentiate into neurons (Kim et al. 2014; Hahn et al. 2013). 5-hmC is specifically enriched in the bodies of neurogenesis-related genes and is associated with their upregulation during differentiation (Kim et al. 2014). In embryonic mouse forebrain, 5-hmC is increased at promoters and gene bodies during neuronal differentiation (Hahn et al. 2013). TET2 and TET3 are expressed more highly than TET1 in the cortex and not surprisingly, are upregulated during neuronal differentiation. Overexpression or deficiency of these TET proteins can alter the timing of neurogenesis or cause inappropriate cellular clustering within the layers of the cortex, respectively. Although not as critical, TET1 also mediates embryonic neurogenesis and loss of TET1 in mice also alters the timing of neuronal production to a degree (Kim et al. 2016).

Hydroxymethylation plays a significant role in adult neuronal function. 5-hmC increases in the brain postnatally and continues to increase with aging (Wen et al. 2014; Szulwach et al. 2011; Kraus et al. 2015). It is particularly enriched in Purkinje cells of the cerebellum (Kriaucionis and Heintz 2009) and in the cortex (Kraus et al. 2012). Compared to 5-mC, it is still only 25% as abundant overall in the brain (Wagner et al. 2015). 5-hmC shows specific enrichment in genes related to synaptic function (Khare et al. 2012). In vitro studies have shown that TET1 is necessary for neurogenesis and progenitor cell proliferation in the adult brain (Zhang et al. 2013). TET1-deficient mice show reduced proliferation of neural progenitor cells (NPCs) in the hippocampus, and in vitro NPCs lacking TET1 have downregulated neurogenesis-related genes (Zhang et al. 2013). Although TET1 global knockout mice display normal brain morphology, they exhibit memory impairments as adults (Zhang et al. 2013; Rudenko et al. 2013). TET2 deficiency impairs the balance of proliferation and differentiation in mouse adult NSCs (Li et al. 2017b). TET2 is required for the increase in 5-hmC that occurs during differentiation of adult NSCs, and it controls critical neurogenic gene transcription (Li et al. 2017b). Mice that have a hippocampal deficiency of TET2 exhibit reductions in 5-hmC and cognitive impairments, both of which are rescued by replacement of TET2 (Gontier et al. 2018).

On the other hand, 5-hmC enrichment and TET enzyme expression can be regulated by neuronal activity (Kaas et al. 2013; Kremer et al. 2018). Fear conditioning in mice causes TET1 to be downregulated for several hours, and both overexpression or knockout of TET1 impairs the formation of long-term memories (Kaas et al. 2013; Rudenko et al. 2013). In primary cortical neurons, TET3 is upregulated in response to neuronal activity (Li et al. 2014). TET3 is upregulated in the cortex following fear extinction in mice (a behavioral paradigm in which fear is "un-learned") (Li et al. 2014). In summary, 5-hydroxymethylcytosine and its machinery have critical roles in the adult brain, including neuronal differentiation and regulating cognitive processes.

### 3.3 Methylation and the Aging Brain

Aging cells enter into a senescent state, whereby the cell cycle is halted (for dividing cells) and progressive phenotypic changes take place, beginning with alterations in genomic methylation, the formation of heterochromatic foci and increased DNA damage (van Deursen 2014; Baker and Petersen 2018). These changes have significant effects on gene expression, such as upregulating immune factors like pro-inflammatory molecules. The overall immune response becomes gradually elevated in the aging brain (Cribbs et al. 2012), forming the "senescence-associated secretory phenotype" (Rodier et al. 2009; Baker and Petersen 2018). Microglial cells, the resident macrophages of the brain, also become activated and produce local inflammation (Samorajski 1976; Norden and Godbout 2013). With age, DNA base-excision repair is reduced, impairing genomic integrity (Gan et al. 2012). Due to an increase in double-stranded DNA breaks, the DNA damage response pathway becomes activated and sustained (Sedelnikova et al. 2004; Rodier et al. 2009). Impairments in the antioxidant pathway lead to increased oxidative stress (Droge and Schipper 2007). Consequently, signaling pathways, metabolism, synapses and neuronal circuitry all become impaired. In the hippocampus especially, neurons display mitochondrial dysfunction and dysregulated calcium homeostasis (Pandya et al. 2016; Gant et al. 2006). Regions of the cerebral cortex begin to show a reduction of synapses (Adams 1987). After the age of 60 the brain gradually begins to reduce in weight and volume due to cell loss (Samorajski 1976). Cell loss is particularly prominent in the superior temporal gyrus and precentral gyrus regions of the cerebral cortex (Samorajski 1976). Collectively, these progressive changes cause a gradual decline in cognitive function that is associated with normal aging (Poddar et al. 2018).

The widely-held opinion has been that 5-mC globally decreases with aging and that this hypomethylation may be involved in age-related disease (Unnikrishnan et al. 2018). However, this is a grossly oversimplified statement, as the dynamics of 5-mC throughout aging depend on the tissue, the gene and (within the brain) the specific brain region. In many tissues, global 5-mC does decrease with age (Unnikrishnan et al. 2018; Wilson et al. 1987; Hoal-van Helden and van Helden 1989) and this often occurs in repetitive elements (Jintaridth and Mutirangura 2010). Yet, these findings seem dependent on the method used to quantify 5-mC (Unnikrishnan et al. 2018). In the brain, 5-mC dynamics throughout aging are region-specific. One study found no difference in the level of 5-mC in the hippocampi between 2-month old (young) and 22-month old (old) mice (Chen et al. 2012), which was corroborated by another study that examined 3, 12 and 24-month old mice of both sexes (Hadad et al. 2016). This same study also demonstrated no change in hippocampal DNMT or TET enzymes with increasing age. Certain genes involved in neuronal function show age-specific changes in methylation that may lead to the cognitive decline observed in aging mammals. Methylation of *Egr1*, an immediate early gene that plays an important role in synaptic plasticity in the hippocampal dentate gyrus, shows increased promoter methylation and reduced transcription in aged rats as compared to young adult rats (Penner et al. 2016). In the prefrontal cortex however, a number of genes involved in synaptic plasticity exhibit increased methylation and reduced expression in aged versus younger rats (Ianov et al. 2017). Thus, although 5-mC does not appear to change globally with aging, brain region-specific or loci-specific 5-mC changes are certainly involved in the aging process of the brain and age-associated diseases.

Aging is associated with gradual reductions in the length and integrity of telomeres (Bekaert et al. 2005; Rizvi et al. 2014). Methylation levels are high in the subtelomeric regions directly adjacent to the telomeres, and this methylation supports a heterochromatic state that confers genomic stability (Ng et al. 2009). DNMTs appear to play a protective role by both controlling telomere length and preventing excessive telomeric recombination (Gonzalo et al. 2006).

Opposite to 5-mC, 5-hmC has been shown to increase with age in the brain and peripheral tissues (Wagner et al. 2015). In specific brain regions of mice, such as the cerebellum and hippocampus, 5-hmC increases from postnatal day 7 (P7) to 1 year of age (Szulwach et al. 2011). In the tree shrew, another mammal, 5-hmC is also elevated from P10 to 2 years of age in the prefrontal, parietal and occipital cortices, the hippocampus and cerebellum (Wei et al. 2017). In mice, 5-hmC increases in the hippocampus from 12 months to 22 months of age (Chen et al. 2012), although this result was not confirmed by another research group using the same quantification method (Hadad et al. 2016).

# 4 Methylation and Neurodegenerative Diseases

## *4.1 Alzheimer's Disease*

Alzheimer's disease (AD) is the leading cause of dementia and is characterized by progressive and profound memory loss and impairments in cognitive functioning, for which there is no cure (Scheltens et al. 2016). The neurological hallmarks of this disease include β-amyloid deposits and plaques, and neurofibrillary tangles that build up within the brain tissue, driven by the proteins β-amyloid and Tau, respectively (Goedert et al. 1991; Gorevic et al. 1986). These pathological changes lead to neuronal loss in multiple brain areas (Braak and Braak 1991). The most common form is late-onset AD, which occurs after the age of 65, while early onset, familial AD occurs before age 65 and is more strongly associated with genetic mutations (Tanzi 1999). In general, late-onset AD is marked by 6 stages of neuronal pathology (Braak and Braak 1991). β-amyloid deposits are usually the first presentation, followed by neurofibrillary changes. While the majority of β-amyloid deposits tend to appear in the cortex, neurofibrillary tangles begin accumulating in the transentorhinal cortex (in the temporal lobe), and then move into the hippocampus and the cerebral cortex (Braak and Braak 1991; Calderon-Garciduenas and Charles 2017). Not all AD cases present the exact same pattern of β-amyloid deposits and/or neurofibrillary changes. It is unknown what exactly causes AD to develop, and which markers are causal versus symptomatic of the disease.

It is estimated that at least 80% of AD cases involve a genetic component, as evidenced by family and twin studies (Heston 1989). Only 5% of AD cases are early-onset, and most of these rare cases are strongly associated with mutations in amyloid precursor protein (APP), presenilin 1 (PSEN1) and presenilin 2 (PSEN2) (Tanzi et al. 1996). For the more common late-onset, sporadic AD, the genetic component is less clear. Polymorphisms in allele 4 of the apolipoprotein E gene (APOEε4) may contribute to over half of the late-onset cases and it is considered to be the main known genetic risk factor (Tanzi 2012). Genome-wide association studies (GWAS) have uncovered at least 11 other genes that show polymorphisms linked to AD (Tanzi 2012). Genetic variants alone however, do not fully account for the development of AD. An extensive amount of research has focused on the association of environmental factors (such as diet) in the development of the disease (Dosunmu et al. 2007). Because epigenetic modifications alter gene regulation as a result of environment or experience, methylation and hydroxymethylation have been targeted as possible risk factors in late-onset, sporadic AD (Bihaqi et al. 2012).

Postmortem studies of AD human brain tissue have revealed significant alterations in DNA methylation patterns, depending on the brain region examined. For example, both 5-mC and 5-hmC are overall significantly reduced in brains of people with AD as compared with age-matched controls with no dementia (Chouliaras et al. 2013). Furthermore, these methylation and hydroxymethylation levels strongly correlate with the build-up of β-amyloid plaques within the same brain region. Depletion of S-adenosylmethionine in vitro and in mouse models increases the proteins APP, PSEN1 and β-site amyloid precursor protein cleaving enzyme 1 (BACE-1) (all major players in AD) while decreasing β-amyloid accumulation (Do Carmo et al. 2016). When examining specific brain regions however, the evidence is contradictory. In one study, 5-mC was decreased in the entorhinal cortex, a region which shows the earliest signs of neurofibrillary changes in people with AD (Mastroeni et al. 2010). However, two other studies reported no change of 5-mC in the entorhinal cortex (Condliffe et al. 2014; Lashley et al. 2015). There is similar contradictory evidence for changes in the level of 5-hmC, with one study showing no change (Lashley et al. 2015) and another showing decreases (Condliffe et al. 2014) in the entorhinal cortex. 5-mC and 5-hmC were increased in the medial frontal and temporal gyri of AD patients in a study that controlled for age, gender, postmortem delay and tissue storage time (Coppieters et al. 2014). Interestingly, 5-mC and 5-hmC, as well as TET1 enzyme levels, are increased in the hippocampal/parahippocampal gyrus in both preclinical and late-stage AD patients (van der Flier et al. 2011), giving support to the possibility that aberrant DNA methylation facilitates the progression of, or perhaps even plays a causal role in AD. Discordant disease phenotypes among monozygotic twins also provide evidence for an epigenetic role in AD development (Ketelaar et al. 2012). One report describes an individual with AD who had reduced 5-mC in the anterior temporal neocortex and superior frontal gyrus (regions highly vulnerable to the disease) compared to his healthy, monozygotic twin (Mastroeni et al. 2009). Although there are discrepancies in the scientific literature (possibly due to differences in age, cause of death, postmortem delay, environmental, or unknown comorbidities that were unaccounted for), it is clear that the DNA modification landscapes, including 5-mC and 5-hmC, are altered in AD and could play causal roles in disease progression even before the onset of AD pathology.

The first indication that specific AD-related genes might be affected by methylation or hydroxymethylation was reported in the 1990s using a methylation-sensitive restriction enzyme to digest the *APP* gene from the temporal lobe brain tissue (West et al. 1995). These results showed an altered methylation pattern of *APP* in the AD patient. Another gene, triggering receptor expressed on myeloid cells 2 (*TREM2*) has been identified as a rare variant for AD with a large effect size and has anti-inflammatory functions (Guerreiro et al. 2013; Hamerman et al. 2006). Hypermethylation of *TREM2*, upstream of the transcription start site, has been reported on three separate occasions by a research group using two different methods of quantification (Smith et al. 2016). Hypermethylated *TREM2* was seen most strongly in the superior temporal gyrus of AD human brains as compared to control brains. Modified cytosines in neurons specifically (excluding other brain cells, such as glia or astrocytes) have been analyzed from human AD or control age-matched brain tissues (Mano et al. 2017). In this study, differentially methylated regions (DMRs) were identified in the breast cancer Type 1 susceptibility protein (*BRCA1*) gene promoter that were specifically hypomethylated in AD neurons. These results were confirmed with pyrosequencing, further finding that *BRCA1* methylation levels correlated with the level of *APOEε4* alleles, indicating a possible mechanism by which methylation may contribute to β-amyloid pathology. This coincided with increased mRNA and protein expression of BRCA1 that was present in the hippocampal and entorhinal cortex of AD brains, but not of the control brains. Interestingly, although *BRCA1* is a DNA repair gene, this protein was mostly present in the cytoplasm of AD brains, suggesting its dysfunction.

Greater methylation on the β-site amyloid precursor protein cleaving enzyme 1 (*BACE-1*) gene promoter is associated with β-amyloid load in AD brains (Do Carmo et al. 2016). In fact, even at the level of a single CpG site, hypermethylation was associated with reduced β-amyloid in those AD patients. Data from two AD mouse models (APP/Psen1 and 3xTg-AD) are consistent with human data that show hypermethylation of the genes thromboxane A2 receptor (*Tbxa2r*), sorbin and SH3 domain containing 3 (*Sorbs3*) and spectrin beta 4 (*Sptbn4*) are correlated with reduced gene and protein expression in frontal cortex brain tissue (Sanchez-Mut et al. 2013). Blocking the thromboxane receptor has been shown reduce AD pathologies in mice (Lauretti et al. 2015). *Sorbs3* hypermethylation in AD brains has also been previously reported (Siegmund et al. 2007) and may be a normal, age-dependent change that is accelerated with AD. *Sorbs3* encodes the protein vinexin, which has roles in cell signaling, gene expression and synaptic functions (Ito et al. 2007).

Mouse models with mutations in various AD-related genes have identified clues to possible mechanisms by which methylation could regulate AD pathogenesis. However, it is important to bear in mind that transgenic mice only recapitulate key aspects of the disease and not the complete pathology. For example, J20 mice, which have a mutation in *APP*, exhibit plaques but no neurofibrillary tangles (Lardenoije et al. 2018). In these mice, 5-mC and the ratio of 5-mC:5-hmC decreases with age in the hippocampal CA3 region, but there is no change in 5-hmC amount. Interestingly, both 5-mC and 5-hmC levels were negatively correlated with the plaque amount in the dentate gyrus of the hippocampus (Lardenoije et al. 2018). The 3xTg-AD mouse model has mutations in *Psen1*, *App* and *Tau* genes and exhibits both plaques and neurofibrillary tangles. In these mice, 5-mC increases with age (from 3 to 17 months) in the dentate gyrus and in the CA1 and CA2 hippocampal regions, but it does not correlate with plaque accumulation (Lardenoije et al. 2018). Neither the J20 nor the 3xTg-Ad mouse models show any age-related change in 5-hmC. In contrast to this, a different study did find that 3xTg-AD mice show an age-related increase in 5-hmC in multiple brain regions, but found no correlation with the plaque accumulation (Cadena-del-Castillo et al. 2014). McGill-Thy1-APPTg mice (which express human *APP* with several mutations) have reduced global 5-mC in the cortex and hippocampus which correlate with the β-amyloid load (Do Carmo et al. 2016). The Tg5xFAD mouse (which highly expresses mutant forms of *APP* and *PSEN1*) exhibits an increase in global neuronal 5-mC, as well as increases in *Dnmt3a* expression (Grinan-Ferre et al. 2016). An important consideration is that 5-mC and 5-hmC were measured by multiple methods across these studies, such as immunoreactivity, LUMA, ELISA and Dot-Blot, in addition to using different mouse lines that express differing genetic risk factors for AD.

We note several interesting points beyond the mere association of methylation and neuropathologies in AD. Importantly, evidence suggests that epigenetic changes may initiate AD, given that they sometimes occur in presymptomatic patients and in AD mouse models prior to the development of cognitive dysfunction or neuronal pathology. The methylation level of *BIN1* and other genes have been shown to be significantly associated with the plaque amount in AD patients—curiously, these same effect sizes between methylation levels and β-amyloid plaques were observed in elderly people with no cognitive impairment, suggesting that altered methylation pattern might be a cause, rather than an effect, of the disease (De Jager et al. 2014). In McGill-Thy1-APPTg mice, early increases in the β-amyloid peptide within neurons (which occurs prior to the β-amyloid pathology and cognitive impairment) is associated with both global hypomethylation in the hippocampus and cortex, and *Bace-1* promoter hypomethylation in the cortex. Further, S-adenosylmethionine (SAM) administration could rescue the global and *Bace-1* hypomethylation, reverse the cognitive impairments and reduce extracellular and intraneuronal β-amyloid pathology in the transgenic mice (Do Carmo et al. 2016). In vitro studies have demonstrated a role of β-amyloid in driving hypomethylation. For example, murine cerebral endothelial cells treated with β-amyloid have reduced global methylation (Chen et al. 2009). On the other hand, restoring methylation has been shown to reduce β-amyloid. Specifically, SAM administration to human neuroblastoma cells inhibits promoter demethylation of *PSEN1* and significantly reduces β-amyloid load (Scarpa et al. 2003).

AD is a disease with multiple genetic components and complex risk factors that make untangling its developmental mechanisms difficult. Methylation studies aimed at AD genes should be carefully designed to focus on the "low-hanging fruit" of potential therapeutic targets, and on loci that are likely to be involved in the majority of AD cases. Given the evidence of certain methylation changes occurring prior to disease onset, more efforts should be made to understand the potential causal role of 5-mC and 5-hmC in AD.

## *4.2 Parkinson's Disease*

Parkinson's disease (PD) is the second most prevalent neurodegenerative disease after Alzheimer's (de Lau and Breteler 2006; Bertram and Tanzi 2005). PD is characterized by cytoplasmic aggregation of the α-synuclein protein (which creates Lewy bodies) and selective loss of dopaminergic neurons, both of which occur in the substantia nigra pars compacta of the midbrain (Wakabayashi et al. 2000; Braak et al. 2004; Gibb and Lees 1991). These pathologies lead to movement disorders including resting tremors, slow movement, rigidity, postural instability and akinesia (Erro and Stamelou 2017), and numerous other non-motor comorbidities such as constipation, sleep dysfunction, depression and cognitive impairments (Titova et al. 2017; Albers et al. 2017). The substantia nigra is a part of the basal ganglia, a group of neuronal networks that provide input to other brain regions to facilitate movement, especially by inhibiting other movements at the same time (Mink 1996). α-synuclein is a protein involved in synaptic vesicles and is typically found at the presynaptic terminal of neurons (not in the cytoplasm), which suggests its dysfunction that it accumulates in the cytoplasm (Burre et al. 2010; Wakabayashi et al. 1992). A disturbing fact is that by the time Parkinsonian symptoms appear, the neurodegeneration and loss of dopaminergic neurons is significant and irreversible (Cheng et al. 2010). Because the neuronal pathology occurs years before symptom onset, understanding early biomarkers has become key in early diagnosis (DeKosky and Marek 2003).

Like AD, the majority of PD cases are sporadic (Lill 2016). GWAS studies have confirmed loci on the genes α-synuclein (*SNCA*) and microtubule-associated protein tau (*MAPT*) as primary susceptibility factors for sporadic PD (Billingsley et al. 2018; Edwards et al. 2010; Pihlstrom et al. 2018). A large CpG island is present in the human *SNCA* gene that covers the transcription start sites, the promoter region and intron 1. Multiple studies support that hypomethylation of this CpG region may lead to PD pathology. For example, postmortem substantia nigra tissue from humans with sporadic PD, as well as in vitro neuronal culture experiments, show an association between *SNCA* intron 1 hypomethylation and increased *SNCA* expression (Jowaed et al. 2010; Matsumoto et al. 2010). However, a later study found no difference in the *SNCA* methylation in this brain region between PD and control (although these findings may be due in part to differences in the DNA isolation process prior to sequencing) (Guhathakurta et al. 2017b). A different study reported that PD individuals had lower *SNCA* promoter methylation, which was associated with a specific variant of *SNCA* in both blood and cerebral cortex tissue (Pihlstrom et al. 2015). Similar findings of *SNCA* methylation in blood and frontal cortex brain tissue from PD individuals have been reported, as well (Masliah et al. 2013). Other studies have failed to find differential *SNCA* methylation in PD blood cells (Song et al. 2014; Richter et al. 2012). Of note, only the human *SNCA* gene shows this many CpG sites, as compared to mice or rats (Jowaed et al. 2010; Guhathakurta et al. 2017a), underscoring the limitations of rodent data in this specific area of PD epigenetic research.

A key finding is that levodopa (L-DOPA; the primary drug used to treat Parkinsonian symptoms) increases *SNCA* methylation both in vitro and in vivo (Schmitt et al. 2015). The link between L-DOPA and methylation is unique in that L-DOPA is metabolized in part through O-methylation using SAM as a methyl donor (Sandler 1972). L-DOPA administration results in decreased SAM concentrations in the brain (Liu et al. 2000). Interestingly, administration of SAM into rat brains causes neuronal degeneration (Charlton and Mack 1994) and motor impairments reminiscent of PD clinical symptoms (Charlton and Crowell 1995). It is clear that methylation is involved, likely at many different aspects, of PD pathology, and it is an additional mechanism through which L-DOPA may exert its therapeutic effects. Further, the link between L-DOPA and methylation may yield insights regarding ways to improve the efficacy of this drug.

*MAPT* and *LRRK2* are two other major genetic risk factors that show altered methylation in PD patients (Simon-Sanchez et al. 2009). PD patients have shown greater *MAPT* promoter methylation in the frontal cortex as compared to control individuals (Masliah et al. 2013). Another report however, demonstrated greater *MAPT* methylation in the cerebellum, but reduced methylation in the putamen, a basal ganglia structure near to the substantia nigra (Coupland et al. 2014). *MAPT* methylation in leukocytes and brain tissue also positively correlates with the age of disease onset in PD (Coupland et al. 2014). Although *LRRK2* is increased in sporadic PD brains (Cho et al. 2013) and *LRRK2* overexpression elevates α-synuclein pathology (Lin et al. 2009), no distinct *LRRK2* promoter methylation patterns were seen in leukocytes from individuals with PD as compared to controls (Tan et al. 2014).

DNA methyltransferases play a specific role in PD neuropathology. Nuclear DNMT1 is reduced in the cerebral cortex of PD patients and shows greater cytoplasmic localization as compared to non-PD human brains (Desplats et al. 2011). Consistent with this, global 5-mC is also reduced, as quantified using immunohistochemistry and ELISA (Desplats et al. 2011). Using rat neuronal cell culture, this same group demonstrated that overexpression of α-synuclein causes Dnmt1 to shift its location to the cytoplasm. α-synuclein and Dnmt1 proteins co-immunoprecipitate (when pulling down either protein) both in cell culture and in human brain tissue, suggesting a direct association between these two proteins. Supporting this, lentiviral expression of *DNMT1* partially reverses α-synuclein-induced global hypomethylation. Similar phenomena are observed in transgenic mice that overexpress human α-synuclein under the control of a Thy-1 promoter. These mice show cytoplasmic localization of Dnmt1 and global hypomethylation; but lentiviral delivery of *DNMT1* increases Dnmt1 nuclear localization (Desplats et al. 2011). Additional studies further implicate DNMT activity in PD pathologies. For example, inhibiting DNA methyltransferases by 5-aza-2′-deoxycytidine in dopaminergic neuronal cell lines (human, mouse and rat) increases apoptosis (Wang et al. 2013b). 5-aza-2′-deoxycytidine also increases the degenerative effects of MPP+, 6-OHDA and rotenone (neurotoxins commonly used in PD research) and reduces *SNCA* promoter methylation (Wang et al. 2013b). Finally, several polymorphic variants of the *DNMT3B* gene correlate with PD risk in Brazilian and Chinese populations (Pezzi et al. 2017; Chen et al. 2017).

In summary, a fair amount of research has already been established regarding the role of methylation in PD. Because the neurodegeneration in PD is largely silent until after significant neuronal loss has occurred, biomarkers of early neurodegeneration would be essential to timely diagnosis and treatment. Further research may be able to identify a novel circulating factor in the blood whose methylation status might signal the onset of neurodegeneration, in advance of symptoms onset. Additional research should also be aimed at understanding the dysregulation of DNMT1 and its sequestration in the cytoplasm by α-synuclein.

## 4.3 Amyotrophic Lateral Sclerosis

Amyotrophic lateral sclerosis (ALS; also known as Lou Gehrig's disease) is a devastating disease involving degeneration of the neurons involved in muscle control (Goetz 2000). It is considered a rare disease, as the incidence is ~5 cases per 100,000 individuals in the U.S. (Mehta et al. 2018), with a mean age of onset of around 60 years (ranging from 40 to 70 years) (Govoni et al. 2017). The neurological hallmark of this disease is the aggregation of ubiquitinated proteinaceous inclusions within the cytoplasm of both upper motor neurons of the cerebral cortex and lower motor neurons of the spinal cord (Hardiman et al. 2017; Goetz 2000). This pathology leads to reduced synapses and eventually denervation of muscle (Sasaki and Maruyama 1994). The clinical presentations of ALS are muscle atrophy, weakness, spasticity, cramps, fasciculation (muscle twitching) and dysphagia (difficulty swallowing) (Rowland and Shneider 2001; Wijesekera and Leigh 2009). Most individuals with ALS die within 3–5 years following the diagnosis due to degeneration of the neurons that lead to the muscles required to breathe (Balendra et al. 2014). Neurons in the motor cortex are lost early on in the disease, and at death more than half of the spinal motor neurons are gone (Chio et al. 2014). Astrogliosis is prominent in neurons of the ventral and dorsal horns of the spinal cord (Schiffer et al. 1996). Like Alzheimer's and Parkinson's diseases, only 5–10% of ALS cases are familial while the rest are sporadic (Ajroud-Driss and Siddique 2015). The cause of ALS is unknown, although an expanded hexanucleotide repeat in the chromosome 9 open reading frame 72 (*C9ORF72*) is the most common genetic cause of both familial and sporadic forms (DeJesus-Hernandez et al. 2011).

Spinal cord tissue from ALS patients has elevated global 5-mC levels and even greater elevations in 5-hmC levels (threefold above non-ALS individuals) (Figueroa-Romero et al. 2012). In the motor cortex of the brain, 5-mC immunoreactivity can be observed in neurons from ALS patients, while it is hard to detect in non-ALS individuals (Chestnut et al. 2011). Global methylation levels from blood samples provide conflicting evidence. 5-mC is increased in whole blood from ALS patients when measured by a restriction enzymes assay (Tremolizzo et al. 2014), and similarly ALS individuals from five families carrying a superoxide dismutase 1 (*SOD1*) mutation show increased 5-mC in blood using an ELISA assay (Coppede et al. 2018). However, another study reported no such increase using an ELISA assay (Figueroa-Romero et al. 2012). In general, very few studies on 5-hmC in ALS were found in a search of the literature. Regardless of the dynamics of global 5-mC and 5-hmC in ALS, methylation of specific genes may guide us to new potential therapeutic targets that could delay the progression of ALS.

The most studied loci associated with sporadic ALS is *C9ORF72*, which bears a hexanucleotide repeat expansion (Shatunov et al. 2010; Ahmeti et al. 2013). The expanded repeat of *C9ORF72* can cause a buildup of nuclear RNA foci in individuals that carry the expanded allele (DeJesus-Hernandez et al. 2011)—a pathology that may interfere with mRNA splicing, RNA binding proteins and other mechanisms leading to cellular dysfunction. Approximately 30% of ALS patients with a *C9ORF72* repeat expansion have hypermethylation of this gene's promoter (van Blitterswijk et al. 2012). *C9ORF72* promoter hypermethylation has been associated with less accumulation of protein aggregates in human ALS brains (Liu et al. 2014), suggesting that it may be a compensatory mechanism. Furthermore, the same study reported that inhibiting methylation in human cell lines using 5-aza-2′-deoxycytidine increases the susceptibility of the cells to oxidative and autophagic stress, leading the authors to hypothesize that *C9ORF72* hypermethylation could be protective against neuronal toxicity. Consistent with this, another group found a negative correlation between *C9ORF72* hypermethylation and neuronal loss in the hippocampus and frontal cortex of human brains (McMillan et al. 2015). Unexpectedly, ALS patients with *C9ORF72* hypermethylation also have enrichment of 5-hmC at the *C9ORF72* promoter (Esanov et al. 2016). There are some contrary reports regarding *C9ORF72* hypermethylation, however. One set of monozygotic twins discordant for ALS showed a repeat expansion of *C9ORF72* but no differences in its methylation pattern (Xi et al. 2014). Additionally, carriers of a shorter expansion of *C9ORF72* show less methylation of this gene as compared to carriers with the long expansion (Gijselinck et al. 2016). Thus, it seems that *C9ORF72* promoter methylation may mediate pathology in some ALS individuals with the *C9ORF72* expansion but not in others.

Other candidate risk genes that have been identified for ALS, such as superoxide dismutase 1 (*SOD1*), vascular endothelial growth factor (*VEGF*), TAR DNA-binding protein 43 (*TARDBP*) and angiogenin (*ANG*) have not been reported to have differential methylation in individuals with ALS. Analysis of postmortem lateral frontal cortex tissue from sporadic ALS individuals using ChIP on Chip techniques (combining chromatin immunoprecipitation [ChIP] with whole genome scanning using a gene chip) revealed no methylation differences in any of the four candidate genes describe above (Morahan et al. 2009). Further, analysis of peripheral blood mononuclear cells (PBMCs) and brain tissue from ALS individuals using bisulfite sequencing revealed no differential methylation in the *SOD1* or *VEGF* promoters (Oates and Pamphlett 2007). However, differential methylation has been observed in over 30 CpG sites in various genes involved in calcium dynamics, excitotoxicity and oxidative stress among ALS and control cortex samples (Morahan et al. 2009). Spinal cord tissue from ALS individuals displays over 1000 differentially expressed genes related to inflammatory and immune responses, with many of these overlapping with differentially methylated regions as well (Figueroa-Romero et al. 2012).

Studies of disease-discordant twins can be of great insight to unraveling disease etiology because it controls for genetic variation, leaving epigenetic (e.g., methylation) factors as possible mechanisms. The methylation profile of peripheral blood cells in five sets of monozygotic twins discordant for ALS revealed common differentially methylated CpG sites, including regions involved in glutamate metabolism and GABA signaling (important neurotransmitters required for synaptic transmission in many neural circuits) in the ALS siblings as compared to their unaffected twins (Young et al. 2017). Interestingly, the ALS siblings also demonstrated a more aged methylation profile compared to their non-ALS siblings [assessed by applying a Horvath algorithm to the methylation data; also known as DNAm age (Horvath 2013)] This aged DNAm profile has also been reported in another set of monozygotic twins discordant for ALS, pointing to an interesting link between epigenetic aging mechanisms and ALS (Zhang et al. 2016).

DNA methyltransferases may play a unique role in neuronal cell death in ALS. In humans with ALS, both DNMT1 and DNMT3A are upregulated in both the nucleus and mitochondria of motor cortex neurons (Chestnut et al. 2011). Transfection of DNMT3A into motor neuron-like cell culture (NSC34) causes degeneration of those cells (Chestnut et al. 2011). Interestingly, this forced expression of DNMT3A is localized to the cytoplasm not the nucleus, and frequently localizes to mitochondria. In NSC34 cells that are undergoing apoptosis, both DNMT1 and DNMT3A are significantly elevated. When the cells are induced to undergo apoptosis, inhibiting DNMTs pharmacologically reduces cell death, supporting the hypothesis that DNMTs can drive apoptosis in neuronal cell culture. Spinal cord lesions in mice also upregulate DNMT1 and DNMT3A, with DNMT3A expression localizing in mitochondria and spinal neuron synapses, and 5-mC is also upregulated upon motor neuron apoptosis. Collectively, these reports highlight a possible role for DNMTs in neuronal death occurring in ALS.

In summary, understanding *C9ORF72* methylation mechanisms, contexts and regulation is an area of ALS research that should continue to be deeply explored. Judging from previously published results, it seems that epigenetic analyses of spinal tissue may yield more insight than that of blood samples. Finally, it is clear that the dynamics of DNMTs in motor neuron apoptosis is an area that should be further exploited for potential therapeutic targets against motor neuron loss.

## 4.4 Multiple Sclerosis

Multiple sclerosis (MS) is a chronic inflammatory disease that attacks the central nervous system (CNS) and is the most common cause of neurological disability in younger adults (Smith and McDonald 1999; Compston and Coles 2002; Grigoriadis et al. 2015). MS pathology begins as inflammatory lesions that cause demyelination of neuronal axons, and eventual transection and degeneration of axons (Trapp et al. 1998; Trapp and Nave 2008; Kraft and Wessman 1974). Because MS affects axonal conduction signals throughout the CNS, the symptoms are broad and include numbness, tingling, spasticity, fatigue/weakness, difficulty walking, vision problems, sexual dysfunction and pain (Samkoff and Goodman 2011; Schwendimann 2006). MS is diagnosed by several lesions in the central nervous system (CNS) that are separated by time and space, which are typically observed using MRI (De Angelis et al. 2019). This disease has "silent symptoms," as evidenced by the development of many new lesions that are seen using MRI, with no simultaneous change in clinical symptoms (Isaac et al. 1988). The general course of MS has two stages: (1) relapsing and remitting MS (RRMS) and secondary progressive MS (SPSS). RRMS is characterized by bouts of symptoms (relapses; that can last several months and occur several times a year) followed by a period of recovery, in which remyelination, inflammation reduction, or compensatory mechanisms to axonal function may occur (Smith and McDonald 1999). Fewer than 10–20% of individuals with MS do not follow this typical course, but instead have primary progressive MS (PPMS) without the prolonged bouts of RRMS (Lublin and Reingold 1996; D'Amico et al. 2018). The age of onset for most MS cases is in the 20s and 30s (later for individuals with PPMS) and the symptom course and disease progression can be quite variable. The RRMS phase can last 8–20 years, but typically 50% of individuals with MS are unable to walk 25 years after disease onset (Trapp and Nave 2008).

Unlike the previously discussed neurodegenerative diseases, the heritability of MS is estimated to be 10–20 times greater for siblings of an individual with MS as compared to the general population (Didonna and Jorge 2017; Sadovnick and Baird 1988). The primary genetic risk factor has been localized to the major histocompatibility complex (*MHC*) region on chromosome 6p21.3 (Sawcer et al. 2005). This region encompasses 160 genes, most of which are involved in the immune response (Kalman and Lublin 1999). The human leukocyte antigen (*HLA*) loci within this region is the largest known MS susceptibility factor (Kalman and Lublin 1999; Isobe et al. 2016). Specific *HLA* haplotypes associate with age of disease onset and gender in MS patients (Hensiek et al. 2002; Moutsianas et al. 2015).

A majority of methylation research in MS has been conducted on blood samples, which is important given that MS may be initiated by circulating inflammatory factors that infiltrate the blood brain barrier (Frischer et al. 2009; Dendrou et al. 2015). Methylation patterns vary depending on the blood fraction analyzed [e.g., whole blood, peripheral mononuclear blood cells (PBMCs; containing a mixture of lymphocytes and leukocytes) or specific sub-groups of lymphocytes (Li et al. 2017a)]. Therefore, the scope and specificity of methylation analyses in blood cells is key to forming accurate conclusions. A good illustration of this is the interleukin 2 receptor subunit a (*IL2RA*) gene, which is elevated in PBMCs of MS patients, but shows no change in promoter methylation (Field et al. 2017). However, both *IL2RA* expression and promoter methylation changes are apparent when analyzing isolated T-cells. In PBMCs of MS patients, 5-hmC is decreased globally (Calabrese et al. 2014). Consistent with this, *TET2* is downregulated in PBMCs (as is *DNMT1*), and both of these genes show aberrant promoter methylation in MS patients (Calabrese et al. 2014). PBMCs from individuals with the less common PPMS form of the disease show more differentially methylated sites than do individuals with RRMS (Kulakova et al. 2016). Alu and LINE-1 repetitive elements are hypermethylated in whole blood from MS patients, and this methylation level correlates with a more severe disability score (Neven et al. 2016). In the monocyte fraction of the blood, 19 hypomethylated CpG sites have been found in the exon 2 of the *HLA DRB1*15:01* loci of MS individuals (Kular et al. 2018). Interestingly, individuals with MS that carried two copies of this allele showed less methylation of *HLA DRB1*15:01* than those with one or no copies of the allele (Kular et al. 2018). When analyzing just CD8+ T cells from patients in the RRMS disease phase, significantly more hypermethylated CpG sites were observed as compared to controls (Bos et al. 2015).

The methylation profile of CD8+ T cells is starkly different from that of CD4+ T cells, with very little overlap (which is not surprising given the different functions of these two cell types in immunity (Maltby et al. 2015; Bos et al. 2015). CD4+ T cells show differentially methylated regions in the *HLA-DRB* loci of individuals with RRMS (Graves et al. 2014), a finding which has been replicated several times by the same group (Maltby et al. 2017). In MS patients taking the anti-inflammatory drug dimethyl fumarate, CD4+ T cells show over 900 differentially methylated positions, most of which are hypermethylated, as compared to MS patients not taking the drug (Maltby et al. 2018). The question has been posed whether methylation status of certain genes could be used as a distinguishing factor between MS individuals who were relapsing versus MS individuals who were in remittance. Eight genes were analyzed that were involved in CD4+/CD8+ T cell differentiation, oligodendrocyte differentiation or neuroinflammation (Sokratous et al. 2018). However, methylation status differed only between MS versus healthy individuals on the genes runt-related transcription factor 3 (*RUNX3*; which regulates CD4/CD8 T cell differentiation) and cyclin-dependent kinase inhibitor 2A (*CDKN2A*; which is indirectly involved in the regulation of oligodendrocyte apoptosis).

Oligodendrocytes, which make up the myelin sheath around neuronal axons, become repeatedly destroyed and regenerated during RRMS (Lee et al. 2015). DNA methylation is known to be a critical component of oligodendrocyte proliferation in brain development (Moyon and Casaccia 2017), but it is also important to the differentiation of oligodendrocyte progenitors after demyelination has occurred in the adult brain (Moyon et al. 2017). 5-mC levels are elevated in oligodendrocytes during remyelination of mice that have been given spinal cord lesions (Moyon et al. 2017). Using an inducible knockout strategy in mice, this same study identified Dnmt1 as being critical to the early phase of remyelination in lesioned mice, and

Dnmt3a critical to the later phase. An analysis of lesion-free regions of MS brains showed hypermethylation and reduced expression of genes Bcl2-like protein 2 (*BCL2L2*) and n-myc downstream-regulated gene 1 (*NDRG1*), which play a role in the survival of oligodendrocytes (Huynh et al. 2014).

Environmental factors such as vitamin D deficiency and smoking also confer major risk in MS development (Michel 2018). T cells from RRMS patients show hypermethylation of an alternative promoter at exon 1C of the vitamin D receptor (*VDR*) gene, which surprisingly is associated with upregulation of the receptor, rather than the expected downregulation typically associated with promoter methylation (Ayuso et al. 2017). Another known environmental MS risk factor, smoking, is known to alter methylation patterns and does so most strongly in those individuals who are *HLA* haplotype carriers (Marabita et al. 2017). It is possible that the effect of smoking load on methylation may be increased in MS individuals, although the evidence for this is not robust (Marabita et al. 2017).

In conclusion, MS is a disease characterized by complex changes in the immune system, both in the periphery and the brain. DNA methylation should be a major focus of MS research, given the strong environmental risk factors that are already known to cause aberrant gene methylation. The focus on methylation in blood cells is important to MS research, given the myriad of circulating immune factors, and that many of these factors can infiltrate the brain in MS. The number of potential questions that can be asked regarding the role of methylation in MS pathology are limitless, and the continuous published research in this area is reflective of that. Understanding the methylation dynamics in inflammatory factors of AD patients will also greatly affect our understanding of other diseases that have major inflammatory components.

## 4.5 *Huntington's Disease*

Huntington's disease (HD) is the most common monogenic neurological disease and is caused by an autosomal dominant expanded repeat of the trinucleotide CAG in the huntingtin (*HTT*) gene (Sun et al. 2017; Gusella and MacDonald 1995). This expansion results in an excessively long polyglutamine sequence on the N-terminus of the HTT protein, which becomes fragmented and forms nuclear aggregates, conferring neuronal toxicity (DiFiglia et al. 1997). The primary clinical symptoms of HD are motor dysfunction (specifically involuntary movements and impaired voluntary movements), cognitive impairments, and frequent psychiatric comorbidities (Ghosh and Tabrizi 2018; Roos 2010). Individuals with HD have between 35 and 55 copies of the repeat (Bates et al. 2015). Greater than 55 CAG copies results in juvenile HD, in which the disease develops before the age of 20, and is almost always inherited through the father (Farrer et al. 1992; Gordon 2003). Typical disease diagnosis occurs between 30 and 50 years of age (Roos 2010). Medium spiny GABAergic neurons in the striatum (a part of the basal ganglia) are particularly vulnerable to the disease (Sieradzan and Mann 2001), especially those expressing adenosine $A_{2A}$ receptors ($A_{2A}$R), which show alterations before symptom onset (Glass et al. 2000; Orru et al. 2011). HD is a progressive and fatal disease (Bates et al. 2015).

While the cause of HD is known to be completely genetic, understanding methylation patterns that correspond to neuronal changes in the HD brain could still be advantageous to therapeutic development. Global hydroxymethylation dynamics have been studied at various ages of an HD mouse model. In the YAC128 transgenic mouse (which contains 128 CAG repeats under control of human endogenous regulatory elements), deep sequencing of 5-hmC-enriched DNA showed no differences in the overall 5-hmC landscape in the striatum and cortex (Wang et al. 2013a). However, there were 747 differentially hydroxymethylated regions (DhMRs; mostly downregulated) that were related to gene pathways such as GABA/glutamate receptor signaling, synaptic long-term potentiation and axonal guidance signaling. Importantly, global 5-hmC was reduced in YAC128 mice (compared to wildtype) by 6 weeks of age, which is well before the onset of disease pathology. Although still lower than wildtype mice, hydroxymethylation in YAC128 mice shows an age-dependent increase until 3 months, in the striatum and hippocampus (Wang et al. 2013a). Thus, decreases in 5-hmC may be potentially useful as an HD biomarker. Surprisingly, despite lower 5-hmC observed in 3-month old HD mice, age-associated reductions in heterochromatin are enhanced in more aged YAC128 mice (Park et al. 2017).

The reports on gene-specific methylation changes in HD are underwhelming. An analysis of whole blood methylation between symptomatic HD, presymptomatic HD and non-HD individuals did not yield robust findings, perhaps due to lack of power (Zadel et al. 2018). Other studies have been consistent with this, in that no differential methylation has been observed in the blood (Hamzeiy et al. 2018) nor the cortex of HD individuals as compared to control (De Souza et al. 2016). An in vitro study of striatal cells expressing polyglutamine-expanded HTT identified a number of differentially methylated regions (most of which were low in CpG sites), using reduced representation bisulfite sequencing (Ng et al. 2013). Only SRY-box 2 (*SOX2*), *PAX6* and nestin (*NES*) (all involved in neurogenesis and differentiation) showed increased methylation of regulatory regions as well as reduced expression. The *D4S95* loci is closely linked to the *HTT* gene, and methylation of this loci shows a high degree of variability (Wasmuth et al. 1988; Pritchard et al. 1989). It has been proposed to use the methylation status of *D4S95* as a predictor of HD (Theilmann et al. 1989), although no further literature on this has been found. Another study has examined whether *D4S95* methylation correlates with age of onset, but discovered that it only correlates with age, in general (Reik et al. 1993).

Humans with HD, even at the earliest stage of the disease, show a decrease in adenosine $A_{2A}$ receptor ($A_{2A}$R) protein in the putamen, a part of the striatum. Consistently, an increase of methylation at exon 1e of the $A_{2A}R$ gene was observed and validated using MeDIP and bisulfite sequencing in these tissues (Villar-Menendez et al. 2013). These findings were recapitulated in two mouse models of HD, the R/61 mouse (containing 145 CAG repeats in the HTT gene) and the R6/2 mouse (115 repeats). Although no change in $A_{2A}R$ promoter methylation was observed in either mouse model, exon m2 showed reduced methylation in

30-week old R6/1 mice, but not in the younger diseased mice. In the R6/2 mice, exon 2 showed reduced hydroxymethylation.

Methylation may also regulate DNA damage and repair in HD. In patient-derived HD cell lines, the *APEX1* gene (involved in DNA repair) is downregulated and shows promoter hypermethylation as compared to control cells (Mollica et al. 2016). Inhibiting DNA methyltransferases rescues *APEX1* gene expression and further, induces stability of the trinucleotide repeats in dividing cells. Similar to individuals with ALS, individuals with HD also show an accelerated epigenetic aging in the frontal and parietal lobes and cingulate cortex regions of the brain (Horvath et al. 2016).

Overall, there does not seem to be a huge consensus in the literature regarding a specific role for DNA methylation or hydroxymethylation in HD. Because HD has a known monogenic cause, the emphasis on methylation research in this area is not as strong as for other neurodegenerative diseases that are strongly affected by environmental factors. Nonetheless, 5-hmC may be a likely biomarker to target in order to achieve an early diagnosis (Table 1).

## 5 Conclusions and Future Research

Neurodegenerative diseases vary greatly in their pathologies, course of progression and impact on quality of life, although each is devastating in its own way. While the cause of Huntington's disease may be clearly known, the instigating factors for other diseases have not been identified. Further, even though we understand that β-amyloid is responsible for creating much of the pathology in AD, it is not clear what causes β-amyloid dysfunction, or how to delay or prevent it. Unraveling the DNA methylation and hydroxymethylation dynamics in neurodegenerative diseases adds another dimension of possible answers regarding causal, mediating, and symptomatic factors in Alzheimer's and other diseases.

A major focus in neurodegenerative disease research is the use of new models that can produce results that are more translatable to humans. The majority of non-human neurodegenerative research to date has relied on mouse models. Despite the strong genetic similarity between mice and humans, therapies that are effective in mouse models often fail to show the same effect in human patients, especially for Alzheimer's disease (Li et al. 2013; Anand et al. 2017; Godyn et al. 2016). Even though there are conserved epigenetic loci between mice and humans that associate with Alzheimer's disease (Sanchez-Mut et al. 2013; Gjoneska et al. 2015), there are apparently other unknown significant differences that prevent drug efficacy from translating to humans. One contributing factor is that mouse models of Alzheimer's disease are unable to entirely recapitulate the disease pathology, typically only exhibiting several but not all major presentations (Van Dam and De Deyn 2011; Li et al. 2013; Esquerda-Canals et al. 2017). Performing studies on tissue that is genetically human (either human tissue sample or human-derived iPSCs) has become a key focus. Brain organoids (small spheroids of neurons that harbor properties of various brain regions) grown from iPSCs will be of great significance

**Table 1** A summary of the neurodegenerative diseases and their associated major methylation changes that are discussed in this chapter

| Disease | Primary symptoms | Pathology | Candidate genes | Distinct methylation changes | References |
|---|---|---|---|---|---|
| Alzheimer's disease | Profound memory loss, cognitive impairment | β-amyloid plaques, neurofibrillary tangles | *APOEε4*, presenilin-1/2, β-site amyloid precursor protein cleaving enzyme 1 | Hippocampal 5-mC, 5-hmC and TET1 enzymes increased in pre-clinical and late-stage AD patients. | Van der Flier et al. (2011) |
| | | | | *BRCA* promoter hypomethylation correlates with β-amyloid load. | Mano et al. (2017) |
| | | | | Hypermethylation of *BACE-1* associated with β-amyloid load. | Do Carmo et al. (2016) |
| | | | | β-amyloid plaque and *BIN1* methylation changes in both AD and non-symptomatic patients. | De Jager et al. (2014) |
| | | | | *TREM2* hypermethylation in superior temporal gyrus. | Smith et al. (2016) |
| Parkinson's disease | Resting tremors, low movement, rigidity, akinesia, postural instability | Selective loss of dopaminergic neurons in the substantia nigra | α-synuclein (SNCA), *MAPT*, *LRRK2* | *SNCA* Intron 1 hypomethylation associated with increased *SNCA* expression. | Jowaed et al. (2010), Matsumoto et al. (2010) Pihlstrom et al. (2015) |
| | | | | L-DOPA increases *SNCA* methylation. | Schmitt et al. (2015) |
| | | | | *MAPT* methylation correlates with age of disease onset. | Coupland et al. (2014) |
| | | | | DNMT1 cytoplasmic localization and associates with α-synuclein. | Desplats et al. (2011) |
| | | | | *DNMT3b* variants correlate with PD risk in some populations. | Pezzi et al. (2017), Chen et al. (2017) |
| Amyotrophic lateral sclerosis | Muscle weakness, atrophy | Degeneration of upper/lower motor neurons | Hexanucleotide repeat expansion of *C9ORF72* | Elevated 5-mC and 5-hmC levels in spinal cord. | Figueroa-Romero et al. (2012) |
| | | | | 5-mC detected in motor cortex of ALS but not control. | Chestnut et al. (2011) |
| | | | | 30% of ALS patients with *C9ORF72* expansion have *C9ORF72* promoter hypermethylation. | Van Blitterswijk et al. (2012) |
| | | | | *C9ORF72* hypermethylation associated with reduced protein aggregates, negatively correlated with neuron loss. | Liu et al. (2014), McMillan et al. (2015) |
| | | | | Elevated *C9ORF72* promoter 5-hmC in people with *C9ORF72* hypermethylation | Esanov et al. (2016) |
| | | | | DNMT1 and DNMT3a upregulated in motor cortex neurons (may drive cell death). | Chestnut et al. (2011) |

| | | | | | |
|---|---|---|---|---|---|
| Multiple sclerosis | Numbness, tingling, fatigue, weakness | Chronic CNS inflammation, demyelination | *HLA-DRB1* loci on chromosome 6p21.3 | CD8+ and CD4+ T cells have distinct methylation profiles. | Maltby et al. (2015), Bos et al. (2015) |
| | | | | 5-hmC decreased in PBMCs; *TET2* and *DNMT1* downregulated with altered promoter methylation. | Calabrese et al. (2014) |
| | | | | Alu and LINE-1 elements hypermethylated, correlated with disability score. | Neven et al. (2016) |
| | | | | 5-mC increased in oligodendrocytes of spinal lesions in mice. | Moyon et al. (2017) |
| | | | | Vitamin D receptor upregulated and promoter hypermethylated in T cells. | Ayuso et al. (2017) |
| Huntington's disease | Motor dysfunction, cognitive impairments, psychiatric disorders | Expanded CAG repeat of huntingtin (HTT) gene | Huntingtin; *D4S95* loci | *A2AR* hypermethylation in striatal neurons. | Villar-Menendez et al. (2013) |
| | | | | Reduced global 5-hmC and enhanced age-associated chromatin relaxation | Park et al. (2017) |

to the field of neurodegenerative disease (Wang 2018). These 3-dimensional organoids can be cultured from human fibroblasts (either healthy or diseased humans), and can provide more complex neuronal material than standard cell culture. In another example, the pig has recently been highlighted as a more relevant model of Huntington's disease as compared to smaller mammals, such as mice (Yan et al. 2018). Transgenic HD pigs display a more severe Huntington's disease phenotype than do mice, and exhibit other HD symptoms such as breathing difficulties. These new models may yield greater translatability into effective therapies for humans.

This review has highlighted several key findings. First is the novel finding that amyotrophic lateral sclerosis and Huntington's disease show accelerated epigenetic age (DNAm age). Because many of the age-related brain changes are similar to neurodegenerative alterations (albeit at a milder level), it is critical that we understand the factors that can switch the normal progression of brain aging into neurodegenerative disease. DNAm age would be an important analysis to carry out for all of the neurodegenerative diseases, and not just ALS and HD. Secondly, there is a unique interaction between methylation and the primary Parkinson's disease drug, L-DOPA, in that the drug's metabolism requires a methyl donor from SAM and thus impacts endogenous DNA methylation dynamics. This interaction could be further explored to develop secondary drugs that may enhance the efficacy of L-DOPA for PD symptoms. Third, given the strong risk that environmental factors like vitamin D deficiency and smoking confer upon the development of multiple sclerosis, aberrant de novo methylation and the stability of that methylation, should be thoroughly investigated with regards to these factors.

In summary, exploring the roles of DNA methylation and hydroxymethylation in neurodegenerative diseases provides another dimension through which more complex questions regarding mechanism can be addressed, and especially it can supply us with insight into how environmental factors can predispose certain individuals to such diseases.

## References

Adams I (1987) Comparison of synaptic changes in the precentral and postcentral cerebral cortex of aging humans: a quantitative ultrastructural study. Neurobiol Aging 8:203–212

Ahmeti KB, Ajroud-Driss S, Al-Chalabi A et al (2013) Age of onset of amyotrophic lateral sclerosis is modulated by a locus on 1p34.1. Neurobiol Aging 34:357 e357–357 e319

Ajroud-Driss S, Siddique T (2015) Sporadic and hereditary amyotrophic lateral sclerosis (ALS). Biochim Biophys Acta 1852:679–684

Albers JA, Chand P, Anch AM (2017) Multifactorial sleep disturbance in Parkinson's disease. Sleep Med 35:41–48

Anand A, Patience AA, Sharma N et al (2017) The present and future of pharmacotherapy of Alzheimer's disease: a comprehensive review. Eur J Pharmacol 815:364–375

Ayuso T, Aznar P, Soriano L et al (2017) Vitamin D receptor gene is epigenetically altered and transcriptionally up-regulated in multiple sclerosis. PLoS One 12:e0174726

Baker DJ, Petersen RC (2018) Cellular senescence in brain aging and neurodegenerative diseases: evidence and perspectives. J Clin Invest 128:1208–1216

Balendra R, Jones A, Jivraj N et al (2014) Estimating clinical stage of amyotrophic lateral sclerosis from the ALS Functional Rating Scale. Amyotroph Lateral Scler Frontotemporal Degener 15:279–284

Ballestar E, Wolffe AP (2001) Methyl-CpG-binding proteins. Targeting specific gene repression. Eur J Biochem 268:1–6

Bashtrykov P, Jankevicius G, Smarandache A et al (2012) Specificity of Dnmt1 for methylation of hemimethylated CpG sites resides in its catalytic domain. Chem Biol 19:572–578

Bates GP, Dorsey R, Gusella JF et al (2015) Huntington disease. Nat Rev Dis Primers 1:15005

Bekaert S, De Meyer T, Van Oostveldt P (2005) Telomere attrition as ageing biomarker. Anticancer Res 25:3011–3021

Bertram L, Tanzi RE (2005) The genetic epidemiology of neurodegenerative disease. J Clin Invest 115:1449–1457

Bihaqi SW, Schumacher A, Maloney B et al (2012) Do epigenetic pathways initiate late onset Alzheimer disease (LOAD): towards a new paradigm. Curr Alzheimer Res 9:574–588

Billingsley KJ, Bandres-Ciga S, Saez-Atienzar S et al (2018) Genetic risk factors in Parkinson's disease. Cell Tissue Res 373:9–20

Biniszkiewicz D, Gribnau J, Ramsahoye B et al (2002) Dnmt1 overexpression causes genomic hypermethylation, loss of imprinting, and embryonic lethality. Mol Cell Biol 22:2124–2135

Bogdanovic O, Veenstra GJ (2009) DNA methylation and methyl-CpG binding proteins: developmental requirements and function. Chromosoma 118:549–565

Bos SD, Page CM, Andreassen BK et al (2015) Genome-wide DNA methylation profiles indicate CD8+ T cell hypermethylation in multiple sclerosis. PLoS One 10:e0117403

Braak H, Braak E (1991) Neuropathological stageing of Alzheimer-related changes. Acta Neuropathol 82:239–259

Braak H, Ghebremedhin E, Rub U et al (2004) Stages in the development of Parkinson's disease-related pathology. Cell Tissue Res 318:121–134

Brandeis M, Ariel M, Cedar H (1993) Dynamics of DNA methylation during development. Bioessays 15:709–713

Burre J, Sharma M, Tsetsenis T et al (2010) Alpha-synuclein promotes SNARE-complex assembly in vivo and in vitro. Science 329:1663–1667

Cadena-del-Castillo C, Valdes-Quezada C, Carmona-Aldana F et al (2014) Age-dependent increment of hydroxymethylation in the brain cortex in the triple-transgenic mouse model of Alzheimer's disease. J Alzheimers Dis 41:845–854

Calabrese R, Valentini E, Ciccarone F et al (2014) TET2 gene expression and 5-hydroxymethylcytosine level in multiple sclerosis peripheral blood cells. Biochim Biophys Acta 1842:1130–1136

Calderon-Garciduenas ALD, Charles (2017) Alzheimer disease. In: Kovacs GA, Alafuzoff I (ed) Neuropathology, vol 145. Handbook of clinical neurology, vol 3rd series. Elsevier, Virginia Commonwealth University

Carlson LL, Page AW, Bestor TH (1992) Properties and localization of DNA methyltransferase in preimplantation mouse embryos: implications for genomic imprinting. Genes Dev 6:2536–2541

Charlton CG, Crowell B Jr (1995) Striatal dopamine depletion, tremors, and hypokinesia following the intracranial injection of S-adenosylmethionine: a possible role of hypermethylation in parkinsonism. Mol Chem Neuropathol 26:269–284

Charlton CG, Mack J (1994) Substantia nigra degeneration and tyrosine hydroxylase depletion caused by excess S-adenosylmethionine in the rat brain. Support for an excess methylation hypothesis for parkinsonism. Mol Neurobiol 9:149–161

Chen KL, Wang SS, Yang YY et al (2009) The epigenetic effects of amyloid-beta(1-40) on global DNA and neprilysin genes in murine cerebral endothelial cells. Biochem Biophys Res Commun 378:57–61

Chen H, Dzitoyeva S, Manev H (2012) Effect of aging on 5-hydroxymethylcytosine in the mouse hippocampus. Restor Neurol Neurosci 30:237–245

Chen X, Xiao Y, Wei L et al (2017) Association of DNMT3b gene variants with sporadic Parkinson's disease in a Chinese Han population. J Gene Med 19:360–365

Cheng HC, Ulane CM, Burke RE (2010) Clinical progression in Parkinson disease and the neurobiology of axons. Ann Neurol 67:715–725

Chestnut BA, Chang Q, Price A et al (2011) Epigenetic regulation of motor neuron cell death through DNA methylation. J Neurosci 31:16619–16636

Chio A, Pagani M, Agosta F et al (2014) Neuroimaging in amyotrophic lateral sclerosis: insights into structural and functional changes. Lancet Neurol 13:1228–1240

Cho HJ, Liu G, Jin SM et al (2013) MicroRNA-205 regulates the expression of Parkinson's disease-related leucine-rich repeat kinase 2 protein. Hum Mol Genet 22:608–620

Chouliaras L, Mastroeni D, Delvaux E et al (2013) Consistent decrease in global DNA methylation and hydroxymethylation in the hippocampus of Alzheimer's disease patients. Neurobiol Aging 34:2091–2099

Choy JS, Wei S, Lee JY et al (2010) DNA methylation increases nucleosome compaction and rigidity. J Am Chem Soc 132:1782–1783

Collins AL, Levenson JM, Vilaythong AP et al (2004) Mild overexpression of MeCP2 causes a progressive neurological disorder in mice. Hum Mol Genet 13:2679–2689

Compston A, Coles A (2002) Multiple sclerosis. Lancet 359:1221–1231

Condliffe D, Wong A, Troakes C et al (2014) Cross-region reduction in 5-hydroxymethylcytosine in Alzheimer's disease brain. Neurobiol Aging 35:1850–1854

Condorelli DF, Dell'Albani P, Conticello SG et al (1997) A neural-specific hypomethylated domain in the 5' flanking region of the glial fibrillary acidic protein gene. Dev Neurosci 19:446–456

Coppede F, Stoccoro A, Mosca L et al (2018) Increase in DNA methylation in patients with amyotrophic lateral sclerosis carriers of not fully penetrant SOD1 mutations. Amyotroph Lateral Scler Frontotemporal Degener 19:93–101

Coppieters N, Dieriks BV, Lill C et al (2014) Global changes in DNA methylation and hydroxymethylation in Alzheimer's disease human brain. Neurobiol Aging 35:1334–1344

Coupland KG, Mellick GD, Silburn PA et al (2014) DNA methylation of the MAPT gene in Parkinson's disease cohorts and modulation by vitamin E in vitro. Mov Disord 29:1606–1614

Cribbs DH, Berchtold NC, Perreau V et al (2012) Extensive innate immune gene activation accompanies brain aging, increasing vulnerability to cognitive decline and neurodegeneration: a microarray study. J Neuroinflammation 9:179

D'Amico E, Patti F, Zanghi A et al (2018) Late-onset and young-onset relapsing-remitting multiple sclerosis: evidence from a retrospective long-term follow-up study. Eur J Neurol 25 (12):1425–1431

De Angelis F, Brownlee WJ, Chard DT et al (2019) New MS diagnostic criteria in practice. Pract Neurol 19(1):64–67

De Jager PL, Srivastava G, Lunnon K et al (2014) Alzheimer's disease: early alterations in brain DNA methylation at ANK1, BIN1, RHBDF2 and other loci. Nat Neurosci 17:1156–1163

de Lau LM, Breteler MM (2006) Epidemiology of Parkinson's disease. Lancet Neurol 5:525–535

De Souza RA, Islam SA, McEwen LM et al (2016) DNA methylation profiling in human Huntington's disease brain. Hum Mol Genet 25:2013–2030

DeJesus-Hernandez M, Mackenzie IR, Boeve BF et al (2011) Expanded GGGGCC hexanucleotide repeat in noncoding region of C9ORF72 causes chromosome 9p-linked FTD and ALS. Neuron 72:245–256

DeKosky ST, Marek K (2003) Looking backward to move forward: early detection of neurodegenerative disorders. Science 302:830–834

Dendrou CA, Fugger L, Friese MA (2015) Immunopathology of multiple sclerosis. Nat Rev Immunol 15:545–558

Desplats P, Spencer B, Coffee E et al (2011) Alpha-synuclein sequesters Dnmt1 from the nucleus: a novel mechanism for epigenetic alterations in Lewy body diseases. J Biol Chem 286:9031–9037

Didonna AO, Jorge R (2017) The genetics of multiple sclerosis. In: Zagon IS, McLaughlin PJ (eds) Multiple sclerosis: perspectives in treatment and pathogenesis. Codon Publications, Brisbane (AU)

DiFiglia M, Sapp E, Chase KO et al (1997) Aggregation of huntingtin in neuronal intranuclear inclusions and dystrophic neurites in brain. Science 277:1990–1993

Do Carmo S, Hanzel CE, Jacobs ML et al (2016) Rescue of early bace-1 and global DNA demethylation by S-adenosylmethionine reduces amyloid pathology and improves cognition in an Alzheimer's model. Sci Rep 6:34051

Doskocil J, Sorm F (1962) Distribution of 5-methylcytosine in pyrimidine sequences of deoxyribonucleic acids. Biochim Biophys Acta 55:953–959

Dosunmu R, Wu J, Basha MR et al (2007) Environmental and dietary risk factors in Alzheimer's disease. Expert Rev Neurother 7:887–900

Droge W, Schipper HM (2007) Oxidative stress and aberrant signaling in aging and cognitive decline. Aging Cell 6:361–370

Edwards TL, Scott WK, Almonte C et al (2010) Genome-wide association study confirms SNPs in SNCA and the MAPT region as common risk factors for Parkinson disease. Ann Hum Genet 74:97–109

Ehrlich M, Gama-Sosa MA, Huang LH et al (1982) Amount and distribution of 5-methylcytosine in human DNA from different types of tissues of cells. Nucleic Acids Res 10:2709–2721

Erro R, Stamelou M (2017) The motor syndrome of Parkinson's disease. Int Rev Neurobiol 132:25–32

Esanov R, Belle KC, van Blitterswijk M et al (2016) C9orf72 promoter hypermethylation is reduced while hydroxymethylation is acquired during reprogramming of ALS patient cells. Exp Neurol 277:171–177

Esquerda-Canals G, Montoliu-Gaya L, Guell-Bosch J et al (2017) Mouse models of Alzheimer's disease. J Alzheimers Dis 57:1171–1183

Fan G, Martinowich K, Chin MH et al (2005) DNA methylation controls the timing of astrogliogenesis through regulation of JAK-STAT signaling. Development 132:3345–3356

Farrer LA, Cupples LA, Kiely DK et al (1992) Inverse relationship between age at onset of Huntington disease and paternal age suggests involvement of genetic imprinting. Am J Hum Genet 50:528–535

Fasolino M, Zhou Z (2017) The crucial role of DNA methylation and MeCP2 in neuronal function. Genes (Basel) 8

Feng J, Zhou Y, Campbell SL et al (2010) Dnmt1 and Dnmt3a maintain DNA methylation and regulate synaptic function in adult forebrain neurons. Nat Neurosci 13:423–430

Ficz G, Branco MR, Seisenberger S et al (2011) Dynamic regulation of 5-hydroxymethylcytosine in mouse ES cells and during differentiation. Nature 473:398–402

Field J, Fox A, Jordan MA et al (2017) Interleukin-2 receptor-alpha proximal promoter hypomethylation is associated with multiple sclerosis. Genes Immun 18:59–66

Figueroa-Romero C, Hur J, Bender DE et al (2012) Identification of epigenetically altered genes in sporadic amyotrophic lateral sclerosis. PLoS One 7:e52672

Frischer JM, Bramow S, Dal-Bianco A et al (2009) The relation between inflammation and neurodegeneration in multiple sclerosis brains. Brain 132:1175–1189

Fuks F, Hurd PJ, Wolf D et al (2003) The methyl-CpG-binding protein MeCP2 links DNA methylation to histone methylation. J Biol Chem 278:4035–4040

Gan W, Nie B, Shi F et al (2012) Age-dependent increases in the oxidative damage of DNA, RNA, and their metabolites in normal and senescence-accelerated mice analyzed by LC-MS/MS: urinary 8-oxoguanosine as a novel biomarker of aging. Free Radic Biol Med 52:1700–1707

Gant JC, Sama MM, Landfield PW et al (2006) Early and simultaneous emergence of multiple hippocampal biomarkers of aging is mediated by Ca2+-induced Ca2+ release. J Neurosci 26:3482–3490

Ghosh R, Tabrizi SJ (2018) Clinical features of Huntington's disease. Adv Exp Med Biol 1049:1–28

Gibb WR, Lees AJ (1991) Anatomy, pigmentation, ventral and dorsal subpopulations of the substantia nigra, and differential cell death in Parkinson's disease. J Neurol Neurosurg Psychiatry 54:388–396
Gijselinck I, Van Mossevelde S, van der Zee J et al (2016) The C9orf72 repeat size correlates with onset age of disease, DNA methylation and transcriptional downregulation of the promoter. Mol Psychiatry 21:1112–1124
Gjoneska E, Pfenning AR, Mathys H et al (2015) Conserved epigenomic signals in mice and humans reveal immune basis of Alzheimer's disease. Nature 518:365–369
Glass M, Dragunow M, Faull RL (2000) The pattern of neurodegeneration in Huntington's disease: a comparative study of cannabinoid, dopamine, adenosine and GABA(A) receptor alterations in the human basal ganglia in Huntington's disease. Neuroscience 97:505–519
Globisch D, Munzel M, Muller M et al (2010) Tissue distribution of 5-hydroxymethylcytosine and search for active demethylation intermediates. PLoS One 5:e15367
Godyn J, Jonczyk J, Panek D et al (2016) Therapeutic strategies for Alzheimer's disease in clinical trials. Pharmacol Rep 68:127–138
Goedert M, Sisodia SS, Price DL (1991) Neurofibrillary tangles and beta-amyloid deposits in Alzheimer's disease. Curr Opin Neurobiol 1:441–447
Goetz CG (2000) Amyotrophic lateral sclerosis: early contributions of Jean-Martin Charcot. Muscle Nerve 23:336–343
Golshani P, Hutnick L, Schweizer F et al (2005) Conditional Dnmt1 deletion in dorsal forebrain disrupts development of somatosensory barrel cortex and thalamocortical long-term potentiation. Thalamus Relat Syst 3:227–233
Gontier G, Iyer M, Shea JM et al (2018) Tet2 rescues age-related regenerative decline and enhances cognitive function in the adult mouse brain. Cell Rep 22:1974–1981
Gonzalo S, Jaco I, Fraga MF et al (2006) DNA methyltransferases control telomere length and telomere recombination in mammalian cells. Nat Cell Biol 8:416–424
Gordon N (2003) Huntington's disease of early onset or juvenile Huntington's disease. Hosp Med 64:576–580
Gorevic PD, Goni F, Pons-Estel B et al (1986) Isolation and partial characterization of neurofibrillary tangles and amyloid plaque core in Alzheimer's disease: immunohistological studies. J Neuropathol Exp Neurol 45:647–664
Goto K, Numata M, Komura JI et al (1994) Expression of DNA methyltransferase gene in mature and immature neurons as well as proliferating cells in mice. Differentiation 56:39–44
Govoni V, Della Coletta E, Cesnik E et al (2017) Can the age at onset give a clue to the pathogenesis of ALS? Acta Neurol Belg 117:221–227
Graves MC, Benton M, Lea RA et al (2014) Methylation differences at the HLA-DRB1 locus in CD4+ T-Cells are associated with multiple sclerosis. Mult Scler 20:1033–1041
Grigoriadis N, van Pesch V, Paradig MSG (2015) A basic overview of multiple sclerosis immunopathology. Eur J Neurol 22(Suppl 2):3–13
Grinan-Ferre C, Sarroca S, Ivanova A et al (2016) Epigenetic mechanisms underlying cognitive impairment and Alzheimer disease hallmarks in 5XFAD mice. Aging (Albany NY) 8:664–684
Guerreiro R, Wojtas A, Bras J et al (2013) TREM2 variants in Alzheimer's disease. N Engl J Med 368:117–127
Guhathakurta S, Bok E, Evangelista BA et al (2017a) Deregulation of alpha-synuclein in Parkinson's disease: Insight from epigenetic structure and transcriptional regulation of SNCA. Prog Neurobiol 154:21–36
Guhathakurta S, Evangelista BA, Ghosh S et al (2017b) Hypomethylation of intron1 of alpha-synuclein gene does not correlate with Parkinson's disease. Mol Brain 10:6
Guo JU, Ma DK, Mo H et al (2011) Neuronal activity modifies the DNA methylation landscape in the adult brain. Nat Neurosci 14:1345–1351
Guo JU, Su Y, Shin JH et al (2014) Distribution, recognition and regulation of non-CpG methylation in the adult mammalian brain. Nat Neurosci 17:215–222
Gusella JF, MacDonald ME (1995) Huntington's disease. Semin Cell Biol 6:21–28

Hadad N, Masser DR, Logan S et al (2016) Absence of genomic hypomethylation or regulation of cytosine-modifying enzymes with aging in male and female mice. Epigenetics Chromatin 9:30
Hahn MA, Qiu R, Wu X et al (2013) Dynamics of 5-hydroxymethylcytosine and chromatin marks in Mammalian neurogenesis. Cell Rep 3:291–300
Hamerman JA, Jarjoura JR, Humphrey MB et al (2006) Cutting edge: inhibition of TLR and FcR responses in macrophages by triggering receptor expressed on myeloid cells (TREM)-2 and DAP12. J Immunol 177:2051–2055
Hamzeiy H, Savas D, Tunca C et al (2018) Elevated global DNA methylation is not exclusive to amyotrophic lateral sclerosis and is also observed in spinocerebellar ataxia types 1 and 2. Neurodegener Dis 18:38–48
Hardiman O, Al-Chalabi A, Chio A et al (2017) Amyotrophic lateral sclerosis. Nat Rev Dis Primers 3:17085
Hendrich B, Guy J, Ramsahoye B et al (2001) Closely related proteins MBD2 and MBD3 play distinctive but interacting roles in mouse development. Genes Dev 15:710–723
Hensiek AE, Sawcer SJ, Feakes R et al (2002) HLA-DR 15 is associated with female sex and younger age at diagnosis in multiple sclerosis. J Neurol Neurosurg Psychiatry 72:184–187
Heston LL (1989) Family studies in Alzheimer's disease. Prog Clin Biol Res 317:195–200
Hoal-van Helden EG, van Helden PD (1989) Age-related methylation changes in DNA may reflect the proliferative potential of organs. Mutat Res 219:263–266
Horvath S (2013) DNA methylation age of human tissues and cell types. Genome Biol 14:R115
Horvath S, Langfelder P, Kwak S et al (2016) Huntington's disease accelerates epigenetic aging of human brain and disrupts DNA methylation levels. Aging (Albany NY) 8:1485–1512
Hotchkiss RD (1948) The quantitative separation of purines, pyrimidines, and nucleosides by paper chromatography. J Biol Chem 175:315–332
Howell CY, Bestor TH, Ding F et al (2001) Genomic imprinting disrupted by a maternal effect mutation in the Dnmt1 gene. Cell 104:829–838
Hu L, Lu J, Cheng J et al (2015) Structural insight into substrate preference for TET-mediated oxidation. Nature 527:118–122
Huang Y, Chavez L, Chang X et al (2014) Distinct roles of the methylcytosine oxidases Tet1 and Tet2 in mouse embryonic stem cells. Proc Natl Acad Sci U S A 111:1361–1366
Hutnick LK, Golshani P, Namihira M et al (2009) DNA hypomethylation restricted to the murine forebrain induces cortical degeneration and impairs postnatal neuronal maturation. Hum Mol Genet 18:2875–2888
Huynh JL, Garg P, Thin TH et al (2014) Epigenome-wide differences in pathology-free regions of multiple sclerosis-affected brains. Nat Neurosci 17:121–130
Ianov L, Riva A, Kumar A et al (2017) DNA methylation of synaptic genes in the prefrontal cortex is associated with aging and age-related cognitive impairment. Front Aging Neurosci 9:249
Inano K, Suetake I, Ueda T et al (2000) Maintenance-type DNA methyltransferase is highly expressed in post-mitotic neurons and localized in the cytoplasmic compartment. J Biochem 128:315–321
Ip JPK, Mellios N, Sur M (2018) Rett syndrome: insights into genetic, molecular and circuit mechanisms. Nat Rev Neurosci 19:368–382
Isaac C, Li DK, Genton M et al (1988) Multiple sclerosis: a serial study using MRI in relapsing patients. Neurology 38:1511–1515
Isobe N, Keshavan A, Gourraud PA et al (2016) Association of HLA genetic risk burden with disease phenotypes in multiple sclerosis. JAMA Neurol 73:795–802
Ito H, Usuda N, Atsuzawa K et al (2007) Phosphorylation by extracellular signal-regulated kinase of a multidomain adaptor protein, vinexin, at synapses. J Neurochem 100:545–554
Ito S, Shen L, Dai Q et al (2011) Tet proteins can convert 5-methylcytosine to 5-formylcytosine and 5-carboxylcytosine. Science 333:1300–1303
Jeltsch A, Jurkowska RZ (2014) New concepts in DNA methylation. Trends Biochem Sci 39:310–318

Jin SG, Zhang ZM, Dunwell TL et al (2016) Tet3 reads 5-carboxylcytosine through its CXXC domain and is a potential guardian against neurodegeneration. Cell Rep 14:493–505

Jintaridth P, Mutirangura A (2010) Distinctive patterns of age-dependent hypomethylation in interspersed repetitive sequences. Physiol Genomics 41:194–200

Jones PL, Veenstra GJ, Wade PA et al (1998) Methylated DNA and MeCP2 recruit histone deacetylase to repress transcription. Nat Genet 19:187–191

Jowaed A, Schmitt I, Kaut O et al (2010) Methylation regulates alpha-synuclein expression and is decreased in Parkinson's disease patients' brains. J Neurosci 30:6355–6359

Jung BP, Zhang G, Nitsch R et al (2003) Differential expression of methyl CpG-binding domain containing factor MBD3 in the developing and adult rat brain. J Neurobiol 55:220–232

Kaas GA, Zhong C, Eason DE et al (2013) TET1 controls CNS 5-methylcytosine hydroxylation, active DNA demethylation, gene transcription, and memory formation. Neuron 79:1086–1093

Kalman B, Lublin FD (1999) The genetics of multiple sclerosis. A review. Biomed Pharmacother 53:358–370

Keshet I, Yisraeli J, Cedar H (1985) Effect of regional DNA methylation on gene expression. Proc Natl Acad Sci U S A 82:2560–2564

Ketelaar ME, Hofstra EM, Hayden MR (2012) What monozygotic twins discordant for phenotype illustrate about mechanisms influencing genetic forms of neurodegeneration. Clin Genet 81:325–333

Khare T, Pai S, Koncevicius K et al (2012) 5-hmC in the brain is abundant in synaptic genes and shows differences at the exon-intron boundary. Nat Struct Mol Biol 19:1037–1043

Kim M, Park YK, Kang TW et al (2014) Dynamic changes in DNA methylation and hydroxymethylation when hES cells undergo differentiation toward a neuronal lineage. Hum Mol Genet 23:657–667

Kim H, Jang WY, Kang MC et al (2016) TET1 contributes to neurogenesis onset time during fetal brain development in mice. Biochem Biophys Res Commun 471:437–443

Kinney SM, Chin HG, Vaisvila R et al (2011) Tissue-specific distribution and dynamic changes of 5-hydroxymethylcytosine in mammalian genomes. J Biol Chem 286:24685–24693

Kraft AM, Wessman HC (1974) Pathology and etiology in multiple sclerosis: a review. Phys Ther 54:716–720

Kraus TF, Globisch D, Wagner M et al (2012) Low values of 5-hydroxymethylcytosine (5hmC), the "sixth base," are associated with anaplasia in human brain tumors. Int J Cancer 131:1577–1590

Kraus TF, Guibourt V, Kretzschmar HA (2015) 5-Hydroxymethylcytosine, the "Sixth Base", during brain development and ageing. J Neural Transm (Vienna) 122:1035–1043

Kremer EA, Gaur N, Lee MA et al (2018) Interplay between TETs and microRNAs in the adult brain for memory formation. Sci Rep 8:1678

Kriaucionis S, Heintz N (2009) The nuclear DNA base 5-hydroxymethylcytosine is present in Purkinje neurons and the brain. Science 324:929–930

Kulakova OG, Kabilov MR, Danilova LV et al (2016) Whole-genome DNA methylation analysis of peripheral blood mononuclear cells in multiple sclerosis patients with different disease courses. Acta Nat 8:103–110

Kular L, Liu Y, Ruhrmann S et al (2018) DNA methylation as a mediator of HLA-DRB1*15:01 and a protective variant in multiple sclerosis. Nat Commun 9:2397

Kumar S, Cheng X, Klimasauskas S et al (1994) The DNA (cytosine-5) methyltransferases. Nucleic Acids Res 22:1–10

Langner KD, Vardimon L, Renz D et al (1984) DNA methylation of three 5' C-C-G-G 3' sites in the promoter and 5' region inactivate the E2a gene of adenovirus type 2. Proc Natl Acad Sci U S A 81:2950–2954

Lardenoije R, van den Hove DLA, Havermans M et al (2018) Age-related epigenetic changes in hippocampal subregions of four animal models of Alzheimer's disease. Mol Cell Neurosci 86:1–15

Lashley T, Gami P, Valizadeh N et al (2015) Alterations in global DNA methylation and hydroxymethylation are not detected in Alzheimer's disease. Neuropathol Appl Neurobiol 41:497–506

Lauretti E, Di Meco A, Chu J et al (2015) Modulation of AD neuropathology and memory impairments by the isoprostane F2alpha is mediated by the thromboxane receptor. Neurobiol Aging 36:812–820

Lee JY, Biemond M, Petratos S (2015) Axonal degeneration in multiple sclerosis: defining therapeutic targets by identifying the causes of pathology. Neurodegener Dis Manag 5:527–548

Li W, Liu M (2011) Distribution of 5-hydroxymethylcytosine in different human tissues. J Nucleic Acids 2011:870726

Li E, Bestor TH, Jaenisch R (1992) Targeted mutation of the DNA methyltransferase gene results in embryonic lethality. Cell 69:915–926

Li C, Ebrahimi A, Schluesener H (2013) Drug pipeline in neurodegeneration based on transgenic mice models of Alzheimer's disease. Ageing Res Rev 12:116–140

Li X, Wei W, Zhao QY et al (2014) Neocortical Tet3-mediated accumulation of 5-hydroxymethylcytosine promotes rapid behavioral adaptation. Proc Natl Acad Sci U S A 111:7120–7125

Li X, Xiao B, Chen XS (2017a) DNA methylation: a new player in multiple sclerosis. Mol Neurobiol 54:4049–4059

Li X, Yao B, Chen L et al (2017b) Ten-eleven translocation 2 interacts with forkhead box O3 and regulates adult neurogenesis. Nat Commun 8:15903

Lill CM (2016) Genetics of Parkinson's disease. Mol Cell Probes 30:386–396

Lin X, Parisiadou L, Gu XL et al (2009) Leucine-rich repeat kinase 2 regulates the progression of neuropathology induced by Parkinson's-disease-related mutant alpha-synuclein. Neuron 64:807–827

Liu XX, Wilson K, Charlton CG (2000) Effects of L-dopa treatment on methylation in mouse brain: implications for the side effects of L-dopa. Life Sci 66:2277–2288

Liu EY, Russ J, Wu K et al (2014) C9orf72 hypermethylation protects against repeat expansion-associated pathology in ALS/FTD. Acta Neuropathol 128:525–541

Lublin FD, Reingold SC (1996) Defining the clinical course of multiple sclerosis: results of an international survey. National Multiple Sclerosis Society (USA) Advisory Committee on Clinical Trials of New Agents in Multiple Sclerosis. Neurology 46:907–911

Maddox SA, Watts CS, Schafe GE (2014) DNA methyltransferase activity is required for memory-related neural plasticity in the lateral amygdala. Neurobiol Learn Mem 107:93–100

Maiti A, Drohat AC (2011) Thymine DNA glycosylase can rapidly excise 5-formylcytosine and 5-carboxylcytosine: potential implications for active demethylation of CpG sites. J Biol Chem 286:35334–35338

Maltby VE, Graves MC, Lea RA et al (2015) Genome-wide DNA methylation profiling of CD8+ T cells shows a distinct epigenetic signature to CD4+ T cells in multiple sclerosis patients. Clin Epigenetics 7:118

Maltby VE, Lea RA, Sanders KA et al (2017) Differential methylation at MHC in CD4(+) T cells is associated with multiple sclerosis independently of HLA-DRB1. Clin Epigenetics 9:71

Maltby VE, Lea RA, Ribbons KA et al (2018) DNA methylation changes in CD4(+) T cells isolated from multiple sclerosis patients on dimethyl fumarate. Mult Scler J Exp Transl Clin 4:2055217318787826

Mano T, Nagata K, Nonaka T et al (2017) Neuron-specific methylome analysis reveals epigenetic regulation and tau-related dysfunction of BRCA1 in Alzheimer's disease. Proc Natl Acad Sci U S A 114:E9645–E9654

Marabita F, Almgren M, Sjoholm LK et al (2017) Smoking induces DNA methylation changes in Multiple Sclerosis patients with exposure-response relationship. Sci Rep 7:14589

Martynoga B, Drechsel D, Guillemot F (2012) Molecular control of neurogenesis: a view from the mammalian cerebral cortex. Cold Spring Harb Perspect Biol 4

Masliah E, Dumaop W, Galasko D et al (2013) Distinctive patterns of DNA methylation associated with Parkinson disease: identification of concordant epigenetic changes in brain and peripheral blood leukocytes. Epigenetics 8:1030–1038
Mastroeni D, McKee A, Grover A et al (2009) Epigenetic differences in cortical neurons from a pair of monozygotic twins discordant for Alzheimer's disease. PLoS One 4:e6617
Mastroeni D, Grover A, Delvaux E et al (2010) Epigenetic changes in Alzheimer's disease: decrements in DNA methylation. Neurobiol Aging 31:2025–2037
Matsumoto L, Takuma H, Tamaoka A et al (2010) CpG demethylation enhances alpha-synuclein expression and affects the pathogenesis of Parkinson's disease. PLoS One 5:e15522
McGraw CM, Samaco RC, Zoghbi HY (2011) Adult neural function requires MeCP2. Science 333:186
McMillan CT, Russ J, Wood EM et al (2015) C9orf72 promoter hypermethylation is neuroprotective: neuroimaging and neuropathologic evidence. Neurology 84:1622–1630
Meadows JP, Guzman-Karlsson MC, Phillips S et al (2016) Dynamic DNA methylation regulates neuronal intrinsic membrane excitability. Sci Signal 9:ra83
Mehta P, Kaye W, Raymond J et al (2018) Prevalence of amyotrophic lateral sclerosis - United States, 2014. MMWR Morb Mortal Wkly Rep 67:216–218
Mellen M, Ayata P, Dewell S et al (2012) MeCP2 binds to 5hmC enriched within active genes and accessible chromatin in the nervous system. Cell 151:1417–1430
Messerschmidt DM, Knowles BB, Solter D (2014) DNA methylation dynamics during epigenetic reprogramming in the germline and preimplantation embryos. Genes Dev 28:812–828
Michel L (2018) Environmental factors in the development of multiple sclerosis. Rev Neurol (Paris) 174:372–377
Miller CA, Sweatt JD (2007) Covalent modification of DNA regulates memory formation. Neuron 53:857–869
Mink JW (1996) The basal ganglia: focused selection and inhibition of competing motor programs. Prog Neurobiol 50:381–425
Mo A, Mukamel EA, Davis FP et al (2015) Epigenomic signatures of neuronal diversity in the mammalian brain. Neuron 86:1369–1384
Mollica PA, Reid JA, Ogle RC et al (2016) DNA methylation leads to DNA repair gene down-regulation and trinucleotide repeat expansion in patient-derived huntington disease cells. Am J Pathol 186:1967–1976
Morahan JM, Yu B, Trent RJ et al (2009) A genome-wide analysis of brain DNA methylation identifies new candidate genes for sporadic amyotrophic lateral sclerosis. Amyotroph Lateral Scler 10:418–429
Moutsianas L, Jostins L, Beecham AH et al (2015) Class II HLA interactions modulate genetic risk for multiple sclerosis. Nat Genet 47:1107–1113
Moyon S, Casaccia P (2017) DNA methylation in oligodendroglial cells during developmental myelination and in disease. Neurogenesis (Austin) 4:e1270381
Moyon S, Ma D, Huynh JL et al (2017) Efficient remyelination requires DNA methylation. eNeuro 4
Nan X, Ng HH, Johnson CA et al (1998) Transcriptional repression by the methyl-CpG-binding protein MeCP2 involves a histone deacetylase complex. Nature 393:386–389
Neven KY, Piola M, Angelici L et al (2016) Repetitive element hypermethylation in multiple sclerosis patients. BMC Genet 17:84
Ng LJ, Cropley JE, Pickett HA et al (2009) Telomerase activity is associated with an increase in DNA methylation at the proximal subtelomere and a reduction in telomeric transcription. Nucleic Acids Res 37:1152–1159
Ng CW, Yildirim F, Yap YS et al (2013) Extensive changes in DNA methylation are associated with expression of mutant huntingtin. Proc Natl Acad Sci U S A 110:2354–2359
Nichol K, Pearson CE (2002) CpG methylation modifies the genetic stability of cloned repeat sequences. Genome Res 12:1246–1256
Norden DM, Godbout JP (2013) Review: microglia of the aged brain: primed to be activated and resistant to regulation. Neuropathol Appl Neurobiol 39:19–34

Oates N, Pamphlett R (2007) An epigenetic analysis of SOD1 and VEGF in ALS. Amyotroph Lateral Scler 8:83–86
Okano M, Xie S, Li E (1998) Cloning and characterization of a family of novel mammalian DNA (cytosine-5) methyltransferases. Nat Genet 19:219–220
Okano M, Bell DW, Haber DA et al (1999) DNA methyltransferases Dnmt3a and Dnmt3b are essential for de novo methylation and mammalian development. Cell 99:247–257
Orru M, Zanoveli JM, Quiroz C et al (2011) Functional changes in postsynaptic adenosine A (2A) receptors during early stages of a rat model of Huntington disease. Exp Neurol 232:76–80
Pandya JD, Royland JE, MacPhail RC et al (2016) Age- and brain region-specific differences in mitochondrial bioenergetics in Brown Norway rats. Neurobiol Aging 42:25–34
Park M, Min B, Jeon K et al (2017) Age-associated chromatin relaxation is enhanced in Huntington's disease mice. Aging (Albany NY) 9:803–822
Penner MR, Parrish RR, Hoang LT et al (2016) Age-related changes in Egr1 transcription and DNA methylation within the hippocampus. Hippocampus 26:1008–1020
Pensold D, Zimmer G (2018) Single-cell transcriptomics reveals regulators of neuronal migration and maturation during brain development. J Exp Neurosci 12:1179069518760783
Pensold D, Symmank J, Hahn A et al (2017) The DNA methyltransferase 1 (DNMT1) controls the shape and dynamics of migrating POA-derived interneurons fated for the murine cerebral cortex. Cereb Cortex 27:5696–5714
Pezzi JC, de Bem CM, da Rocha TJ et al (2017) Association between DNA methyltransferase gene polymorphism and Parkinson's disease. Neurosci Lett 639:146–150
Pihlstrom L, Berge V, Rengmark A et al (2015) Parkinson's disease correlates with promoter methylation in the alpha-synuclein gene. Mov Disord 30:577–580
Pihlstrom L, Blauwendraat C, Cappelletti C et al (2018) A comprehensive analysis of SNCA-related genetic risk in sporadic parkinson disease. Ann Neurol 84:117–129
Poddar J, Pradhan M, Ganguly G et al (2018) Biochemical deficits and cognitive decline in brain aging: Intervention by dietary supplements. J Chem Neuroanat 95:70–80
Pohodich AE, Zoghbi HY (2015) Rett syndrome: disruption of epigenetic control of postnatal neurological functions. Hum Mol Genet 24:R10–R16
Pritchard CA, Cox DR, Myers RM (1989) Methylation at the Huntington disease-linked D4S95 locus. Am J Hum Genet 45:335–336
Reik W, Maher ER, Morrison PJ et al (1993) Age at onset in Huntington's disease and methylation at D4S95. J Med Genet 30:185–188
Richter J, Appenzeller S, Ammerpohl O et al (2012) No evidence for differential methylation of alpha-synuclein in leukocyte DNA of Parkinson's disease patients. Mov Disord 27:590–591
Rizvi S, Raza ST, Mahdi F (2014) Telomere length variations in aging and age-related diseases. Curr Aging Sci 7:161–167
Robertson KD, Uzvolgyi E, Liang G et al (1999) The human DNA methyltransferases (DNMTs) 1, 3a and 3b: coordinate mRNA expression in normal tissues and overexpression in tumors. Nucleic Acids Res 27:2291–2298
Rodier F, Coppe JP, Patil CK et al (2009) Persistent DNA damage signalling triggers senescence-associated inflammatory cytokine secretion. Nat Cell Biol 11:973–979
Roos RA (2010) Huntington's disease: a clinical review. Orphanet J Rare Dis 5:40
Roost MS, Slieker RC, Bialecka M et al (2017) DNA methylation and transcriptional trajectories during human development and reprogramming of isogenic pluripotent stem cells. Nat Commun 8:908
Rowland LP, Shneider NA (2001) Amyotrophic lateral sclerosis. N Engl J Med 344:1688–1700
Rudenko A, Dawlaty MM, Seo J et al (2013) Tet1 is critical for neuronal activity-regulated gene expression and memory extinction. Neuron 79:1109–1122
Ruzov A, Tsenkina Y, Serio A et al (2011) Lineage-specific distribution of high levels of genomic 5-hydroxymethylcytosine in mammalian development. Cell Res 21:1332–1342
Sadovnick AD, Baird PA (1988) The familial nature of multiple sclerosis: age-corrected empiric recurrence risks for children and siblings of patients. Neurology 38:990–991

Samkoff LM, Goodman AD (2011) Symptomatic management in multiple sclerosis. Neurol Clin 29:449–463
Samorajski T (1976) How the human brain responds to aging. J Am Geriatr Soc 24:4–11
Sanchez-Mut JV, Aso E, Panayotis N et al (2013) DNA methylation map of mouse and human brain identifies target genes in Alzheimer's disease. Brain 136:3018–3027
Sandler M (1972) L-dopa metabolism in man (with special reference to parkinsonism). Sci Basis Med Annu Rev:161–173
Sanosaka T, Imamura T, Hamazaki N et al (2017) DNA methylome analysis identifies transcription factor-based epigenomic signatures of multilineage competence in neural stem/progenitor cells. Cell Rep 20:2992–3003
Santiago M, Antunes C, Guedes M et al (2014) TET enzymes and DNA hydroxymethylation in neural development and function - how critical are they? Genomics 104:334–340
Sarraf SA, Stancheva I (2004) Methyl-CpG binding protein MBD1 couples histone H3 methylation at lysine 9 by SETDB1 to DNA replication and chromatin assembly. Mol Cell 15:595–605
Sasaki S, Maruyama S (1994) Synapse loss in anterior horn neurons in amyotrophic lateral sclerosis. Acta Neuropathol 88:222–227
Sawcer S, Ban M, Maranian M et al (2005) A high-density screen for linkage in multiple sclerosis. Am J Hum Genet 77:454–467
Scarpa S, Fuso A, D'Anselmi F et al (2003) Presenilin 1 gene silencing by S-adenosylmethionine: a treatment for Alzheimer disease? FEBS Lett 541:145–148
Scheltens P, Blennow K, Breteler MM et al (2016) Alzheimer's disease. Lancet 388:505–517
Schiffer D, Cordera S, Cavalla P et al (1996) Reactive astrogliosis of the spinal cord in amyotrophic lateral sclerosis. J Neurol Sci 139(Suppl):27–33
Schmitt I, Kaut O, Khazneh H et al (2015) L-dopa increases alpha-synuclein DNA methylation in Parkinson's disease patients in vivo and in vitro. Mov Disord 30:1794–1801
Schwendimann RN (2006) Treatment of symptoms in multiple sclerosis. Neurol Res 28:306–315
Sedelnikova OA, Horikawa I, Zimonjic DB et al (2004) Senescing human cells and ageing mice accumulate DNA lesions with unrepairable double-strand breaks. Nat Cell Biol 6:168–170
Shatunov A, Mok K, Newhouse S et al (2010) Chromosome 9p21 in sporadic amyotrophic lateral sclerosis in the UK and seven other countries: a genome-wide association study. Lancet Neurol 9:986–994
Siegmund KD, Connor CM, Campan M et al (2007) DNA methylation in the human cerebral cortex is dynamically regulated throughout the life span and involves differentiated neurons. PLoS One 2:e895
Sieradzan KA, Mann DM (2001) The selective vulnerability of nerve cells in Huntington's disease. Neuropathol Appl Neurobiol 27:1–21
Simon-Sanchez J, Schulte C, Bras JM et al (2009) Genome-wide association study reveals genetic risk underlying Parkinson's disease. Nat Genet 41:1308–1312
Singer J, Roberts-Ems J, Riggs AD (1979) Methylation of mouse liver DNA studied by means of the restriction enzymes msp I and hpa II. Science 203:1019–1021
Smith KJ, McDonald WI (1999) The pathophysiology of multiple sclerosis: the mechanisms underlying the production of symptoms and the natural history of the disease. Philos Trans R Soc Lond Ser B Biol Sci 354:1649–1673
Smith AR, Smith RG, Condliffe D et al (2016) Increased DNA methylation near TREM2 is consistently seen in the superior temporal gyrus in Alzheimer's disease brain. Neurobiol Aging 47:35–40
Sokratous M, Dardiotis E, Bellou E et al (2018) CpG Island methylation patterns in relapsing-remitting multiple sclerosis. J Mol Neurosci 64:478–484
Song Y, Ding H, Yang J et al (2014) Pyrosequencing analysis of SNCA methylation levels in leukocytes from Parkinson's disease patients. Neurosci Lett 569:85–88
Stewart KR, Veselovska L, Kelsey G (2016) Establishment and functions of DNA methylation in the germline. Epigenomics 8:1399–1413

Stroud H, Feng S, Morey Kinney S et al (2011) 5-Hydroxymethylcytosine is associated with enhancers and gene bodies in human embryonic stem cells. Genome Biol 12:R54
Sun YM, Zhang YB, Wu ZY (2017) Huntington's disease: relationship between phenotype and genotype. Mol Neurobiol 54:342–348
Sweatt JD (2016) Dynamic DNA methylation controls glutamate receptor trafficking and synaptic scaling. J Neurochem 137:312–330
Szulwach KE, Li X, Li Y et al (2011) 5-hmC-mediated epigenetic dynamics during postnatal neurodevelopment and aging. Nat Neurosci 14:1607–1616
Szwagierczak A, Bultmann S, Schmidt CS et al (2010) Sensitive enzymatic quantification of 5-hydroxymethylcytosine in genomic DNA. Nucleic Acids Res 38:e181
Tahiliani M, Koh KP, Shen Y et al (2009) Conversion of 5-methylcytosine to 5-hydroxymethylcytosine in mammalian DNA by MLL partner TET1. Science 324:930–935
Takizawa T, Nakashima K, Namihira M et al (2001) DNA methylation is a critical cell-intrinsic determinant of astrocyte differentiation in the fetal brain. Dev Cell 1:749–758
Tan YY, Wu L, Zhao ZB et al (2014) Methylation of alpha-synuclein and leucine-rich repeat kinase 2 in leukocyte DNA of Parkinson's disease patients. Parkinsonism Relat Disord 20:308–313
Tanzi RE (1999) A genetic dichotomy model for the inheritance of Alzheimer's disease and common age-related disorders. J Clin Invest 104:1175–1179
Tanzi RE (2012) The genetics of Alzheimer disease. Cold Spring Harb Perspect Med 2
Tanzi RE, Kovacs DM, Kim TW et al (1996) The gene defects responsible for familial Alzheimer's disease. Neurobiol Dis 3:159–168
Teter B, Rozovsky I, Krohn K et al (1996) Methylation of the glial fibrillary acidic protein gene shows novel biphasic changes during brain development. Glia 17:195–205
Theilmann JL, Robbins CA, Hayden MR (1989) Methylation at the D4S95 locus and predictive testing. Am J Hum Genet 45:477–479
Titova N, Qamar MA, Chaudhuri KR (2017) The nonmotor features of Parkinson's disease. Int Rev Neurobiol 132:33–54
Trapp BD, Nave KA (2008) Multiple sclerosis: an immune or neurodegenerative disorder? Annu Rev Neurosci 31:247–269
Trapp BD, Peterson J, Ransohoff RM et al (1998) Axonal transection in the lesions of multiple sclerosis. N Engl J Med 338:278–285
Tremolizzo L, Messina P, Conti E et al (2014) Whole-blood global DNA methylation is increased in amyotrophic lateral sclerosis independently of age of onset. Amyotroph Lateral Scler Frontotemporal Degener 15 98–105
Turek-Plewa J, Jagodzinski PP (2005) The role of mammalian DNA methyltransferases in the regulation of gene expression. Cell Mol Biol Lett 10:631–647
Unnikrishnan A, Hadad N, Masser DR et al (2018) Revisiting the genomic hypomethylation hypothesis of aging. Ann N Y Acad Sci 1418:69–79
Uysal F, Ozturk S, Akkoyunlu G (2017) DNMT1, DNMT3A and DNMT3B proteins are differently expressed in mouse oocytes and early embryos. J Mol Histol 48:417–426
van Blitterswijk M, DeJesus-Hernandez M, Rademakers R (2012) How do C9ORF72 repeat expansions cause amyotrophic lateral sclerosis and frontotemporal dementia: can we learn from other noncoding repeat expansion disorders? Curr Opin Neurol 25:689–700
Van Dam D, De Deyn PP (2011) Animal models in the drug discovery pipeline for Alzheimer's disease. Br J Pharmacol 164:1285–1300
van der Flier WM, Pijnenburg YA, Fox NC et al (2011) Early-onset versus late-onset Alzheimer's disease: the case of the missing APOE varepsilon4 allele. Lancet Neurol 10:280–288
van Deursen JM (2014) The role of senescent cells in ageing. Nature 509:439–446
Villar-Menendez I, Blanch M, Tyebji S et al (2013) Increased 5-methylcytosine and decreased 5-hydroxymethylcytosine levels are associated with reduced striatal A2AR levels in Huntington's disease. NeuroMolecular Med 15:295–309
Waalwijk C, Flavell RA (1978) DNA methylation at a CCGG sequence in the large intron of the rabbit beta-globin gene: tissue-specific variations. Nucleic Acids Res 5:4631–4634

Wagner M, Steinbacher J, Kraus TF et al (2015) Age-dependent levels of 5-methyl-, 5-hydroxymethyl-, and 5-formylcytosine in human and mouse brain tissues. Angew Chem Int Ed Eng 54:12511–12514
Wakabayashi K, Takahashi H, Obata K et al (1992) Immunocytochemical localization of synaptic vesicle-specific protein in Lewy body-containing neurons in Parkinson's disease. Neurosci Lett 138:237–240
Wakabayashi K, Hayashi S, Yoshimoto M et al (2000) NACP/alpha-synuclein-positive filamentous inclusions in astrocytes and oligodendrocytes of Parkinson's disease brains. Acta Neuropathol 99:14–20
Wang H (2018) Modeling neurological diseases with human brain organoids. Front Synaptic Neurosci 10:15
Wang F, Yang Y, Lin X et al (2013a) Genome-wide loss of 5-hmC is a novel epigenetic feature of Huntington's disease. Hum Mol Genet 22:3641–3653
Wang Y, Wang X, Li R et al (2013b) A DNA methyltransferase inhibitor, 5-aza-2'-deoxycytidine, exacerbates neurotoxicity and upregulates Parkinson's disease-related genes in dopaminergic neurons. CNS Neurosci Ther 19:183–190
Wasmuth JJ, Hewitt J, Smith B et al (1988) A highly polymorphic locus very tightly linked to the Huntington's disease gene. Nature 332:734–736
Wei S, Hua HR, Chen QQ et al (2017) Dynamic changes in DNA demethylation in the tree shrew (Tupaia belangeri chinensis) brain during postnatal development and aging. Zool Res 38:96–102
Wen L, Li X, Yan L et al (2014) Whole-genome analysis of 5-hydroxymethylcytosine and 5-methylcytosine at base resolution in the human brain. Genome Biol 15:R49
West RL, Lee JM, Maroun LE (1995) Hypomethylation of the amyloid precursor protein gene in the brain of an Alzheimer's disease patient. J Mol Neurosci 6:141–146
Wijesekera LC, Leigh PN (2009) Amyotrophic lateral sclerosis. Orphanet J Rare Dis 4:3
Wilson VL, Smith RA, Ma S et al (1987) Genomic 5-methyldeoxycytidine decreases with age. J Biol Chem 262:9948–9951
Wu H, Coskun V, Tao J et al (2010) Dnmt3a-dependent nonpromoter DNA methylation facilitates transcription of neurogenic genes. Science 329:444–448
Xi Z, Yunusova Y, van Blitterswijk M et al (2014) Identical twins with the C9orf72 repeat expansion are discordant for ALS. Neurology 83:1476–1478
Xie W, Barr CL, Kim A et al (2012) Base-resolution analyses of sequence and parent-of-origin dependent DNA methylation in the mouse genome. Cell 148:816–831
Yan S, Tu Z, Liu Z et al (2018) A huntingtin knockin pig model recapitulates features of selective neurodegeneration in Huntington's disease. Cell 173:989–1002, e1013
Yoder JA, Walsh CP, Bestor TH (1997) Cytosine methylation and the ecology of intragenomic parasites. Trends Genet 13:335–340
Young PE, Kum Jew S, Buckland ME et al (2017) Epigenetic differences between monozygotic twins discordant for amyotrophic lateral sclerosis (ALS) provide clues to disease pathogenesis. PLoS One 12:e0182638
Zadel M, Maver A, Kovanda A et al (2018) DNA methylation profiles in whole blood of Huntington's disease patients. Front Neurol 9:655
Zhang RR, Cui QY, Murai K et al (2013) Tet1 regulates adult hippocampal neurogenesis and cognition. Cell Stem Cell 13:237–245
Zhang M, Xi Z, Ghani M et al (2016) Genetic and epigenetic study of ALS-discordant identical twins with double mutations in SOD1 and ARHGEF28. J Neurol Neurosurg Psychiatry 87:1268–1270
Zhao X, Ueba T, Christie BR et al (2003) Mice lacking methyl-CpG binding protein 1 have deficits in adult neurogenesis and hippocampal function. Proc Natl Acad Sci U S A 100:6777–6782
Zhou D, Robertson KD (2016) Role of DNA methylation of genomic stability. In: Tollefsbol TO (ed) Translational epigenetics series. Elsevier, London

# Functional Implications of Dynamic DNA Methylation for the Developing, Aging and Diseased Brain

**Geraldine Zimmer-Bensch**

## Contents



**Abstract** Epigenetic mechanisms of gene regulation as the interface between the genome and environment, control diverse processes in development, aging and disease. As proposed by increasing body of evidence, defects in epigenetic remodeling during brain development, function and aging seem central to diverse aspects of the pathophysiology of psychiatric and neurological diseases.

The discovery of active ways of DNA demethylation has paved the way to reconsider the functional implications of DNA methylation in the brain, where dynamic reconfiguration of the DNA methylation landscape has been observed during development and aging. High-throughput studies profiling global DNA methylation and transcriptional changes suggest that DNA methylation-dependent gene regulation is crucially involved in regulating neuronal differentiation and maturation processes, as well as in age-related declines of neuronal function. As DNA methylation and DNA methyltransferases (DNMTs) also influences the histone code, the crosstalk of these two mechanisms of epigenetic gene regulation in neuronal development and function

G. Zimmer-Bensch (✉)
Division of Functional Epigenetics in the Animal Model, Institute for Biology II, RWTH Aachen University, Aachen, Germany
e-mail: zimmer@bio2.rwth-aachen.de



S. Jurga, J. Barciszewski (eds.), *The DNA, RNA, and Histone Methylomes*, RNA Technologies, https://doi.org/10.1007/978-3-030-14792-1_6

has been started to be investigated. Here, an overview is provided about the currently known functional implications dynamic DNA methylation and the crosstalk with histone modifications have in neuronal development and aging, as well as in associated diseases. Further, we discuss the integration and applicability of animal models as tool to gain insights in human brain aging.

**Keywords** DNA methyltransferases · Cortical interneurons · Neuronal migration · Cell death · Neuropsychiatric diseases

# 1 Introduction

Epigenetic mechanisms of gene expression control a variety of processes during development, aging and disease. Similar to histone modifications, DNA methylation catalyzed by DNA methyltransferases (DNMTs) turned out as dynamic epigenetic mark, due to the discovery of active ways of DNA demethylation involving TET-mediated oxidation of 5-methylcytosines. Dynamic DNA methylation is evident during neuronal development and maturation (Lister et al. 2013; Sharma et al. 2008), and seems implicated in regulating adult neuronal functions (Meadows et al. 2015, 2016; Sweatt 2016) as well as age-associated processes (Akbarian et al. 2013; Barter and Foster 2018; Lardenoije et al. 2015). Moreover, altered DNA methylation emerged to be involved in the etiology of neuropsychiatric disorders, including major depressive disorder, autism spectrum disorder and schizophrenia (Akbarian et al. 2013). Besides, dynamic changes in the DNA methylation landscape were observed during brain aging (Barter and Foster 2018; Lardenoije et al. 2015).

In this chapter, we discuss the role of DNA methylation in the developing, aging and diseased brain with focus on the cerebral cortex.

# 2 The Cerebral Cortex

Neuronal circuitries in the six-layered cerebral cortex, the seat of higher cognitive functions in the mammalian brain, are established by excitatory glutamatergic principal neurons and inhibitory gamma-aminobutyric acid (GABA)-expressing interneurons. Excitatory projection neurons adopt layer-specific identities and form specific dendritic and axonal connections. Layer II/III neurons mainly project contra- or ipsilateral to other cortical areas, while neurons of layer V and VI project to subcortical targets, including the thalamus (layer VI neurons), midbrain, hindbrain and spinal cord (layer V neurons) (Merot et al. 2009). The inhibitory GABAergic neurons populate different cortical layers and act as local modulators of excitatory neurons. Although only representing about 20% of the overall neuronal population, inhibitory interneurons are critical for cortical information processing, learning and memory formation (Hensch 2005; Letzkus et al. 2015). Due to their enormous morphological and physiological diversity, inhibitory interneuron subtypes have the capacity to selectively target different sub-cellular compartments of projection neurons (De Marco Garcia et al. 2011), enabling a dynamic inhibition-dependent regulation of input and output processing (Gidon and Segev 2012; Pouille et al. 2013). Parvalbumin (PV)-positive interneurons primarily target the soma and axon initial segments of glutamatergic neurons, while dendritic inhibition is achieved by somatostatin (SST)-expressing interneurons. Vasointenstinal peptide (VIP)-positive interneurons inhibit mainly other cortical interneurons (Druga 2009).

The relevance of inhibitory interneuron function for cortical information processing is reflected by diverse neurologic and psychiatric diseases which involve defects in the cortical inhibition (Marin 2012). For example, deficits of SST-positive cortical interneuron function including impaired GABAergic transmission and decreased *Sst* expression levels are suggested to be implicated in the pathophysiology of schizophrenia (Lin and Sibille 2013; Morris et al. 2008). Defects in SST interneurons were further observed in numerous other human psychiatric and neurological disorders such as major depressive disorder, bipolar disorder, Alzheimer's disease and Parkinson's disease (Lin and Sibille 2013). Various studies already provided evidence that impairments during development contribute to defective inhibition underlying such diseases (Marin 2012). Hence, the correct establishment of the cortical GABAergic system during development is crucial for proper cortical function.

## 3 Cerebral Cortex Development

The formation of the cerebral cortex is a highly sophisticated process requiring the precise interplay of several developmental steps. These include proliferation of neuronal stems cells, differentiation, migration from the proliferative zone to their cortical target layer, axonal and dendritic growth as well as establishment of synaptic contacts.

Cortical projection neurons arise exclusively from progenitors located within the dorsal pallium. From there they migrate radially out to form the different cortical layers in an "inside-out" fashion, with deep layer neurons born first and upper-layer neurons born later, migrating past earlier born ones (Merot et al. 2009).

Neuroepithelial cells (NECs) as the earliest progenitors of the cortex, are organized in a pseudostratified neuroepithelium resulting from the apico-basal movement of their nuclei during cell-cycle progression. After initial expansion of the progenitor pool by symmetric proliferative divisions, they divide asymmetrically generating radial glial cells (RGCs) that are located in the ventricular zone (VZ) und display similar bipolar morphology (Agirman et al. 2017). At the onset of neurogenesis, RGCs divide asymmetrically to generate post-mitotic neurons or intermediate, transient amplifying progenitor cells. These intermediate progenitors delaminate and translocate their cell bodies more basally, forming the subventricular zone and dividing symmetrically to indirectly generate neurons. The transient amplifying progenitors are already present at early stages of neurogenesis and are suggested to contribute to the neuronal production of all cortical layers (Agirman et al. 2017; Merot et al. 2009). In addition to short neural precursor cells (SNPs), outer RGCs (oRGCs) are described in the murine cortex to appear as a minor population, whereas they are proportionally more important in the developing cortex of gyrencephalic mammals contributing to the folding of the cortex (Hansen et al. 2010; Nonaka-Kinoshita et al. 2013). They share common molecular features with RGCs but reside in the outer part of the SVZ lacking basal attachment. Alike RGCs, SNPs reside in the VZ. However, they are transcriptionally distinct from RGCs, lack basal attachment and are programmed to generate neurons via symmetric differentiative divisions (Agirman et al. 2017).

In contrast to excitatory cortical neurons that arise from the cortical proliferative zones, comparatively little is known about progenitor subtypes generating the diverse subsets of inhibitory GABAergic interneurons that are located in spatially distinct domains of the subpallium. These include the medial and caudal ganglionic eminences, abbreviated with MGE and CGE, respectively, as well as the pre-optic area (POA) (Druga 2009). The MGE generates parvalbumin (PV)-positive basket and chandelier cells, as well as Martinotti and multipolar somatostatin (SST)-expressing interneurons (Butt et al. 2005, 2008; Xu et al. 2003), whereby SST-interneuron generation precedes the PV-interneuron generation (Butt et al. 2005, 2008; Inan et al. 2012). The POA contributes to a diverse subset of cortical interneurons, including neuropeptide Y (NPY), reelin, PV, SST, CTIP2 positive interneurons and neurogliaform cells (Gelman et al. 2009, 2011; Symmank et al. 2019). Likewise, the CGE produces a large variety of cortical interneurons including reelin positive cells, vasointestinal peptide (VIP)/calretinin positive bipolar interneurons and VIP/cholecystokinin positive basket cells (Hu et al. 2017; Miyoshi et al. 2010; Murthy et al. 2014; Rubin and Kessaris 2013).

Upon becoming post-mitotic, the different interneuron subsets migrate along particular routes through the basal telencephalon up to the cortex (Corbin and Butt 2011). This long-range tangential migration to cortical target regions represents a critical step. Apart from the initiation of migration by adapting a migratory morphology and the maintenance of their motility throughout the migratory period, the directionality has to be strictly controlled to achieve successful migration to the cortex, to precisely distribute over cortical areas and layers, and to finally integrate appropriately into cortical circuits [reviewed in Metin et al. (2006); Zimmer-Bensch (2018)].

# 4 Dynamic DNA Methylation in Neuronal Development

## 4.1 Key Players of DNA Methylation and Demethylation

DNA methylation is accomplished by DNA methyltransferases that in eukaryotes catalyze the methylation of predominantly cytosines at the fifth carbon of the pyrimidine ring yielding in 5-methylcytosine (5mC). DNA methylation of cytosines that are followed by guanines is called CpG methylation. In brain tissue as well as in human embryonic stem cells non-CpG or CpH methylation (H refers to adenine, thymine or another cytosine) is further prevalent (Guo et al. 2014; Lee et al. 2017; Pinney 2014). DNA methylation can be associated with silencing or activation of transcription, dependent on the methylated genomic regions and the DNA methylation-interacting proteins. DNA methylation can result in blocking the binding of transcription factors or recruiting methyl-binding proteins involved in gene silencing, thereby causing repression of gene transcription (Zhu et al. 2016). In addition to methyl-binding proteins a battery of transcription factors lacking the methyl-binding domain was suggested to interact with methylated DNA through different motifs, whereby the physiological relevance remains to be elucidated (Zhu et al. 2016).

Hypermethylation of CpG sites located in promoter or enhancer regions is often correlated with transcriptional repression (Chodavarapu et al. 2010; Lister et al. 2009). However, a substantial proportion of DNA methylation sites appears to be positively correlated with gene expression. Besides gene body methylation, which can be associated with repression and activation of gene expression (Lister et al. 2013; Yang et al. 2011), methylation upstream of transcriptional start sites can lead to transcriptional activation (Irizarry et al. 2009). Methylated cytosines are also evident in intergenic regions that control the transcription of genes nearby (Jones 2012). In neurons, alterations in CpH methylation were also found to correlate with transcriptional changes (Guo et al. 2014; Lister et al. 2013), emphasizing the gene regulatory potential of CpH methylation.

In the developing and adult nervous system, DNA methylation is achieved by DNMT1, DNMT3a and DNMT3b (Jang et al. 2017). Whereas DNMT1 acts as maintenance enzyme in dividing progenitors due to its high affinity to hemimethylated DNA, DNMT3a and DNMT3b were described as de novo methyltransferases (Jin and Robertson 2013). However, DNMT1 is also expressed in non-dividing post-mitotic neurons (Kadriu et al. 2012), where DNMT1 and DNMT3a can exert partly redundant (Feng et al. 2010) but also distinctive functions (Morris et al. 2016).

The discovery of active ways of DNA demethylation by Ten-eleven translocation (TET) family enzyme- dependent mechanisms (Wu and Zhang 2017) initiated a re-thinking about the functional implications of DNA methylation in post-mitotic and differentiated neurons. In the central nervous system, the DNA methylation landscape is dynamically altered throughout the developmental time course (Lister and Mukamel 2015; Lister et al. 2013), which has been related to cell-type specific development and maturation (Lister and Mukamel 2015; Lister et al. 2013; Mo et al. 2015; Sharma et al. 2016). In the adult brain, dynamic DNA methylation was suggested to be involved in synaptic plasticity and memory formation (Kennedy and Sweatt 2016; Sweatt 2016; Meadows et al. 2015, 2016; Zovkic et al. 2013), while upon aging a shift in CpG methylation and a continuous increase in CpH methylation was described (Ianov et al. 2017).

## 4.2 DNA Methylation and Neurogenesis

The establishment of neuronal circuits relies on the proper generation of its diverse neuronal composites. Neurons are generated by neuronal stem cells, which become progressively restricted to generate the different types of neurons first (neurogenesis) and glia cells afterwards (gliogenesis). In addition to this temporal restriction, a spatial determination occurs early in development mediated by patterning (Kiecker and Lumsden 2005). For example, the excitatory and inhibitory neurons of the cerebral cortex derive from progenitors located in the dorsal and ventral telencephalon, respectively (Hu et al. 2017; Martynoga et al. 2012). The sequential generation of the excitatory neurons fated for the distinct layers of the cerebral cortex relies on progressive fate restriction (Martynoga et al. 2012), whereas progenitors of distinct spatial domains are proposed to give rise to different cortical interneuron subtypes (Hu et al. 2017). Although, diverse transcriptional networks and cascades implicated in interneuron subtype generation are already described, comparatively little is clear yet about the mechanisms of cell fate restriction in cortical interneuron progenitors, which contemporaneously give rise to inhibitory interneurons destined for diverse telencephalon regions (Hu et al. 2017). However, cell fate determination of both, excitatory principal cortical neurons and inhibitory interneurons, is associated with setting up subtype-specific transcriptional programs, directing subsequent developmental steps like migration, targeting and morphological differentiation (Franco and Muller 2013; Hu et al. 2017). Increasing body of evidence proposes a close connection between the epigenetic machinery and such stage- and subtype-specific transcriptional programs during neuronal differentiation. For example, the Nkx-class homeobox transcription factor 2.1 (NKX2-1), which is on top of the hierarchical transcriptional cascade governing development of MGE-derived inhibitory cortical interneurons (Flandin et al. 2010; McKinsey et al. 2013; Nobrega-Pereira et al. 2008; Sandberg et al. 2016; van den Berghe et al. 2013) also affects the epigenome, as significant alterations in histone profiles were observed in *NKX2-1* conditional knockout animals (Sandberg et al. 2016).

Indeed, dynamic temporal changes in DNA methylation patterns have been observed alongside with the sequentially generated neuronal subtypes (Lister and Mukamel 2015; Lister et al. 2013; Mo et al. 2015; Sharma et al. 2016). However, whether the methylome defines cell identity by suppressing alternative fates and thereby promoting a certain lineage, or whether the emergence of particular DNA methylation profiles is a consequence of fate restriction driving subtype-specific developmental programs, is not clear so far.

In support of a role for DNA methylation in cell fate restriction, DNMTs are found widely expressed in neuronal precursors of the central nervous system (Feng et al. 2005). DNMT1 is suggested to be crucial for driving the neuronal fate by inhibiting astroglial differentiation during the neurogenic period. In the spinal cord, *Dnmt1* deficiency at progenitor level causes precocious astroglial differentiation and hypomethylation of genes associated to the gliogenic JAK/STAT pathway (Fan et al. 2005). Likewise, *Dnmt1*-deficiency promotes the differentiation of neuronal stem cells into astrocytes in precursors of the dendate gyrus (Noguchi et al. 2016b). Moreover, TET1 was suggested to contribute to the neurogenesis onset by promoting the expression of neuronal markers (Kim et al. 2016). In contrast to these findings, no indications of cell fate changes were observed upon the loss of *Uhrf1* in neuronal stem cells as determined by RNA-sequencing experiments (Ramesh et al. 2016), acting as important adaptor for DNMTs (Berkyurek et al. 2014). Hence, further research is required to decipher the detailed role of DNA methylation in neuronal progenitors, especially as DNMTs are known to act non-canonically through interactions with histone modifications in developing neurons (Symmank et al. 2018).

## *4.3 DNA Methylation in Post-mitotic Neuronal Development*

Upon becoming post-mitotic, immature neurons migrate to their target regions where they adopt subtype-specific features in regard to morphology, molecular properties, firing and connectivity patterns. In addition to migration and morphological maturation including axonal and dendritic growth, programmed cell death is another crucial aspect of post-mitotic maturation that has to be highly regulated, to remove unconnected neurons and to regulate final neuron number (Southwell et al. 2012).

The establishment of methods for high resolution and large-scale methylome profiling lead to the discovery of highly dynamic DNA methylation reconfiguration during neuronal maturation (Lister and Mukamel 2015; Lister et al. 2013; Mo et al. 2015; Sharma et al. 2016). Thereby, different cell types like glia cells and neurons, but also distinct neuronal subtypes like GABAergic interneurons and glutamatergic projection neurons of the cerebral cortex differ vastly in their DNA methylation profiles (Kozlenkov et al. 2014, 2016; Lister et al. 2013). This points to a role of DNA methylation in cell type-specific maturation programs, whereby cell type-specific DNA methylation patterns seem rather a consequence than the cause of lineage-specification (Sharma et al. 2016).

Many post-mitotic developmental processes require the coordinated remodeling of the cytoskeleton, for example during migration, dendritic and axonal growth, and branching. In migrating cortical interneurons, DNMT1-dependent DNA methylation is suggested to regulate cytoskeleton-associated genes, thereby promoting the migratory morphology required for proper migration (Pensold et al. 2017). DNMT1 target genes were identified by correlative methylome and transcriptome analysis applying MeDIP and RNA sequencing of FACS-enriched embryonic *Dnmt1* wild-type and knockout interneurons (Pensold et al. 2017). Among them *Pak6*, a member of the p21-activated kinases (PAKs), was found up-regulated in expression in *Dnmt1*-deficient cells (Pensold et al. 2017). PAKs are known to be involved in cell survival regulation as well as cytoskeletal rearrangements (Kumar et al. 2017), and PAK6 in particular was already shown to promote neurite complexity in excitatory cortical neurons (Civiero et al. 2015). Consistently, forced expression of PAK6 induced by a PAK6-GFP expression construct caused a multipolar morphology of embryonic interneurons, reminiscent to the phenotype determined for migrating *Dnmt1*-deficient interneurons (Pensold et al. 2017). In contrast, siRNA-mediated *Pak6* depletion reduced neurite complexity and cell death (Pensold et al. 2017). Hence, *Pak6* represents a downstream target of DNMT1-dependent transcriptional repression involved in cytoskeleton and cell death regulation underlying proper cortical interneuron migration (Fig. 1). However, no changes in the DNA methylation level of the *Pak6* gene locus, neither upstream nor downstream was observed in *Dnmt1*-deficient embryonic interneurons (Pensold et al. 2017; Symmank et al. 2018). Hence, DNA methylation-independent actions of DNMT1 likely account for the transcriptional regulation of *Pak6*. Indeed, many genes found altered in expression between *Dnmt1*-deficient and wild-type embryonic interneurons were not in conjunction with respective changes in DNA methylation and vice versa, pointing to non-canonical actions of DNMT1 (Pensold et al. 2017; Symmank et al. 2018).

In addition to DNA-methylation, DNMTs can also interact with histone modifying complexes (Du et al. 2015), thereby modulating transcription. There is evidence that DNA methylation inhibits permissive and supports repressive histone methylation to ensure gene silencing (Hashimshony et al. 2003; Lande-Diner et al. 2007). This can be achieved by direct interactions between DNA methylating and histone modifying enzymes via specific binding domains, which modulate the recruitment of proteins to complexes and the catalytic activity of their binding partners (Clements et al. 2012; Smallwood et al. 2007; Vire et al. 2006). For DNMT1, an interaction with EZH2, the core enzyme of the polycomp repressor complex 2 (PRC2) executing repressive trimethylations on lysine 27 at the N-terminal amino acid tail of histone 3 (H3K27me3) (Margueron and Reinberg 2011), was described to occur in non-neuronal cells (Ning et al. 2015; Purkait et al. 2016; Vire et al. 2006). Moreover, DNMT1 affects H3K27 trimethylation by modulating *Ezh2* expression levels (Purkait et al. 2016; So et al. 2011).

In migrating cortical interneurons, a crucial role of DNMT1-dependent establishment of repressive H3K27me3 marks was suggested to negatively act on *Pak6* gene expression (Symmank et al. 2018) (Fig. 1). Transcriptional repression of *Pak6* is crucial to maintain the migratory morphology and to promote interneuron survival, as determined by knockout and forced expression experiments (Pensold et al. 2017). Inhibition of EZH2, the core enzyme of the PRC2 (Chittock et al. 2017), executing H3K27 trimethylation, causes similar effects on neuronal

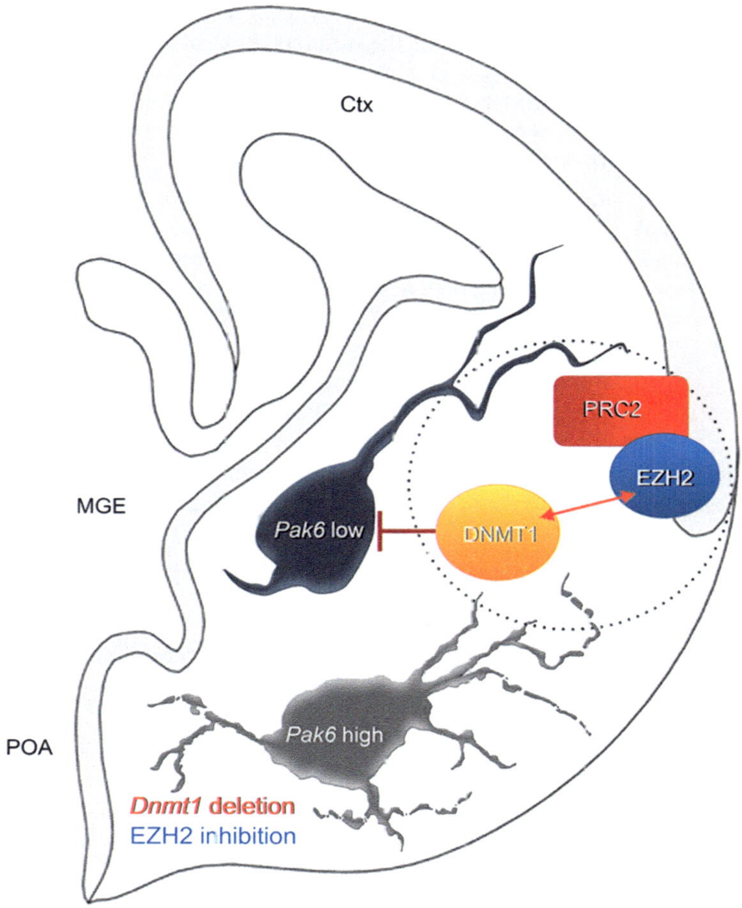


**Fig. 1** Schematic view of a coronal section of one hemisphere of an embryonic mouse brain, illustrating a polarized migrating interneuron (dark blue) and a multipolar, degenerating interneuron (grey). DNMT1 promotes migration and survival by repressing *Pak6* expression, through interactions with EZH2 catalyzing the establishment of repressive H3K27me3 histone marks as the core enzyme of the polycomb repressor complex 2 (PRC2). In turn, *Dnmt1* deletion or EZH2 inhibition cause elevated *Pak6* expression levels and cellular complexity as well as cell death. *POA* preoptic area, *MGE* medial ganglionic eminence, *Ctx* cerebral cortex

complexity (Fig. 1), which are rescued by *Pak6* depletion (Symmank et al. 2018). Thereby, the DNMT1-dependent establishment of H3K27me3 marks were identified to rely on direct interactions of DNMT1 and EZH2 at protein level (Fig. 1) (Symmank et al. 2018).

DNMT1 has already been described to be critical for the post-mitotic maturation of other neuronal subtypes in vitro and in vivo (Chestnut et al. 2011; Fan et al. 2001; Hutnick et al. 2009; Rhee et al. 2012). DNMT1 promotes the morphological maturation and refinement of cortical excitatory neurons (Feng et al. 2010; Hutnick et al. 2009), and is further crucial for the differentiation of dendate gyrus neurons (Noguchi et al. 2016b).

Another common role of DNMTs and DNA methylation during development of diverse neuronal subsets refers to cell death and survival regulation at post-mitotic level (Fan et al. 2001; Noguchi et al. 2016a; Pensold et al. 2017; Rhee et al. 2012). DNA hypomethylation perturbs the survival of neurons of the central nervous system (Fan et al. 2001) including retinal neurons (Rhee et al. 2012). *Dnmt1* deletion caused impaired survival of post-mitotic cortical interneurons (Pensold et al. 2017) and of newly generated hippocampal neurons in adult brains (Noguchi et al. 2015). While for retinal neuron survival DNMT1-dependent DNA methylation was proposed to be required (Rhee et al. 2012), non-canonical actions of DNMT1 through a crosstalk with histone modifications were suggested to contribute to the survival regulation in immature cortical interneurons (Pensold et al. 2017; Symmank et al. 2018).

The relevance of DNA methylation for survival regulation is further sustained by in vitro studies, showing an implication of TET2 function, involved in DNA demethylation, in cortical neuron survival (Mi et al. 2015). Consistently, *Tet1* deletion makes cerebellar granular cells more vulnerable towards oxidative stress-induced neuronal cell death (Xin et al. 2015).

Together, these studies emphasize a crucial role of DNA methylation as well as of non-canonical DNMT actions in post-mitotic neuronal development, including migration, morphological maturation, neuronal survival and cell death regulation.

# 5 DNA Methylation in the Aging Brain

## *5.1 Difficulties in Facing the Neurobiology of Aging*

Aging causes structural, neurochemical and physiological alterations in the brain that lead to behavioral changes, memory decline and cognitive impairments (Rozycka and Liguz-Lecznar 2017). Cognitive aging depends on numerous factors and results in metabolic, hormonal and immune dysregulation, increased oxidative stress and inflammation, altered neurotransmission and reduced neurotrophic support of neurons (Rozycka and Liguz-Lecznar 2017). Thereby, different brain regions and neuronal cell types are distinctively affected by the aging process. In addition to reduced excitability and plasticity (Clark and Taylor 2011), and a decline of the inhibitory function (Cheng and Lin 2013; Shetty and Turner 1998; Stanley and Shetty 2004), a selective vulnerability of particular neuronal subtypes like inhibitory interneurons and GABAergic synapses (Rozycka and Liguz-Lecznar 2017) were observed in particular regions of aged brains.

However, observations like age-related changes in cell numbers differ between selected animal models and humans, and conflicting data even exist for the same species (Flood and Coleman 1988). Due to the important functions GABAergic inhibitory interneurons have in cortical information processing, age-associated defects in inhibitory circuits appear as attractive hypothesis for cognitive decline and age-associated disorders (Rozycka and Liguz-Lecznar 2017). Indeed, several studies found reduced cell numbers of cortical interneuron subtypes across different species and brain regions (summarized in Table 1). Moreover, functional and structural changes of GABAergic synapses appear to occur in aged brains. These include loss of synaptic contacts, decreased neurotransmitter release, reduced post-synaptic responsiveness to neurotransmitters, suggested to contribute to the age-associated cognitive decline (Rozycka and Liguz-Lecznar 2017).

In agreement with reduced neurotransmitter release, major changes in the expression of genes related to neurotransmission and transcriptional repression especially of GABA-related transcripts have been reported for the human prefrontal cortex, which could however not be detected in non-primate mammals (Loerch et al. 2008). In contrast, to this, several studies described changes in transcripts related to GABAergic transmission across different species (summarized in Table 2).

Elevated neuroprotection-related gene expression and diminished expression of genes involved in general synaptic function at least appear as conserved features of mammalian brain aging (Ianov et al. 2016; Jiang et al. 2001; Loerch et al. 2008). Consistently, RNA sequencing of synaptosomes from cerebral cortices of aged mice moreover revealed changes in expression of synaptic transmission-related genes (Chen et al. 2017). Of note, in this study differential expression of diverse long non-coding RNAs were detected between young and old synaptosomes, proposed to be crucial for synaptic physiology.

Due to this heterogeneity in the reported structural, functional and transcriptional alterations in aged brains within one specie and between different species, approaching the functional implications of DNA methylation in brain aging is far from being a simple task. Comparative studies with more stringency in regard to the analysis of particular brain regions and individual cell types achieved by single cell methods enabling parallel single-cell based methylation and transcriptional analysis are needed to determine cell and species-specific age-related changes in DNA methylation and their transcriptional consequences.

Despite conflicting reports, murine models have called increased attention for investigating the neurobiology of aging and age-associated neurodegenerative diseases, due to the rapid evolution of mouse genetics and the comparatively short life span of mice (Bilkei-Gorzo 2014; Jucker and Ingram 1997).

## *5.2 The Implication of DNA Methylation Signatures for Brain Aging*

Although DNA methylation signatures are altered upon aging in human and mouse brains (Lister et al. 2013; Siegmund et al. 2007), apparent region-specific differences impede general conclusions about their functional implications (Kraus et al. 2016;

**Table 1** Summary of studies investigating age-associated alterations in cortical interneuron numbers across species

| Species | Cortical area | Observations | Age of old species | References |
|---|---|---|---|---|
| Human | Ctx/Hc | Unchanged number of PV cells | >65 years | Bu et al. (2003) |
| Human | Visual Ctx and parahippocampal gyrus | Reduced density of CB-immunopositive cells | >65 years | Bu et al. (2003) |
| Human | Auditory Ctx | Reduced density of CCR-immunopositive cells | >65 years | Bu et al. (2003) |
| Cat | Visual Ctx | Reduced GABAergic interneurons | 12 years | Hua et al. (2008) |
| Rat | Hc | Reduced GABAergic interneurons | 26–30 months | Stanley et al. (2012) |
| Rat | Auditory Ctx | Decreased numbers of GAD65- and 67-immunoreactive neurons | 30–35 months | Burianova et al. (2009) |
| Rat | Perirhinal Ctx | No differences in PV- or CR immunoreactivity | 26 months | Moyer et al. (2011) |
| Rat | Auditory Ctx | Reduced CB interneuron numbers | >28 months | Ouda et al. (2012) |
| Rat | Auditory Ctx | Reduced numbers of PV interneurons | >28 months | Ouda et al. (2008) |
| Rat | Somatosensory and motor Ctx | Reduced numbers of PV interneurons | 26 months | Miettinen et al. (1993) |
| Rat | Somatosensory and motor Ctx | Decreased SOM interneurons | 26 months | Miettinen et al. (1993) |
| Rat | Hc | Decreased SOM interneurons | 23 months | Stanley et al. (2012) |
| Rat | Hc | Reduced numbers of CB interneurons | 25–30 months | Potier et al. (2006) |
| Rat | Sensory Ctx | Reduced numbers of VIP interneurons | 20–29 months | Cha et al. (1997) |
| Rat | Frontal, occipital and temporal cortical areas, Hc | Reduced numbers of NPY neurons | 20–29 months | Cha et al. (1997) |
| Rat | Auditory Ctx | Reduced numbers of NPY cells | 25 months | Ouellet and de Villers-Sidani (2014) |
| Rat | Auditory Ctx | Decreased numbers of SOM and PV-interneurons | 25 months | Ouellet and de Villers-Sidani (2014) |

*Ctx* cortex, *Hc* hippocampus

**Table 2** Summary of studies reporting age-related changes in mRNA or protein level of GABA-related genes across different species

| Species | Cortical area | Observations | References |
|---|---|---|---|
| Primates | Ctx, Hc | Reduction of *Sst* mRNA | Hayashi et al. (1997) |
| Rat | Auditory Ctx | Reduced levels of *Gad1* and *Gad2* mRNAs | Ling et al. (2005) |
| Rat | Auditory Ctx | Decrease in the protein levels of GAD65 and GAD67 | Burianova et al. (2009) |
| Monkey | Visual Ctx | Altered GABAergic gene expression | Liao et al. (2016) |
| Rat | Medial PFC | Reduction in GAT-1 | Banuelos et al. (2014) |
| Human | Frontal Ctx | Reduction in GAT-1 | Sundman-Eriksson and Allard (2006) |
| Rat | Hc | Decrease in the VGAT level | Canas et al. (2009) |
| Mouse | Barrel Ctx | Decreased *Vgat* mRNA and VGAT protein levels | Liguz-Lecznar et al. (2015) |

*PFC* prefrontal cortex, *Ctx* cortex, *Hc* hippocampus

Numata et al. 2012). Another general challenge is the correlation of methylation marks with the transcriptional output to elucidate the physiological and biological relevance. Is the changed transcription a consequence of altered DNA methylation or do transcriptional alterations predispose for alterations in DNA methylation signatures?

As described above, the relationship between DNA methylation and expression depends on the genomic localization, with transcriptional potential being shown for DNA methylation within promoter regions, as well as within gene bodies, presumably at enhancer and silencer regions in introns and exons (Clermont et al. 2016; Kulis et al. 2013; Lee et al. 2015; Vinson and Chatterjee 2012). Hereinafter an overview about reported age-related changes in DNA methylation found for particular brain regions, genomic localizations and genes will be provided.

A decrease in CpG methylation upon aging was observed within repetitive sequences, including transposable elements (Ianov et al. 2017). Repressive DNA methylation contributes to genomic stability by preventing transposable elements from translocating in the DNA. Reduced DNA methylation causes increased transposon activity that has been related to diminished neuronal function and memory impairments during aging in Drosophila (Li et al. 2013).

In contrast to reduced methylation levels at CpG sites, non-CpG methylation, which can also causes gene silencing (Guo et al. 2014), continues to increase in the aging brain (Ianov et al. 2017; Lister et al. 2013). Interestingly, for aged cognitively impaired animals, hypermethylation of non-CpGs is enriched for synaptic genes suggesting that de novo methylation of non-CpGs is linked to the decrease in their expression (Ianov et al. 2017).

For activity and synapse-associated genes an increase in promoter methylation has further been reported (Haberman et al. 2012; Keleshian et al. 2013; Penner et al. 2016). In contrast, promoter hypomethylation was detected for immune-related genes and seems associated with increased neuroinflammation (Mangold et al. 2017).

Hypermethylation in gene bodies of synaptic genes in conjunction with decreased expression was further reported for aged animals that display impaired PFC-dependent behavior (Ianov et al. 2017). CpG and non-CpG methylation of gene bodies and intergenic regions of synaptic plasticity genes can be modulated by environmental factors and correlate with respective changes in gene expression (Guo et al. 2011, 2014; Halder et al. 2016). These studies emphasize a potential relevance of gene body methylation of synapse-related gene expression during aging.

Among synapse-related genes found to be differentially methylated and expressed in orbital frontal cortices of aged human brains, many GABA-related genes were identified (McKinney et al. 2015), which is consistent with the age-associated alterations in the cortical GABAergic system observed across different species (Tables 1 and 2).

DNA methylation has been reported to be modulated by neuronal activity in the adult brain, which can be mediated by NMDA receptor activity (Guo et al. 2011; Penner et al. 2016). As many synapse and neuronal activity-related genes are altered in expression upon aging, subsequent physiological changes can act in turn on the DNA methylation landscape. For sure, more work needs to be done to dissect the function of DNA methylation in the aging brain.

Although different studies described DNA methylation as crucial for neuronal survival during development (Hutnick et al. 2009; Pensold et al. 2017; Rhee et al. 2012), evidence for direct survival regulation in the aging brain is still lacking. Support for potential functional implications of DNA methylation in neuronal cell death regulation arouse from patients diagnosed with Alzheimer's Disease, an age-related neurodegenerative disorder. In neurons of postmortem cortical tissue 5mC and 5hmC immunoreactivity was found globally altered compared to age-matched control individuals (Coppieters et al. 2014; Mastroeni et al. 2010). However, the age-related mechanisms that can culminate in neuronal death or neurodegeneration seem very diverse, involving oxidative stress, disturbed calcium homeostasis, chromosomal instability, impaired DNA repair, and the accumulation of nuclear and mitochondrial DNA damage. These can either contribute individually or in combination to age-associated cell death in the central nervous system. DNMT1 was already reported to function coordinately with the DNA damage repair in cancer (Jin and Robertson 2013), whereas potential involvements in regulating neuronal aging-related cell death still remain elusive and require further investigations.

Despite numerous open questions, the current data points to an implication of a drift of DNA methylation upon aging in influencing the regulation of long-term neuronal survival and the vulnerability towards age-associated neurodegenerative disorders.

## 6 DNA Methylation in Neuropsychiatric Diseases

Increasing body of evidence points to an epigenetic component in multifactorial neuropsychiatric disorders, to which genetic and environmental factors contribute. Epigenetic marks, which are sensitive to environmental insult, may account for the yet unexplained individual susceptibility and the variability in the course and etiology of diseases like schizophrenia, major depression disorder and autism.

DNA methylation turns out as a key epigenetic mechanism in major depression disorder (Pishva et al. 2017). Social psychological stress is proposed to cause methylation of genes relevant to the disease (McGowan et al. 2009; Oberlander et al. 2008), and DNA demethylation of neuronal cell death-related genes together with neuronal cell death were described to be associated with major depression disorder [reviewed in Symmank and Zimmer (2017)]. Moreover, DNA methylation of *Bdnf* causing reduced synthesis of BDNF, which is crucial for the development, survival and maintenance of neurons, has been linked to depression (Na et al. 2016; Roth et al. 2011).

A genome-wide methylation study has provided evidence for dysregulated DNA methylation profiles in cortical neurons in Autism Spectrum Disorder, whereby changes in DNA methylation affect genes involved in synaptic, neuronal and GABAergic processes (Nardone et al. 2017).

Altered DNA methylation in GABAergic interneurons seems further to be involved in the pathophysiology of schizophrenia. Increased *Dnmt1* expression and subsequently elevated DNA methylation levels are detected in cortical interneurons of patients diagnosed with schizophrenia (Costa et al. 2007; Ruzicka et al. 2007; Veldic et al. 2004). Site-specific analysis revealed that genes like *Reln* and *Gad1* relevant for GABAergic neurotransmission and interneuron function display elevated levels of DNA methylation (Costa et al. 2007; Ruzicka et al. 2007; Veldic et al. 2004). The altered methylation patterns correlate with reduced expression of these genes suggested to account for impaired interneuron function (Costa et al. 2007; Ruzicka et al. 2007; Veldic et al. 2004). Besides schizophrenia, disruption of GABAergic interneuron functionality has been associated with the pathophysiology of other psychological disorders including autism and epilepsy, whereby defects in cortical interneuron development might be of relevance (Marin 2012; Symmank and Zimmer 2017). In support of this, prenatal stress elevates *Dnmt1* and *Dnmt3a* expression in GABAergic interneurons and induces abnormalities in the DNA methylation network as well as behaviors indicative of a schizophrenia-like phenotype in offspring (Matrisciano et al. 2013).

In addition to the reported transcriptional changes caused by altered DNA methylation, a significant layer-specific loss of inhibitory interneurons was identified in postmortem studies of schizophrenia patients (Benes et al. 1991). In agreement with the cell loss, a death receptor pathway was recently shown to be implicated in the pathology of schizophrenia (Catts and Weickert 2012). However, similar to the ageing brain a direct link between cell death genes and DNA methylation is still lacking in the context of schizophrenia.

The transcriptional regulation by DNA methylation in cortical interneurons in disease-related contexts reported so far mostly refers to genes relevant for brain development and physiology including neuronal activity (Costa et al. 2007; Ruzicka et al. 2007; Veldic et al. 2004). The modulation of signal transmission, synaptic plasticity and membrane excitability by DNMT1 was also reported in cortical excitatory neurons under normal conditions (Feng et al. 2010; Levenson et al. 2006; Meadows et al. 2016). As neuronal activity is closely linked to neuron survival (Pfisterer and Khodosevich 2017; Rozycka and Liguz-Lecznar 2017), cell loss observed in diseased brains could be an indirect consequence of DNMT-dependent DNA methylation of genes involved in synaptic neurotransmission. Elevated *Dnmt1* expression in cortical interneurons is also related to the pathogenesis of mental impairments and psychosis due to neural injury and drug abuse (Guidotti et al. 2011; Lewis 2012; Veldic et al. 2005). Thus, the modulation of DNMT1 expression and function, particular in developing and adult cortical interneurons, appears crucial for proper circuitry and the functionality of the adult cerebral cortex, with potential impact on neuronal survival.

## 7 Conclusive Remarks

Epigenetic mechanisms of gene regulation like DNA methylation emerge as attractive mediators integrating external stimuli into the genome, as they appear sensitive towards environmental insults. The dynamic changes of DNA methylation signatures in the developing, adult and aging brain may account for the yet unexplained individual susceptibility and variability of age-related disorders as well as for neuropsychiatric diseases, which in part are developmental in their origin.

However, the implications of DNA methylation for discrete sub-cellular processes necessitate more detailed research. Besides deciphering cell subtype-specific effects, which can be addressed by innovative single cell sequencing approaches, the correlation with transcriptional changes represents a crucial aspect. Moreover, the crosstalk of DNA methylation with histone modifying mechanisms multiplies the spectrum of potential effects on gene transcription, and needs to be investigated context- and stage-specifically. Apart from that it is important to dissect, how context-dependent target-specificity of DNA methylation and demethylation is achieved during neuronal development and aging, and how environmental stimuli mechanistically act on DNA methylation.

**Acknowledgments** This work was funded by the DFG ZI 1224/4-1.

## References


Agirman G, Broix L, Nguyen L (2017) Cerebral cortex development: an outside-in perspective. FEBS Lett 591:3978–3992

Akbarian S, Beeri MS, Haroutunian V (2013) Epigenetic determinants of healthy and diseased brain aging and cognition. JAMA Neurol 70:711–718

Banuelos C, Beas BS, McQuail JA et al (2014) Prefrontal cortical GABAergic dysfunction contributes to age-related working memory impairment. J Neurosci 34:3457–3466

Barter JD, Foster TC (2018) Aging in the brain: new roles of epigenetics in cognitive decline. Neuroscience 24:516–525

Benes FM, McSparren J, Bird ED et al (1991) Deficits in small interneurons in prefrontal and cingulate cortices of schizophrenic and schizoaffective patients. Arch Gen Psychiatry 48:996–1001

Berkyurek AC, Suetake I, Arita K et al (2014) The DNA methyltransferase Dnmt1 directly interacts with the SET and RING finger-associated (SRA) domain of the multifunctional protein Uhrf1 to facilitate accession of the catalytic center to hemi-methylated DNA. J Biol Chem 289:379–386

Bilkei-Gorzo A (2014) Genetic mouse models of brain ageing and Alzheimer's disease. Pharmacol Ther 142:244–257

Bu J, Sathyendra V, Nagykery N et al (2003) Age-related changes in calbindin-D28k, calretinin, and parvalbumin-immunoreactive neurons in the human cerebral cortex. Exp Neurol 182:220–231

Burianova J, Ouda L, Profant O et al (2009) Age-related changes in GAD levels in the central auditory system of the rat. Exp Gerontol 44:161–169

Butt SJ, Fuccillo M, Nery S et al (2005) The temporal and spatial origins of cortical interneurons predict their physiological subtype. Neuron 48:591–604

Butt SJ, Sousa VH, Fuccillo MV et al (2008) The requirement of Nkx2-1 in the temporal specification of cortical interneuron subtypes. Neuron 59:722–732

Canas PM, Duarte JM, Rodrigues RJ et al (2009) Modification upon aging of the density of presynaptic modulation systems in the hippocampus. Neurobiol Aging 30:1877–1884

Catts VS, Weickert CS (2012) Gene expression analysis implicates a death receptor pathway in schizophrenia pathology. PLoS One 7:e35511

Cha CI, Lee YI, Lee EY et al (1997) Age-related changes of VIP, NPY and somatostatin-immunoreactive neurons in the cerebral cortex of aged rats. Brain Res 753:235–244

Chen BJ, Ueberham U, Mills JD et al (2017) RNA sequencing reveals pronounced changes in the noncoding transcriptome of aging synaptosomes. Neurobiol Aging 56:67–77

Cheng CH, Lin YY (2013) Aging-related decline in somatosensory inhibition of the human cerebral cortex. Exp Brain Res 226:145–152

Chestnut BA, Chang Q, Price A et al (2011) Epigenetic regulation of motor neuron cell death through DNA methylation. J Neurosci 31:16619–16636

Chittock EC, Latwiel S, Miller TC et al (2017) Molecular architecture of polycomb repressive complexes. Biochem Soc Trans 45(1):193–205. https://doi.org/10.1042/bst20160173

Chodavarapu RK, Feng S, Bernatavichute YV et al (2010) Relationship between nucleosome positioning and DNA methylation. Nature 466(7304):388–392. https://doi.org/10.1038/nature09147

Civiero L, Cirnaru MD, Beilina A et al (2015) Leucine-rich repeat kinase 2 interacts with p21-activated kinase 6 to control neurite complexity in mammalian brain. J Neurochem 135(6):1242–1256. https://doi.org/10.1111/jnc.13369

Clark BC, Taylor JL (2011) Age-related changes in motor cortical properties and voluntary activation of skeletal muscle. Curr Aging Sci 4(3):192–199

Clements EG, Mohammad HP, Leadem BR et al (2012) DNMT1 modulates gene expression without its catalytic activity partially through its interactions with histone-modifying enzymes. Nucleic Acids Res 40:4334–4346

Clermont PL, Parolia A, Liu HH et al (2016) DNA methylation at enhancer regions: novel avenues for epigenetic biomarker development. Front Biosci (Landmark edition) 21:430–446

Coppieters N, Dieriks BV, Lill C et al (2014) Global changes in DNA methylation and hydroxymethylation in Alzheimer's disease human brain. Neurobiol Aging 35:1334–1344

Corbin JG, Butt SJ (2011) Developmental mechanisms for the generation of telencephalic interneurons. Dev Neurobiol 71:710–732

Costa E, Dong E, Grayson DR, Guidotti A, Ruzicka W, Veldic M (2007) Reviewing the role of DNA (cytosine-5) methyltransferase overexpression in the cortical GABAergic dysfunction associated with psychosis vulnerability. Epigenetics 2(1):29–36

De Marco Garcia NV, Karayannis T, Fishell G (2011) Neuronal activity is required for the development of specific cortical interneuron subtypes. Nature 472:351–355

Druga R (2009) Neocortical inhibitory system. Folia Biol 55(6):201–217

Du J, Johnson LM, Jacobsen SE et al (2015) DNA methylation pathways and their crosstalk with histone methylation. Nat Rev Mol Cell Biol 16:519–532

Fan G, Beard C, Chen RZ et al (2001) DNA hypomethylation perturbs the function and survival of CNS neurons in postnatal animals. J Neurosci 21:788–797

Fan G, Martinowich K, Chin MH et al (2005) DNA methylation controls the timing of astrogliogenesis through regulation of JAK-STAT signaling. Development (Cambridge, England) 132:3345–3356

Feng J, Chang H, Li E et al (2005) Dynamic expression of de novo DNA methyltransferases Dnmt3a and Dnmt3b in the central nervous system. J Neurosci Res 79:734–746

Feng J, Zhou Y, Campbell SL et al (2010) Dnmt1 and Dnmt3a maintain DNA methylation and regulate synaptic function in adult forebrain neurons. Nat Neurosci 13:423–430

Flandin P, Kimura S, Rubenstein JL (2010) The progenitor zone of the ventral medial ganglionic eminence requires Nkx2-1 to generate most of the globus pallidus but few neocortical interneurons. J Neurosci 30:2812–2823

Flood DG, Coleman PD (1988) Neuron numbers and sizes in aging brain: comparisons of human, monkey, and rodent data. Neurobiol Aging 9:453–463

Franco SJ, Muller U (2013) Shaping our minds: stem and progenitor cell diversity in the mammalian neocortex. Neuron 77:19–34

Gelman DM, Martini FJ, Nobrega-Pereira S et al (2009) The embryonic preoptic area is a novel source of cortical GABAergic interneurons. J Neurosci 29:9380–9389

Gelman D, Griveau A, Dehorter N et al (2011) A wide diversity of cortical GABAergic interneurons derives from the embryonic preoptic area. J Neurosci 31:16570–16580

Gidon A, Segev I (2012) Principles governing the operation of synaptic inhibition in dendrites. Neuron 75:330–341

Guidotti A, Auta J, Chen Y et al (2011) Epigenetic GABAergic targets in schizophrenia and bipolar disorder. Neuropharmacology 60:1007–1016

Guo JU, Ma DK, Mo H et al (2011) Neuronal activity modifies the DNA methylation landscape in the adult brain. Nat Neurosci 14:1345–1351

Guo JU, Su Y, Shin JH et al (2014) Distribution, recognition and regulation of non-CpG methylation in the adult mammalian brain. Nat Neurosci 17:215–222

Haberman RP, Quigley CK, Gallagher M (2012) Characterization of CpG island DNA methylation of impairment-related genes in a rat model of cognitive aging. Epigenetics 7:1008–1019

Halder R, Hennion M, Vidal RO et al (2016) DNA methylation changes in plasticity genes accompany the formation and maintenance of memory. Nat Neurosci 19:102–110

Hansen DV, Lui JH, Parker PR et al (2010) Neurogenic radial glia in the outer subventricular zone of human neocortex. Nature 464:554–561

Hashimshony T, Zhang J, Keshet I et al (2003) The role of DNA methylation in setting up chromatin structure during development. Nat Genet 34:187–192

Hayashi M, Yamashita A, Shimizu K (1997) Somatostatin and brain-derived neurotrophic factor mRNA expression in the primate brain: decreased levels of mRNAs during aging. Brain Res 749:283–289

Hensch TK (2005) Critical period plasticity in local cortical circuits. Nat Rev Neurosci 6:877–888
Hu JS, Vogt D, Sandberg M et al (2017) Cortical interneuron development: a tale of time and space. Development (Cambridge, England) 144:3867–3878
Hua T, Kao C, Sun Q et al (2008) Decreased proportion of GABA neurons accompanies age-related degradation of neuronal function in cat striate cortex. Brain Res Bull 75:119–125
Hutnick LK, Golshani P, Namihira M et al (2009) DNA hypomethylation restricted to the murine forebrain induces cortical degeneration and impairs postnatal neuronal maturation. Hum Mol Genet 18:2875–2888
Ianov L, Rani A, Beas BS et al (2016) Transcription profile of aging and cognition-related genes in the medial prefrontal cortex. Front Aging Neurosci 8:113
Ianov L, Riva A, Kumar A et al (2017) DNA methylation of synaptic genes in the prefrontal cortex is associated with aging and age-related cognitive impairment. Front Aging Neurosci 9:249
Inan M, Welagen J, Anderson SA (2012) Spatial and temporal bias in the mitotic origins of somatostatin- and parvalbumin-expressing interneuron subgroups and the chandelier subtype in the medial ganglionic eminence. Cerebral Cortex (New York, NY) 1991(22):820–827
Irizarry RA, Ladd-Acosta C, Wen B et al (2009) The human colon cancer methylome shows similar hypo- and hypermethylation at conserved tissue-specific CpG island shores. Nat Genet 41:178–186
Jang HS, Shin WJ, Lee JE et al (2017) CpG and non-CpG methylation in epigenetic gene regulation and brain function. Genes 8:148
Jiang CH, Tsien JZ, Schultz PG et al (2001) The effects of aging on gene expression in the hypothalamus and cortex of mice. Proc Natl Acad Sci U S A 98:1930–1934
Jin B, Robertson KD (2013) DNA methyltransferases, DNA damage repair, and cancer. Adv Exp Med Biol 754:3–29
Jones PA (2012) Functions of DNA methylation: islands, start sites, gene bodies and beyond. Nat Rev Genet 13:484–492
Jucker M, Ingram DK (1997) Murine models of brain aging and age-related neurodegenerative diseases. Behav Brain Res 85:1–26
Kadriu B, Guidotti A, Chen Y et al (2012) DNA methyltransferases1 (DNMT1) and 3a (DNMT3a) colocalize with GAD67-positive neurons in the GAD67-GFP mouse brain. J Comp Neurol 520:1951–1964
Keleshian VL, Modi HR, Rapoport SI et al (2013) Aging is associated with altered inflammatory, arachidonic acid cascade, and synaptic markers, influenced by epigenetic modifications, in the human frontal cortex. J Neurochem 125:63–73
Kennedy AJ, Sweatt JD (2016) Drugging the methylome: DNA methylation and memory. Crit Rev Biochem Mol Biol 51:185–194
Kiecker C, Lumsden A (2005) Compartments and their boundaries in vertebrate brain development. Nat Rev Neurosci 6:553–564
Kim H, Jang WY, Kang MC et al (2016) TET1 contributes to neurogenesis onset time during fetal brain development in mice. Biochem Biophys Res Commun 471:437–443
Kozlenkov A, Roussos P, Timashpolsky A et al (2014) Differences in DNA methylation between human neuronal and glial cells are concentrated in enhancers and non-CpG sites. Nucleic Acids Res 42:109–127
Kozlenkov A, Wang M, Roussos P et al (2016) Substantial DNA methylation differences between two major neuronal subtypes in human brain. Nucleic Acids Res 44:2593–2612
Kraus TF, Kilinc S, Steinmaurer M et al (2016) Profiling of methylation and demethylation pathways during brain development and ageing. J Neural Transm (Vienna) 123:189–203
Kulis M, Queiros AC, Beekman R et al (2013) Intragenic DNA methylation in transcriptional regulation, normal differentiation and cancer. Biochim Biophys Acta 1829:1161–1174
Kumar R, Sanawar R, Li X, Li F (2017) Structure, biochemistry, and biology of PAK kinases. Gene 605:20–31
Lande-Diner L, Zhang J, Ben-Porath I et al (2007) Role of DNA methylation in stable gene repression. J Biol Chem 282:12194–12200

Lardenoije R, Iatrou A, Kenis G et al (2015) The epigenetics of aging and neurodegeneration. Prog Neurobiol 131:21–64
Lee SM, Choi WY, Lee J et al (2015) The regulatory mechanisms of intragenic DNA methylation. Epigenomics 7:527–531
Lee JH, Park SJ, Nakai K (2017) Differential landscape of non-CpG methylation in embryonic stem cells and neurons caused by DNMT3s. Sci Rep 7:11295
Letzkus JJ, Wolff SB, Luthi A (2015) Disinhibition, a circuit mechanism for associative learning and memory. Neuron 88:264–276
Levenson JM, Roth TL, Lubin FD, Miller CA, Huang IC, Desai P, Malone LM, Sweatt JD (2006) Evidence that DNA (cytosine-5) methyltransferase regulates synaptic plasticity in the hippocampus. J Biol Chem 281(23):15763–15773
Lewis DA (2012) Cortical circuit dysfunction and cognitive deficits in schizophrenia--implications for preemptive interventions. Eur J Neurosci 35:1871–1878
Li W, Prazak L, Chatterjee N et al (2013) Activation of transposable elements during aging and neuronal decline in Drosophila. Nat Neurosci 16:529–531
Liao C, Han Q, Ma Y et al (2016) Age-related gene expression change of GABAergic system in visual cortex of rhesus macaque. Gene 590:227–233
Liguz-Lecznar M, Lehner M, Kaliszewska A et al (2015) Altered glutamate/GABA equilibrium in aged mice cortex influences cortical plasticity. Brain Struct Funct 220:1681–1693
Lin LC, Sibille E (2013) Reduced brain somatostatin in mood disorders: a common pathophysiological substrate and drug target? Front Pharmacol 4:110
Ling LL, Hughes LF, Caspary DM (2005) Age-related loss of the GABA synthetic enzyme glutamic acid decarboxylase in rat primary auditory cortex. Neuroscience 132:1103–1113
Lister R, Mukamel EA (2015) Turning over DNA methylation in the mind. Front Neurosci 9:252
Lister R, Pelizzola M, Dowen RH et al (2009) Human DNA methylomes at base resolution show widespread epigenomic differences. Nature 462:315–322
Lister R, Mukamel EA, Nery JR et al (2013) Global epigenomic reconfiguration during mammalian brain development. Science 341:1237905
Loerch PM, Lu T, Dakin KA et al (2008) Evolution of the aging brain transcriptome and synaptic regulation. PLoS One 3:e3329
Mangold CA, Masser DR, Stanford DR et al (2017) CNS-wide sexually dimorphic induction of the major histocompatibility complex 1 pathway with aging. J Gerontol A Biol Sci Med Sci 72:16–29
Margueron R, Reinberg D (2011) The polycomb complex PRC2 and its mark in life. Nature 469:343–349
Marin O (2012) Interneuron dysfunction in psychiatric disorders. Nat Rev Neurosci 13:107–120
Martynoga B, Drechsel D, Guillemot F (2012) Molecular control of neurogenesis: a view from the mammalian cerebral cortex. Cold Spring Harb Perspect Biol 4(10)
Mastroeni D, Grover A, Delvaux E et al (2010) Epigenetic changes in Alzheimer's disease: decrements in DNA methylation. Neurobiol Aging 31:2025–2037
Matrisciano F, Tueting P, Dalal I et al (2013) Epigenetic modifications of GABAergic interneurons are associated with the schizophrenia-like phenotype induced by prenatal stress in mice. Neuropharmacology 68:184–194
McGowan PO, Sasaki A, D'Alessio AC et al (2009) Epigenetic regulation of the glucocorticoid receptor in human brain associates with childhood abuse. Nat Neurosci 12:342–348
McKinney BC, Lin CW, Oh H et al (2015) Hypermethylation of BDNF and SST genes in the orbital frontal cortex of older individuals: a putative mechanism for declining gene expression with age. Neuropsychopharmacology 40:2604–2613
McKinsey GL, Lindtner S, Trzcinski B et al (2013) Dlx1&2-dependent expression of Zfhx1b (Sip1, Zeb2) regulates the fate switch between cortical and striatal interneurons. Neuron 77:83–98
Meadows JP, Guzman-Karlsson MC, Phillips S et al (2015) DNA methylation regulates neuronal glutamatergic synaptic scaling. Sci Signal 8(382):ra61

Meadows JP, Guzman-Karlsson MC, Phillips S et al (2016) Dynamic DNA methylation regulates neuronal intrinsic membrane excitability. Sci Signal 9:ra83

Merot Y, Retaux S, Heng JI (2009) Molecular mechanisms of projection neuron production and maturation in the developing cerebral cortex. Semin Cell Dev Biol 20:726–734

Metin C, Baudoin JP, Rakic S et al (2006) Cell and molecular mechanisms involved in the migration of cortical interneurons. Eur J Neurosci 23:894–900

Mi Y, Gao X, Dai J et al (2015) A novel function of TET2 in CNS: sustaining neuronal survival. Int J Mol Sci 16:21846–21857

Miettinen R, Sirvio J, Riekkinen P et al (1993) Neocortical, hippocampal and septal parvalbumin- and somatostatin-containing neurons in young and aged rats: correlation with passive avoidance and water maze performance. Neuroscience 53:367–378

Miyoshi G, Hjerling-Leffler J, Karayannis T et al (2010) Genetic fate mapping reveals that the caudal ganglionic eminence produces a large and diverse population of superficial cortical interneurons. J Neurosci 30:1582–1594

Mo A, Mukamel EA, Davis FP et al (2015) Epigenomic signatures of neuronal diversity in the mammalian brain. Neuron 86:1369–1384

Morris HM, Hashimoto T, Lewis DA (2008) Alterations in somatostatin mRNA expression in the dorsolateral prefrontal cortex of subjects with schizophrenia or schizoaffective disorder. Cereb Cortex 18:1575–1587

Morris MJ, Na ES, Autry AE et al (2016) Impact of DNMT1 and DNMT3a forebrain knockout on depressive- and anxiety like behavior in mice. Neurobiol Learn Mem 135:139–145

Moyer JR Jr, Furtak SC, McGann JP et al (2011) Aging-related changes in calcium-binding proteins in rat perirhinal cortex. Neurobiol Aging 32:1693–1706

Murthy S, Niquille M, Hurni N et al (2014) Serotonin receptor 3A controls interneuron migration into the neocortex. Nat Commun 5:5524

Na KS, Won E, Kang J et al (2016) Brain-derived neurotrophic factor promoter methylation and cortical thickness in recurrent major depressive disorder. Sci Rep 6:21089

Nardone S, Sams DS, Zito A et al (2017) Dysregulation of cortical neuron DNA methylation profile in autism spectrum disorder. Cereb Cortex 27:5739–5754

Ning X, Shi Z, Liu X et al (2015) DNMT1 and EZH2 mediated methylation silences the microRNA-200b/a/429 gene and promotes tumor progression. Cancer Lett 359:198–205

Nobrega-Pereira S, Kessaris N, Du T et al (2008) Postmitotic Nkx2-1 controls the migration of telencephalic interneurons by direct repression of guidance receptors. Neuron 59:733–745

Noguchi H, Kimura A, Murao N et al (2015) Expression of DNMT1 in neural stem/precursor cells is critical for survival of newly generated neurons in the adult hippocampus. Neurosci Res 95:1–11

Noguchi H, Kimura A, Murao N et al (2016a) Prenatal deletion of DNA methyltransferase 1 in neural stem cells impairs neurogenesis and causes anxiety-like behavior in adulthood. Neurogenesis (Austin) 3:e1232679

Noguchi H, Murao N, Kimura A et al (2016b) DNA methyltransferase 1 is indispensable for development of the hippocampal dentate gyrus. J Neurosci 36:6050–6068

Nonaka-Kinoshita M, Reillo I, Artegiani B et al (2013) Regulation of cerebral cortex size and folding by expansion of basal progenitors. EMBO J 32:1817–1828

Numata S, Ye T, Hyde TM et al (2012) DNA methylation signatures in development and aging of the human prefrontal cortex. Am J Hum Genet 90:260–272

Oberlander TF, Weinberg J, Papsdorf M et al (2008) Prenatal exposure to maternal depression, neonatal methylation of human glucocorticoid receptor gene (NR3C1) and infant cortisol stress responses. Epigenetics 3:97–106

Ouda L, Druga R, Syka J (2008) Changes in parvalbumin immunoreactivity with aging in the central auditory system of the rat. Exp Gerontol 43:782–789

Ouda L, Burianova J, Syka J (2012) Age-related changes in calbindin and calretinin immunoreactivity in the central auditory system of the rat. Exp Gerontol 47:497–506

Ouellet L, de Villers-Sidani E (2014) Trajectory of the main GABAergic interneuron populations from early development to old age in the rat primary auditory cortex. Front Neuroanat 8:40

Penner MR, Parrish RR, Hoang LT et al (2016) Age-related changes in Egr1 transcription and DNA methylation within the hippocampus. Hippocampus 26:1008–1020

Pensold D, Symmank J, Hahn A et al (2017) The DNA methyltransferase 1 (DNMT1) controls the shape and dynamics of migrating POA-derived interneurons fated for the murine cerebral cortex. Cereb Cortex 27:5696–5714

Pfisterer U, Khodosevich K (2017) Neuronal survival in the brain: neuron type-specific mechanisms. Cell Death Dis 8:e2643

Pinney SE (2014) Mammalian non-CpG methylation: stem cells and beyond. Biology 3:739–751

Pishva E, Rutten BPF, van den Hove D (2017) DNA methylation in major depressive disorder. Adv Exp Med Biol 978:185–196

Potier B, Jouvenceau A, Epelbaum J et al (2006) Age-related alterations of GABAergic input to CA1 pyramidal neurons and its control by nicotinic acetylcholine receptors in rat hippocampus. Neuroscience 142:187–201

Pouille F, Watkinson O, Scanziani M et al (2013) The contribution of synaptic location to inhibitory gain control in pyramidal cells. Phys Rep 1:e00067

Purkait S, Sharma V, Kumar A et al (2016) Expression of DNA methyltransferases 1 and 3B correlates with EZH2 and this 3-marker epigenetic signature predicts outcome in glioblastomas. Exp Mol Pathol 100:312–320

Ramesh V, Bayam E, Cernilogar FM et al (2016) Loss of Uhrf1 in neural stem cells leads to activation of retroviral elements and delayed neurodegeneration. Genes Dev 30:2199–2212

Rhee KD, Yu J, Zhao CY et al (2012) Dnmt1-dependent DNA methylation is essential for photoreceptor terminal differentiation and retinal neuron survival. Cell Death Dis 3:e427

Roth TL, Zoladz PR, Sweatt JD et al (2011) Epigenetic modification of hippocampal Bdnf DNA in adult rats in an animal model of post-traumatic stress disorder. J Psychiatr Res 45:919–926

Rozycka A, Liguz-Lecznar M (2017) The space where aging acts: focus on the GABAergic synapse. Aging Cell 16:634–643

Rubin AN, Kessaris N (2013) PROX1: a lineage tracer for cortical interneurons originating in the lateral/caudal ganglionic eminence and preoptic area. PLoS One 8:e77339

Ruzicka WB, Zhubi A, Veldic M et al (2007) Selective epigenetic alteration of layer I GABAergic neurons isolated from prefrontal cortex of schizophrenia patients using laser-assisted microdissection. Mol Psychiatry 12:385–397

Sandberg M, Flandin P, Silberberg S et al (2016) Transcriptional networks controlled by NKX2-1 in the development of forebrain GABAergic neurons. Neuron 91:1260–1275

Sharma RP, Tun N, Grayson DR (2008) Depolarization induces downregulation of DNMT1 and DNMT3a in primary cortical cultures. Epigenetics 3:74–80

Sharma A, Klein SS, Barboza L et al (2016) Principles governing DNA methylation during neuronal lineage and subtype specification. J Neurosci 36:1711–1722

Shetty AK, Turner DA (1998) Hippocampal interneurons expressing glutamic acid decarboxylase and calcium-binding proteins decrease with aging in Fischer 344 rats. J Comp Neurol 394:252–269

Siegmund KD, Connor CM, Campan M et al (2007) DNA methylation in the human cerebral cortex is dynamically regulated throughout the life span and involves differentiated neurons. PLoS One 2:e895

Smallwood A, Esteve PO, Pradhan S et al (2007) Functional cooperation between HP1 and DNMT1 mediates gene silencing. Genes Dev 21:1169–1178

So AY, Jung JW, Lee S et al (2011) DNA methyltransferase controls stem cell aging by regulating BMI1 and EZH2 through microRNAs. PLoS One 6:e19503

Southwell DG, Paredes MF, Galvao RP et al (2012) Intrinsically determined cell death of developing cortical interneurons. Nature 491:109–113

Stanley DP, Shetty AK (2004) Aging in the rat hippocampus is associated with widespread reductions in the number of glutamate decarboxylase-67 positive interneurons but not interneuron degeneration. J Neurochem 89:204–216

Stanley EM, Fadel JR, Mott DD (2012) Interneuron loss reduces dendritic inhibition and GABA release in hippocampus of aged rats. Neurobiol Aging 33:431 e431–431 e413
Sundman-Eriksson I, Allard P (2006) Age-correlated decline in [3H]tiagabine binding to GAT-1 in human frontal cortex. Aging Clin Exp Res 18:257–260
Sweatt JD (2016) Dynamic DNA methylation controls glutamate receptor trafficking and synaptic scaling. J Neurochem 137:312–330
Symmank J, Zimmer G (2017) Regulation of neuronal survival by DNA methyltransferases. Neural Regen Res 12(11):1768–1775
Symmank J, Bayer C, Schmidt C et al (2018) DNMT1 modulates interneuron morphology by regulating Pak6 expression through crosstalk with histone modifications. Epigenetics 13:536–556
Symmank J, Gölling V, Gerstmann K, Zimmer G (2019) The transcription factor LHX1 regulates the survival and directed migration of POA-derived cortical interneurons. Cereb Cortex 29(4):1644–1658
van den Berghe V, Stappers E, Vandesande B et al (2013) Directed migration of cortical interneurons depends on the cell-autonomous action of Sip1. Neuron 77:70–82
Veldic M, Caruncho HJ, Liu WS et al (2004) DNA-methyltransferase 1 mRNA is selectively overexpressed in telencephalic GABAergic interneurons of schizophrenia brains. Proc Natl Acad Sci U S A 101:348–353
Veldic M, Guidotti A, Maloku E et al (2005) In psychosis, cortical interneurons overexpress DNA-methyltransferase 1. Proc Natl Acad Sci U S A 102:2152–2157
Vinson C, Chatterjee R (2012) CG methylation. Epigenomics 4:655–663
Vire E, Brenner C, Deplus R et al (2006) The Polycomb group protein EZH2 directly controls DNA methylation. Nature 439:871–874
Wu X, Zhang Y (2017) TET-mediated active DNA demethylation: mechanism, function and beyond. Nat Rev Genet 18:517–534
Xin YJ, Yuan B, Yu B et al (2015) Tet1-mediated DNA demethylation regulates neuronal cell death induced by oxidative stress. Sci Rep 5:7645
Xu Q, de la Cruz E, Anderson SA (2003) Cortical interneuron fate determination: diverse sources for distinct subtypes? Cereb Cortex 13:670–676
Yang J, Ji WY, Qu YR et al (2011) DNA methylation and histone modification relate to RASSF1A gene deletion in laryngeal carcinoma tissues. Zhonghua Er Bi Yan Hou Tou Jing Wai Ke Za Zhi 46:308–312
Zhu H, Wang G, Qian J (2016) Transcription factors as readers and effectors of DNA methylation. Nat Rev Genet 17:551–565
Zimmer-Bensch G (2018) Diverse facets of cortical interneuron migration regulation – implications of neuronal activity and epigenetics. Brain Res 1700:160–169
Zovkic IB, Guzman-Karlsson MC, Sweatt JD (2013) Epigenetic regulation of memory formation and maintenance. Learn Mem 20:61–74

# The Methylome of Bipolar Disorder: Evidence from Human and Animal Studies

**Consuelo Walss-Bass and Gabriel R. Fries**

## Contents



**Abstract** Bipolar disorder is a chronic and often severe psychiatric disorder with a complex multifactorial heritability. While it is known that the pathophysiology of bipolar disorder involves the interaction of several genetic variants, each of small effect size, current molecular studies have failed to explain the high heritability of bipolar disorder based only on single nucleotide polymorphisms. Based on the known key role of the environment in modulating the risk of bipolar disorder, epigenetic mechanisms have been proposed as mediators of gene-environment interactions in this disorder. In particular, several studies have identified DNA methylation alterations that interact with susceptibility-conferring genotypes to modulate the expression of neurodevelopment genes, ultimately contributing to bipolar disorder pathogenesis and/or progression. This chapter aims to review recent genome-wide findings of alterations in DNA methylation in brain and blood of patients with bipolar disorder and in relevant animal models. In addition, it discusses the potential clinical implications, limitations, and future directions of the field of bipolar disorder epigenetics.

**Keywords** Bipolar disorder · Epigenetics · Genetics · DNA methylation · Brain · Blood · Biomarker · Neurodevelopment

---

C. Walss-Bass (✉) · G. R. Fries
Department of Psychiatry and Behavioral Sciences, McGovern Medical School, The University of Texas Health Science Center at Houston (UTHealth), Houston, TX, USA
e-mail: Consuelo.WalssBass@uth.tmc.edu

 
S. Jurga, J. Barciszewski (eds.), *The DNA, RNA, and Histone Methylomes*, RNA Technologies, https://doi.org/10.1007/978-3-030-14792-1_7

# 1 Bipolar Disorder as an Epigenetic Neurodevelopmental Disorder

Bipolar disorder is a chronic, recurrent, and often severe psychiatric disorder that affects more than 1% of the world's population (Grande et al. 2016). According to the Diagnostic and Statistical Manual of Mental Disorders, fifth Edition, bipolar disorder can be divided into different diagnostic types based on longitudinal course and severity of mood disturbance. For instance, bipolar disorder type I is characterized by the occurrence of at least one manic episode, and bipolar disorder type II requires at least one hypomanic episode and one major depressive episode for its diagnosis (Association 2013). Given the high occurrence of bipolar disorder and medical comorbidities, in addition to the typical prevalence of cognitive and functional impairments, bipolar disorder has been ranked among the main causes of disability among young people (Grande et al. 2016). In addition, only a small to moderate fraction of bipolar disorder patients adequately respond to the medications currently available, with evidence of high heterogeneity in overall responsiveness among patients (Garnham et al. 2007; Routhieaux et al. 2018). The study of bipolar disorder's pathophysiology is, therefore, of uttermost importance in the search for novel and more efficacious medications that can significantly reduce the burden inflicted upon patients and improve their quality of lives.

The efforts in identification of causative genes in bipolar disorder, although to date have largely focused on genetic studies, have in recent years turned towards understanding how environmental influences on gene function modulate risk for psychopathology. The fact that genome-wide association studies (GWAS) have explained only a small fraction of bipolar disorder's heritability (Gershon et al. 2011; Kerner 2015), which is estimated to be between 60% and 80% from twin and adoptions studies (McGuffin et al. 2003; Kieseppa et al. 2004), argues for epigenetic modifications such as DNA methylation to account for at least part of the heritability. This idea is supported by our current knowledge of how epigenetic control of gene expression, particularly via DNA methylation, can be altered by environmental influences, is highly regulated throughout development, and may be inherited (Fries et al. 2016). Indeed, the establishment and maintenance of methylation *loci* is crucial for central nervous system differentiation and for regulation of brain processes, such as synaptic plasticity, learning, and memory (Grayson and Guidotti 2013). A plethora of studies in preclinical models and in humans have suggested that DNA methylation alterations can be induced early in life and sometimes maintained into adulthood (Jawahar et al. 2015; Mitchell et al. 2016), meeting the criteria for a neurodevelopmental marker influenced by environmental exposures. This dynamic epigenetic regulation of gene expression during development provides a biological mechanism for the premise that bipolar disorder is a neurodevelopmental disorder in which the interaction of inherited genetic susceptibility and epigenetic processes modulates brain plasticity during development, leading to brain functional abnormalities and manifestation of the disorder. This neurodevelopmental hypothesis is further supported by other studies, as well (Buoli et al. 2017; Muhleisen et al. 2018).

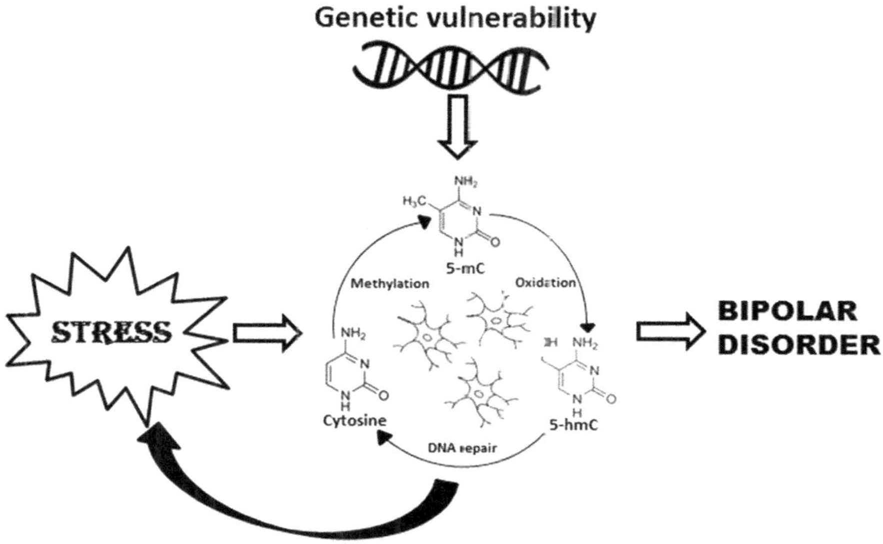


**Fig. 1** Gene environment interacations early in life lead to alterations of epigenetic processes in brain cells that are worsened by further exposure to stress, eventually leading to brain dysfuction and manifestation of bipolar disorder

Accounting for both genetic and epigenetic modifications, as well as a role for environmental insults, a model for bipolar disorder development would be as follows: inherited genetic mutations in stress response pathways lead to an inability to respond appropriately to environmental insults early in life, causing epigenetic modifications and biological alterations that lead to disruption of normal cell function in the brain. These early abnormalities then initiate a feedback loop of increased sensitivity to stress, such that exposure to further stress causes additional and more severe cellular abnormalities, eventually leading to global brain system dysfunction and manifestation of disease symptoms (Fig. 1) (Fries et al. 2012; Walss-Bass et al. 2018). An overall understanding of the genomic and environmental factors influencing disease manifestation could lead to an early detection of brain and behavioral abnormalities, and importantly, may lead to identification of specific treatments that could alter disease course or reduce severity.

## 2 Modulation of Brain Function by DNA Methylation in Bipolar Disorder

Studies showing brain structural abnormalities in individuals with first-episodes of psychosis or mania support the hypothesis of bipolar disorder as a neurodevelopmental disorder (Fornito et al. 2007, 2009). In further epigenetic support of this idea, using a genome-scale hairpin bisulfite sequencing approach in human and mouse frontal cortices at different developmental stages, it has been demonstrated that cell-specific methylation increases dramatically during early stages of brain development and the methylated *loci* are enriched for GWAS variants associated with neurological disorders including bipolar disorder, suggesting that early brain methylation may be altered in this disorder (Sun et al. 2016). Of interest, overall hypomethylation has been observed in frontal cortex of bipolar disorder subjects compared to healthy controls, preferentially targeting the terminal ends of the chromosomes (Xiao et al. 2014).

Strengthening the aforementioned hypothesis of bipolar disorder as a neurodevelopmental disorder, differentially methylated genes identified in postmortem brain studies of bipolar disorder play important roles in neuronal development (Table 1). In a genome-wide analysis of twins discordant for bipolar disorder, a study identified *ST6GALNAC1*, which codes for a protein that transfers sialic acid to O-linked N-acetylgalactosamine residues, to be hypomethylated in peripheral blood from affected twins as well as in an independent sample of postmortem brain tissue (Dempster et al. 2011). Another study observed hypomethylation of the catechol-O-methyl transferase gene (*COMT*) promoter region in the frontal lobe of patients, together with a tendency for the enrichment of the Val allele of the *COMT* VAL158Met polymorphism (Abdolmaleky et al. 2006). Using a genome-wide approach in postmortem brain, a study identified differential methylation in the frontal cortex of genes involved in brain development, glutamatergic and GABAergic neurotransmission, mitochondrial function and stress response, all pathways that are implicated in BD (Mill et al. 2008). One of the strongest findings in this study was hypomethylation of the HLA complex group 9 (*HCG9*), which was confirmed in a subsequent study (Kaminsky et al. 2012).

In regards to which specific environmental stressors lead to DNA methylation changes that modulate risk for psychopathology, several lines of evidence show the harmful effect of early life stress, particularly childhood maltreatment, on the course of bipolar disorder leading to earlier age of onset and greater symptom severity compared to patients without such history (Levandowski and Grassi-Oliveira 2018). In a landmark study, researchers from the University of Montreal were the first to show that early life experiences influence behavior and mental health across the lifespan (Fish et al. 2004). Using a rodent model, the authors showed that lack of maternal care leads to alternations in methylation of hippocampal *NGFI-A*, a gene involved in stress response via the glucocorticoid receptor pathway. Another study on the association between childhood trauma and genetic variants of the gene coding for the FK506-binding protein 51 (FKBP51, which is also involved in regulation of stress response via the glucocorticoid receptor pathway) showed that childhood trauma can lead to a genotype-specific DNA methylation change of this gene, ultimately determining changes in the response to stress in adults (Klengel et al. 2013). These studies were the first to empirically demonstrate the role of DNA methylation in mediating the interaction between gene and environment. In a more recent study using methylome analyses in human and non-human models of early life stress, Ankyrin-3, a gene strongly associated with bipolar disorder by GWAS,

**Table 1** Summary of genome-wide methylation studies in bipolar disorder

| Source/type of tissue | Finding | Study |
|---|---|---|
| HEK293 cells | Valproate induces histone acetylation and activated DNA demethylation in the same gene systems | Milutinovic et al. (2007) |
| Human frontal cortex | Differential methylation of genes involved in brain development, glutamatergic and GABAergic neurotransmission, mitochondrial function and stress response | Mill et al. (2008) |
| Lymphoblasts | Altered methylation of *SMS* and *PPIEL* in BD twins compared to control twins | Kuratomi et al. (2008) |
| Rat primary astrocytes | Lamotrigine induced no changes in global DNA methylation | Perisic et al. (2010) |
| Mouse embryonic stem cells | Lithium treatment resulted in hypomethylation of Igf2, Igf2r, and H19 in mouse embryonic stem cells | Popkie et al. (2010) |
| Human blood and postmortem brain | *ST6GALNAC1* hypomethylation | Dempster et al. (2011) |
| Lymphoblasts and postmortem brain | Hypomethylation of SLC6A4 | Sugawara et al. (2011) |
| Neuroblastoma cell line | Carbamazapine induced hypermethylation of 64 genes and hypomethylation of 14 genes | Asai et al. (2013) |
| Human cerebellum | Enrichment of cis regulatory loci on DNA methylation among top BD susceptibility variants | Gamazon et al. (2013) |
| Human brain methylation and GWAS datasets | Allele-specific methylation of genes in ion channel pathways | Chuang et al. (2013) |
| Human blood and postmortem brain | SNPs that influence DNA methylation are more likely to be located in microRNA binding sites | Smith et al. (2014) |
| Human cerebellum | Differentially expressed genes with an aberrant methylation pattern | Chen et al. (2014) |
| Lymphoblasts | Decreased global methylation in lithium responders | Huzayyin et al. (2014) |
| Human frontal cortex | Overall hypomethylation in BD | Xiao et al. (2014) |
| Human blood cells | Thousands of differentially methylated regions, preferentially located in promoters, 3′-UTR and 5′-UTR of genes | Li et al. (2015) |
| Human blood cells | Hypomethylation of a locus near CYP11A1 | Sabunciyan et al. (2015) |
| Hippocampus | Circuit-specific DNA methylation changes | Ruzicka et al. (2015) |
| Dorsolateral prefrontal cortex | Differentially methylated regions are distributed preferentially across miRNA introns | Zhao et al. (2015) |
| Rat hippocampus | Lithium and valproate cause common epigenetic effects on the leptin receptor gene in rats | Lee et al. (2015) |
| Human and mouse frontal cortex | Cell-subset specific methylation across development was enriched for BD GWAS variants | Sun et al. (2016) |
| Rats, non-human primates and human | *Ank3* methylation as a marker for early-life stress and vulnerability to psychiatric disorders | Luoni et al. (2016) |

(continued)

**Table 1** (continued)

| Source/type of tissue | Finding | Study |
|---|---|---|
| Human blood cells | Valproate induced a significant alteration in overall methylation in BD patients | Houtepen et al. (2016) |
| Dorsolateral prefrontal cortex | SNPs associated with gene expression, DNA methylation and histone acetylation | Ng et al. (2017) |
| Human white blood cells | Altered methylation and expression of genes related to the glucocorticoid receptor pathway in offspring of BD patients | Fries et al. (2017b) |
| Hippocampal GABAergic interneurons | Differentially methylated regions associated with multiple zinc finger genes and WNT signaling factors | Ruzicka et al. (2018) |
| Human blood cells | Hypomethylation of *FAM63* and an intergenetic region on chromosome 16. Hypermethylation of *TBC1D22A*. | Sugawara et al. (2018) |

*Ank3* ankyrin 3; *BD* bipolar disorder; *CYP11A1* cytochrome P450, family 11, subfamily A, polypeptide 1; *FAM63* family with sequence similarity 63, member B; *GWAS* genome-wide association studies; *PPIEL* peptidylprolyl isomerase E-like; *SLC6A4* solute carrier family 6 member 4; *SMS* spermine synthase; *SNP* single nucleotide polymorphism; *TBC1D22A* TBC1 domain family member 22A; *UTR* untranslated region

was identified as a molecular marker of early-life stress and vulnerability to psychiatric disorders (Luoni et al. 2016). Altogether, these studies suggest a possible mechanism for epigenetic modifications in mediating gene-environment interactions and susceptibility to mood disorders.

## 3 Comprehensive Integration Analyses of Genomic Data

The complexity of the bipolar disorder phenotype, together with mostly small sample sizes (particularly in postmortem brain), has limited the ability to identify gene methylation differences that achieve genome-wide significance ($p$-value $<1 \times 10^{-8}$ after correction for multiple comparisons). Integration of multiple 'omic' datasets (i.e., transcriptome, methylome, genome) obtained from the same individuals, followed by pathway analysis, has recently been demonstrated as a powerful strategy to overcome sample size limitations and disease complexity as specific genes may be associated with the disorder in more than one dataset, and several genes may be identified to be part of specific biological pathways or gene networks, thus enhancing the ability to explore the underlying molecular mechanisms for bipolar disorder. For example, a study utilizing allele-specific methylation (ASM), where specific genetic variants are known to influence brain DNA methylation, identified ion channel related pathways associated with bipolar disorder in two Caucasian populations (Chuang et al. 2013). A study exploring the effects of bipolar disorder susceptibility variants previously identified by GWAS on gene expression and DNA methylation in human cerebellum samples found an enrichment of *cis* regulatory loci on mRNA expression and DNA methylation among the top susceptibility variants, and SNPs that regulate gene expression are different from those that regulate methylation of the same gene. Further, the use of this information to reduce the number of variants studied enhanced the ability to detect significant associations with bipolar disorder (Gamazon et al. 2013). Another study applied quantitative trait locus (xQTL) analyses to integrate RNA sequence, DNA methylation, and histone acetylation data from the dorsolateral prefrontal cortex of 411 older adults, prioritizing by cell type. The study identified SNPs that significantly associated with alterations at all three levels and demonstrated that SNP effects on RNA expression are fully mediated by epigenetic features in 9% of the identified loci (Ng et al. 2017). A study integrating genome-wide methylation and expression data obtained from the cerebellum of patients with schizophrenia and bipolar disorder identified differentially expressed genes with an aberrant methylation pattern including phosphoinositide-3-kinase, regulatory subunit 1 (*PIK3R1*), butyrophilin, subfamily 3, member A3 (*BTN3A3*), nescient helix-loop-helix 1 (*NHLH1*), and solute carrier family 16, member 7 (*SLC16A7*) in patients with psychosis (Chen et al. 2014). By performing an integrative analysis of methylation and RNA sequencing data on brain samples from subjects with psychosis and controls, a study found that differentially methylated regions (DMRs) were distributed preferentially across introns that were enriched for regulatory elements such as enhancers, as well as introns that overlapped with microRNAs, thus providing novel mechanisms by which DNA methylation may mediate gene expression changes (Zhao et al. 2015). Similarly, a previous study had observed that SNPs that influence DNA methylation were more likely to be located in microRNA binding sites across populations with different ancestries and developmental stages, in both blood and brain tissue. Of interest, this study found that, compared with a random group of SNPs, those influencing DNA methylation were overrepresented among SNPs previously associated with bipolar disorder (Smith et al. 2014).

## 4 Blood as a Proxy of Brain

Although of high relevance and significance to directly inform neuropathophysiological mechanisms involved in bipolar disorder, postmortem studies are inherently limited due to technical issues (postmortem interval, comorbidities, and the influence of the cause of death on biological findings, to name a few) as well as the impossibility of linking findings with longitudinal assessments in large samples of patients. Because of this, several studies have attempted to identify disease mechanisms using peripheral tissues easily available from living subjects, such as blood.

Because DNA methylation has been shown to be highly cell type- and tissue-specific (with tissue differences being considered one of the largest contributors to variability in the human DNA methylome) (Farre et al. 2015), a few studies have attempted to identify potential overlaps between blood and brain. While initial studies found strong correlations between the mean methylation levels in blood and brain tissue in a between-subjects design (Davies et al. 2012; Horvath et al. 2012), weak to moderate correlations have been reported in within-subject designs. Using paired blood and temporal lobe biopsy from 12 epilepsy patients, around 7.9% of CpG sites were found to be highly correlated between tissues (Walton et al. 2016). Another study has estimated that 9.7% of CpGs measured in blood correlate with brain regions, with CpG sites with a higher between-subject variability presenting stronger between-tissue correlations (Edgar et al. 2017). Overall, these results suggest that a small fraction of CpGs are highly correlated and can be successfully used as proxy of brain tissues. The use of blood is also supported by evidence of significant overlap in detected methylation quantitative trait loci (mQTLs) between both tissues (Smith et al. 2014). In this sense, adequate approaches in blood epigenome-wide studies of brain-related disorders, such as bipolar disorder, may include filtering the initial CpGs to those with high between-tissue correlations, or following up significant hits using available datasets of correlation between tissues, such as BECon (Edgar et al. 2017), EpiBrain (Bediaga et al. 2017), and the 'Blood Brain DNA Methylation Comparison Tool' (Hannon et al. 2015).

Based on the premise that some blood findings may reliably inform of brain mechanisms and can thus be used for the discovery of clinically-useful biomarkers, several studies have investigated blood methylation alterations in bipolar disorder. While a few have identified no statistical significant alterations (corrected for multiple testing) possibly due to low statistical power (Walker et al. 2016a, b; Fries et al. 2017a), other groups have identified significant hits that may be informative of bipolar disorder's epigenetic mechanisms. For instance, altered methylation levels of upstream regions of spermine synthase (*SMS*) and peptidylprolyl isomerase E-like (*PPIEL*) have been reported in bipolar disorder twins compared to control twins (Kuratomi et al. 2008), although this has not been replicated in other case-control cohorts. Other differentially methylated findings in blood from bipolar disorder patients (not including studies focusing on candidate genes) include hypomethylation of a locus near the cytochrome P450, family 11, subfamily A, polypeptide 1 (*CYP11A1*) (Sabunciyan et al. 2015), hypomethylation of family with sequence similarity 63, member B (*FAM63*) (Sugawara et al. 2018), hypomethylation of an intergenic region on chromosome 16 (Sugawara et al. 2018), hypermethylation of TBC1 domain family member 22A (*TBC1D22A*) (Sugawara et al. 2018), hypermethylation of solute carrier family 6 member 4 (*SLC6A4*) (Sugawara et al. 2011), and also thousands of regions located in promoters, 3′-UTR and 5′-UTRs of genes (Li et al. 2015). Alterations in blood of youth at high risk for bipolar disorder, such as offspring of parents with bipolar disorder, have also been identified in genes related to the glucocorticoid receptor signaling pathway (Fries et al. 2017b). In addition, different degrees of suicidal behavior within bipolar disorder patients has been associated with altered methylation of the 5′-UTR of membrane palmitoylated protein 4 (*MPP4*), the intron 3 of TRE2/BUB2/CDC16 domain family member 16 (*TBC1D16*), and exon 1 of nucleoporin 133 (*NUP133*) (Jeremian et al. 2017). Whether these findings are related to alterations in brain tissue, influenced by medication use, or are causal alterations involved in the onset of illness remains to be elucidated.

## 5 The Effects of Psychotropic Medications on DNA Methylation

Pharmacological treatment of bipolar disorder includes mood stabilizers (e.g. lithium, valproate, lamotrigine, and carbamazepine), antidepressants and atypical antipsychotics (Jann 2014), with mood stabilizers being the most common type of medication prescribed. Although widely used, their mechanisms of action remain poorly understood, particularly downstream of receptor binding and intracellular signaling. Drug-induced epigenetic changes are attractive read-outs for potential long-term effects of medications. Lithium and valproate cause common epigenetic effects on the leptin receptor gene in rats (Lee et al. 2015). Lithium treatment resulted in hypomethylation of Igf2 (which encodes a protein involved in cell proliferation, differentiation and survival), Igf2r, and H19 (regulation of cell proliferation) in mouse embryonic stem cells (Popkie et al. 2010), as well as of the BDNF gene in peripheral blood mononuclear cells (PBMCs) from patients (D'Addario et al. 2012; Dell'Osso et al. 2014) and in rat hippocampal neurons (Dwivedi and Zhang 2014). Lithium caused global 5-methylcytosine (5mC) levels to be reduced in lymphoblastoid cell lines from relatives of bipolar disorder patients (Huzayyin et al. 2014) compared to controls. Further, lithium caused decreased global methylation in bipolar disorder patients who responded to lithium (Huzayyin et al. 2014).

Valproate, an anticonvulsant used as a mood stabilizer and to control impulsive behavior in bipolar disorder patients, has been shown to be a potent inhibitor of histone deacetylases (HDACs) (Phiel et al. 2001), which are believed to directly influence DNA methylation (Dobosy and Selker 2001). In line with this, a study found that valproate induced histone acetylation and activated DNA demethylation in the same gene systems in HEK293 cells, suggesting that valproate might cause demethylation of genes through histone acetylation (Milutinovic et al. 2007). Moreover, in human peripheral blood cells, treatment with valproate has been found to reduce BDNF promoter methylation (D'Addario et al. 2012; Dell'Osso et al. 2014), as well as p21 (Aizawa and Yamamuro 2015), RELN (Dong et al. 2008), and glutamate type I transporter (GLT-1) genes (Perisic et al. 2010), and to induce a significant alteration in overall methylation in BD patients (Houtepen et al. 2016). In regards to other mood stabilizers, carbamazapine has been shown to induce hypermethylation of 64 genes and hypomethylation of 14 genes in a neuroblastoma cell line (Asai et al. 2013), while lamotrigine induced no changes in global DNA methylation (Perisic et al. 2010). Altogether, these findings suggest drug-specific effects of mood stabilizers on DNA methylation and warrant further analysis in different cohorts, although common effects on the methylation of specific groups of genes have been reported (Asai et al. 2013). Similar epigenetic effects have also been suggested to play a role in the mechanism of action of antidepressants (Menke and Binder 2014) and antipsychotics (Guidotti and Grayson, 2014; Houtepen et al. 2016), which supports the idea that the reversal of bipolar disorder symptoms induced by different medications involves modulation of gene expression via DNA methylation.

## 6 Future Directions

As discussed, given the knowledge that DNA methylation is tissue- and cell-specific, studies of the role of this epigenetic modification in brain disorders should consider differences in methylation status across specific brain regions and cell types. Although still in its infancy, high-throughput single cell analysis of DNA methylation is rapidly becoming the gold standard in analysis of brain DNA methylation alterations in bipolar disorder and other psychiatric disorders. A study using Illumina Human Methylation 450 BeadChips and laser capture microdissection to obtain DNA methylation data from hippocampal *stratum oriens* GABAergic interneurons from eight control, eight schizophrenia, and eight bipolar disorder subjects, identified distinct DNA methylomes among phenotypically similar populations of GABAergic interneurons, where 11 highly significant differentially methylated regions were associated with multiple zinc finger of the cerebellum gene family members and WNT signaling factors, and a greater number of differentially methylated regions were identified in bipolar disorder cases than in schizophrenia or controls (Ruzicka et al. 2018). This study follows a similar previous study by the same group where circuit-specific DNA methylation changes were identified in a subset of GAD1 regulatory network genes in the hippocampus of schizophrenia and bipolar disorder subjects (Ruzicka et al. 2015).

In addition to single-cell analyses, future studies investigating the DNA methylome in bipolar disorder will involve assessment of both methylation and hydroxymethylation. 5-methylcytosine (5mC) can be enzymatically modified to 5-hydroxymethylcytosine (5hmC) by the ten-eleven translocation (TET) protein family, and may act as an intermediate in an active, replication-independent DNA demethylation process (Hahn et al. 2014). In addition, 5hmC can be a strong inhibitor of the DNA methyltransferase (DNMT) maintenance reaction catalyzed by DNMT1, leading to passive DNA demethylation over subsequence replication cycles. Of particular interest in bipolar disorder studies, 5hmC is reported to be highly enriched and particularly stable in the brain (Hahn et al. 2014; Cheng et al. 2015; Madrid et al. 2016), suggesting it as a distinct epigenetic mark with a characteristic function independent of the DNA demethylation process. In fact, 5hmC has been shown to interact with several chromatin binding proteins (Cheng et al. 2015) and can have opposing effects on gene expression compared to 5mC (Klungland and Robertson, 2017; Ponnaluri et al. 2017). In this sense, it is essential to be able to distinguish 5mC from 5hmC in epigenomic studies and accurately detect and quantitate the levels of 5hmC at a single-base resolution and in specific cell types. The cell type-specific epigenetic landscapes might ultimately determine the selective vulnerability of specific cells to neurodevelopmental or environmental insults that could culminate in manifestation of bipolar disorder (Fig. 1).

State-of-the-art sequencing-based technologies are starting to unravel novel methylation alterations in the human genome, which may be of future interest for the field of psychiatric epigenetics. Non-CpG methylation sites have been proposed to play key roles in neuronal cells (Guo et al. 2014) and have not yet been properly explored in human studies. Moreover, the DNA modification $N^6$-methyladenine (6 mA), which has been shown to be highly prevalent in prokaryotes, has recently been reported in the human genome (Xiao et al. 2018) and is thought to be involved in the response to environmental stress (Yao et al. 2017). Overall, these recent findings suggest a plethora of novel and fairly understudied epigenetic mechanisms that may significantly improve our understanding of bipolar disorder and provide targets for the development of new therapeutics.

In regards to new therapeutics, the advances in knowledge of DNA methylation alterations in bipolar disorder point towards the use of methylation inhibitors as a potential novel approach. Preclinical studies in bipolar disorder animal models show promising results (Sales et al. 2011; Sales and Joca 2016), and modulation of DNMT has been shown to be involved in the mechanism of action of antidepressants and the improvement of depression symptoms (Gassen et al. 2015; Zimmermann et al. 2012). Several inhibitors of DNMT have been identified and are being tested in a number of medical conditions, particularly different types of cancer (Erdmann et al. 2015). However, the non-specificity of the DNMT inhibitors to unique genomic loci, particularly those specifically relevant to bipolar disorder, needs to be taken into consideration before this approach can be used in the clinic. Further studies on targeting specific bipolar disorder-related genes for the modulation of symptoms are warranted. Moreover, because methylation is known to determine and maintain different cell phenotypes within an organism, the potential negative side effects of these drugs needs to be considered, including potential carcinogenic effects. In summary, the ability to target methylation at specific loci rather than simply inhibiting non-specific enzymes will likely determine the success of methylation inhibitors in treatment of bipolar disorder.

## References

Abdolmaleky HM, Cheng KH, Faraone SV et al (2006) Hypomethylation of MB-COMT promoter is a major risk factor for schizophrenia and bipolar disorder. Hum Mol Genet 15:3132–3145

Aizawa S, Yamamuro Y (2015) Valproate administration to mice increases hippocampal p21 expression by altering genomic DNA methylation. Neuroreport 26:915–920

Asai T, Bundo M, Sugawara H et al (2013) Effect of mood stabilizers on DNA methylation in human neuroblastoma cells. Int J Neuropsychopharmacol 16:2285–2294

Association AP (2013) Diagnostic and statistical manual of mental disorders, DSM-5, 5th edn. American Psychiatric Association, Arlington, VA

Bediaga NG, Elcoroaristiazabal X, Calvo B et al (2017) Blood samples as a surrogate for brain samples in methylation studies. EC Neurol 5:74–90

Buoli M, Serati M, Caldiroli A et al (2017) Neurodevelopmental versus neurodegenerative model of schizophrenia and bipolar disorder: comparison with physiological brain development and aging. Psychiat Danub 29:24–27

Chen C, Zhang C, Cheng L et al (2014) Correlation between DNA methylation and gene expression in the brains of patients with bipolar disorder and schizophrenia. Bipolar Disord 16:790–799

Cheng Y, Bernstein A, Chen D, Jin P (2015) 5-Hydroxymethylcytosine: a new player in brain disorders? Exp Neurol 268:3–9

Chuang LC, Kao CF, Shih WL et al (2013) Pathway analysis using information from allele-specific gene methylation in genome-wide association studies for bipolar disorder. PLoS One 8:e53092
D'Addario C, Dell'Osso B, Palazzo MC et al (2012) Selective DNA methylation of BDNF promoter in bipolar disorder: differences among patients with BDI and BDII. Neuropsychopharmacology 37:1647–1655
Davies MN, Volta M, Pidsley R et al (2012) Functional annotation of the human brain methylome identifies tissue-specific epigenetic variation across brain and blood. Genome Biol 13:R43
Dell'Osso B, D'Addario C, Carlotta Palazzo M et al (2014) Epigenetic modulation of BDNF gene: differences in DNA methylation between unipolar and bipolar patients. J Affect Disord 166:330–333
Dempster EL, Pidsley R, Schalkwyk LC et al (2011) Disease-associated epigenetic changes in monozygotic twins discordant for schizophrenia and bipolar disorder. Hum Mol Genet 20:4786–4796
Dobosy JR, Selker EU (2001) Emerging connections between DNA methylation and histone acetylation. Cell Mol Life Sci 58:721–727
Dong E, Nelson M, Grayson DR et al (2008) Clozapine and sulpiride but not haloperidol or olanzapine activate brain DNA demethylation. Proc Natl Acad Sci USA 105:13614–13619
Dwivedi T, Zhang H (2014) Lithium-induced neuroprotection is associated with epigenetic modification of specific BDNF gene promoter and altered expression of apoptotic-regulatory proteins. Front Neurosci 8:457
Edgar RD, Jones MJ, Robinson WP et al (2017) An empirically driven data reduction method on the human 450K methylation array to remove tissue specific non-variable CpGs. Clin Epigenetics 9:11
Erdmann A, Halby L, Fahy J et al (2015) Targeting DNA methylation with small molecules: what's next? J Med Chem 58:2569–2583
Farre P, Jones MJ, Meaney MJ et al (2015) Concordant and discordant DNA methylation signatures of aging in human blood and brain. Epigenetics Chromatin 8:19
Fish EW, Shahrokh D, Bagot R et al (2004) Epigenetic programming of stress responses through variations in maternal care. Ann N Y Acad Sci 1036:167–180
Fornito A, Malhi GS, Lagopoulos J et al (2007) In vivo evidence for early neurodevelopmental anomaly of the anterior cingulate cortex in bipolar disorder. Acta Psychiatr Scand 116:467–472
Fornito A, Yucel M, Wood SJ et al (2009) Anterior cingulate cortex abnormalities associated with a first psychotic episode in bipolar disorder. Br J Psychiatry 194:426–433
Fries GR, Pfaffenseller B, Stertz L et al (2012) Staging and neuroprogression in bipolar disorder. Curr Psychiatry Rep 14:667–675
Fries GR, Walss-Bass C, Soares JC et al (2016) Non-genetic transgenerational transmission of bipolar disorder: targeting DNA methyltransferases. Mol Psychiatry 21:1653–1654
Fries GR, Bauer IE, Scaini G et al (2017a) Accelerated epigenetic aging and mitochondrial DNA copy number in bipolar disorder. Transl Psychiatry 7:1283
Fries GR, Quevedo J, Zeni CP et al (2017b) Integrated transcriptome and methylome analysis in youth at high risk for bipolar disorder: a preliminary analysis. Transl Psychiatry 7:e1059
Gamazon ER, Badner JA, Cheng L et al (2013) Enrichment of cis-regulatory gene expression SNPs and methylation quantitative trait loci among bipolar disorder susceptibility variants. Mol Psychiatry 18:340–346
Garnham J, Munro A, Slaney C et al (2007) Prophylactic treatment response in bipolar disorder: results of a naturalistic observation study. Affect Disord 104:185–190
Gassen NC, Fries GR, Zannas AS et al (2015) Chaperoning epigenetics: FKBP51 decreases the activity of DNMT1 and mediates epigenetic effects of the antidepressant paroxetine. Sci Signal 8:ra119
Gershon ES, Alliey-Rodriguez N, Liu C (2011) After GWAS: searching for genetic risk for schizophrenia and bipolar disorder. Am J Psychiatry 168:253–256
Grande I, Berk M, Birmaher B et al (2016) Bipolar disorder. Lancet (London, England) 387:1561–1572

Grayson DR, Guidotti A (2013) The dynamics of DNA methylation in schizophrenia and related psychiatric disorders. Neuropsychopharmacology 38:138–166
Guidotti A, Grayson DR (2014) DNA methylation and demethylation as targets for antipsychotic therapy. Dialogues Clin Neurosci 16:419–429
Guo JU, Su Y, Shin JH et al (2014) Distribution, recognition and regulation of non-CpG methylation in the adult mammalian brain. Nat Neurosci 17:215–222
Hahn MA, Szabo PE, Pfeifer GP (2014) 5-Hydroxymethylcytosine: a stable or transient DNA modification? Genomics 104:314–323
Hannon E, Lunnon K, Schalkwyk L et al (2015) Interindividual methylomic variation across blood, cortex, and cerebellum: implications for epigenetic studies of neurological and neuropsychiatric phenotypes. Epigenetics 10:1024–1032
Horvath S, Zhang Y, Langfelder P et al (2012) Aging effects on DNA methylation modules in human brain and blood tissue. Genome Biol 13:R97
Houtepen LC, van Bergen AH, Vinkers CH et al (2016) DNA methylation signatures of mood stabilizers and antipsychotics in bipolar disorder. Epigenomics 8:197–208
Huzayyin AA, Andreazza AC, Turecki G et al (2014) Decreased global methylation in patients with bipolar disorder who respond to lithium. Int J Neuropsychopharmacol 17:561–569
Jann MW (2014) Diagnosis and treatment of bipolar disorders in adults: a review of the evidence on pharmacologic treatments. Am Health Drug Benefits 7:489–499
Jawahar MC, Murgatroyd C, Harrison EL et al (2015) Epigenetic alterations following early postnatal stress: a review on novel aetiological mechanisms of common psychiatric disorders. Clin Epigenetics 7:122
Jeremian R, Chen YA, De Luca V et al (2017) Investigation of correlations between DNA methylation, suicidal behavior and aging. Bipolar Disord 19:32–40
Kaminsky Z, Tochigi M, Jia P et al (2012) A multi-tissue analysis identifies HLA complex group 9 gene methylation differences in bipolar disorder. Mol Psychiatry 17:728–740
Kerner B (2015) Toward a deeper understanding of the genetics of bipolar disorder. Front Psych 6:105
Kieseppa T, Partonen T, Haukka J et al (2004) High concordance of bipolar I disorder in a nationwide sample of twins. Am J Psychiatry 161:1814–1821
Klengel T, Mehta D, Anacker C et al (2013) Allele-specific FKBP5 DNA demethylation mediates gene-childhood trauma interactions. Nat Neurosci 16:33–41
Klungland A, Robertson AB (2017) Oxidized C5-methyl cytosine bases in DNA: 5-hydroxymethylcytosine; 5-formylcytosine; and 5-carboxycytosine. Free Radic Biol Med 107:62–68
Kuratomi G, Iwamoto K, Bundo M et al (2008) Aberrant DNA methylation associated with bipolar disorder identified from discordant monozygotic twins. Mol Psychiatry 13:429–441
Lee RS, Pirooznia M, Guintivano J et al (2015) Search for common targets of lithium and valproic acid identifies novel epigenetic effects of lithium on the rat leptin receptor gene. Transl Psychiatry 5:e600
Levandowski ML, Grassi-Oliveira R (2018) Influence of early childhood trauma on risk for bipolar disorder. In: Soares JC, Walss-Bass C, Brambilla P (eds) Bipolar disorder vulnerability: perspectives from pediatric and high risk populations. Elsevier, London, UK
Li Y, Camarillo C, Xu J et al (2015) Genome-wide methylome analyses reveal novel epigenetic regulation patterns in schizophrenia and bipolar disorder. BioMed Res Inter 2015:201587
Luoni A, Massart R, Nieratschker V et al (2016) Ankyrin-3 as a molecular marker of early-life stress and vulnerability to psychiatric disorders. Transl Psychiatry 6:e943
Madrid A, Papale LA, Alisch RS (2016) New hope: the emerging role of 5-hydroxymethylcytosine in mental health and disease. Epigenomics 8:981–991
McGuffin P, Rijsdijk F, Andrew M et al (2003) The heritability of bipolar affective disorder and the genetic relationship to unipolar depression. Arch Gen Psychiatry 60:497–502
Menke A, Binder EB (2014) Epigenetic alterations in depression and antidepressant treatment. Dialogues Clin Neurosci 16(3):395–404

Mill J, Tang T, Kaminsky Z et al (2008) Epigenomic profiling reveals DNA-methylation changes associated with major psychosis. Am J Hum Genet 82:696–711

Milutinovic S, D'Alessio AC, Detich N et al (2007) Valproate induces widespread epigenetic reprogramming which involves demethylation of specific genes. Carcinogenesis 28:560–571

Mitchell C, Schneper LM, Notterman DA (2016) DNA methylation, early life environment, and health outcomes. Pediatr Res 79:212–219

Muhleisen TW, Reinbold CS, Forstner AJ et al (2018) Gene set enrichment analysis and expression pattern exploration implicate an involvement of neurodevelopmental processes in bipolar disorder. Affect Disord 228:20–25

Ng B, White CC, Klein HU et al (2017) An xQTL map integrates the genetic architecture of the human brain's transcriptome and epigenome. Nat Neurosci 20:1418–1426

Perisic T, Zimmermann N, Kirmeier T et al (2010) Valproate and amitriptyline exert common and divergent influences on global and gene promoter-specific chromatin modifications in rat primary astrocytes. Neuropsychopharmacology 35:792–805

Phiel CJ, Zhang F, Huang EY et al (2001) Histone deacetylase is a direct target of valproic acid, a potent anticonvulsant, mood stabilizer, and teratogen. J Biol Chem 276:36734–36741

Ponnaluri VK, Ehrlich KC, Zhang G et al (2017) Association of 5-hydroxymethylation and 5-methylation of DNA cytosine with tissue-specific gene expression. Epigenetics 12:123–138

Popkie AP, Zeidner LC, Albrecht AM et al (2010) Phosphatidylinositol 3-kinase (PI3K) signaling via glycogen synthase kinase-3 (Gsk-3) regulates DNA methylation of imprinted loci. J Biol Chem 285:41337–41347

Routhieaux M, Keels J, Tillery EE (2018) The use of pharmacogenetic testing in patients with schizophrenia or bipolar disorder: a systematic review. Ment Health Clin 8:294–302

Ruzicka WB, Subburaju S, Benes FM (2015) Circuit- and diagnosis-specific DNA methylation changes at γ-aminobutyric acid-related genes in postmortem human hippocampus in schizophrenia and bipolar disorder. JAMA Psychiat 72:541–551

Ruzicka WB, Subburaju S, Coyle JT et al (2018) Location matters: distinct DNA methylation patterns in GABAergic interneuronal populations from separate microcircuits within the human hippocampus. Hum Mol Genet 27:254–265

Sabunciyan S, Maher B, Bahn S et al (2015) Association of DNA methylation with acute mania and inflammatory markers. PLoS One 10:e0132001

Sales AJ, Joca SR (2016) Effects of DNA methylation inhibitors and conventional antidepressants on mice behaviour and brain DNA methylation levels. Acta Neuropsychiatry 28:11–22

Sales AJ, Biojone C, Terceti MS et al (2011) Antidepressant-like effect induced by systemic and intra-hippocampal administration of DNA methylation inhibitors. Br J Pharmacol 164:1711–1721

Smith AK, Kilaru V, Kocak M et al (2014) Methylation quantitative trait loci (meQTLs) are consistently detected across ancestry, developmental stage, and tissue type. BMC Genomics 15:145

Sugawara H, Iwamoto K, Bundo M et al (2011) Hypermethylation of serotonin transporter gene in bipolar disorder detected by epigenome analysis of discordant monozygotic twins. Transl Psychiatry 1:e24

Sugawara H, Murata Y, Ikegame T et al (2018) DNA methylation analyses of the candidate genes identified by a methylome-wide association study revealed common epigenetic alterations in schizophrenia and bipolar disorder. Psychiatry Clin Neurosci 72:245–254

Sun MA, Sun Z, Wu X et al (2016) Mammalian brain development is accompanied by a dramatic increase in bipolar DNA methylation. Sci Rep 6:32298

Walker RM, Christoforou AN, McCartney DL et al (2016a) DNA methylation in a Scottish family multiply affected by bipolar disorder and major depressive disorder. Clin Epigenetics 8:5

Walker RM, Sussmann JE, Whalley HC et al (2016b) Preliminary assessment of pre-morbid DNA methylation in individuals at high genetic risk of mood disorders. Bipolar Disord 18:410–422

Walss-Bass C, Soares JC, Brambilla P (2018) Summary and integration of current findings: a model for bipolar disorder development. In: Soares J, Walss-Bass C, Brambilla P (eds) Bipolar disorder vulnerability: perspectives from pediatric and high risk populations. Elsevier, London, UK

Walton E, Hass J, Liu J, Ehrlich S et al (2016) Correspondence of DNA methylation between blood and brain tissue and its application to schizophrenia research. Schizophr Bull 42:406–414

Xiao Y, Camarillo C, Ping Y et al (2014) The DNA methylome and transcriptome of different brain regions in schizophrenia and bipolar disorder. PLoS One 9:e95875

Xiao CL, Zhu S, He M et al (2018) N(6)-methyladenine DNA modification in the human genome. Mol Cell 71:306–318.e307

Yao B, Cheng Y, Wang Z et al (2017) DNA N6-methyladenine is dynamically regulated in the mouse brain following environmental stress. Nat Commun 8:1122

Zhao H, Xu J, Pang L et al (2015) Genome-wide DNA methylome reveals the dysfunction of intronic microRNAs in major psychosis. BMC Med Genet 8:62

Zimmermann N, Zschocke J, Perisic T et al (2012) Antidepressants inhibit DNA methyltransferase 1 through reducing G9a levels. Biochem J 448:93–102

# DNA Methylation in Multiple Sclerosis

**Lara Kular and Maja Jagodic**

## Contents



**Abstract** Multiple Sclerosis (MS) is a leading cause of lifelong disability in young adults. The disease strikes individuals in their most productive years with incurable and progressive course that results in development of fatigue and accumulation of physical and cognitive disability. MS is characterized by autoimmune destruction of the myelin and subsequent neurodegeneration. This chronic disease of the central nervous system is likely triggered by environmental factors such as smoking, lack of sun exposure/vitamin D deficiency and infection in genetically predisposed individuals, the strongest influence coming from *HLA-DRB1* variants within the HLA class II locus. However, the mechanisms underlying susceptibility to MS are still puzzling and specific clinical translations are lacking. Emerging evidence suggests the implication of epigenetic mechanisms such as DNA methylation in the pathogenesis of MS. In this chapter, we aimed to review findings from DNA methylation studies in MS and discuss their clinical relevance. We first present a critical overview of the outcomes of DNA methylation studies in immune cells and brain tissue from MS patients. We then discuss emerging evidence supporting a role of DNA

L. Kular (✉) · M. Jagodic
Department of Clinical Neuroscience, Center for Molecular Medicine, Karolinska Institutet, Stockholm, Sweden
e-mail: lara.kular@ki.se



S. Jurga, J. Barciszewski (eds.), *The DNA, RNA, and Histone Methylomes*, RNA Technologies, https://doi.org/10.1007/978-3-030-14792-1_8

methylation in mediating the effect from the major genetic risk variant *HLA-DRB1*15:01* and environmental risk factors, smoking and vitamin D deficiency, in MS. We also describe the potential of DNA methylation-based biomarkers and therapies for precision medicine in MS. We expect that the encouraging findings from DNA methylation studies in MS might open new avenues for a better understanding and treatment MS patients.

**Keywords** Multiple sclerosis · DNA methylation · Immune cells · Brain · HLA · Smoking · Vitamin D · Biomarkers · Therapy

## 1 Introduction

Multiple Sclerosis (MS) is a chronic inflammatory disease of the central nervous system (CNS) characterized by autoimmune destruction of myelin and subsequent neuronal death (Compston and Coles 2008). MS is the leading cause on non-traumatic disability among young adults worldwide, affecting up to 200/100,000 individuals in Northern populations with nearly 70% of patients being women (female to male ratio ranging from 2:1 to 3:1) (Trojano et al. 2012; Bezzini and Battaglia 2017). At diagnosis (between 20 and 40 years of age), the majority of MS patients (80–90%) present with a relapsing-remitting (RRMS) form of disease characterized by repeated and transient episodes of neurological symptoms (relapse) followed by complete or partial recovery (remission) (Compston and Coles 2008). Current treatments are effective only in the early inflammatory RRMS stage, but they target broadly the immune system and pose serious safety concerns (Soelberg Sorensen 2017). Most RRMS patients (80%) will eventually convert to a secondary progressive stage (SPMS) with persistent neuronal loss and continuous accumulation of disability. Additionally, in ~10% of cases, patients will manifest a primary progressive form of MS (PPMS) already from onset. MS pathology is believed to be initiated by disruption of the blood-brain-carrier (BBB) and infiltration of peripheral immune cells into the CNS, resulting in confined areas of inflammatory demyelination and axonal injury, called lesions or plaques, which continuously arise in the CNS (Compston and Coles 2008). Variation of clinical symptoms between patients and during disease course is conditioned by anatomical localization and severity of the lesions and range from sensory, motor and visual deficit to fatigue and cognitive impairment. Thus, MS is a highly heterogeneous incurable chronic disease leading not only to personal debilitation but also to considerable economic and societal burden (Brundin et al. 2017).

Even though the exact cause of MS remains unknown, disease likely results from a complex interplay between genetic and environmental risk factors. The genetic basis of MS was first demonstrated in familial studies with an overall recurrence risk for monozygotic (MZ) twins of 18.2%, which significantly differs from the one for dizygotic twins (4.2%) and siblings (2.7%), and an overall genetic heritability estimated at 54% (O'Gorman et al. 2013). The first genetic association was established in the 1970s with the Human Leukocyte Antigen (HLA) region on chromosome 6p21 (Jersild et al. 1975) and was later refined to the haplotype *HLA-DRB5*0101–HLA-DRB1*1501–HLA-DQA1*0102–HLA-DQB1*0602* (Fogdell et al. 1995) encoding HLA class II molecules involved in regulation of immune processes. Over the past decade, genome-wide association studies (GWAS) conducted in large case-control cohorts have enabled the identification of single nucleotide polymorphism (SNP) associated with the risk of developing disease at the population level. These collective efforts have made a breakthrough in decoding the genetic architecture of MS risk by identifying >200 MS-associated loci, with the strongest influence coming from the aforementioned *HLA-DRB1*15:01* variant (odd ratio, OR~3) (International Multiple Sclerosis Genetics et al. 2011, 2013). Yet, the functional interpretation of MS causal variants remains challenging as most of them are located in non-coding regions of the genome (Farh et al. 2015). Furthermore, a more complex pattern of inheritance is likely driven by parent-of-origin effects where the risk depends on whether the allele is inherited from the mother or the father (Ebers et al. 2004; Chao et al. 2009). Collectively, genetic data converge on a polygenic model of the risk of developing MS, with one locus conferring moderate effect and many loci of small effects. They further indicate that a limited part of the disease heritability can be explained by genetic variants, with population-based studies estimating significantly lower sibling relative risk compared to family studies (O'Gorman et al. 2013; Westerlind et al. 2014). This gap suggests a 'hidden' heritability which, together with a yet-unexplained rise in incidence of MS during the last decades, may be explained by non-genetic processes such as gene-environment interactions. Accordingly, vast epidemiological data support a role of environmental exposures and lifestyle habits in disease susceptibility. Compelling body of evidence associates tobacco smoking, Epstein-Barr virus (EBV)-mediated infectious mononucleosis, low vitamin D and sun exposure as well as obesity with susceptibility to develop MS (Olsson et al. 2017). Other environmental and lifestyle factors such as night shift work, seasonal change, alcohol and diet have also been reported to affect MS risk and warrant replication (Olsson et al. 2017). Interestingly, for most of the identified environmental factors such as sun exposure/vitamin D deficiency, mononucleosis, night shift work or high BMI, the childhood-adolescence period seems to represent a specific window of susceptibility in the risk to develop MS (Olsson et al. 2017). Moreover, gene-environment interactions have been shown to contribute to risk, as evidenced for interaction between smoking or EBV and MS-associated *HLA* factors (Hedstrom et al. 2011; Xiao et al. 2015). Jointly, known genetic and environmental factors and their interactions can explain a substantial fraction of disease risk (van der Mei et al. 2016). Yet, the mechanisms underpinning disease initiation and progression are poorly annotated and robust prognostic tools and more specific and potent therapies are lacking, thus posing major challenges for an efficient care of MS patients.

The low concordance rate of MS in MZ twins together with parent-of-origin effects, 'hidden' heritability and long-term impact of environmental risk factors suggest involvement of epigenetic mechanisms in disease pathogenesis. Epigenetics refers to mitotically (and meiotically) heritable changes in gene expression that do not entail variation in the DNA sequence. Epigenetic processes are therefore primarily of non-genetic origin and cell type-specific, with non-shared environmental influence accounting for most of the variance (Busche et al. 2015). Epigenetic mechanisms refer to biochemical modifications of the genome, such as DNA methylation and histone posttranslational modifications, and their regulatory effects on chromatin dynamics and transcription. DNA methylation, the deposition of a methyl group to cytosine, mostly in the context of a CpG dinucleotide, is by far the most studied epigenetic modification in clinical studies of MS. Because DNA methylation inhibits gene expression when associated to promoter region of genes, hypermethylation in this region is considered as a mark of transcriptional repression. De novo deposition and maintenance of methylation are orchestrated by DNA methyltransferases DNMT3A/B and DNMT1, respectively, while active demethylation is catalyzed by members of the ten–eleven translocations (TETs) family of enzymes. Notably, the implication of epigenetics in MS is further supported by the identification of MS-associated genetic variation and transcriptional changes of genes encoding members of the DNA methylation machinery (Calabrese et al. 2014; Andlauer et al. 2016; Fagone et al. 2016). Moreover, early dysregulation of methionine metabolism (an essential metabolite in the methyl group transfer to DNA) has been recently proposed to impact DNA methylation patterns in MS as well (Singhal et al. 2018). Overall, DNA methylation dynamics is responsive to the environment and can lead to stable and heritable but reversible changes in gene-regulatory networks. Thus, DNA methylation studies represent a promising approach for improved understanding of MS pathogenesis and therapeutic opportunities. In this chapter, we will review the studies reporting DNA methylation alterations in MS patients and discuss the clinical translations of these findings.

## 2 DNA Methylation Studies in MS

### *2.1 DNA Methylation in Peripheral Immune Cells*

Findings from immunological, genetic and histopathological studies of patients with MS have revealed a crucial role of immune cells in the pathogenesis of MS. The strongest genetic influence comes from the *HLA* class II region, which encodes essential molecules for antigen presentation by antigen presenting cells (APCs, such as macrophages) and antigen recognition by pathogenic T helper (Th) cells. MS is regarded as $CD4^+$ Th cell-driven disease with predominant Th1- and Th17-mediated proinflammatory processes. Accordingly, experimental autoimmune encephalomyelitis (EAE), an MS-like animal model, can be induced by passive transfer of activated CNS antigen specific $CD4^+$ T cells (Ben-Nun et al. 1981) and pharmacological treatment of RRMS patients with inhibitor of lymphocytes migration showed efficacy in reducing inflammation and disease activity (Polman et al. 2006; Kappos et al. 2010). In that context, DNA methylation studies in MS have aimed to explore the molecular mechanisms underlying MS immunopathogenesis by profiling whole blood, blood peripheral mononuclear cells (PBMCs) and sorted $CD4^+$ and $CD8^+$ T cells, the majority of them in case-control cohorts. Details about the cohorts and main findings are described in Table 1.

Studies addressing global methylation exploit the fact that methylation measured at repetitive elements, such as Alu repeats and long interspersed nucleotide elements (LINE-1), jointly representing one third of the entire genome, can serve as a surrogate of total genomic methyl cytosine. Increased global DNA methylation in blood cells and sera has been reported in RRMS patients compared to controls (Neven et al. 2016; Pinto-Medel et al. 2017; Dunaeva et al. 2018). LINE-1 methylation further correlates with either motor disability, measured as Expanded Disability Status Scale (EDSS) status (Neven et al. 2016) or IFN-treatment duration (Pinto-Medel et al. 2017). Given that loss of global methylation typically leads to chromosomal instability, loss of imprinting and activation of transposable element, these data suggest an increased genome stability in MS patients. However, such interpretation should be considered with caution as global methylation methods do not reveal locus-specific changes. This is important in light of the consequences of the locus-specific genome instability on reactivation of specific endogenous retroviruses, which has been observed in MS patients (Morris et al. 2018).

Gene-candidate approaches have examined DNA methylation at *a priori* selected candidate genes involved in inflammatory processes and/or MS susceptibility. They have focused on promoter regions and thus inferred the putative impact of the observed changes on transcription. Results from promoter profiling of eight neuroinflammatory genes in whole blood from RRMS patients and healthy donors showed increased methylation levels at *RUNX3*, *CDKN2A*, *SOCS1*, and *NEUROG1* genes implicated in neuroglial and T cell differentiation, most of them being reported as dysregulated in MS patients (Sokratous et al. 2018). The negative regulator of proinflammatory signaling *SHP-1* gene displayed hypermethylation correlating with reduced transcript levels in peripheral blood leukocytes from MS patients compared to controls (Kumagai et al. 2012). Interestingly, studies have shown that T cells from RRMS patients exhibit hypomethylation of CpGs in the two previously identified MS risk loci, vitamin D receptor (*VDR*) and IL2 receptor (*IL2RA*) genes compared to controls (Ayuso et al. 2017; Field et al. 2017). Lower methylation levels at these regions correlate with MS-specific increased *IL2RA* and *VDR* expression in T cells and blood leukocytes, respectively. Of note, *IL2RA* expression by $CD4^+$ T cells has been shown to be regulated by vitamin D, further supporting its relevance in MS susceptibility (Berge et al. 2016). Finally, we have shown that $CD4^+$ T cells from RRMS patients exhibit hypomethylation of the *VMP1/MIR21* locus compared to SPMS patients and healthy controls. Lower methylation associated with upregulation of the microRNA (miRNA) *miR-21* and concomitant downregulation of its target genes, important in cell apoptosis and proliferation, in $CD4^+$ T cells (Ruhrmann et al. 2018). This data highlights an interplay between epigenetic mechanisms where DNA methylation changes at restricted CpGs of a miRNA can lead to perturbed expression of multiple genes involved in immune processes. Thus, even though most of these studies are biased towards pre-selected candidates, they reveal functionally relevant changes in methylation which could contribute to enhanced inflammation in MS. Of special notice, similar changes of some candidate genes, such as *RUNX3*, *NEUROG1*,

**Table 1** DNA methylation studies in Multiple Sclerosis

| Tissue | Cohort (F/M) | Meth | Findings[a] | Reference |
|---|---|---|---|---|
| Global methylation approaches | | | | |
| WB | 51 RRMS (38/13), 137 HC (73/37) | Pyroseq | Global hypermethylation (Alu, LINE-1, SAT-α) in MS. Correlation between Alu, LINE-1 DNA methylation and EDSS score but not MS course. | Neven et al. (2016) |
| Serum | 24 untreated RRMS (17/7), 24 HC (19/5) | BS-seq, MSP | Cell-free circulating DNA displays hypermethylation of L1PA2 sub-family of LINE-1 fragments. | Dunaeva et al. (2018) |
| Buffy coat | 54 untreated RRMS (36/18), 36 IFN-treated RRMS (21/15), 25 HC (14/11) | LINE-1 assay | Slight global hypermethylation in MS. Negative correlation with duration of IFNbeta treatment | Pinto-Medel et al. (2017) |
| Candidate-gene DNA methylation approaches | | | | |
| WB | 50 pairs of MZ twins discordant in MS (35/15) | MSP | No difference of *MHC2TA* pIV promoter methylation between discordant MZ twin pairs. | Ramagopalan et al. (2009) |
| WB | Benign cohort: 48 RRMS (EDSS ≤ 3). Malignant cohort: 20 PPMS (EDSS > 6) | Pyroseq | No difference at *HLA-DRB1*1501* and *HLA-DRB5* methylation in benign vs. malignant MS. | Handel et al. (2010) |
| Buffy coat | 7 PPMS, 50 RRMS, 12 SPMS (49/20), 19 HC (10/9) | Cloning BS-seq | Hypermethylation of *SHP-1* promoter 2 in MS vs. HC. No correlation with MS clinical parameters. | Kumagai et al. (2012) |
| PBMCs | 39 RRMS, 1 SPMS (32/8), 40 HC (30/10) | EpiTyper, dot blot | Downregulation of *TET2* and *DNMT1* gene expression in MS. Hypermethylated CpGs in *TET2*. Reduced global 5hmC level and slightly increased global 5mC in MS. | Calabrese et al. (2014) |
| PBMCs | 31 RRMS, 1 SPMS (22/10), 30 HC (15/15) | Cloning BS-seq | *PADI2* hypomethylated promoter correlation with upregulated gene in MS. No correlation with clinical parameters. | Calabrese et al. (2012) |

(continued)

**Table 1** (continued)

| Tissue | Cohort (F/M) | Meth | Findings[a] | Reference |
|---|---|---|---|---|
| T cells, PBL | 23 RRMS (14/9), 12 HC (8/4) | Cloning BS-seq | *VDR* alternative promoter hypermethylation in MS, no correlation with clinical parameters. *VDR* mRNA upregulation in PBLs of MS. | Ayuso et al. (2017) |
| WB, PBMCs, NAWM | PBMCs: 28 RRMS, 10 HC, WB: 14 MS, 14 HC, NAWM: 8 MS, 6 HC | EpiTyper, 450K | No difference in *IL2RA* methylation in PBMCs and NAWM in MS vs. HC. After mixed-tissue deconvolution: 1 hypomethylated T cell-specific DMP at *IL2RA* promoter correlating with increased *IL2RA* expression in T cells in MS vs. HC. | Field et al. (2017) |
| $CD4^+$ T | DC: 12 RRMS (9/3), 8 SPMS (4/4), 12 HC (8/4). VC: 30 RRMS (22/8) 11 SPMS (8/3), 12 HC (5/7), 9 INDC (7/2) | 450K, pyroseq | Hypomethylation of *VMP1/MIR21* locus in RRMS (compared to HC and SPMS) and association with lower miR-21 expression. | Ruhrmann et al. (2018) |
| WB | 66 RRMS (33 rel, 33 rem, 44/22), 33 HC (22/11) | MS-MLPA | Hypermethylation of *RUNX3, CDKN2A, SOCS1, and NEUROG1* in MS vs. HC. No difference inbetween relapse vs. remission. | Sokratous et al. (2018) |
| Genome-wide DNA methylation approaches | | | | |
| $CD4^+$ T | 2 RRMS, 1 SPMS pairs of discordant MZ twins (2/1) | RRBS | Two common DMPs (*TMEM1, PEX14*) between two twin pairs. | Baranzini et al. (2010) |
| $CD4^+$ T, $CD8^+$ T | 30 treated RRMS (26/4), 28 HC (15/13) | 450K | $CD4^+$: 74 DMPs (35 genes) in RRMs vs HC: 19 *HLA*-DMPs, 55 non-*HLA* DMPs. Correlation of *HLA-DRB1* DNA methylation with *HLA-DRB1*1501* haplotype. $CD8^+$: 79 non-MHC DMPs (51 genes). No overlap with $CD4^+$ T cells. | Graves et al. (2014), Maltby et al. (2015) |

(continued)

**Table 1** (continued)

| Tissue | Cohort (F/M) | Meth | Findings[a] | Reference |
|---|---|---|---|---|
| WB, CD4⁺ T, CD8⁺ T | 16 RRMS (16/0), 14 HC (14/0) | 450K | No genome-wide DMPs. Nominally significant CpGs: predominant hypermethylation in CD8⁺ specifically. Two common DMPs (at *TMEM48 and APC2*) in CD4⁺, CD8⁺ T cells and WB. No difference between different disease duration. | Bos et al. (2015) |
| PBMCs | 14 RRMS (9/5), 8 PPMS (6/2), 8 HC (6/2) | 450K | 136 DMPs between RRMS, PPMS and HC: 30 DMPs (17 genes) in RRMS vs. HC, 67 DMPs (25 genes) in PPMS vs. HC and 51 DMPs (22 genes) in PPMS vs. RRMS, respectively. Most PPMS-DMPs are hypermethylated. | Kulakova et al. (2016) |
| CD4⁺ T | 28 untreated RRMS (28/0), 22 HC (22/0) | 450K | 153 genes with DMRs: *HLA-DRB1* hypomethylated, *SNORD1A*, *SHTN1*, *MZB1* and *TNF* displayed DMRs at TSS region. | Maltby et al. (2017) |
| WB | Selected cohort: 50 MS (19 current, 9 past, 22 never-smoker, 50/0); Broad cohort: 132 MS (33 current, 34 past, 65 never-smoker, 90/42), 135 HC (34 current, 31 past, 70 never-smoker, 100/35). | 450K, pyroseq | Effect of smoking dependent on smoking load and time since cessation. 58 DMPs (29 genes) in current vs. never-smokers with MS, including 8 unreported DMPs. Reversible changes with time post cessation. *AHRR* gene: correlation with expression in PBMCs. Effect of smoking load interacts with MS disease. | Marabita et al. (2017) |
| WB, CD4⁺ T, CD8⁺ T, CD14⁺, | WB: 140 MS (98/42), 139 HC (104/35); CD14⁺: 23 MS (15/8), 13 HC (9/4), CD4⁺: 21 MS (14/7), 12 HC | 450K, pyroseq, BS-seq, RNAseq, GWAS | Hypomethylation of *HLA-DRB1* exon 2 mediate genetic risk from *HLA-DRB1*15:01* and a novel protective | Kular et al. (2018) |

(continued)

**Table 1** (continued)

| Tissue | Cohort (F/M) | Meth | Findings[a] | Reference |
|---|---|---|---|---|
| $CD19^+$ B, PBMCs | (8/4), $CD8^+$: 15 MS (6/9), 14 HC (9/5), $CD19^+$: 17 MS (9/8), 12 HC (6/6). | | variant through change of *HLA-DRB1* expression. | |
| $CD4^+$ T | 7 (4/3) baseline and 6 month after DMF treatment | EPIC | 974 DMPs after treatment, 97% hypermethylated | Maltby et al. (2018) |
| $CD8^+$ T, $CD4^+$ T | Combined cohorts including samples from (Bos et al. 2015; Maltby et al. 2015, 2017): $CD4^+$: 94 RRMS (94/0), 94 HC (94/0), $CD8^+$: 68 RRMS (68/0), 57 HC (57/0) | 450K, EPIC | No genome-wide DMPs in $CD8^+$ but CpGs hypermethylated in MS. DMR: 2 intragenic DMR in both $CD4^+$ and $CD8^+$: *HLA-DRB1* (hypomethylated), *SLFN12* (hypermethylated). DMR in $CD4^+$ only: intergenic (*MOG/ZFP57* and downstream *SLFN12*), *NINJ2* (hypermethylated). Correlation with expression in WB. | Rhead et al. (2018) |

[a]Of note, described are findings deemed significant by the original study using originally reported criteria for significance, which vary widely

*WB* whole blood, *PBMCs* peripheral blood mononuclear cells, *PBL* peripheral blood leukocytes, *MS* Multiple Sclerosis, *RRMS* relapsing-remitting MS, *SPMS* secondary progressive MS, *PPMS* primary progressive MS, *HC* healthy controls, *INDC* inflammatory neurological disease control, *F/M* female/male number, *EDSS* expanded disability status scale, *MZ* monozygotic, *DMF* dimethylfumarate, *DMP* differentially methylated position, *DMR* differentially methylated region, *GWAS* genome-wide association study, *RRBS* reduced representation bisulfite sequencing, *450K* Infinium HumanMethylation450 BeadChip, *pyroseq* pyrosequencing, *BS-seq* bisulfite cloning-sequencing, *MSP* methyl sensitive PCR, vs. versus

*MIR21* or *IL2RA*, could be identified in genome-wide investigations of specific cell types, described below (Bos et al. 2015).

The recent progress in genome-wide methylation analyses has advanced the field beyond candidate gene approaches by enabling investigation of the methylome landscape of patients. A seminal genome-wide study has investigated the $CD4^+$ T methylome of three MZ twin pairs discordant for MS and found only two common DMPs between twin pairs, which is not unexpected given a small and heterogeneous cohort (Baranzini et al. 2010). The development of array-based technologies such as Infinium HumanMethylation BeadChip arrays has allowed cost- and time-effective DNA methylation profiling of blood cells from case-control cohorts, the large majority comparing RRMS patients with healthy controls. These epigenome-wide association studies, referred to as EWAS, have revealed that epigenetic alterations occur at multiple loci throughout the genome of immune cells, reporting detailed mapping of differentially methylated single CpG positions (DMPs) or regions (DMRs), their effect sizes (represented by Δβ-values) and directionality (Table 1). However, EWAS conducted in Norwegian and Australian cohorts have yielded various results to date, discrepancies that can be further reflected by the low overlap between studies (Bos et al. 2015). Indeed, no common DMPs were observed between the three existing studies focusing on CD4$^+$ T cells from RRMS compared to controls (Graves et al. 2014; Bos et al. 2015; Maltby et al. 2017). Overall, three annotated genes, *HLA-DRB1* and *HLA-DRB6* genes from the MS risk *HLA* class II locus as well as the RNA-editing *ADARB2* gene, overlap between the studies, although with different reported CpGs (Fig. 1a). Among common DMPs reported by at least two studies, 12 of them displayed consistent changes (i.e. same directionality) (Fig. 1b). They map to immune genes, i.e. *HLA* class II genes (*HLA-DRB1, HLA-DRB5, HLA-DRB6*), TGF-β induced gene *TGFBI,* as well as the ribosomal kinase *RPS6KA2* gene and the protein-ubiquitin ligase *FBXO27* gene. In CD8$^+$ T cells, while only one intergenic DMP overlaps between the two existing studies (Bos et al. 2015; Maltby et al. 2015), 17 common genes were found to exhibit at least one differentially methylated CpG, although at different locations, in the two studies (Fig. 1b). They are involved in phagocytosis (*MEGF10, BAI1*), cell proliferation (*CDKN1C, CAMTA1*), cell migration (*IGF2BP1, CDC42BPB*), iron sequestering (*FTL*) and xenobiotic metabolism (*UGT1A10*). Recently, joint analysis of the combined Norwegian and Australian data (Bos et al. 2015; Maltby et al. 2015, 2017) reported five significant DMRs in CD4$^+$ and CD8$^+$ T cells from MS patients compared to controls (Rhead et al. 2018). Two intragenic DMRs, that map *HLA-DRB1* and *SLFN12* genes, were found hypomethylated and hypermethylated, respectively in RRMS patients compared to controls in both CD4$^+$ and CD8$^+$ T cells, while an intergenic DMR located between *MOG* and *ZFP57* genes was specific to CD4$^+$ T cells. Additionally, two DMRs in *NINJ2* gene and downstream *SLFN12* locus were identified in CD4$^+$ T cells only when comparing treatment-naïve MS patients and controls. Methylation differences at *HLA-DRB1*, *NINJ2* and *SLFN12* genes associated with changes in expression in whole blood from MS patients compared to controls.

Despite disparities between studies (see explanations in Sect. 2.3), interesting findings point to two MS-related features. Predominant genome-wide hypermethylation could be observed in CD8$^+$ T cells, specifically (i.e. not CD4$^+$ T cells), from RRMS patients compared to controls (Bos et al. 2015; Rhead et al. 2018), in PBMCs from PPMS patients, specifically, compared to RRMS patients and healthy controls (Kulakova et al. 2016) and in CD4$^+$ T cell from MS patients after pharmacological treatment with the MS-drug dimethylfumarate (Maltby et al. 2018). One can speculate that dysregulation of DNA methylation machinery, such as DNMT and TET enzymes could have such a global impact on DNA methylation. Moreover, we and others have found striking hypomethylation of the *HLA-DRB1* gene (Graves et al. 2014; Maltby et al. 2017; Kular et al. 2018; Rhead et al. 2018), the relevance of such altered methylation at the major MS risk locus will be further described in Sect. 3.1 of this chapter.

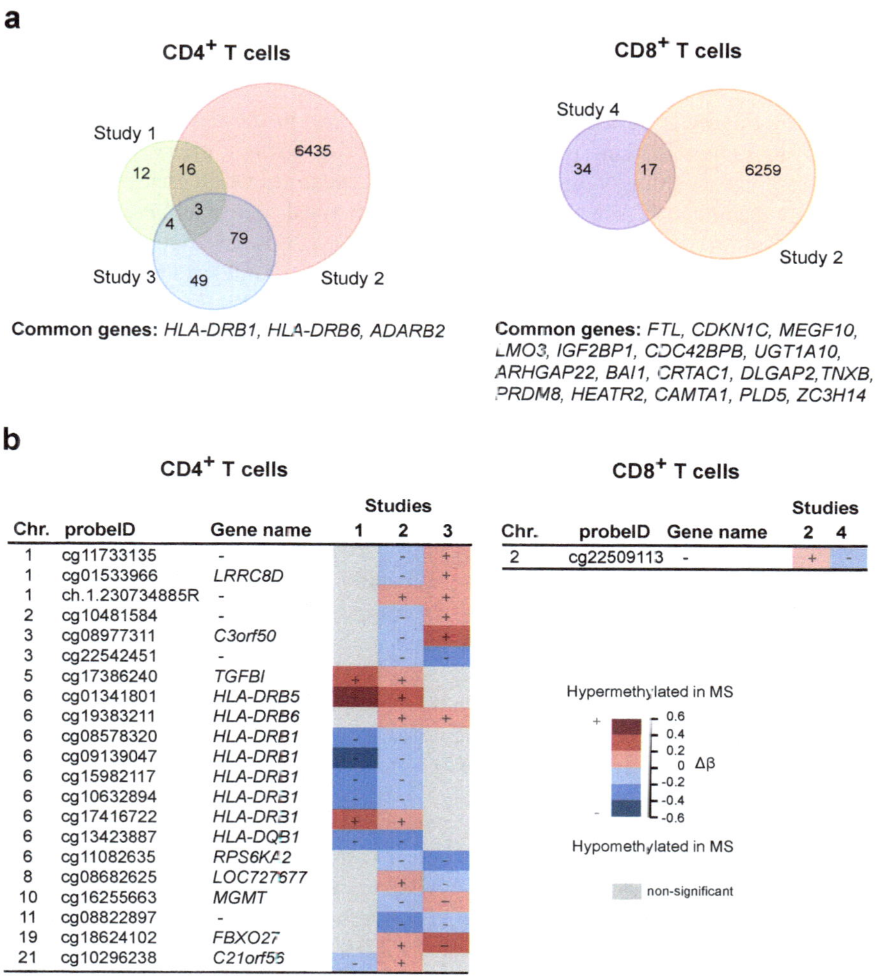


**Fig. 1** Overlap between DNA methylation studies in immune cell type-specific in MS. (**a**) Venn-diagram illustrating the number of differentially methylated genes between MS patients and healthy controls in studies profiling $CD4^+$ (left panel) and $CD8^+$ (right panel) T cells. The names of overlapping genes between all cell type-specific studies appear below the diagram. (**b**) Heatmap of differentially methylated CpGs between MS patients and healthy controls reported in at least two studies profiling $CD4^+$ (left) and $CD8^+$ (right) T cells. The color gradient represents the direction of change ($\Delta\beta$-value), with blue and red being hypomethylated (−) and hypermethylated (+), respectively, in MS patients compared to controls. Studies 1, 2, 3 and 4 correspond to $CD4^+$ (Graves et al. 2014), $CD4^+$ and $CD8^+$ (Bos et al. 2015), $CD4^+$ (Maltby et al. 2017) and $CD8^+$ (Maltby et al. 2015), respectively. Note that comparisons were conducted on the reported DMPs with varying significance thresholds: genome-wide significance (False Discovery rate or Benjamini-Hochberg-adjusted p-value $<0.05$, $\Delta\beta \geq \pm\ 0.1$) in studies 1, 3 and 4 and nominal significance (p-value $<0.05$) for study 2

## 2.2 DNA Methylation in Post-Mortem Brain

Neuroimaging and histopathological studies of the CNS white matter in MS patients have distinguished a variety of lesion types differing in the degree of leukocytes infiltration, demyelination, remyelination and neuro-axonal injury. Importantly, areas from the seemingly unaffected normal appearing white matter (NAWM) frequently manifest diffuse pathology along with focal abnormalities even in the absence of infiltrating leukocytes (Barnett and Prineas 2004; Henderson et al. 2009; van Horssen et al. 2012; Burm et al. 2016). NAWM lack of integrity has been further associated with neurological disability (Dineen et al. 2009; Francis et al. 2014; Meijer et al. 2016). Additionally, recent evidence suggests that neurodegenerative processes might start earlier in life (Chard et al. 2002; Hagstrom et al. 2017; Tortorella et al. 2018), without prior demyelination (DeLuca et al. 2006) and in the grey matter as well (Geurts and Barkhof 2008; Calabrese et al. 2010). Overall, brain atrophy is the strongest predictor of disability, its impact becoming apparent later in life when the neurological reserves are likely exhausted from long-standing compensatory mechanisms. Due to the limited accessibility of the target organ in MS, the molecular mechanisms underlying the neuropathology of MS remain elusive. Given that DNA methylation is chemically stable, studies have so far relied on observation in post-mortem brain tissue, composed of mixed cell populations.

The first study examining DNA methylation in the MS brain was a candidate gene study of *PADI2* encoding a citrunillating enzyme. This study was motivated by the observation of elevated citrunillated myelin basic protein (MBP) in NAWM of MS patients, which is suggested to contribute to myelin destabilization in MS. The authors found hypomethylation of *PADI2* promoter, which associated with increased levels of PADI2 enzyme and citrunillated MBP in NAWM from 12 MS patients compared to white matter samples from 19 non-MS controls, i.e. non-neurological controls and patients with other neurological diseases (Mastronardi et al. 2007). Altogether, epigenetic dysregulation of PADI2 enzyme might participate, at least in part, in aberrant citrunillation of MBP and subsequent myelin breakdown. Interestingly, hypomethylation and upregulation of *PADI2* gene could also be observed in peripheral blood from MS patients compared to controls, suggesting that some changes occurring in the brain can be detected in blood (Calabrese et al. 2012).

However, alteration of *PADI2* gene could not be identified in recent genome-wide DNA methylation studies comparing MS-NAWM versus (vs.) control white matter (Huynh et al. 2014) or demyelinated vs. myelinated hippocampi from MS patients (Chomyk et al. 2017) (Table 2). Epigenome-wide profiling of NAWM from MS patients instead revealed numerous, albeit subtle, changes clustering on 539 DMRs throughout the genome (Huynh et al. 2014). Interestingly, hypomethylated CpGs occurred predominantly within promoters, i.e. transcription starting sit (TSS) $\pm$ 2000 bp, and were depleted from gene bodies (intron, exon), the latter being enriched in hypermethylated CpGs. Gene ontology annotation of DMR-related genes further showed that hypomethylated DMRs affected genes associated with immune responses while hypermethylated DMRs are enriched in genes involved in general cell functions

**Table 2** Genome-wide DNA methylation studies in *post-mortem* brain tissue from MS patients

| Reference | Huynh et al. (2014) | Chomyk et al. (2017) |
|---|---|---|
| Brain tissue | Frontal lobe | Hippocampus |
| Analysis | MS-NAWM vs. NNC | MS-demyelinated vs. MS-myelinated |
| Cohort (F/M) | DC: 28 NAWM-MS (3 RRMS, 17 SPMS, 7 PPMS, 17 /11), 19 NNC (7/12); VC: 10 MS (SPMS, 7/3), 20 NNC (14/6) | 8 myelinated MS (6 SPMS, 2 PPMS, 5/3), 7 demyelinated MS (6 SPMS, 1 PPMS, 5/2) |
| Method | 450K, EpiTyper, RNA-seq | 450K, ELISA, IHC, RT-qPCR |
| Findings | – 220 hypomethylated DMRs (1235 CpGs)<br>– 319 hypermethylated DMRs (1292 CpGs)<br>– at oligodendrocyte-specific genes (*BCL2L2, HAGHL, NDRG1, CTSZ, LGMN*).<br>– correlation with expression change of a fraction of corresponding genes. | – 144 DMPs (75 genes)<br>– 62 hypermethylated DMPs<br>– 82 hypomethylated DMPs<br>– at astrocytic and neuronal genes<br>– Promoter (TSS)-DMPs at *AKNA, EBPL, HERC6, SFRP1, NHLH2, PLCH1, TMEM132B* and *WDR81* correlated with expression changes. |

*MS* Multiple Sclerosis, *RRMS* relapsing-remitting MS, *SPMS* secondary progressive MS, *PPMS* primary progressive MS, *NNC* non-neurological disease control, *F/M* female/male number, *NAWM* normal appearing white matter, *DC* discovery cohort, *VC* validation cohort, *DMP* differentially methylated position, *DMR* differentially methylated region, *450K* Infinium HumanMethylation450 BeadChip, *TSS* transcription starting site, *IHC* immunohistochemistry, *TSS* transcription starting site, vs. versus

and oligodendrocyte-related processes. Some of the changes could be further associated with gene expression differences from RNA-seq data in MS NAWM compared to control samples. Of note, the transcriptional changes did not necessarily anti-correlate with DMR direction of change, probably due to differences in DMR location and cellular origin in bulk tissue.

A recent study has investigated DNA methylation changes following hippocampus demyelination in MS (Chomyk et al. 2017). Comparison of demyelinated vs. myelinated hippocampi from MS patients led to the identification of 144 hypomethylated and hypermethylated DMPs with large changes ($\Delta\beta > 20\%$). These changes coincide with differential expression of DNA methylation enzymatic machinery, with significant upregulation of methylating enzymes (DNMT1, DNMT3A/B) concomitant with downregulation of demethylating TET enzymes in the MS hippocampus following demyelination. This finding together with the predominant detection of DNMTs and TETs proteins in hippocampal neurons compared to other cell types, imply that methylation patterns likely differ in a locus- and, importantly, cell type-specific manner, which might not be reflected in bulk tissue analysis. In line with this, the 75 genes harboring DMPs had been previously reported to be expressed by multiple brain cell types, i.e. microglia, oligodendrocyte, astrocytes and neurons, with however, an overrepresentation of astrocytic- and neuronal-specific genes. Nevertheless, changes occurring at promoter-related sequences (TSS1500, 16 genes) could further associate with anti-correlated transcriptional changes of the corresponding genes (Table 2), most of them being involved in immune or neuronal processes.

These genome-wide characterizations in blood immune cells and bulk post-mortem brain tissue of MS patients unravel DNA methylation changes at genes involved in immune and nervous processes and set the stage for future studies in larger and more homogenous cohorts.

## 2.3 *Methodological Considerations in Clinical DNA Methylation Studies*

We will here review several biases that might impair proper interpretation of DNA methylation changes, particularly in the clinical context.

### 2.3.1 Cohort and Sample Heterogeneity

The varying results between studies examining blood immune cells in MS underscores the context-dependent nature of epigenetic marks where DNA methylation is highly sensitive to sample and cohort heterogeneity. Therefore, cohort characteristics, such as genetic background, disease course and sub-type, age and sex, largely influence the outcome of the analyses. A variety of other confounders associated to treatment history, lifestyle habits or environmental exposures such as smoking likely affect the methylome and could account for some of the signal detected in DNA methylation studies as well.

Cell type-specificity of DNA methylation represents another challenge in data interpretation, as different cell proportions from mixed blood and brain tissue might drive the observed DNA methylation changes and therefore mask 'true' changes. Accordingly, DNA methylation changes in cell types sorted from the same case-control cohorts display little overlap (Graves et al. 2014; Bos et al. 2015; Maltby et al. 2015). Undoubtedly, cell type sorting prior to analysis offers the most optimal strategy to account for tissue heterogeneity. This approach is feasible in blood samples but it is confined only to the most abundant cell types and cell sorting is rather limited in *post-mortem* brain tissue, due to reliance on only a restricted number of cell type-specific nuclear antibodies (Yeung et al. 2014). Therefore, in most brain studies, spatial and cellular heterogeneity is therefore lost. In that context, progress in emerging fields such as single-cell methylomics (Smallwood et al. 2014) together with in situ *DNA* methylation analysis (Shiura et al. 2014) and spatial 'omics' (Stahl et al. 2016) will eventually provide useful tools to complement bulk methylome by mapping a molecular atlas at a single-cell level. Alternatively, analytical strategies correcting for confounders could aid in deciphering biologically relevant DNA methylation signals, as exemplified in two aforementioned studies in which accounting for age and blood cell proportions strengthened or even enabled the identification of the *MIR21* and *IL2RA* loci, respectively, in CD4$^+$ T cells from MS patients (Field et al. 2017; Ruhrmann et al. 2018). It is noteworthy that the current lack of reference methylomes from distinct human brain cell types hinders the use of reference-based cellular deconvolution from brain tissue-generated data, contrasting with its common use in blood DNA methylation analysis (Titus et al. 2017). This challenge can be partly overcome by the use of unsupervised methods base on reference-free algorithms (Titus et al. 2017). Additionally, further work is needed in order to characterize changes in other immune cell types than CD4$^+$ and CD8$^+$ T cells, e.g. APCs such as monocytes and B cells which are believed to play pivotal roles in MS. This is supported by the considerable benefit RRMS patients gained after treatment with monoclonal antibody depleting B cells (Hauser et al. 2008). Finally, additional heterogeneity come from sub-cell types from immune (e.g. Th1, Th17, Th2, Treg CD4$^+$ T cells) and nervous (e.g. excitatory glutamatergic vs. inhibitory GABAergic neurons) cell populations, which are known to exert different functions. Optimization of genome-wide methylation profiling for sample with low input will undeniably aid in capturing this variety of changes occurring in MS. Altogether, these issues highlight the need to take into account cell and cohort heterogeneity in EWAS design prior to and during downstream analysis.

### 2.3.2 Methodological Challenges

Genome-wide technologies have facilitated EWAS analysis by promising unbiased 'hypothesis-free' approach to comprehensively characterize variations associated to complex diseases. Among the technologies available, the utilization of cost-effective array platforms such as Infinium HumanMethylation BeadChips represents the best compromise for DNA methylation analysis in clinical samples. Despite their extensive use, methylation arrays pose several limitations (Barker et al. 2018), some being inherent to the technology itself, other to the commonly employed bisulfite (BS) treatment of DNA prior to sample hybridization. Indeed, even though these so-called "genome-wide" Illumina 450K and EPIC arrays cover ~99% of the Refseq genes together with some well-known intergenic regulatory regions, they target each gene with few probes and overall annotate only a fraction, i.e. 1.7% and 3%, respectively, of the total CpGs present in the genome. An additional bias in array-based methods is the role of SNPs in the pre-design probe locations (Chen et al. 2013), causing differences in binding to certain alleles for a given gene. Moreover, the most widely-used DNA methylation studies (array included) rely on the bisulfite (BS) treatment of genomic DNA, which converts unmethylated cytosines to thymine, leaving methylated cytosine unchanged. Signals from each base are subsequently used to estimate the proportion of methylated vs. unmethylated CpGs. Yet, by changing most cytosines at non-CpG sites, BS conversion drastically reduces the complexity of the genome to three bases, therefore hampering proper exploration of highly polymorphic loci. This is of particular importance in the HLA region since it is a highly polymorphic region with a complex pattern of linkage disequilibrium and the high similarity between the proximal *HLA* class II genes. The challenge to study such locus can be illustrated by our effort to validate BS array-generated DMR at

*HLA-DRB1* using multiple methods (detailed in Sect. 3.1.1 of this chapter) (Kular et al. 2018). Additionally, conventional BS-based arrays do not allow distinction of true CpG methylation (5mC) from its antagonist CpG hydroxymethylation (5hmC), nor do they cover sufficient probes targeting non-CpGs, both non-canonical modifications being highly prevalent in the human brain, especially neurons. The impact of mixed 5mC/5hmC signals in *post-mortem* brain samples could be technically circumvented by the use of BS and oxidative BS side-by-side treatments prior to hybridization on the array (Stewart et al. 2015). Besides technical limitations, heterogeneity in analytical approaches, computational pipelines and statistical approaches play an evident role in the reported outcome. As seen earlier, studies have invariably favored either DMP or DMR analysis, reporting significant changes at nominal or adjusted P-value (with or without $\Delta\beta$ cutoff) and using different normalization strategies, thus making comparison between studies challenging.

### 2.3.3 Biological Relevance

Inferring the functional consequence of methylation changes is still very demanding given our partial knowledge of the impact of such changes. Data interpretation likely builds on an integrative model incorporating a variety of parameters such as the location (promoter, intragenic, intergenic), the nature (single vs. contiguous CpGs) and the amplitude of change. Importantly, DNA methylation acts in concert with histone posttranslational modifications and chromatin conformational regulators, these interactions shaping gene regulation in responses to internal (genetic) and external (environmental) influences. Therefore superimposing information from additional molecular layers, ideally transcriptional and organizational, appears crucial for proper data interpretation. These can be derived from publicly available databases integrating genetic architecture of the human epigenome and transcriptome, such as ENCODE, Roadmap and Blueprint epigenomes (Bujold et al. 2016), Genotype-Tissue Expression (GTEx) (Gamazon et al. 2018) or the brain-specific xQTL (Ng et al. 2017) platforms.

The biological relevance of epigenetics in complex human diseases such as MS is further complicated by the difficulty to infer causality between epigenetic marks and pathological processes. Further work is necessary to define the nature of the interactions between the genome, methylome and exposome. Indeed, DNA methylation can be triggered by genetic, environmental and stochastic cues and impact proper functioning of virtually all immune or CNS cell types implicated in MS. As such, DNA methylation could be a cause or consequence of disease, act independently or in mediation of risk factors. To address this issue, the use of analytical strategies, namely causal inference testing and Mendelian randomization (described in the next section) or methodological approaches such as longitudinal cohorts, could undeniably aid in elucidating the epigenetic contribution in MS disease. At the tissue and cellular level, the use of emerging methodologies such as CRISPR-dCas9-based epigenome-editing (Pulecio et al. 2017) (described in Sect. 3.3.2 of this chapter) in combination with adequate experimental design in animal and cellular models will certainly assist in the quest for biological relevance of identified epigenetic changes.

# 3 Exploiting DNA Methylation for a Better Understanding and Treatment of MS Patients

GWAS have revealed that the genetic architecture of MS is polygenic and related to more than 200 common variants. The identification of various environmental risk factors further increases the complexity of the risk for developing MS. Similarly, EWAS are continuously increasing the catalog of putatively relevant candidate loci associating with the disease. A key challenge now is to place the identified variants, exposures and methylation alterations in the context of pathological mechanisms. This further underscores the need for investigation of causal alleles in the relevant tissue/cell type and under specific environmental conditions. In that context, since DNA methylation integrates signals from both genetic and environmental influences, it can be regarded to act at different levels of genetic predisposition to disease: in an additive manner, in synergy or as a mediator of genetic risk (Fig. 2). In this section, we will describe the potential roles of DNA methylation in mediating risk for MS.

## *3.1 DNA Methylation as a Mediator of Genetic Risk in MS*

### 3.1.1 DNA Methylation Mediates Risk from the Major MS Risk *HLA-DRB1*15:01*

Integrated approaches combining the multiple layers of the interplay between genetic and epigenetic factors in gene regulation have shown that, overall, a substantial fraction of the methylome is controlled by the DNA sequence (Liu et al. 2014; Chen et al. 2016). This dependence, referred to as methylation quantitative trait locus (meQTL), can occur by local SNPs disrupting the CpG site or by proximal or distal SNPs affecting epigenetic status in *cis* or *trans*, respectively, through long-range physical and functional interactions. Therefore, a genetic-epigenetic paradigm appears instrumental to understand how risk variants could shape individuals into susceptibility for MS. Importantly, DNA methylation changes at the major MS risk gene, *HLA-DRB1,* have been consistently identified in blood immune cells and seem to partially dependent on carriage of the risk *HLA-DRB1*15:01* variant (Graves et al. 2014). We have investigated whether DNA methylation mediates effect of genetic variation in MS by integrating genome-wide genotype data with epigenome-wide data in case-control cohorts (Kular et al. 2018). A summary of the results is depicted in Fig. 3. Using Illumina 450K arrays for methylation profiling in $CD14^+$ monocytes sorted from blood of MS patients and healthy controls, we found that monocytes of the risk *HLA-DRB1*15:01* carriers display a considerably lower methylation at 19 CpGs of a DMR encompassing the exon 2 of the *HLA-DRB1* gene (Fig. 3a). Validation of *HLA-DRB1*-specific DNA methylation changes using BS-pyrosequencing confirmed hypomethylation in *HLA-DRB1*15:01* carriers,

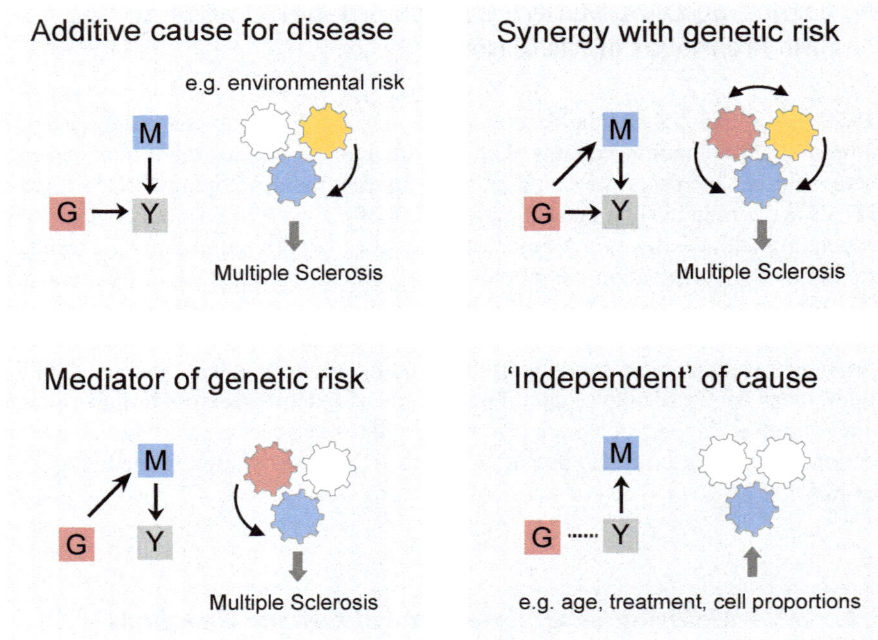


**Fig. 2** Potential roles of DNA methylation in MS. Interplay between DNA methylation (M), genetic risk (G) and environmental risk exposure in phenotypic outcome (Y), depicted by blue, red and yellow wheels, respectively. DNA methylation can act in addition, in synergy, as a mediator of genetic risk or be seemingly 'independent' of any cause. The latter occur for example if DNA methylation changes are driven by cohort or sample confounders (e.g. age, sex or treatment history, cell proportions)

which could be further correlated with higher *HLA-DRB1* expression in monocytes from risk carriers compared to heterozygotes and non-carriers. We next tested whether the *HLA-DRB1*15:01* variant specifically, i.e. compared to other *HLA-DRB1* alleles, drives the observed differences using allele-specific DNA methylation and expression analyses. Single-strand BS-DNA cloning and sequencing of a fragment encompassing the DMR sequence in homozygous *HLA-DRB1*15:01* individuals confirmed the unmethylated status of more than 52 CpGs harboring exon 2 of the gene in *HLA-DRB1*15:01*. The use of methyl-sensitive restriction enzyme followed by allele-specific qPCR, which alleviates biases induced by BS conversion and pre-designed probes from commercial arrays, established *HLA-DRB1**15:01 as the sole hypomethylated variant compared to the most common *HLA-DRB1* alleles. We next functionally tested whether intragenic methylation change at *HLA-DRB1* can actively impact gene expression or rather be consequence of transcriptional activity in the locus, as reported for other genes (Mendizabal et al. 2017; Neri et al. 2017). Results revealed that PBMCs treated with a demethylating agent exhibit increased *HLA-DRB1* expression and the *HLA-DRB1* DMR sequence displays methylation-sensitive enhancer properties using an in-vitro reporter system. Taken

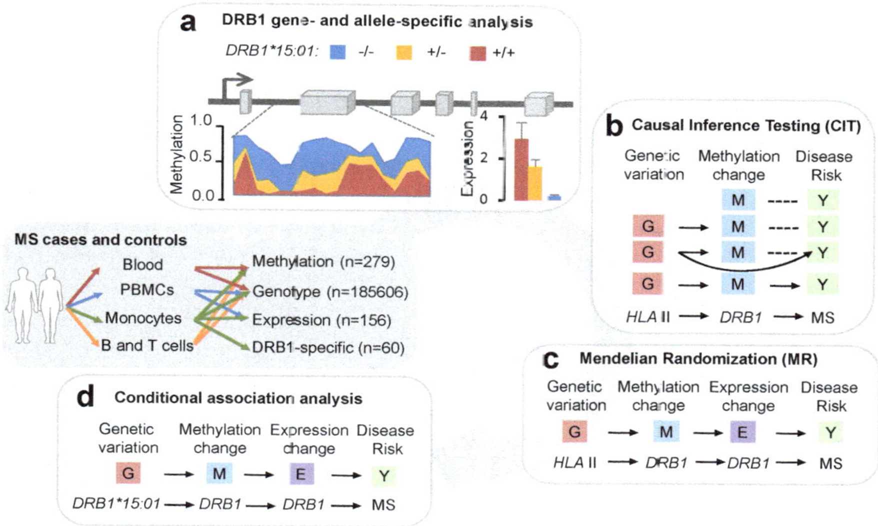


**Fig. 3** DNA methylation as a mediator of genetic risk in MS. (**a**) Genome-wide, locus-specific, functional methylation analyses in monocytes revealed that *HLA-DRB1*15:01* is unmethylated and pre-dominantly expressed allele. Causal inference methods such as causal inference testing (CIT) with the different analytical steps described in (**b**) and Mendelian Randomization (MR) with Egger's regression (**c**) established significant causation between *HLA-DRB1* methylation, expression and MS risk. (**d**) Genetic association identified *HLA-DRB1*15:01* and a novel protective *HLA* variant as interacting variants acting through the changes in *HLA-DRB1* DNA methylation and gene expression to modulate risk of MS

together, these findings strongly suggest that DNA methylation in *HLA-DRB1* can mediate the risk of MS. To formally test this hypothesis we performed genome-wide mediation analysis using a large case-control cohort with genotype and methylation data and applied Causal Inference Test (CIT) to establish the mediation (Fig. 3b). CIT analysis identified 50 genetic variants that predispose for MS through methylation changes, primarily in the same *HLA-DRB1* exon 2 region identified in monocytes. We then addressed the functional impact of genetically-controlled methylation at exon 2 of *HLA-DRB1* on transcription by carrying out two-sample Mendelian Randomization (MR) and MR-Egger's regression. Findings corroborated a causal relationship between methylation at the *HLA-DRB1* DMR and *HLA-DRB1* gene expression, with MR-Steiger test validating this directionality (Fig. 3c). Finally, association analyses in four large cohorts (14,259 cases and 171,347 controls), conditioning for all known MS variants in the *HLA* locus, confirmed that the main effect comes from *HLA-DRB1*15:01* (Fig. 3d), as observed in monocytes, but also identified a novel protective MS variant (rs9267649) counteracting the effect of *HLA-DRB1*15:01* on methylation and expression.

This study highlights the importance of integrating multi-layered data to explore the molecular mechanisms underlying risk variants and to further identify new disease-modifying variants that might escape detection by conventional genetic studies. It also raises the novel hypothesis that methylation-mediated regulation of expression of the HLA class II molecules, in addition to conformational changes of the protein, mediates the risk of MS. The HLA class II molecules present specific antigens, mainly derived from extracellular pathogens, on the surface of APCs. To do so, *HLA-DRB1*-encoded molecules are constantly being synthetized and remain ready to accept peptides, only translocation to the surface requiring peptide binding. In line with this, higher *HLA-DRB1* transcript levels observed in risk carriers would increase the probability of them binding to the peptide and presenting it in higher amounts on the surface. In the context of MS pathogenesis, this is most likely the case for specific APCs presenting MS autoantigenic peptide(s). The exposed peptides are then recognized by $CD4^+$ T lymphocytes, leading to a complex cascade of specific immune responses driving autoimmunity against CNS myelin. In this conceptual framework, it is tempting to speculate about the potential of DNA methylation as a mediator of cellular plasticity in MS.

### 3.1.2 DNA Methylation as a Mediator of Cellular Plasticity in MS?

Normal body functions during development and homeostasis and its aberrant expression in the case of disease involve changes in phenotypic plasticity. In this context, 'susceptibility' genes may in reality act as 'plasticity' genes, rendering some individuals more responsive than others to external (stochastic, environmental) factors (Belsky et al. 2009). This notion applied to development was already suggested by C. Waddington, pioneer in epigenetics, as for him, it should be possible to alter the degree of flexibility-inflexibility by selecting appropriate genotypes (Waddington 1959). Important work conducted in animal models of phenotypic plasticity in the context of genetic homogeneity has further evidenced the existence of a third source of phenotypic variability, neither genetic nor environmental, by demonstrating the contribution of DNA methylation changes in the varying phenotypes, even in parthenogenetic species upon environmental stability (Kucharski et al. 2008; Vogt et al. 2008). These studies are consistent with observations from MZ twin studies (Kaminsky et al. 2009) and collectively imply that epigenetic processes are likely pervasive guarantors of plasticity, from organismal to cellular and molecular levels. An altered epigenetic plasticity and chromatin dynamics could conceivably underpin pathogenic processes as well, as illustrated by altered DNA methylation variability in cancer (Hansen et al. 2011), aging (Cheung et al. 2018) or autoimmune diseases such as Rheumatoid Arthritis (Webster et al. 2018). Undoubtedly, low or high plasticity at specific genes are both likely to alter cellular phenotype in a locus-, cell type- and context-dependent manner, thus conferring adaptive or maladaptive response under specific circumstances. Whether genome-wide DNA methylation variability is affected in Multiple Sclerosis remains an open question warranting further investigation. Nevertheless, a shift from a risk model to a plasticity paradigm might aid in capturing the complexity underlying disease susceptibility at the individual level. Accordingly, genetic inheritance of risk from *HLA-DRB1*15:01* might rely not only on transmission of the impact of genetic variation on structural specificity of *HLA-DRB1* protein for MS autoantigen, but also on transmission of certain plasticity.

From an evolutionary perspective, exposure to pathogens is believed to be one of the strongest selective forces in human evolution, largely contributing to the remarkable genetic diversity found in the *HLA* locus. This phenomenon can be regarded as part of the classical genetic assimilation of an adaptive acquired trait. Interestingly, a recent study has found that, unlike other common *HLA* genes (e.g. *HLA-DQB1*), *HLA-DRB1* locus does not display any biologically meaningful pathogen group-specific bias (Pierini and Lenz 2018), indicating that specific *HLA-DRB1* alleles might have been selected by specific pathogens. Additionally, HLA-DR molecules have been associated to presentation of intracellular endogenous antigens as well, such as following autophagy of intracellular component (Munz 2016) or in the case of viral infection (Martin and Carrington 2005), including MS risk EBV infection (Paludan et al. 2005). Altogether, these findings suggest that the typical 'divergent allele advantage' in this locus (translating to a better chance for efficient immunosurveillance due to sequence diversity between alleles in heterozygous individuals) has probably not played a substantial role in *HLA-DRB1* allele selection, this locus has likely evolved under advantageous selection by specific pathogens.

In line with this, the remarkable sequence variation of *HLA-DRB1* gene mirroring the extended pathogen recognition repertoire maps to the HLA peptide-binding groove encoded by exon 2 of *HLA-DRB1* gene. It is noteworthy that the very same locus also harbors the identified hypomethylated DMR mediating risk via enhanced *HLA-DRB*15:01* gene expression. One can hypothesize that by buffering the molecular stochasticity and subsequent intrinsic noise at the DNA methylation (Smallwood et al. 2014) and transcriptional level (Elowitz et al. 2002; Chang et al. 2008; Kellogg and Tay 2015), *HLA-DRB1*15:01* variant is poised to a latent activation state. This suggests that the additional source of plasticity conferred by *HLA-DRB1*15:01* through DNA hypomethylation at exon 2 has been selectively favored over the years, probably due its potential to drive efficient immune response. At the cellular level in MS disease, because *HLA-DRB1*15:01* mean expression appears constitutively set at a high level, this plasticity might however elicit maladaptive responses upon stochastic events observed in MS-dependent context, such as autoantigen peptide encounter. This further supports the match-mismatch hypothesis underlying adaptive and maladaptive response in the case of a mismatch between one's predicted versus actual environment. Further work will be essential to decipher whether epigenetic-mediated cellular plasticity plays a role in autoimmune processes in MS.

### 3.2 DNA Methylation as a Mediator of Environmental Risk Factors

More than 10 environmental exposures or lifestyle habits have been suggested to increase susceptibility to MS. However, few studies have put focus on the mechanisms underlying these effects. Recent evidence indicates that DNA methylation could play a role in mediating the effect of two of the most established risk factors, cigarette smoking and lack of sun exposure/vitamin D, in disease pathogenesis. The main findings are illustrated in Fig. 4.

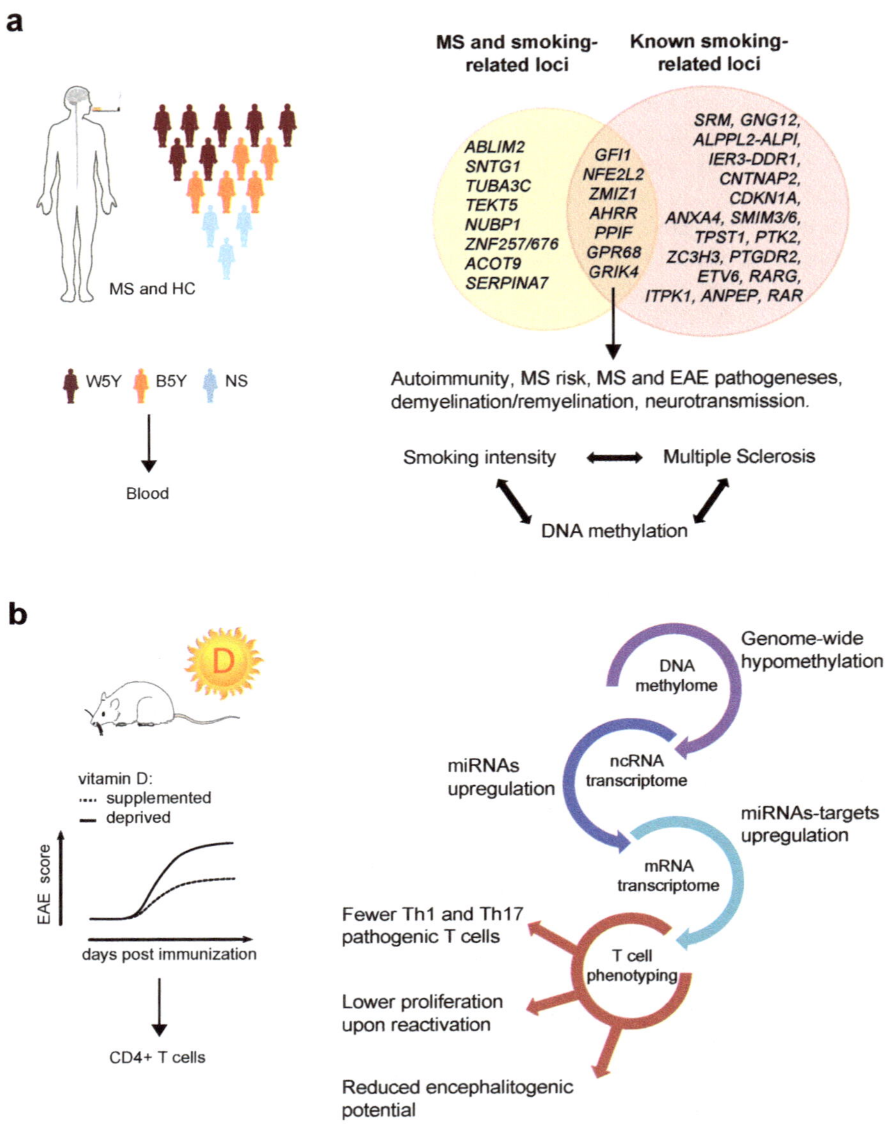


**Fig. 4** DNA methylation as a mediator of environmental risk in MS. (**a**). DNA methylation changes underlying cigarette smoking in blood samples from case-control cohort with individuals categorized regarding the time post smoking cessation: within 5 years (W5Y) smoker (dark red), beyond 5 years (B5Y) smoker (orange) or never smoker (light blue). Altered CpGs implicate known smoking-related loci as well as novel loci in MS patients, with the smoking load having an enhanced impact in patients compared to controls. (**b**). DNA methylation changes mediating protective effect of vitamin D in $CD4^+$ T cells from juvenile rats vitamin D-supplemented or -deprived in an animal MS model, experimental autoimmune encephalomyelitis (EAE). Genome-wide changes correlate with dysregulation of microRNA (miRNAs) and their target genes, concomitant with differential phenotypes of $CD4^+$ T cells and protection against EAE in vitamin D-supplemented rats

### 3.2.1 DNA Methylation and Smoking in MS

Both active and passive smoking have been associated with increased risk, progression and disability in MS (Olsson et al. 2017). Pathogenic mechanisms underpinning smoke exposure likely involve lung irritation and inflammation rather than systemic nicotine exposure itself (Hedstrom et al. 2009, 2013a, b). In line with this, unspecific lung irritation due to other toxic compounds has been reported to increase disease severity (Bergamaschi et al. 2018; Jeanjean et al. 2018). Local immune responses resulting from lung irritation would trigger peripheral immune dysregulation and further promote autoimmune reactions. Both duration and intensity of smoking increase MS risk independently, and unlike other MS risk factor, the effect is not associated to a particular age of exposure (Hedstrom et al. 2009). Interestingly, the impact of smoking is reversible as it remains up to 5 years after cessation and is nullified a decade after cessation (Hedstrom et al. 2013a, b). Studies have examined interaction between the strongest genetic risk, i.e. carriage of the risk allele *HLA-DRB1*15:01* and absence of the protective *HLA-A*02* allele, and cigarette smoking. Results showed that both active and passive smokers, carrying both *HLA* risk factors, display considerably increased risk to develop MS compared to non-smokers, e.g. nearly 14-fold for active smoker compared to a fivefold increase in non-smokers (Hedstrom et al. 2011, 2014). More recently, *NAT1* gene involved in metabolism of smoke compounds emerged as a putative genetic modifier of tobacco smoke exposure in MS susceptibility (Briggs et al. 2014). However, even though several hypotheses implicate impact of pro-inflammatory processes occurring locally in the lungs, the underlying mechanisms supporting gene-environment interactions in MS immunopathogenesis are still elusive.

We recently explored the impact of smoking on blood DNA methylation profiles in cohorts of MS patients by comparing individuals among MS risk categories: less than 5 years after cessation (within 5 year (W5Y)-smokers), more than 5 years after cessation (beyond 5 years (B5Y)-past smokers) and never smokers (NS) (Marabita et al. 2017). Expectedly, comparison of methylome signatures in MS patients revealed that the majority of differences could be observed between W5Y-smokers and NS groups. A large fraction (84%) of the 58 identified DMPs (mapping 38 loci) were found hypomethylated after smoking, most of them corresponding to CpGs known to be affected by smoking in healthy individuals (Gao et al. 2015). In addition to these established loci, eight potentially novel smoking-associated DMPs were found in the context of MS. Overall, the identified DMPs locate in regulatory regions of genes that have been implicated in MS and/or EAE pathogenesis (Fig. 4a). Importantly, hypomethylation of CpGs of *AHRR* gene could be associated with increased *AHRR* expression in PBMCs from smoker compared to non-smoker MS patients. The aryl-hydrocarbon receptor (AHR) repressor *AHRR* gene was first described to encode a competitive repressor of AHR activity involved in xenobiotic detoxification, e.g. under smoke exposure, and was later implicated as a negative regulator of inflammation and aberrant proliferation. One possible interpretation would speculate that the observed hypomethylation and upregulation of *AHRR* gene in smokers could reflect compensatory mechanisms to AHR-mediated toxicity of compounds from cigarette smoke. Another interpretation links *AHRR* dysregulation to a specific cell type involved in MS pathogenesis, where increased *AHRR* levels might contribute to impaired homeostasis in smokers. In support of this hypothesis, smoking-related DMPs found in MS patients were enriched in hematopoietic stem and progenitor cells (HSPC)-specific regulatory regions. Interestingly, a response to Natalizumab treatment in MS patients has been associated with HSPC mobilization (Mattoscio et al. 2015).

Of note, the identified DNA methylation changes appear more pronounced in a cohort considered to be at high risk, i.e. composed of only MS female patients with both MS genetic risk (*HLA-DRB1*15:01*$^{+}$, *HLA-A2*$^{-/-}$) and higher smoking load, suggesting a modifying effect of any of these variables at the DNA methylation level. A putative interaction between genetic risk and smoking-related DNA methylation change is consistent with the increased risk observed in *HLA-DRB1*15:01*$^{+}$ MS smokers mentioned above. This hypothesis is further supported by the significant effect that the intensity of smoking has on DNA methylation in W5Y-smoker as well. Moreover, the effect of smoking is reversible, as methylation levels reached the ones observed in MS patients who have quit smoking more than 5 years prior to blood sampling or who have never smoked. Jointly, these findings are in accordance with observations from epidemiological studies on the impact of smoking in MS. Interestingly, while the disease status (RRMS, SPMS) did not have a significant impact on DNA methylation, the presence of MS disease per se exacerbated the effect of smoking load, inflating the extent of hypomethylation. This novel finding about a modifier effect of disease on the impact of smoking intensity is of high relevance both for clinicians and MS patients.

### 3.2.2 Vitamin D and DNA Methylation in MS

The first evidence of the effect of environmental factor in MS susceptibility arose from the observation of latitude-gradient effect on MS prevalence (Koch-Henriksen and Sorensen 2010; Simpson et al. 2011). The effect was later attributed to lack of sun exposure, low vitamin D levels or a combination of both, as the major source of vitamin D originates from skin exposure to UV-radiation and to a lesser extent from dietary intake. Additionally, polymorphism in genes involved vitamin D metabolism such as the MS risk *CYP27B1* locus (Sundqvist et al. 2010) have been associated to vitamin D levels as well (Bahrami et al. 2018). The age of exposure was refined to childhood/adolescence, with migration studies showing that moving to country of high latitude before adolescence, and not at adulthood, is likely responsible for greater individual risk to develop MS (Gale and Martyn 1995). In line with this, low vitamin D levels detected before the age of 20 years old associate with increased MS risk (Munger et al. 2006) and vitamin D levels in adolescent, but not adult, rats affect EAE incidence and course via immunomodulatory actions (Adzemovic et al. 2013). These findings jointly support recommendation for prophylactic vitamin supplementation of adolescents in prevention of MS risk. Overall, lack of sun exposure and/or hypovitaminosis D were shown to affect MS susceptibility, disease activity, disability and progression (Olsson et al. 2017). Interestingly, a vitamin D-dependent regulation of *HLA-DRB1*15:01* variant involving functional interaction at vitamin D response element (VDRE), has been reported in-vitro (Ramagopalan et al. 2009). Evidence of a long-term impact of vitamin D/lack of sun exposure, remaining after the period of exposure, further reinforces the hypothesis of a role of epigenetics in mediating the effect. Even though the interaction between vitamin D cognate receptor VDR and transcription factors and histone posttranslational modifiers is well established (Lu et al. 2018), the impact of vitamin D on DNA methylation is less documented.

Using several 'omic' approaches in rodents, we recently investigated the mechanisms underlying the protective effect of vitamin D supplementation, compared to deprivation, on EAE in juvenile rats (Zeitelhofer et al. 2017). Genome-wide DNA methylation profiling of $CD4^+$ T cells displayed subtle but widespread methylation changes upon vitamin D supplementation compared to deprivation. Notably, the vast majority of DMRs were found hypomethylated, probably due to reduced expression levels of all DNMTs in these cells. Moreover, a large fraction of the identified DMRs affected the expression of corresponding genes, which were suggested to be proximal mediators of VDR signaling. Among them, hypomethylation and concomitant upregulation of small non-coding RNAs was associated to subsequent modulation of their target genes (Fig. 4b). Expectedly, altered genes were enriched in pathways related to T cell activation and differentiation. Accordingly, $CD4^+$ T cells presented with reduced ability to differentiate into Th1 and Th17 cells, to proliferate and importantly, to exert encephalitogenic effect. This study thus provides functional evidence that vitamin D affects the pathogenic potential of $CD4^+$ T cells directly via DNA methylation changes. Results contrast with findings from another study where global methylation in $CD4^+$ T cells, assessed at LINE-1 sequence, was found increased in EAE adult mice following vitamin D treatment (Moore et al. 2018). This effect was associated with increased methyltransferase *BHMT1* gene involved in methionine epigenetic metabolism and with a transition from encephalitogenic $CD4^+$ T cell to regulatory Treg cell population. The differences between studies can be explained by the large disparities in experimental conditions or by differential mechanisms of vitamin D in euchromatin and heterochromatin compartments. Limited studies have reported association between DNA methylation changes and vitamin D status in humans (Bahrami et al. 2018) and future research is needed to confirm these findings in MS.

Altogether these data support the relevance of studying DNA methylation in understanding the interaction between the exposome and the genome, and how this interaction may thus affect risk to develop MS.

### 3.3 DNA Methylation as Biomarker and Putative Therapeutic Target

Despite advance in identifying the molecular mechanisms underpinning MS disease and progress in development of potent immunomodulatory drugs for early stages of disease, no treatment cure disease, leaving MS patients with constant progression of disability. Moreover, valid biomarkers are still lacking for disease phenotyping and prediction of treatment response and disease progression. In this context, the remarkable properties of stability and reversibility of DNA methylation offer unprecedented perspective for improved biomarker and therapies.

### 3.3.1 DNA Methylation as Biomarker

Monitoring and prognostic methods in MS are primarily based on neuroimaging methods such as Magnetic Resonance Imaging (MRI) detecting brain burden, i.e. atrophy and lesional damage which correlate with long-term disability. On the other hand, the use of biomolecular biomarkers such as DNA methylation has the potential to reflect ongoing rather than delayed events occurring in the brain of MS patients. Correlation of some DNA methylation changes between post-mortem brain and peripheral blood, as observed for *PADI2* gene (Calabrese et al. 2012, 2014), advocates the use of DNA methylation as supplementary information compared to existing molecular biomarkers. In line with this, profiling of DNA methylation in cell-free blood samples has been shown to accurately capture cell type-specific signature of dying cells from peripheral organ, such as oligodendrocyte degeneration in RRMS patients (Lehmann-Werman et al. 2016). Additionally, the use of locus-specific DNA methylation patterns in $CD4^+$ T cells, such as *IL17A* and *FOXP3* genes, allows for accurate estimation of Th lineage commitment and imbalance, thus representing interesting phenotyping tool (Janson et al. 2011). Methylation patterns of cell-free plasma DNA have been shown to potentially serve as a discriminatory biomarker of relapse vs. remission for RRMS patients (Liggett et al. 2010). Finally, novel approaches based on droplet/digital assay (Yu et al. 2018) or the generation of methylation biosignatures from multiplexed DNA methylation profiles might represent new strategies for early, specific and quantitative detection of DNA methylation-based biomarkers in MS patients.

### 3.3.2 DNA Methylation as Therapeutic Target

The plastic nature of DNA methylation marks makes them attractive target for pharmacological therapy. The impact of DNMT inhibitors such as 5-aza-2′deoxycytabine (5-aza, known as decitabine) has been investigated in the context of EAE. Decitabine, an FDA-approved hypomethylating agent, is a chemical analog of cytidine that incorporates into replicating DNA where it irreversibly blocks DNMT1 activity and leads to loss of methylation in a cell division-dependent manner. Several studies have demonstrated protective effect of 5-aza treatment in EAE and consistently associated protection with a direct effect on Th1 and Th17 pathogenic cells (Chan et al. 2014; Mangano et al. 2014; Wang et al. 2017). Both prophylactic and therapeutic administration of 5-aza resulted in amelioration of EAE clinical score and histological hallmarks, i.e. reduced lymphocyte infiltration in the CNS and demyelination. Interestingly, differential effects were observed depending on the dose and duration of 5-aza treatment, with a chronically administered low 5-aza dose promoting polarization of T cells into a beneficial Treg phenotype (Chan et al. 2014; Mangano et al. 2014) while high acute 5-aza treatment acting primarily on T effector (Teff) cell proliferation (Wang et al. 2017). Indeed, low doses of 5-aza result in higher number of circulating and infiltrating Treg cells, concomitant with decrease of Th1 and Th17 pathogenic T cell populations. Thymic Treg display enhanced immunosuppressive activity, inhibiting proliferation and lowering the activation potential of Teff cells (Chan et al. 2014). These effects were attributed to predominant hypomethylation of Treg-specific loci such as the *FOXP3* gene (Mangano et al. 2014). In contrast, the infiltrating Treg compartment was unchanged when treating animal with higher 5-aza dose (Wang et al. 2017), this acute administration could instead inhibit activation of the CNS-resident macrophage/microglia and proliferation of encephalitogenic T cells. Unexpectedly, inhibition of T cell proliferation was further shown to be mediated by a TET2 action on cell cycle-related genes. While 5-aza treatment of $CD4^+$ T cells leads to increased TET2 and TET3 and reduced TET1 expression, knock-down of TET2, specifically, can partially abrogate 5-aza effect on proliferation by restoring the expression of key cell cycle inhibitors (p15, p16 and p21 genes) to basal level. TET2 mediation of 5-aza effect operated through direct binding of TET2 to the promoters of these genes (Wang et al. 2017). Thus, 5-aza can favor promoter demethylation not only by inhibiting DNMT1 but also by promoting TET expression in certain cell types. It seems however that the use of global epigenetic modifiers such as DNMT inhibitors for clinical purpose is limited, such global manipulation eliciting broad effects on the methylome. Additionally, as previously mentioned, 5-aza action relies on cell division, thereby restricting its impacts on proliferating cells only, as observed for immune cells, and can exert serious neurotoxic effect (Wang et al. 2013). These deleterious effects therefore hamper its potential for chronic neurodegenerative diseases such as MS.

Novel approaches based on targeted-epigenetic therapy are expected to overcome the global effect of DNMT inhibitors. Such epigenome-editing strategies aim at correcting deleterious DNA methylation changes while leaving homeostatic marks unaltered and would therefore set the stage toward precision and personalized medicine. Among the epigenome editing tools available in animal experimentation, the recently developed CRISPR-dCas9 system appears a promising strategy for targeted epigenetic therapy (Pulecio et al. 2017). CRISPR-dCas9 design builds on the ability of the CRISPR-based system to induce stable locus-specific changes in DNA methylation (Klann et al. 2017). A short single guide RNA (sgRNA) is used to deliver a catalytically inactive Cas9 (dCas9) fused to the catalytic domain of methyltransferase (Dnmt3a) or demethylase (Tet1) to modulate methylation at a specific locus without modifying the DNA sequence. To this aim, a chimera comprising catalytic domains of Dnmt3a and Dnmt3l has been shown to induce more robust degree of methylation compared to Dnmt3a alone (Stepper et al. 2017). Additionally, it has been recently proposed that fusing Dnmt3a-Dnmt3l with the Krüppel-associated box repressor domain (or other repressors) can achieve stable loss of expression that is resistant to external activation stimuli (Amabile et al. 2016). Conversely, CRISPR-dCas9 construct tethering Tet1 catalytic domain, successfully used for targeted demethylation of specific loci, can be utilized to demethylate genes and release them from repression, as demonstrated in-vivo (Liu et al. 2016). Therefore, the attractiveness of targeted epigenetic therapy would rely on its locus-specific, long-lasting, albeit reversible, action, thereby possibly limiting continuous or repeated administration in patients. However, the therapeutic potential of CRISPR-dCas9 epimodifier system is still at early pre-clinical stage of development and additional concerns related to toxicity and safety, with possible off-targets and poor tissue-specificity, warrant further investigation. In that regards, novel molecular design and delivery approaches, e.g. using natural or synthetic carrier nanoparticles (Lu et al. 2014) to either pathogenic immune cells or nervous cell types, would provide complementary tools for specific delivery and efficiency of the epimodifier system.

## 4 Conclusion and Future Perspectives

Multiple Sclerosis is a clinically heterogeneous disease that affects young individuals and results in progressive debilitating disability. Neurologists face daily challenges in the care of MS patients, not the least with variable efficacy and side effects of MS medications and unpredictable disease progression. The MS paradigm proposes that environmental exposures operate on susceptible genetic background to cause disease. The potential of DNA methylation studies to aid in better understanding and treating disease relies on its remarkable properties of sensitivity, stability and reversibility. A growing body of evidence from EWAS supports alteration of DNA methylation levels at specific loci of the genome involved in immune and nervous processes in affected patients compared to controls. However, methylome characterization in MS is still in its infancy and thus warrant further investigation in order to overcome current challenges related to cohort and samples heterogeneity, methodological limitations and biological interpretation of the data. Nevertheless, emerging findings suggest that DNA methylation could mediate genetic and environmental risk in MS and further support DNA methylation as a mediator of phenotypic plasticity driving disease development and progression. In addition, DNA methylation may provide potential biomarkers for phenotypic profiling and prognosis of MS patients. Future perspectives also include the use of DNA methylation as therapeutic targets based on targeted-epigenetic therapy. Overall, DNA methylation could serve as a molecular substratum for precision medicine and personalized care of MS patients. Future research incorporating advanced epigenetic methods in large and homogeneous cohorts appears essential to further integrate encouraging scientific findings to a clinical perspective.

## References

Adzemovic MZ, Zeitelhofer M, Hochmeister S et al (2013) Efficacy of vitamin D in treating multiple sclerosis-like neuroinflammation depends on developmental stage. Exp Neurol 249:39–48

Amabile A, Migliara A, Capasso P et al (2016) Inheritable silencing of endogenous genes by hit-and-run targeted epigenetic editing. Cell 167(1):219–232 e14

Andlauer TF, Buck D, Antony G et al (2016) Novel multiple sclerosis susceptibility loci implicated in epigenetic regulation. Sci Adv 2:e1501678

Ayuso T, Aznar P, Soriano L et al (2017) Vitamin D receptor gene is epigenetically altered and transcriptionally up-regulated in multiple sclerosis. PLoS One 12:e0174726

Bahrami A, Sadeghnia HR, Tabatabaeizadeh SA et al (2018) Genetic and epigenetic factors influencing vitamin D status. J Cell Physiol 233:4033–4043

Baranzini SE, Mudge J, van Velkinburgh JC et al (2010) Genome, epigenome and RNA sequences of monozygotic twins discordant for multiple sclerosis. Nature 464:1351–1356

Barker ED, Roberts S, Walton E (2018) Hidden hypotheses in 'hypothesis-free' genome-wide epigenetic associations. Curr Opin Psychol 27:13–17

Barnett MH, Prineas JW (2004) Relapsing and remitting multiple sclerosis: pathology of the newly forming lesion. Ann Neurol 55:458–468

Belsky J, Jonassaint C, Pluess M et al (2009) Vulnerability genes or plasticity genes? Mol Psychiatry 14:746–754

Ben-Nun A, Wekerle H, Cohen IR (1981) The rapid isolation of clonable antigen-specific T lymphocyte lines capable of mediating autoimmune encephalomyelitis. Eur J Immunol 11:195–199

Bergamaschi R, Cortese A, Pichiecchio A et al (2018) Air pollution is associated to the multiple sclerosis inflammatory activity as measured by brain MRI. Mult Scler 24:1578–1584

Berge T, Leikfoss IS, Brorson IS et al (2016) The multiple sclerosis susceptibility genes TAGAP and IL2RA are regulated by vitamin D in CD4+ T cells. Genes Immun 17:118–127

Bezzini D, Battaglia MA (2017) Multiple sclerosis epidemiology in Europe. Adv Exp Med Biol 958:141–159

Bos SD, Page CM, Andreassen BK et al (2015) Genome-wide DNA methylation profiles indicate CD8+ T cell hypermethylation in multiple sclerosis. PLoS One 10:e0117403

Briggs FB, Acuna B, Shen L et al (2014) Smoking and risk of multiple sclerosis: evidence of modification by NAT1 variants. Epidemiology 25:605–614

Brundin L, Kobelt G, Berg J et al (2017) New insights into the burden and costs of multiple sclerosis in Europe: results for Sweden. Mult Scler 23:179–191

Bujold D, Morais DAL, Gauthier C et al (2016) The international human epigenome consortium data portal. Cell Syst 3:496–499.e2

Burm SM, Peferoen LA, Zuiderwijk-Sick EA et al (2016) Expression of IL-1beta in rhesus EAE and MS lesions is mainly induced in the CNS itself. J Neuroinflammation 13:138

Busche S, Shao X, Caron M et al (2015) Population whole-genome bisulfite sequencing across two tissues highlights the environment as the principal source of human methylome variation. Genome Biol 16:290

Calabrese M, Filippi M, Gallo P (2010) Cortical lesions in multiple sclerosis. Nat Rev Neurol 6:438–444

Calabrese R, Zampieri M, Mechelli R et al (2012) Methylation-dependent PAD2 upregulation in multiple sclerosis peripheral blood. Mult Scler 18:299–304

Calabrese R, Valentini E, Ciccarone F et al (2014) TET2 gene expression and 5-hydroxymethylcytosine level in multiple sclerosis peripheral blood cells. Biochim Biophys Acta 1842:1130–1136

Chan MW, Chang CB, Tung CH et al (2014) Low-dose 5-aza-2′-deoxycytidine pretreatment inhibits experimental autoimmune encephalomyelitis by induction of regulatory T cells. Mol Med 20:248–256

Chang HH, Hemberg M, Barahona M et al (2008) Transcriptome-wide noise controls lineage choice in mammalian progenitor cells. Nature 453:544–547

Chao MJ, Ramagopalan SV, Herrera BM et al (2009) Epigenetics in multiple sclerosis susceptibility: difference in transgenerational risk localizes to the major histocompatibility complex. Hum Mol Genet 18:261–266

Chard DT, Griffin CM, Parker GJ et al (2002) Brain atrophy in clinically early relapsing-remitting multiple sclerosis. Brain 125:327–337
Chen YA, Lemire M, Choufani S et al (2013) Discovery of cross-reactive probes and polymorphic CpGs in the Illumina Infinium HumanMethylation450 microarray. Epigenetics 8:203–209
Chen L, Ge B, Casale FP et al (2016) Genetic drivers of epigenetic and transcriptional variation in human immune cells. Cell 167:1398–1414.e24
Cheung P, Vallania F, Warsinske HC et al (2018) Single-cell chromatin modification profiling reveals increased epigenetic variations with aging. Cell 173:1385–1397e.14
Chomyk AM, Volsko C, Tripathi A et al (2017) DNA methylation in demyelinated multiple sclerosis hippocampus. Sci Rep 7:8696
Compston A, Coles A (2008) Multiple sclerosis. Lancet 372:1502–1517
DeLuca GC, Williams K, Evangelou N et al (2006) The contribution of demyelination to axonal loss in multiple sclerosis. Brain 129:1507–1516
Dineen RA, Vilisaar J, Hlinka J et al (2009) Disconnection as a mechanism for cognitive dysfunction in multiple sclerosis. Brain 132:239–249
Dunaeva M, Derksen M, Pruijn GJM (2018) LINE-1 hypermethylation in serum cell-free DNA of relapsing remitting multiple sclerosis patients. Mol Neurobiol 55:4681–4688
Ebers GC, Sadovnick AD, Dyment DA et al (2004) Parent-of-origin effect in multiple sclerosis: observations in half-siblings. Lancet 363:1773–1774
Elowitz MB, Levine AJ, Siggia ED et al (2002) Stochastic gene expression in a single cell. Science 297:1183–1186
Fagone P, Mangano K, Di Marco R et al (2016) Expression of DNA methylation genes in secondary progressive multiple sclerosis. J Neuroimmunol 290:66–69
Farh KK, Marson A, Zhu J et al (2015) Genetic and epigenetic fine mapping of causal autoimmune disease variants. Nature 518:337–343
Field J, Fox A, Jordan MA et al (2017) Interleukin-2 receptor-alpha proximal promoter hypomethylation is associated with multiple sclerosis. Genes Immun 18:59–66
Fogdell A, Hillert J, Sachs C et al (1995) The multiple sclerosis- and narcolepsy-associated HLA class II haplotype includes the DRB5*0101 allele. Tissue Antigens 46:333–336
Francis PL, Chia TL, Jakubovic R et al (2014) Extensive white matter dysfunction in cognitively impaired patients with secondary-progressive multiple sclerosis. AJNR Am J Neuroradiol 35:1910–1915
Gale CR, Martyn CN (1995) Migrant studies in multiple sclerosis. Prog Neurobiol 47:425–448
Gamazon ER, Segre AV, van de Bunt M et al (2018) Using an atlas of gene regulation across 44 human tissues to inform complex disease- and trait-associated variation. Nat Genet 50:956–967
Gao X, Jia M, Zhang Y et al (2015) DNA methylation changes of whole blood cells in response to active smoking exposure in adults: a systematic review of DNA methylation studies. Clin Epigenetics 7:113
Geurts JJ, Barkhof F (2008) Grey matter pathology in multiple sclerosis. Lancet Neurol 7:841–851
Graves MC, Benton M, Lea RA et al (2014) Methylation differences at the HLA-DRB1 locus in CD4+ T-cells are associated with multiple sclerosis. Mult Scler 20:1033–1041
Hagstrom IT, Schneider R, Bellenberg B et al (2017) Relevance of early cervical cord volume loss in the disease evolution of clinically isolated syndrome and early multiple sclerosis: a 2-year follow-up study. J Neurol 264:1402–1412
Handel AE, De Luca GC, Morahan J et al (2010) No evidence for an effect of DNA methylation on multiple sclerosis severity at HLA-DRB1*15 or HLA-DRB5. J Neuroimmunol 223:120–123
Hansen KD, Timp W, Bravo HC et al (2011) Increased methylation variation in epigenetic domains across cancer types. Nat Genet 43:768–775
Hauser SL, Waubant E, Arnold DL et al (2008) B-cell depletion with rituximab in relapsing-remitting multiple sclerosis. N Engl J Med 358:676–688
Hedstrom AK, Baarnhielm M, Olsson T et al (2009) Tobacco smoking, but not Swedish snuff use, increases the risk of multiple sclerosis. Neurology 73:696–701

Hedstrom AK, Sundqvist E, Baarnhielm M et al (2011) Smoking and two human leukocyte antigen genes interact to increase the risk for multiple sclerosis. Brain 134:653–664
Hedstrom AK, Hillert J, Olsson T et al (2013a) Nicotine might have a protective effect in the etiology of multiple sclerosis. Mult Scler 19:1009–1013
Hedstrom AK, Hillert J, Olsson T et al (2013b) Smoking and multiple sclerosis susceptibility. Eur J Epidemiol 28:867–874
Hedstrom AK, Bomfim IL, Barcellos LF et al (2014) Interaction between passive smoking and two HLA genes with regard to multiple sclerosis risk. Int J Epidemiol 43:1791–1798
Henderson AP, Barnett MH, Parratt JD et al (2009) Multiple sclerosis: distribution of inflammatory cells in newly forming lesions. Ann Neurol 66:739–753
Huynh JL, Garg P, Thin TH et al (2014) Epigenome-wide differences in pathology-free regions of multiple sclerosis-affected brains. Nat Neurosci 17:121–130
International Multiple Sclerosis Genetics C, Wellcome Trust Case Control C, Sawcer S et al (2011) Genetic risk and a primary role for cell-mediated immune mechanisms in multiple sclerosis. Nature 476:214–219
International Multiple Sclerosis Genetics C, Beecham AH, Patsopoulos NA et al (2013) Analysis of immune-related loci identifies 48 new susceptibility variants for multiple sclerosis. Nat Genet 45:1353–1360
Janson PC, Linton LB, Bergman EA et al (2011) Profiling of CD4+ T cells with epigenetic immune lineage analysis. J Immunol 186:92–102
Jeanjean M, Bind MA, Roux J et al (2018) Ozone, NO2 and PM10 are associated with the occurrence of multiple sclerosis relapses. Evidence from seasonal multi-pollutant analyses. Environ Res 163:43–52
Jersild C, Dupont B, Fog T et al (1975) Histocompatibility determinants in multiple sclerosis. Transplant Rev 22:148–163
Kaminsky ZA, Tang T, Wang SC et al (2009) DNA methylation profiles in monozygotic and dizygotic twins. Nat Genet 41:240–245
Kappos L, Radue EW, O'Connor P et al (2010) A placebo-controlled trial of oral fingolimod in relapsing multiple sclerosis. N Engl J Med 362:387–401
Kellogg RA, Tay S (2015) Noise facilitates transcriptional control under dynamic inputs. Cell 160:381–392
Klann TS, Black JB, Chellappan M et al (2017) CRISPR-Cas9 epigenome editing enables high-throughput screening for functional regulatory elements in the human genome. Nat Biotechnol 35:561–568
Koch-Henriksen N, Sorensen PS (2010) The changing demographic pattern of multiple sclerosis epidemiology. Lancet Neurol 9:520–532
Kucharski R, Maleszka J, Foret S et al (2008) Nutritional control of reproductive status in honeybees via DNA methylation. Science 319:1827–1830
Kulakova OG, Kabilov MR, Danilova LV et al (2016) Whole-genome DNA methylation analysis of peripheral blood mononuclear cells in multiple sclerosis patients with different disease courses. Acta Nat 8:103–110
Kular L, Liu Y, Ruhrmann S et al (2018) DNA methylation as a mediator of HLA-DRB1*15:01 and a protective variant in multiple sclerosis. Nat Commun 9:2397
Kumagai C, Kalman B, Middleton FA et al (2012) Increased promoter methylation of the immune regulatory gene SHP-1 in leukocytes of multiple sclerosis subjects. J Neuroimmunol 246:51–57
Lehmann-Werman R, Neiman D, Zemmour H et al (2016) Identification of tissue-specific cell death using methylation patterns of circulating DNA. Proc Natl Acad Sci U S A 113:E1826–E1834
Liggett T, Melnikov A, Tilwalli S et al (2010) Methylation patterns of cell-free plasma DNA in relapsing-remitting multiple sclerosis. J Neurol Sci 290:16–21
Liu Y, Li X, Aryee MJ et al (2014) GeMes, clusters of DNA methylation under genetic control, can inform genetic and epigenetic analysis of disease. Am J Hum Genet 94:485–495
Liu XS, Wu H, Ji X et al (2016) Editing DNA methylation in the mammalian genome. Cell 167 (233–247):e217

Lu CT, Zhao YZ, Wong HL et al (2014) Current approaches to enhance CNS delivery of drugs across the brain barriers. Int J Nanomedicine 9:2241–2257

Lu M, Taylor BV, Korner H (2018) Genomic effects of the vitamin D receptor: potentially the link between vitamin D, immune cells, and multiple sclerosis. Front Immunol 9:477

Maltby VE, Graves MC, Lea RA et al (2015) Genome-wide DNA methylation profiling of CD8+ T cells shows a distinct epigenetic signature to CD4+ T cells in multiple sclerosis patients. Clin Epigenetics 7:118

Maltby VE, Lea RA, Sanders KA et al (2017) Differential methylation at MHC in CD4(+) T cells is associated with multiple sclerosis independently of HLA-DRB1. Clin Epigenetics 9:71

Maltby VE, Lea RA, Ribbons KA et al (2018) DNA methylation changes in CD4(+) T cells isolated from multiple sclerosis patients on dimethyl fumarate. Mult Scler J Exp Transl Clin 4:2055217318787826

Mangano K, Fagone P, Bendtzen K et al (2014) Hypomethylating agent 5-aza-2′-deoxycytidine (DAC) ameliorates multiple sclerosis in mouse models. J Cell Physiol 229:1918–1925

Marabita F, Almgren M, Sjoholm LK et al (2017) Smoking induces DNA methylation changes in multiple sclerosis patients with exposure-response relationship. Sci Rep 7:14589

Martin MP, Carrington M (2005) Immunogenetics of viral infections. Curr Opin Immunol 17:510–516

Mastronardi FG, Noor A, Wood DD et al (2007) Peptidyl argininedeiminase 2 CpG island in multiple sclerosis white matter is hypomethylated. J Neurosci Res 85:2006–2016

Mattoscio M, Nicholas R, Sormani MP et al (2015) Hematopoietic mobilization: potential biomarker of response to natalizumab in multiple sclerosis. Neurology 84:1473–1482

Meijer KA, Muhlert N, Cercignani M et al (2016) White matter tract abnormalities are associated with cognitive dysfunction in secondary progressive multiple sclerosis. Mult Scler 22:1429–1437

Mendizabal I, Zeng J, Keller TE et al (2017) Body-hypomethylated human genes harbor extensive intragenic transcriptional activity and are prone to cancer-associated dysregulation. Nucleic Acids Res 45:4390–4400

Moore JR, Hubler SL, Nelson CD et al (2018) 1,25-dihydroxyvitamin D3 increases the methionine cycle, CD4(+) T cell DNA methylation and Helios(+)Foxp3(+) T regulatory cells to reverse autoimmune neurodegenerative disease. J Neuroimmunol 324:100–114

Morris G, Maes M, Murdjeva M et al (2018) Do human endogenous retroviruses contribute to multiple sclerosis, and if so, how? Mol Neurobiol:1–16. https://doi.org/10.1007/s12035-018-1255-x

Munger KL, Levin LI, Hollis BW et al (2006) Serum 25-hydroxyvitamin D levels and risk of multiple sclerosis. JAMA 296:2832–2838

Munz C (2016) Autophagy beyond intracellular MHC class II antigen presentation. Trends Immunol 37:755–763

Neri F, Rapelli S, Krepelova A et al (2017) Intragenic DNA methylation prevents spurious transcription initiation. Nature 543:72–77

Neven KY, Piola M, Angelici L et al (2016) Repetitive element hypermethylation in multiple sclerosis patients. BMC Genet 17:84

Ng B, White CC, Klein HU et al (2017) An xQTL map integrates the genetic architecture of the human brain's transcriptome and epigenome. Nat Neurosci 20:1418–1426

O'Gorman C, Lin R, Stankovich J et al (2013) Modelling genetic susceptibility to multiple sclerosis with family data. Neuroepidemiology 40:1–12

Olsson T, Barcellos LF, Alfredsson L (2017) Interactions between genetic, lifestyle and environmental risk factors for multiple sclerosis. Nat Rev Neurol 13:25–36

Paludan C, Schmid D, Landthaler M et al (2005) Endogenous MHC class II processing of a viral nuclear antigen after autophagy. Science 307:593–596

Pierini F, Lenz TL (2018) Divergent allele advantage at human MHC genes: signatures of past and ongoing selection. Mol Biol Evol 35(9):2145–2158. https://doi.org/10.1093/molbev/msy116

Pinto-Medel MJ, Oliver-Martos B, Urbaneja-Romero P et al (2017) Global methylation correlates with clinical status in multiple sclerosis patients in the first year of IFNbeta treatment. Sci Rep 7:8727

Polman CH, O'Connor PW, Havrdova E et al (2006) A randomized, placebo-controlled trial of natalizumab for relapsing multiple sclerosis. N Engl J Med 354:899–910

Pulecio J, Verma N, Mejia-Ramirez E et al (2017) CRISPR/Cas9-based engineering of the epigenome. Cell Stem Cell 21:431–447

Ramagopalan SV, Maugeri NJ, Handunnetthi L et al (2009) Expression of the multiple sclerosis-associated MHC class II allele HLA-DRB1*1501 is regulated by vitamin D. PLoS Genet 5: e1000369

Rhead B, Brorson IS, Berge T et al (2018) Increased DNA methylation of SLFN12 in CD4+ and CD8+ T cells from multiple sclerosis patients. PLoS One 13:e0206511

Ruhrmann S, Ewing E, Piket E et al (2018) Hypermethylation of MIR21 in CD4+ T cells from patients with relapsing-remitting multiple sclerosis associates with lower miRNA-21 levels and concomitant up-regulation of its target genes. Mult Scler 24:1288–1300

Shiura H, Okamoto A, Sasaki H et al (2014) Whole-mount MeFISH: a novel technique for simultaneous visualization of specific DNA methylation and protein/RNA expression. PLoS One 9:e95750

Simpson S Jr, Blizzard L, Otahal P et al (2011) Latitude is significantly associated with the prevalence of multiple sclerosis: a meta-analysis. J Neurol Neurosurg Psychiatry 82:1132–1141

Singhal NK, Freeman E, Arning E et al (2018) Dysregulation of methionine metabolism in multiple sclerosis. Neurochem Int 112:1–4

Smallwood SA, Lee HJ, Angermueller C et al (2014) Single-cell genome-wide bisulfite sequencing for assessing epigenetic heterogeneity. Nat Methods 11:817–820

Soelberg Sorensen P (2017) Safety concerns and risk management of multiple sclerosis therapies. Acta Neurol Scand 136:168–186

Sokratous M, Dardiotis E, Bellou E et al (2018) CpG island methylation patterns in relapsing-remitting multiple sclerosis. J Mol Neurosci 64:478–484

Stahl PL, Salmen F, Vickovic S et al (2016) Visualization and analysis of gene expression in tissue sections by spatial transcriptomics. Science 353:78–82

Stepper P, Kungulovski G, Jurkowska RZ et al (2017) Efficient targeted DNA methylation with chimeric dCas9-Dnmt3a-Dnmt3L methyltransferase. Nucleic Acids Res 45:1703–1713

Stewart SK, Morris TJ, Guilhamon P et al (2015) oxBS-450K: a method for analysing hydroxymethylation using 450K BeadChips. Methods 72:9–15

Sundqvist E, Baarnhielm M, Alfredsson L et al (2010) Confirmation of association between multiple sclerosis and CYP27B1. Eur J Hum Genet 18:1349–1352

Titus AJ, Gallimore RM, Salas LA et al (2017) Cell-type deconvolution from DNA methylation: a review of recent applications. Hum Mol Genet 26:R216–R224

Tortorella C, Direnzo V, Ruggieri M et al (2018) Cerebrospinal fluid neurofilament light levels mark grey matter volume in clinically isolated syndrome suggestive of multiple sclerosis. Mult Scler 24:1039–1045

Trojano M, Lucchese G, Graziano G et al (2012) Geographical variations in sex ratio trends over time in multiple sclerosis. PLoS One 7:e48078

van der Mei I, Lucas RM, Taylor BV et al (2016) Population attributable fractions and joint effects of key risk factors for multiple sclerosis. Mult Scler 22:461–469

van Horssen J, Singh S, van der Pol S et al (2012) Clusters of activated microglia in normal-appearing white matter show signs of innate immune activation. J Neuroinflammation 9:156

Vogt G, Huber M, Thiemann M et al (2008) Production of different phenotypes from the same genotype in the same environment by developmental variation. J Exp Biol 211:510–523

Waddington CH (1959) Canalization of development and genetic assimilation of acquired characters. Nature 183:1654–1655

Wang Y, Wang X, Li R et al (2013) A DNA methyltransferase inhibitor, 5-aza-2′-deoxycytidine, exacerbates neurotoxicity and upregulates Parkinson's disease-related genes in dopaminergic neurons. CNS Neurosci Ther 19:183–190

Wang X, Wang J, Yu Y et al (2017) Decitabine inhibits T cell proliferation via a novel TET2-dependent mechanism and exerts potent protective effect in mouse auto- and allo-immunity models. Oncotarget 8:56802–56815

Webster AP, Plant D, Ecker S et al (2018) Increased DNA methylation variability in rheumatoid arthritis-discordant monozygotic twins. Genome Med 10:64

Westerlind H, Ramanujam R, Uvehag D et al (2014) Modest familial risks for multiple sclerosis: a registry-based study of the population of Sweden. Brain 137:770–778

Xiao D, Ye X, Zhang N et al (2015) A meta-analysis of interaction between Epstein-Barr virus and HLA-DRB1*1501 on risk of multiple sclerosis. Sci Rep 5:18083

Yeung MS, Zdunek S, Bergmann O et al (2014) Dynamics of oligodendrocyte generation and myelination in the human brain. Cell 159:766–774

Yu M, Heinzerling TJ, Grady WM (2018) DNA methylation analysis using droplet digital PCR. Methods Mol Biol 1768:363–383

Zeitelhofer M, Adzemovic MZ, Gomez-Cabrero D et al (2017) Functional genomics analysis of vitamin D effects on CD4+ T cells in vivo in experimental autoimmune encephalomyelitis. Proc Natl Acad Sci U S A 114:E1678–E1687

# Early Life Stress and DNA Methylation

**Annakarina Mundorf and Nadja Freund**

## Contents



**Abstract** DNA methylation and demethylation can be influenced by several environmental factors including diet, smoking, drug consumption, parental behavior and stress. Given that methylation changes can lead to altered gene transcription their impact can be enormous. Therefore, it is very important to understand the processes and underlying factors influencing methylation. Changes in DNA methylation that occur early during development induce altered gene expression that can affect the development of the brain and other organs right from the beginning. Stress during early development is linked to an increased risk for psychiatric and physiological disorders and altered DNA methylation could be the mediating factor. Whether the addition or the removal of methyl groups is linked to psychiatric outcome depends on several factors like the specific gene and the exposure. There are different approaches to investigate this relationship and to identify risk genes. Some groups focus on the mediating effect of gene methylation on early life stress exposure and psychiatric outcome. Another approach is the study of gene methylation in adults

A. Mundorf · N. Freund (✉)
Research Division Experimental and Molecular Psychiatry, Department of Psychiatry, Psychotherapy and Preventive Medicine, LWL University Hospital, Ruhr-University Bochum, Bochum, Germany
e-mail: nadja.freund@rub.de

 
S. Jurga, J. Barciszewski (eds.), *The DNA, RNA, and Histone Methylomes*, RNA Technologies, https://doi.org/10.1007/978-3-030-14792-1_9

with already diagnosed psychiatric disorders. Others investigated the reversible effect of psychotherapy on gene methylation in patients. Only a few studies correlate gene methylation in healthy adults with subclinical symptoms.

The following chapter will first give a brief introduction on environmental influences, DNA methylation and increased risk for the development of psychiatric disorders. It will then summarize findings in human and animal studies on early life stress, gene methylation and stress-related psychiatric disorders. At the end of the chapter, we will give an outlook on possible clinical applications.

**Keywords** Psychopathology · Lifestyle · Glucocorticoid receptor · Animal models · Maternal separation · Gene expression · Behavior

# 1 Environmental Influences on DNA Methylation

DNA methylation is either the spontaneous alteration of gene expression or the reaction to environmental influences, a post-replication modification by adding or removing a methyl group at a CpG side (Jaenisch and Bird 2003). It often occurs during the development of the organism and therefore changes are carried on through mitosis. In general, differences in gene expression are a mechanism that allows heterogeneity between genetically homologous cells. Environmental factors are known to influence DNA methylation patterns, thereby altering gene function and expression (Razin and Riggs 1980; Razin and Cedar 1991) and even causing disorders. Epigenetic modifications have been associated with various diseases like cancer, cardiovascular and neurodegenerative diseases as well as psychiatric disorders (Santos-Rebouças and Pimentel 2007; Mahgoub and Monteggia 2013). Therefore, understanding epigenetic processes and influences on the organism is an important step in clinical research. In cancer research the field of cancer epigenetics is rapidly growing and epigenetic therapy is making promising progress (Sharma et al. 2010). Driven by the success in cancer, other fields including psychiatry are now starting to also investigate epigenetics involved in disease processes.

## *1.1 Lifestyle*

Not only severe life events but several environmental factors can influence DNA methylation. An individual's lifestyle, physical activities, nutrition, alcohol consumption, smoking or stress are known to induce DNA methylation changes (Alegría-Torres et al. 2011; Lim and Song 2012). Physical activity e.g. has been associated with higher methylation of the *LINE-1* gene which is linked to a reduced risk of ischemic heart disease and stroke in elderly (Alegría-Torres et al. 2011).

The abuse of substances including cocaine, opioids or alcohol is well-known for altering the DNA methylation state of specific genes (Nielsen et al. 2012). Altered gene expression that arises as consequence might then be responsible for dysfunction in reward signaling, craving, and relapse leading to addiction and relapse (Nielsen et al. 2012). Even recent alcohol intake (i.e. consumption in the past 6 month) is known to change gene methylation in healthy participants (investigated in blood cells), leading to altered gene expression (Philibert et al. 2012; Liu et al. 2018). Thereby methylation changes depend on drinking frequency and indicate that higher amounts of consumed alcohol might lead to stronger DNA methylation changes (Philibert et al. 2012). Smoking can also effect DNA methylation in a dose- and time-dependent manner (Philibert et al. 2014). Altered DNA methylation induced by smoking affects gene regulation and is hypothesized to increase vulnerability to other diseases (Philibert et al. 2014).

Not only the amount but also timing of exposure or consumption is crucial. Vulnerable windows for environmental influences are especially periods of neural growth like early childhood or adolescence (Andersen and Teicher 2008; Freund et al. 2013). Therefore, lifestyle of a pregnant woman can even have long term consequences for the unborn child. DNA methylation changes induced by prenatal exposure might alter gene expression patterns even before crucial brain structures or pathways are fully developed (Lucassen et al. 2013) and induce long-lasting DNA methylation changes which e.g. increase the risk to develop psychiatric disorders later in life (Knopik et al. 2019). An unhealthy diet during pregnancy e.g. can affect DNA methylation of insulin-like growth factor 2 gene (*IGF2*) of the offspring and increase the risk to develop attention-deficit/hyperactivity disorder (ADHD) symptoms (Rijlaarsdam et al. 2017). Perinatal malnutrition also affects hippocampal growth via altered DNA methylation and increase the risk to develop psychiatric disorders (Lucassen et al. 2013).

## 1.2 Stress

Stress is known to affect an individual at many different levels. Influences on the immune and cardiovascular system as well as the brain can induce various diseases (McEwen and Stellar 1993; Romeo 2016). Specifically, during development, like early childhood and adolescence, the brain is vulnerable to stress exposure resulting in an increased risk to develop mental disorders consequently. Severe stress exposure is linked to several psychiatric disorders including major depressive disorder (MDD), bipolar disorder, posttraumatic stress disorder (PTSD) and schizophrenia (Vinkers et al. 2015; Matosin et al. 2017). Changes induced by stress exposure leading to the onset of a psychiatric disorder might be mediated by epigenetic alteration (Klengel et al. 2014; Vinkers et al. 2015; Matosin et al. 2017; Barker et al. 2017). Thereby, timing of stress exposure is critical as mentioned above. Consequences on the individual's development and health depend if stress exposure occurred prenatally, in early childhood or in adolescence (Matosin et al. 2017).

A study investigating DNA methylation changes (in buccal cells) in 15-year-old adolescents reported altered methylation after experiencing a stressful childhood (Essex et al. 2013). Interestingly, they found different time windows for maternal and paternal stress impacting the child's DNA methylation. Maternal stressors in infancy and paternal stressors in preschool years (parental stressors being e.g. financial stress, parenting stress, depression) were predictive for several DNA methylation changes (Essex et al. 2013). Methylation changes following stress exposure have been reported on several genes. The glucocorticoid receptor gene (*NR3C1*), the serotonin transporter gene (*SLC6A4*) and *FKBP5* (a gene involved in the immune system and interaction with the glucocorticoid receptor), however, seem to be mainly affected (Vinkers et al. 2015).

# 2 Findings in Human and Animal Studies Regarding Early Life Stress, Gene Methylation, and Psychiatric Disorders

## *2.1 Human Studies*

Early life stress (ELS) affects DNA methylation patterns (Vinkers et al. 2015) and ample evidence indicates that ELS increases the risk for several psychiatric disorders (Teicher et al. 2003). Therefore, it is of special interest to link the reported methylation changes after ELS to psychopathology. Indeed methylation of the serotonin transporter gene *SLC6A4* has been reported to be altered in patients with bipolar disorder, MDD, PTSD, schizophrenia, ADHD and obesity (Palma-Gudiel and Fañanás 2017). Similarly, MDD and PTSD have been linked to methylation changes of the genes *NR3C* and *FKBP5* while increased methylation of *NR3C1* changes have also been reported in borderline personality disorder and patients with bipolar disorder show increased methylation on *FKBP5* (Argentieri et al. 2017).

Identifying methylation changes that mediate psychopathology after ELS, however, entails some challenges. Different aspects of life like smoking, nutrition, and parental behavior can influence DNA methylation. Therefore, it is hard to identify if altered DNA methylation is caused by a specific ELS and not being influenced by other environmental factors. Cause and consequence of methylation changes and psychopathology are hard to identify. Nevertheless, first studies try to identify the direct link between ELS, methylation and psychopathology. Specifically for the glucocorticoid receptor, there are some hints that this link exists (Smart et al. 2015). Radtke and colleagues e.g. assessed the occurrence of lifetime childhood maltreatment in 46 individuals, measured the methylation of the glucocorticoid receptor in blood samples and conducted a structured interview to evaluate psychological wellbeing. They found that number of events of maltreatment correlated positively with methylation and symptoms of borderline personality disorder and depression also positively correlated with methylation (Radtke et al. 2015). Longitudinal, population-based studies also support the role of DNA methylation in mediating psychopathology after ELS (Barker et al. 2017). One example is the Avon Longitudinal Study of Parents and Children (ALSPAC) which is an ongoing epidemiological study of children and parents gathering psychological and physiological information e.g. whole-genome methylation data at different time points (Boyd et al. 2013; Fraser et al. 2013). With an integrated data resource for epigenomic studies (called ARIES) longitudinal, population-based DNA methylation profiling with a great number of subjects (N = 1018) is possible (Relton et al. 2015). Out of this complex study, multiple smaller studies have been conducted investigating the role of DNA methylation in connecting ELS and psychopathology. It has been found that methylation of the oxytocin receptor gene (*OXTR*) is associated with higher prenatal risk exposure and with higher callous-unemotional traits in adolescence (Cecil et al. 2014). *OXTR* methylation could therefore mediate callous-unemotional traits, a risk factor for developing early-onset conduct problems (CP).

In another epigenome-wide analysis, cord blood DNA methylation of children from the ALSPAC has been examined to investigate if altered DNA methylation could serve as a biomarker to detect early-onset CP (Cecil et al. 2018). Specifically, they used trajectories of CP to search for DNA methylation alterations between early-onset versus low CP and analyzed if early exposure influences DNA methylation. For example, maternal smoking showed a strong correlation with increased DNA methylation of *MGLL*, a gene encoding for a protein involved in pain perception, and was associated with early-onset CP in late childhood (Cecil et al. 2018).

Apart from the timing of stress exposure (prenatal, early life, adolescence) most epigenome-wide studies differ in (*I*) investigated cell types (umbilical cord blood, leukocytes, neonatal cord blood, blood, buccal cells, serum, brain tissue of suicide cases), (*II*) trauma type being chosen as ELS (childhood sexual or physical abuse, neglect, maltreatment, parental death/loss, institutionalized children), (*III*) technique being used for methylation analysis and (*IV*) the way of analyzing the results (genome-wide or candidate gene approach) (Vinkers et al. 2015). This high variability in the study design complicates validation of results and therewith, might weaken the potential role of DNA methylation as a mediator or biomarker for psychopathologies (Barker et al. 2017).

Another obstacle in finding methylation markers that link ELS and psychopathology is that ELS affects the brain and leads to neuroanatomical changes like volume reduction, of e.g. the hippocampus, the amygdala, the prefrontal cortex, and the corpus callosum (Hart and Rubia 2012; Frodl and O'Keane 2013). Changes in DNA methylation in peripheral tissue, however, do not reflect on the brain as DNA methylation is tissue specific (Bakulski et al. 2016). Studies in human post-mortem brain tissue are rare and limited (Vinkers et al. 2015). Animal models might help to overcome this obstacle. They are an excellent tool to investigate early life experience in a very controlled manner and link its effect on methylation changes in brain tissue to behavioural consequences.

## 2.2 Animal Studies

A well-established animal model for ELS is maternal separation (MS) (Freund et al. 2013; Nieto et al. 2016). Briefly, the pups are separated from their mother for several hours each day during their early childhood. Affected brain regions and behavioral phenotype following MS are comparable to findings in humans who experienced ELS (Teicher et al. 2003). Following MS, multiple DNA methylation changes in genes (frequently relevant in brain maturation or plasticity) in different brain regions (e.g. hippocampus, hypothalamus, amygdala, prefrontal cortex) have been found (Fumagalli et al. 2007; Nieto et al. 2016). Regarding the psychopathology, conclusions can be drawn by investigating the behavior of the animal. As well as in humans, disorders are first characterized by alterations in the normal behavior. Therewith, increased anxiety, depressive-like, hedonic or anhedonic behavior can be measured as well as impaired memory, stress coping and impulsiveness. Standardized assays have been developed and validated to investigate psychiatric psychopathologies in rodents (Nestler and Hyman 2010). For example, by using the natural aversion of rodents towards open spaces, altered anxiety behavior can be assessed by exposing them to a large open field (open field test).

Investigating effects of ELS on DNA methylation, an increased methylation of *Nr3c1* as shown in humans has also been confirmed in the hippocampus of maternally separated mice (Kember et al. 2012). These animals furthermore showed increased methylation of the gene for vasopressin *Avp* and decreased methylation of *Nr4a1*, a growth factor gen (Kember et al. 2012). Changes in methylation status were furthermore accompanied by increased anxiety behavior and an increased corticosterone response to stress (Kember et al. 2012). In the hypothalamus MS resulted in decreased methylation of the corticotrophin gen *Crh* (Chen et al. 2012) and *Avp* (Murgatroyd et al. 2009). Another benefit of animal studies is the analysis of DNA methylation and mRNA expression at once. The decreased methylation of *Avp* resulted in increased vasopressin expression and animals showing alterations in stress coping and memory. Interestingly, a vasopressin receptor agonist was able to reverse the behavioral changes (Murgatroyd et al. 2009) reinforcing the important role of *Avp* in stress coping and memory behavior. Synapsin gen *Syn1* was more methylated in the amygdala after MS resulting in lower mRNA expression (Park et al. 2014). In the same region increased methylation of *Ntsr1* (encoding for the neurotensin reporter 1) induced enhanced fear conditioning and reduced gene expression of *Ntsr1* (Toda et al. 2014). Microinjection of a neurotensin receptor agonist or antagonist was able to de- or increase fear conditioning, respectively (Toda et al. 2014). All studies reinforce the mediating role of methylation. In the pituitary gland, MS caused decreased methylation of the Proopiomelanocortin gen *Pomc* (a gene involved in stress and immune modulation) that was accompanied by an increase in mRNA expression (Wu et al. 2014). Furthermore, the reduction of reward seeking after MS was accompanied by hypermethylation of the Dopamine D1 receptor in the nucleus accumbens and a consequent decrease of mRNA as well as protein levels (Sasagawa et al. 2017). These studies confirm the potential role of

DNA methylation in mediating the development of psychopathologies after experiencing ELS.

First hints for treatment options were shown in animal models as well. Adding a methyl donor to the animals diet improved anxiety and depressive-like behavior after MS (McCoy et al. 2016; Paternain et al. 2016). As the methyl donor enriched diet was given when the MS animals were adults the behavioral alterations induced by MS are still reversible later on. The preventive effects of an enriched diet has been suggested in humans as well (Rijlaarsdam et al. 2017).

But not only separation from the mother, even low maternal care can influence DNA methylation. Disrupted maternal care increased the expression of the DNA methyltransferase in the offspring's amygdala leading to an increase in total methylation and increased anxiety. Abusive maternal behavior increased methylation of the *Bdnf* gene, coding for a neurotrohic factor, as well as decreased its mRNA expression in the prefrontal cortex (Roth et al. 2009). This fact is in line with results found in human patients with borderline personality disorder showing increased *BDNF* methylation (Thomas et al. 2018, see Sect. 3). In the animal model, a potential treatment option was already discovered. Infusing the grown up rats with a DNA methylation inhibitor resulted in decreased methylation of the *Bdnf* gene and normalized mRNA expression (Roth et al. 2009). Increased maternal care, on the other hand, can have beneficial effects. Increased licking and grooming of the pups decreased the methylation of the glucocorticoid gene in the offspring's hippocampus (Weaver et al. 2004). This alteration was persistent into adulthood and might influence the stress response via the hypothalamic-pituitary-adrenal axis. So, early maternal care might have a high effect on the stress response of the offspring by affecting brain plasticity.

The findings in animal studies emphasize the relevance of DNA methylation changes that disrupt brain maturation, especially, when the methylation changes are induced during vulnerable windows. Nevertheless, most studies only investigate the effects of ELS on DNA methylation without considering the consequences on psychopathology (behavior). In addition, apart from the glucocorticoid receptor, human studies point to different genes compared to animal studies. A recently conducted evaluation of human and animal studies on ELS and DNA methylation found that results match in less than 50% between animal models and studies with humans (Watamura and Roth 2018). One explanation for this finding could be that, preclinical (animal) and clinical (human) studies are not well-aligned and often differ in time of stress exposure, type of stressor, timing of tissue sampling and tissue investigated (Nieto et al. 2016). Cross-species, multi-tissue studies to investigate the effect of ELS on DNA methylation help to identify stable effects that are comparable across species.

## 2.3 Cross Species Studies

Nieratschker and colleagues compared DNA methylation of the *MORC family CW-type zinc finger 1* (*MORC1*) gene from (*I*) human cord blood following prenatal stress, (*II*) the prefrontal cortex tissue of adult rats that had been exposed to prenatal stress, and (*III*) blood cells of adolescent nonhuman primates after maternal separation with (*IV*) matched non-stressed control groups (Nieratschker et al. 2014). They found reduced methylation of *MORC1* in all tissues of all species. Then, they went one step further and performed a gene-based case-control analysis with data from a previous genome-wide association study with blood from major depression patients. Interestingly, specific gene variants of *MORC1* were associated with major depressive disorder (Nieratschker et al. 2014).

Similarly, methylation profiles in post-mortem hippocampus samples of suicide victims with and without a history of severe childhood abuse were studied (Suderman et al. 2012). The findings were then compared with results in hippocampus samples of rats that received low or high maternal care. Interestingly, they found high similarities of methylation of the glucocorticoid receptor gene after ELS in humans and rats indicating cross-species regulatory mechanisms that are conserved. As mentioned, altered methylation of the glucocorticoid receptor gene after ELS has already been reported in human blood (Radtke et al. 2015; Smart et al. 2015) and rat hippocampus (Weaver et al. 2004). So far, this alteration has been linked with increased risk of MDD and stress related disorders like PTSD. Therewith, it seems possible that methylation of the glucocorticoid receptor gene might rather play a role in stress reaction than in the development of a specific disorder. Even though cross species studies hold promise there are still rare and more studies are urgently needed.

## 3 Possible Clinical Applications of Gene Methylation: An Outlook

As stated in the beginning, there are different approaches to investigate the effects of ELS and DNA methylation on psychopathology. We reported that DNA methylation changes have been found when comparing humans and animals with or without stress experience or between patients and healthy controls.

Another approach is to correlate symptoms and methylation patterns in population wide studies. In a recent study with healthy adults (30 women and 30 men), we found a significant association between increased *MORC1* promoter methylation in buccal cells and increased self-depression scoring according to the Beck Depression Inventory (Mundorf et al. 2018). Increased *MORC1* methylation could potentially serve as an early detection marker for depressive symptoms. Similarly, methylation patterns of several genes might act as biomarker for early detection of psychiatric disorders. One advantage of early detection is that our epigenetic pattern including DNA methylation is changeable. As mentioned above, drugs can influence our DNA methylation signature. Therefore it is no surprise, that pharmacological treatment alters DNA methylation (Boks et al. 2012; Ovenden et al. 2018). But even psychotherapy can alter DNA methylation. The serotonin transporter as well as *FKBP5* showed changes in methylation pattern after psychotherapy (Roberts et al. 2014, 2019). Even more interesting, stronger changes in methylation pattern are associated with a greater response to the therapy. Therefore, DNA methylation could potentially serve as marker for therapy outcome.

A different but interesting approach is the study of gene methylation in adults with already diagnosed psychiatric disorders. Especially, as early stress exposure is linked to an increased risk of a psychiatric disorder and altered DNA methylation. Identifying risk genes in patients will facilitate early diagnosis.

In a recent clinical study in borderline personality disorder patients (85% women) and matched controls, *BDNF* promoter methylation was investigated in buccal and blood cells (Thomas et al. 2018). Interestingly, the researchers found a significantly increased *BDNF* promoter methylation in the patients but only in saliva samples. Moreover, they investigated the effects of psychotherapy on *BDNF* methylation in the patients which decreased significantly after 12-month treatment (Thomas et al. 2018). More than half of the borderline patients declared having experienced childhood traumata. These findings seems to be in line with previous findings in animal studies where an increased methylation of *Bdnf* in the prefrontal cortex was found after experiencing abusive maternal care (Roth et al. 2009). In the animal study, treating the grown-up rats with a DNA methylation inhibitor resulted in decreased methylation of the *Bdnf* gene and normalized mRNA expression giving hope for new and effective treatment options of borderline personality disorder.

As mentioned above (see Sect. 1.1) alcohol consumption can alter DNA methylation. Whether alcohol consumption induces DNA methylation changes leading to addiction or DNA methylation changes leading to more consumption and therewith addictive behavior is unknown. For a better understanding, both scenarios have to be analyzed. In a study with alcohol-dependent patients (49 male patients; mean age $49.14 \pm 10.47$ years), the patients showed decreased methylation of *GDAP1*, a gene, that has so far only been associated with the neurological disorder Charcot-Marie-Tooth disease. Furthermore the analyses conducted in blood cells revealed that the degree of hypomethylation was associated with increased alcohol dependence (Brückmann et al. 2016). Methylation of *GDAP1* might therefore serve as a biomarker for disease severity. In addition, the comparison of *GDAP1* methylation before and after 3 weeks of an alcohol treatment program showed an increase of methylation after treatment (Brückmann et al. 2016). Other studies confirmed, that an altered DNA methylation signature in blood cells could enable clinical diagnosis of heavy alcohol consumption which is difficult to access in patients (Liu et al. 2018). As heavy alcohol consumption (and dependence) is frequent in patients with psychiatric disorders knowing the DNA methylation signature will facilitate differential diagnosis by DNA methylation.

Even though the recent findings in DNA methylation predicting psychopathology or treatment outcome seems promising, the application of DNA methylation as biomarker posses challenges. Most studies use case-control or cross-sectional methods and only a few apply a prospective approach. Therefore, it is still unclear whether gene methylation is a predisposing factor for diseases or a consequence of pathology (Argentieri et al. 2017).

## 4 Conclusions

In conclusion, human and animal studies report an effect of ELS on DNA methylation. In addition, ELS and altered DNA methylation have been linked to different psychiatric disorders. Therefore, altered DNA methylation might be the missing link between stress exposure and the development of psychopathology. So far, most studies use different designs or investigate different genes, making it difficult to validate results. Moreover, animal and human studies are not well aligned, and more cross-species studies are needed. Nevertheless, DNA methylation has the potential for an easy-to-apply biomarker not only to facilite diagnosis but also as an early detection marker of symptoms. Findings like the reversal effect of psychotherapy on DNA methylation or the successful treatment with a methyl donor put DNA methylation in the position of a possible clinical treatment target.

## References

Alegría-Torres JA, Baccarelli A, Bollati V (2011) Epigenetics and lifestyle. Epigenomics 3:267–277

Andersen SL, Teicher MH (2008) Stress, sensitive periods and maturational events in adolescent depression. Trends Neurosci 31:183–191

Argentieri MA, Nagarajan S, Seddighzadeh B et al (2017) Epigenetic pathways in human disease: the impact of DNA methylation on stress-related pathogenesis and current challenges in biomarker development. EBioMedicine 18:327–350

Bakulski KM, Halladay A, Hu VW et al (2016) Epigenetic research in neuropsychiatric disorders: the "tissue issue". Curr Behav Neurosci Rep 3:264–274

Barker ED, Walton E, Cecil CAM (2017) Annual Research Review: DNA methylation as a mediator in the association between risk exposure and child and adolescent psychopathology. J Child Psychol Psychiatry 59:303–322

Boks MP, de Jong NM, Kas MJH et al (2012) Current status and future prospects for epigenetic psychopharmacology. Epigenetics 7:20–28

Boyd A, Golding J, Macleod J et al (2013) Cohort profile: The 'Children of the 90s'—the index offspring of the Avon Longitudinal Study of Parents and Children. Int J Epidemiol 42:111–127

Brückmann C, Di Santo A, Karle KN et al (2016) Validation of differential GDAP1 DNA methylation in alcohol dependence and its potential function as a biomarker for disease severity and therapy outcome. Epigenetics 11:456–463

Cecil CAM, Lysenko LJ, Jaffee SR et al (2014) Environmental risk, oxytocin receptor gene (OXTR) methylation and youth callous-unemotional traits: a 13-year longitudinal study. Mol Psychiatry 19:1071–1077

Cecil CAM, Walton E, Jaffee SR et al (2018) Neonatal DNA methylation and early-onset conduct problems: a genome-wide, prospective study. Dev Psychopathol 30:383–397

Chen J, Evans AN, Liu Y et al (2012) Maternal deprivation in rats is associated with corticotrophin-releasing hormone (CRH) promoter hypomethylation and enhances CRH transcriptional responses to stress in adulthood. J Neuroendocrinol 24:1055–1064

Essex MJ, Boyce WT, Hertzman C et al (2013) Epigenetic vestiges of early developmental adversity: childhood stress exposure and DNA methylation in adolescence. Child Dev 84:58–75

Fraser A, Macdonald-Wallis C, Tilling K et al (2013) Cohort profile: the Avon Longitudinal Study of parents and children: ALSPAC mothers cohort. Int J Epidemiol 42:97–110

Freund N, Thompson BS, Denormandie J et al (2013) Windows of vulnerability: maternal separation, age, and fluoxetine on adolescent depressive-like behavior in rats. Neuroscience 249:88–97
Frodl T, O'Keane V (2013) How does the brain deal with cumulative stress? A review with focus on developmental stress, HPA axis function and hippocampal structure in humans. Neurobiol Dis 52:24–37
Fumagalli F, Molteni R, Racagni G et al (2007) Stress during development: Impact on neuroplasticity and relevance to psychopathology. Prog Neurobiol 81:197–217
Hart H, Rubia K (2012) Neuroimaging of child abuse: a critical review. Front Hum Neurosci 6:52
Jaenisch R, Bird A (2003) Epigenetic regulation of gene expression: how the genome integrates intrinsic and environmental signals. Nat Genet 33(Suppl):245–254
Kember RL, Dempster EL, Lee THA et al (2012) Maternal separation is associated with strain-specific responses to stress and epigenetic alterations to Nr3c1, Avp, and Nr4a1 in mouse. Brain Behav 2:455–467
Klengel T, Pape J, Binder EB et al (2014) The role of DNA methylation in stress-related psychiatric disorders. Neuropharmacology 80:115–132
Knopik VS, Marceau K, Bidwell LC et al (2019) Prenatal substance exposure and offspring development: Does DNA methylation play a role? Neurotoxicol Teratol 71:50–63. https://doi.org/10.1016/j.ntt.2018.01.009
Lim U, Song M-A (2012) Dietary and lifestyle factors of DNA methylation. Methods Mol Biol Clifton NJ 863:359–376
Liu C, Marioni RE, Hedman ÅK et al (2018) A DNA methylation biomarker of alcohol consumption. Mol Psychiatry 23:422–433
Lucassen PJ, Naninck EFG, van Goudoever JB et al (2013) Perinatal programming of adult hippocampal structure and function; emerging roles of stress, nutrition and epigenetics. Trends Neurosci 36:621–631
Mahgoub M, Monteggia LM (2013) Epigenetics and psychiatry. Neurother J Am Soc Exp Neurother 10:734–741
Matosin N, Cruceanu C, Binder EB (2017) Preclinical and clinical evidence of DNA methylation changes in response to trauma and chronic stress. Chronic Stress Thousand Oaks Calif 1. https://doi.org/10.1177/2470547017710764
McCoy CR, Rana S, Stringfellow SA et al (2016) Neonatal maternal separation stress elicits lasting DNA methylation changes in the hippocampus of stress-reactive Wistar Kyoto rats. Eur J Neurosci 44:2829–2845
McEwen BS, Stellar E (1993) Stress and the individual. Mechanisms leading to disease. Arch Intern Med 153:2093–2101
Mundorf A, Schmitz J, Güntürkün O et al (2018) Methylation of MORC1: a possible biomarker for depression? J Psychiatr Res 103:208–211
Murgatroyd C, Patchev AV, Wu Y et al (2009) Dynamic DNA methylation programs persistent adverse effects of early-life stress. Nat Neurosci 12:1559–1566
Nestler EJ, Hyman SE (2010) Animal models of neuropsychiatric disorders. Nat Neurosci 13:1161–1169
Nielsen DA, Utrankar A, Reyes JA et al (2012) Epigenetics of drug abuse: predisposition or response. Pharmacogenomics 13:1149–1160
Nieratschker V, Massart R, Gilles M et al (2014) MORC1 exhibits cross-species differential methylation in association with early life stress as well as genome-wide association with MDD. Transl Psychiatry 4:e429
Nieto SJ, Patriquin MA, Nielsen DA, Kosten TA (2016) Don't worry; be informed about the epigenetics of anxiety. Pharmacol Biochem Behav 146–147:60–72
Ovenden ES, McGregor NW, Emsley RA et al (2018) DNA methylation and antipsychotic treatment mechanisms in schizophrenia: progress and future directions. Prog Neuro-Psychopharmacol Biol Psychiatry 81:38–49
Palma-Gudiel H, Fañanás L (2017) An integrative review of methylation at the serotonin transporter gene and its dialogue with environmental risk factors, psychopathology and 5-HTTLPR. Neurosci Biobehav Rev 72:190–209

Park H-J, Kim S-K, Kang W-S et al (2014) Increased activation of synapsin 1 and mitogen-activated protein kinases/extracellular signal-regulated kinase in the amygdala of maternal separation rats. CNS Neurosci Ther 20:172–181

Paternain L, Martisova E, Campión J et al (2016) Methyl donor supplementation in rats reverses the deleterious effect of maternal separation on depression-like behaviour. Behav Brain Res 299:51–58

Philibert RA, Plume JM, Gibbons FX et al (2012) The impact of recent alcohol use on genome wide DNA methylation signatures. Front Genet 3:54

Philibert RA, Beach SRH, Brody GH (2014) The DNA methylation signature of smoking: an archetype for the identification of biomarkers for behavioral illness. Neb Symp Motiv 61:109–127

Radtke KM, Schauer M, Gunter HM et al (2015) Epigenetic modifications of the glucocorticoid receptor gene are associated with the vulnerability to psychopathology in childhood maltreatment. Transl Psychiatry 5:e571

Razin A, Cedar H (1991) DNA methylation and gene expression. Microbiol Rev 55:451–458

Razin A, Riggs AD (1980) DNA methylation and gene function. Science 210:604–610

Relton CL, Gaunt T, McArdle W et al (2015) Data resource profile: accessible resource for integrated epigenomic studies (ARIES). Int J Epidemiol 44:1181–1190

Rijlaarsdam J, Cecil CAM, Walton E et al (2017) Prenatal unhealthy diet, insulin-like growth factor 2 gene (IGF2) methylation, and attention deficit hyperactivity disorder symptoms in youth with early-onset conduct problems. J Child Psychol Psychiatry 58:19–27

Roberts S, Lester KJ, Hudson JL, Rapee RM, Creswell C, Cooper PJ, Thirlwall KJ, Coleman JRI, Breen G, Wong CCY, Eley TC (2014) Serotonin tranporter methylation and response to cognitive behaviour therapy in children with anxiety disorders. Transl Psychiatry 4(9):e444

Roberts S, Keers R, Breen G, Coleman JRI, Jöhren P, Kepa A, Lester KJ, Margraf J, Scheider S, Teismann T, Wannemüller A, Eley TC, Wong CCY (2019) DNA methylation of FKBP5 and response to exposure-based psychological therapy. Am J Med Genet B Neuropsychiatr Genet 180(2):150–158

Romeo RD (2016) The impact of stress on the structure of the adolescent brain: implications for adolescent mental health. Brain Res 1654:185–191

Roth TL, Lubin FD, Funk AJ, Sweatt JD (2009) Lasting epigenetic influence of early-life adversity on the BDNF gene. Biol Psychiatry 65:760–769

Santos-Rebouças CB, Pimentel MMG (2007) Implication of abnormal epigenetic patterns for human diseases. Eur J Hum Genet 15:10–17

Sasagawa T, Horii-Hayashi N, Okuda A et al (2017) Long-term effects of maternal separation coupled with social isolation on reward seeking and changes in dopamine D1 receptor expression in the nucleus accumbens via DNA methylation in mice. Neurosci Lett 641:33–39

Sharma S, Kelly TK, Jones PA (2010) Epigenetics in cancer. Carcinogenesis 31:27–36

Smart C, Strathdee G, Watson S et al (2015) Early life trauma, depression and the glucocorticoid receptor gene – an epigenetic perspective. Psychol Med 45:3393–3410

Suderman M, McGowan PO, Sasaki A et al (2012) Conserved epigenetic sensitivity to early life experience in the rat and human hippocampus. Proc Natl Acad Sci U S A 109(Suppl 2):17266–17272

Teicher MH, Andersen SL, Polcari A et al (2003) The neurobiological consequences of early stress and childhood maltreatment. Neurosci Biobehav Rev 27:33–44

Thomas M, Knoblich N, Wallisch A et al (2018) Increased BDNF methylation in saliva, but not blood, of patients with borderline personality disorder. Clin Epigenetics 10:109

Toda H, Boku S, Nakagawa S et al (2014) Maternal separation enhances conditioned fear and decreases the mRNA levels of the neurotensin receptor 1 gene with hypermethylation of this gene in the rat amygdala. PLoS One 9:e97421

Vinkers CH, Kalafateli AL, Rutten BPF et al (2015) Traumatic stress and human DNA methylation: a critical review. Epigenomics 7:593–608

Watamura SE, Roth TL (2018) Looking back and moving forward: evaluating and advancing translation from animal models to human studies of early life stress and DNA methylation. Dev Psychobiol. https://doi.org/10.1002/dev.21796

Weaver ICG, Cervoni N, Champagne FA et al (2004) Epigenetic programming by maternal behavior. Nat Neurosci 7:847–854

Wu Y, Patchev AV, Daniel G et al (2014) Early-life stress reduces DNA methylation of the Pomc gene in male mice. Endocrinology 155:1751–1762

# Regulation of 5-Hydroxymethylcytosine Distribution by the TET Enzymes

**John Z. Cao, Anastasia E. Hains, and Lucy A. Godley**

## Contents



---

Authors "John Z. Cao" and "Anastasia E. Hains" contributed equally to this chapter.

---

J. Z. Cao
Committee on Cancer Biology, University of Chicago, Chicago, IL, USA

A. E. Hains
Program on Molecular Pathogenesis and Molecular Medicine, University of Chicago, Chicago, IL, USA

L. A. Godley (✉)
Committee on Cancer Biology, University of Chicago, Chicago, IL, USA

Section of Hematology/Oncology, Department of Medicine, University of Chicago, Chicago, IL, USA
e-mail: lgodley@medicine.bsd.uchicago.edu



S. Jurga, J. Barciszewski (eds.), *The DNA, RNA, and Histone Methylomes*, RNA Technologies, https://doi.org/10.1007/978-3-030-14792-1_10

**Abstract** 5-Hydroxymethylcytosine (5-hmC) has a critical role in the epigenetic regulation of many cellular processes, including differentiation, cytokine responses, and memory formation. This modified base is generated by the TET enzymes through oxidizing 5-methylcytosine and can be further oxidized to 5-formylcytosine and 5-carboxylcytosine. The level and distribution of 5-hmC in the genome is tightly controlled through transcription of *TET* genes, post-translational modifications on the TET enzymes, and various interacting partners of TET enzymes. Deregulation of 5-hmC is often found in pathological conditions and plays crucial roles in disease progression. This chapter will discuss recent findings regarding 5-hmC functions, TET proteins, and their gene regulations, as well as their relationships to cancer.

**Keywords** TET enzymes · 5-Hydroxymethylcytosine · 5-Methylcytosine · CpG · Demethylation

# 1 Introduction

In mammalian DNA, cytosine bases found within the context of a 5′-CpG-3′ dinucleotide are substrates for covalent modification to 5-methylcytosine (5-mC) by the DNA methyltransferases DNMT1, DNMT3A, and DNMT3B, which then can be oxidized sequentially by Ten-Eleven Translocation methylcytosine dioxygenases TET1, TET2, and TET3 to produce 5-hydroxymethylcytosine (5-hmC), 5-formylcytosine (5-fC), and 5-carboxylcytosine (5-caC) (Wu and Zhang 2017) (Fig. 1a). With the discovery of TET catalytic activity (Iyer et al. 2009; Tahiliani et al. 2009), there have been great advances in understanding the role of 5-hmC in various biological processes. Emerging evidence indicates that 5-hmC plays important roles in many cellular processes, including differentiation, cytokine responses, and memory formation (Cao 2016; Vogel Ciernia and LaSalle 2016; An et al. 2017). In this chapter, we discuss the regulation of TET enzymes and their roles in modulating the 5-hmC landscape as well as the impact of TET and 5-hmC deregulation in human disease.

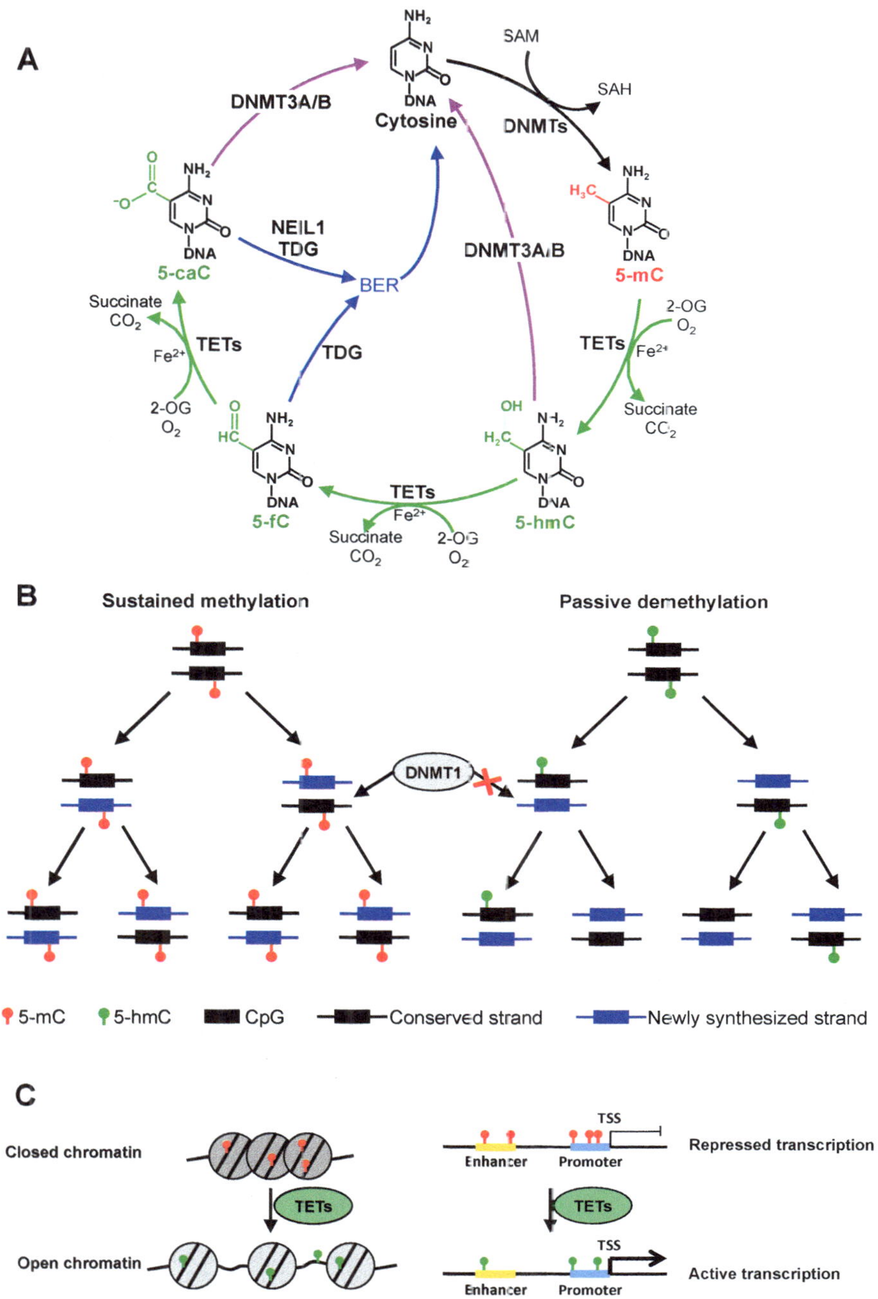


**Fig. 1** Regulation and functions of 5-hmC. (**a**) Biochemical pathways that control cytosine modifications. Unmodified cytosine is methylated to 5-mC by DNMTs using SAM as methyl

# 2 TET and 5-hmC Functions

## 2.1 Mammalian Cytosine Modifications

In mammalian genomes, cytosines present within 5′-CpG-3′ dinucleotides can be methylated to 5-mC by DNMTs using S-adenosylmethionine (SAM) as the methyl group donor. DNMT1 has optimal activity in methylating hemimethylated cytosines and thus functions mainly to maintain DNA methylation patterns after DNA replication (Lee et al. 2014). In contrast, DNMT3A and DNMT3B function mainly as *de novo* methyltransferases (Lee et al. 2014) (Fig. 1b). Many studies published since the 1980s have established that 5-mC is a base that regulates chromatin function: When found within gene bodies, 5-mC is associated with gene expression (Jjingo et al. 2012; Jones 2012; Varley et al. 2013), and when concentrated in CpG islands that overlap gene promoters, the base represses gene transcription, a phenomenon commonly observed in cancer cells (Feinberg and Vogelstein 1983; Gama-Sosa et al. 1983; Baylin et al. 1986; Feinberg and Tycko 2004).

The first oxidized product of 5-mC, 5-hmC, was originally identified as a specialized base in T2, T4, and T6 bacteriophages (Wyatt and Cohen 1953). Although the presence of 5-hmC was also observed in mammalian DNA, it was believed that this base was the result of oxidative damage (Tardy-Planechaud et al. 1997; Valinluck and Sowers 2007). 5-hmC did not receive major attention as a chromatin regulating base until it was discovered that 5-hmC is the oxidized product of TET dioxygenase activity using 5-mC as a substrate (Iyer et al. 2009; Tahiliani et al. 2009). Since then, multiple 5-hmC specific assays have been developed (Table 1), which have facilitated the elucidation of 5-hmC functions.

---

**Fig. 1** (continued) donor. TETs oxidize 5-mC to 5-hmC, 5-fC, and 5-caC stepwise using 2-OG and $O_2$ as co-substrates and $Fe^{2+}$ as a co-factor (green arrows). 5-hmC and 5-caC can be converted to unmodified cytosine in vitro by direct removal of 5-modifications through DNMT3A/B (purple arrows). 5-fC and 5-caC can be excised by TDG and/or NEIL1 and replaced with unmodified cytosine through base-excision repair (blue arrows). (**b**) 5-hmC promotes passive demethylation through replication. Newly synthesized DNA contains no modified cytosines. When the CpG dinucleotide on the old strand is methylated, DNMT1 can bind and methylate the complementary CpG dinucleotide on the newly synthesized strand, maintaining methylation over DNA replication and cell division. When the CpG dinucleotide in the old strand is hydroxymethylated, however, if DNMT1 fails to bind, the newly synthesized strand will remain unmodified, leading to loss of modification through multiple rounds of DNA replication. (**c**) Stable 5-hmC promotes open chromatin and gene transcription. Abbreviations: *2-OG* 2-oxoglutarate, *5-mC* 5-methylcytosine, *5-hmC* 5-hydroxymethylcytosine, *5-fC* 5-formylcytosine, *5-caC* 5-carboxylcytosine, *5-mod* 5-modified cytosine, *BER* base-excision repair, *SAM* S-adenosyl methionine, *SAH* S-adenosyl homocysteine, *TSS* transcription start site

**Table 1** Commonly used methods that specifically detect 5-hmC

| Method | Name | Description | Resolution | Limitation | References |
|---|---|---|---|---|---|
| Mass spectrometry | HPLC-MS/MS | Digested nucleoside separated by liquid chromatography and measured by tandem mass-spectrometry | Global | Cannot detect local cytosine modifications | Fernandez et al. (2018) |
| Affinity | hMe-SEAL | Streptavidin pull-down of biotin-labeled 5-hmC in fragmented DNA | 50–100 bp | Requires expensive specialized reagents | Song et al. (2011) |
| Affinity | hMe-DIP | Immunoprecipitation of fragmented DNA with 5-hmC specific antibody | 50–100 bp | Antibody specificity | Nestor and Meehan (2014) |
| Affinity | JBP1-seq | JBP1 affinity pull-down of glycosylated 5-hmC in fragmented DNA | 50–100 bp | Requires specialized beads | Robertson et al. (2012) |
| Redox | TAB-seq | TET-mediated oxidation of 5-mC to 5-caC, while 5-hmC is protected by glycosylation, followed by bisulfite treatment | Single-base | Expensive enzyme | Yu et al. (2012) |
| Redox | oxBS-seq | Selective oxidation of 5-hmC to 5-fC followed by bisulfite treatment | Single-base | Requires parallel bisulfite sequencing | Booth et al. (2013) |
| Restriction enzymes | MspJ1 family | Digest DNA near modified cytosine except glycosylated 5-hmC | Single-base | Sequence specificity of the enzyme | Cohen-Karni et al. (2011) |
| Restriction enzymes | PvuRts1I family | 5-hmC specific restriction enzyme | Single-base | Sequence specificity of the enzyme | Wang et al. (2011) |

## 2.2 *Roles of 5-hmC*

5-hmC is a relatively rare base in mammalian genomes, and its abundance is highly tissue dependent. For example, 5-hmC consists of 0.01–0.3% of all cytosine species in myeloid cells (Ko et al. 2010; Madzo et al. 2014); constitutes around 1% of all cytosines in cerebral tissue (Kriaucionis and Heintz 2009; Wen et al. 2014; Field et al. 2015); and is present at levels as high as 5% of all cytosines in undifferentiated embryonic stem (ES) cells (Tahiliani et al. 2009). Currently, two major functions of 5-hmC are proposed: the base is a component of the demethylation process, and the base can also function as a stable epigenetic mark with distinct roles in chromatin structure and transcriptional regulation.

5-hmC facilitates demethylation both within passive and active demethylation pathways (Fig. 1). When DNA is replicated, a newly inserted cytosine on the daughter strand is unmodified at first. If the cytosine base within the parent 5′-CpG-3′ dinucleotide is methylated, then DNMT1 methylates the cytosine within the newly synthesized strand since its highest activity is on hemi-methylated CpG dinucleotides, and DNMT1 sits at the replication fork, thereby maintaining the DNA methylation pattern (Fig. 1b). However, if the cytosine base within the parent DNA strand is hydroxymethylated, then maintenance of this base in the face of DNA replication requires two enzymes, first a DNMT to produce 5-mC, and then a TET to convert that base to 5-hmC. Therefore, in the face of DNA replication, if a DNMT fails to recognize a 5-hmC, that base will result in an unmodified cytosine in the daughter strand. This process is referred to as passive demethylation (Valinluck and Sowers 2007; Seiler et al. 2018). A more active and cell-cycle independent demethylation pathway involves further oxidation of 5-hmC by TETs to 5-fC and 5-caC (Ito et al. 2011), both of which can be replaced by unmodified cytosines through base-excision repair (BER) by thymine-DNA glucosylase (TDG) (He et al. 2011; Maiti and Drohat 2011; Coey et al. 2016; Pidugu et al. 2016) (Fig. 1a). 5-caC, but not 5-fC, can also undergo BER by Nei-like 1 (NEIL1) excision (Slyvka et al. 2017) (Fig. 1a). In addition to BER-mediated demethylation, some studies have also found that DNMT3A/3B have the ability to remove the 5-modification from 5-hmC and 5-caC directly, but not from 5-mC or 5-fC, resulting in unmodified cytosine (Chen et al. 2012; Liutkevičiūtė et al. 2014) (Fig. 1a).

5-hmC can also function as a stable and distinct epigenetic mark associated with active gene transcription. Many reports have found that 5-hmC is strongly associated with promoters, enhancers, and transcription factor binding sites (Madzo et al. 2014; Mariani et al. 2014; Vasanthakumar and Godley 2015) (Fig. 1c). Similarly, 5-hmC was shown to be associated with active transcription and histone modifications that mark open chromatin (Ficz et al. 2011; Szulwach et al. 2011; Madzo et al. 2014; Lin et al. 2017) (Fig. 1c). Evidence for the role of 5-hmC as a stable and actively maintained epigenetic mark is most apparent in the context of cell differentiation, where gene expression and chromatin structure undergo major changes. For example, in the course of erythropoiesis, 5-hmC at certain genomic loci remains highly enriched despite a global decrease in total 5-hmC level and multiple rounds of DNA replication (Madzo et al. 2014). To maintain 5-hmC during consecutive rounds of DNA replication, both DNMTs and TETs must be involved, indicating a functional role of 5-hmC at these loci.

In addition to its role in regulating gene expression and chromatin structure, recent studies have revealed potential involvement of 5-hmC in DNA damage repair. During murine embryonic development, *Tet1* loss-of-function oocytes have a much higher rate of defective meiosis and unresolved double strand breaks compared to those with wild-type *Tet1* (Yamaguchi et al. 2012). In addition, DNA damage in mouse Purkinje cells leads to increased 5-hmC levels (Jiang et al. 2015). Further studies in mouse ES cells showed that knock-out (KO) of all three *Tet* genes led to a high level of mitotic defects, as measured by the presence of chromatin fragments in anaphase, which resulted from a higher sensitivity to replication stress and delayed DNA repair (Kafer et al. 2016). In human cancer cell lines, 5-hmC foci co-localize with DNA damage foci marked by γH2AX and 53BP1 upon induced DNA damage (Kafer et al. 2016). When *TET2* is knocked-down with shRNA, 5-hmC foci no longer form, suggesting that TET2 is directly involved the DNA damage response

(Kafer et al. 2016). It is possible that 5-hmC promotes local chromatin remodeling or serves as an epigenetic mark to recruit additional DNA damage repair machinery (Kafer et al. 2016).

## 2.3 *TET Protein Structure and Function*

*TET* genes are present in all metazoan genomes. In mammals and other gnathostomata, the *TET* gene underwent triplication to generate *TET1*, *TET2*, and *TET3*. During this process, *TET2* likely underwent a local chromosome inversion causing the 5′ end of the inversion to create a distinct gene, *CXXC4* (or *IDAX*) (Iyer et al. 2009; Ko et al. 2013). All three *TET* genes are transcribed and translated into catalytically functional TET enzymes.

All three TET proteins share a C-terminal catalytic domain, whereas TET1 and TET3 share an N-terminal CXXC domain (Fig. 2a). The catalytic domains consist of a cysteine-rich domain and a double-stranded β-helix (DSBH) domain, together forming a globular structure with the core DNA-binding DSBH in the center (Hu et al. 2013, 2015) (Fig. 2b). TET catalytic domains specifically recognize and bind CpGs and flip the modified cytosine base into the catalytic cavity, with no preference for flanking sequences (Hu et al. 2013) (Fig. 2b). This base flipping mechanism appears to be conserved evolutionally, as the same mechanism is found in the *Naegleria gruberi* Tet enzyme (Hashimoto et al. 2014, 2015).

TET proteins are 2-oxoglutarate (2OG) and oxygen ($O_2$) dependent dioxygenases that use Fe(II) as a co-factor. TET catalytic domains share high homology to the previously discovered J-binding proteins (JBP1/2) (Iyer et al. 2009). JBP1/2 are dioxygenases in kinetoplastids that catalyze the oxidation of thymine at the methyl group to carboxyuracil, which is the first catalytic step towards base J. Based on this homology, the catalytic function of oxidizing 5-mC to 5-hmC was identified for the TETs (Tahiliani et al. 2009). Additional studies found that TETs can oxidize 5-hmC further to 5-fC and 5-caC in a stepwise fashion (He et al. 2011; Ito et al. 2011). In mammalian cells, the levels of 5-fC and 5-caC are 2–3 orders of magnitude lower than those of 5-hmC (Ito et al. 2011).

Enzyme kinetic studies showed that the $K_m$ values of TETs for oxygen, Fe(II), and 2OG are about 30 μM, 4 μM, and 60 μM, respectively (Laukka et al. 2015). This study also showed that TETs bind modified CpGs (5-mC, 5-hmC, or 5-fC) at around 100 nM. At a minimum, TET enzymes are two- to threefold more efficient at oxidizing 5-mC compared to 5-hmC or 5-fC due to differences in the conformation of the active sites (Hu et al. 2015), which leads to faster 5-hmC creation than its removal through further oxidation. The catalytic differences, together with TDG/NEIL1 mediated removal of 5-fC and 5-caC, may explain the relative abundance of 5-hmC over 5-fC and 5-caC.

The affinity of TETs for oxygen is much higher than that of the HIF prolyl-hydroxylases, which are 2OG/$O_2$ dependent dioxygenases like TETs that hydroxylate hypoxia-inducible factor α (HIFα) to induce HIFα degradation (Laukka et al.

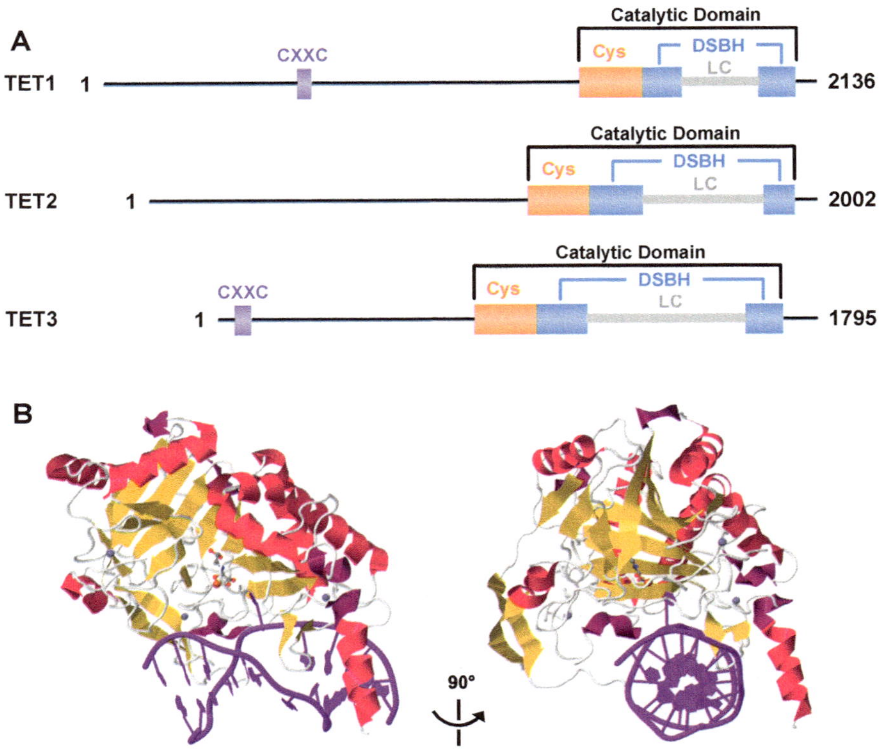


**Fig. 2** TET protein domains and binding with DNA. (**a**) Human TET proteins. All three TETs have conserved C-terminal catalytic domain, which consists of a cysteine-rich domain (Cys) and two double-stranded beta-helix (DSBH) DNA binding domains. The two DSBH domains are separated by a low complexity (LC) insert. Both TET1 and TET3 have a CpG binding CXXC domain near the N-terminus. During evolution, the sequences encoding the CXXC domain of TET2 became an independent gene: *CXXC4/IDAX*. (**b**) The crystal structure of TET2:DNA binding (Hu et al. 2013). The TET2 catalytic domain forms a double-stranded beta-helix (DSBH, yellow) structure that binds DNA modified CpG (purple). The DSBH structure is stabilized by several surrounding alpha-helices (red). The modified cytosine base is flipped out from the DNA double-helix and inserted in the active site along with 2-oxoglutarate and oxygen. Protein Data Bank (PDB) ID: 4NM6 (Hu, L., Li, Z., Cheng, J., Rao, Q., Gong, W., Liu, M., Shi, Y.G., Zhu, J., Wang, P., and Xu, Y. (2013). Crystal structure of TET2-DNA complex: insight into TET-mediated 5mC oxidation. Cell 155, 1545–1555.), image created with Jmol (Jmol: an open-source Java viewer for chemical structures in 3D. http://www.jmol.org/)

2015). Although both of these enzyme families use $O_2$ as a substrate, the higher affinity for $O_2$ in TETs suggests that the TETs may remain functional in hypoxic environments. Later studies showed that TETs retain catalytic activity under physiological hypoxia, at $O_2$ concentrations at which the HIF prolyl-hydroxylases no longer function. The TETs may act with less efficiency or may lose their activities under the pathological hypoxia found in solid tumors (Thienpont et al. 2016).

2OG is produced from isocitrate by the isocitrate dehydrogenases (IDH1/2) in the TCA cycle. It is usually abundant in cells and is not limiting for TET activities. However, 2OG analogs can accumulate in pathological conditions as a result of mutations in genes encoding TCA cycle enzymes (Pollard et al. 2005; Dang et al. 2009). The IDH1 R100/R132 and IDH2 R140/R172 missense mutations commonly found in cancers catalyze the production of R-2-hydroxyglutarate (R-2-HG) (Dang et al. 2009; Losman and Kaelin 2013). R-2-HG competes with 2OG for TET binding, but cannot be used in the oxidation reaction and thus acts as an inhibitor of TETs (Xu et al. 2011a). Consistent with this finding, glioblastomas and leukemias with these *IDH* mutations display global hypermethylation (Figueroa et al. 2010; Ko et al. 2010; Turcan et al. 2012). In addition, *IDH* and *TET* mutations are almost always mutually exclusive in leukemias, strongly suggesting that the two enzymes function in the same cellular processes (Gaidzik et al. 2012; Patel et al. 2012). Two additional TCA cycle metabolites, fumarate and succinate, were found to accumulate in solid tumors with mutated fumarate hydratase (*FH*) or succinate dehydrogenase (*SDH*) (Pollard et al. 2005). Fumarate and succinate are also able to inhibit TET activity by competing with 2OG for TET binding (Laukka et al. 2015). This was demonstrated in SK-N-BE(2) neuroblastoma cells, where 5-hmC levels decreased 10–40% upon fumarate or succinate treatment (Laukka et al. 2015).

TET activity is enhanced by ascorbate (vitamin C) by maintaining the Fe co-factor in a reduced (II) state (Kuiper and Vissers 2014). Ascorbate is essential for the function of collagen prolyl-4-hydroxylase, another 2OG/$O_2$ dependent dioxygenase similar to the TETs. Ascorbate deficiency causes failure of prolyl hydroxylation and leads to scurvy (Gorres and Raines 2010). The addition of ascorbate to cultured mouse ES cells or mouse embryonic fibroblasts leads to a rapid increase in 5-hmC levels that is Tet dependent (Blaschke et al. 2013; Minor et al. 2013; Yin et al. 2013). Other studies showed demethylation in ascorbate treated human ES cells and mouse embryonic fibroblasts as well as significant changes in the transcriptome (Chung et al. 2010; Blaschke et al. 2013; Chen et al. 2013).

## 3 Normal Transcription of *TET* Genes and Its Impact on the 5-hmC Landscape

Highly regulated expression of *TET* genes is essential for proper maintenance of the 5-hmC landscape in steady state or epigenetic reprogramming. Despite the functional redundancy, expression of each *TET* gene is distinct depending on the developmental context and tissue type, and disrupted expression patterns can give rise to pathogenic phenotypes. Here, we highlight the regulation of *TET* expression and its impact on 5-hmC by examining ES cells and adult neuronal cells, two cell types where 5-hmC is highly abundant and undergoes dynamic changes. It should be noted that the majority of embryonic development studies were done in murine systems, but there are supporting data obtained from studies in human cell lines. The species differences are denoted by mouse *Tet* and human *TET*.

## *3.1* TET *Transcription and TET Protein Regulation of 5-hmC in Embryonic Development*

Embryonic development is characterized by extensive epigenetic remodeling. Immediately after fertilization, 5-mC in the paternal zygotic genome is oxidized to 5-hmC and is further converted to unmodified cytosine through the demethylation pathways (Fig. 1a) (Mayer et al. 2000; Oswald et al. 2000). Mammalian ES cells have high levels of 5-hmC, which is enriched in enhancers, gene bodies, exons, transcription start sites, and 5′-untranslated regions of genes (Tahiliani et al. 2009; Pastor et al. 2011; Stroud et al. 2011; Wu and Zhang 2011). It is also enriched in the promoters of genes that feature both H3K27me3 and H3K4me3 marks, which are associated with open chromatin (Pastor et al. 2011; Stroud et al. 2011; Wu and Zhang 2011). These enrichments are often associated with elevated gene transcription (Ficz et al. 2011). Furthermore, throughout the development of the embryo, there are dynamic gains and losses of 5-hmC at specific regions of the genome, especially in the primordial germ cells (PGCs). Murine PGCs undergo their own demethylation events, between e9.5 and e10.5, in which there is global erasure of 5-mC (Hajkova et al. 2002; Seki et al. 2005). The TET enzymes are critical for 5-hmC changes in embryonic development and PGCs, as loss of all three *TET* genes is lethal and loss of one or combinations of two has distinct phenotypes (Ficz et al. 2011; Wu and Zhang 2011). During epigenetic remodeling of the developing embryo, expression of *TET1*, *TET2*, and *TET3* is tightly controlled.

## *3.2 Normal* Tet1 *Transcription in ES Cells and PGCs*

In mammalian ES cells, normal *TET1* expression is very high and plays a role in maintaining the pluripotent state (Koh et al. 2011). *TET1* expression in ES cells is promoted by canonical ES factors, such as OCT3/4, MYC, and NANOG, and is reduced when these factors are absent (Koh et al. 2011; Neri et al. 2015). *TET1* expression decreases as cells differentiate (Neri et al. 2015). The decrease in expression is mediated by the polycomb complex, which epigenetically silences the transcription start sites of *TET1* through deposition of the repressive histone mark, H3K27me (Neri et al. 2015). Although it is known that *Tet1* is expressed as early as e9.25 in PGCs, it is not known what transcription factors control expression of *Tet1* (Vincent et al. 2013).

Truncated *TET1* RNAs have been identified in mouse and human embryonic systems. In mouse ES cells, there is a *Tet1* isoform switch after e14.5. Full length

*Tet1* is expressed in early embryos, ES cells, and PGCs, but only a *Tet1* RNA that does not code for part of the N-terminus, including the CXXC domain (*Tet1s*), is expressed in adult tissues (Zhang et al. 2016). Despite having a lower chromatin affinity, Tet1s is still capable of binding DNA and its binding pattern is similar to that of full-length Tet1 (Zhang et al. 2016). Nonetheless, when homozygous *Tet1s* male mice were crossed with WT female mice, the resulting pups featured higher levels of perinatal lethality and lower birth weights compared to pups from WT crosses (Zhang et al. 2016). These experiments indicate that full-length Tet1 is important for normal embryonic development in mice, but not necessary in adult mice. The regulation of 5-hmC by Tet1s has yet to be studied in mice. A similar mechanism has been hypothesized in human systems: full-length TET1 is present in ES cells, and an alternate truncated protein that lack the CXXC domain is present in adult tissues (Good et al. 2017). This smaller isoform of TET1 has been implicated in many cancers, including breast, uterine, and ovarian cancers (Good et al. 2017). This isoform is catalytically active, but does not have as great an impact on gene expression as the full-length TET1 (Good et al. 2017). The mechanism behind this is unclear, although it has been shown that the TET1 short-isoform binds outside of CpG islands, a distinction from full-length TET1 (Good et al. 2017).

### 3.3 *Loss of* Tet1 *Disrupts 5-hmC Levels and Distribution in ES Cells and PGCs*

Loss of *Tet1* in ES cells has a measurable impact on total levels and distribution of 5-hmC. When *Tet1* was knocked down in mouse ES cells, there is a decrease of about 35–50% of total 5-hmC levels (Tahiliani et al. 2009; Dawlaty et al. 2011; Hon et al. 2014). Correspondingly, there is an increase in total 5-mC levels (Ito et al. 2010; Dawlaty et al. 2011). Loss of 5-hmC in *Tet1* knock-down (KD) ES cells has been localized to promoters, gene bodies, and a few intergenic regions (Wu and Zhang 2011; Huang et al. 2014). Similarly, *Tet1* KD in PGCs results in an increase in global 5-mC levels and a 45% reduction in 5-hmC (Yamaguchi et al. 2012).

In addition to aberrant cytosine modification patterns, loss of *Tet1* in ES cells and PGCs leads to deregulation of gene expression (Dawlaty et al. 2011). Specifically, there is reduced expression of pluripotent factors *Nanog*, *Oct4*, and *Sox2*, as well as other genes associated with pluripotency (Ito et al. 2010; Ficz et al. 2011). In PGCs, loss of *Tet1* was found to reduce expression of some meiotic genes (Yamaguchi et al. 2012).

Loss of *Tet1* in ES cells and PGCs results in reduced growth and up-regulation of differentiation-associated genes (Ito et al. 2010; Pastor et al. 2011). When mouse ES cells that lack *Tet1* undergo differentiation, there is upregulation of genes encoding primary trophectoderm factors (*Cdx2*, *Eomes*, and *Elf5*) and downregulation of genes encoding neuroectoderm factors (*Pax6* and *NeuroD2*) (Ito et al. 2010; Dawlaty et al. 2011; Koh et al. 2011; Xu et al. 2011b). Further, as mentioned previously, depletion of

*Tet1* in PGCs leads to defective meiotic prophase that includes DSBs and formation of univalent chromosomes (Yamaguchi et al. 2012).

*Tet1* KO mouse ES cells are able to differentiate and form living mice (Dawlaty et al. 2011). However, a number of defects have been reported in mice that lack *Tet1* (Dawlaty et al. 2011; Rudenko et al. 2013; Zhang et al. 2013; Kumar et al. 2015). Some *Tet1* deficient mice have a smaller body size compared to WT (Dawlaty et al. 2011). Mice that lack *Tet1* have memory extinction impairment and abnormally enhanced hippocampal long-term depression, indicating that the downregulation of neuroectoderm factors impacts normal brain function (Rudenko et al. 2013). Others have reported impaired spatial memory, object location memory, and threat recognition memory in *Tet1* deficient mice (Zhang et al. 2013; Kumar et al. 2015). The meiosis defects mentioned above in mouse PGCs lead to loss of oocytes and consequent decrease in mouse litter size (Yamaguchi et al. 2012).

## 3.4 *Loss of* Tet2 *Transcription and Its Impact in ES Cells and PGCs*

*Tet2* is expressed about fivefold less than *Tet1* in ES cells, and its expression has been shown to be under the control of Oct4 (Koh et al. 2011; Sohni et al. 2015). Loss of *Tet2* results in about a 50–90% reduction in 5-hmC (Koh et al. 2011; Hon et al. 2014). This loss of 5-hmC in *Tet2* deficient cells is localized to the gene bodies and exon-intron boundaries of highly expressed genes (Huang et al. 2014). An additional report found *Tet2* deficient cells to have increased 5-hmC around the *Nanog* promoter, which is believed to be the result of compensatory *Tet1* expression (Huang et al. 2014; Langlois et al. 2014). Similar to *Tet1* deficient cells, there is an increase in 5-mC in reaction to the loss of *Tet2*, but specifically at enhancers under this condition (Hon et al. 2014).

Overall, loss of *Tet2* has a less dramatic impact on downstream gene expression than the loss of *Tet1* (Koh et al. 2011; Huang et al. 2014). Broadly, there is delayed gene induction in *Tet2* deficient ES cells during differentiation (Hon et al. 2014). In one study, there was only a modest increase in neuroectoderm markers *Pax6*, *NeuroD1*, *Lefty1*, and *Lefty2,* the opposite effect of *Tet1* KD (Koh et al. 2011). When *Tet2* deficient cells were allowed to differentiate, there was a bias towards neuroectoderm differentiation in one mouse and one human model (Koh et al. 2011; Langlois et al. 2014). However, according to another group, loss of *Tet2* has no impact on gene expression or ES cell morphology (Ito et al. 2010). Differentiation of *Tet2* deficient cells resulted in a decrease in both number and clonogenic capability of hematopoietic progenitor cells (Langlois et al. 2014). In PGCs, *Tet2* is heterogeneously expressed at e9.5 and e10.5 and is generally regarded as redundant with *Tet1* (Hackett et al. 2013; Vincent et al. 2013). When *Tet1* and *Tet2* are silenced together, demethylation of genes in PGCs is inhibited (Hackett et al. 2013).

The evidence for the roles of *Tet2* in ES cells is not always consistent across studies. First, as previously stated, the loss of *Tet2* has highly variable impacts on the level of 5-hmC (50–90% reduction) (Koh et al. 2011; Hon et al. 2014). There are also differing reports on the impact of *Tet2* on gene expression. Loss of *Tet2* promoted the expression of neuroectoderm markers in two studies (Koh et al. 2011; Langlois et al. 2014), but two other groups found that loss of *Tet2* negatively regulated *Lefty1* and had no impact on global gene expression (Ito et al. 2010; Hon et al. 2014). In addition, *Tet2* is frequently studied within the context of *Tet1* loss in embryonic development (Ficz et al. 2011; Koh et al. 2011; Vincent et al. 2013). Few studies have looked at the role of *Tet2* independent of *Tet1* in ES cells and PGCs.

## 3.5 Tet3 *Transcription and Its Epigenetic Role in the Zygote*

*Tet3* expression is essential for normal 5-hmC epigenetic remodeling of the zygote, but is nearly undetectable during very early embryonic development before the gene gets re-expressed in specific adult tissues (Gu et al. 2011; Koh et al. 2011; Wossidlo et al. 2011). *Tet3* expression is high in oocytes and zygotes and is significantly decreased in two-cell embryos (Gu et al. 2011; Wossidlo et al. 2011). When *Tet3* is knocked down, the conversion of 5-mC to 5-hmC in the paternal DNA is abrogated, and there is little 5-hmC (Gu et al. 2011; Wossidlo et al. 2011). Consequently, the demethylation of key paternal embryonic genes (*Oct4*, *Lemd1*, and *Nanog*) is impaired (Gu et al. 2011). The pattern of 5-hmC enrichment has yet to be investigated at other regions of the genome. As a result, the *Oct4* gene activation is delayed in the early embryo (Gu et al. 2011). The loss of *Tet3* from the maternal genome causes developmental failure in mouse embryos, due to retention of paternal 5-mC (Gu et al. 2011).

Although *Tet3* expression is low in ES cells, it increases during neuronal differentiation (Li et al. 2015b). *Tet3* KO cells are capable of expressing neuronal differentiation markers but often undergo apoptosis (Li et al. 2015b). *Tet3* is also critical for *Xenopus* eye and neural development (Xu et al. 2012).

## 3.6 *Loss of Multiple* Tets *Has Serious Functional Consequences in Embryonic Development*

Double and triple KO (DKO and TKO) of *TET* genes in embryonic systems has dramatic effects on 5-hmC maintenance and cell phenotype. When *Tet1* and *Tet2* are silenced together, there is loss of 75–80% of 5-hmC in ES cells, greater than those of the effects of either single KD (Koh et al. 2011). *Tet3* is upregulated in the double silenced cells, presumably to compensate the loss of *Tet1* and *Tet2* (Koh et al. 2011). In *Tet1*/*Tet2* DKO cells, the changes in gene expression are similar to *Tet1* KD alone, suggesting that *Tet1* has a dominant role on transcription over *Tet2* in ES cells (Koh et al. 2011). During differentiation of mouse embryos, *Tet1/Tet2* DKO cells remain depleted of 5-hmC, and there are developmental defects in chimeric embryos consistent with perinatal lethality (Dawlaty et al. 2013). The defects *Tet1/Tet2* DKO embryos exhibit include exencephaly, hemorrhages in the head, or profound growth retardation. However, some *Tet1/Tet2* DKO mice are viable (Dawlaty et al. 2013). Although both sexes of the DKO mice are fertile, females have smaller ovaries and reduced fertility (Dawlaty et al. 2013).

*Tet1/Tet3* DKO in embryonic systems also has severe biological consequences, including abnormal mitosis and nuclear blebbing in the two-cell stage (Kang et al. 2015). Although not well studied in ES cells, gestation of *Tet1/Tet3* DKO mouse embryos had been assessed in one study (Kang et al. 2015). In these embryos, 5-hmC was undetectable by antibody staining, and only about 70% of them survived to eight-cell embryos. By e3.5, 60% of the remaining embryos had aberrant *Nanog* expression. During the entire process, the transcriptomes of the embryos featured very high variability. Ultimately, embryos were not viable beyond e10.5.

*tet2/tet3* DKO has only been examined in the context of embryonic development in zebrafish (Li et al. 2015a). In this model, tet2 and tet3 are the predominant 5-hmC regulators in embryos. *tet2/tet3* DKO resulted in a 30-fold loss of 5-hmC, although the distribution of 5-hmC was not characterized. These embryos featured abnormal differentiation, including altered brain morphology, smaller eyes, and a lack of a definitive hematopoietic stem cell population, and ultimately did not survive beyond the larval stage. Gene expression was not assessed, so it is not known how loss of *tet2* and *tet3* impacted the transcriptome of these animals.

*Tet1/Tet2/Tet3* TKO is invariably lethal and leads to complete depletion of 5-hmC in ES cells (Dawlaty et al. 2014; Hu et al. 2014; Lu et al. 2014). Promoters and enhancers are hypermethylated, and many genes are deregulated (Lu et al. 2014). However, studies disagree about whether the majority of genes are upregulated or downregulated in TKO cells (Dawlaty et al. 2014; Lu et al. 2014). One study found that when the TKO cells underwent differentiation, there was reduced expression of mesodermal and endodermal markers (Dawlaty et al. 2014). In addition, the cells formed fewer differentiated structures. When injected into embryos, there was a lower incidence of chimeric embryos formed from TKO cells compared to controls, indicating that the *Tet*s are essential for normal embryonic development.

## *3.7 Cocaine Stimulation Remodels the 5-hmC Landscape Through* Tet1 *Expression*

The brain is particularly enriched with high levels of 5-hmC (Kriaucionis and Heintz 2009; Globisch et al. 2010), where the base is thought to play a major role in neuronal adaptation to stimuli, which requires epigenetic remodeling. Once considered static epigenetic marks, we now appreciate that covalent cytosine modifications in the brain are reversible and dynamic, with frequent remodeling of 5-mC and 5-hmC, due in part to *TET1* expression (Reik 2007; Guo et al. 2011a, b). Changes to the 5-hmC landscape occur in response to neuronal activity, indicating specific stimuli have the potential to remodel 5-hmC level and distribution (Guo et al. 2011a). One such stimulus that has been studied within the context of brain tissue is cocaine exposure (Feng et al. 2015). This study found repeated cocaine administration resulted in downregulation of *Tet1* expression and increased 5-hmC levels within the nucleus accumbens of mice. The increases of 5-hmC were enriched around the enhancers and gene bodies of genes known to be involved in drug addiction. The downregulation of *Tet1* and altered 5-hmC appeared to enhance behavioral responses to cocaine. Nonetheless, the mechanism of how *Tet*s are regulated in the brain is yet to be identified. Despite being a tissue enriched in 5-hmC, very little is known about the expression and function of the *Tets* in this system.

Epigenetic reprogramming of the 5-hmC landscape is highly dependent on rigorous regulation of expression from the *TET* genes. It is clear that each TET enzyme has a unique role in each process, and the loss of one enzyme can only be partially compensated by the other two, indicating non-overlapping functions.

# 4 Mutations of *TET* Genes in Cancers

Epigenetic abnormalities such as global hypomethylation and site-specific hypermethylation have been known to be characteristics of cancers since the 1980s (Feinberg and Vogelstein 1983; Gama-Sosa et al. 1983; Baylin et al. 1986; Feinberg and Tycko 2004). Since the discovery of TET function, the role of 5-hmC in pathologic conditions has become a new focus in epigenetics. However, many techniques used to study 5-mC, including sodium bisulfite treatment and several methylation-sensitive restriction enzymes, cannot distinguish 5-mC from 5-hmC. For this reason, many interpretations of 'methylcytosine' should be more properly described as '5-mC or 5-hmC'. The recent development of 5-hmC specific techniques (Table 1), such as TAB-seq (Yu et al. 2012) and hMe-SEAL (Song et al. 2011), allow specific detection of 5-hmC. Through the use of these new techniques, widespread 5-hmC alterations were found in blood and solid tumor cancers (Vasanthakumar and Godley 2015). Here, we will discuss some of the most common causes of 5-hmC deregulation in cancer, including mutations of *TET*s and changes in *TET* expression.

## *4.1* TET1 *Mutations and Transcriptional Changes in Cancer*

The *TET1* gene was originally discovered as a fusion partner of *KMT2A* (*MLL*) in cases of t(10;11)(q22;23) acute myeloid leukemia (AML) (Ono et al. 2002; Lorsbach et al. 2003). *MLL* encodes a histone methyltransferase, and leukemia-specific chromosomal fusions involving this gene frequently result in production of MLL-fusion proteins. The *MLL-TET1* fusion is found in about 0.3% of all *MLL*-fusion leukemias, resulting in the fusion of the N-terminus of MLL with the C-terminal catalytic domain of TET1 (Lee et al. 2013; Meyer et al. 2013). The *MLL-TET1* fusion has also been found in some acute lymphoblastic leukemia (ALL) cases and has the capability to drive a phenotypic switch from ALL to AML (Burmeister et al. 2009; Ittel et al. 2013). *MLL-TET1* fusion causes translocation of the TET1 catalytic domain, which is normally absent in myeloid systems, therefore promoting AML progression (Huang et al. 2013). In fact, *TET1* is a direct target of all MLL-fusion proteins and is found to be significantly upregulated in MLL-rearranged leukemia samples compared to normal bone marrow samples (Huang et al. 2013). This alteration results in increased global level of 5-hmC (Huang et al. 2013, 2016). However, genomic enrichment of 5-hmC in MLL-rearranged leukemia cells has yet to be studied.

*TET1* mutations not involving chromosomal fusion are rare in hematopoietic malignancies, especially compared to *TET2* mutations. A study of 83 patients revealed only 2 cases with *TET1* mutations in myelodysplastic syndrome (MDS) or chronic myelomonocytic leukemia (CMML) patients (Lasho et al. 2018) (Fig. 3), but it remains unclear whether these are driver or passenger mutations.

The role of TET1 in the pathogenicity of cancers is usually linked to its gene expression. The 5-hmC levels in solid tumors are often low compared to surrounding normal tissue in solid tumors, including liver, brain, kidney, lung, colorectal, and gastric cancers (Kudo et al. 2012; Vasanthakumar and Godley 2015). Knock-down

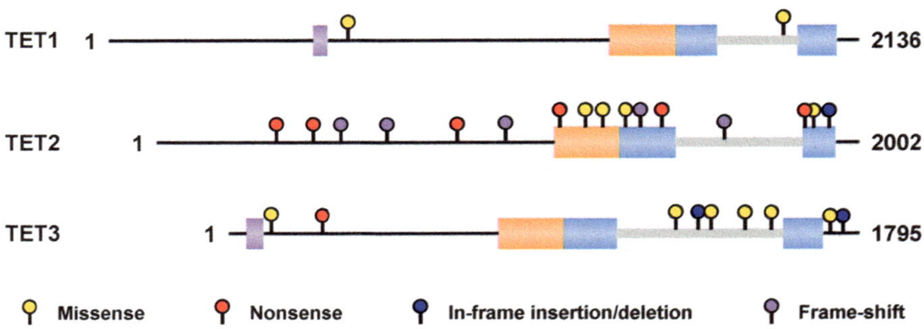


**Fig. 3** Mutation effects in TET proteins. *TET1* and *TET3* are rarely mutated in human diseases. Most of the mutations identified in *TET1* and *TET3* are missense mutations or in-frame insertion/deletion (indel), leading to single amino acid substitutions in the protein. All of these mutations occur outside of the core catalytic domain or the CXXC domain, likely maintaining the catalytic function of the protein. A notable exception is a nonsense mutation in *TET3*, which truncates the protein before the catalytic domain, leading to complete loss of the catalytic domain in the protein. Numerous *TET2* mutations have been described over the last decades, most of which lead to loss of function in TET2 protein. In contrast to *TET1* and *TET3* mutations, *TET2* missense mutations often directly impact the catalytic domain and cause impaired catalytic activities. Nonsense and frame-shift mutations are commonly found throughout the entire coding sequence, also leading to loss of TET2 catalytic activity

of *Tet1* alone in 3T3 fibroblasts is sufficient to replicate the 5-hmC loss seen in tumors, but not enough to induce malignant transformation (Kudo et al. 2012). Although many of the earlier studies on 5-hmC did not provide mechanistic links between 5-hmC alterations and pathogenic phenotypes, more recent studies began to provide better insight into the roles of 5-hmC regulation in cancers.

For example, we have recently gained insight in how TET1 is involved in hypoxic responses in certain solid tumors. When *MYCN*-amplified neuroblastoma cell lines are exposed to hypoxia, 5-hmC levels increase (Mariani et al. 2014). This result was surprising given, as discussed earlier in this chapter, TETs rely on oxygen for activity. The increased 5-hmC was the result of hypoxia-induced expression of *TET1*. In hypoxia, 5-hmC was enriched along the gene bodies of direct targets of HIF-1, the master regulator of hypoxia. These HIF-1 targets are known as canonical hypoxic response genes. Upon loss of HIF-1α, *TET1* expression no longer increased, and there was no induction of 5-hmC under hypoxic conditions. Loss of hypoxia induced-*TET1* resulted in decreased expression from canonical hypoxic response genes in comparison to hypoxic controls, demonstrating TET1 facilitates expression from canonical hypoxic response genes in hypoxia (Mariani et al. 2014).

## *4.2* TET2 *Mutations and Transcriptional Changes in Cancer*

*TET2* is the most commonly mutated gene among the three *TET* genes (Fig. 3). In hematologic malignancies, *TET2* mutations are found in 20–30% of AML, 20–30% of MDS or myeloproliferative neoplasm, and up to 58% of CMML (Delhommeau et al. 2009; Jankowska et al. 2009; Langemeijer et al. 2009; Yamazaki et al. 2012, 2015; Itzykson et al. 2013). *TET2* mutations are also found in other types of cancers at lower frequencies: 12% of T-cell ALL (Quivoron et al. 2011), 7% of clear cell renal cell carcinoma (clear cell RCC) (Sato et al. 2013), 12% of stage IV non-small cell lung cancer (Jin et al. 2016), and in metastatic prostate cancer (Nickerson et al. 2013). Copy number alterations of *TET2* are found in about 5.6% of various hematologic malignancies, 70% of which also have cytogenetic alterations (Bacher et al. 2012). Notably, most in-frame deletions and missense mutations occur in exons that code for the C-terminal catalytic domain, whereas nonsense and frameshift deletions can occur along the entire gene (Fig. 3) (Langemeijer et al. 2009; Euba et al. 2012). Most of the known mutations result in truncated protein or impaired catalytic functions (Langemeijer et al. 2009; Euba et al. 2012). One study found that many of the leukemia associated missense mutations in the catalytic domain of TET2 reduced the affinity of the enzyme for Fe(II) and 2OG and greatly reduce its activity (Laukka et al. 2015).

The effect of *TET2* mutation on 5-hmC and 5-mC has been studied, but not thoroughly. It is clear that *TET2* mutated leukemia cells have globally decreased 5-hmC (Ko et al. 2010; Madzo et al. 2014), but the mechanism behind this loss is not understood. Early studies of *TET2* mutated samples found no global difference in 5-mC, but showed site-specific hyper- and hypomethylation (Ko et al. 2010). In contrast, other studies have shown global hypermethylation in leukemic cells (Madzo et al. 2014). It should be noted that Ko et al. measured global methylation using the Illumina Infinium 27K array that probes 27,578 selected CpG sites, whereas Madzo et al. used mass spectrometry that measure all nucleotides in the genome, which may explain the discrepancy in these findings. Curiously, in this study, distribution of 5-hmC in *TET2*-mutated leukemia is dramatically different from normal, which is not explained solely by the global decrease of 5-hmC. Although most 5-hmC peaks decrease correspondingly with the global decrease of 5-hmC, certain sites gain high density of 5-hmC not seen in normal control. This redistribution process suggests that *TET1* or *TET3* may compensate for the loss of *TET2,* resulting in skewed 5-hmC distribution. Lastly, in CMML cells, *TET2* mutation-induced cytosine modification changes, especially hyper-modifications, are enriched in enhancer regions and negatively correlate with gene expression changes (Meldi et al. 2015; Yamazaki et al. 2015). Note that these studies used bisulfite sequencing based methods to characterize genome-wide cytosine modification changes, which cannot distinguish 5-mC from 5-hmC. Using modification status from as few as 14 sites, a support vector machine model was able to achieve 79% accuracy in predicting patient response to decitabine, a hypomethylating agent commonly used to treat CMML (Meldi et al. 2015). This result suggests that the changes in cytosine modifications have prognostic value to risk-stratify patients at diagnosis.

In addition to *TET2* mutations, it is becoming increasingly clear that *TET2* expression is a target of microRNAs in blood cancers (Cheng et al. 2013; Song et al. 2013). *TET2* has been found to be downregulated by miR-22, miR-29b, miR-125a, miR-26, miR-101, and miR-520d. 5-hmC levels were found to be decreased when *TET2* was down-regulated by miR-22, miR-29b, or miR-125a. Some of these microRNAs were also shown to regulate *TET1* and *TET3*, but an effect on 5-hmC level or distribution was not demonstrated (Cheng et al. 2013). Expression of miR-22, miR-29b, or miR-125a increases the oncogenic traits of cells, including increased replating ability of hematopoietic stem cells. In addition, mouse models with overexpression of these miRs featured development of MDS, and splenomegaly (Cheng et al. 2013; Song et al. 2013).

Although *TET2* has been investigated mostly in the context of blood cancers, there have been some instances of abnormal *TET2* expression in solid tumors. Low 5-hmC levels correlate with down-regulated *TET2* expression in oral cell carcinoma and esophageal squamous cell carcinoma (ESCC) (Jäwert et al. 2013; Shi et al. 2016). In both tumors, low 5-hmC was associated with low *TET2* expression. ESCC tumors have significantly decreased *TET2* and *TET3* expression compared to a normal tissue control (Shi et al. 2016). It is unknown how downregulation of *TET2* and low 5-hmC impact tumor oncogenicity, but loss of 5-hmC is an unfavorable prognostic factor in ESCC (Shi et al. 2016).

## 4.3 TET3 *Mutations and Transcriptional Changes in Cancer*

*TET3* mutations are rare compared to *TET2* mutations in cancer. A previous study with a cohort of 408 leukemia patients did not find any *TET3* mutations (Abdel-Wahab et al. 2009). More recently, two loss-of-function mutations were identified in *TET3* in the bone marrow of 28 CMML patients, both coexisting with *TET2* mutations (Merlevede et al. 2016) (Fig. 3). In addition, another study identified seven *TET3* mutations in a study of 83 leukemia patients (Lasho et al. 2018) (Fig. 3). None of the newly identified mutations co-occurred with *TET2* mutations. Six of these mutations were found in CMML patients (16% of all CMML patients), two of which were found in the same patient, and one of which co-occurred with a previously mentioned *TET1* mutation (Lasho et al. 2018). These findings suggest that *TET3* mutations preferentially occur in CMML, although the significance and consequences of *TET3* mutations remain to be studied.

There are emerging data that *TET3* expression is associated with an aggressive phenotype in RCC and head and neck cancer (Chen et al. 2017a; Misawa et al. 2018). High expression of *TET3* in RCC is associated with poor survival, but the effect on the 5-hmC landscape is unknown (Chen et al. 2017a). *TET3* is expressed at low levels in head and neck cancer, but perhaps more interestingly, the cytosine modification status of its promoter is negatively associated with disease survival (Misawa et al. 2018). The mechanism behind this finding is not yet known.

## 4.4 *Deregulated Expression of Multiple* TETs *in Cancer*

Many other cancers with low 5-hmC levels feature low expression or no change in expression of all three *TET* enzymes, indicating a more complex mechanism behind the aberrant 5-hmC landscape. Low levels of 5-hmC have been identified in brain, breast, liver, lung, pancreatic, prostate, and skin tumors (Jin et al. 2011; Lian et al. 2012; Yang et al. 2013). Breast, liver, and skin cancers all demonstrate low expression of all three *TET* enzymes (Lian et al. 2012; Yang et al. 2013; Chen et al. 2017b). The mechanism of low *TET* expression has only been explored in liver cancer. Low 5-hmC correlated with metastasis and poor prognosis of hepatocellular carcinoma (HCC) (Chen et al. 2017b). Expression from all three *TET* enzymes was downregulated by elevated expression of miR-29a, which in turn, resulted in decreased 5-hmC and increased 5-mC levels. The promoter of tumor suppressor gene *SOCS1* was found to be hypermethylated, and cell motility was increased.

In contrast, there has been no reported change in expression in any of the *TETs* in lung and brain cancer (Jin et al. 2011). Tumors that arise from pancreatic and prostate tissue have featured low levels of 5-hmC, however *TET* expression level has yet to be investigated.

Correcting altered transcription from *TET* genes is an attractive goal because it has the potential to 'reset' the 5-hmC landscape to pre-cancerous status. Further work to identify mechanisms of *TET* transcription regulation is required to achieve this goal.

# 5 Post-Translational Modifications (PTMs) of TET Enzymes

TET proteins are subject to an assortment of PTMs, many of which are poorly understood functionally. However, TET PTMs might reconcile a major discrepancy in studies concerning the DNA binding properties of TETs. There is conflicting research into the binding of TET to DNA: some studies have found that TET does not bind DNA at all, and others have found TETs can bind sequences containing unmodified, methylated, and hydroxymethylated cytosine (Zhang et al. 2010; Frauer et al. 2011b; Xu et al. 2011b). Preliminary research indicates that PTMs play a role in stability, activity, and localization of the TET enzymes. Considering the role of TETs in the pathogenesis of diseases, it is likely the role of TET PTMs and their effects on any aberrant activity will become a target for more studies. Known PTMs of the TET proteins include phosphorylation, O-GlcNAcylation (O-GlcNAc), acetylation, and ubiquitination (Fig. 4) (Bauer et al. 2015; Nakagawa et al. 2015; Zhang et al. 2017).

## *5.1 Phosphorylation of TET Proteins*

A study performed in mouse and human systems indicated phosphorylation of the three TET proteins largely occurs in the N-terminus (Fig. 4a) (Bauer et al. 2015). This modification can occur singly or in tandem with another modification. For example, in TET2, phospho-Ser-23 is found only with phosphor-Ser-15. The kinase (s) responsible for the phosphorylation modifications on the TET proteins are largely unknown at this time. Recent studies are starting to reveal the pathways regulating TET2 phosphorylation. For example, a specific site on TET2, Ser-99, was found to be phosphorylated by AMP-activated kinase (AMPK) (Wu et al. 2018). TET2 with this phosphorylation mark was found to have a longer half-life than TET2 without this mark. A Ser-99-Asp phospho-mimic TET2 was found to increase 5-hmC levels when expressed. However, distribution of 5-hmC was not investigated. Further research on TET2 phosphorylation will shed light on other kinases that modulate TET2 activity and stability. However, because modifications are enriched in the non-catalytic domains, it is hypothesized that TET phosphorylation may also play a role in interaction with TET binding partners (Bauer et al. 2015; Liu et al. 2018).

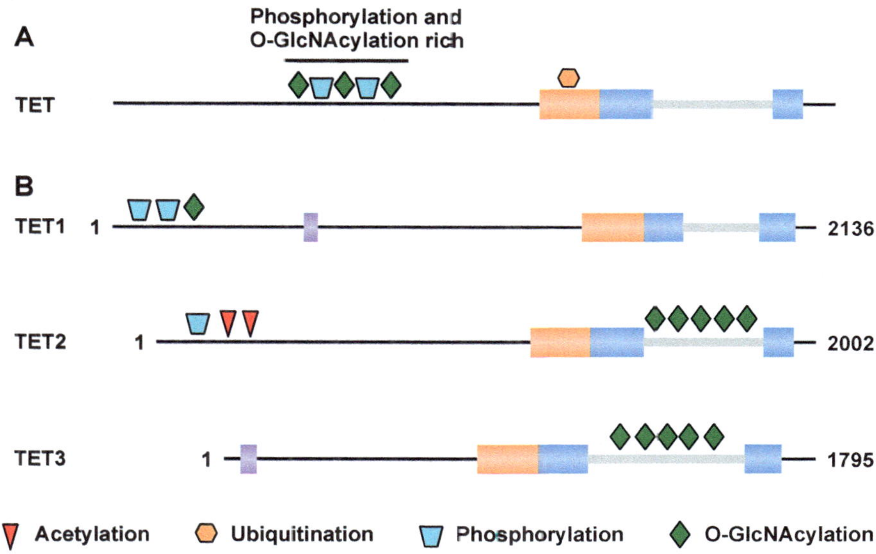


**Fig. 4** General distribution of TET PTMs. (**a**) Schematic representation of the common TET structure featuring modifications that are consistent across all three TET enzymes. A phosphorylation and O-GlcNAcylation-rich region is comprised of a large number of phosphate and O-GlcNAc marks that are variably distributed. A monoubiquitination modification is also found in the cysteine-rich domain of all three TET enzymes. (**b**) General distribution of modifications unique to each TET protein. Colors: Red: Acetylation, Orange: Ubiquitination, Blue: Phosphorylation, Green: O-GlcNAcylation

## *5.2 O-GlcNAc Modifications Facilitate TET Protein Stability, Activity, and Localization*

O-GlcNAc and phosphorylation modifications are occasionally in competition with each other (Bauer et al. 2015). Specifically, Ser-97 and Ser-374 of TET2 and Ser-362 and Ser-557 of TET3 could be phosphorylated or O-GlcNAcylated. Overall, some TET phosphorylation marks are stable, such as Ser-950 of TET1, but some are lost when *OGT* is highly expressed, such as Ser-2016 of TET1. *OGT* encodes the enzyme *O*-linked *N*-Acetylglucosamine Transferase (OGT) which is responsible for the addition of an O-GlcNAc group to serine and threonine. Similar to phosphorylation modifications, O-GlcNAc modifications are generally localized to the N-terminus and the low complexity insert (Fig. 4a, b). However, there is some distinction of the O-GlcNAc modification patterns between the three TET proteins (Fig. 4b). TET1 features fewer O-GlcNAc modifications than TET2 or TET3, with most found in the N-terminus and hardly any in the C-terminal region. In addition, O-GlcNAc modifications of TET1 have been shown to be highly dynamic (Shi et al. 2013; Bauer et al. 2015). TET2 and TET3 display most O-GlcNAc in the low complexity insert region in one comprehensive study (Bauer et al. 2015). In addition, the first 350 amino acids of TET3 are unmodified. In mouse and human systems, there appears to be crosstalk between the phosphorylation and the O-GlcNAc modifications of the TET proteins. Although increase in *OGT* expression is associated with decreased phosphorylation in TETs, phosphate and O-GlcNAc groups are rarely found on the same residue. Instead of direct competition, it appears that neighboring residues are engaging in crosstalk as several modifications of the same type often occur in close proximity. There have been many proposed functions of TET O-GlcNAc modifications (Vella et al. 2013; Bauer et al. 2015; Hrit et al. 2018). Several studies have demonstrated that this modification increases Tet1 protein stability and activity (Vella et al. 2013; Hrit et al. 2018). Others have implicated O-GlcNAc groups in mediating binding of TET partners and localization (Bauer et al. 2015). Studies in mouse embryonic development have shown that loss of O-GlcNAc modifications on Tet3 resulted in nuclear rather than cytosolic subcellular localization of the protein and substantially increased global 5-hmC levels (Zhang et al. 2014). Interestingly, O-GlcNAcylation of Tet1 and Tet2 had no impact on their respective subcellular localizations (Zhang et al. 2014).

## *5.3 Acetylation Has Been Implicated in the Stability of TET2*

Acetylation has been identified as a modification of only TET2 in human cell lines. One study found TET2 is acetylated by the epigenetic modifier p300 (Zhang et al. 2017). TET2 is acetylated at multiple lysines in the N-terminus, and the presence of two (K110 and K111) are necessary for p300-mediated acetylation (Fig. 4b). Although there is no accumulation of TET2 protein, global 5-hmC level increases in the presence of acetylated TET2, indicating that acetylation of TET2 enhances its enzymatic activity. The acetylation marks were also found to inhibit polyubiquitination of TET2, therefore promoting its stability. Lastly, TET2 acetylation has also been implicated in potential binding of DNMT1.

## *5.4 TETs Are Monoubiquitinated*

Monoubiquitination of all three TET proteins occurs on a lysine, mediated by VprBP, in the highly conserved cysteine-rich domain (Fig. 4a) (Nakagawa et al. 2015). In vivo ubiquitination assays with TET proteins yielded a single band on a gel, indicating addition of a one ubiquitination group. The monoubiquitination occurs on K1589 in TET1, K1299 in TET2, and K859 in TET3. Loss of monoubiquitination prevents binding of TETs to DNA and therefore results in an overall loss of 5-hmC in in vitro and in vivo experiments. Several oncogenic mutations of TET2 disrupt its ubiquitination, indicating this modification is important for normal TET2 activity.

## 6 Proteins Interact Directly with Modified Cytosine and TET Binding Proteins

Stable epigenetic marks interact with epigenetic "reader" proteins that mediate their downstream effects. Although there is some overlap between binding partners, 5-mC, 5-hmC, and TETs each have their own unique set of interactors.

### *6.1 Proteins Recognize, Bind, and Read 5-mC and 5-hmC*

Identified readers of 5-mC all contain the conserved methyl-binding domain, but not all proteins that feature a methyl-binding domain bind 5-mC (Hendrich and Tweedie 2003). Most proteins that read 5-mC are thought to reinforce the repression of genes with 5-mC rich promoters (Boyes and Bird 1991; Nan et al. 1997; Hendrich and Bird 1998). In fact, it is hypothesized that the ability to bind 5-mC may be cell type specific (Iurlaro et al. 2013). Therefore, this section will only discuss the most universal 5-mC readers. The very first proteins identified that read 5-mC were MeCP1 and MeCP2 (Meehan et al. 1989; Lewis et al. 1992). Although it was originally thought that 5-mC prevents transcription exclusively by directly blocking the binding of transcription factors, it is now known that MeCP1, MeCP2, and MBD1/2/4, are all indirect mediators of the biological function of the 5-mC mark by recruiting chromatin remodeling machinery such as NuRD complex to downregulate gene expression (Boyes and Bird 1991; Nan et al. 1997; Hendrich and Bird 1998; Hendrich and Tweedie 2003; Pan et al. 2017).

Researchers have tried to identify proteins that mediate the biological function of 5-hmC using a variety of screening methods (Iurlaro et al. 2013; Spruijt et al. 2013). Unfortunately, many studies have been unable to replicate results presented by other groups. This may be due to a lack of rigorous testing across multiple model systems, but also may indicate many 5-hmC readers are cell type specific, implying a more complex mechanism at work than simple hydroxymethylcytosine recognition (Iurlaro et al. 2013). Some proposed 5-hmC readers include RBM14, PRP8, RPL26, MSH6, PNKP, Wdr76, and Thy28. However, each of these has only been identified in a single study, and the findings remain to be validated (Hendrich and Bird 1998; Frauer et al. 2011a; Yildirim et al. 2011; Mellén et al. 2012; Iurlaro et al. 2013; Spruijt et al. 2013). Only three proteins, NP95/UHRF1, MeCP2, and MBD3, have been confirmed in more than one study (Sharif et al. 2007; Arita et al. 2008; Frauer et al. 2011a; Yildirim et al. 2011; Mellén et al. 2012; Baubec et al. 2013; Iurlaro et al. 2013; Spruijt et al. 2013). NP95/UHRF1 has been identified as a 5-hmC reader in three studies in two cell-free assays and in mouse ES cells (Frauer et al. 2011a; Iurlaro et al. 2013; Spruijt et al. 2013). Uhrf1 is traditionally recognized as a binder of hemimethylated DNA and functions in DNA methylation maintenance, but has lately been characterized as a reader of both 5-mC and 5-hmC (Sharif et al. 2007; Arita et al. 2008; Frauer et al. 2011a). The functional consequences of NP95/UHRF1 binding 5-hmC have yet to be examined in vivo (Frauer et al. 2011a). MeCP2, already established as a 5-mC binding protein, has also demonstrated 5-hmC binding in two studies in neural tissues and mouse ES cells (Mellén et al. 2012; Spruijt et al. 2013). It is unclear how the MeCP2:5-mC binding profile differs from MeCP2:5-hmC binding profile or whether the function of this protein depends on its ligand. However, a mutated version of *MeCP2* is expressed in Rett Syndrome: R133C. Experiments with electrophoretic mobility shift assays indicated that this MeCP2 protein is incapable of binding 5-hmC (Mellén et al. 2012). This indicates that readers of 5-hmC may have an essential role in normal neuronal phenotype. Finally, MBD3 has been identified as a weak binder of 5-hmC in two studies (Yildirim et al. 2011; Mellén et al. 2012). One study that indicated MDB3 as a 5-hmC reader also indicated Tet1:Mbd3 complexes bind 5-hmC rich regions of the genome in ES cells, and identified Mbd3 as both a recruiter of Tet1 and an effector that mediates gene expression (Yildirim et al. 2011). However, other studies have found that MBD3 binds the same genomic regions independent of the presence of 5-mC and 5-hmC, contradicting its proposed function as a 5-hmC reader (Iurlaro et al. 2013; Spruijt et al. 2013). This also potentially complicates the findings of the relationship between Tet1 and Mdb3 (Baubec et al. 2013; Iurlaro et al. 2013; Spruijt et al. 2013). The significance of MBD3:5-hmC binding is under debate.

Overall, although many potential proteins that bind 5-mC and 5-hmC have been identified, the function of many of these interactions is mostly unknown. Future studies of the biological functions of these interactions are needed to better understand how cytosine modifications regulate the genome.

## 6.2 Some Binding Partners Interact with All Three TETs

Current research on TET binding proteins shows that these binding partners can modify TET activity, localization, and stability (Fig. 5). Many of the identified proteins are not yet fully characterized and may have functions outside of their relationship with TETs.

The interaction between Tet and Ogt has been studied extensively. Independent of the O-GlcNAcylation activity, stable Tet:Ogt interaction promotes Tet activity and regulates Tet genomic localization. OGT binds TET1 through a conserved C-terminal domain, and is thought to form a larger complex with at least SIN3A (Vella et al. 2013; Hrit et al. 2018). Ogt and Tet1 colocalized at unmethylated CpG-rich promoters (Vella et al. 2013). Additional studies have found that loss of Ogt protein resulted in abnormal Tet1 localization in ES cells, indicating Ogt is important for Tet1 binding of some genomic sites (Shi et al. 2013). Multiple groups have reported that the loss of Tet1:Ogt interaction also results in reduced 5-hmC level and expression from Tet1-activated genes (Shi et al. 2013; Hrit et al. 2018). Ultimately, these experiments indicate that the role of Tet1:Ogt binding is to increase its activity and recruit it to genomic sites (Fig. 5a) (Shi et al. 2013; Hrit et al. 2018). TET2/3:OGT binding does not appear to have a significant impact on the 5-hmC

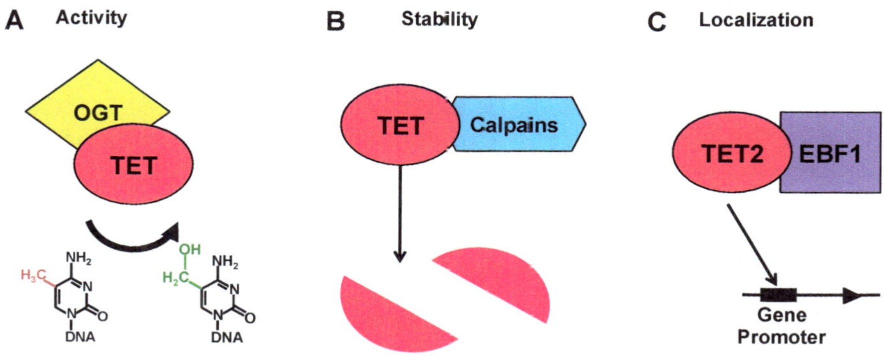


**Fig. 5** TET binding partners impact TET activity, localization, and stability. (**a**) Binding of OGT increases TET activity when bound to DNA. (**b**) Calpains bind and degrade TETs. (**c**) EBF1 brings TET2 to promoters of genes, where it hydroxylates 5-mC to 5-hmC

epigenetic landscape, instead playing a scaffolding role for OGT (Deplus et al. 2013; Ito et al. 2014).

Tet stability appears to be regulated, at least in part, by calpains in mouse ES cells (Wang and Zhang 2014). Calpains 1 and 2 regulate TET stability via direct binding and cleavage of TET enzymes (Fig. 5b) (Wang and Zhang 2014). This mechanism of TET protein degradation is predominantly found in certain stages of development when *TET* expression is high (Wang and Zhang 2014). Tet1 and Tet2 are degraded by calpain1 in mouse ES cells, whereas Tet3 is degraded by calpain2 during ES cell differentiation (Wang and Zhang 2014). Loss of calpains resulted in twofold higher 5-hmC levels, downregulation of *Cdx2* and *Eomes,* and upregulation of *Ngn2* and *Pax6* in ES cells (Wang and Zhang 2014). Therefore, TET:calpain complexes are required for normal 5-hmC regulation and gene expression in ES cells.

TDG, MBD4, SMUG1, NEIL1, NEIL2, NEIL3, PARP1, LIG3, and XRCC1 are part of the BER pathway and interact with TET during this process (Müller et al. 2014). Their colocalization with TETs is thought to facilitate removal of 5-caC and repair of the site. However, a role for direct binding between TET and each specific BER protein has yet to be described (Müller et al. 2014).

Beyond the shared interactors, the TET proteins have been shown to have binding partners unique to only one or two of them. These unique partners may explain the non-overlapping functions of TETs and provide further insight to the regulation of 5-hmC.

## 6.3 *TET1 Binding Proteins*

The function of most described interactors is regulating the localization of TET1 within the genome. The binding partners of TET1 are chromatin-binding proteins:

transcription factors and epigenetic modifiers. They are thought to function at different levels of TET1 localization, from promoting TET1 binding to DNA to recruiting TET1 to target regions in the genome.

Proteins that are thought to facilitate TET1 binding to DNA include MeCP2, HDAC1/6/7, EZH2, mSin3A, PCNA, and LSD1 (Cartron et al. 2013). All of these proteins were found to bind TET1 directly and each other, thus stabilizing the TET1-DNA interaction. Further experimentation revealed that, in addition to this function, TET1:EZH2 and TET1:SIN3A are thought to recruit TET1 to specific genomic elements (Chandru et al. 2018). It should be stated that many proteins labeled as 'recruiters' of TET1 have not been thoroughly tested on a genomic scale and are usually tested only at selected targets (Williams et al. 2011; Cartron et al. 2013). TET1:EZH2 and TET1:SIN3A have been implicated in the regulation of *HOXD12* and *NES1* respectively. Loss of the complexes resulted in increased methylation at each promoter and, therefore, decreased expression. Another study looked at the impact of Tet1:Sin3a complex on global genomic targets in mouse ES cells and showed that Sin3a recruited Tet1 to demethylate the promoter of *Lefty1* (Zhu et al. 2018). Loss of Sin3a resulted in altered localization of Tet1 and deregulation of 111 genes. However, the level and distribution of 5-hmC was not assessed.

Other identified proteins that act to recruit TET1 to genomic sites include PCNA and the Mbd3/NuRD complex (Williams et al. 2011; Yildirim et al. 2011; Cartron et al. 2013; Costa et al. 2013). Complexes containing TET1:PCNA are thought to play a role in the regulation of *MUC2* and *RRMI*, as experiments identified both proteins bound to these genes through chromatin-immunoprecipitation (ChIP) with α-PCNA antibody and sequential ChIP with α-TET1 antibody.

## *6.4 TET2 Binding Proteins*

Identified binding partners of TET2 promote its localization and stabilization. Two factors that recruit TET2 to genomic sites are EBF1 and WT1 (Fig. 5c). EBF1 and TET2 are part of the same complex, which localize to hypermethylated promoters in the context of *IDH* mutations in chondrosarcoma (Fig. 5c) (Guilhamon et al. 2013). It is hypothesized that TET2:EBF1 complexes act to target tissue specific demethylation sites, but further research is required to identify changes in the 5-hmC landscape and impacts on gene expression. The relationship between WT1 and TET2 was discovered in hematopoietic cells, which commonly feature mutually exclusive mutations in *WT1* and *TET2* in AML (Rampal et al. 2014). The TET2: WT1 complexes are believed to play a role in maintaining normal 5-hmC levels and localization in hematopoietic cells. Another study showed that TET2 occupied the promoters of WT1 target genes, and this occupancy is lost when *WT1* expression was depleted, indicating WT1 recruited TET2 to genomic sites (Wang et al. 2015). Further, mutation of either WT1 or TET2 resulted in lower total 5-hmC levels and aberrant distribution at enhancers, gene bodies, and distal regulatory elements. Conversely, when *WT1* and *TET2* were coexpressed in HEK293T cells, 5-hmC levels increased at promoters and transcription start sites of WT1 targets. When WT1 was depleted, 5-hmC enrichment at WT1 target genes was lost. When *IDH, WT1*, and *TET2* were mutated separately, 5-hmC changes in the gene body and distal regulatory regions had a positive correlation with gene expression changes (Rampal et al. 2014). When *TET2* was ectopically expressed in HEK293T cells, there was increased gene expression from WT1 targets, which was abrogated upon loss of TET2 catalytic activity (Wang et al. 2015). Finally, proliferation of leukemia cells was stimulated by the loss of either TET2 or WT1 but the loss of both was not additive, indicating that TET2 and WT1 function in the same pathway that controls cell proliferation. Taken together, evidence suggests that the TET2:WT1 complex acts to maintain 5-hmC and normal gene expression, and that loss of either protein results in an oncogenic phenotype (Rampal et al. 2014; Wang et al. 2015).

Idax and DNMT1 have been found to be involved in TET2 protein degradation and stability (Ko et al. 2010; Rampal et al. 2014; Wang et al. 2015; Zhang et al. 2017). A study shows that Idax and Tet2 bind to each other, and the coexpression of *Idax* and *Tet2* in HEK293T cells led to a decrease of Tet2 protein and a decrease in 5-hmC, but only when the DNA binding domain of Idax was present (Ko et al. 2013). This evidence suggests that Idax binds DNA and Tet2 then recruits an unknown degradation complex to target Tet2, which results in decreased 5-hmC. Binding of TET2 to DNMT1 occurs after the acetylation of TET2, and appears to promote protein stability (Zhang et al. 2017). The impact of TET2:DNMT1 on 5-hmC has yet to be explored, but it appears to prevent abnormal methylation of DNA. The location of 5-hmC loss and any impact on gene expression have yet to be investigated.

## 6.5 TET3 Binding Proteins

The known binding partners of Tet3, Rest and WT1, are believed to be involved in TET3 genomic localization (Rampal et al. 2014). The Tet3:Rest complex forms during murine retina maturation, which requires the formation of a complex neural network (Perera et al. 2015). Rest binds a Tet3 isoform that lacks the CXXC domain, indicating it may serve as a DNA targeting factor for Tet3. Coexpression of *Tet3* and *Rest* leads to a dramatic increase in 5-hmC levels, more than overexpression of *Tet3* alone. Rest target genes were enriched in a subset of genes that gained 5-hmC, suggesting a synergistic interaction. Measurement of global protein levels indicated that the protein levels of many Rest gene targets were significantly increased when *Tet3* was overexpressed. However, this increase was lost when Rest was not present, meaning both Tet3 and Rest are required for the observed phenotype. Finally, TET3:WT1 forms a complex following the depletion of TET2 in hematopoietic models, but this is considered a compensatory mechanism for the loss of TET2 (Rampal et al. 2014).

# 7 Conclusions

The 5-hmC epigenetic landscape is modulated by the regulation of TET enzymes at the genetic, transcription, and protein levels. As discussed above, the activity of TET enzymes is tightly controlled at each of these levels. However, much of our knowledge is incomplete. More research is needed to understand the mechanisms that control *TET* gene expression and TET PTMs. In addition, although many TET binding partners have been identified, more research is required to determine their functions and mechanisms of action. Loss or change of any of these regulation mechanisms causes TET deregulation, which can alter the 5-hmC landscape and play a pathogenic role in disease. For this reason, study of the TET enzymes, in particular how they are regulated and impact the 5-hmC landscape, will provide answers to questions about the role of the epigenetic state in health and disease.

# References

Abdel-Wahab O, Mullally A, Hedvat C et al (2009) Genetic characterization of TET1, TET2, and TET3 alterations in myeloid malignancies. Blood 114:144–147

An J, Rao A, Ko M (2017) TET family dioxygenases and DNA demethylation in stem cells and cancers. Exp Mol Med 49:e323

Arita K, Ariyoshi M, Tochio H et al (2008) Recognition of hemi-methylated DNA by the SRA protein UHRF1 by a base-flipping mechanism. Nature 455:818–821

Bacher U, Weissmann S, Kohlmann A et al (2012) TET2 deletions are a recurrent but rare phenomenon in myeloid malignancies and are frequently accompanied by TET2 mutations on the remaining allele. Br J Haematol 156:67–75

Baubec T, Ivánek R, Lienert F et al (2013) Methylation-dependent and -independent genomic targeting principles of the MBD protein family. Cell 153:480–492

Bauer C, Göbel K, Nagaraj N et al (2015) Phosphorylation of TET proteins is regulated via O-GlcNAcylation by the O-linked N-acetylglucosamine transferase (OGT). J Biol Chem 290: 4801–4812

Baylin SB, Höppener JW, de Bustros A et al (1986) DNA methylation patterns of the calcitonin gene in human lung cancers and lymphomas. Canc Res 46:2917–2922

Blaschke K, Ebata KT, Karimi MM et al (2013) Vitamin C induces Tet-dependent DNA demethylation and a blastocyst-like state in ES cells. Nature 500:222–226

Booth MJ, Ost TW, Beraldi D et al (2013) Oxidative bisulfite sequencing of 5-methylcytosine and 5-hydroxymethylcytosine. Nat Protocol 8:1841–1851

Boyes J, Bird A (1991) DNA methylation inhibits transcription indirectly via a methyl-CpG binding protein. Cell 64:1123–1134

Burmeister T, Meyer C, Schwartz S et al (2009) The MLL recombinome of adult CD10-negative B-cell precursor acute lymphoblastic leukemia: results from the GMALL study group. Blood 113:4011–4015

Cao X (2016) Self-regulation and cross-regulation of pattern-recognition receptor signalling in health and disease. Nat Rev Immunol 16:35–50

Cartron P-FF, Nadaradjane A, Lepape F et al (2013) Identification of TET1 partners that control its DNA-demethylating function. Gene Canc 4:235–241

Chandru A, Bate N, Vuister GW et al (2018) Sin3A recruits Tet1 to the PAH1 domain via a highly conserved Sin3-interaction domain. Sci Rep 8:14689

Chen C-CC, Wang K-YY, Shen C-KJK (2012) The mammalian de novo DNA methyltransferases DNMT3A and DNMT3B are also DNA 5-hydroxymethylcytosine dehydroxymethylases. J Biol Chem 287:33116–33121
Chen J, Guo L, Zhang L et al (2013) Vitamin C modulates TET1 function during somatic cell reprogramming. Nat Genet 45:1504–1509
Chen D, Maruschke M, Hakenberg O et al (2017a) TOP2A, HELLS, ATAD2, and TET3 are novel prognostic markers in renal cell carcinoma. Urology 102:e1-265–e1e267
Chen Q, Yin D, Zhang Y et al (2017b) MicroRNA-29a induces loss of 5-hydroxymethylcytosine and promotes metastasis of hepatocellular carcinoma through a TET-SOCS1-MMP9 signaling axis. Cell Death Dis 8:e2906
Cheng J, Guo S, Chen S et al (2013) An extensive network of TET2-targeting MicroRNAs regulates malignant hematopoiesis. Cell Rep 5:471–481
Chung TL, Brena RM, Kolle G et al (2010) Vitamin C promotes widespread yet specific DNA demethylation of the epigenome in human embryonic stem cells. Stem Cells 28:1848–1855
Coey CT, Malik SS, Pidugu LS et al (2016) Structural basis of damage recognition by thymine DNA glycosylase: key roles for N-terminal residues. Nucleic Acids Res 44:10248–10258
Cohen-Karni D, Xu D, Apone L et al (2011) The MspJI family of modification-dependent restriction endonucleases for epigenetic studies. Proc Natl Acad Sci U S A 108:11040–11045
Costa Y, Ding J, Theunissen TW et al (2013) NANOG-dependent function of TET1 and TET2 in establishment of pluripotency. Nature 495:370–374
Dang L, White DW, Gross S et al (2009) Cancer-associated IDH1 mutations produce 2-hydroxyglutarate. Nature 462:739–744
Dawlaty MM, Ganz K, Powell BE et al (2011) Tet1 is dispensable for maintaining pluripotency and its loss is compatible with embryonic and postnatal development. Cell Stem Cell 9:166–175
Dawlaty MM, Breiling A, Le T et al (2013) Combined deficiency of Tet1 and Tet2 causes epigenetic abnormalities but is compatible with postnatal development. Dev Cell 24:310–323
Dawlaty MM, Breiling A, Le T et al (2014) Loss of Tet enzymes compromises proper differentiation of embryonic stem cells. Dev Cell 29:102–111
Delhommeau F, Dupont S, Della Valle V et al (2009) Mutation in TET2 in myeloid cancers. N Engl J Med 360:2289–2301
Deplus R, Delatte B, Schwinn MK et al (2013) TET2 and TET3 regulate GlcNAcylation and H3K4 methylation through OGT and SET1/COMPASS. EMBO J 32:645–655
Euba B, Vizmanos JLL, García-Granero M et al (2012) A meta-analysis of TET2 mutations shows a distinct distribution pattern in de novo acute myeloid leukemia and chronic myelomonocytic leukemia. Leuk Lymphoma 53:1230–1233
Feinberg AP, Tycko B (2004) The history of cancer epigenetics. Nat Rev Canc 4:143–153
Feinberg AP, Vogelstein B (1983) Hypomethylation distinguishes genes of some human cancers from their normal counterparts. Nature 301:89–92
Feng J, Shao N, Szulwach KE et al (2015) Role of Tet1 and 5-hydroxymethylcytosine in cocaine action. Nat Neurosci 18:536–544
Fernandez AF, Valledor L, Vallejo F et al (2018) Quantification of global DNA methylation levels by mass spectrometry. Meth Mol Biol 1708:49–58
Ficz G, Branco MR, Seisenberger S et al (2011) Dynamic regulation of 5-hydroxymethylcytosine in mouse ES cells and during differentiation. Nature 473:398–402
Field SF, Beraldi D, Bachman M et al (2015) Accurate measurement of 5-methylcytosine and 5-hydroxymethylcytosine in human cerebellum DNA by oxidative bisulfite on an array (OxBS-array). PLoS One 10:e0118202
Figueroa ME, Abdel-Wahab O, Lu C et al (2010) Leukemic IDH1 and IDH2 mutations result in a hypermethylation phenotype, disrupt TET2 function, and impair hematopoietic differentiation. Canc Cell 18:553–567
Frauer C, Hoffmann T, Bultmann S et al (2011a) Recognition of 5-hydroxymethylcytosine by the Uhrf1 SRA domain. PLoS One 6:e21306

Frauer C, Rottach A, Meilinger D et al (2011b) Different binding properties and function of CXXC zinc finger domains in Dnmt1 and Tet1. PLoS One 6:e16627
Gaidzik VI, Paschka P, Späth D et al (2012) TET2 mutations in acute myeloid leukemia (AML): results from a comprehensive genetic and clinical analysis of the AML study group. J Clin Oncol 30:1350–1357
Gama-Sosa MA, Slagel VA, Trewyn RW et al (1983) The 5-methylcytosine content of DNA from human tumors. Nucleic Acids Res 11:6883–6894
Globisch D, Münzel M, Müller M et al (2010) Tissue distribution of 5-hydroxymethylcytosine and search for active demethylation intermediates. PLoS One 5:e15367
Good CR, Madzo J, Patel B et al (2017) A novel isoform of TET1 that lacks a CXXC domain is overexpressed in cancer. Nucleic Acids Res 45:8269–8281
Gorres KL, Raines RT (2010) Prolyl 4-hydroxylase. Crit Rev Biochem Mol Biol 45:106–124
Gu T-PP, Guo F, Yang H et al (2011) The role of Tet3 DNA dioxygenase in epigenetic reprogramming by oocytes. Nature 477:606–610
Guilhamon P, Eskandarpour M, Halai D et al (2013) Meta-analysis of IDH-mutant cancers identifies EBF1 as an interaction partner for TET2. Nat Comm 4:2166
Guo JU, Ma DK, Mo H et al (2011a) Neuronal activity modifies the DNA methylation landscape in the adult brain. Nat Neurosci 14:1345–1351
Guo JU, Su Y, Zhong C et al (2011b) Hydroxylation of 5-methylcytosine by TET1 promotes active DNA demethylation in the adult brain. Cell 145:423–434
Hackett JA, Sengupta R, Zylicz JJ et al (2013) Germline DNA demethylation dynamics and imprint erasure through 5-hydroxymethylcytosine. Science 339:448–452
Hajkova P, Erhardt S, Lane N et al (2002) Epigenetic reprogramming in mouse primordial germ cells. Mech Dev 117:15–23
Hashimoto H, Pais JE, Zhang X et al (2014) Structure of a Naegleria Tet-like dioxygenase in complex with 5-methylcytosine DNA. Nature 506:391–395
Hashimoto H, Pais JE, Dai N et al (2015) Structure of Naegleria Tet-like dioxygenase (NgTet1) in complexes with a reaction intermediate 5-hydroxymethylcytosine DNA. Nucleic Acids Res 43: 10713–10721
He Y-FF, Li B-ZZ, Li Z et al (2011) Tet-mediated formation of 5-carboxylcytosine and its excision by TDG in mammalian DNA. Science 333:1303–1307
Hendrich B, Bird A (1998) Identification and characterization of a family of mammalian methyl-CpG binding proteins. Mol Cell Biol 18:6538–6547
Hendrich B, Tweedie S (2003) The methyl-CpG binding domain and the evolving role of DNA methylation in animals. Trends Genet 19:269–277
Hon GC, Song C-XX, Du T et al (2014) 5mC oxidation by Tet2 modulates enhancer activity and timing of transcriptome reprogramming during differentiation. Mol Cell 56:286–297
Hrit J, Goodrich L, Li C et al (2018) OGT binds a conserved C-terminal domain of TET1 to regulate TET1 activity and function in development. eLife 7:e34870
Hu L, Li Z, Cheng J et al (2013) Crystal structure of TET2-DNA complex: insight into TET-mediated 5mC oxidation. Cell 155:1545–1555
Hu X, Zhang L, Mao S-QQ et al (2014) Tet and TDG mediate DNA demethylation essential for mesenchymal-to-epithelial transition in somatic cell reprogramming. Cell Stem Cell 14: 512–522
Hu L, Lu J, Cheng J et al (2015) Structural insight into substrate preference for TET-mediated oxidation. Nature 527:118–122
Huang H, Jiang X, Li Z et al (2013) TET1 plays an essential oncogenic role in MLL-rearranged leukemia. Proc Natl Acad Sci U S A 110:11994–11999
Huang Y, Chavez L, Chang X et al (2014) Distinct roles of the methylcytosine oxidases Tet1 and Tet2 in mouse embryonic stem cells. Proc Natl Acad Sci U S A 111:1361–1366
Huang H, Jiang X, Wang J et al (2016) Identification of MLL-fusion/MYC⊣miR-26⊣TET1 signaling circuit in MLL-rearranged leukemia. Canc Lett 372:157–165

Ito S, D'Alessio AC, Taranova OV et al (2010) Role of Tet proteins in 5mC to 5hmC conversion, ES-cell self-renewal and inner cell mass specification. Nature 466:1129–1133
Ito S, Shen L, Dai Q et al (2011) Tet proteins can convert 5-methylcytosine to 5-formylcytosine and 5-carboxylcytosine. Science 333:1300–1303
Ito R, Katsura S, Shimada H et al (2014) TET3-OGT interaction increases the stability and the presence of OGT in chromatin. Genes Cells 19:52–65
Ittel A, Jeandidier E, Helias C et al (2013) First description of the t(10;11)(q22;q23)/MLL-TET1 translocation in a T-cell lymphoblastic lymphoma, with subsequent lineage switch to acute myelomonocytic myeloid leukemia. Haematologica 98:e166–e168
Itzykson R, Kosmider O, Renneville A et al (2013) Prognostic score including gene mutations in chronic myelomonocytic leukemia. J Clin Oncol 31:2428–2436
Iurlaro M, Ficz G, Oxley D et al (2013) A screen for hydroxymethylcytosine and formylcytosine binding proteins suggests functions in transcription and chromatin regulation. Genome Biol 14: R119
Iyer LM, Tahiliani M, Rao A et al (2009) Prediction of novel families of enzymes involved in oxidative and other complex modifications of bases in nucleic acids. Cell Cycle 8:1698–1710
Jankowska AM, Szpurka H, Tiu RV et al (2009) Loss of heterozygosity 4q24 and TET2 mutations associated with myelodysplastic/myeloproliferative neoplasms. Blood 113:6403–6410
Jäwert F, Hasséus B, Kjeller G et al (2013) Loss of 5-hydroxymethylcytosine and TET2 in oral squamous cell carcinoma. Anticancer Res 33:4325–4328
Jiang D, Zhang Y, Hart RP et al (2015) Alteration in 5-hydroxymethylcytosine-mediated epigenetic regulation leads to Purkinje cell vulnerability in ATM deficiency. Brain 138:3520–3536
Jin S-GG, Jiang Y, Qiu R et al (2011) 5-Hydroxymethylcytosine is strongly depleted in human cancers but its levels do not correlate with IDH1 mutations. Canc Res 71:7360–7365
Jin Y, Shao Y, Shi X et al (2016) Mutational profiling of non-small-cell lung cancer patients resistant to first-generation EGFR tyrosine kinase inhibitors using next generation sequencing. Oncotarget 7:61755–61763
Jjingo D, Conley AB, Yi SV et al (2012) On the presence and role of human gene-body DNA methylation. Oncotarget 3:462–474
Jones PA (2012) Functions of DNA methylation: islands, start sites, gene bodies and beyond. Nat Rev Genet 13:484–492
Kafer GR, Li X, Horii T et al (2016) 5-Hydroxymethylcytosine marks sites of DNA damage and promotes genome stability. Cell Rep 14:1283–1292
Kang J, Lienhard M, Pastor WA et al (2015) Simultaneous deletion of the methylcytosine oxidases Tet1 and Tet3 increases transcriptome variability in early embryogenesis. Proc Natl Acad Sci U S A 112:45
Ko M, Huang Y, Jankowska AM et al (2010) Impaired hydroxylation of 5-methylcytosine in myeloid cancers with mutant TET2. Nature 468:839–843
Ko M, An J, Bandukwala HS et al (2013) Modulation of TET2 expression and 5-methylcytosine oxidation by the CXXC domain protein IDAX. Nature 497:122–126
Koh KP, Yabuuchi A, Rao S et al (2011) Tet1 and Tet2 regulate 5-hydroxymethylcytosine production and cell lineage specification in mouse embryonic stem cells. Cell Stem Cell 8: 200–213
Kriaucionis S, Heintz N (2009) The nuclear DNA base 5-hydroxymethylcytosine is present in Purkinje neurons and the brain. Science 324:929–930
Kudo Y, Tateishi K, Yamamoto K et al (2012) Loss of 5-hydroxymethylcytosine is accompanied with malignant cellular transformation. Canc Sci 103:670–676
Kuiper C, Vissers MC (2014) Ascorbate as a co-factor for Fe- and 2-oxoglutarate dependent dioxygenases: physiological activity in tumor growth and progression. Front Oncol 4:359
Kumar D, Aggarwal M, Kaas GA et al (2015) Tet1 oxidase regulates neuronal gene transcription, active DNA hydroxy-methylation, object location memory, and threat recognition memory. Neuroepigenetics 4:12–27

Langemeijer SM, Kuiper RP, Berends M et al (2009) Acquired mutations in TET2 are common in myelodysplastic syndromes. Nat Genet 41:838–842
Langlois T, da Costa Reis Monte-Mor B, Lenglet G et al (2014) TET2 deficiency inhibits mesoderm and hematopoietic differentiation in human embryonic stem cells. Stem Cells 32:2084–2097
Lasho TL, Vallapureddy R, Finke CM et al (2018) Infrequent occurrence of TET1, TET3, and ASXL2 mutations in myelodysplastic/myeloproliferative neoplasms. Blood Canc J 8:32
Laukka T, Mariani CJ, Ihantola T et al (2015) Fumarate and succinate regulate expression of hypoxia-inducible genes via TET enzymes. J Biol Chem 291:4256–4265
Lee S-GG, Cho SY, Kim MJ et al (2013) Genomic breakpoints and clinical features of MLL-TET1 rearrangement in acute leukemias. Haematologica 98:7
Lee HJ, Hore TA, Reik W (2014) Reprogramming the methylome: erasing memory and creating diversity. Cell Stem Cell 14:710–719
Lewis JD, Meehan RR, Henzel WJ et al (1992) Purification, sequence, and cellular localization of a novel chromosomal protein that binds to methylated DNA. Cell 69:905–914
Li C, Lan Y, Schwartz-Orbach L et al (2015a) Overlapping requirements for Tet2 and Tet3 in normal development and hematopoietic stem cell emergence. Cell Rep 12:1133–1143
Li T, Yang D, Li J et al (2015b) Critical role of Tet3 in neural progenitor cell maintenance and terminal differentiation. Mol Neurobiol 51:142–154
Lian CG, Xu Y, Ceol C et al (2012) Loss of 5-hydroxymethylcytosine is an epigenetic hallmark of melanoma. Cell 150:1135–1146
Lin IHH, Chen Y-FF, Hsu M-TT (2017) Correlated 5-hydroxymethylcytosine (5hmC) and gene expression profiles underpin gene and organ-specific epigenetic regulation in adult mouse brain and liver. PLoS One 12:e0170779
Liu D, Li G, Zuo Y (2018) Function determinants of TET proteins: the arrangements of sequence motifs with specific codes. Brief Bioinform
Liutkevičiūtė Z, Kriukienė E, Ličytė J et al (2014) Direct decarboxylation of 5-carboxylcytosine by DNA C5-methyltransferases. J Am Chem Soc 136:5884–5887
Lorsbach RB, Moore J, Mathew S et al (2003) TET1, a member of a novel protein family, is fused to MLL in acute myeloid leukemia containing the t(10;11)(q22;q23). Leukemia 17:637–641
Losman J-AA, Kaelin WG (2013) What a difference a hydroxyl makes: mutant IDH, (R)-2-hydroxyglutarate, and cancer. Gene Dev 27:836–852
Lu F, Liu Y, Jiang L et al (2014) Role of Tet proteins in enhancer activity and telomere elongation. Gene Dev 28:2103–2119
Madzo J, Liu H, Rodriguez A et al (2014) Hydroxymethylation at gene regulatory regions directs stem/early progenitor cell commitment during erythropoiesis. Cell Rep 6:231–244
Maiti A, Drohat AC (2011) Thymine DNA glycosylase can rapidly excise 5-formylcytosine and 5-carboxylcytosine: potential implications for active demethylation of CpG sites. J Biol Chem 286:35334–35338
Mariani CJ, Vasanthakumar A, Madzo J et al (2014) TET1-mediated hydroxymethylation facilitates hypoxic gene induction in neuroblastoma. Cell Rep 7:1343–1352
Mayer W, Niveleau A, Walter J et al (2000) Demethylation of the zygotic paternal genome. Nature 403:501–502
Meehan RR, Lewis JD, McKay S et al (1989) Identification of a mammalian protein that binds specifically to DNA containing methylated CpGs. Cell 58:499–507
Meldi K, Qin T, Buchi F et al (2015) Specific molecular signatures predict decitabine response in chronic myelomonocytic leukemia. J Clin Investig 125:1857–1872
Mellén M, Ayata P, Dewell S et al (2012) MeCP2 binds to 5hmC enriched within active genes and accessible chromatin in the nervous system. Cell 151:1417–1430
Merlevede J, Droin N, Qin T et al (2016) Mutation allele burden remains unchanged in chronic myelomonocytic leukaemia responding to hypomethylating agents. Nat Commun 7: 10767
Meyer C, Hofmann J, Burmeister T et al (2013) The MLL recombinome of acute leukemias in 2013. Leukemia 27:2165–2176

Minor EA, Young JI, Wang G (2013) Ascorbate induces ten-eleven translocation (Tet) methylcytosine dioxygenase-mediated generation of 5-hydroxymethylcytosine. J Biol Chem 288: 13669–13674
Misawa K, Imai A, Mochizuki D et al (2018) Association of TET3 epigenetic inactivation with head and neck cancer. Oncotarget 9:24480–24493
Müller U, Bauer C, Siegl M et al (2014) TET-mediated oxidation of methylcytosine causes TDG or NEIL glycosylase dependent gene reactivation. Nucleic Acids Res 42:8592–8604
Nakagawa T, Lv L, Nakagawa M et al (2015) CRL4(VprBP) E3 ligase promotes monoubiquitylation and chromatin binding of TET dioxygenases. Mol Cell 57:247–260
Nan X, Campoy FJ, Bird A (1997) MeCP2 is a transcriptional repressor with abundant binding sites in genomic chromatin. Cell 88:471–481
Neri F, Incarnato D, Krepelova A et al (2015) TET1 is controlled by pluripotency-associated factors in ESCs and downmodulated by PRC2 in differentiated cells and tissues. Nucleic Acids Res 43:6814–6826
Nestor CE, Meehan RR (2014) Hydroxymethylated DNA immunoprecipitation (hmeDIP). Meth Mol Biol 1094:259–267
Nickerson ML, Im KM, Misner KJ et al (2013) Somatic alterations contributing to metastasis of a castration-resistant prostate cancer. Hum Mutat 34:1231–1241
Ono R, Taki T, Taketani T et al (2002) LCX, leukemia-associated protein with a CXXC domain, is fused to MLL in acute myeloid leukemia with trilineage dysplasia having t(10;11)(q22;q23). Canc Res 62:4075–4080
Oswald J, Engemann S, Lane N et al (2000) Active demethylation of the paternal genome in the mouse zygote. Curr Biol 10:475–478
Pan H, Bilinovich SM, Kaur P et al (2017) CpG and methylation-dependent DNA binding and dynamics of the methylcytosine binding domain 2 protein at the single-molecule level. Nucleic Acids Res 45:9164–9177
Pastor WA, Pape UJ, Huang Y et al (2011) Genome-wide mapping of 5-hydroxymethylcytosine in embryonic stem cells. Nature 473:394–397
Patel JP, Gönen M, Figueroa ME et al (2012) Prognostic relevance of integrated genetic profiling in acute myeloid leukemia. N Engl J Med 366:1079–1089
Perera A, Eisen D, Wagner M et al (2015) TET3 is recruited by REST for context-specific hydroxymethylation and induction of gene expression. Cell Rep 11:283–294
Pidugu LS, Flowers JW, Coey CT et al (2016) Structural basis for excision of 5-formylcytosine by thymine DNA glycosylase. Biochemistry 55:6205–6208
Pollard PJ, Brière JJ, Alam NA et al (2005) Accumulation of Krebs cycle intermediates and overexpression of HIF1alpha in tumours which result from germline FH and SDH mutations. Hum Mol Genet 14:2231–2239
Quivoron C, Couronné L, Valle DV et al (2011) TET2 inactivation results in pleiotropic hematopoietic abnormalities in mouse and is a recurrent event during human lymphomagenesis. Canc Cell 20:25–38
Rampal R, Alkalin A, Madzo J et al (2014) DNA hydroxymethylation profiling reveals that WT1 mutations result in loss of TET2 function in acute myeloid leukemia. Cell Rep 9:1841–1855
Reik W (2007) Stability and flexibility of epigenetic gene regulation in mammalian development. Nature 447:425–432
Robertson AB, Dahl JA, Ougland R et al (2012) Pull-down of 5-hydroxymethylcytosine DNA using JBP1-coated magnetic beads. Nat Protocol 7:340–350
Rudenko A, Dawlaty MM, Seo J et al (2013) Tet1 is critical for neuronal activity-regulated gene expression and memory extinction. Neuron 79:1109–1122
Sato Y, Yoshizato T, Shiraishi Y et al (2013) Integrated molecular analysis of clear-cell renal cell carcinoma. Nat Genet 45:860–867
Seiler CL, Fernandez J, Koerperich Z et al (2018) Maintenance DNA methyltransferase activity in the presence of oxidized forms of 5-methylcytosine: structural basis for ten eleven translocation-mediated DNA demethylation. Biochemistry 57:6061–6069

Seki Y, Hayashi K, Itoh K et al (2005) Extensive and orderly reprogramming of genome-wide chromatin modifications associated with specification and early development of germ cells in mice. Dev Biol 278:440–458

Sharif J, Muto M, Takebayashi S-i et al (2007) The SRA protein Np95 mediates epigenetic inheritance by recruiting Dnmt1 to methylated DNA. Nature 450:908–912

Shi F-TT, Kim H, Lu W et al (2013) Ten-eleven translocation 1 (Tet1) is regulated by O-linked N-acetylglucosamine transferase (Ogt) for target gene repression in mouse embryonic stem cells. J Biol Chem 288:20776–20784

Shi X, Yu Y, Luo M et al (2016) Loss of 5-hydroxymethylcytosine is an independent unfavorable prognostic factor for esophageal squamous cell carcinoma. PLoS One 11:e0153100

Slyvka A, Mierzejewska K, Bochtler M (2017) Nei-like 1 (NEIL1) excises 5-carboxylcytosine directly and stimulates TDG-mediated 5-formyl and 5-carboxylcytosine excision. Sci Rep 7: 9001

Sohni A, Bartoccetti M, Khoueiry R et al (2015) Dynamic switching of active promoter and enhancer domains regulates Tet1 and Tet2 expression during cell state transitions between pluripotency and differentiation. Mol Cell Biol 35:1026–1042

Song C-XX, Szulwach KE, Fu Y et al (2011) Selective chemical labeling reveals the genome-wide distribution of 5-hydroxymethylcytosine. Nat Biotechnol 29:68–72

Song SJ, Ito K, Ala U et al (2013) The oncogenic microRNA miR-22 targets the TET2 tumor suppressor to promote hematopoietic stem cell self-renewal and transformation. Cell Stem Cell 13:87–101

Spruijt CG, Gnerlich F, Smits AH et al (2013) Dynamic readers for 5-(hydroxy)methylcytosine and its oxidized derivatives. Cell 152:1146–1159

Stroud H, Feng S, Morey Kinney S et al (2011) 5-Hydroxymethylcytosine is associated with enhancers and gene bodies in human embryonic stem cells. Genome Biol 12:R54

Szulwach KE, Li X, Li Y et al (2011) Integrating 5-hydroxymethylcytosine into the epigenomic landscape of human embryonic stem cells. PLoS Genet 7:e1002154

Tahiliani M, Koh KP, Shen Y et al (2009) Conversion of 5-methylcytosine to 5-hydroxymethylcytosine in mammalian DNA by MLL partner TET1. Science 324:930–935

Tardy-Planechaud S, Fujimoto J, Lin SS et al (1997) Solid phase synthesis and restriction endonuclease cleavage of oligodeoxynucleotides containing 5-(hydroxymethyl)-cytosine. Nucleic Acids Res 25:553–559

Thienpont B, Steinbacher J, Zhao H et al (2016) Tumour hypoxia causes DNA hypermethylation by reducing TET activity. Nature 537:63–68

Turcan S, Rohle D, Goenka A et al (2012) IDH1 mutation is sufficient to establish the glioma hypermethylator phenotype. Nature 483:479–483

Valinluck V, Sowers LC (2007) Endogenous cytosine damage products alter the site selectivity of human DNA maintenance methyltransferase DNMT1. Canc Res 67:946–950

Varley KE, Gertz J, Bowling KM et al (2013) Dynamic DNA methylation across diverse human cell lines and tissues. Genome Res 23:555–567

Vasanthakumar A, Godley LA (2015) 5-Hydroxymethylcytosine in cancer: significance in diagnosis and therapy. Canc Genet 208:167–177

Vella P, Scelfo A, Jammula S et al (2013) Tet proteins connect the O-linked N-acetylglucosamine transferase Ogt to chromatin in embryonic stem cells. Mol Cell 49:645–656

Vincent JJ, Huang Y, Chen P-YY et al (2013) Stage-specific roles for tet1 and tet2 in DNA demethylation in primordial germ cells. Cell Stem Cell 12:470–478

Vogel Ciernia A, LaSalle J (2016) The landscape of DNA methylation amid a perfect storm of autism aetiologies. Nat Rev Neurosci 17:411–423

Wang Y, Zhang Y (2014) Regulation of TET protein stability by calpains. Cell Rep 6:278–284

Wang H, Guan S, Quimby A et al (2011) Comparative characterization of the PvuRts1I family of restriction enzymes and their application in mapping genomic 5-hydroxymethylcytosine. Nucleic Acids Res 39:9294–9305

Wang Y, Xiao M, Chen X et al (2015) WT1 recruits TET2 to regulate its target gene expression and suppress leukemia cell proliferation. Mol Cell 57:662–673

Wen L, Li X, Yan L et al (2014) Whole-genome analysis of 5-hydroxymethylcytosine and 5-methylcytosine at base resolution in the human brain. Genome Biol 15:R49

Williams K, Christensen J, Pedersen MT et al (2011) TET1 and hydroxymethylcytosine in transcription and DNA methylation fidelity. Nature 473:343–348

Wossidlo M, Nakamura T, Lepikhov K et al (2011) 5-Hydroxymethylcytosine in the mammalian zygote is linked with epigenetic reprogramming. Nat Commun 2:241

Wu H, Zhang Y (2011) Tet1 and 5-hydroxymethylation: a genome-wide view in mouse embryonic stem cells. Cell Cycle 10:2428–2436

Wu X, Zhang Y (2017) TET-mediated active DNA demethylation: mechanism, function and beyond. Nat Rev Genet 18:517–534

Wu D, Hu D, Chen H et al (2018) Glucose-regulated phosphorylation of TET2 by AMPK reveals a pathway linking diabetes to cancer. Nature 559:637–641

Wyatt GR, Cohen SS (1953) The bases of the nucleic acids of some bacterial and animal viruses: the occurrence of 5-hydroxymethylcytosine. Biochem J 55:774–782

Xu W, Yang H, Liu Y et al (2011a) Oncometabolite 2-hydroxyglutarate is a competitive inhibitor of α-ketoglutarate-dependent dioxygenases. Canc Cell 19:17–30

Xu Y, Wu F, Tan L et al (2011b) Genome-wide regulation of 5hmC, 5mC, and gene expression by Tet1 hydroxylase in mouse embryonic stem cells. Mol Cell 42:451–464

Xu Y, Xu C, Kato A et al (2012) Tet3 CXXC domain and dioxygenase activity cooperatively regulate key genes for Xenopus eye and neural development. Cell 151:1200–1213

Yamaguchi S, Hong K, Liu R et al (2012) Tet1 controls meiosis by regulating meiotic gene expression. Nature 492:443–447

Yamazaki J, Taby R, Vasanthakumar A et al (2012) Effects of TET2 mutations on DNA methylation in chronic myelomonocytic leukemia. Epigenetics 7:201–207

Yamazaki J, Jelinek J, Lu Y et al (2015) TET2 mutations affect non-CpG island DNA methylation at enhancers and transcription factor-binding sites in chronic myelomonocytic leukemia. Canc Res 75:2833–2843

Yang H, Liu Y, Bai F et al (2013) Tumor development is associated with decrease of TET gene expression and 5-methylcytosine hydroxylation. Oncogene 32:663–669

Yildirim O, Li R, Hung J-HH et al (2011) Mbd3/NURD complex regulates expression of 5-hydroxymethylcytosine marked genes in embryonic stem cells. Cell 147:1498–1510

Yin R, Mao SQ, Zhao B et al (2013) Ascorbic acid enhances Tet-mediated 5-methylcytosine oxidation and promotes DNA demethylation in mammals. J Am Chem Soc 135:10396–10403

Yu M, Hon GC, Szulwach KE et al (2012) Tet-assisted bisulfite sequencing of 5-hydroxymethylcytosine. Nat Protocol 7:2159–2170

Zhang H, Zhang X, Clark E et al (2010) TET1 is a DNA-binding protein that modulates DNA methylation and gene transcription via hydroxylation of 5-methylcytosine. Cell Res 20: 1390–1393

Zhang R-RR, Cui Q-YY, Murai K et al (2013) Tet1 regulates adult hippocampal neurogenesis and cognition. Cell Stem Cell 13:237–245

Zhang Q, Liu X, Gao W et al (2014) Differential regulation of the ten-eleven translocation (TET) family of dioxygenases by O-linked β-N-acetylglucosamine transferase (OGT). J Biol Chem 289:5986–5996

Zhang W, Xia W, Wang Q et al (2016) Isoform switch of TET1 regulates DNA demethylation and mouse development. Mol Cell 64:1062–1073

Zhang YW, Wang Z, Xie W et al (2017) Acetylation enhances TET2 function in protecting against abnormal DNA methylation during oxidative stress. Mol Cell 65:323–335

Zhu F, Zhu Q, Ye D et al (2018) Sin3a-Tet1 interaction activates gene transcription and is required for embryonic stem cell pluripotency. Nucleic Acids Res 46:6026–6040

# Epigenetic Alterations: The Relation Between Occupational Exposure and Biological Effects in Humans

**Vivian Silva Kahl, Mónica Cappetta, and Juliana Da Silva**

## Contents



**Abstract** Exposome encompass the totality of human environmental exposures, providing a lifelong exposure history and complementing the genome. One of its domains is a specific external environment, which includes occupational exposure. Over the last decades, several publications have shown the higher incidences of exposure-related diseases and its relationship with DNA damage. However, there is a body of evidence that genetic variants cannot fully explain the variability in the risk of chronic diseases initiation and development, leaving a potential role the interaction between environmental and genetic factors. A key phenomenon are epigenetic modifications, heritable changes in gene expression that occur without changes in DNA sequence and play an important role in identifying mechanisms of xenobiotic-induced non-genotoxic carcinogenesis. Recently studies with occupational exposure individuals have shown substantial epigenetic alterations as effect of work-related activity with several xenobiotics, such as benzene, solvent, styrene, heavy metals, and mixtures of chemicals. Exposure to occupational toxicants may contribute to

V. S. Kahl (✉)
Telomere Length Regulation Unit, Children's Medical Research Institute (CMRI), Westmead, NSW, Australia
e-mail: vkahl@cmri.org.au

M. Cappetta
Laboratorio de Epidemiología Genética, Departamento de Genética, Facultad de Medicina, Universidad de la República, Montevideo, Uruguay

J. Da Silva
Laboratory of Toxicological Genetics, Post-graduate Program in Cellular and Molecular Biology Applied to Health, Lutheran University of Brazil (ULBRA), Canoas, Brazil



S. Jurga, J. Barciszewski (eds.), *The DNA, RNA, and Histone Methylomes*, RNA Technologies, https://doi.org/10.1007/978-3-030-14792-1_11

arising of adverse birth outcomes, neurological and other multifactorial diseases, and increased risk of cancer, and there is evidence that epigenetic aspects intermediate their effects in human health. In the current chapter, we review recently discoveries in the field of occupational exposure, health effects, and the interaction of epigenetic factors for such outcomes. The solid identification of key genetic and/or epigenetic events involved in chemical occupational-related carcinogenesis is a relevant step towards improvement of biomarkers to evaluate exposure, predict biological effects, and prevent adverse health consequences.

**Keywords** Occupational exposure · Metals · Organic compounds · Complex mixtures · Biomarkers of exposure

## 1 Introduction

Exposome encompass the totality of human environmental exposures, providing a lifelong exposure history and complementing the genome, being a new paradigm for studying the sum of environmental causes of diseases. Environmental research and public health aspects currently face several challenges such as air and water, as well as industrial pollution, which are particularly of prevalent concern in developing economies (Holland 2017). One of the major domains of exposome is the evaluation of an individual's external environment, which encompasses the increasing in global warming, widespread use of chemicals including pesticides and heavy metals, as much as other endocrinal disruptors, and major changes in nutrition and lifestyle of modern society, such as smoking and drinking habits, hormone-based medicines, high-fat foods and low fibers intake (Faisandier et al. 2011; Holland 2017; Siroux et al. 2016).

Individual's exposome is defined as the total of many exposure factors that comprehend the lifetime of such individual, including exposure to chemicals, radiation, environmental agents, nutritional patterns, stress, among others. Their health behavior, physical activities routine and their microbiome profile are components of the exposome. Specially, occupational exposure is a major issue as regard public health, as the proper identification of hazards and prevention of new threats to health may help in minimizing concerns (Holland 2017). In the past decades, some pathologies have been linked to different occupational exposures, with the main findings for respiratory tract (Gaffney and Christiani 2015), endocrinology diseases (Silins and Högberg 2011), cardiovascular impairments (Fang et al. 2010; Sekhotha et al. 2016), and risk of cancer (Alvanja and Bonner 2012; Charbotel et al. 2014; Fritisch et al. 2015).

Biomarkers are then a powerful tool for occupational health risk assessment. They are generally divided in three main classes for human studies: biomarkers of exposure, of susceptibility and of effect. While the first ones involve measurements of metabolites, mainly compounds and reflects internal and biologically effect dose, the second indicate an often-constitutive ability of an individual to respond to a given exposure (Schulte and Hauser 2012). Biomarkers of effect compose the majority of occupational studies as, in general, workers are exposed to mixtures of agents. Therefore, those biomarkers help to identify both active components of the mixtures and consequences of specific mixtures exposures. A sub-class of biomarkers of effect is called biomarkers of early disease, which comprehends tests more closely indicative of a plain clinical effect (Silins and Högberg 2011).

Genetics is considered the main player in phenotype, therefore biomarkers of susceptibility represent a substantial knowledge for occupational risk assessment, as they include polymorphisms of specific genes associated with metabolism and detoxification of chemical material in the organism (Schulte and Hauser 2012). Genetic differences in metabolism can have an effect on population level, rather than in individual level, and may result in different effects for a given exposure. However, it is known that DNA sequence alone (i.e., genetic variation) cannot fully explain the observed phenotypic traits. Mutations in several genes are a distinctive feature of cancer cells and support the knowledge that cancer arises through the accumulation of irreversible DNA damage, and act in a 'genotoxic' manner. Despite this, there is a group of carcinogens that induce cancer via non-genotoxic mechanisms. Thus, other determinants of phenotype variation should be considered, and these include epigenetic modifications related to environmental exposure (Ravegnini et al. 2015; Meehan et al. 2018).

Epigenetic mechanisms, such as DNA, RNA and histones modifications, and microRNAs, have been shown to be potential links between the genetic and environmental exposure, which can be determinant to health and disease development. Epigenetics investigates heritable changes in gene expression without modifications in DNA sequence itself and, unlike genetics, they could be reversible. Particularly, epigenetic modifications can alter genome expression and function under exogenous influence (Baccarelli and Bollati 2009; Holland 2017). In the current chapter, the most recently discoveries in the field of occupational exposure health effects and the possible interaction of some epigenetic factors for such outcomes will be reviewed. The solid identification of key genetic and/or epigenetic events involved in chemical occupational-related carcinogenesis is a relevant step towards improvement of biomarkers to evaluate exposure, predict biological effects, and prevent adverse health consequences.

## 2 Metals

The genetic toxicity aspect of metals has been extensively studied, demonstrating that many common metals in human routine can cause DNA damage (Bal et al. 2011). Lately, its effect on epigenome has been shown through several in vitro and in vivo studies, along with epidemiological research as well.

Table 1 shows the main results observed in different studies as regard epigenetic markers in occupational exposure to metals. Arsenic (As) compounds are important environmental carcinogens that affect DNA methylation status in cell (Cheng et al.

**Table 1** Epigenetic alterations in occupational exposure to metals

| Exposure | Country | N. of participants | Results[a] | References |
|---|---|---|---|---|
| Arsenic (As) | China | *Exposed*: 43 As trioxide producers (plant 1), 36 workers who stopped producing 85 days previous (plant 2). *Control*: 24 individuals never exposed | ↓ 5 LincRNAs in workers for plant 1. Significant higher base modifications of three exons of *p53* in workers from both plants. Several correlations between different exon base modifications of *p53* and expressions of LincRNAs. | Wen et al. (2016) |
| Chromium (Cr) | China | *Exposed*: 115 chromate producing facility workers. *Control:* 60 non-exposed local residents | ↓ global DNA methylation. RBC-Cr levels negatively associated with global DNA methylation. Folate positively associated with global DNA methylation for both groups. | Wang et al. (2012) |
| | China | *Exposed*: 29 chrome plating workers. *Control*: 29 non-exposed to Cr matched individuals | methylation of *MT-TF* and *MT-RNR1*. Negatively correlation found for Cr levels and *MT-TF* and *MT-RNR1* gene methylation. CpG sites in *MT-TF* and *MT-RNR1* negatively associated with Cr level. | Linqing et al. (2016) |
| | China | *Exposed*: 87 blue-collar workers from a chromate factory. *Control*: 30 office workers from the same factory | Cr levels in exposed workers positively correlated with: methylation level of CpG sites in DNA repair genes (*MGMT* and *HOGG1*) and with CpG sites in *RAD51* gene. | Hu et al. (2018) |
| Lead (Pb) | China | *Exposed*: 103 battery plant workers. *Control:* age- and gender-matched 103 healthy volunteers | ↓ methylation of *ALAD* CpG. Individuals with methylated *ALAD* had increased risk of lead poisoning. | Li et al. (2011) |
| | China | *Exposed*: 53 battery plant workers. *Control*: 57 healthy individuals matched by age and gender, smoking status and alcohol consumption | ↓ methylation of *LINE-1.* Lower methylation levels as higher Pb blood levels. | Li et al. (2013) |
| | China | *Exposed:* 1130 battery factories; top 10% with highest blood lead level (BLL) and bottom 10% | ↓ expression of three *miRNAs* in high Pb exposure and high BLL workers: *miR-520c-3p*, | Xu et al. (2017) |

(continued)

**Table 1** (continued)

| Exposure | Country | N. of participants | Results[a] | References |
|---|---|---|---|---|
| | | with lowest BLL defined as high and minimal lead-exposure groups | *miR-211* and *miR-148a,* and ↑ *miR-572* expression. | |
| Mercury (Hg) | United States | *Exposed:* 41 dentists (36 males, 5 females). *Control*: 90 non-dentists (28 males and 62 females). | *LINE-1* methylation positively correlated with age. Trend of *SEPP1* hypomethylation with increasing Hg hair levels, significant among males, for both groups. Trend remaining when for dentists only. | Goodrich et al. (2013) |
| Nickel (Ni) | China | *Exposed*: phase 1–30 flash smelting workshop where Ni is processed; phase 2—additional 15 subjects occupationally exposed to Ni. *Control*: phase 1–60 maintenance and office workers; phase 2–15 additional subjects from same place; all frequency-matched by age and smoking habits | ↑ H3K4me3 and ↓ H3K9me2. H3K4me3 and H3K9ac were positively and negatively associated with urinary Ni, respectively. H3K4me3, H3K9me2 and H3K9ac histone modifications were relatively stable over time. | Arita et al. (2012) |
| | China | *Exposed*: 140 nickel-smelting workers divided in seven groups according to age and years of work. *Control*: 140 office workers age-matched | ↑ levels of H3K4me3 and ↓ levels of H3K27me3. ↑ H3K4me3 level was the highest in the 30+ service length subgroup. ↓ H3K27me3 levels associated with years of exposure. | Ma et al. (2015) |
| Nickel (Ni), arsenic (As) and iron (Fe) | Italy | *Exposed*: 63 male steel production plant (pre- and post-exposure in a given week) | ↑ H3K4me3 and H3K9ac in association with years of employment; ↑ H3K4me3 increased in association with air levels of Ni, As and Fe; Cumulative exposure to the three agents was positively correlated with H3K4me3 and H3K9ac | Cantone et al. (2011) |

[a]Describes results statistically significant (unless stated otherwise) for exposure groups in relation to control or baseline groups (unless stated otherwise)
*RBC* red blood cells

2012) and are classified in group 1 of IARC (Beyersmann and Hartwig 2008). As has been shown to cause different types of cancer via exposure to contaminated drinking water and/or air pathway (Tchounwou et al. 2003). Oxidative methylation and reduction reactions of As lead to the generation of methylated metabolites, which are excreted in urine. A Chinese study evaluated As trioxide producers and observed higher levels of urinary As (Wen et al. 2016). The toxicity of As is more frequently related to its trivalent state, the As trioxide, due to its ability to bind thiol groups in various cellular components. The exposed group from Wen et al. (2016) study was composed by workers currently exposed and workers who had stopped producing 85 days previously to the study begins. Regardless, both subgroups showed significant base modifications of exons 5, 6, 7 and 8 of *p53* tumor suppressor gene (Wen et al. 2016). DNA microarray study has found up-regulation of various oncogenes after As exposure, but not for *p53*, while another report show that human lung adenocarcinoma cell lines exposed to As had dose-response hypermethylation of *p53* promoter (reviewed by Cheng et al. 2012). The tumor suppressor p53 plays a key role in the induction of apoptosis and cell cycle arrest in response to a variety of genotoxic stressors, preventing the propagation of damaged cells. Gene-specific methylation of both *p53* and *p16* has been associated with As exposure in different occupational and environmental settings. The p16 is a cyclin-dependent kinase inhibitor that cycle-regulates cells senescence through induction of inflammatory markers (Fischer et al. 2013). Several lines of evidence suggest that As compounds genotoxicity is mediated by increased levels of reactive oxygen species (ROS) and they have also been shown to inhibit DNA repair, mainly thorough nucleotide excision repair (NER) (Beyersmann and Hartwig 2008).

Oxidative stress is a major pathway through which metal compounds can cause DNA damage and epigenetic imbalances in humans. Pentavalent chromium reacts with isolated DNA to produce 8-oxo-7,8-dihydro-2′-deoxyguanosine (8-OHdG), the more relevant marker of guanines oxidation. Significantly higher urinary 8-OHdG and increased DNA damage (Hu et al. 2018; Wang et al. 2012), evaluated through Comet assay (Wang et al. 2012) and micronucleus frequency (Hu et al. 2018), were observed in chromate producing facility workers. Altered global methylation of DNA, and methylation of DNA repair genes and specific mtDNA genes were observed in chromate exposed workers in three different studies conducted in China (Hu et al. 2018; Wang et al. 2012; Linqing et al. 2016). Chromium (Cr) has been shown to reduce in-vitro H3 phosphorylation and trimethylation, as well as various acetylation marks in H3 and H4. These changes inhibit RNA polymerase II recruitment and transcription initiation. Thus, epigenetic mechanisms might be a central target of chromium toxicity and inhibition of these mechanisms reduces the capacity of cells to respond to environmental hazards. Long-term exposure to chromium may cause a significant increase in histone deacetylation, which would lead to histone methylation in specific positions involved in gene repression and silencing and to subsequent DNA hypermethylation, which would soon be converted into a complete and efficient state of gene silencing (Schnekenburger et al. 2007). Exposure to different Cr compounds has been extensively related with incidence of respiratory cancer in human and animal models (Beyersmann and Hartwig 2008). In this context, the severity of symptoms of chronic obstructive pulmonary disease (COPD) was correlated with a reduction in histone deacetylase activity (HDAC) in lung cancer and alveolar macrophages (Schottenfeld and Beebe-Dimmer 2006). Increased proliferation kinetics and the interaction of hydroxyl radicals with DNA increase the likelihood of DNA structural and transcriptional errors. A key enzymatic function of HDAC is the inhibition or modulation of production of proinflammatory cytokines and matrix metalloproteinases by macrophages (O'Sullivan et al. 2010). What is more, regulation of the $p16^{ink4a}$ tumor suppressor gene appears to be a major target of chromium toxicity. Cigarette smoking is a major source of coexposure to chromium and B[a]P, and several studies have reported the association between aberrant p16 methylation and smoking (Jarmalaite et al. 2003; Sun et al. 2015). Hypermethylation of the $p16^{ink4a}$ promoter has also been found in one-third of chromate workers with a history of exposure for 15 years or more who developed lung cancer (Kondo et al. 2006).

The disruption of DNA repair mechanism seems to be relevant as per toxicity of metal compounds to humans. Similar to Cr, lead (Pb) highly interacts with both NER and base excision repair (BER), being considered a genotoxic agent. The occupational exposure to Pb shows increased in MN frequencies accompanied by influence of polymorphisms in genes involved in DNA repair, such as OGG1 and XRCC4 (García-Lestón et al. 2012). This metal also seems to perturb telomere replication, leading to chromosomal abnormalities, including, telomere loss (Pottier et al. 2013). Nonhomologous end joining (NHEJ) is the predominant form of repair of unprotected DNA ends in mammalian cells and involves proteins that are well characterized, including Ku70, Ku80, DNA-PKcs, LIG4, and XRCC4. NHEJ is also involved in fusion of telomeres as a result of deficiency in TRF2, one of the components of shelterin complex in telomeres (Murnare 2011). Battery factor workers is the largest studied group of individuals occupationally exposed to Pb, for which all the metal levels are significantly higher than control groups (García-Lestón et al. 2012; Li et al. 2011, 2013; Xu et al. 2017). The *ALAD* gene exists in two forms, *ALAD1* and *ALAD2*, and codes for the enzyme ALAD that catalyzes the second step of heme synthesis and may modify lead toxicokinetics and exert impact on individual susceptibility to lead poisoning. Methylated *ALAD* was observed in Chinese battery factory workers and was associated with increased risk of lead poisoning (Li et al. 2011). In other group of workers occupationally exposed to Pb, hypomethylation of *LINE-1* was significantly different from control individuals and inversely associated with Pb levels (Li et al. 2013). Methylation of *LINE-1* helps maintain genomic stability and integrity, while loss of methylation in *LINE-1* may result in higher chances of mitotic recombination (Cheung et al. 2009). Dysregulation of miRNA interferes in translation of their mRNA. In a large study where 1130 battery workers were stratified as per their Pb blood level (BLL), the top 10% of BLL showed lower expression of three miRNAs. Their functional analysis showed a network involved in cellular process, such as apoptosis, phagocytosis and cell proliferation, potentially mediating pathways related to different by Pb exposure (Xu et al. 2017).

Industrial and commercial uses are the main source of exposure to metals for humans. This aspect is not different for nickel (Ni), a transition metal used in industry along with other metals to form alloys to produce jewelry, household equipment, cooking utensils, coins, orthopedic implants, among others. Although Ni is a known carcinogen, it is non-mutagenic or weakly mutagenic in rodent assays (Costa et al. 2005), but its carcinogenicity is thought to be associated with its ability to exacerbate epigenetic modifications (Arita et al. 2012). Two different studies conducted in China with nickel-smelting workers evidenced nickel's high influence over histone modifications (Arita et al. 2012; Ma et al. 2015). While all histone modifications occur during normal cellular development and processes, dysregulation of the balance of appropriate histone modifications can lead to disease. Histone modifications are of particular interest because histone dynamics play a role in the toxic potential of the chemicals by influencing both transcriptional activity and DNA repair mechanisms (Chappel et al. 2016). Ni induces transcriptional repression of genes involved in homology-dependent DNA double-strand break repair and mismatch repair and Ni accumulation in lung tissues is associated with increased p53 risk mutation in lung cancer patients (Scanlon et al. 2017). Higher levels of H3K4me3 in workers than in the control group were seen in both studies with nickel-smelting occupational exposure, being positively associated with urinary Ni levels (Arita et al. 2012) and with years of exposure to Ni (Ma et al. 2015). Interestingly, an Italian group of male steel production plant, occupationally exposed to nickel, arsenic and iron, had also increased levels of H3K4me3 associated with years of employment and with air levels of the three metals (Cantone et al. 2011). Each of those studies also observed lower levels of H3K9me2 (Arita et al. 2012) and H3K27me3 (Ma et al. 2015), and higher levels of H3K9ac (Cantone et al. 2011) in exposed workers. The redox activity of some nickel compounds is related to histone biding. Furthermore, Ni-induced methylation of H3K9 histones are considered repressing modifications, leading to disruption of transcription factors' access to DNA and silencing of telomeric marker genes (reviewed by Maxwell and Salnikow 2004). The accumulation of the number of subchromosomal regions with allelic imbalance extending to the telomeres is a genomic marker of impaired DNA repair and DNA-damaging agents (Birkbak et al. 2012).

Dental amalgams are a route of exposure to mercury (Hg) by dentists worldwide. Even though, because Hg is also available to humans through consumption of seafood, not many studies are published specific as regard occupational exposure to this metal. Male and female dentists were evaluated in USA and their occupation was predictor of higher Hg blood levels (Goodrich et al. 2013). Although not differences for global DNA methylation were observed, *LINE-1* methylation was positively correlated with age, while a trend of *SEPP1* hypomethylation was observed for male dentists (Goodrich et al. 2013). The *SEPP1* gene is important for Hg toxicokinetics and protection against its toxicity through direct binding, besides its well-established antioxidant activity (Chen et al. 2006). Mercury has been shown to be genotoxic and cause damage to neuronal, cardiovascular and renal systems, besides showing substantial epigenetic modifications in mice offspring (reviewed by Cheng et al. 2012).

## 3 Organic Compounds

Benzene, toluene, xylene, solvents and carbons remains the environmental chemicals highly used in industrial setting worldwide. They are found in gas stations, in leather products, sports equipment manufacturing, outdoor air, among other workplaces. Professions related to vehicle traffic and petrochemical production are the top ones as regard benzene occupational exposure (Carugno et al. 2012; Byun et al. 2013; Jamebozorgi et al. 2018), as summarized in Table 2. Hypermethylation of the *p15* gene was observed in petrochemical male workers (Carugno et al. 2012; Jamebozorgi et al. 2018), bus drivers, gas station attendants and police officers (Carugno et al. 2012). Benzene exposure has been consistently associated with acute myelogenous leukemia (AML) and although the straight forward mechanism has not been fully understood, aberrant DNA methylation patterns, including global hypomethylation, gene-specific hypermethylation or hypomethylation and loss of imprinting, are commonly observed in AML tissues (Rinsky et al. 2002). Mitochondrial DNA copy number (mtDNAcn) was significantly higher and associated with *LINE-1* hypomethylation in an Italian study comprising individuals working somehow with vehicle traffic (Carugno et al. 2012). Curiously, the aberrant methylation observed for *p15* gene in petrochemical workers was not associated with age or smoking status, neither with DNA damage parameters, such as micronucleus (MN), nucleoplasmic bridges (NPB), and nuclear buds (NBUD) (Jamebozorgi et al. 2018). Additionally, another Italian study did not find any difference on mtDNA methylation between gas-station workers with high- and low-exposure to benzene (Byun et al. 2013).

House builders and decorators represent another group of workers with risk of exposure to benzene in their workplace (Table 2), but mainly to a mixture of benzene, toluene and xylene (BTX). This mixture is used as a solvent and the concentration of its components may vary broadly. In a Chinese study comprehending decorators and painters, BTX levels were significantly higher in those workers compared to control group, but even higher in painters (Sha et al. 2014). The increased levels of BTX were associated with lower mRNA expression of genes involved in genome's methylation pattern maintenance. Cytokinesis-block micronucleus parameters assay did not show differences in this study (Sha et al. 2014). Benzene, toluene and xylene are monocyclic aromatic hydrocarbon compounds and when evaluated separately, it is assumed that toluene and xylene may interact determining benzene toxicity by influencing its toxicokinetics (ATSDR 2004). Exposure to a mix of organic solvents, including chloroform, was also associated with global DNA hypermethylation in a study conducted with 128 pharmaceutical plant workers in Belgium (Godderis et al. 2012). A group of chronic toxic encephalopathy (CTE) patients was also included in this study. Since CTE is a neurobehavioral disorder associated with solvent exposure, authors aimed to explore if DNA methylation patterns could play a role in this disease development and prognostic. Although CTE patients had longer exposure to mix of organic solvents than the pharmaceutical workers, their global DNA methylation patterns were

**Table 2** Epigenetic alterations in occupational exposure to organic compounds

| Exposure | Country | N. of participants | Results[a] | References |
|---|---|---|---|---|
| Benzene | Italy | *Exposed*: 153 bus drivers, 78 gas station attendants, 77 police officers, 33 petrochemical workers; *control*: 178 individuals from the same region of study | ↑ mtDNAcn for exposed groups; Interquartile range increase in exposure associated with increase in mtDNAcn; mtDNAcn associated with *LINE-1* hypomethylation and *p15* hypermethylation. | Carugno et al. (2012) |
| | Italy | *Exposed*: 20 gas-station attendants with high exposure to $PM_1$ ($\geq 7.6\ \mu g/m^3$) and 20 with low-exposure to $PM_1$ ($\leq 3.8\ \mu g/m^3$) | No effects on mtDNA methylation (*MT-TF*, *MT-RNR1* and D-loop control region) No association of mtDNA methylation with benzene levels | Byun et al. (2013) |
| | Iran | *Exposed:* petrochemical male workers (40) exposed to <1 mm of benzene. *Control:* 31 office workers | ↑ DNA methylation rate in *p15*$^{INK4b}$ gene in exposed individuals. No association between methylation and frequency of MN, NPB and NBUD in peripheral blood lymphocytes. | Jamebozorgi et al. (2018) |
| Toluene | Korea | *Exposed*: 14 short-term (<6.4 year. of exposure) and 14 long-term (>6.4 year. of exposure). *Control*: 14 non-exposed | 631 genes upregulated and 263 downregulated in the short-term exposure. 662 genes upregulated and 260 downregulated in the long-term exposure, some overlapping short-term; Cell survival, immune and nerve systems functions associated with upregulated genes. | Hong et al. (2016) |
| Benzene, toluene and xylene (BTX) | China | *Exposed*: 132 decorators and 129 painters. *Control*: 130 non-exposed workers frequency matched by sex and age | ↓ expression of *PARP1, DNMTs* and *MBD2*, and ↓ PARP activity. ↓ *PARP1, DNMT1, DNMT3a, DNMT3b* and *MBD2* mRNA expression was correlated with increased BTX levels. | Sha et al. (2014) |
| Mixture of organic | Belgium | *Exposed*: 128 pharmaceutical plant workers. | Global DNA hypermethylation | Godderis et al. (2012) |

(continued)

**Table 2** (continued)

| Exposure | Country | N. of participants | Results[a] | References |
|---|---|---|---|---|
| solvents (mainly chloroform) | | *Control*: 41 healthy individuals | associated with solvent exposure<br>*GSTP1* polymorphism significantly associated with global DNA methylation. | |
| 1,6-hexamethylene diisocyanate (HDI) | United States | *Exposed*: 20 automotive spray-painters based on stratified HDI | Two methylated CpG sites from genes LPHN3 and SCARA5 were associated with urine HDI levels and creatinine.<br>Thirty methylated CpG sites from 28 different genes associated with HDI inhalation and skin exposure. | Nylander-French et al. (2014) |
| Carbon nanotubes | Belgium | *Exposed:* 24 multi-wall carbon nanotubes (MWCNT) workers.<br>*Control*: age-matched 43 office workers from the same company | No differences in global DNA methylation (5-mC), hydroxymethylation (5-hmC) and *LINE-1* methylation between groups.<br>5-mC and 5-hmC were positively correlated between them.<br>Gene-specific methylation in MWCNT group: DNMT1, HDAC4, NPAT/ATM and SKI. | Ghosh et al. (2017) |

[a]Describes results statistically significant (unless stated otherwise) for exposure groups in relation to control or baseline groups (unless stated otherwise)
*CBMN* cytokinesis-block micronucleus assay, *NBUD* nuclear buds, *NPB* nucleoplasmic bridges

similar to the control group (Godderis et al. 2012). Solvent components may rise the occurrence of ROS and cytotoxicity, which explain the association observed between *GSTP1* polymorphism and global DNA methylation in workers (Godderis et al. 2012).

Table 2 also summarizes results observed for some other solvents. Toluene itself accumulates in tissues, including parts of the brain with high lipid content (Tas et al. 2011). Exposure to toluene has been shown to affect gene expression. A pilot study conducted with individuals short-term and long-term (<6.4 and >6.4 years of exposure) occupationally exposed to this chemical evaluated the variation in gene expression and occurrence of methylation (Hong et al. 2016). Authors found that 26 genes were upregulated and hypomethylated, while 32 genes were downregulated and hypermethylated. The upregulated genes, such as MAPK1, TGFB1, TNFs and

ACHE, were mainly associated with cell survival, nervous and immune systems pathways, suggesting that they can help predict the effects of time-dependent toluene exposure (Hong et al. 2016). A previous study showed that footwear workers exposed to solvent-based adhesives had increased DNA damage than control individuals and workers exposed to water-based adhesives (Heuser et al. 2005). The main solvent used by footwear workers was toluene, a chemical that can induce oxidative stress (Martínez-Alfaro et al. 2010). As much as toluene, 1,6-hexamethylene diisocyanate (HDI), is absorbed by human body through inhalation and skin exposure. This chemical is commonly found in automotive spray dye workplaces. Through system biology approach, Nylander-French et al. (2014) analyzed inter-individual differences for automotive spray-painters based on stratified HDI-exposure levels, as regard CpG DNA methylation interactions with blood and urine markers. Two methylated CpG sites in *LPHN3* and *SCARA5* were associated with urine HDI levels and creatinine (Nylander-French et al. 2014). LPHN3 belongs to a family of proteins that function in both cell adhesion and signal transduction and is a binding partner of ubiquitin, suggesting a role in protein ubiquitination (Boucard et al. 2014). The *SCARA5* gene codes for proteins that work in recognition of host defense by initiating immune system (Jiang et al. 2006). Although *SCARA5* is in chromosome 8 and, along with other genes, is implicated in frequent copy number variation, it has not been related to specific diseases. On the other hand, *LPHN3* gene is reported to be associated with cognitive disabilities (Nylander-French et al. 2014).

Apart from benzene, current knowledge on organic compound toxicity is controversial as regard genotoxicity and human long-term effects. Carbon is one of the major organic compounds that humans are exposed to, in both environmentally and occupationally settings (Table 2). No difference in global DNA methylation, hydroxymethylation and *LINE-1* methylation was observed between a group of multi-wall carbon nanotubes (MWCNT) and a control group (Ghosh et al. 2017). However, the MWCNT group showed gene-specific methylation, such as *DNMT1*, *HDAC4*, *NPAT/ATM* and *SKI* genes. While *DNMT1* plays a role in epigenetic regulation itself, *NPAT/ATM* codes for proteins involved in DNA repair and cell cycle pathways (Blackford and Jackson 2017). The *SKI* gene is considered a potential TGF-β repressor (Zhang et al. 2017). Taken together, such data shows that occupational exposure to carbon do not alter global DNA methylation but modifies gene-specific methylation towards cellular process highly important to genomic stability.

## 4 Complex Mixtures

The evaluation of occupational exposure to complex mixtures is a challenge itself since such types of exposure have many constituents in common and people can be exposed to more than one of those mixtures at the same period (Manno et al. 2009). Work environments are hardly composed by only one chemical, therefore, biomarkers of exposure to mixtures can be a strategic tool to understand risks and prevent diseases outburst. Several studies approaching epigenetic effects in individuals occupationally exposed to complex mixtures are shown in Table 3.

Brazilians coal miners showed global DNA hypermethylation when compared to a control group (De Souza et al. 2018). Coal is a mixture of several chemicals, mainly inorganic elements and polycyclic aromatic hydrocarbons (PAHs), many of which have mutagenic and carcinogenic effects (Léon et al. 2007). Although there was no correlation of DNA methylation with the other parameters analyzed in the study, authors also observed shorter telomere length for coal miners. DNA methylation regulates and determines transcription, chromatin structure, chromosome integrity, and genomic imprinting. Aberrant DNA methylation can lead to disruption of any or all of these processes and may contribute to carcinogenesis, which is also highly associated with telomeric imbalance. PAHs are known for producing DNA adducts leading to genomic instability, as shown in a mouse model orally exposed to coal tar (Long et al. 2016). Different studies approaching the single components of coal demonstrated their epigenetic alterations in several study models as well (reviewed by De Souza et al. 2018). Polycyclic aromatic hydrocarbons (PAH) are one of the most studied groups of xenobiotics to which people are exposed in their workplaces. Since PAHs are generated in different industries, researchers have high interest in it. Regardless which occupation, all studies with PAHs in Table 3 showed significantly higher levels of PAHs markers in individuals occupationally exposed (Alegría-Torres et al. 2013; Alhamdow et al. 2018; Duan et al. 2013; Pavanello et al. 2009; Yang et al. 2012). Three different studies performed with coke-oven workers observed higher indexes of DNA damage parameters, such as MN and comet cells (Duan et al. 2013; Pavanello et al. 2009; Yang et al. 2012). Methylation of *p53* gene and hypermethylation of 22 CpG sites in *p16* gene were determinant in MN increase (Pavanello et al. 2009; Yang et al. 2012), while hypomethylation of *MGMT* was correlated with higher MN frequency (Duan et al. 2013). Differently from Brazilian coal miners, who are exposed to a mixture of chemicals, Sweden chimney sweeps and creosote-exposed males showed no difference in telomere length when compared to control group (Alhamdow et al. 2018). However, those individuals presented hypomethylation of *AHRR* and *F2RL3* genes, the former only for creosote-exposed ones. A cohort follow-up study pointed hypomethylation of the *F2RL3* gene as a potent predictor of incidence and mortality of lung cancer (Zhang et al. 2015a). *AHRR* hypomethylation status is also considered a predictor of lung cancer risk, in addition to be linked to lymphoblastic anemia (de Smith et al. 2017; Zhang et al. 2016a). Occupational exposure to PAHs alters patterns of global DNA methylation in several markers of this status, such as *MGMT, LINE-1* and *Alu*. It is also interesting that disturbance of methylation provoked by exposure to PAH interferes with other parameters: *Alu* methylation is negatively associated with TNF-α (Alegría-Torres et al. 2013) and positively correlated with MN (Pavanello et al. 2009), *LINE-1* is inversely associated with comet cells and MN frequency (Duan et al. 2013) and positively with MN only (Pavanello et al. 2009). Pavanello et al. (2009) also showed hypomethylation of tumor suppressors genes *p53* and *HIC1*, that synergizes in tumor suppression (Guerardel et al. 2001). The *p53* gene is

**Table 3** Epigenetic alterations in occupational exposure to complex mixtures

| Exposure | Country | N. of participants | Results[a] | References |
|---|---|---|---|---|
| Coal | Brazil | *Exposed:* 55 coal miners.<br>*Control:*<br>27 non-exposed from same region | ↑ global DNA methylation.<br>No correlations between global DNA methylation with comet assay, MN, oxidative stress and inorganic elements. | De Souza et al. (2018) |
| Diesel engine exhaust (DEE) | China | *Exposed*: 117 male DDE-exposed workers from a diesel engine manufacturing plant.<br>*Control*: 112 male non-exposed | ↓ methylation of DDR-related genes (*p16, RASSF1A* and *MGMT*).<br>*p16, RASSF1A, MGMT* and *LINE-1* methylation levels negative correlated with CBMN indexes. | Zhang et al. (2016b) |
| | China | *Exposed:* 20 truck drivers exposed to high traffic-derived elemental carbon (EC; $\geq$16.6 μg/m$^3$) and 20 with low-exposure to EC ($\leq$16.1 μg/m$^3$).<br>*Control:* age-matched 20 office workers | No effects on mtDNA methylation (MT-TF, MT-RNR1 and D-loop control region). | Byun et al. (2013) |
| Hair dye and hair waving products | Sweden | *Exposed*: 295 hairdressers.<br>*Control*:<br>92 non-hairdressers | ↓ frequency of *CDKN2A* methylation | Li et al. (2016) |
| Particulate matter (PM) | Italy | *Exposed*: 63 steel production plant workers (baseline: first day of a workweek before shift starts X postexposure: fourth day of the week) | ↑ *miR-222* and *miR-21* expression in postexposure samples. No correlation of miRNA expression with any personal and demographic characteristics. *miR-222* expression positively associated with lead levels. *miR-146a* expression negatively associated with lead and cadmium. | Bollati et al. (2010) |
| | Italy | *Exposed*: 63 steel production plant workers (baseline: first day of a workweek before shift | ↓ *hTERT* expression in post-exposure but not dose-dependent with PM.<br>No differences for CpG | Dioni et al. (2011) |

(continued)

**Table 3** (continued)

| Exposure | Country | N. of participants | Results[a] | References |
|---|---|---|---|---|
| | | starts X postexposure: fourth day of the week) | sites in *hTERT* promoter. | |
| | Italy | *Exposed*: 40 steel workers exposed to $PM_1$ (20 high-, 20 low-exposure) | ↑ methylation of *MT-TF* and *MT-RNR1* in the 20-high exposure group. | Byun et al. (2013) |
| | United States | *Exposed*: 38 male boilermaker welders in high-exposure welding day and low-exposure welding day (pre-shift and post-shift) | $PM_{2.5}$ was associated with hypermethylation of *iNOS* promoter gene. Years of work were associated with *iNOS* hypermethylation. | Kile et al. (2013) |
| Polycyclic aromatic hydrocarbons (PAH) | Mexico | *Exposed*: 39 male brick manufacturers (pre- and post-exposure in a given week) | 1-hydroxypyrene (1-OHP) urine concentration negatively associated with *IL-12* and *p53* DNA methylation; negative association trend observed for *TNF-alpha* and *Alu* methylation | Alegría-Torres et al. (2013) |
| | Sweden | *Exposed*: 151 chimney sweeps and 19 creosote-exposed male workers. *Controls*: 152 healthy men | ↓ methylation of *AHRR* CpG sites. ↓ methylation of *F2RL3* in creosote-exposed. | Alhamdow et al. (2018) |
| | China | *Exposed*: 82 coke-oven workers. *Control*: 62 unexposed workers age and gender matched | ↓ methylation of *LINE-1* and *MGMT*. *LINE-1*, *MGMT* and its hot CpG site-specific methylation negatively correlated with 1-OHP. *LINE-1* methylation inversely associated with comet cells and micronucleus frequency. ↑ MN in *MGMT* hypomethylation individuals | Duan et al. (2013) |
| | Poland | *Exposed*: 49 non-smoking coke-oven workers. *Controls*: 43 non-exposed, matched by gender, age and smoke status | ↑ methylation of *LINE-1* and *Alu* sequences. ↓ methylation of *p53* and *H1C1*. | Pavanello et al. (2009) |

(continued)

**Table 3** (continued)

| Exposure | Country | N. of participants | Results[a] | References |
|---|---|---|---|---|
| | China | *Exposed:* 69 male coke-oven workers. *Control:* 59 male non-exposed workers | ↑ *p16[INKa]* expression. Hypermethylation of 22 CpG sites in *p16[INKa]*. Hypermethylated CpG sites positively correlated with 1-OHP and CBMN parameters | Yang et al. (2012) |
| Pesticides | Brazil | *Exposed*: 137 male soybean farmers. *Control:* 83 male non-farmers | ↑ global DNA methylation. Positive correlation between MN and global DNA methylation | Benedetti et al. (2018) |
| | Mexico | *Exposed:* 127 urban pesticide sprayers (moderate- and high-exposure). *Control*: 63 non-exposed | ↓ *LINE-1* methylation in exposed group, but higher for high-exposure compared to moderate-exposure. ↓ CpG1 *LINE-1* methylation in both exposed groups compared to control. ↓ CpG2 and CpG3 *LINE-1* methylation in moderate-exposed group compared to control. *LINE-1* methylation associated with alcohol consumption in high-exposure group | Benitez-Trinidad et al. (2018) |
| | Brazil | *Exposed*: 56 tobacco farmers. *Control*: 74 unexposed individuals from the same region | ↓ global methylation ↑ *p16* methylation associated with shortest telomeres. | Kahl et al. (2018a) |
| | Brazil | *Exposed*: 40 tobacco farmers. *Control*: 40 individuals non-exposed matched by gender and age | ↓ global methylation | Kahl et al. (2018b) |
| | Netherlands | *Exposed:* 108 low-exposure and 61 high- exposure. *Control:* 1392 non-exposed subjects | ↑ DNA methylation in 4 CpGs for women in high-exposure group and ↓ DNA methylation in one CpG. High pesticide exposure individuals showed differential | van der Plaat et al. (2018) |

(continued)

**Table 3** (continued)

| Exposure | Country | N. of participants | Results[a] | References |
|---|---|---|---|---|
| | | | DNA methylation of 31 CpG sites annotated to 29 genes; 20 of those found in subjects with airway obstruction; Seven of the 31 CpG sites were associated with modified gene expression levels | |
| Volatile organic compounds (VOC) | Mexico | *Exposed*: 40 shoe factory workers (LS) and 36 gas station attendants (GS). *Control*: 66 university employees | ↑ promoter methylation in *TOP2A* (compare to control group), *SOD1* and *TNF-alpha* (compare to both control and GS group) genes in LS group. Correlation between *GSTP1* promoter methylation and iNOS and COX-2 methylation in LS group. Both LS and GS groups had ethylbenzene levels correlated with *TOP2A* methylation | Jiménez-Garza et al. (2018) |
| | Korea | *Exposed:* 128 workers from dockyards. *Control:* 41 unexposed individuals from different working areas | Identification of deregulated miRNAs: 467 for toluene, 221 for xylene and 695 ethylbenzene exposures | Song and Ryu (2015) |
| Welding fumes/respirable dust | | *Exposed*: 101 men welders. *Control*: 127 non-welders men | No significant difference of *APC* methylation in the fully model. | Li et al. (2015) |

[a]Describes results statistically significant (unless stated otherwise) for exposure groups in relation to control or baseline groups (unless stated otherwise)
*CBMN* cytokinesis-block micronucleus assay, *MN* micronucleus

the best characterized B[*a*]P (a marker of PAH exposure) mutagenic target and, together with tumor suppressor *p16*, is frequently epigenetically altered in smoking PAH-associated lung cancer (Risch and Plass 2008).

Epidemiological approaches have consistently linked both short- and long-term exposure to particulate matter (PM) with increased morbidity and mortality (Anderson et al. 2012). The two major mechanisms by which PMs act on human body are through increased inflammation and oxidative stress. A study conducted with boilermakers welders demonstrated $PM_{2.5}$ levels associated with hypermethylation of promoter region of *iNOS* gene (Kile et al. 2013), which is involved in production of nitric oxide and plays an important role in cardiovascular health (Tsutsui et al. 2010). Authors highlight that increased in *iNOS* DNA methylation may have been produced by systemic inflammation from inhaling fine particulate (Kile et al. 2013). Two Italian studies evaluated the same group of production plant steel workers but looking into different aspects of epigenetics and genotoxicity (Bollati et al. 2010; Dioni et al. 2011). There was increase in expression for *miR-222* and *miR-21* (Bollati et al. 2010), increase in telomere length and decrease in *hTERT* expression (Dioni et al. 2011) in the post-exposure period. Methylation of CpG-rich sequences of the hTERT promoter is involved in hTERT expression (Guilleret et al. 2002). Telomere length was also positively associated with $PM_{10}$ and $PM_1$ levels (Dioni et al. 2011). Telomeres shorten in each cell division due the end-replication problem and are, therefore, markers of cellular senescence (O'Sullivan and Karlseder 2010). The end-replication problem can be overcome, and telomeres can be maintained, by telomerase: a core enzyme composed by a reverse transcriptase catalytic component (hTERT) and an RNA component (hTR). Telomerase is suppressed to undetectable levels in human somatic cells but can be reactivated in majority of cancer cells to counteract telomere shortening (Reddel 2014) and is positively regulated by hTERT. The decreased *hTERT* expression would not explain the increase in telomere length also observed in this group (Dioni et al. 2011). However, there is a high possibility that those individuals are activating the alternative telomere lengthening (ALT), based on homologous recombinant that overcomes telomere trimming, and is present in 15% of cancers (Reddel 2014). Interestingly, PM is a major toxic component of air pollution that has been associated in epidemiological investigations with increased mortality from several outcomes, including lung cancer (Brook et al. 2004). In concordance on how ALT may be triggering the telomere length elongation on those individuals, a study demonstrated that the lung carcinoma cell line SK-LU-1 is telomerase negative and presents ALT (Bryan et al. 1997).

Diesel engine exhaust (DEE) is a mixture of several chemicals, among which are elemental carbon, PM and PAHs. Recently, DEE was reevaluated by IARC and included as a known carcinogen for humans (Benbrahim-Tallaa et al. 2012; Zhang et al. 2016b). For exposed populations, the genotoxic effects of DEE are determined not only by DNA damage induced, but also by DNA damage response (DDR). No effects of elemental carbon were observed in mtDNA methylation parameters in 20 truck drivers in China, regardless being exposed to high or low levels of it (Byun et al. 2013). On the other hand, 117 Chinese DDE-exposed workers from a diesel engine manufacturing plant showed exacerbated levels of $PM_{2.5}$, elemental carbon and six urinary PAHs, when compared to 112 control individuals (Zhang et al. 2016b). A previous work with same group of DDE-exposed individuals demonstrated higher levels of CBMN parameters, indicating DNA damage related to the exposure (Zhang et al. 2015b). Afterwards, authors observed a negative correlation of CBMN parameters and methylation of *p16, LINE-1, MGMT* and *RASSF1A* genes. While *LINE-1* represents global DNA methylation of repetitive elements, the other three genes are DDR-related ones, and were found hypomethylated in this study (Zhang et al. 2016b). The hypomethylation of those genes can lead to their increased expression levels and consequent activation of DDR. Hypomethylation is also typically associated with higher transcriptional levels, which leads to cell-cycle arrest that facilitates DNA damage repair in the case of *p16* and *RASSF1A* and directly strengthen DDR in the case of *MGMT*, mitigating genomic instability.

A study performed with samples (benign and malignant biopsies, blood and saliva) from individuals with impalpable lesions of breast cancer also observed higher methylation of *CDKN2A* gene (Delmonico et al. 2015), that encodes for p16$^{INK4a}$ protein. Among the malignant samples, ATM, a serine/threonine kinase that is recruited and activated by DNA double-strand breaks leading to DDR, was the most hypermethylated in lesions, followed by p16$^{INK4a}$ in blood and saliva samples (Delmonico et al. 2015). In this aspect, *CDKN2A* is a key factor in cell cycle regulation and its hypermethylation has been found associated with bladder cancer. Working as hairdressers has been associated with increased bladder cancer risk, particularly due to aromatic amines in hair dyes and oxidative hair dying, waving and bleaching products (Bolt and Golka 2007; IARC 2010). Hairdressers showed shorter telomere length in comparison with non-hairdressers in a study performed in 2016 in Sweden (Li et al. 2016). Hair waving was associated with less frequent *CDKN2A* methylation and with the shortest telomeres observed in studied group. Authors highlight that the methylation patterns found were not expected as per the literature states. Nevertheless, the high content of oxidative chemicals in the hairdressers' work environment may lead to complex and controversial results (Li et al. 2016).

Welding fumes have been categorized as possible human carcinogen into Group 2B (IARC 1990) and studies show increased risk for lung cancer in welders. Several million people worldwide are occupationally exposed to welding fumes (Antonini 2003), which means exposure to combustion-derived products, such as metal oxide particles (Leonard et al. 2010). A group of 101 Sweden male welders were recently evaluated as per oxidative stress, telomere length and DNA methylation patterns in relation to their occupational exposure (Li et al. 2015). Because telomere length is highly affected by oxidative stress, mainly 8-OHdG generation, authors measured telomere length in studied population through qPCR. They did not observe any significant difference between control and exposed groups, and similar result was found for 8-OHdG measurement through liquid chromatography (Li et al. 2015). However, every working year was associated with shorter telomeres and hypermethylation of *MGMT* in time-dependent manner, indicating a possible cumulative effect of welding fumes. Among the five investigated genes, only *APC* had higher methylation frequency in welders. The tumor suppressor *APC* gene regulates the Wnt signaling pathway, which plays an important role in cell growth regulation (Sparks et al. 1998) and has been found hypermethylated in serum and plasma of lung cancer patients (Li et al. 2015).

Human health effects ranging from neurotoxicity to cancer have been reported in cases of chronic exposure to volatile organic compounds (VOC). Exposure to VOC occurs through both environmental (motor vehicle exhaust) and industrial (solvents) sources (Masiol et al. 2014) Due to its composition, including mainly but not only benzene, toluene, ethylbenzene and xylene, VOCs are compounds hard to evaluate as per occupational exposure. Increased levels of six VOCs were found in shoe factory workers (LS group) and of two VOCs in gas station attendants (GS group) in a recent Mexican study (Jiménez-Garza et al. 2018). In both LS and GS groups, authors observed hypermethylation of *TOP2A* promoter, a gene that encodes for TOP2A, an enzyme that catalyzes the breaking and rejoining of DNA strands, playing a critical role in DNA replication, recombination, chromosome separation and condensation, and gene transcription (Nitiss 2009). The LS group, when compared to GS and control groups, also showed *SOD1* and *TNF-α* promoter hypermethylation, demonstrating an effect of oxidative stress and inflammation. In agreement, the *GSTP1* promoter methylation frequency was correlated with both *iNOS* and *COX-2* methylation (Jiménez-Garza et al. 2018). *TNF-α, iNOS* and *COX-2* are genes associated with inflammation parameters, while *GSTP1* and *SOD1* are linked to oxidative stress response. Cellular events for cancer development have been characterized by ongoing oxidative stress that may lead to inflammation (Imbesi et al. 2013). Taken together that promoter methylation correlates negatively with gene expression, several miRNAs were identified as deregulated in another group of VOC-exposed workers. In 128 dockyards employees, over 450 and 220 miRNAs were deregulated for toluene and xylene, respectively (Song and Ryu 2015). Interestingly, the highest number of deregulated miRNAs, 695, was found for ethylbenzene among dockyards workers (Song and Ryu 2015), while in the Mexican study, both LS and GS groups had ethylbenzene levels strongly correlated with *TOP2A* methylation status (Jiménez-Garza et al. 2018).

Pesticides are one of the most xenobiotics concerns worldwide. General population can be exposed to low concentrations of those chemicals through contamination of air, water and food, while high exposures are associated with occupational exposure, such as production, packaging and application of these compounds. A majority of 39 out of 46 studies reported positive findings as regard pesticide occupational exposure and CBMN parameters (Bolognesi and Holland 2016). Three different Brazilian studies evaluated epigenetic parameters in farmworkers in soybean and tobacco fields. Soybean farmers showed increased global DNA methylation, positively correlated with MN frequency (Benedetti et al. 2018). In this same work, authors observed higher levels of DNA damage and cell death parameters through buccal micronucleus Cytome assay (BMCyt) and Comet assay, also associated with oxidized guanines (Benedetti et al. 2018). On the other hand, tobacco farmers, who are exposed to pesticides and nicotine (natural pesticide), showed global DNA hypomethylation in two studies (Kahl et al. 2018a, b), showing that the epigenetic mechanism is different for farmers working in those two crops. While acute effects of pesticides are widely known, chronic effects are still largely under speculation. For some pesticides, mechanisms such as endocrine disruption are hypothesized. Additionally, it had been suggested that health effects observed are related to specific mutagenic effects of particular pesticides (Mostafalou and Abdollahi 2017), which may explain differences between soybean and tobacco farmers. Moreover, Brazilian tobacco farmers had shorter telomeres and increased DNA damage (Kahl et al. 2018a, b), which was associated with years of exposure (Kahl et al. 2018a). Hypermethylation of tumor suppressor *p16* was positively correlated with the shortest telomeres among tobacco farmers (Kahl et al. 2018a). $p16^{INK4a}$ contributes to the p53-independent response to telomere damage (Jacobs and De Lange 2004), suggesting that in tobacco farmers, the hypermethylation of *p16* in shortest telomeres may be a response to pesticide-induced oxidative stress (Mostafalou and Abdollahi 2017).

Global dose-dependent DNA hypermethylation was observed in Indian adults with chronic arsenic exposure, but this effect was modified by folate (Pilsner et al. 2007). Interestingly, the *MTHFR C677T* polymorphism, the most common gene in folate metabolism, influenced in both DNA damage and telomere length in tobacco farmers (Kahl et al. 2018b), suggesting that DNA methylation is dependent of methyl availability, interfering with genomic stability. Global DNA methylation in oncogenes or genes that favors apoptosis resistance argue in the addition of methionine, choline, folate and vitamin B12 as methyl donors to both prevent and limit cancer aggressiveness (Hervouet et al. 2013). Hypomethylation of *LINE-1* repetitive element was observed in a group of urban pesticide sprayers, showing differences for gender, as men had higher *LINE-1* methylation when compared to women. The exposed group was also evaluated as per level of exposure, as high- and moderate-exposure. The high-exposure subgroup had exacerbated *LINE-1* hypomethylation when compared to moderate-exposure (Benitez-Trinidad et al. 2018). Similarly, high-exposure to pesticides in a Netherland study showed differential DNA methylation patterns for 31 CpG sites annotated for 29 genes; 20 of those found in subjects with airway obstruction (van der Plaat et al. 2018), which is a common human chronic disease associated with pesticide exposure (Mostafalou and Abdollahi 2017). Those data suggest that levels of exposure are related to human response to pesticide exposure, including epigenetic modifications. Tobacco and soybean farmers were not divided according to levels of exposure, mainly because they are all exposed in long-term periods (Benedetti et al. 2018; Kahl et al. 2018a, b). Occupational long-term exposure to pesticides is associated with birth defects, reproductive problems, diabetes, respiratory diseases, amyotrophic lateral sclerosis, cognitive impairments, and cancer (Mostafalou and Abdollahi 2017). Growing progress has been made in the recognition of epigenetic modifications in pesticide-exposure approach leading to chronic diseases.

## 5 Discussion and Conclusion

Faisandier et al. (2011) establish that, in occupational field, the exposome is considered the network of occupational health problems, sharing components of the set of occupational exposure. This view helps building basis for coherent discussion towards the development of networks for monitoring occupational exposure situations, including its varied origins, several effects and biomarkers of exposure usefulness (Faisandier et al. 2011). Adverse health effects due to exposure have been estimated for up to 75% of global noncommunicable diseases (NCDs). Environmental exposures are ranked as top risk factors for chronic disease mortality (WHO 2010). "Environmental epigenetics" is a term that refers to how environmental exposures affect epigenetic changes (Reamon-Buettner et al. 2008). Life experiences, habits, and the environment shape who individuals are by virtue of their impact on epigenome and health. A great example is identical twins that, although they share the same genome and are superficially phenotypically similar, they are unique individuals with definable differences. These differences result from distinct gene expression influenced by epigenetic factors. Behavior, nutrition, and exposure to toxins and pollutants are among the lifestyle factors known to be associated with epigenetic modifications (Tiffon 2018).

Environmental toxicants can alter epigenetic regulatory processes, and mediate specific mechanisms of toxicity and responses. Growing evidence suggests that at least fifteen environmental chemicals may lead directly to diseases via epigenetic mechanism-regulated gene expression changes (Hou et al. 2012). Environmental and occupational factors induce epigenetic alterations that can contribute to the onset of NCDs, of which cancer is the most prevalent. Because these epigenetic changes are small, potentially cumulative, and may develop in long-term periods, there is a difficulty in establishing a direct relationship of cause-effect among occupational factors, epigenetic changes and diseases arising (Baccarelli and Bollati 2009; Ravegnini et al. 2015). On the other hand, literature also suggests that epigenetic modifications play a major role in human complexes diseases, particularly cancer. Carcinogenicity is now considered to develop as an epigenetic disease the same as it is considering a genetic disease. There is massive understanding of the contribution of epigenetic events in the initiation, promotion and progression of different types of cancers, mainly through silencing of tumors suppressor genes and/or activation of proto-oncogenes (Jones and Baylin 2002). Importantly, DNA methylation and apoptosis resistance are characteristics of cancer cells. Proteins related to apoptosis are considered to counteract the oncogenic Wnt signaling pathway and the inactivation of this gene may increase cancer development and progression (Hervouet et al. 2013). Many cancers are characterized by global DNA hypomethylation, previously associated with chromosomal instability and, paradoxically, with both gene-specific hypo- and hypermethylation (Esteller 2008). Knowledge on heritable changes in gene expression that result from epigenetic events is of increasing relevance in the development of strategies for prevention, early diagnosis and treatment of cancer. In addition, non-genotoxic carcinogens are a group of chemical agents that are known to cause tumors without directly damaging DNA. Evidences suggest that these compounds can lead to prominent epigenomic alterations in tissues that are targets for carcinogenesis as a result of exposure (Koturbash et al. 2011).

In animal studies, several chemicals have been reported to induce transgenerational phenotypic effects (Baccarelli and Bollati 2009; Chen et al. 2016) and humans studies have also shown that males can pass their traits acquired during lifetime as regard changes in dietary intake, chemical exposure, stress, or trauma, onto their offspring (Chen et al. 2016). Transgenerational transmission of chemically-induced epigenetic modifications have been considered the probable mechanism for these effects (Baccarelli and Bollati 2009; Tran and Miyake 2017). In fact, both paternal and maternal exposure to environmental xenobiotics during gametogenesis or gestation has been shown to be responsible for the offspring's epigenome. In some cases, the potential for persistent transgenerational modification of the epigenome may also inform on parental germ cell exposures. Exposure to toxicants during fetal life and exposure of germ cells, possibly at a specific developmental stage, can induce heritable epigenetic changes. Epigenetic mechanisms can underlie the effects of in utero and early life exposures on adult health, as these fetal exposure to epigenetically-active chemicals can lead to health effects later in life, even independently of environmental and/or occupational risk factors in adulthood (Baccarelli and Bollati 2009; Tran and Miyake 2017).

Epigenetic parameters have been currently reported as a robust tool for studying carcinogenesis of occupational settings (Ziech et al. 2010). Therefore, they represent a class of biomarkers with great potential in the identification of exposure status, damage response, and/or disease state. The incorporation of such parameters in chemical safety assessments still depends on characterization of the epigenetic alterations induced by xenobiotics (Holland 2017; Koturbash et al. 2011). Epigenetic modifications are relatively stable and influenced by environment. Exposure to different classes of xenobiotics, such as metals, organic compounds and complexes mixtures, may lead to epigenome alterations, as seen in this chapter. Experimental, clinical and epidemiological studies have increased the current knowledge of mechanisms of action by which such chemical compounds can modify gene expression. Taken together, the evidence outlined in this chapter demonstrate that epigenome can be regulated by systematic factors, i.e., in response to environmental changes (Hou et al. 2012).

Both genetic and epigenetic responses of organisms to environmental factors, including chemical exposures, influence adaptation, susceptibility to toxicity and biodiversity. In model organisms, it is established that epigenetic alterations, including changes to the methylome, can create a memory of the received signal. Thus, it is proposed that epigenetic "foot-printing" of organisms could identify classes of chemical contaminants to which they have been exposed throughout their lifetime. However, a better understanding is necessary to decide which epigenetic alterations are most informative, which can take to an effective use of epigenetic endpoints as markers of exposure. Specifically, additional studies are needed to characterize the relationship between epigenetic alterations and toxicity phenotypes, and the epigenetic-specific dose-response (Baccarelli and Bollati 2009; Hou et al. 2012) and how, ultimately, toxicant exposure affects the composition and differentiation status of cell types in a given tissue (Meehan et al. 2018).

Overall, studies of aberrant DNA methylation represented the most commonly studied epigenetic feature, followed by changes in the expression of noncoding RNAs, and finally histone modifications. Knowledge on heritable changes in gene expression that result from epigenetic events is of increasing relevance in the development of strategies for prevention, early diagnosis and treatment of cancer. Today, epigenetics is one of the most exciting fields of biological sciences, as it is involved in occupational and environmental exposures related programming and transgenerational effects. Risks arising from some NCDs, like pneumoconiosis, cancers and allergies, are predictable and preventable. Because most of epigenetic changes are also reversible, there is growing field for developing personalized preventive medicine (Hou et al. 2012; Meehan et al. 2018). Preventive strategies, such as exposure reduction, and pharmacological, dietary and/or lifestyle interventions may take an important part in future epigenomic research. Progress in these areas will require development and adaptation of new technologies, as much as interdisciplinary research, including toxicology, bioinformatics, epigenomics and data generation (Holland 2017; Hou et al. 2012; Meehan et al. 2018). Consequently, preventive action could lead to decreasing disease morbidity and mortality arising from many occupational-related diseases that are of major public concern.

## References


Alegría-Torres JA, Barretta F, Batres-Esquivel LE et al (2013) Epigenetic markers of exposure to polycyclic aromatic hydrocarbons in Mexican brickmakers: a pilot study. Chemosphere 91:475–480

Alhamdow A, Lindh C, Hagberg J et al (2018) DNA methylation of the cancer-related genes *F2RL3* and *AHRR* is associated with occupational exposure to polycyclic aromatic hydrocarbons. Carcinogenesis 39:869–878

Alvanja MC, Bonner MR (2012) Occupational pesticide exposure and cancer risk: a review. J Toxicol Environ Health B Crit Rev 15:238–263

Anderson JO, Thundiyil JG, Stolbach A (2012) Clearing the air: a review of the effects of particulate matter air pollution on human health. J Med Toxicol 8:166–175

Antonini JM (2003) Health effects of welding. Crit Rev Toxicol 33:61–103

Arita A, Niu J, Qu Q et al (2012) Global levels of histone modifications in peripheral blood mononuclear cells of subjects with exposure to nickel. Environ Health Perspect 120:198–203

ATSDR (Agency for Toxic Substances and Disease Registry) (2004) Interaction profile for: Benzene, Toluene, Ethylbenzene, and Xylenes (BTEX). US Department of Health and Human Services; Public Health Service, Atlanta

Baccarelli A, Bollati V (2009) Epigenetics and environmental chemicals. Curr Opin Pediatr 21:243–251

Bal W, Protas AM, Kasprzak KS (2011) Genotoxicity of metal ions: chemical insights. Met Ions Life Sci 8:319–373

Benbrahim-Tallaa L, Baan RA, Grosse Y et al (2012) Carcinogenicity of diesel engine and gasoline-engine exhausts and some nitroarenes. Lancet Oncol 13:663–664

Benedetti D, Alderete BL, de Souza CT et al (2018) DNA damage and epigenetic alterations in soybean farmers exposed to complex mixtures of pesticides. Mutagenesis 33:87–95

Benitez-Trinidad AB, Medina-Díaz IM, Bernal-Hernández YY et al (2018) Relationship between LINE-1 methylation pattern and pesticide exposure in urban sprayers. Food Chem Toxicol 113:125–133

Beyersmann D, Hartwig A (2008) Carcinogenic metal compounds: recent insight into molecular and cellular mechanisms. Arch Toxicol 82:493–512

Birkbak NJ, Wang ZC, Kim JY et al (2012) Telomeric allelic imbalance indicates defective DNA repair and sensitivity to DNA-damaging agents. Cancer Discov 2:366–375

Blackford AN, Jackson SP (2017) ATM, ATR, and DNA-PK: the trinity at the heart of the DNA damage response. Mol Cell 66:801–817

Bollati S, Pesatori AC, Dioni L et al (2010) Shorter telomere length in peripheral blood lymphocytes of workers exposed to polycyclic aromatic hydrocarbons. Carcinogenesis 31:216–221

Bolognesi C, Holland N (2016) The use of lymphocyte cytokinesis-block micronucleus assay for monitoring pesticide-exposed populations. Mutat Res 770:183–203

Bolt HM, Golka K (2007) The debate on carcinogenicity of permanent hair dyes: new insights. Crit Rev Toxicol 37:521–536

Boucard AA, Maxeiner S, Sudhof TC (2014) Latrophilins function as heterophilic cell-adhesion molecules by binding to teneurins: regulation by alternative splicing. J Biol Chem 289:387–402
Brook RD, Franklin B, Cascio W et al (2004) Air pollution and cardiovascular disease: a statement for healthcare professionals from the expert panel on population and prevention science of the American heart association. Circulation 109:2655–2671
Bryan TM, Englezou A, Dalla-Pozza L et al (1997) Evidence for an alternative mechanism for maintaining telomere length in human tumors and tumor-derived cell lines. Nature Med 3:1271–1274
Byun HM, Panni T, Motta V et al (2013) Effects of airbone pollutants on mitochondrial DNA methylation. Part Fibre Toxicol 10:18
Cantone L, Nordio F, Hou L et al (2011) Inhalable metal-rich air particles and histone H3K4 dimethylation and H3K9 acetylation in a cross-sectional study of steel workers. Environ Health Perspect 119:964–969
Carugno M, Pesatori AC, Dioni L et al (2012) Increased mitochondrial DNA copy number in occupations associated with low-dose benzene exposure. Environ Health Perspect 120:210–215
Chappel G, Pogribny IP, Guyton KZ et al (2016) Epigenetic alterations induced by genotoxic occupational and environmental human chemical carcinogens: a systematic literature review. Mutat Res Rev Mutat Res 768:27–45
Charbotel B, Fervers B, Droz JP (2014) Occupational exposures in rare cancers: a critical review of the literature. Crit Rev Oncol Hematol 90:99–134
Chen C, Yu H, Zhao J et al (2006) The roles of serum selenium and selenoproteins on mercury toxicity in environmental and occupational exposure. Environ Health Perspec 114:297–301
Chen Q, Yan W, Duan E (2016) Epigenetic inheritance of acquired traits through sperm RNAs and sperm RNA modifications. Nat Rev Genet 17:733–743
Cheng TF, Choudhuri S, Muldoon-Jacobs K (2012) Epigenetic target of some toxicologically relevant metals: a review of the literature. J Appl Toxicol 32:643–653
Cheung HH, Lee TL, Rennert OM et al (2009) DNA methylation of cancer genome. Birth Defects Res C Embryo Today 87:335–350
Costa M, Davidson TL, Chen H et al (2005) Nickel carcinogenesis: epigenetics and hypoxia signaling. Mutat Res 592:79–88
De Smith AJ, Kaur M, Gonseth S et al (2017) Correlates of prenatal and early-life tobacco smoke exposure and frequency of common gene deletions in childhood acute lymphoblastic leukemia. Cancer Res 77:1674–1683
De Souza MR, Kahl VFS, Rohr P et al (2018) Shorter telomere length and DNA hypermethylation in peripheral blood of coal workers. Mutat Res 836:36–41
Delmonico L, Moreira AS, Franco MF et al (2015) CDKN2A(p14(ARF)/p16(INK4a)) and ATM promoter methylation in patients with impalpable breast lesions. Hum Pathol 46:1540–1547
Dioni L, Hoxha M, Nordio F et al (2011) Effects of short-term exposure to inhalate particulate matter on telomere length, telomerase expression, and telomerase methylation in steel workers. Environ Health Perspect 119:622–627
Duan H, He Z, Ma J et al (2013) Global and MGMT promoter hypomethylation independently associated with genomic stability of lymphocytes in subjects exposed to high-dose polycyclic aromatic hydrocarbon. Arch Toxicol 87:2013–2022
Esteller M (2008) Epigenetics in cancer. N Engl J Med 358:1148–1159
Faisandier L, Bonneterre V, De Gaudemaris R et al (2011) Occupational exposome: a network-based approach for characterizing occupational health problems. J Biomed Inform 44:545–552
Fang SC, Cassidy A, Christiani DC (2010) A systematic review of occupational exposure to particulate matter and cardiovascular disease. Int J Environ Res Public Health 7:1773–1806
Fischer BM, Wong JK, Degan S et al (2013) Increased expression of senescence markers in cystic fibrosis airways. Am J Physiol Lung Cell Mol Physiol 304:L394–L400
Fritisch L, Benke G, Risch HA et al (2015) Occupational exposure to N-nitrosamines and pesticides and risk of pancreatic cancer. Occup Environ Med 72:678–683

Gaffney A, Christiani DC (2015) Gene-environment interaction from international cohorts: impact on development and evolution of occupational and environmental lung and airway disease. Semin Respir Crit Care Med 36:347–357

García-Lestón J, Roma-Torres J, Vilares M et al (2012) Genotoxic effects of occupational exposure to lead and influence of polymorphisms in genes involved in lead toxicokinetics and in DNA repair. Environ Int 43:29–36

Ghosh M, Öner D, Poels K et al (2017) Changes in DNA methylation induced by multi-walled carbon nanotube exposure in the workplace. Nanotoxicology 11:1195–1210

Godderis L, De Raedt K, Tabish AM et al (2012) Epigenetic changes in lymphocytes of solvent-exposed individuals. Epigenomics 4:269–277

Goodrich JM, Basu N, Franzblau A et al (2013) Mercury biomarkers and DNA methylation among Michigan dental professionals. Environ Mol Mutagen 54:195–203

Guerardel C, Deltour S, Pinte S et al (2001) Identification in the human candidate tumor suppressor gene HIC-1 of a new major alternative TATA-less promoter positively regulated by p53. J Biol Chem 276:3078–3089

Guilleret I, Yan P, Grange F et al (2002) Hypermethylation of the human telomerase catalytic subunit (hTERT) gene correlates with telomerase activity. Int J Cancer 101:335–341

Hervouet E, Cheray M, Vallette FM et al (2013) DNA methylation and apoptosis resistance in cancer cells. Cell 2(3):545–573

Heuser VD, Andrade VM, da Silva J et al (2005) Comparison of genetic damage in Brazilian footwear-workers exposed to solvent-based or water-based adhesive. Mutat Res 583:85–94

Holland N (2017) Future of environmental research in the age of epigenomics and exposomics. Rev Environ Health 32:45–54

Hong JY, Yu SY, Kim SY et al (2016) Association analysis of toluene exposure time with high-throughput mRNA expressions and methylation patterns using *in vivo* samples. Environ Res 146:59–64

Hou L, Zhang X, Wang D et al (2012) Environmental chemical exposures and human epigenetics. Int J Epidemiol 41:79–105

Hu G, Li P, Cui X et al (2018) Cr(VI)-induced methylation and down-regulation of DNA repair genes and its association with markers of genetic damage in workers and 16HBE cells. Environ Poll 238:833–843

IARC (1990) International agency for research on cancer. Chromium, nickel and welding. IARC Monogr Eval Carcinog Risks Hum 49:447–525

IARC (2010) International agency for research on cancer. Somearomatic amines, organic dyes, and related exposures. IARC Monogr Eval Carcinog Risks Hum 99:499–658

Imbesi S, Musolino C, Allegra A (2013) Oxidative stress in oncohematologic diseases: an update. Expert Rev Hematol 6:317–325

Jacobs JJL, de Lange T (2004) Significant role for p16$^{INK4a}$ in p53-independent telomere-directed senescence. Curr Biol 14:2302–2308

Jamebozorgi I, Majidizadeh T, Pouryaghoub G et al (2018) Aberrant DNA methylation of two tumor suppressor genes, p14ARF and p15INK4b, after chronic occupational exposure to low level of benzene. Int J Occup Environ Med 9:145–151

Jarmalaite S, Kannio A, Anttila S et al (2003) Aberrant p16 promoter methylation in smokers and former smokers with nonsmall cell lung cancer. Int J Cancer 106:913–918

Jiang Y, Oliver P, Davies KE et al (2006) Identification and characterization of murine SCARA5, a novel class A scavenger receptor that is expressed by populations of epithelial cells. J Biol Chem 281:11834–11845

Jiménez-Garza O, Guo L, Byun HM et al (2018) Aberrant promoter methylation in genes related to hematopoietic malignancy in workers exposed to a VOC mixture. Toxicol Appl Pharmacol 339:65–72

Jones PA, Baylin SB (2002) The fundamental role of epigenetic events in cancer. Nat Rev Genet 3:415–428

Kahl VFS, Dhillon VS, Simon D et al (2018a) Chronic occupational exposure endured by tobacco farmers from Brazil and association with DNA damage. Mutagenesis 33:119–128

Kahl VFS, Dhillon V, Fenech M et al (2018b) Occupational exposure to pesticides in tobacco fields: the integrated evaluation of nutritional intake and susceptibility on genomic and epigenetic instability. Oxid Med Cell Longev 2018:7017423

Kile ML, Fang S, Baccarelli AA et al (2013) A panel study of occupational exposure to fine particulate matter and changes in DNA methylation over a single workday and years worked in boilermaker welders. Environ Health 12:47

Kondo K, Takahashi Y, Hirose Y et al (2006) The reduced expression and aberrant methylation of p16(INK4a) in chromate workers with lung cancer. Lung Cancer 53:295–302

Koturbash I, Beland FA, Pogribny IP (2011) Role of epigenetic events in chemical carcinogenesis—a justification for incorporating epigenetic evaluations in cancer risk assessment. Toxicol Mech Methods 21:289–297

Léon G, Perez LE, Linares JC et al (2007) Genotoxic effects in wild rodents (*Rattus rattus* and *Mus musculus*) in an open coal mining area. Mutat Res 630:42–49

Leonard SS, Chen BT, Stone SG (2010) Comparison of stainless and mild steel welding fumes in generation of reactive oxygen species. Part Fibre Toxicol 7:32

Li C, Xu M, Wang S et al (2011) Lead exposure suppressed *ALAD* transcription by increasing methylation level of the promoter CpG islands. Toxicol Lett 203:48–53

Li C, Yang X, Xu M et al (2013) Epigenetic marker (LINE-1 promoter) methylation level was associated with occupational lead exposure. Clin Toxicol 51:225–229

Li H, Hedmer M, Wodjacz T et al (2015) Oxidative stress, telomere shortening, and DNA methylation in relation to low-to-moderate occupational exposure to welding fumes. Environ Mol Mutagen 56:684–693

Li H, Akerman G, Lidén C et al (2016) Alterations of telomere length and DNA methylation in hairdressers: a cross-sectional study. Environ Mol Mutagen 57:159–167

Linqing Y, Bo X, Xueqin Y et al (2016) Mitochondrial DNA hypomethylation in chrome plating workers. Toxicol Lett 243:1–6

Long AS, Watson M, Arlt VM et al (2016) Oral exposure to commercially available coal tar-based pavement sealcoat induces murine genetic damage and mutations. Environ Mol Mutagen 57:535–545

Ma L, Bai Y, Pu H et al (2015) Histone methylation in nickel-smelting industrial workers. PLoS One 10:e0140339

Manno M, Viau C, Cocker J et al (2009) Biomonitoring for occupational health risk assessment (BOHRA). Toxicol Lett 192:3–16

Martínez-Alfaro M, Cárabez-Trejo A, Gallegos-Corona MA et al (2010) Thinner inhalation effects on oxidative stress and DNA repair in a rat model of abuse. J Appl Toxicol 30:226–232

Masiol M, Agostinelli C, Formenton G (2014) Thirteen years of air pollution hourly monitoring in a large city: potential sources, trends, cycles and effects of car-free days. Sci Total Environ 494-495:84–96

Maxwell P, Salnikow K (2004) HIF-1, an oxygen and metal responsive transcription factor. Cancer Biol Ther 3:29–35

Meehan RR, Thomson JP, Lentini A et al (2018) DNA methylation as a genomic marker of exposure to chemical and environmental agents. Curr Opinion Chem Biol 45:48–56

Mostafalou S, Abdollahi M (2017) Pesticides: an update of human exposure and toxicity. Arch Toxicol 91:549–599

Murnare JP (2011) Telomere dysfunction and chromosome instability. Mutat Res 730:28–36

Nitiss JL (2009) DNA topoisomerase II and its growing repertoire of biological functions. Nat Rev Cancer 9:327–337

Nylander-French LA, Wu MC, French JE et al (2014) DNA methylation modifies urine biomarker levels in 1,6-hexamethylene diisocyanate exposed workers: a pilot study. Toxicol Lett 231:217–226

O'Sullivan RJ, Karlseder J (2010) Telomeres: protecting chromosomes against genome instability. Nature Rev 11:171–181
O'Sullivan RJ, Kubicek S, Schreiber SL et al (2010) Reduced histone biosynthesis and chromatin changes arising from a damage signal at telomeres. Nat Struct Mol Biol 17:1218–1225
Pavanello S, Bollati V, Pesatori AC et al (2009) Global and gene-specific promoter methylation changes are related to *anti*-B[*a*]PDE-DNA adduct levels and influence micronuclei levels in polycyclic aromatic hydrocarbon-exposed individuals. Int J Cancer 125:1692–1697
Pilsner JR, Liu X, Ahsan H et al (2007) Genomic methylation of peripheral blood leukocyte DNA: influences of arsenic and folate in Bangladesh adults. Am J Clin Nutr 86:1179–1186
Pottier G, Viau M, Ricoul M et al (2013) Lead exposure induces telomere stability in human cells. PLoS One 8:e67501
Ravegnini G, Sammarini G, Hrelia P et al (2015) Key genetic and epigenetic mechanisms in chemical carcinogenesis. Toxicol Sci 148:2–13
Reamon-Buettner SM, Mutschler V, Borlak J (2008) The next innovation cycle in toxicogenomics: environmental epigenetics. Mutat Res 659:158–165
Reddel R (2014) Telomere maintenance mechanism in cancer: clinical implications. Curr Pharm Des 20:6361–6374
Rinsky RA, Hornung RW, Silver SR et al (2002) Benzene exposure and hematopoietic mortality: a long-term epidemiologic risk assessment. Am J Ind Med 42:474–480
Risch A, Plass C (2008) Lung cancer epigenetics and genetics. Int J Cancer 123:1–7
Scanlon SE, Scanlon CD, Hegan DC et al (2017) Nickel induces transcriptional down-regulation of DNA repair pathways in tumorigenic and non-tumorigenic lung cells. Carcinogenesis 38:627–637
Schnekenburger M, Talaska G, Puga A (2007) Chromium cross-links histone deacetylase 1-DNA methyltransferase 1 complexes to chromatin, inhibiting histone-remodeling marks critical for transcriptional activation. Mol Cell Biol 27:7089–7101
Schottenfeld D, Beebe-Dimmer J (2006) Chronic inflammation: a common and important factor in the pathogenesis of neoplasia. CA Cancer J Clin 56:69–83
Schulte PA, Hauser JE (2012) The use of biomarkers in occupational health research, practice, and policy. Toxicol Lett 213:91–99
Sekhotha MM, Monyeki KD, Sibuyi ME (2016) Exposure to agrochemicals and cardiovascular disease: a review. Int J Environ Res Public Health 13:229
Sha Y, Zhou W, Yang Z et al (2014) Changes in poly(ADP-ribosyl)ation patterns in workers exposed to BTX. PLoS One 9:e106146
Silins I, Högberg J (2011) Combined toxic exposures and human health: biomarkers of exposure and effect. Int J Environ Res Public Health 8:629–647
Siroux V, Agier L, Slama R (2016) The exposome concept: a challenge and a potential driver for environmental health research. Eur Respir Rev 25:124–129
Song MK, Ryu JC (2015) Blood miRNAs as sensitive and specific biological indicators of environmental and occupational exposure to volatile organic compound (VOC). Int J Hyg Environ Health 218:590–602
Sparks AB, Morin PJ, Vogelstein B (1998) Mutational analysis of the APC/beta catenin/tcf pathway in colorectal cancer. Cancer Res 58:1130–1134
Sun R, Liu J, Wang B et al (2015) Deoxyribonucleic acid (DNA) methyltransferase contributes to p16 promoter CpG island methylation in lung adenocarcinoma with smoking. Int J Clin Exp Med 8:15773–15779
Tas U, Ogeturk M, etal MS (2011) Hepatotoxic activity of toluene inhalation and protective role of melatonin. Toxicol Ind Health 27:465–473
Tchounwou PB, Patiolla AK, Centeno JA (2003) Carcinogenic and systemic health effects associated with arsenic exposure – a critical review. Toxicol Pathol 31:575–588
Tiffon C (2018) The impact of nutrition and environmental epigenetics on human health and disease. Int J Mol Sci 19(11):E3425

Tran NQV, Miyake K (2017) Neurodevelopmental disorders and environmental toxicants: epigenetics as an underlying mechanism. Int J Genomics 2017:7526592
Tsutsui M, Shimokawa H, Otsuji Y et al (2010) Pathophysiological relevance of NO signaling in the cardiovascular system: novel insight from mice lacking all NO synthases. Pharmacol Ther 128:499–508
van der Plaat DA, Jong K, de Viries M et al (2018) Occupational exposure to pesticides is associated with differential DNA methylation. Occup Environ Med 75:427–435
Wang TC, YS S, Wang H et al (2012) Oxidative DNA damage and global DNA hypomethylation are related to folate deficiency in chromate manufacturing workers. J Hazard Mater 213-214:440–446
Wen W, Lu L, He Y et al (2016) LincRNAs and base modifications of *p53* induced by arsenic methylation in workers. Chem Biol Interact 246:1–10
WHO (World Health Organziation) (2010) Global status report on noncommunicable diseases. WHO, Geneva, p 176
Xu M, Yu Z, Hu F et al (2017) Identification of differential plasma miRNA profiles in Chinese workers with occupational lead exposure. Biosci Rep 37:BSR20171111
Yang P, Ma J, Zhang B et al (2012) CpG site-specific hypermethylation of $p16^{INK4\alpha}$ in peripheral blood lymphocytes of PAH-exposed workers. Cancer Epidemiol Biom Prev 21:182–190
Zhang Y, Schottker B, Ordóñez-Mena J et al (2015a) F2RL3 methylation, lung cancer incidence and mortality. Int J Cancer 137:1739–1748
Zhang X, Duan H, Gao F et al (2015b) Increased micronucleus, nucleoplasmic bridge, nuclear bud frequencies in the peripheral blood lymphocytes of diesel engine exhaust-exposed workers. Toxicol Sci 143:408–417
Zhang Y, Elgizouli M, Schottker B et al (2016a) Smoking-associated DNA methylation markers predict lung cancer incidence. Clin Epigenetics 8:12
Zhang Z, Li J, He Z et al (2016b) Associations between DNA methylation in DNA damage response-related genes and cytokinesis-block micronucleus cytome index in diesel engine exhaust-exposed workers. Arch Toxicol 90:1997–2008
Zhang S, Takaku M, Zou L et al (2017) Reversing SKI-SMAD4-mediated suppression is essential for $T_H17$ cell differentiation. Nature 551:105–109
Ziech D, Franco R, Pappa A (2010) The role of epigenetics in environmental and occupational. Chem Biol Interact 188:340–349

# DNA Methylation: Biological Implications and Modulation of Its Aberrant Dysregulation

**Alessia Lucidi, Daniela Tomaselli, Dante Rotili, and Antonello Mai**

## Contents



**Abstract** The alteration of the DNA methylation pattern is often related to the onset of diseases based on epigenetic dysregulation, primarily cancer. In this scenery the development of DNA methyltransferase inhibitors is one of the most attractive challenges for anticancer therapy. The present chapter proposes a comprehensive classification of the DNA methyltransferase inhibitors known in literature, on the basis of their natural or synthetic nature and their mechanism of action.

**Keywords** Epigenetics · DNA methyltransferase · DNMT inhibitors · Anticancer therapy · Tumor suppressor genes

Authors "Alessia Lucidi" and "Daniela Tomaselli" have contributed equally to this chapter.

A. Lucidi · D. Tomaselli · D. Rotili · A. Mai (✉)
Department of Chemistry and Technologies of Drugs, Sapienza University of Rome, Roma, Italy
e-mail: antonello.mai@uniroma1.it



S. Jurga, J. Barciszewski (eds.), *The DNA, RNA, and Histone Methylomes*, RNA Technologies, https://doi.org/10.1007/978-3-030-14792-1_12

# 1 Introduction About Epigenetics and DNA Methylation

In eukaryotic cells the genetic information is enclosed in DNA molecules which are packaged in a dynamic structure known as chromatin. Chromatin results from the physical association between DNA double helix and histone proteins and is in constant balance between its "open" and transcriptionally active form called euchromatin, and its "closed" and transcriptionally repressed one, known as heterochromatin. The basic unit of this dynamic structure, known as nucleosome, consists of a histone core (an octamer resulting of two tetramers, each composed by histones H2A, H2B, H3 and H4) around which DNA is wrapped approximately with 147 bp (base pair). Although all the cells of the same organism contain the same genetic information, the phenotype results different and tissues-specific thanking to epigenetics, which acts as the first director of the correlation between genotype and phenotype. Therefore, it is widely known that gene expression is intimately regulated by epigenetic mechanisms which response to environmental exposure (early life experience, stress, etc.) (Berger et al. 2009).

Over the time, however, the meaning of the term "epigenetics" has continuously evolved until it reached a definition coined by Arthur D. Riggs: "The study of mitotically and/or meiotically heritable changes in gene function that cannot be explained by changes in DNA sequence" (Berger et al. 2009). The accessibility of the transcription machinery to DNA is modulated by several epigenetic mechanisms, among which the most significant are the histone modifications and the DNA methylation (Fig. 1). Proteins involved in chromatin remodeling resulted able to reversibly add, remove, or bind such covalent modifications (for this reason they are classified into "writers", "erasers", or "readers", respectively), offering the possibility to pharmacologically revert these effects. In humans, a wide variety of histone modifications have been identified such as methylation, acetylation, phosphorylation, ubiquitination, and sumoylation.

DNA methylation, that we can describe as the most stable epigenetic mark known in humans (Reik 2007), occurs at the C5 of the cytosine residue mainly in the context of 5′-CpG-3′ (5′-cytosine-phosphate-guanine-3′) dinucleotide, and is the result of the catalytic activity of a specific class of enzymes known as DNA methyltransferases (DNMTs), which use *S*-adenosyl-L-methionine (SAM) as the co-substrate (Jurkowska et al. 2011). While methylated CpG sites are randomly distributed among the genome (approximately 70–80% of CpG dinucleotides resulted methylated and therefore not transcribed such as the CpG-rich satellites and retroelements), the unmethylated CpG sites are localized in the so-called "CpG island", mainly identifiable into the promoter region of more than half of all human genes (Ehrlich 2002). DNA methylation acts, generally, as a repressive mark as it is able to arrest transcriptional initiation, either by recruiting methyl-binding proteins (MBPs) (inducing a repressed chromatin condition), or by preventing the binding of specific transcription factors. So, if the promoter results hypermethylated the transcription of the corresponding gene is repressed. This epigenetic modification can be propagated during cell division, proving that epigenetic regulation could be able to modulate

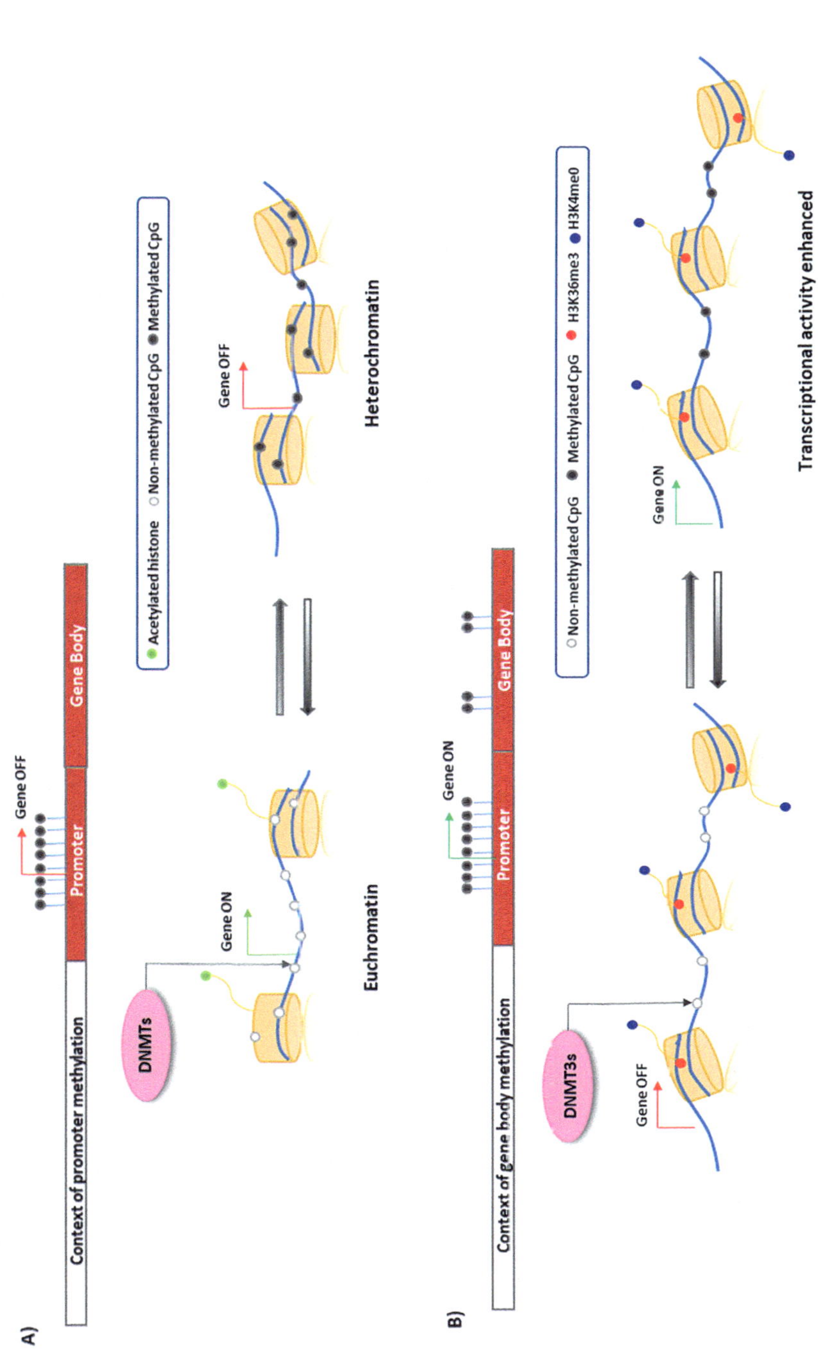


**Fig. 1** Scheme of dual function of DNA methylation in gene regulation. Epigenetics modifications resulted able to modulate gene expression in a context- and site-specific mode: **(a)** Hypermethylation of gene promoter induce gene inactivation; **(b)** In cancer, gene body CpGs are preferential sites for de novo methylation and is associated with gene expression

gene expression even when the signal upstream of the process is no longer present (Auclair and Weber 2012).

The repressed chromatin condition induced by the recruitment of MBPs is especially due to the involvement of some histone methyltransferases (HMTs) and/or histone deacetylases (HDACs) enzymes (that are able to remove methyl units and acetyl group from histone tails, respectively) that allowed the binding of chromatin silencer such as the heterochromatin protein 1 (HP1) inducing a transcriptionally inactive state (Bannister et al. 2001).

An alteration of the methylation pattern is responsible for the onset of different pathological conditions and this is particularly evident in cancer in which both aberrant hypermethylation of tumor-suppressor genes (or non-coding RNAs) and genome-wide hypomethylation occur (Lujambio et al. 2010; Hon et al. 2012).

## 2 DNMT Enzymes

Mammals DNMTs (Fig. 2) are enclosed in two functionally and structurally distinct families that result able to establish and maintain DNA methylation patterns: while DNMT3 family results able to establish the basic CpG methylation level, DNMT1 maintains the DNA methylation state during the sensitive phase of chromosome replication and repair. Until today, three different isoforms of the "de novo" DNMT3

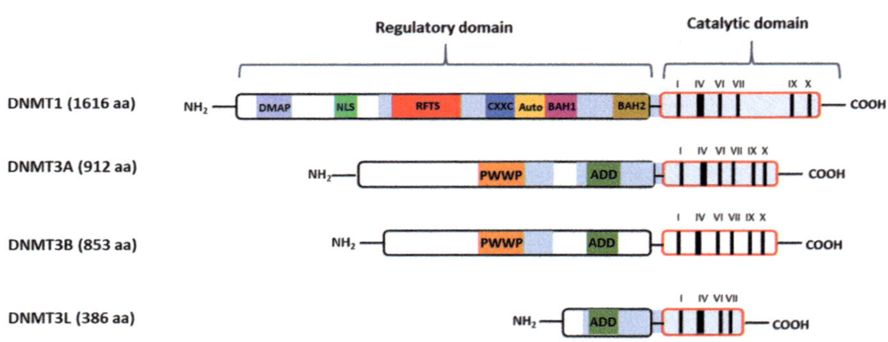


**Fig. 2** Schematic representation of the structure of mammalian DNMT isoforms domain. The portions represented in blue refer to domains structurally solved. N-terminal domain of DNMT3A and 3B has a ProTrpTrpPro (PWWP) domain for activity regulation and nucleosome recognition and ATRX-DNMT3-DNMT3L (ADD, in green) (Guo et al. 2014; Baubec et al. 2015). I, IV, VI, IX and X are conserved motifs of the *C*-terminal catalytic domain (red box). The regulatory domain of DNMT1 comprises a nuclear localization signal (NLS, in green), a cysteine-rich (CXXC, in blue) DNA binding domain and bromo-adjacent homology domains 1 and 2 (BAH1 in purple, BAH2 in brown), DNA methyltransferase associated protein (DMAP, in turquoise) and replication *foci* targeting sequence (RFTS, in red). The auto-inhibitory linker (Auto, in yellow) between BAH1 and CXXC prevents de novo methylation (Song et al. 2010). DNMT3L is a catalytically inactive isoform of DNMTs that showed only an ADD domain on the *N*-terminal portion

subfamily are known: two active methyltransferases, DNMT3A and 3B (that catalyze the methylation of unmethylated and methylated DNA at an equal rate), and one regulatory component, DNMT3-like protein (DNMT3L). DNMT3L, which we can described as a coactivator, takes part in specific crosstalk with chromatin remodeling actors, in fact, since it lacks the typical essential motifs useful for cosubstrate and DNA binding, the catalytic activity is not present (Jia et al. 2007). Furthermore, another isoform is known, named DNMT2, which showed a minimal DNMT activity in vitro but acts as an RNA methyltransferase toward tRNA molecules and for this reason is also called TRDMT1. DNMT1 and DNMT3s are structurally characterized by a *C*-terminal catalytic domain and a *N*-terminal regulatory motif (Jurkowska et al. 2011). While the *C*-terminal catalytic domain, that showed six highly conserved motifs, plays an essential role in DNA binding and in the transfer of the methyl unit, the *N*-terminal domain, interacting with several specific proteins and DNA molecules, is able to regulate its activity driving their localization to chromatin. The binding of the cosubstrate is allowed by specific residues in motif X and I, motifs IV, VI and VIII are involved in the catalysis reaction and in the target base binding and motif IX is essential for DNA recognition. As previously reported, DNMT3L lacks motifs IX and X that resulted essential for DNA and cosubstrate binding (Ruthenburg et al. 2007).

## 2.1 DNMT1

The first DNMT to be characterized and purified was DNMT1. In somatic cells, DNMT1 is the most abundant enzyme among the DNMTs and its structure is characterized by 1616 amino acids and exists in three different isoforms: DNMT1s, DNMT1o, DNMT1p typical of somatic cells, occyte and pachytene, respectively (Dhe-paganon et al. 2011; Giraldo et al. 2013). Next to the center core of the *N*-terminal domain we can found a region, present in all the mammalian cytosine methyltransferases, rich in cystein residues that is able to bind zinc ions, as well as the CpG binding proteins and MBD1. As previously mentioned, DNMT1 due to its major affinity for hemimethylated DNA (about 10–40 folds), is renamed "maintenance methyltransferase", allowing the methylation of the newly synthesized strands after DNA replication. Recent studies showed that probably the well accepted site-specific maintenance methylation pattern needs to be reinterpreted, in fact, this model is based on two pillars: (1) the methylation level of a single CpG site is inherited in a stable way; (2) an enzyme capable of carrying out this task precisely exists. Despite the fundamental importance of such process at the cellular level, this is not sufficient to guarantee the perfect reproduction of the methylation status of 56 million CpG sites during the numerous DNA replication processes, thus leading to a revision of the classical model before exposed: (1) DNA methylation should be considered generally not-site specific, and (2) the maintenance DNMTs enzymes are not able to specifically copy the genome wide methylation patters. Additionally, in vitro Dnmt1 acts as "de novo" methyltransferase too (Goyal et al. 2006) and different studies confirmed this activity also in vivo, in fact, its overexpression in cells seems to be associated with the methylation of some previously un-methylated sites. DNMT1 covers an essential role in somatic cells during developmental stages, as it ensures cell survival and proliferation. In fact, different evidences demonstrated that the disruption of Dnmt1 gene in mice can induce a strong embryonic lethality correlated to a wide genome demethylation. In addition, in non-cancer cells, knock-down or knockout condition of DNMT1 arises in severe mitotic defects (Chen et al. 2007), chromosomal instability, apoptosis (Takashima et al. 2009) and tumorigenesis. DNMT1 can be considered a promising therapeutic target because the disruption of DNMT1 in cancer cells leads to cells differentiation blocking tumor growth and cells invasiveness (Chik and Szyf 2011). DNMT1 allows the maintenance of DNA methylation state in two different ways: (1) interacting directly with the replication fork; (2) or through the interaction with UHRF1 (ubiquitin-like, containing PHD and RING finger domains 1). DNMT1 is transported in correspondence of the replication *foci* into the nucleus of somatic cells during the beginning of the S phase through the nuclear location signal (NLS) thanks to its PCNA (Proliferating Cell Nuclear Antigen) binding domain (PBD) and to the specific sequence located within its *N*-terminal domain (Targeting Replication Foci: TRF). UHRF1, showing a marked affinity for hemimetylated CpG sites recruits DNMT1 in correspondence of these. The occurrence of an error with a frequency of 5% per cell division and per CpG site gives to cells some flexibility for decisive changes in their methylation pattern. This aspect is controlled and restored by DNMT3A and 3B that resulted able to assist DNMT1 during cells replication (Jones and Liang 2009).

## 2.2 The DNMT3 Family

To date two different functional DNMT3 are known, DNMT3A and DNMT3B, which are able to catalyze the "de novo" methylation of CpG dinucleotides during development. The third DNMT3, DNMT3L, instead, lacks the enzymatic activity typical of this class, acting as a regulatory factor of germ cells. DNMT3A covers an essential role in the imprinting phenomena resulting able to induce the methylation mainly at the pericentromeric regions of DNA playing an essential role in the setting of methylation state in maturing gametes (Smallwood et al. 2011) and in postnatal somatic stem cell (Wu et al. 2010; Challen et al. 2012) while DNMT3B appears to acts at centromeric regions (Jones and Liang 2009) resulting essential during implantation, in later embryonic stages and differentiated cells (Borgel et al. 2010). Currently, DNMT3L is regarded a DNMT3A cofactor able to interact with different histone modifying proteins, thus acting as a linker between histone modifications and DNA methylation. While the structure at the level of the *C*-terminal region of DNMT3A and DNMT3B (which overlap for the 84%) is close to those of DNMT1, the *N*-terminal region appears to be half the length of DNMT1 and it is characterized by: (1) a domain rich in cysteine residues located in the *N*-terminal region near the GK-linker capable of binding the zinc ion, which acts as a transcriptional repressor; (2) A PWWP sequence rich in Trp and Pro located in the *N*-terminal of the domain rich in cytosine residues of 161 aminoacids, which appears to be fundamental in directing DNMT3B on metaphase chromosomes and DNMT3A over pericentric heterochromatin.

## 2.3 Mechanism of Action

The mechanism of action of DNMT enzymes can be summarized in the catalysis of the transfer of a methyl unit from the co-substrate SAM (that allosterically modulates the activity of DNMTs) to the C5 of cytosine through a covalent catalytic reaction. Three are the key protagonists that are involved in the methylation program: DNMT enzyme, the co-substrate SAM (responsible for the transport of monocarboxylic units at the lower oxidation state), and, obviously, DNA. What energetically allowed the entrance of the cytosine in the catalytic pocket of the enzyme is: (1) the substrate-enzyme interaction (that is favored by the dissolution of the hydrogen bond between the cytosine and its coupled base); (2) the simultaneous establishment of the π-interactions with the adjacent base; (3) the nucleophilicity of the cytosine.

Once the enzyme has recognized the CpG site which has to be methylated, it is able to pull out the deoxycytidine directing it to the catalytic pocket through a process known as "base flipping". The next step involves the binding of a DNMTs cysteine (Cys1226 for hDMT1, Cys711 in hDNMT3A and Cys652 in hDNMT3B) to the C6 of the cytosine forming the corresponding enamine, permitted by the transient protonation of the nitrogen atom in the third position (N3) promoted by the glutamate residue belonging to the VI motif of DNMT enzyme, thus breaking the aromaticity of pyrimidine ring. Subsequently, the transfer of a methyl group to the C5 is allowed by its delocalized electrons that provide the nucleophilic attack on the methyl unit of SAM. Lastly, the release of the cysteine through a β-elimination reaction (permitted by the deprotonation of the C5 by a basic aminoacidic residue specific for each DNMT isoform) restore the aromaticity of the cytosine releasing a molecule of S-Adenosyl-L-Homocysteine (SAH) and the DNMT enzyme which results ready for a new catalytic cycle (Fig. 3) (Xu et al. 2010; Lopez et al. 2016). As previously described, the added methyl unit allows the recruitment of several regulatory proteins.

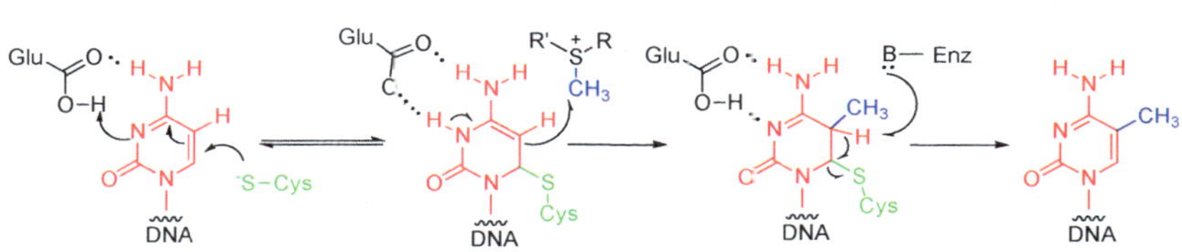


**Fig. 3** DNA cytosine methylation mechanism catalyzed by DNMT enzymes

## 2.4 Physiological Implication

DNA methylation is one of the more characterized epigenetic changes as a consequence of the important role that it covers in the physio-pathological field, in particular, it plays a fundamental role in genomic imprinting, embryonic development, X-chromosome inactivation, silencing of retrotransposon, repetitive elements, DNA repair etc. (Miller 2010).

It is already known that the methylation of CpG islands of promoter regions is often correlated to the silencing of the expression of the gene involved, while only a few evidence showed a weak correlation between the methylation of CpG site located in the gene coding sequence localized outside the first exon and the gene expression (Luczak and Jagodziński 2006).

The physiological role typical of the methylation process and DNMT enzymes has been deeply studied in mouse models since in 1992 Li et al. were able to report the first Dnmt1-knockout mouse model (Li et al. 1992). In mice, knock-out of *Dnmt1* and *Dnmt3b* are both responsible of the induction of embryonic lethality while the loss of *Dnmt3a* alters postnatal development (Li et al. 1992; Takashima et al. 2009).

The methylation mechanism that occur during the embryonic development dynamically reprograms this phase in a tissue-specific mode, in fact, the lack of the expression of Dnmt genes in mice, induces their death in the first 8 days of pregnancy (Sulewska et al. 2007).

Obviously, changes in the levels of the methylation state occur physiologically during development and differentiation of mammals cells. For example, in the zygote stage variations of the DNA methylation pattern and several dynamic histone modifications take place, while a global demethylation occurs within the early embryo. After the implantation, instead, a significant de novo methylation happened at the level of most CpG sites and, during the gastrulation, a tissue-specific demethylation occurs. To conclude, we have a sex-dependent de novo DNA methylation process during gametogenesis with a decrease in the DNA methylation pattern during postembryonic life state. As it happens in vitro in ageing cells, a slow decrease in the DNA methylation level has been observed during postembryonic phase (Xu et al. 2009).

## 2.5 DNMT and Disease

DNA methylation is one of the main guardians of cell integrity and gene expression, protecting cells from mutation through the repression of repeated and transposable elements (Esteller 2007). In this complex framework, overexpression of DNMT enzymes or mutation in their structures lead to the establishment of different disease, first of all cancer. Numerous pathologies have been linked to epigenetic disruptions, particularly diseases that are influenced by the environment. Being involved in CNS development and ageing, DNA methylation controls the age-related cognitive functions such as the memory process and it has recently been proved to have an important role in the pathophysiology of mood disorders (major depression disorder, bipolar disorder, schizophrenia) (Higuchi et al. 2011), but especially in Alzheimer's disease. Indeed, Fuso et al. reported that alterations of DNA methylation metabolism (both DNA methyltransferase and demethylase activities) were associated with the overexpression of presenilin1 (PSEN1), a gene involved in the production of amyloid β peptide whose deposits in intraneuronal spaces are one of the key features of Alzheimer's disease (Fuso et al. 2011). The treatment with DNMT inhibitors such as azacytidine, zebularine and procainamide proved to be effective in schizophrenia in which the overexpression of DNMT1 causes the hypermethylation and consequently the repression of RELN and GAD67 genes (Veldic et al. 2003; Kundakovic et al. 2007; Karsli-Ceppioglu 2016), but also in depression to counteract the stress mediated hypermethylation in the hippocampal region, associated with low levels of the neuroprotective protein brain-derived neurotrophic factor (BDNF) (Tian et al. 2009).

DNA methylation is clearly involved in auto-immune diseases (López-Pedrera et al. 2012), and in particular its role in systemic lupus erythematosus (SLE), systemic sclerosis (SSc), Rheumatoid arthritis (RA), Sjögren syndrome, cryopyrin-associated periodic syndromes (CAPS) and familial Mediterranean fever (FMF) was deeply investigated (Ballestar and Li 2017). In SLE a global hypomethylation was found to occur mainly due to the downregulation of DNMT1 especially in T-cells, further confirmed by the evidence that procainamide, a DNMT inhibitor specific for DNMT1, induced the onset of SLE in syngeneic mice. Obviously, defects in DNMT1 activity result in the overexpression of genes encoding for inflammatory cytokines (IL10, IL13) (Zhao et al. 2010) and surface antigens (CD11a, CD70) (Li et al. 2010), manifested as T-cells autoreactivity. DNA hypomethylation was also observed in RA in various cell types such as T and B lymphocytes, peripheral blood mononuclear cells (PBMCs) and synovial fibroblast, the last being the most important cell line for the pathogenesis of RA (Sun et al. 2016).

Several evidences in the last decade proved that the dysregulation of DNA methylation activity is strictly associated even with some genetic disorders such as Fragile X syndrome (FRAXA) and Facioscapulohumeral muscular dystrophy (FSHD) (Robertson 2005) but also with cardiovascular diseases (Bressler et al. 2011; Sacconi et al. 2012). By the way, in hypertrophic cardiomyopathy (HCM) the alteration of gene expression was observed. In particular, the treatment of HL-1 cardiomyocytes with 5-azacytidine induced the reactivation of *Myh7*, *Myh7b*, *Bnp*, *Gata4*, and *Nfatc1-a* genes (Fang et al. 2015).

Recently an important role of DNA methylation in the energy balance was also revealed. In the adipocytes of obese subjects DNMT1 is responsible for the methylation and the consequent repression of the adiponectin promoter region, being the adiponectin the hormone which controls the glucose blood levels and fatty acids breakdown (Kim et al. 2015).

Aberrant DNA methylation patterns have been extensively described in numerous cancers. It has been shown that cancer cells display a global hypomethylation and at the same time a hypermethylation of the CpG sites correspondent to gene promoter regions. On one hand, DNA hypomethylation leads to chromosomal instability, since repeated sequences are no longer methylated; on the other hand, DNA hypermethylation leads to the transcriptional repression of TSGs. Genes that acquire hypermethylation in regulatory regions are involved in a variety of important cellular pathways. For instance, two cell cycle-related genes, $p16^{INK4a}$ (CDKN2A) and $p15^{INK4a}$ (CDKN2B), undergo DNA methylation-mediated silencing in different types of cancers. They are both involved in the control of G1 progression, acting as cyclin-dependent kinase inhibitors and are important tumor suppressors. Various genes associated with DNA repair processes are also hypermethylated in tumor tissues, thus confirming the fact that epigenetic events may promote classical genetic alteration such as mutations. Defective DNA mismatch repair is observed in a significant portion of colorectal cancers and is associated with the hypermethylation of MLH1 (Gausachs et al. 2012). Moreover, the transcriptional repression of hypermethylated genes involved in cell adhesion, such CDH13 (H-cadherin) may lead to invasion and/or metastasis, and thereby tumor progression (Kim et al. 2005). Some of the genes hypermethylated in cancer are connected with cancer cell survival since they have pro-apoptotic functions, as in the case of the death-associated protein kinase 1 (DAPK1), which mediates interferon-induced apoptosis (Michie et al. 2010), or of the target of methylation-induced silencing (TMS1), a pro-apoptotic caspase domain protein responsible for the activation of apoptotic signaling pathways (Gordian et al. 2009). However, Suzuki et al., by investigating DNA methylation pattern of gastrointestinal cancer cells, concluded that DNA hypomethylation has more oncogenic weight than DNA hypermethylation (Suzuki et al. 2006). No information is known about the initial mechanism which causes the global hypomethylation observed in human tumors, but probably, DNA demethylation of CpG sites occurs as a consequence of a gradual age-dependent failure to guard methylation replication faithfulness, rather than due to a pathological defect. This "wear and tear" model proves the existence of cellular mechanisms responsible of the automatic preservation of methyl-deoxycytidine replication, even if not with 100% fidelity. These errors, by affecting the methylation of sensitive pericentromeric repetitive sequences may influence the genetic alterations, thus leading to mistakes in recombination and chromosome replication (Gaudet 2003). As mentioned above, the reversibility of DNA methylation represents an interesting strategy in oncology. Hence, the use of specific DNMTi might reactivate TSGs and induce the reprogramming of cancer cells, leading to their proliferation arrest and ultimately to their death.

## 3 DNMT Inhibitors (DNMTi)

The potential reversibility of the methylation mark covers an important therapeutic role because DNMTi represent powerful tools to correct and reverse it pharmacologically. A lot of DNMTi have been reported in literature which are grouped in two different families: (1) the most studied nucleoside analogs; (2) the non-nucleoside inhibitors, structurally variable according to their mechanism of action (Lopez et al. 2016).

### 3.1 Nucleoside Inhibitors

To date, only two drugs among the known DNMTi received FDA (Food and Drug Administration) approval for clinical use in hematological malignance such as acute myeloid leukemia (AML), chronic myelomonocytic leukemia (CMML) and myelodysplastic syndrome (MDS): 5-azacytidine (**1**, azacytidine, 5azaC, Vidaza®, Fig. 4, Table 1) and decitabine (**2**, 5-aza-2′-deoxycytidine, 5azadC, Dacogene®, Fig. 4, Table 1), two suicide inhibitors (Jones and Taylor 1980; Christman 2002; Wijermans et al. 2008; Issa and Kantarjian 2009; Gros et al. 2012).

Unfortunately, the chemical instability (due to the deamination of cytidine and deoxycytidine and to the opening of triazine ring), the high toxicity and low specificity (both linked to their mechanism of action), and the poor bioavailability of these molecules have limited their use, encouraging researchers towards new solutions (Christman 2002; Brueckner et al. 2007).

In fact, it is well-known that the methylation process is typical for both normal and cancer cells, so the achievement of the inhibition selectivity is a hard challenge.

However, the different thresholds for promoter activation and the higher proliferation activity that characterized cancer cells, suggested that the latter could be more affected by the treatment than normal cells (Pechalrieu et al. 2017).

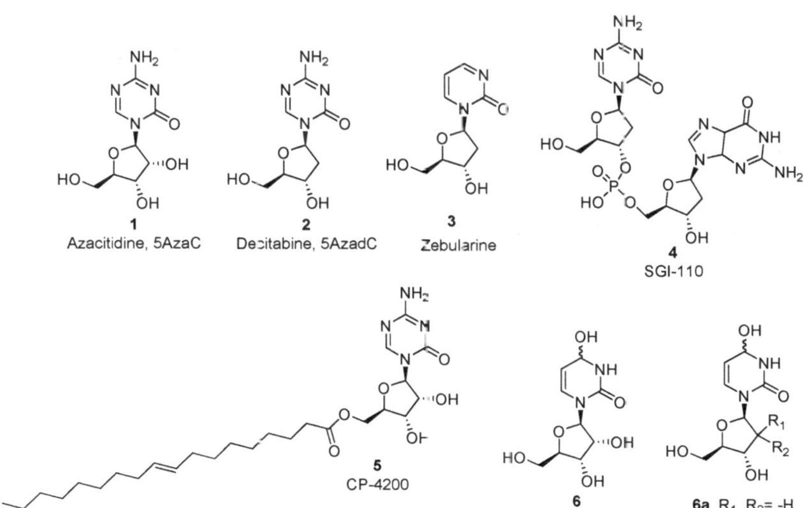


**Fig. 4** Structure of DNMT nucleoside inhibitors and THU derivatives (CDA inhibitors)

**Table 1** Nucleoside and non-nucleoside DNMTi evaluated in clinical trials as monotherapy

| Drugs | Type of DNMTi | Induced effects | Type of cancer treated | Clinical trials |
|---|---|---|---|---|
| **1** 5azaC | Nucleoside | DNA demethylation, differentiation of leukemic cell lines, reduction of tumor volume | AML, MDS | FDA approved |
| | | | Solid tumors | Phase I |
| **2** 5azadC | Nucleoside | DNA demethylation, DNMT depletion, growth inhibition, antineoplastic activity | AML, MDS, CMML | FDA approved |
| | | | Solid tumors | Phase I |
| **4** SGI-110 | Nucleoside | DNA demethylation, antitumor activity | Solid and hematological tumors | Phase I, II, III |
| **30** Hydralazine | Non-nucleoside | Not yet known | Ovarian and brain tumors | Phase III |
| **32** Disulfiram | Non-nucleoside | DNA demethylation, reduction of tumor progression | Prostate cancer | Study phase not applicable |

The nucleoside analogues effected their activity only after being incorporated (as a result of triple phosphorylation by cellular kinases) into DNA strands (in particular 5azaC **1** needs to be converted first in its deoxyribose analogue) by the DNA polymerases (in place of deoxycytidine), forming, in this way, an irreversible covalent complex with DNMT enzymes and inducing the proteasomal degradation of the latter (Ghoshal et al. 2005).

Both these compounds are characterized by the presence of a nitrogen atom in position 5 instead of the carbon C1, and are linked to a ribose (5azaC **1**) or a deoxyribose (5azadC **2**). To carry out their pharmacological action these molecules need to be incorporated into the DNA during the S phase (replication), entailing, in this way, a small specificity towards the more rapidly proliferating cells, the cancer cells. Once integrated, these analogues of cytosine are recognized by DNMTs acting as suicide substrates: after the formation of the complex, unlike what happens with cytosine residues, the β-elimination reaction is not allowed as the consequence of the presence of the nitrogen atom in position 5 (Fig. 5) (Fahy et al. 2012).

In cancer cells, 5azaC **1** and 5azadC **2** induced the re-expression of TSG, resulting in apoptosis and cell cycle arrest, and prompting, on the whole, reprogramming of the cell (Lopez et al. 2016). In addition, these molecules also determined cytotoxicity at high doses due to the formation of DNA-DNMT adducts, being both the effects responsible for the anti-tumor activity (Flotho et al. 2009). For the reason just exposed, these compounds are administered in low doses with the aim to further exploit the de-methylating activity over the cytotoxic one (Klimasauskas et al. 1994). In prostate cancer, 5azaC **1** acts as a chemosensitizer (Festuccia et al. 2009) reactivating the expression of the phosphatase and tensin homolog deleted on chromosome ten (PTEN), a TSG involved in cell signaling. PTEN acts on the

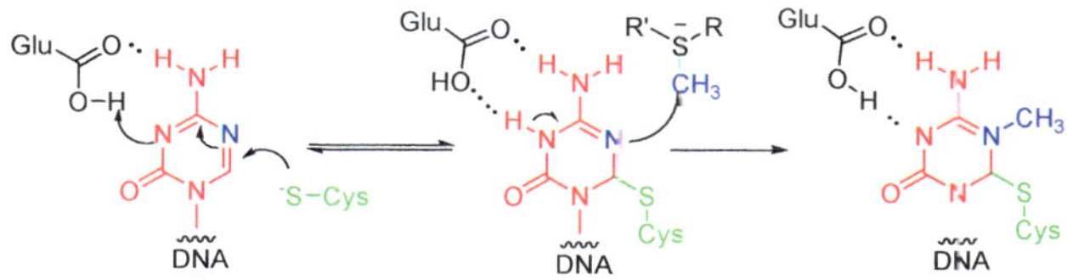


2
5azadC

irreversibie covalent complex

**Fig. 5** Mechanism of action of DNMTi that act as suicide substrates

phosphoinositide-3-kinase (PI3K) which in turn regulates the activation of Akt, a protein kinase B involved in the cellular growth and in the apoptotic resistance.

Furthermore, the ribose analogues of this class of DNMTi are also integrated into RNA residues interfering with the protein synthesis. This last aspect showed the reason why 5azadC **2** resulted more effective than 5azaC **1** with less adverse reactions.

Despite their marked effectiveness, their mechanism of action involves the occurrence of various adverse reactions, first of all dose-limiting neutropenia and renal toxicity, while long term-safety, as well as carcinogenic and mutagenic potential of these compounds are subjects of debate (Yang et al. 2003; Issa and Kantarjian 2009).

In addition, as previously reported, other limiting factors related to the application of these molecules have been highlighted beyond the lack of selectivity over the different DNMTs isoforms, such as their metabolic/chemical instability and their low bioavailability (Derissen et al. 2013). For example, under physiological pH and temperature condition, these compounds undergo degradation in several products as a consequence of anomerization, deformylation, deamination reaction (by cytidine deaminase, CDA) and hydrolytic opening of the triazine ring (Weber et al. 2007).

Evidence of this aspect is the plasmatic half-life related to subcutaneous administration for 5azadC **2** and 5azaC **1** (about 10–35 min and 41 min respectively) (Marcucci et al. 2005; Karahoca and Momparler 2013).

To overcome the main limits related to the use of 5azaC **1** and 5azadC **2**, others nucleoside analogues were developed. Zebularine (**3**, Fig. 4, Table 1), is a first-generation nucleoside inhibitor and in particular a cytidine analogue in which the amino group in position -4 of the pyrimidine ring is absent. **3** showed the capability to hit not only DNMTs but also cytidine deaminase following oral administration (Lemaire et al. 2009).

Despite Zebularine **3** was characterized by a lower cytotoxicity and greater stability than 5azaC **1** and 5azadC **2**, it requires higher doses of administration to achieve the same demethylation effect in cells. This condition was explained considering a different mechanism of action for **3** that did not requires the formation of an irreversible covalent complex with DNMTs but a reversible one characterized by a slow dissociation constant (Flotho et al. 2009; van Bemmel et al. 2009; Champion et al. 2010). In clear cell renocarcinoma (ccRCC) the hypermethylation of the von Hippel-Lindau (VHL) gene promoter causes the accumulation of the hypoxia inducible factor (HIF) which in turn induces angiogenesis. In this context, as reported by

Alleman et al., Zebularine **3** and 5azadC **2**, by reactivating the gene expression of VHL, exhibited a marked angiostatic effect (Alleman et al. 2004) even strengthened by the transcriptional activation of different angiogenesis inhibiting genes such as TSP1, JUNB and IGFBP3 (Li et al. 1999; Gravina et al. 2010). However, the clinical use of Zebularine **3** as single agent is hindered because its inefficient metabolic activation.

Another approach identified with the aim to improve the bioavailability and stability of **1** and **2** was the design of prodrugs. The most interesting and promising among this series was the prodrug of **2**, the CpG nucleotide analog called SGI-110 (**4**, Fig. 4) (Yoo et al. 2007). The new aspect related to SGI-110 **4** is its capability to resist to the action of CDA enzymes that inactivate 5azaC **1** and 5azadC **2**. In addition, **4** showed a better cell penetration than **1** and **2** and a cytotoxicity and aqueous solubility comparable to **2**. To date, SGI-110 **4** is in phase II for hepatocellular carcinoma (NCT01752933) in phase II/III clinical evaluation for AML (NCT02348289), in combination with immunotherapy for metastatic melanoma treatment in Europe (EUdract 2015-001329-17) and with different chemotherapy drugs for ovarian cancer (NCT01696032) (Table 1) (Chuang et al. 2010).

In addition, another strategy was based on the functionalization of cytidine analogues with different fatty acids in order to overcome the drug resistance due to poor cellular uptake. Among them, CP-4200 (**5**, Fig. 4), a lipophilic ester of 5azaC **1** was developed with the purpose of exploit a different drug transport mechanism within cells (compared to the conventional nucleoside transport systems which is often lacking as a mechanism of resistance to the treatment) improving the activity against orthotropic acute lymphoblastic leukemia (ALL) mouse model.

**5** is able to deplete DNMT1, suggesting that, despite the additional metabolic activation step necessary as a pro-drug, it ensures the sufficient concentration of its active metabolite in cancer cells (Breistøl et al. 1999; Bergman et al. 2004; Brueckner et al. 2010).

Since demethylating agents require a long treatment time and repeated and daily doses, the development of oral formulations could induce a large increase in patient compliance. Oral formulations of 5azaC **1** and 5azadC **2** are noteworthy because they are in phase I and II of clinical trials alone or in association for the treatment of both solid and hematological tumors. The most interesting is the Celgene Corporation Product, CC-486, an oral formulation of 5azaC **1,** that resulted effective against MDS and CML (phase I) (Tables 1 and 2) (Garcia-Manero et al. 2011).

## *3.2 Non-Nucleoside Inhibitors*

Although the use of nucleoside inhibitors as single agents in hematological cancers is already approved, in solid tumors their application as therapeutic agents is still controversial because the toxicity issue limits their administration at high doses. Moreover, there is a significant need for DNMT inhibitors that do not rely on DNA incorporation to exhibit their activity. A recent study using a panel of human cancer

**Table 2** Most promising combination therapies based on DNMTi

| Type of drug | Combined drugs | Type of cancer and relative phase of the study |
|---|---|---|
| Nucleoside DNMTi + CDAi | **2** 5azadC + E-7727 | MDS (I) (II) |
| Nucleoside DNMTi + Chemo-drugs | **1** 5azaC + Ara-C | AML (II) and other hematological tumors |
| | **1** 5azaC/**2**azadC + ATRA + VPA | AML (II) |
| | CC-486 + abraxane/gemcitabine | Pancreatic cancer (I) |
| | CC-486 + carboplatin/ABI-007 | Refractory or relapse solid tumors (I) |
| Nucleoside DNMTi + Immuno-drugs | **2** 5azadC + plerixafor + interferon α-2b | Metastatic solid tumors (I)<br>Renal cell carcinoma (II)<br>Melanoma (I)(II) |
| | **2** 5azadC + plerixafor + panitumumab | *Wild*-type metastatic colorectal-cancer (I) |
| Non-nucleoside DNMTi + Immuno-drugs | **37** MG98 + Interferon | Advanced renal carcinoma (I) |
| Nucleoside DNMTi + other Epi-drugs | **1** 5azaC/**2** 5azadC + SAHA | Hematological tumors (I)(II) |
| | **1** 5azaC + Mocetinostat/Entinostat | Hematological tumors (I)(II) |
| | **1** 5azaC + Entinostat | Lung cancer (I), Breast cancer (II) metastatic colorectal cancer (II) |
| | **2** 5azadC + VPA | MDS (II) and AML (II) |
| | **2** 5azadC + Panobinostat | Breast cancer (I) (II) |

cell lines showed a 1000-fold variability in 5azadC **2** potency, attributed to differential incorporation of 5azadC **2** into DNA. As DNA hypermethylation correlates with clinical benefit of 5azadC **2** in myelodysplastic syndrome (MDS) (Shen et al. 2010), it is possible that some element of MDS resistance to 5azaC **1** and 5azadC **2** stems from the ability of tumor cells to block the import, phosphorylation, and incorporation of the drug into DNA. Non-nucleoside inhibitors could bypass these hurdles by targeting DNMTs directly. They exhibit a wide structural diversity and can be divided in the following groups.

### 3.2.1 Natural Compounds: Flavonoids, Psammaplin A, Curcumin, Laccaic Acid, Nanaomycin A and Resveratrol

Flavonoids are polyphenolic compounds, mainly extracted from plant, and have considerable interest due to their potential antiviral, anti-inflammatory and antitumor activity. One of well-known polyphenol molecule is Genistein (**7**, Fig. 6). It was first isolated from Dyer's Broom (*Genista tinctoria*) in 1899 by Perkin and Newbury, then Baker and Robinson characterized and first synthesized the isoflavone nucleus in 1928 (Walter 1941). Although Genistein **7** has been considered a phytoestrogenic

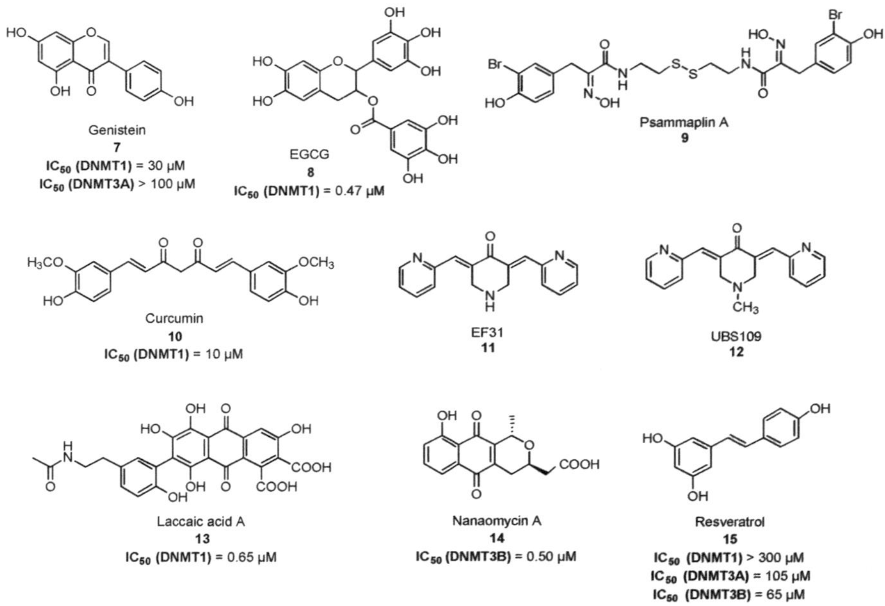


**Fig. 6** Natural compounds, synthetic analogues as DNMT inhibitors and relative $IC_{50}$ where available

molecule, its role as a potential anticancer agent has been recently evidenced because it showed in vitro activity against diverse enzymes such as topoisomerase I or II, histidine kinase, tyrosine kinases and protooncogene HER-2. Another important studied flavonoid is (-) epigallocatechin-3-O-gallate (EGCG, **8**, Fig. 6), a major component of green tea extracts. EGCG **8** is generally considered the biologically most active compound in vitro. The changes in the activities of various protein kinases, growth factors, and transcription factors represent a common mechanism involved in cellular effects of tea polyphenols. In addition to induce the alteration of intracellular signaling by activating cellular receptors, it was shown that EGCG can enter within cells and directly interact with their molecular targets (Yang et al. 2009). Recently, Genistein **7** and EGCG **8** have been reported as enzymatic and cellular DNMT inhibitors: EGCG **8** directly inhibited DNMTs activity and partially reversed RAR-β methylation status while Genistein **7** (20–50 mmol/L) is a DNMT inhibitor noncompetitive with SAM but shows competitive and noncompetitive inhibitory activity with the substrate poly(dI-dC). Genistein **7** is a weaker DNMT inhibitor than EGCG, but it results more effective in reactivating the expression of normally hypermethylated genes, probably because its higher stability in cell culture medium respect to EGCG **8**, allows to reach higher intracellular concentrations. The demethylating activity can be enhanced by extending the treatment period with EGCG **8** or Genistein **7** or by their association therapy with a HDAC inhibitor (Yang et al. 2008).

The supposed mechanism of action of flavonoids consists on the enzymatic inhibition of the DNA methylation due to an increased formation of SAH during the catechol-*O*-methyltransferase-mediated *O*-methylation of this flavonoids. In comparison, the strong inhibitory effect of EGCG **8** on DNMT-mediated DNA methylation is independent of its own methylation and is largely due to its direct inhibition of the DNMTs (Lee et al. 2005). These two flavonoids are now regarded more as chemo-preventive drugs than actual treatment drugs, even though their metabolisms drastically decrease their bioavailability, hence lowering their potential activity.

Psammaplin A (**9**, Fig. 6) is a natural compound extracted for the first time in 1987 from a sponge, the *Psammaplin Aplysilia*. This dimer of two derivatives of 3-bromotyrosine was synthesized primary in 1992 by Hoshino et al. and shows both antibacterial and antitumor properties (Hoshino et al. 1992). Psammaplin A **9** was described as inhibitor of almost ten different enzymes, such as leucine aminopeptidase, DNA gyrase, topoisomerase II, HDACs and DNMTs (Piña et al. 2003). It has been shown that several natural derivatives of Psammaplin A **9** have an activity against DNMT1 and HDACs. Moreover, Psammaplin A **9** shows antiproliferative properties also on MDA-MB-435 and A549 cell lines (breast and lung cancer, respectively) with promising results, as proved by the inhibition of cell growth at low doses ($IC_{50} = 2\ \mu M$). Nevertheless, no DNA demethylation has been highlighted in HCT116 cells (Godert et al. 2006) and, in a recent study, Psammaplin A **9** exhibited no activity on DNMT1 at the concentrations of 30 and 120 μM (Baud et al. 2012).

Molecular docking of the interaction between curcumin (**10**, Fig. 6) and DNMT1 suggested that such compound covalently blocks the catalytic thiolate of DNMT1 to exert its inhibitory effect on DNA methylation. This was validated by showing that curcumin **10** inhibits the activity of the CpG methyltransferase *M. SssI* with an $IC_{50}$ of 30 nM, but no inhibitory activity of hexahydrocurcumin up to 100 μM was found to occur (Liu et al. 2009). Finally, curcumin **10** at doses higher than 3 mM induced a decrease of global DNA demethylation of leukemia MV4-11 cells. However, recent studies showed that curcumin does not exhibit a demethylating activity, which suggested that curcumin **10** has no relevant DNMT inhibition (Medina-Franco et al. 2011). The fact that the number of studies showing positive effects of curcumin **10** is much higher than that showing negative effects, it may just indicate that there are more researchers evaluating the beneficial effects of curcumin **10** than evaluating its toxicity. Future research is needed to establish the benefit-risk profile of curcumin **10**. Recently, the activity of novel curcumin **10** analogues EF31 (**11**, Fig. 6) and UBS109 (**12**, Fig. 6) as demethylating agents were investigated, by Nagaraju et al., in two pancreatic cell lines (Nagaraju et al. 2013). They selected the pancreatic cell lines based on baseline level of DNA methylation. MiaPaCa-2 cells have high levels of DNA methylation associated with loss of tumor suppressor genes SPARC, *p16* and E-cadherin. In contrast, PANC-1 cells have low levels of baseline DNA methylation and expression of SPARC, *p16*, and E-cadherin. The EF31 **11** and UBS109 **12** showed several folds more potency than curcumin and inhibit DNMT1, NF-κB and HSP90. In addition, EF31 **11** and UBS109 **12** have better solubility and bioavailability. The unique properties of EF31 **11** and UBS109 **12** make them promising therapeutic agents better than the parent compound curcumin **10** (Nagaraju et al. 2013).

In 2013, the natural anthraquinone Laccaic acid A (**13**, Fig. 6) was found to behave as a DNA-competitive inhibitor of DNMT1 with an $IC_{50}$ of 650 ± 40 nM, displaying a slight selectivity for this isoform respect to DNMT3A and *M.SssI* methyltransferase and effectiveness in reactivating some genes involved in breast cancer invasiveness (Fagan et al. 2013).

Nanaomycin A (**14**, Fig. 6) is a quinone antibiotic isolated from a *Streptomyces* strain, in 1975. Its DNMT1 potential inhibitory properties have been deduced after a virtual in silico screening, (Kuck et al. 2010b) but exhibited no activity when evaluated on DNMT1 biochemical assay. During further investigations, Nanaomycin A **14** demonstrated a selective inhibition of DNMT3B. Subsequent cellular characterization was undertaken, specifically cytotoxicity evaluations and DNA methylation levels examination on three cancer cell lines (A549, HCT116 and HL60, lung, colon and leukemia cell lines, respectively). Molecular docking calculation using a homology model of DNMT3B resulted in a possible model of Nanaomycin A **14** into the catalytic domain. However, despite the effects observed on the enzyme and a weak but significant demethylation of the RASSF1A promoter region, the authors concluded that DNMT3B inhibition is not the only mechanism of action of Nanaomycin A **14**, but rather it may hit other cellular targets (Kuck et al. 2010a).

Resveratrol (**15**, Fig. 6) is a natural compound endowed with a large pharmacological profile being able to manifest anticancer, anti-inflammatory, cardioprotective and antioxidant activities. Even DNMTs are targeted by Resveratrol as proved by Aldawsari et al, who reported for the first time the inhibition data ($IC_{50}$ vs DNMT1 > 300 μM, $IC_{50}$ vs DNMT3A = 105 μM, $IC_{50}$ vs DNMT3B = 65 μM) (Aldawsari et al. 2016), thus supporting its effect in decreasing DNMTs expression levels (Qin et al. 2014).

### 3.2.2 Synthetic Compounds

DNA Binders

Procainamide (**16a**, Fig. 7) is a cardiovascular drug with apparent activity against DNMT in cells (Segura-Pacheco et al. 2003). Like hydralazine, however, the DNA methylation inhibition of procainamide **16a** in cells has been disputed (Brueckner et al. 2007). Procaine (**16b**, Fig. 6), a closely related molecule, has also been reported to inhibit DNA methylation in MCF-7 breast cancer cells and to restore the transcription of RARβ2 by demethylating its promoter in the same cell line (Villar-Garea et al. 2003). The chemical evolution of these two compounds are identified by Castellano et al. and consists of oxazoline and isooxazoline constrained analogues of procainamide, among which the most potent against DNMTs are the derivatives **17** and **18** (Fig. 7), both endowed with high micromolar inhibitory activity against DNMT1 of 150 μM (Castellano et al. 2008, 2011).

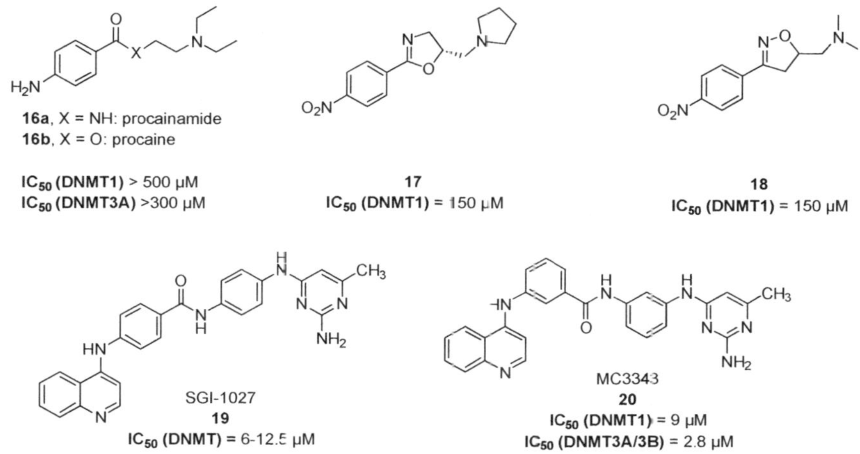


**Fig. 7** Structure of DNA binder inhibitors and relative $IC_{50}$ values

SGI-1027 (**19**, Fig. 7), a quinoline-based compound, has demonstrated inhibitory activity against DNMT1, DNMT3A, and DNMT3B in biochemical assays and resulted in decreased methylation at tumor suppressor gene CpG islands and corresponding gene upregulation (Datta et al. 2009). Quinolinium bisquaternary salts related to SGI-1027 **19** are known to bind reversibly but strongly in the minor groove of DNA (Adams et al. 2005). The non quaternized, weakly basic compound SGI-1027 **19** also binds reversibly but much less strongly to DNA, is indefinitely stable in aqueous solution, is highly lipophilic and has a low polar surface suggesting a good distribution and cell uptake abilities. SGI-1027 **19** inhibits all the different DNMT isoforms ($IC_{50} = 6–12.5$ μM) due to the highly-conserved motifs I and X (involved in the recognition of the SAM cosubstrate) and even if it has been considered a SAM competitive inhibitor, in the last years it has been classified as a DNA-binder (Datta et al. 2009; Gros et al. 2015). A medicinal chemistry optimization performed by Valente et al. led to the identification of the *meta-meta* analogue of SGI1027 **19**, called MC3343 (**20**, Fig. 7), which exhibited a higher potency and selectivity than its analogue ($IC_{50}$ vs DNMT1 $= 9 \pm 1$ μM, $IC_{50}$ vs DNMT3A/3B $= 2.8 \pm 0.2$ μM), and showed also to be effective in several cancer cell lines (RAJI, U-937, PC-3, MDA-MB-231) still retaining an appreciable toxicity profile (Valente et al. 2014).

## Substrate and SAM-Cosubstrate Competitors

RG-108 (**21**, Fig. 8) is a small-molecule inhibitor of DNMT1 that was discovered using a computational screening approach (Pawel Siedlecki et al. 2005). The compound blocked DNA methyltransferase activity in vitro, and in cells inhibited the methylation of tumor-suppressor gene promoters, but not the DNA in centromeric

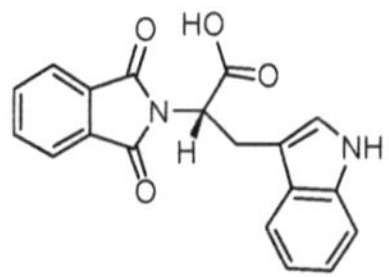

RG-108
**21**
$IC_{50}$ **(DNMT1)** = 390 μM
$IC_{50}$ **(DNMT3A)** = 315 μM

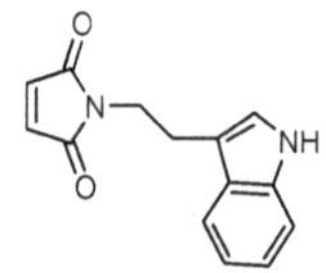

RG119-1
**22**
$IC_{50}$ **(DNMT1)** = 19 μM
$IC_{50}$ **(DNMT3A)** = 3.2 μM

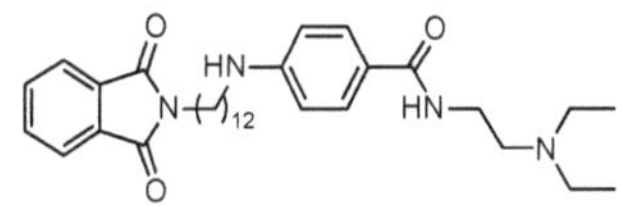

**23**
$IC_{50}$ **(DNMT1)** = 4.9 μM
$IC_{50}$ **(DNMT3A)** = 8.2 μM

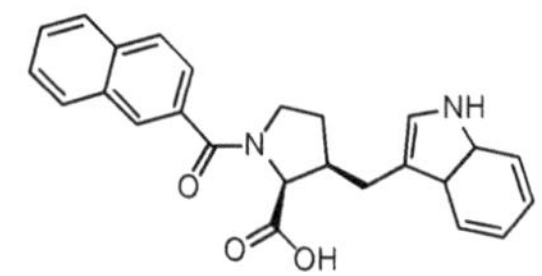

**24**
$IC_{50}$ **(DNMT1)** = 50 μM

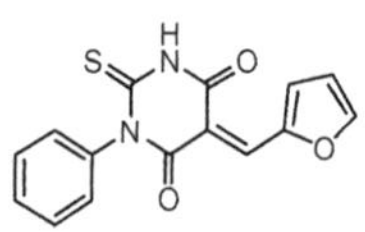

SID49645275
**25**
$IC_{50}$ **(DNMT1)** = 0.811 μM

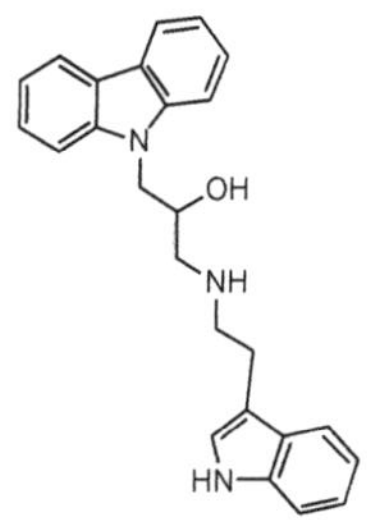

DC_05
**26**
$IC_{50}$ **(DNMT1)** = 10.3 μM

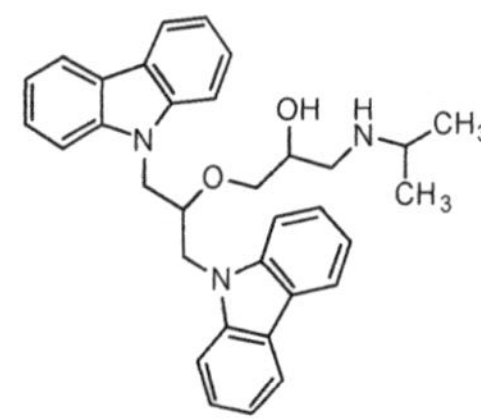

DC_517
**27**
$IC_{50}$ **(DNMT1)** = 1.7 μM

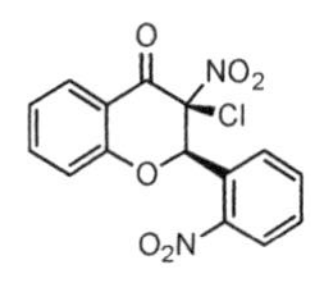

**28**
$IC_{50}$ **(DNMT3A/3L)** = 0.37-0.49 μM

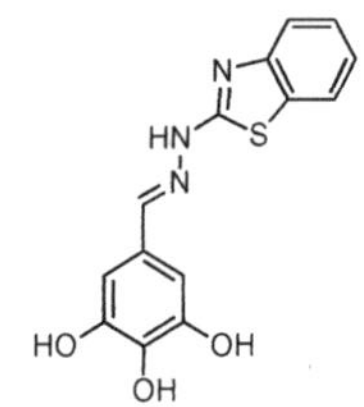

**29**
$IC_{50}$ **(DNMT3A)** = 1.6 μM

**Fig. 8** Substrate and SAM competitors DNMT inhibitors and relative $IC_{50}$ values

satellite elements, suggesting a context-dependent DNA methylation inhibition (Brueckner et al. 2005). RG-108 **21**, unlike the nucleoside analogs 5azaC **1**, 5azadC **2**, zebularine **3**, procaine **16b**, and EGCG **8**, did not demonstrate cytotoxic or genotoxic effects on cells even at high concentrations (Stresemann et al. 2006), but its high hydrophobicity limited a broader in vivo application of the compound (Mai and Altucci 2009). However, subsequent investigations, pointed out that RG-108 **21** behaves as a weak inhibitor of DNMTs with $IC_{50}$ values in the high micromolar range in comparison with the 115 nM value initially reported by the authors (Suzuki et al. 2010; Asgatay et al. 2014).

Rondelet and coworkers published an interesting study about the identification of a novel maleimide analogue of RG-108 **21**, named RG119-1 (**22**, Fig. 8). This compound is not only able to inhibit both DNMT1 and DNMT3A with $IC_{50}$ values of 19 and 3.2 μM respectively acting as a SAM-competitive inhibitor but proved to be effective also in M14K and H28 lung cancer cells (Grégoire et al. 2017).

In 2012, Halby et al. developed an interesting approach consisting in the conjugation of procainamide **16a** with phthalimide of RG-108 **21**. The result was the identification of compound **23** (Fig. 8), characterized by a 12-carbon linker connecting the pharmacophoric moieties, behaving as a substrate competitor with an $IC_{50}$ of 4.9 ± 1.3 and 8.2 ± 1.9 μM against DNMT1 and DNMT3A/3L respectively (Halby et al. 2012).

Starting again from RG-108 **21**, in 2014, Asgatay et al. published a work in which a systematic structural optimization was described, having as final result, the identification of a SAM-competitive constrained analogue of RG-108 **21**, here reported as compound **24** (Fig. 8). It is a proline analogue of the tryptophan with a naphthoyl moiety linked to the proline amino-terminal, displaying a selective inhibitory activity against DNMT1 ($IC_{50}$ = 50 ± 14 μM) (Asgatay et al. 2014).

An innovative fluorescence-based high throughput screening described by Ye and Stivers in 2010, allowed to identify SID49645275 **25** (Fig. 8) (Ye and Stivers 2010), a thioxodihydropyrimidine-based compound, exhibiting a potent and selective inhibitory activity against DNMT1 ($IC_{50}$ vs DNMT1 = 811 nM). Since no experimental evidence were known about the binding mode of this new derivative, Medina-Franco and coworkers set up a docking analysis for SID49645275 **25**, in which they found the best pose score when the compound binds into the substrate binding pocket (Medina-franco and Yee 2013).

Chen et al., in 2014, reported a work describing the identification of new DNMTi through a docking-based virtual screening approach. The investigation allowed to individuate two compounds DC_05 **26** (Fig. 8) and DC_517 **27** (Fig. 8), characterized by a carbazole scaffold, never associated to the DNMT inhibition before, which selectively inhibit DNMT with $IC_{50}$ of 10.3 and 1.7 μM respectively. These two derivatives bind within the catalytic site of DNMT1, in such a way to locate the carbazolyl moiety in the SAM binding site and the indolyl portion in the cytidine pocket. Moreover, since the compounds have a chiral center, the authors studied also the activity of DC_05 **26** and DC_517 **27** enantiomers in comparison with the correspondent racemates, but the results suggested that the chirality does not play an important role for the inhibitory activity nor for the binding mode of the compounds. When evaluated in HCT116 and Capan-1 cells, DC_05 **26** and DC_517 **27** exhibited antiproliferative activity in the low micromolar range, DC_517 23 being also able to induce a dose dependent apoptotic effect in HCT116 cells (Chen et al. 2014).

In 2011, Ceccaldi et al. reported the screening of a library of flavonoid derivatives, among which compound **28** (Fig. 8) proved to be the most potent DNMT3A/3L inhibitor with an $IC_{50}$ value of 0.37 ± 0.05 μM and 0.49 ± 0.09 μM by using HTS FluoMet and $^{3}$H-SAM assay respectively. This 3-chloro-3-nitroflavanone inhibits DNMT3A/3L with a SAM competitive mechanism of action (Ceccaldi et al. 2011).

Recently, a novel hydrazone-gallate based compound **29** was identified as a selective inhibitor of DNMT3A with an $IC_{50}$ of 1.6 ± 0.6 μM which binds into the catalytic site of the enzyme. Even if **29** is not cytotoxic against KG-1, MDA-MB-231 and HCT116, it resulted able to determine the reactivation of the luciferase gene at 5 and 10 μM, thus suggesting that it does not act as a promiscuous binder as typically the gallates do, but that the in vivo effects are the results of the DNMT3A inhibition only (Erdmann et al. 2016).

### DNMT Inhibitors with Unknown Mechanism of Action

Hydralazine (**30**, Fig. 9) is another cardiovascular drug that was shown to inhibit DNA methylation in cells (Segura-Pacheco et al. 2003), although not in an in vitro biochemical assay (Deng et al. 2003). Despite its frequent use for numerous years, its mechanism remains unknown. Indeed, hydralazine **30** induces the erythematosus lupus in treated patients and this autoimmune disease is associated with a hypomethylation of T cells, confirmed on T cells in culture (Brueckner et al. 2007). The DNMT inhibitory activity of hydralazine **30** is controversial, as a subsequent study was unable to reproduce DNMT inhibitory activity in cells (Chuang 2005). Clinical studies of hydralazine **30** in combination with a HDACi (valproate) in MDS are undergoing, currently in phase II (Candelaria et al. 2011). Moreover, hydralazine **30** is currently tested in phase III on patients that have developed brain or ovarian tumors (Table 2) (Zambrano et al. 2005). Therefore, hydralazine **30** is a promising molecule in anti-cancer treatments.

Through a Scintillation Proximity-based high throughput Assay, Kilgore and colleagues discovered the 4-chloro-*N*-(4-hydroxynaphthalen-1-yl)-3-nitrobenzenesulfonamide SW155246 (**31**, Fig. 9) as a novel compound able to potently inhibit hDNMT1 ($IC_{50}$ vs hDNMT1 = 1.2 μM) with a moderate selectivity respect to the murine and human DNMT3A ($IC_{50}$ vs mDNMT3A = 38 μM). When evaluated in HeLa and A549 cancer cell lines, SW155246 **31** exhibited cell growth inhibition, and demonstrated to decrease the global methylation levels in HeLa cells and to reactivate the expression of RASSF1A tumor suppressor gene, normally methylated on its promoter in A549 cells. Moreover, this molecule proved to be not effective in decreasing DNMT1 protein levels nor in generating reactive oxygen species (ROS), thus providing an evidence of its ability to target specifically only the

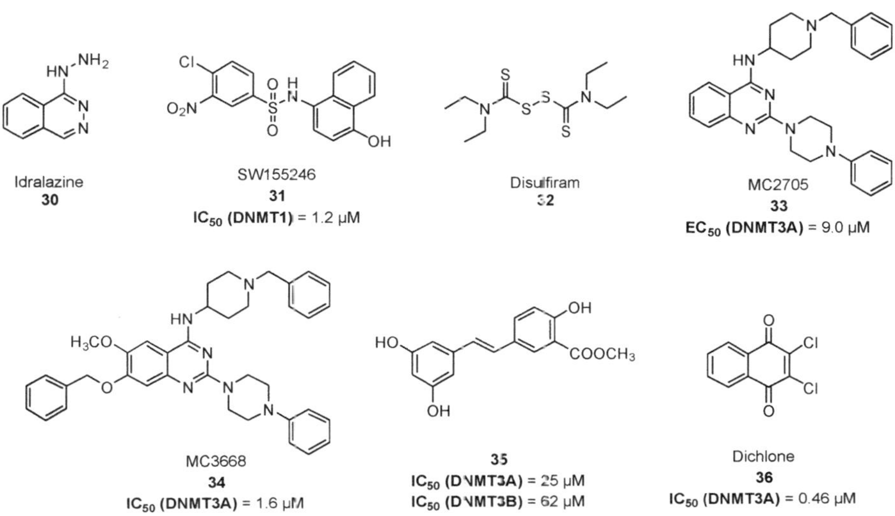


**Fig. 9** DNMT inhibitors with unknown mechanism of action and relative $IC_{50}$ value where available

enzymatic activity of DNMT1, even if the mechanism of action is not known (Kilgore et al. 2013).

Disulfiram (**32**, Fig. 9) is a known chemical agent used for the treatment of the alcohol dependence, which realizes its pharmacological effect by inhibiting the alcohol dehydrogenase. Its capability to interact with thiol groups prompted Lin et al. to test this compound on DNMTs also, thus achieving promising results. Indeed, disulfiram **32** inhibits the methyltransferase activity of DNMTs of 95% at the concentration of 200 μM and, above all, it is able to decrease the global methylation levels of both PC3 and CWR22Rv1 prostate cancer cells, showing also the ability to reactivate the expression of APC and RARβ genes in CWR22Rv1 and C4-2B malignant cells (Lin et al. 2011). Moreover, since disulfiram **32** manifests antiproliferative activity in different prostate cancer cell lines, it was evaluated in a clinical trial for its effect on recurrent prostate cancers (Table 2) (Fahy et al. 2012).

Starting from the G9a/GLP inhibitors BIX01294, Rotili et al. prepared a series of 6,7-desmethoxy quinazolines by maintaining the 1-benzyl-4-piperidinylamino moiety (typical of BIX01294) at the C4 position of the quinazoline ring and changing the substituent at the C2, spanning from differently sized cyclic amines (containing or not heteroatoms and/or lipophilic groups) to open-chain amines. Among the newly synthesized derivatives, compound MC2705 (**33**, Fig. 9), equipped with the *N*-phenyl-piperazine moiety at the C2, inhibited potently and selectively DNMT3A with an $IC_{50}$ value of $9 \pm 2$ μM. Docking studies on this compound show that probably it behaves as a SAM-competitive inhibitor, even if no competition studies have been performed to confirm this hypothesis (Rotili et al. 2014). A series of analogues of MC2705 **33** has been consequently developed and in particular, among these, the compound MC3668, here reported as **34** (Fig. 9), displayed an increased inhibitory potency against human DNMT3A compared with the parent derivative ($IC_{50}$ of 1.6 ± 0.4 μM). Interestingly, it resulted highly effective against drug sensitive 3D7 as well as multidrug resistant Dd2 and W2 *P. Falciparum* lines, with $IC_{50}$ values in the medium nanomolar range ($IC_{50}$ of 34.0 ± 18.6, 81.3 ± 44.3, 39.9 ± 9.9 nM respectively), but no clear evidence has been possible to achieve about the correspondence between the DNMT3A inhibition and the antimalarial effect (Bouchut et al. 2019).

In 2016, Aldawsari and coworkers reported the identification of a novel series of small molecules based on the hybridization of salicylate and resveratrol, with the most potent compound being the stilbene derivative **35** (Fig. 9). It exhibited $IC_{50}$ values of 25 and 62 μM against DNMT3A and DNMT3B respectively and displayed antiproliferative activity in the medium micromolar range in HT-29 colorectal cells, HepG2 liver cells and SK-BR-3 breast cancer cells (Aldawsari et al. 2016).

Finally, the fungicide dichlone **36** (Fig. 9) is reported to be a potent DNMT3A/3L inhibitor with an $IC_{50}$ of 0.46 μM and an appreciable antiproliferative activity against DU145 prostate cancer cells as demonstrated by the $TC_{50}$ value of 8 μM (Ceccaldi et al. 2013).

### Oligonucleotides

Another interesting approach is the development of oligonucleotides able to bind the DNMTs replacing the DNA sequence subjected of methylation. Examples of these are small RNA molecules rich of guanines which maximize the interaction with DNMTs (Pradhan et al. 2016), oligonucleotides containing DNMT nucleoside inhibitors replacing the cytosine of a CpG dinucleotide (Sledziewski et al. 2015), and microRNA like miR29b (Garzon et al. 2009).

The most promising oligonucleotide synthesized until now, MG98 **37** (TTCATGTCAGCCAAGGCCAC), entered in phase I clinical trials for the treatment of AML or MDS, but, unfortunately, the higher toxicity and the lack of efficacy induced the failure of the trials. With the aim to overcome these limits, MG98 **37** has been submitted to other studies based on specific combination with other agents such as those deriving from the association of MG98 **37** and interferon (IFN) for the treatment of advanced renal carcinoma. In this study, the combination MG98 **37**/IFN showed an interesting safety profile but low efficacy (Phase I, only 1 patient on 19 showed partial response) (Amato 2007; Klisovic et al. 2008; Amato et al. 2012; Aguilera et al. 2017).

# 4 Novel Inhibitory Strategies

Although the large number and different chemical type of non-nucleoside DNMT inhibitors, except hydralazine **30** and MG98 **37**, none of them has been studied for clinical application due to their low inhibitory activity in vivo. Hence, it is urgent to identify new inhibitory strategies.

Arimondo's group employed an alternative approach, never experimented before, for the design of DNMTi. They supposed that the design of compounds able to resemble the transition state of the DNMTs catalytic site could be a good strategy to obtain effective inhibition. Precisely, the catalytic site of DNMT1 and DNMT3A has three main protagonists: the cosubstrate SAM, the cytidine of DNA strand, that is the substrate, and the catalytic cysteine residue of the enzyme that triggers the reaction mechanism. Following a systematic medicinal chemistry optimization, the quinazoline was chosen as basal scaffold for the similarity with the adenine moiety of SAM, the insertion of hydrophobic moieties at the C-4 of the quinazoline resulted useful for the DNMT inhibition (Saavedra et al. 2009), and the quinoline ring, attached to the position 7 through constrained linkers, was selected to mimic the cytidine moiety. Among the resulting series of derivatives, described in both a patent and an article (Halby and Arimondo 2015; Halby et al. 2017), compounds **38** (Fig. 10) and **39** (Fig. 10) displayed the best results in terms of enzymatic and biological activity. These compounds present as cytosine mimicking group the quinoline nucleus, whose nitrogen atom must be in *para* position respect to the 2-aminoethyl-piperidinemethanol moiety that works as a linker between the quinoline and the C-7. At the position C4 of the basal scaffold they are equipped with the

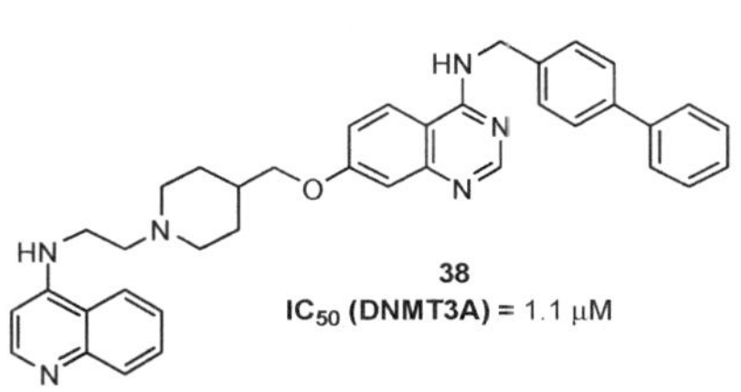

**38**
**$IC_{50}$ (DNMT3A)** = 1.1 μM

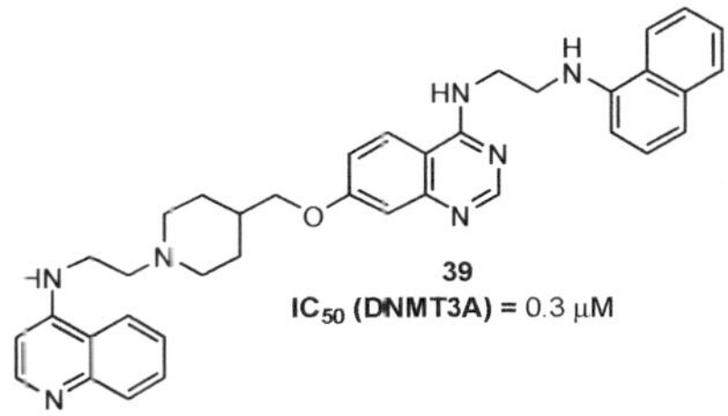

**39**
**$IC_{50}$ (DNMT3A)** = 0.3 μM

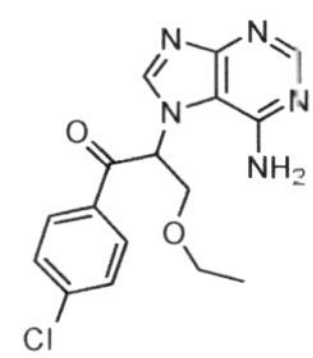

**40**
**$IC_{50}$ (DNMT3Acat)** = 2.[illegible] μM

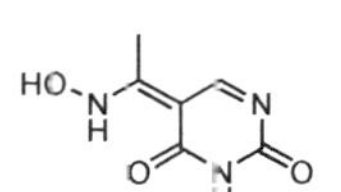

NSC232003
**41**

**Fig. 10** Structures of DNMTi developed through new inhibitory strategy and relative $IC_{50}$ value where available

biphenyl and the 4-(aminoethyl)naphthalene functions respectively. While derivative **39** ($IC_{50}$ vs DNMT3A of 0.3 μM) could not be screened because of its low solubility, compound **38** ($IC_{50}$ vs DNMT3A of 1.1 μM) showed to induce the re-expression of the luciferase reporter gene by demethylating at 52% the luciferase promoter, to determine the 87% of chromatin opening, and to reactivate the expression of *CDKN2A* in HCT116 cells in a time dependent way.

It is worth mentioning the discovery of a substrate competitive DNMT3A inhibitor, here called compound **40** (Fig. 10), which manifested an $IC_{50}$ of 2.1 μM against this isoform. In DMSO, it rapidly coverts in a Michael acceptor, thus being able to form a covalent bond with the catalytic cysteine; however, this chemical property makes compound **40** too little stable, so much to not allow its biological characterization (Erdmann et al. 2015b).

In 2016 Myrianthopoulos et al. published a work in which they reported a new strategy for the modulation of DNA methylation consisting in the identification of a PPI disruptor of DNMT1/UHRF1, called NSC232003 (**41**, Fig. 10). It is an uracil derivative bearing an oxime group at the C3 which binds the 5mC binding cavity of UHRF1 SRA domain and proved to decrease the 5mC content in MCF7 of 50% at the concentration of 15 μM (Myrianthopoulos et al. 2016).

## *4.1 Dual DNMT and HDAC or G9a Inhibitors*

In the last few years the pharmaceutical approach is moving toward the design and synthesis of molecules resulting from the chemical association of two different entities, each of which is capable of carrying out its own pharmacological action. As previously reported, the synergic effect related to the simultaneous inhibition of DNMT and HDAC enzymes, encouraged the design of hybrid molecules able to hit both the targets at the same time (Robertson et al. 2000; Rountree et al. 2000; Griffiths and Gore 2008; Hatada 2010; Pathania et al. 2016). Starting from the structure of the SAM competitive DNMTs inhibitor NSC-319745 **42** (Kuck et al. 2010b; Kabro et al. 2013), a series of hydroxamic acid derivatives was designed allowing the chelation of zinc ion and the consequent inhibition of HDAC enzymes without losing activity towards DNMTs. Among these, the most effective was **43** (Fig. 11), obtained as a result of different chemical modifications such as the substitution of 5-chloro-2-methoxybenzoic group with the 2-chloro-5-(trifluoromethyl)-phenyl (that is known to be a moiety able to improve the activity against DNMTs) (Kabro et al. 2013), the introduction of a *N*-hydroxycinnamide residue (typical of panobinostat and belinostat, two well-known HDACi), as well as, as already mentioned, the conversion of the carboxylic acid portion to the hydroxamic one. **43** showed $IC_{50}$ value in the nanomolar range against HDAC1 with a strong cytotoxic effect against human cancer cells K562 and U937. In addition, **43** increased the acetylation state of H4K8 and H3K9, inducing the re-expression of *p16* (through *p16* CpG island demethylation) and remarkable apoptosis in U937 cell lines (Yuan et al. 2017). Based on the evidence that the

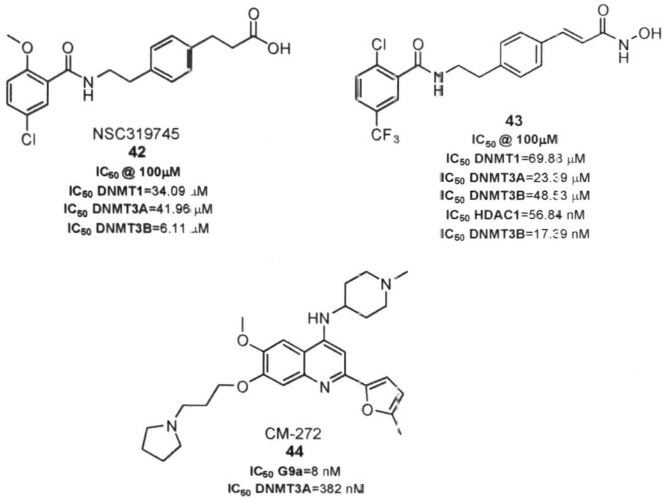


**Fig. 11** Structure of dual DNMT/HDAC or G9a Inhibitors and relative $IC_{50}$ value

histone H3K9-methyltransferase G9a physically interacts with DNMT1 to coordinate DNA and histone methylation during cell division (Estéve et al. 2006), a dual G9a/DNMT quinoline derivatives was individuated. CM-272 (**44**, Fig. 11) exhibited an $IC_{50}$ value in the nanomolar range ($IC_{50}$ vs G9a = 8 nM, $IC_{50}$ vs DNMT1 = 382 nM and $IC_{50}$ vs DNMT3A = 85 nM) with interesting pharmacokinetic and toxicity profiles. When tested in vivo, this compound proved to increase the overall survival of different hematological cancer cell lines like CEMO-1, MV-4-11 and OCI-Ly10 by inducing immunogenic cell death as a consequence of the activation of the type I IFN pathway determined by the inhibition of the methyltransferase activity of G9a and DNMTs. CM-272 **44** probably compete with the substrate or binds within the H3 binding site of DNMT enzymes, thus reducing as a final effect both DNA and H3K9me2 methylation levels (José-Enériz et al. 2017).

## 5 Combination Therapies

Despite once of the most commonly adopted approaches to hit non-responsive tumors consists in the increasing of radiation or chemotherapeutics dosage, in addition to the potential desired higher pharmacological response, also the development of side effects and resistance occurs. Currently, pharmacological research is moving toward the identification of agents able to induce a synergic effect if administered simultaneously. DNMTi show a pleiotropic effect in cancer cell (such as cytostatic, cytotoxic, apoptotic, differentiating and anti-angiogenic activity), hitting different key aspects typical of the tumor, suggesting the potential related to their use alone or in combination (Table 2) (Gravina et al. 2010).

Preclinical models highlighted the effectiveness of the association therapy derived from the co-administration of the cytosine mimicking derivatives and specific inhibitors of CDA such as tetrahydrouridine (THU) (**6**, Fig. 4). Given the lack of this evidence in superior preclinical models (for example monkeys) due to the low bioavailability of **6**, fluorinated THU derivatives were developed such as **6b** (Fig. 4), which showed better pharmacokinetic and pharmacodynamic profile compared to THU **6** (Ferraris et al. 2014). In addition, another fluorinated compound, knows as E-7727 (the exact chemical structure is not yet disclosed) is in phase I/II clinical trial in combination with 5azadC **2** (NCT02103478) for the treatment of MDS (Erdmann et al. 2015a). Several recent studies highlighted the capability of low doses of 5azaC **1** or 5azadC **2** to act synergistically with different cytotoxic agents overcoming the major limit of in chemotherapeutic treatment, the unavoidable chemoresistance (Cameron et al. 1999; Kristensen et al. 2009; Gravina et al. 2010). In particular the combination of 5azadC **2** with cytarabine (Ara-C) resulted in a phase II clinical trial for the treatment of older AML patients (NCT01829503). In addition, among the most promising association therapies it is worth mentioning those deriving from the association of the all-*trans* retinoic acid (ATRA) and 5azaC **1** or 5azadC **2**. Moreover, the triple association between 5azaC **1** or 5azadC **2**, ATRA and the HDACi valproic acid (VPA), has been evaluated resulting in phase II clinical trial for the treatment of AML (Xiang et al. 2014). A long pretreatment with low doses of 5azaC **1** is known to increase sensitivity to Doxorubicine treatment in chemoresistant-patients affected by diffuse large B-cell lymphoma (DLBCL) thanks to the capability of 5azaC **1** to induce a cellular reprogramming without toxicity in vivo (Clozel et al. 2013). The pretreatment (21–28 days) with CC-486 followed by the administration of first line chemotherapies (such as Gemcitabine or Abraxane) is in phase II clinical trial for the treatment of pancreatic cancer (NCT01845805). Another interesting example was represented by the association resulting from oral-5azaC CC486 and ABI-007 (albumin-bound formulation of paclitaxel) or carboplatin, now in phase I clinical trial for refractory or relapsed solid tumors. The association between 5azaC **1** and Camptothecin in human hepatoma cell line SMMC77221 increases the sensitization to the cytotoxic treatment (Ding et al. 2009). In addition, 5azadC **2** was tested in clinical trial to evaluate its capability to reprogram gene expression in AML in combination with Plerixafor, an inhibitor of stromal cell derived factor 1α (SDF-1α) and: 1 interferon α-2b, to stimulate immune response in renal cell carcinoma (phase II, NCT00561912), melanoma (phase I/II, NCT00791271) or metastatic solid tumor (phase I, NCT00701298); 2) Panitumumab (an anti-EGFR agent) in KRAS wild-type metastatic colorectal cancer to overcome resistance due to EGFR blocking agents (phase I, NCT00879385).

Furthermore, several studies showed the role of 5azadC **2** as radiation sensitizers (thanks to G2/M arrest and gene upregulation) in SaOS2, HOS, and U2OS osteosarcoma cells (Li et al. 2014) and MDA-MB-231 and MDA-MB-435 breast cancer cell lines (Wang et al. 2013) as well as lung cancer A549 and glioblastoma U373MG cells (Kim et al. 2012).

Several promising combinations were obtained from co-administration of epi-drugs. For example, since the treatment based on HDACi Trichostatin A (TSA) as a single agent resulted unable to induce the re-expression of *p15, p16 MLH1* and *TIMP3* genes, this goal is reached through a pretreatment with low doses of 5azadC **2** in colorectal carcinoma cell line (Cameron et al. 1999). In addition, the association of another FDA-approved HDACi, Vorinostat (SAHA) with 5azaC **1** or 5azadC **2** resulted in phase I/II clinical trials for the treatment of different blood tumors and in vivo studies showing a hopeful synergic mechanism in colon carcinoma cell lines (Yang et al. 2012). Worth of note are combinations between 5azaC **1** and other HDACi such as Mocetinostat, Entinostat, that resulted in phase I/II clinical trials for the treatment of hematological cancer (Fandy et al. 2009). In particular, the association of 5azaC **1** with Entinostat could be useful for the treatment of some solid tumors such as metastatic colorectal cancer (NCT01105377), advanced non-small cell lung cancer (NCT01886573) (Juergens et al. 2011) and advanced breast cancer (NCT01349959). The combination based on 5azadC **2** and the HDACi valproic acid (VPA) results in phase II clinical trials (NCT00414310) for the treatment of MDS and AML (Yang et al. 2005). In addition, the association composed by 5azadC **2** and Panobinostat (HDACi) was selected for the treatment of triple negative metastatic breast cancer tamoxifen/trastuzumab-resistant, resulting able to induce the re-expression of estrogen factors, restoring the efficiency of the treatment (phase I/II, NCT01194908).

To conclude, treatment with DNMTi induces an increase in the pharmacological response of cells to other anticancer drugs (such as immunomodulatory, chemotherapies agents or other epi-drugs) in both hematological and solid tumors thanks to their capability to restore cell functions hitting the system as a whole (Cameron et al. 1999; Azad et al. 2013; Ahuja et al. 2014, 2016).

## 6 Conclusion

Over the last two decades, numerous evidences have shown the importance of the role of epigenetic modifications in physio-pathological processes and in particular in the onset of cancer. The current use of agents able to reverse the epigenetic alterations has become a cornerstone of cancer research. In this framework, DNMTi can be considered a promising weapon to fight tumor due to their capability to induce the modulation of growth and differentiation, cell cycle arrest and apoptosis. To date, the only agents that received the approval for clinical use by FDA (5azaC **1** and 5azadC **2**) or results in advanced clinical trials (SGI-110 **4**) are covalent inhibitors of DNMTs. While this kind of inhibitors induced the maximum desirable inhibitory effect, their mechanism of action is also characterized by significant adverse reactions due to the lack of selectivity of action in terms of both cells type and DNMT isoforms. To overcome these limits, the oncological research is moving toward the design of non-nucleoside inhibitors that could guarantee a more selective therapeutic action. Unfortunately, none of the non-nucleoside inhibitors reported in literature reaches the clinical trials (with the exception of hydralazine **30** and MG98 **37**) due to their weak demethylation activity in cells as a consequence, first of all, of their poor solubility and cell penetration (Erdmann et al. 2015a). Based on the well-accepted concept which describes cancer as a multi-factorial disease, the idea of using combination therapies is very promising in particularly because DNMTi offers greatly improved access to DNA-protein complex for radiation or cytotoxic agents.

Therefore, the goal of the imminent future is, first of all, the design of powerful and selective non-nucleoside inhibitors able to induce a marked cell demethylation in order to better understand the role of each specific DNMT isoform with the aim to individuate the more favorable strategy to target DNA methylation.

## References

Adams A, Leong C, Denny WA et al (2005) Structures of two minor-groove-binding quinolinium quaternary salts complexed with d(CGCGAATTCGCG)2 at 1.6 and 1.8 Å resolution. Acta Crystallogr Sect D Biol Crystallogr 61:1348–1353

Aguilera OC, Depreux P, Halby L et al (2017) DNA methylation targeting: the DNMT/HMT crosstalk challenge. Biomolecules 7:1–21

Ahuja N, Easwaran H, Baylin SB (2014) Harnessing the potential of epigenetic therapy to target solid tumors. J Clin Invest 124:56–63

Ahuja N, Sharma AR, Baylin SB (2016) Epigenetic therapeutics: a new weapon in the war against cancer. Annu Rev Med 67:73–89

Aldawsari FS, Aguayo-ortiz R, Kapilashrami K et al (2016) Resveratrol-salicylate derivatives as selective DNMT3 inhibitors and anticancer agents. J Enzyme Inhib Med Chem 31:695–703

Alleman WG, Tabios RL, Chandramouli GVR et al (2004) The in vitro and in vivo effects of re-expressing methylated von Hippel-Lindau tumor suppressor gene in clear cell renal carcinoma with 5-Aza-2-deoxycytidine. Clin Cancer Res 10:7011–7021

Amato RJ (2007) Inhibition of DNA methylation by antisense oligonucleotide MG98 as cancer therapy. Clin Genitourin Cancer 5:422–426

Amato RJ, Stephenson J, Hotte S et al (2012) MG98, a second-generation DNMT1 inhibitor, in the treatment of advanced. Renal Cell Carcinoma 2:415–421

Asgatay S, Champion C, Marloie G et al (2014) Synthesis and evaluation of analogues of N-Phthaloyl-L-tryptophan RG108 as inhibitors of DNA methyltransferase 1. J Med Chem 57:421–434

Auclair G, Weber M (2012) Mechanisms of DNA methylation and demethylation in mammals. Biochimie 94:2202–2211

Azad N, Zahnow CA, Rudin CM et al (2013) The future of epigenetic therapy in solid tumours-lessons from the past. Nat Rev Clin Oncol 10:256–266

Ballestar E, Li T (2017) New insights into the epigenetics of inflammatory rheumatic diseases. Nat Rev Rheumatol 10:593–605

Bannister AJ, Zegerman P, Partridge JF et al (2001) Selective recognition of methylated lysine 9 on histone H3 by the HP1 chromo domain. Nature 410:120–124

Baubec T, Colombo DF, Wirbelauer C et al (2015) Genomic profiling of DNA methyltransferases reveals a role for DNMT3B in genic methylation. Nature 520:243–247

Baud MGJ, Leiser T, Haus P et al (2012) Defining the mechanism of action and enzymatic selectivity of psammaplin A against its epigenetic targets. J Med Chem 55:1731–1750

Berger SL, Kouzarides T, Shiekhattar R et al (2009) An operational definition of epigenetics. Genes Dev 23:781–783

Bergman AM, Kuiper CM, Noordhuis P et al (2004) Antiproliferative activity and mechanism of action of fatty acid derivatives of gemcitabine in leukemia and solid tumor cell lines and in human xenografts. Nucleosides Nucleotides Nucleic Acids 23:1329–1333

Borgel J, Guibert S, Li Y et al (2010) Targets and dynamics of promoter DNA methylation during early mouse development. Nat Genet 42:1093–1100

Bouchut A, Rotili D, Pierrot C et al (2019) Identification of novel quinazoline derivatives as potent antiplasmodial agents. Eur J Med Chem 161:277–291

Breistøl K, Balzarini J, Sandvold ML et al (1999) Antitumor activity of P-4055 (elaidic acid-cytarabine) compared to cytarabine in metastatic and s.c. human tumor xenograft models. Cancer Res 59:2944–2949

Bressler J, Shimmin LC, Boerwinkle E et al (2011) Global DNA methylation and risk of subclinical atherosclerosis in young adults: the Pathobiological Determinants of Atherosclerosis in Youth (PDAY) study. Atherosclerosis 219:958–962

Brueckner B, Boy RG, Siedlecki P et al (2005) Epigenetic reactivation of tumor suppressor genes by a novel small-molecule inhibitor of human DNA methyltransferases. Cancer Res 14:6305–6311

Brueckner B, Kuck D, Lyko F (2007) DNA methyltransferase inhibitors for cancer therapy. Cancer J 13:17–22

Brueckner B, Rius M, Markelova MR et al (2010) Delivery of 5-azacytidine to human cancer cells by elaidic acid esterification increases therapeutic drug efficacy. Mol Cancer Ther 9:1256–1264

Cameron EE, Bachman KE, Myöhänen S et al (1999) Synergy of demethylation and histone deacetylase inhibition in the re-expression of genes silenced in cancer. Nat Genet 21:103–107

Candelaria M, Herrera A, Labardini J et al (2011) Hydralazine and magnesium valproate as epigenetic treatment for myelodysplastic syndrome. Preliminary results of a phase-II trial. Ann Hematol 90:379–387

Castellano S, Kuck D, Sala M et al (2008) Constrained analogues of procaine as novel small molecule inhibitors of DNA methyltransferase-1. J Med Chem 51:2321–2325

Castellano S, Kuck D, Viviano M et al (2011) Synthesis and biochemical evaluation of Δ 2 -Isoxazoline derivatives as DNA methyltransferase 1 inhibitors. J Med Chem 54:7663–7677

Ceccaldi A, Rajavelu A, Champion C et al (2011) C5-DNA methyltransferase inhibitors: from screening to effects on zebrafish embryo development. Chembiochem 12:1337–1345

Ceccaldi A, Rajavelu A, Ragozin S et al (2013) Identification of novel inhibitors of DNA methylation by screening of a chemical library. ACS Chem Biol 8:543–548

Challen GA, Sun D, Jeong M et al (2012) Dnmt3a is essential for hematopoietic stem cell differentiation. Nat Genet 44:23–31

Champion C, Guianvarc'h D, Sénamaud-Beaufort C et al (2010) Mechanistic insights on the inhibition of C5 DNA methyltransferases by zebularine. PLoS One 5(8):e12388. https://doi.org/10.1371/journal.pone.0012388

Chen T, Hevi S, Gay F et al (2007) Complete inactivation of DNMT1 leads to mitotic catastrophe in human cancer cells. Nat Genet 39:391–396

Chen S, Wang Y, Zhou W et al (2014) Identifying novel selective non-nucleoside DNA methyltransferase 1 inhibitors through docking-based virtual screening. J Med Chem 57 (21):9028–9041

Chik F, Szyf M (2011) Effects of specific DNMT gene depletion on cancer cell transformation and breast cancer cell invasion; toward selective DNMT inhibitors. Carcinogenesis 32:224–232

Christman JK (2002) 5-Azacytidine and 5-aza-2′-deoxycytidine as inhibitors of DNA methylation: mechanistic studies and their implications for cancer therapy. Oncogene 21:5483–5495

Chuang JC (2005) Comparison of biological effects of non-nucleoside DNA methylation inhibitors versus 5-aza-2′-deoxycytidine. Mol Cancer Ther 4:1515–1520

Chuang JC, Warner SL, Vollmer D et al (2010) S110, a 5-aza-2′-deoxycytidine-containing dinucleotide, is an effective DNA methylation inhibitor in vivo and can reduce tumor growth. Mol Cancer Ther 9:1443–1450

Clozel T, Yang SN, Elstrom RL et al (2013) Mechanism-based epigenetic chemosensitization therapy of diffuse large B-cell lymphoma. Cancer Discov 3:1002–1019
Datta J, Ghoshal K, Denny WA et al (2009) A new class of quinoline-based DNA hypomethylating agents reactivates tumor suppressor genes by blocking DNA methyltransferase 1 activity and inducing its degradation. Cancer Res 69:4277–4285
Deng C, Lu Q, Zhang Z et al (2003) Hydralazine may induce autoimmunity by inhibiting extracellular signal - regulated kinase pathway signaling. Arthritis Rheum 48:746–756
Derissen EJB, Beijnen JH, Schellens JHM (2013) Concise drug review: azacitidine and decitabine. Oncologist 18:619–624
Dhe-paganon S, Syeda F, Park L (2011) Review article: DNA methyl transferase 1: regulatory mechanisms and im-plications in health and disease. Int J Biochem Mol Biol 2:58–66
Ding L, Qiu L, Zhang J et al (2009) Camptothecin-induced cell proliferation inhibition and apoptosis enhanced by DNA methyltransferase inhibitor, 5-aza-2′-deoxycytidine. Biol Pharm Bull 32:1105–1108
Ehrlich M (2002) DNA methylation in cancer: too much, but also too little. Oncogene 21:5400–5413
Erdmann A, Halby L, Fahy J et al (2015a) Targeting DNA methylation with small molecules: what's next? J Med Chem 58:2569–2583
Erdmann A, Menon Y, Gros C et al (2015b) Design and synthesis of new non nucleoside inhibitors of DNMT3A. Bioorg Med Chem 23:5946–5953
Erdmann A, Menon Y, Gros C et al (2016) Identification and optimization of hydrazone-gallate derivatives as specific inhibitors of DNA methyltransferase 3A. Future Med Chem 8 (4):373–380. https://doi.org/10.1016/B978-0-12-544952-6.50001-0
Esteller M (2007) Cancer epigenomics: DNA methylomes and histone-modification maps. Nat Rev Genet 8:286–298
Estéve P-O, Chin HG, Smallwood A et al (2006) Direct interaction between DNMT1 and G9a coordinates DNA and histone methylation during replication. Genes Dev 20:3089–3103
Fagan RL, Cryderman DE, Kopelovich L et al (2013) Laccaic acid A is a direct, DNA-competitive inhibitor of DNA. J Biol Chem 288:23858–23867
Fahy J, Jeltsch A, Arimondo PB (2012) DNA methyltransferase inhibitors in cancer: a chemical and therapeutic patent overview and selected clinical studies. Expert Opin Ther Pat 22(12):1427–1442
Fandy TE, Herman JG, Kerns P et al (2009) Early epigenetic changes and DNA damage do not predict clinical response in an overlapping schedule of 5-azacytidine and entinostat in patients with myeloid malignancies. Blood 114:2764–2773
Fang X, Robinson J, Wang-hu J et al (2015) cAMP induces hypertrophy and alters DNA methylation in HL-1 cardiomyocytes. Am J Physiol Cell Physiol 6:425–436
Ferraris D, Duvall B, Delahanty G et al (2014) Design, synthesis, and pharmacological evaluation of fluorinated tetrahydrouridine derivatives as inhibitors of cytidine deaminase. J Med Chem 57:2582–2588
Festuccia C, Gravina GL, Alessandro AMD et al (2009) Azacitidine improves antitumor effects of docetaxel and cisplatin in aggressive prostate cancer models. Endocr Relat Cancer 16 (2):401–413
Flotho C, Claus R, Batz C et al (2009) The DNA methyltransferase inhibitors azacitidine, decitabine and zebularine exert differential effects on cancer gene expression in acute myeloid leukemia cells. Leukemia 23:1019–1028
Fuso A, Nicolia V, Cavallaro RA et al (2011) DNA methylase and demethylase activities are modulated by one-carbon metabolism in Alzheimer's disease models. J Nutr Biochem 22:242–251
Garcia-Manero G, Gore SD, Cogle C et al (2011) Phase I study of oral azacitidine in myelodysplastic syndromes, chronic myelomonocytic leukemia, and acute myeloid leukemia. J Clin Oncol 29:2521–2527
Garzon R, Heaphy CEA, Havelange V et al (2009) MicroRNA 29b functions in acute myeloid leukemia. Blood 114:5331–5342

Gaudet F (2003) Induction of tumors in mice by genomic hypomethylation. Science 300:489–492
Gausachs M, Mur P, Corral J et al (2012) MLH1 promoter hypermethylation in the analytical algorithm of Lynch syndrome: a cost-effectiveness study. Eur J Hum Genet 20(7):762
Ghoshal K, Datta J, Majumder S et al (2005) 5-Aza-deoxycytidine induces selective degradation of DNA methyltransferase 1 by a proteasomal pathway that requires the KEN box, bromo-adjacent homology domain, and nuclear localization signal. Mol Cell Biol 25:4727–4741
Giraldo AM, Decourcy K, Ball SF et al (2013) Gene expression of Dnmt1 isoforms. Cell Reprogram 15:309–321
Godert AM, Angelino N, Woloszynska-Read A et al (2006) An improved synthesis of psammaplin A. Bioorganic Med Chem Lett 16:3330–3333
Gordian E, Ramachandran K, Singal R (2009) Methylation mediated silencing of TMS1 in breast cancer and its potential contribution to docetaxel cytotoxicity. Anticancer Res 29:3207–3210
Goyal R, Reinhardt R, Jeltsch A (2006) Accuracy of DNA methylation pattern preservation by the Dnmt1 methyltransferase. Nucleic Acids Res 34:1182–1188
Gravina GL, Festuccia C, Popov VM et al (2010) Biological rationale for the use of DNA methyltransferase inhibitors as new strategy for modulation of tumor response to chemotherapy and radiation. Mol Cancer 9:305
Grégoire R, Fleury L, Céline F et al (2017) Inhibition studies of DNA methyltransferases by maleimide derivatives of RG108 as non-nucleoside inhibitors. Future Med Chem 9:533–536
Griffiths EA, Gore SD (2008) DNA methyltransferase and histone deacetylase inhibitors in the treatment of myelodysplastic syndromes. Semin Hematol 45:23–30
Gros C, Fahy J, Halby L et al (2012) DNA methylation inhibitors in cancer: recent and future approaches. Biochimie 94:2280–2296
Gros C, Fleury L, Nahoum V et al (2015) New insights on the mechanism of quinoline-based DNA methyltransferase inhibitors. J Biol Chem 290:6293–6302
Guo X, Wang L, Li J et al (2014) Structural insight into autoinhibition and histone H3-induced activation of DNMT3A. Nature 517:640–644
Halby L, Arimondo PB (2015) Quinazoline derivatives and their use as DNA methyltransferase inhibitors. WO2015040169A1
Halby L, Champion C, Sénamaud-Beaufort C et al (2012) Rapid synthesis of new DNMT inhibitors derivatives of procainamide. Chembiochem 13:157–165
Halby L, Menon Y, Rilova E et al (2017) Rational design of bisubstrate-type analogues as inhibitors of DNA methyltransferases in cancer cells. J Med Chem 60:4665–4679
Hatada I (2010) The epigenomics of cancer. In: Cho WC (ed) An omics perspective on cancer research. Springer, Dordrecht
Higuchi F, Uchida S, Yamagata H et al (2011) State-dependent changes in the expression of DNA methyltransferases in mood disorder patients. J Psychiatr Res 45:1295–1130
Hon GC, Hawkins RD, Caballero OL et al (2012) Global DNA hypomethylation coupled to repressive chromatin domain formation and gene silencing in breast cancer. Genome Res 22:246–258
Hoshino O, Murakata M, Yamada K (1992) Synthesis of a bromotyrosine derived C02H. Bioorg Med Chem Lett 2:1561–1562
Issa JPJ, Kantarjian HM (2009) Targeting DNA methylation. Clin Cancer Res 15:3938–3946
Jia D, Jurkowska RZ, Zhang X et al (2007) Structure of Dnmt3a bound to Dnmt3L suggests a model for de novo DNA methylation. Nature 449:248–251
Jones PA, Liang G (2009) Rethinking how DNA methylation patterns are maintained. Nat Rev Genet 10:805–811
Jones PA, Taylor SM (1980) Cellular differentiation, cytidine analogs and DNA methylation. Cell 20:85–93
José-Enériz ES, Agirre X, Rabal O et al (2017) Discovery of first-in-class reversible dual small molecule inhibitors against G9a and DNMTs in hematological malignancies. Nat Commun 8:15424. https://doi.org/10.1038/ncomms15424

Juergens RA, Wrangle J, Vendetti FP et al (2011) Combination epigenetic therapy has efficacy in patients with refractory advanced non-small cell lung cancer. Cancer Discov 1:598–607
Jurkowska RZ, Jurkowski TP, Jeltsch A (2011) Structure and function of mammalian DNA methyltransferases. Chembiochem 12:206–222
Kabro A, Lachance H, Marcoux-Archambault I et al (2013) Preparation of phenylethylbenzamide derivatives as modulators of DNMT3 activity. Medchemcomm 4:1562–1570
Karahoca M, Momparler RL (2013) Pharmacokinetic and pharmacodynamic analysis of 5-aza-2-′-deoxycytidine (decitabine) in the design of its dose-schedule for cancer therapy. Clin Epigenetics 5:3
Karsli-Ceppioglu S (2016) Epigenetic mechanisms in psychiatric diseases and epigenetic therapy. Drug Dev Res 77(7):407–413
Kilgore JA, Du X, Melito L et al (2013) Identification of DNMT1 selective antagonists using a novel scintillation proximity assay. J Biol Chem 288:19673–19684
Kim JS, Han J, Shim YM et al (2005) Aberrant methylation of H-cadherin (CDH13) promoter is associated with tumor progression in primary nonsmall cell lung carcinoma. Cancer 104:1825–1833
Kim HJ, Kim JH, Chie EK et al (2012) DNMT (DNA methyltransferase) inhibitors radiosensitize human cancer cells by suppressing DNA repair activity. Radiat Oncol 7:39
Kim AY, Park YJ, Pan X et al (2015) Obesity-induced DNA hypermethylation of the adiponectin gene mediates insulin resistance. Nat Commun 6:1–11
Klimasauskas S, Kumar S, Roberts RJ et al (1994) Hhal methyltransferase flips its target base out of the DNA helix. Cell 76:357–369
Klisovic RB, Stock W, Cataland S et al (2008) A phase I biological study of MG98, an oligodeoxynucleotide antisense to DNA methyltransferase 1, in patients with high-risk myelodysplasia and acute myeloid leukemia. Clin Cancer Res 14:2444–2449
Kristensen LS, Nielsen HM, Hansen LL (2009) Epigenetics and cancer treatment. Eur J Pharmacol 625:131–142
Kuck D, Caulfield T, Lyko F et al (2010a) Nanaomycin A selectively inhibits DNMT3B and reactivates silenced tumor suppressor genes in human cancer cells. Mol Cancer Ther 9:3015–3023
Kuck D, Singh N, Lyko F et al (2010b) Novel and selective DNA methyltransferase inhibitors: docking-based virtual screening and experimental evaluation. Bioorg Med Chem 18:822–829
Kundakovic M, Chen Y, Costa E, Grayson DR (2007) DNA methyltransferase inhibitors coordinately induce expression of the human reelin and glutamic acid decarboxylase 67 genes. Mol Pharmacol 71:644–653
Lee W, Shim J, Zhu B (2005) Mechanisms for the inhibition of DNA methyltransferases by tea catechins and bioflavonoids. Mol Pharmacol 68:1018–1030
Lemaire M, Momparler LF, Raynal NJM et al (2009) Inhibition of cytidine deaminase by zebularine enhances the antineoplastic action of 5-aza-2-deoxycytidine. Cancer Chemother Pharmacol 63:411–416
Li E, Bestor TH, Jaenisch R (1992) Targeted mutation of the DNA methyltransferase gene results in embryonic lethality. Cell 69:915–926
Li Q, Ahuja N, Burger PC et al (1999) Methylation and silencing of the Thrombospondin-1 promoter in human cancer. Oncogene 18(21):3284–3289
Li Y, Zhao M, Yin H et al (2010) Overexpression of the growth arrest and DNA damage – induced 45 gene contributes to autoimmunity by promoting DNA demethylation in lupus T cells. Arthritis Rheum 62:1438–1447
Li Y, Geng PL, Jiang W et al (2014) Enhancement of radiosensitivity by 5-Aza-CdR through activation of G2/M checkpoint response and apoptosis in osteosarcoma cells. Tumor Biol 35:4831–4839
Lin J, Haffner MC, Zhang Y et al (2011) Disulfiram is a DNA demethylating agent and inhibits prostate cancer cell growth. Prostate 71:333–343

Liu Z, Xie Z, Jones W et al (2009) Curcumin is a potent DNA hypomethylation agent. Bioorg Med Chem Lett 19:706–709

Lopez M, Halby L, Arimondo PB (2016) DNA methyltransferase inhibitors: development and applications. In: Jeltsch A, Jurkowska RZ (eds) DNA methyltransferases - role and function. Springer, Cham

López-Pedrera C, Pérez-Sánchez C, Ramos-Casals M et al (2012) Cardiovascular risk in systemic autoimmune diseases: epigenetic mechanisms of immune regulatory functions. Clin Dev Immunol 2012:974648. https://doi.org/10.1155/2012/974648

Luczak MW, Jagodziński PP (2006) The role of DNA methylation in cancer development. Folia Histochem Cytobiol 44:143–154

Lujambio A, Portela A, Liz J et al (2010) CpG island hypermethylation-associated silencing of non-coding RNAs transcribed from ultraconserved regions in human cancer. Oncogene 29:6390–6401

Mai A, Altucci L (2009) Epi-drugs to fight cancer: from chemistry to cancer treatment, the road ahead. Int J Biochem Cell Biol 41:199–213

Marcucci G, Silverman L, Eller M et al (2005) Bioavailability of azacitidine subcutaneous versus intravenous in patients with the myelodysplastic syndromes. J Clin Pharmacol 45:597–602

Medina-franco JL, Yee J (2013) Docking of a novel DNA methyltransferase inhibitor identified from high- throughput screening: insights to unveil inhibitors in chemical databases. Mol Divers 17(2):337–344

Medina-Franco JL, Lopez-Vallejo F, Kuck D et al (2011) Natural products as DNA methyltransferase inhibitors: a computer-aided discovery approach. Mol Divers 15:293–304

Michie AM, McCaig AM, Nakagawa R et al (2010) Death-associated protein kinase (DAPK) and signal transduction: regulation in cancer. FEBS J 277:74–80

Miller G (2010) The seductive allure of behavioral epigenetics. Science 329:24–27

Myrianthopoulos V, Cartron PF, Klimašauskas S et al (2016) Tandem virtual screening targeting the SRA domain of UHRF1 identifies a novel chemical tool modulating DNA methylation. Eur J Med Chem 114:390–396. https://doi.org/10.1016/j.ejmech.2016.02.043

Nagaraju GP, Zhu S, Wen J et al (2013) Novel synthetic curcumin analogues EF31 and UBS109 are potent DNA hypomethylating agents in pancreatic cancer. Cancer Lett 341:195–203

Pathania R, Ramachandran S, Mariappan G et al (2016) Combined inhibition of DNMT and HDAC blocks the tumorigenicity of cancer stem-like cells and attenuates mammary tumor growth. Cancer Res 76:3224–3235

Pechalrieu D, Etievant C, Arimondo PB (2017) DNA methyltransferase inhibitors in cancer: from pharmacology to translational studies. Biochem Pharmacol 129:1–13

Piña IC, Gautschi JT, Wang G-Y-S et al (2003) Psammaplins from the sponge *Pseudoceratina purpurea*: inhibition of both histone deacetylase and DNA methyltransferase. J Org Chem 68:3866–3873

Pradhan S, Esteve P-O, Zhang G (2016) Dnmt inhibitors. Patent US20160272977A1

Qin W, Zhang K, Clarke K et al (2014) Methylation and miRNA effects of resveratrol on mammary tumors vs. normal tissue. Nutr Cancer 66:270–277

Reik W (2007) Stability and flexibility of epigenetic gene regulation in mammalian development. Nature 447:425–432

Robertson KD (2005) DNA methylation and human disease. Nat Rev Genet 6:597–610

Robertson KD, Ait-Si-Ali S, Yokochi T et al (2000) DNMT1 forms a complex with RB, E2F1 and HDAC1 and represses transcription from E2F-responsive promoters. Nat Genet 25:338–342

Rondelet G, Wouters J (2017) Human DNA (cytosine-5)-methyltransferases: a functional and structural perspective for epigenetic cancer therapy. Biochimie 139:137–147

Rotili D, Tarantino D, Marrocco B et al (2014) Properly substituted analogues of BIX-01294 lose inhibition of G9a histone methyltransferase and gain selective anti-DNA methyltransferase 3A activity. PLoS One 9:1–9

Rountree MR, Bachman KE, Baylin SB (2000) DNMT1 binds HDAC2 and a new co-repressor, DMAP1, to form a complex at replication foci. Nat Genet 25:269–277

Ruthenburg AJ, Li H, Patel DJ, David Allis C (2007) Multivalent engagement of chromatin modifications by linked binding modules. Nat Rev Mol Cell Biol 8(12):983–994
Saavedra OM, Isakovic L, Llewellyn DB et al (2009) SAR around (l)-S-adenosyl-l-homocysteine, an inhibitor of human DNA methyltransferase (DNMT) enzymes. Bioorganic Med Chem Lett 19:2747–2751
Sacconi S, Camaño P, de Greef JC et al (2012) Patients with a phenotype consistent with facioscapulohumeral muscular dystrophy display genetic and epigenetic heterogeneity. J Med Genet 49:41–46
Segura-Pacheco B, Trejo-Becerril C, Perez-Cardenas E et al (2003) Reactivation of tumor suppressor genes by the cardiovascular drugs hydralazine and procainamide and their potential use in cancer therapy. Clin Cancer Res 9:1596–1603
Shen L, Kantarjian H, Guo Y et al (2010) DNA methylation predicts survival and response to therapy in patients with myelodysplastic syndromes. J Clin Oncol 28:605–613
Siedlecki P, Garcia Boy R, Musch T et al (2005) Discovery of two novel, small-molecule inhibitors of DNA methylation. J Med Chem 49:678–683
Sledziewski A, Devos T, Kole R (2015) Oligonucleotide, inhibitors of DNA methyltransferase and their use in treating disease. Patent US20150167004A1
Smallwood SA, Tomizawa SI, Krueger F et al (2011) Dynamic CpG island methylation landscape in oocytes and preimplantation embryos. Nat Genet 43:811–814
Song J, Rechkoblit O, Bestor TH et al (2010) Structure of DNMT1-DNA complex reveals a role for autoinhibition in maintenance DNA methylation. Science 331:1036–1040
Stresemann C, Brueckner B, Musch T et al (2006) Functional diversity of DNA methyltransferase inhibitors in human cancer cell lines. Cancer Res 66:2794–2800
Sulewska A, Niklińska W, Kozłowski M et al (2007) DNA methylation in states of cell physiology and pathology. Folia Histochem Cytobiol 45:149–158
Sun B, Hu L, Luo Z et al (2016) DNA methylation perspectives in the pathogenesis of autoimmune diseases. Clin Immunol 164:21–27
Suzuki K, Suzuki I, Leodolter A et al (2006) Global DNA demethylation in gastrointestinal cancer is age dependent and precedes genomic damage. Cancer Cell 9:199–207
Suzuki T, Tanaka R, Hamada S et al (2010) Design, synthesis, inhibitory activity, and binding mode study of novel DNA methyltransferase 1 inhibitors. Bioorg Med Chem Lett 20:1124–1127
Takashima S, Takehashi M, Lee J et al (2009) Abnormal DNA methyltransferase expression in mouse germline stem cells results in spermatogenic defects. Biol Reprod 81:155–164
Tian F, Hu X-Z, Wu X et al (2009) Dynamic chromatin remodeling events in hippocampal neurons are associated with NMDA receptor-mediated activation of Bdnf gene promoter 1. J Neurochem 109(5):1375–1388
Valente S, Liu Y, Schnekenburger M et al (2014) Selective non-nucleoside inhibitors of human DNA methyltransferases active in cancer including in cancer stem cells. J Med Chem 57:701–713
van Bemmel DM, Brank AS, Eritja R et al (2009) DNA (Cytosine-C5) methyltransferase inhibition by oligodeoxyribonucleotides containing 2-(1H)-pyrimidinone (zebularine aglycon) at the enzymatic target site. Biochem Pharmacol 78:633–641
Veldic M, Caruncho HJ, Liu WS et al (2003) DNA-methyltransferase 1 mRNA is selectively overexpressed in telencephalic GABAergic interneurons of schizophrenia brains. Proc Natl Acad Sci USA 1:348–353
Villar-Garea A, Fraga MF, Espada J et al (2003) Procaine is a DNA-demethylating agent with growth-inhibitory effects in human cancer cells. Cancer Res 63:4984–4989
Walter ED (1941) Genistin (an isoflavone glucoside) and its aglucone, genistein, from soybeans. J Am Chem Soc 63:3273–3276
Wang L, Zhang Y, Li R et al (2013) 5-aza-2′-Deoxycytidine enhances the radiosensitivity of breast cancer cells. Cancer Biother Radiopharm 28:34–44
Weber M, Hellmann I, Stadler MB et al (2007) Distribution, silencing potential and evolutionary impact of promoter DNA methylation in the human genome. Nat Genet 39:457–466
Wijermans PW, Ruter B, Baer MR et al (2008) Efficacy of decitabine in the treatment of patients with chronic myelomonocytic leukemia (CMML). Leuk Res 32:587–591

Wu H, Coskun V, Tao J et al (2010) Dnmt3a-dependent nonpromoter DNA methylation facilitates transcription of neurogenic genes. Science 329:444–447

Xiang L, Dong W, Wang R et al (2014) All-trans retinoic acid enhances the effect of 5-aza-2-′-deoxycytidine on p16INK4a demethylation, and the two drugs synergistically activate retinoic acid receptor β gene expression in the human erythroleukemia K562 cell line. Oncol Lett 8:117–122

Xu J, Watts JA, Pope SD et al (2009) Transcriptional competence and the active marking of tissue-specific enhancers by defined transcription factors in embryonic and induced pluripotent stem cells. Genes Dev 23:2824–2838

Xu F, Mao C, Ding Y et al (2010) Molecular and enzymatic profiles of mammalian DNA methyltransferases: structures and targets for drugs. Curr Med Chem 17:4052–4071

Yang AS, Estedo MR, Garcia-Manero G et al (2003) Comment on "Chromosomal instability and tumors promoted by DNA hypomethylation" and "Induction of tumors in mice by genomic hypomethylation". Science 302:1153

Yang H, Hoshino K, Sanchez-Gonzalez B et al (2005) Antileukemia activity of the combination of 5-aza-2′-deoxycytidine with valproic acid. Leuk Res 29:739–748

Yang CS, Fang M, Lambert JD et al (2008) Reversal of hypermethylation and reactivation of genes by dietary polyphenolic compounds. Nutr Rev 66:18–20

Yang CS, Wang X, Lu G et al (2009) Cancer prevention by tea: animal studies, molecular mechanisms and human relevance. Nat Rev Cancer 9:429–439

Yang D, Torres CM, Bardhan K et al (2012) Decitabine and vorinostat cooperate to sensitize colon carcinoma cells to fas ligand-induced apoptosis in vitro and tumor suppression in vivo. J Immunol 188:4441–4449

Ye Y, Stivers JT (2010) Fluorescence-based high-throughput assay for human DNA. Anal Biochem 401:168–172

Yoo CB, Jeong S, Egger G et al (2007) Delivery of 5-aza-2′-deoxycytidine to cells using oligodeoxynucleotides. Cancer Res 67:6400–6408

Yuan Z, Sun Q, Li D et al (2017) Design, synthesis and anticancer potential of NSC-319745 hydroxamic acid derivatives as DNMT and HDAC inhibitors. Eur J Med Chem 134:281–292

Zambrano P, Segura-Pacheco B, Perez-Cardenas E et al (2005) A phase I study of hydralazine to demethylate and reactivate the expression of tumor suppressor genes. BMC Cancer 5:44

Zhao M, Tang J, Gao F et al (2010) Hypomethylation of IL10 and IL13 promoters in CD4 + T cells of patients with systemic lupus erythematosus. J Biomed Biotechnol 2010:1–9

# Functions and Dynamics of Methylation in Eukaryotic mRNA

**Mingjia Chen and Claus-Peter Witte**

## Contents



**Abstract** Eukaryotic messenger RNA (mRNA) contains non-canonical nucleosides, which are modified mostly by methylation. Although some modifications are known for decades, advances in high-throughput sequencing and mass spectrometric techniques now have allowed to elucidate transcriptome wide methylation patterns. The discovery of methyltransferases that write and demethylases that erase methylations in a sequence-specific manner, as well as reader proteins that recognize these modifications leading to a specific biological response, has triggered wide

M. Chen (✉)
Department of Molecular Nutrition and Biochemistry of Plants, Institute of Plant Nutrition, Leibniz University Hannover, Hannover, Germany

College of Life Sciences, Nanjing Agricultural University, Nanjing, P.R. China
e-mail: mjchen@njau.edu.cn

C.-P. Witte (✉)
Department of Molecular Nutrition and Biochemistry of Plants, Institute of Plant Nutrition, Leibniz University Hannover, Hannover, Germany
e-mail: cpwitte@pflern.uni-hannover.de

 
S. Jurga, J. Barciszewski (eds.), *The DNA, RNA, and Histone Methylomes*, RNA Technologies, https://doi.org/10.1007/978-3-030-14792-1_13

attention converting the research field of mRNA methylation into a current hotspot in molecular biology.

Most research has focussed on $N^6$-methyladenosine ($m^6A$), which is the most abundant modification in eukaryotic mRNA. Therefore, this overview has a focus on $m^6A$ summarizing the current knowledge on how specific $m^6A$ patterns are generated and how they are recognized and translated into biological outputs like alternative splicing, altered transcript stability, or modified translational activity of mRNAs. The distribution patterns of other methylations in mRNA, such as $N^1$-methyladenosine ($m^1A$), 5-methylcytidine ($m^5C$) and 5-hydroxymethylcytidine ($hm^5C$) have also been mapped in recent years. We review the current knowledge regarding these and other minor eukaryotic mRNA methylations and provide an outlook suggesting potential future research directions.

**Keywords** Methyltransferase · Demethylase · $N^6$-mAMP · 2′-O-methylation ($N_m$) · 7-methylguannosine ($m^7G$) · 3-methylcytidine ($m^3C$) · $N^4$-acetylcytidine ($ac^4C$) · YTH family proteins · $N^6$-methyl AMP deaminase (MAPDA)

## 1 Introduction

In addition to the four canonical nucleosides adenosine (A), guanosine (G), cytosine (C) and uridine (U), over 150 chemical nucleoside modifications have been discovered in RNA species of all kingdoms of life. Comparatively few modified nucleosides, mostly altered by methylation, are known in eukaryotic messenger RNA (mRNA), including $N^6$-methyladenosine ($m^6A$), 5-methylcytidine ($m^5C$), $N^1$-methyladenosine ($m^1A$), 2′-O-ribose methylation of nucleosides with a canonical base (Nm), $N^7$-methylguanosine ($m^7G$), 3-methylcytidine ($m^3C$), and $N^4$-acetylcytidine ($ac^4C$) (Fig. 1). Although the latter is an acetylation, we will also review the very recent discovery of this modification in eukaryotic mRNA. Some of these modifications are known to be reversible and have essential roles in regulating gene expression. In this chapter, we review the function and dynamics of these eukaryotic mRNA methylations, focusing on the most abundant mRNA methylation, $m^6A$, as well as summarizing the current knowledge concerning other methylations.

## 2 The Functions and Dynamics of $N^6$-methyladenosine ($m^6A$)

$N^6$-methyladenosine is the most prevalent (0.4–0.6% of adenosine is modified in mammalian and plant mRNA) and currently best studied base modification present in the mRNA of all eukaryotes, including yeast (Bodi et al. 2010), plants (Zhong et al. 2008), flies (Hongay and Orr-Weaver 2011) and mammals (Horowitz et al. 1984; Schibler et al. 1977; Yoon et al. 2017). Although it was already discovered in

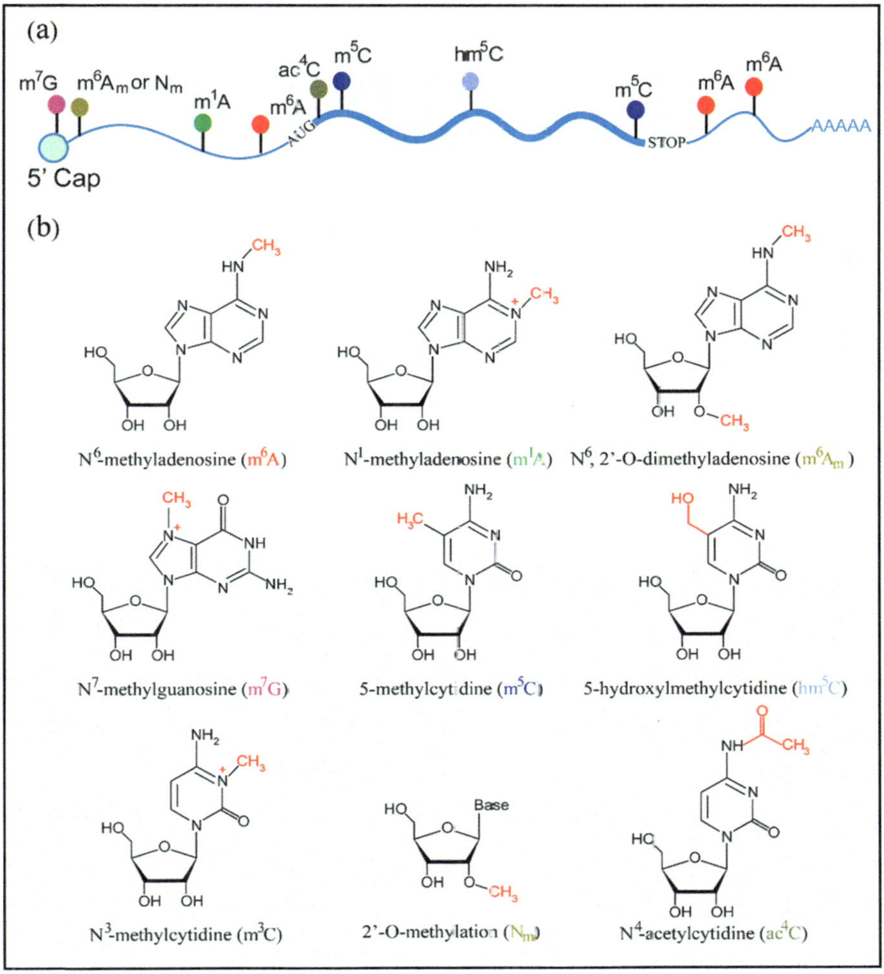


**Fig. 1** Currently known nucleoside methylations, hydroxymethylations, and acetylations in eukaryotic mRNA. (**a**) The typical positions of the respective modified nucleosides in eukaryotic mRNA. Note that the typical distribution of m$^3$C is not yet known. (**b**) The chemical structures of methylated, hydroxymethylated, and acetylated nucleosides found in mRNA

mammalian cells in 1974 (Desrosiers et al. 1974; Perry and Kelley 1974), transcriptome-wide distribution of m$^6$A was poorly characterized until 2012, when it was mapped on the whole transcriptome for the first time by antibody-mediated immunoprecipitation coupled with high-throughput sequencing (Dominissini et al. 2012). In human, over 12,000 m$^6$A sites were unveiled in more than 7000 transcripts, clustered around stop codons, 3′ untranslated regions (3′ UTRs), and within long internal exons (Fig. 1). The modification occurs at the consensus RNA motif RRm$^6$ACH (R = G or A; H = U or A or C) (Dominissini et al. 2012; Meyer et al. 2012). The dynamics and molecular functions of m$^6$A as a novel post-transcriptional

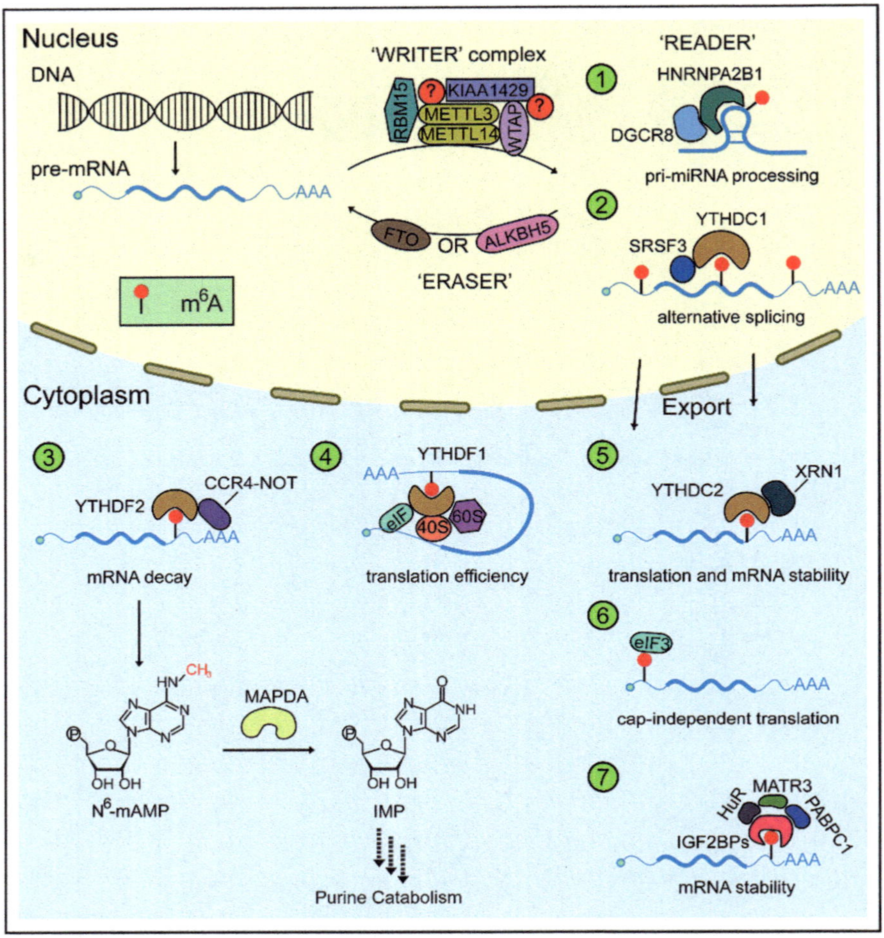


**Fig. 2** The dynamics and functions of m$^6$A in eukaryotic mRNA. m$^6$A formation is catalyzed by the writer protein complex at specific sites in pre-mRNA. Marks can be removed by easer proteins. Reader proteins recognize m$^6$A sites and affect various processes including pri-miRNA processing, alternative splicing, translation, and mRNA stability (①–⑦). The N$^6$-mAMP released from m$^6$A-RNA turnover is catabolized by MAPDA to the canonical nucleotide inosine monophosphate (IMP). All abbreviations for proteins are explained in the text

regulator of gene expression have been established since 2011, when the first m$^6$A demethylase FTO (fat mass and obesity-associated protein) was identified in mammalian cells (Jia et al. 2011). In the nucleus, m$^6$A is installed and removed by so called 'writer' and 'eraser' proteins, respectively. The mature m$^6$A mRNAs are decoded by 'reader' proteins to mediate various biological processes, such as alternative splicing, mRNA translation, and mRNA stability. N$^6$-mAMP is released from these methylated mRNAs upon their degradation, and is catabolized by N$^6$-methyl AMP deaminase (MAPDA) to inosine monophosphate (IMP), which is a metabolite of the canonical purine nucleotide metabolism (Fig. 2).

## 2.1 The Writer Complex: $m^6A$ Methyltransferase Complex

The installation of $m^6A$ occurs in the nucleus by a methyltransferase complex conserved in eukaryotes (Fig. 1) with an approximate molecular weight of 875 kDa (Bokar et al. 1994). The complex contains two core components, Methyltransferase Like 3 (METTL3) (Bokar et al. 1997), and METTL14 (Liu et al. 2014; Wang et al. 2016b), and several accessory proteins like Wilms Tumor 1 Associated Protein (WTAP) (Ping et al. 2014), and RNA Binding Motif Protein 15 (RBM15) (Patil et al. 2016), and KIAA1429 (Schwartz et al. 2014).

In the search for proteins responsible for the $m^6A$ formation, a 70-kDa fraction from partial protein purification was characterized, which exhibited S-adenosylmethionine (SAM)-binding activity (Bokar et al. 1994). Purification of the corresponding recombinant protein identified a $m^6A$ methyltransferase, originally termed MT-A70 but later renamed METTL3. Mutation of the *METTL3* ortholog in Arabidopsis (here called adenosine methylase A, MTA) leads to the complete loss of $m^6A$ in mRNA resulting in failure of the developing embryo to progress past the globular stage (Zhong et al. 2008). Also in other organisms METTL3 has critical functions. In *Saccharomyces cerevisiae*, it is involved in the regulation of cell fate decision during sporulation and meiosis (Agarwala et al. 2012; Bodi et al. 2010; Clancy et al. 2002), while in *Drosophila melanogaster* it modulates neuronal functions and sex determination through alternative splicing of the *Sxl* gene (Haussmann et al. 2016; Lence et al. 2016). METTL3 localizes in the nuclear speckles in plant and mammalian cells.

METTL3 belongs to a larger family of SAM-dependent methyltransferases that is highly conserved in mammals (Schapira 2016). METTL14, another family member, shares 43% identity with METTL3 (Liu et al. 2014). Interestingly, three independent crystallization studies demonstrated that METTL14 has no methyltransferase activity as it lacks a SAM-binding domain (Śledź and Jinek 2016; Wang et al. 2016a, b). It is now known, that METTL3 and METTL14 form a stable heterodimer in the $m^6A$ methyltransferase complex. Crystal structures of the METTL3 and METTL14 complex showed that only METTL3 binds SAM while METTL14 appears to assist in RNA binding in concert with METTL3 and enhances the enzymatic activity of METTL3 (Wang et al. 2016b). Abrogation of *METTL14* in mouse decreases $m^6A$ abundance in mRNAs of embryonic stem cells, which blocks stem cell self-renewal and differentiation leading to early embryonic lethality (Geula et al. 2015; Wang et al. 2014a). The Arabidopsis ortholog of METTL14, MTB (adenosine methylase B), is also a core member of the methyltransferase complex. Its absence results in an arrest of embryonic development at the globular stage and reduces the $m^6A/A$ ratio from 1.6% to 0.75% in mRNA (Růžička et al. 2017).

While characterizing the Arabidopsis $m^6A$ methyltransferase MTA, a yeast two-hybrid screen for MTA-binding proteins was performed, which resulted in the identification of FIP37 (FKBP12-interacting protein of 37 kDa), the plant otholog of human WTAP (Zhong et al. 2008). Subsequent studies highlighted the functional importance of WTAP. It interacts with METTL3 and METTL14, and is required for their localization into nuclear speckles, which are enriched with pre-mRNA processing factors and are the place for m$^6$A formation in vivo (Ping et al. 2014). WTAP depletion causes a strong reduction of the RNA-binding capability of METTL3 and loss of METTL3 and METTL14 localization from nuclear speckles (Ping et al. 2014). Therefore, WTAP is considered to be an adaptor protein connecting the m$^6$A methyltransferase complex to nuclear speckles.

Recently, RBM15 and its paralogue RBM15B were identified in mammalian cells as accessory components of the m$^6$A methylation complex (Patil et al. 2016). They were shown to recruit the m$^6$A methylation core complex to specific sites in mRNA. Co-immunoprecipitation analyses revealed that RBM15 and RBM15B interact with METTL3 in the presence of WTAP. Importantly, knockdown of *RBM15* and *RBM15B* led to a significant reduction of m$^6$A in mRNA, indicating that they are indeed functional components of the m$^6$A methyltransferase complex. These results are also supported by data from Drosophila where Spenito, the RBM15 ortholog in fly, is required for m$^6$A formation (Lence et al. 2016). Interestingly, RBM15 and RBM15B bind to uridine-rich regions in mRNAs, which are in proximity to sites that are methylated to m$^6$A (Patil et al. 2016), suggesting that these proteins guide the m$^6$A methyltransferase complex to appropriate sites on the mRNA. Additional guide proteins may exist, because not all m$^6$A methylation sites are found near uridine-rich regions, and RBM15 and RBM15B-mediated methylations represent only a subset of m$^6$A marks in mRNA.

Two independent proteomic screens revealed that KIAA1429, also known as VIRMA (vir like m$^6$A methyltransferase associated), associates with the core components of the m$^6$A methyltransferase complex in mammalian cells (Schwartz et al. 2014; Yue et al. 2018). Similarly, Virilizer, the Drosophola ortholog of VIRMA, interacts with Fl(2)d (female lethal2), the ortholog of WTAP, and regulates alternative splicing of pre-mRNAs involved in sex determination in flies (Ortega et al. 2003). The depletion of VIRMA leads to about four-fold reduction of m$^6$A, demonstrating that it is required for the full methylation pattern in mammals (Schwartz et al. 2014). Very recently it was reported that VIRMA guides the core components of the m$^6$A methyltransferase complex to specific sites on mRNAs (Yue et al. 2018).

In addition to the above components, other members of m$^6$A methyltransferase complex, such as the E3 ubiquitin ligase HAKAI (Horiuchi et al. 2013; Růžička et al. 2017), may also guide the methyltransferase complex to specific methylation sites.

## 2.2 The Erasers: m$^6$A Demethylases

An important advance in the m$^6$A field was the identification of the first m$^6$A demethylation enzyme, fat mass and obesity-associated protein (FTO) (Jia et al. 2011). Soon later, another m$^6$A demethylase, α-ketoglutarate-dependent dioxygenase alkB homolog 5 (ALKBH5), was identified (Zheng et al. 2013). The discovery of enzymatic activities that remove m$^6$A marks indicated that these RNA modifications are dynamically regulated (Fig. 2).

FTO belongs to the non-heme Fe(II)- and α-ketoglutarate-dependent dioxygenase AlkB family of proteins, and was initially characterized as an enzyme demethylating $m^3T$ and $m^3U$ in single-stranded DNA (ssDNA) and single-stranded RNA (ssRNA) in vitro (Gerken et al. 2007; Jia et al. 2008), although both observed activities were much lower than those of the other AlkB-family proteins (Lee et al. 2005). In 2011, Jia et al. found that FTO could efficiently demethylate $m^6A$ in mRNA both in vitro and in vivo (Jia et al. 2011). During the oxidation of $m^6A$ to A, FTO generates $N^6$-hydroxymethyladenosine and $N^6$-formyladenosine as intermediate products. The depletion of *FTO* in mammalian cells induced a significant increase of $m^6A$ content in mRNA. Interestingly, another study demonstrated that FTO also exhibited demethylation activity towards $N^6$, 2′-O-dimethyladenosine ($m^6A_m$), which is exclusively located at the +1 position downstream of the 7-methylguanosine cap (Fig. 1) in many mRNAs (Mauer et al. 2017). By selectively demethylating $m^6A_m$, FTO reduces the stability of $m^6A_m$-containing mRNAs. A very recent study comprehensively investigated the cellular distribution and the substrates of FTO in vivo (Wei et al. 2018a). The authors demonstrated that FTO preferentially demethylates internal $m^6A$ of pre-mRNAs in the nucleus as well as the cap $m^6A_m$ and $m^6A$ in mRNA in the cytosol. Additionally, internal $m^6A$ in U6 RNA, internal and cap $m^6A_m$ in small nuclear RNAs (snRNAs) and $N^1$-methyladenosine ($m^1A$) in tRNA can be demethylated by FTO.

Another $m^6A$ demethylase, ALKBH5, was initially identified in a biochemical screen testing the demethylation activity of recombinant human AlkB homologs towards $m^6A$-labelled ssRNA substrates (Zheng et al. 2013). As FTO, it is a demethylase that can reverse $m^6A$ methylation by oxidation. The depletion of *ALKBH5* in mammalian cells resulted in approximately 9% increase of the $m^6A$ level in mRNA, whereas a 29% decrease was observed by its overexpression. Similar to the $m^6A$ methyltransferase METTL3, ALKBH5 is located in nuclear speckles, suggesting that it is involved in pre-mRNA processing. In addition, the knockdown of *ALKBH5* leads to a dramatic increase of cytoplasmic mRNA amounts as a result of accelerated nuclear mRNA export (Zheng et al. 2013).

In plants, *ALKBH10B*, a homolog of mammalian *ALKBH5*, influences floral transition. *ALKBH10B* mutation leads to global $m^6A$ hypermethylation and reduces the stability of its target transcripts. Mutant plants exhibit late flowering, producing approximately seven extra leaves before the first bud bloomes (Duan et al. 2017). Furthermore, Arabidopsis ALKBH9B has been reported to be an additional plant $m^6A$ demethylase in vitro. It co-localizes with siRNA bodies and associates with processing bodies (P-bodies) in the cytoplasm, implying its involvement in mRNA silencing and mRNA decay (Martínez-Pérez et al. 2017).

## 2.3 *The Readers: $m^6A$ Recognition Proteins*

$N^6$-methyladenosine in mRNA can directly affect the secondary structure by altering RNA-RNA base pairing, as well as RNA-protein interactions of the transcripts. It can influence every step of an mRNA's lifetime, including splicing, export, translation and degradation (Roundtree et al. 2017). The major mechanism by which $m^6A$ exerts its function is by recruiting specific RNA-binding proteins, the so-called reader proteins. To date, two types of reader proteins for $m^6A$ recognition were identified. The first type binds directly to the $m^6A$ base and includes the YTH family proteins (Hsu et al. 2017; Li et al. 2017b; Shi et al. 2017; Wang et al. 2014b, 2015; Xiao et al. 2016), the eIF3 (Meyer et al. 2015), and the IGF2BPs (Huang et al. 2018). The second type, for example the reader protein HNRNPA2B1 (Alarcón et al. 2015), selectively recognizes an $m^6A$-mediated secondary structure in the mRNA (Fig. 2).

The definition of the YTH family proteins began when an approximately 140 amino acid domain of the human splicing factor YT521-B was shown to have RNA binding capacity (Hartmann et al. 1999). Basic Local Alignment Search Tool (BLAST) analyses revealed that this domain is highly conserved in YT521-B homologs as well as in other proteins of eukaryotes. Accordingly, the YT521-B homologs were called YTH domain-containing family proteins (YTHDFs) while other proteins, which otherwise are not homologous to YT521-B, were called YTH domain-containing proteins (YTHDCs). Together, these proteins are named YTH (for YT521-B homology) family proteins (Stoilov et al. 2002).

In vertebrates, there are three YTHDFs called YTHDF1, YTHDF2, and YTHDF3. YTHDF2 is the first $m^6A$ reader protein identified from mammalian cells (Wang et al. 2014b). Over 3000 cellular RNAs with a conserved core motif of G($m^6A$)C were identified as the targets of YTHDF2. Upon YTHDF2 binding such mRNAs are transferred from the translatable pool to P-bodies, where mRNA decay occurs. Further research revealed that YTHDF2 recruited the CCR4-NOT deadenylase complex (Fig. 2 ③) through a direct interaction and then mediated the degradation of $m^6A$-containing RNAs (Du et al. 2016). The knockdown of *YTHDF2* results in a prolonged lifetime and increased expression of its target transcripts (Wang et al. 2014b). A homolog of YTHDF2, YTHDF1 can also recognize $m^6A$ on mRNA. In contrast to YTHDF2, YTHDF1 actively promotes protein synthesis by interacting with the translation machinery, including the 40S ribosome, the 60S ribosome, and the translation initiation factor eIF3 (Fig. 2 ④). The knockdown of *YTHDF1* leads to reduced translation of its target transcripts (Wang et al. 2014b). YTHDF3, interacts with both YTHDF1 and YTHDF2 and affects the RNA binding specificity of each of them (Li et al. 2017b; Shi et al. 2017). YTHDF3 promotes translation in synergy with YTHDF1 and accelerates $m^6A$-containing mRNA decay through the interaction with YTHDF2. All three YTHDFs are located in the cytoplasm, where they together regulate the cytoplasmic metabolism of $m^6A$ mRNA.

YTHDCs have two members, named YTHDC1 and YTHDC2. Among the YTH family proteins, YTHDC1 is the only member located in the nucleus (Fig. 2 ②) where it binds the $m^6A$-containing pre-mRNAs and regulates their splicing (Xiao et al. 2016). Protein-protein interaction and PAR-CLIP (photoactivatable ribonucleoside crosslinking and immunoprecipitation) sequencing revealed that YTHDC1 interacts with the serine/arginine-rich splicing factors SRSF3 and SRSF10. SRSF3 binds maturing mRNA to promote exon inclusion whereas SRSF10 facilitates exon skipping. YTHDC1 targets m$^6$A in exonic regions, where it recruits SRSF3 but blocks SRSF10-mRNA binding to modulate alternative splicing of pre-mRNA (Xiao et al. 2016). Another study showed that YTHDC1 also binds the m$^6$A residues on lncRNA *XIST*, which in total contains more than 78 m$^6$A sites, to trigger transcriptional repression on the X chromosome (Patil et al. 2016).

The cytoplasmic YTHDC2 has the highest molecular weight of all YTH family proteins (~160 kDa vs ~60 kDa for other members). Biochemical analyses and CLIP-seq (crosslink immunoprecipitation sequencing) indicated that YTHDC2 selectively binds m$^6$A mRNA at its consensus motif, GGm$^6$ACU, which is frequently found around the stop codon (Fig. 2 ⑤). The knockout of *YTHDC2* in mouse leads to an increase in target mRNA abundance and a decrease in translation efficiency, which in turn affects spermatogenesis and results in both male and female infertility (Hsu et al. 2017). YTHDC2 contains not only a YTH domain but also several other domains, including an R3H domain (a RNA binding domain with an R-($X_3$)-H motif), a DEAH-box helicase core domain, two ankyrin repeat (ANK) domains, and an HELICc helicase domain (Kretschmer et al. 2018; Meyer and Jaffrey 2017). The presence of many domains in YTHDC2 implies that this protein may have multiple functions. Subsequent studies demonstrated that YTHDC2 is also an RNA-induced ATPase with 3′→5′ RNA helicase activity, and it interacts with the 5′→3′ exoribonuclease XRN1 via the ankyrin repeats, which are strategically located between the two helicase domains of YTHDC2 (Wojtas et al. 2017). In addition, the R3H domain contributes together with the YTH domain to the selective binding of YTHDC2 to m$^6$A-containing RNA (Kretschmer et al. 2018).

Thirteen YTH domain-containing proteins (ECT1-12 and CPSF30) were identified in *A. thaliana* (Li et al. 2014). Among them, *ECT2* is the most highly and most widely expressed member. It binds the m$^6$A residues, which are enriched in the 3′ UTRs of target mRNAs (Scutenaire et al. 2018; Wei et al. 2018b). Subcellular localization and mRNA-seq analyses suggested that ECT2 regulates 3′UTR processing in the nucleus and its binding to the mRNA promotes transcript stability in the cytoplasm (Wei et al. 2018b). Depletion of *ECT2* leads to increased trichome branching (Arribas-Hernández et al. 2018; Scutenaire et al. 2018; Wei et al. 2018b), which is due to the accelerated degradation of three ECT2-binding trichome morphogenesis related transcripts (Wei et al. 2018b). Furthermore, disruption of *ECT2* and *ECT3* together results in a delayed emergence of the first true leaves (Arribas-Hernández et al. 2018).

Eukaryotic initiation factor 3 (eIF3) is an m$^6$A reader without a YTH domain (Meyer et al. 2015). Normally, the cap-binding protein eIF4 is necessary for translation initiation (Jackson et al. 2010). However, transcripts containing m$^6$A in the 5′UTR do not require eIF4, but recruit eIF3 (Fig. 2 ⑥) and other initiation factors for their translation. Therefore, 5′UTR m$^6$A promotes the translation of certain mRNAs (Meyer et al. 2015). It appears that eIF3 is involved in two distinct types of translational enhancement via m$^6$A. Either eIF3 binds directly to the m$^6$A site in the 5′UTR (Fig. 2 ⑥) (Meyer et al. 2015) or YTHDF1 recognizes m$^6$A residues close to the stop codon and then interacts and transfers eIF3 to the 5′UTR of the transcript (Fig. 2 ④) In this latter scenario eIF3 itself is technically not a reader protein, because the reader function is exerted by YTHDF1 (Wang et al. 2015).

Recently, the insulin-like growth factor 2 mRNA-binding proteins (IGF2BPs, including IGF2BP1-3) were identified as an additional family of $m^6A$ readers in mammals (Huang et al. 2018). All of them contain four tandem KH (K homology) domains, which are conserved ssRNA-binding domains required for $m^6A$ recognition. IGF2BPs target thousands of cytoplasmic $m^6A$-bearing mRNAs through the consensus GG($m^6A$)C motif and promote their stability. Further investigation revealed that several mRNA-stabilizing proteins, including Hu antigen R (HuR), matrin 3 (MATR3), and poly(A)-binding protein cytoplasmic 1 (PABPC1), interact with IGF2BPs. Co-localization of HuR and IGF2BPs (Fig. 2 ⑦) was observed in P-bodies, suggesting that they work together to protect their targets from degradation. Upon heat shock, IGF2BPs are recruited to stress granules possibly to protect mRNAs from harmful conditions (Huang et al. 2018).

Heterogeneous nuclear ribonucleoprotein A2/B1 (HNRNPA2B1), containing an RNA recognition motif (RRM) domain, has been reported to be a nuclear $m^6A$ reader in the regulation of microRNA processing (Alarcón et al. 2015). Depletion of *HNRNPA2B1* leads to an accumulation of pri-miRNA transcripts and to a reduction of mature microRNAs, implying that this reader protein operates in the processing of pri-miRNAs to pre-miRNAs in the nucleus. Further biochemical analyses demonstrated that HNRNPA2B1 influences the microRNA processing via recruiting the microprocessor machinery protein DGCR8 (Fig. 2 ①). In addition, disruption of *HNRNPA2B1* affects alternative splicing patterns similarly to *METTL3* mutation. This suggests that the absence of $m^6A$ marks or the inability to read them has similar effects on splicing (Alarcón et al. 2015). Intriguingly, a structure-based study indicated that HNRNPA2B1 does not directly bind $m^6A$ (Wu et al. 2018), unlike the other reader proteins. In structural studies of the other reader proteins, aromatic cage-like surfaces were identified as key $m^6A$-specificity elements. However, such an aromatic pocket is not found in the RRM domain of HNRNPA2B1. By contrast, HNRNPA2B1 exhibits higher binding affinity towards non-methylated RNA substrates (Wu et al. 2018). Another study suggested that HNRNPA2B1 may selectively bind $m^6A$-containing RNAs through an $m^6A$-switch mode (Liu et al. 2015; Zhou and Pan 2018), in which $m^6A$ decreases the stability of Watson-Crick base-pairing in RNA loops altering the local structure of the transcripts thereby facilitating the interaction of HNRNPA2B1 with the RNA (Fig. 2 ①).

## 2.4 The Metabolic Fate of the $m^6A$ Mark

Quantitatively little alterations of $m^6A$ patterns are observed after mRNAs are exported to the cytoplasm (Ke et al. 2017). Hence, $N^6$-methyl-AMP ($N^6$-mAMP) will be released with other nucleotides from $m^6A$-containing mRNA turnover at the end of its lifetime. Metabolite analysis showed that $N^6$-mAMP does not accumulate in aging Arabidopsis leaf tissue, suggesting that a catabolic pathway for $N^6$-mAMP exists. We recently identified an $N^6$-mAMP deaminase (MAPDA), which is widely conserved in eukaryotes. This enzyme catalyzes the hydrolysis of $N^6$-mAMP to inosine monophosphate (IMP) and methylamine in the cytoplasm (Chen et al. 2018) and thus represents the endpoint of m$^6$A modification (Fig. 2). Mutation of *MAPDA* in *A. thaliana* or depletion in human HeLa cells both lead to a significant increase of N$^6$-mAMP content confirming the function of MAPDA in vivo. Also some N$^6$-mATP accumulates in *mapda* plants, which in vitro can be incorporated into newly synthesized RNAs. Therefore, we investigated whether m$^6$A abundance is altered in *mapda* plants but observed that possible changes of m$^6$A content in vivo were too small to be reliably quantified above the natural background of m$^6$A in RNA (Chen et al. 2018). We speculated that the RNA is protected not only by MAPDA but also by additional molecular filters from random incorporation of N$^6$-mATP. Cytoplasmic adenylate kinases, which exhibit a strong substrate preference for AMP over N$^6$-mAMP were tentatively identified as such additional filters (Chen et al. 2018).

# 3 Other mRNA Methylations and Acetylations in Eukaryotes

## 3.1 *5-methylcytidine (m$^5$C)*

DNA methylation at the 5′ position of cytosine has been known as an epigenetic maker for decades (Zhang et al. 2018). Interestingly, 5-methylcytidine (m$^5$C) also occurs in mRNA (Cui et al. 2017; David et al. 2017) and contributes to the epigenetic regulation of various biological processes (Motorin et al. 2009; Yang et al. 2017). Over 10000 m$^5$C sites were mapped in human mRNAs and other non-coding RNAs, and most of them were identified in untranslated regions, particularly at the binding sites for Argonaute proteins (Squires et al. 2012) and immediately downstream of translation initiation sites (Yang et al. 2017). In plants, 6045 m$^5$C peaks were detected in 4465 transcripts, and these marks are mainly located in the coding sequences of mRNAs with low translation activity immediately after the start codon and before the stop codon (Fig. 1) (Cui et al. 2017). The m$^5$C modification in mRNA is mainly introduced by the tRNA m$^5$C methyltransferase NOL1/NOP2/Sun domain family member 2 (NSUN2) and specifically bound by the mRNA export factor ALYREF in mammals (Yang et al. 2017). ALYREF, together with NSUN2, regulates the nuclear-cytoplasmic shuttling of the m$^5$C-containing mRNAs. In plants, abrogation of the m$^5$C methyltransferase reduces cell division in the root apical meristem, causing shorter primary roots and enhanced sensitivity to oxidative stress (David et al. 2017). Disruption of all four *ALY* genes in Arabidopsis (ALYs are homologs of ALYREF in mammals), results in nuclear mRNA accumulation. Phenotypically, vegetative and reproductive defects are observed, including severely reduced leave and root growth, altered flower morphology, as well as reduced seed production (Pfaff et al. 2018).

Furthermore, a small portion of m$^5$C in RNA (~0.02% m$^5$C) can be oxidized to 5-hydroxymethylcytosine (hm$^5$C) by the ten-eleven translocation (Tet)-family enzymes in Drosophila and mammalian cells (Delatte et al. 2016; Fu et al. 2014). Tet-deficient flies have decreased RNA hydroxymethylation and suffer from impaired brain development. Notably, $hm^5C$ was also present in RNA of *Caenorhabditis elegans* and *A. thaliana*, which do not contain $hm^5dC$ in their DNA and lack *Tet* homologs in their genomes, implying a non-Tet mediated mechanism for $hm^5C$ formation in RNA (Huber et al. 2015). Transcriptome-wide mapping in Drosophila revealed that $hm^5C$ occurs in the transcripts of many genes at the consensus motif of 'UCCUC' and is more abundant in coding sequences (Delatte et al. 2016). The function of $hm^5C$ in RNA, however, remains largely unknown.

## 3.2 *$N^1$-methyladenosine ($m^1A$)*

Besides $m^6A$, also a methylation at the $N^1$ position of adenosine ($m^1A$) occurs in eukaryotic mRNA (Dominissini et al. 2016; Li et al. 2016). However, $m^1A$ is less abundant with an $m^1A/A$ ratio of about 0.02% in mRNA of human cells. The modification is enriched in the 5′ untranslated region (Fig. 1) of mRNA transcripts (Li et al. 2016) and around the start codon upstream of the first splice site (Dominissini et al. 2016). $m^1A$ can rearrange to $m^6A$ under alkaline conditions at elevated temperatures (Dimroth rearrangement) (Macon and Wolfenden 1968), therefore, it might be erased in the detection process leading to its underestimation and the introduction of noise in $m^6A$ measurements. In 2016, Dominissini and colleagues identified more than 7000 $m^1A$ modification sites in human mRNA (Dominissini et al. 2016), while Li and colleagues detected 901 $m^1A$ modification sites originating from 887 transcripts, encoded by 600 human genes (Li et al. 2016). Later, a $m^1A$-seq study reaching single-base resolution reported only 7 $m^1A$ nucleotides at internal sites of cytosolic mRNAs and 5 $m^1A$ modifications in the mitochondrial mRNA (Safra et al. 2017). Interestingly, Li and colleagues, who previously found 901 $m^1A$ modification sites, also developed a single-nucleotide resolution mapping method for $m^1A$ profiling at same time, and they identified 473 sites in cytosolic mRNA and 22 $m^1A$ sites from mitochondrial genes (Li et al. 2017a). It appears that the actual number of $m^1A$ modifications fluctuates either for biological or technical reasons.

The tRNA methyltransferase complex TRMT6/61A is required for a part of the $m^1A$ marks in mRNA (Li et al. 2017a). The RNA demethylase ALKBH3 can remove $m^1A$. The $m^1A$ profile is dynamic in response to stimuli, such as nutrient starvation and heat shock (Li et al. 2016). A very recent study demonstrated that YTHDF1-3 and YTHDC1, but not YTHDC2, bind directly to $m^1A$ in RNA (Dai et al. 2018). However, the mechanism of $m^1A$ installation in RNA, including writer protein(s) other than TRMT6/61A and their preferred sequence motives, as well as the functional roles of this mRNA methylation have yet to be elucidated.

### 3.3 2′-O-methylation ($N_m$)

Ribose methylation in mRNA usually occurs at the 2′O position (2′-O-methylation, $N_m$) in the second or the third base adjacent to the 5′ cap (Schibler and Perry 1977), and these are installed by the Cap-specific mRNA (nucleoside-2′-O-)-methyltransferase 1 (CMTR1) (Belanger et al. 2010) and CMTR2 (Werner et al. 2011), respectively, in mammals. $N_m$ might be involved in the discrimination of self and non-self mRNAs (Daffis et al. 2010). A portion of $N_m$ modifications also bear $m^6A$ methylations to form $m^6A_m$ (Linder et al. 2015). $m^6A_m$-containing transcripts are more stable than others, because the modification prevents DCP2-mediated decapping and microRNA-induced mRNA decay (Mauer et al. 2017). In vertebrates, the methylation of the $N^6$ position generating $m^6A_m$ is introduced by PCIF1 (Phosphorylated C-terminal domain Interacting Factor 1), which interacts with the Ser5-phosphorylated C-terminal domain of RNA polymerase II, as cap-specific adenosine methyltransferase (CAPAM) complex (Akichika et al. 2018). Mutating *PCIF1* by CRISPR-Cas9 in human cells completely abrogates the $N^6$ methylation on $A_m$ in the cap structure. The $m^6A_m$ modification was restored when knockout cells were rescued by plasmid-encoded PCIF1 (Akichika et al. 2018). Dynamically, the $m^6A$ modification of $m^6A_m$ can be removed by the demethylase FTO (Mauer et al. 2017).

2′-O-methylation is also present at the 3′ end of microRNAs, thus protecting microRNA from 3′→5′ degradation. Such modification is installed in plants by the methyltransferase HUA-ENHANCER-1 (HEN1) (Yu et al. 2005).

### 3.4 7-methylguanosine ($m^7G$)

In addition to mRNA methylations listed above, there are a variety of other less abundant modifications in mRNA. Recently, 7-methylguanosine ($m^7G$), which was previously thought to only exist in the 5′ cap structure of mRNA, was mapped at internal positions of mRNA (Chu et al. 2018). In this research, the authors developed a novel analytical method to differentiate internal $m^7G$ from that present in the 5′ cap. In detail, by comparing the digestion properties of different nucleases, namely S1 nuclease and phosphodiesterase I, they found that phosphodiesterase I can digest both cap and internal $m^7G$ containing RNA while the S1 nuclease is only able to digest RNA with internal $m^7G$ (Chu et al. 2018). High resolution mass spectrometry analyses coupled with differential nuclease treatments of mRNA revealed the existence of internal $m^7G$ in mRNA in human cells, rats, as well as in plants. In plants, $m^7G$ in mRNA occurs at a higher frequency (about $3.5 \times 10^{-5}$–$5.5 \times 10^{-5}$ $m^7G/G$) than in mammals (about $0.5 \times 10^{-5}$–$1.5 \times 10^{-5}$ $m^7G/G$). The abundance of $m^7G$ in mRNA was elevated in rice when plants suffered environmental stress, such exposure to the higher concentrations of cadmium, which is widely distributed in soil (Chu et al. 2018).

### 3.5 3-methylcytidine ($m^3C$)

The rare modification 3-methylcytidine ($m^3C$, 0.004% of cytosine) was detected in mRNA from mouse liver and human cells recently (Xu et al. 2017). Apart from mRNA, $m^3C$ is also present and comparatively abundant in tRNA (about 1.7% $m^3C$/C) but is absent in rRNA of mammalian cells. Dramatic reduction of $m^3C$ abundance in mRNA was observed in *mettl8* knockout mice, indicating that its formation in mRNA is mediated by METTL8 (Xu et al. 2017). It was also shown that METTL2 and METTL6 contribute $m^3C$ formation in certain tRNAs (Xu et al. 2017). Mice lacking either METTL2 or METTL6 or METTL8, showed no developmental defects, and growth rates of human cells, that were mutated in *METTL2* and *METTL8* as well as knockdown for *METTL6*, were not significantly altered (Xu et al. 2017). Up to date, little is known about the distribution, dynamics, and function of this newly identified mRNA modification.

### 3.6 $N^4$-acetylcytidine ($ac^4C$)

A very recent publication reported the identification of a novel mRNA modification, $N^4$-acetylcytidine ($ac^4C$), occurring with an abundance of approximately 0.2% $ac^4C$/C in the mRNA of mammalian cells (Arango et al. 2018). $ac^4C$ immunoprecipitation coupled with the next-generation sequencing (acRIP-seq) revealed that the majority of acetylated transcripts possess one to two $ac^4C$ modification site(s), and that the distribution of $ac^4C$ generally displays a 5′ positional bias, especially clustering proximal to translation start sites within coding sequences. Acetyltransferase NAT10, previously determined as a protein acetyltransferase for α-tubulin, histones, and p53, is responsible for $ac^4C$ installation in mRNA. Its mutation resulted in approximately 80% to 90% reduction of $ac^4C$ modification in mRNA. Analyses of mRNA half-lives indicated that the NAT10-mediated mRNA acetylation increases the stability of target transcripts. Further biochemical studies showed that $ac^4C$ also promotes the efficiency of translation (Arango et al. 2018).

## 4 Concluding Remarks and Future Perspectives

The discovery of mRNA methylation revealed a new layer of epigenetic regulation. RNA methylation shares many of the characteristics with the well-known epigenetic DNA and histone modifications, such as the reversibility and the regulatory effects on gene expression. In the nucleus, writer proteins install methylations at specific sites in mRNA according to consensus RNA motives or structures. Eraser proteins can dynamically remove modification marks, and they are hypothesized to fine tune the precise methylation pattern maybe depending on environmental cues. In accordance, only a sub fraction of the consensus $m^6A$ RNA methylation motif ($RRm^6ACH$) in mammals is actually methylated (Dominissini et al. 2012), which might be controlled by $m^6A$ demethylases. $m^6A$ marks in some pre-mRNAs can be recognized by nuclear reader proteins to mediate differential splicing. Mature mRNAs are exported to the cytosol, where their methylation pattern is decoded by different cytoplasmic reader proteins. These readers fine tune several biological processes, such as mRNA decay and translation. Upon RNA turnover, methylated nucleotides are released together with canonical ones. Special enzymes for the catabolism of modified nucleotides will often be required to remove the modifications thereby facilitating the entry of the altered nucleotides into the general nucleotide metabolism.

Today, most of our mechanistic understanding of mRNA methylation is derived from studies on $m^6A$, and little is known about the other methylations, including $m^5C$, $m^1A$, $N_m$, $m^7G$, $m^3C$, and $ac^4C$. Substantial efforts investigating these methylations (acetylations) will still be necessary. It will be required (1) to find enzymes with the respective writer activities and (2) to map their respective binding sites, and (3) to elucidate the biological processes these methylations are involved in, and (4) to describe the metabolic fate of the corresponding methylated nucleotides released upon degradation of the modified mRNA. For all mRNA methylations, including $m^6A$, the large amount of information gained from high-throughput modification mapping approaches is highly valuable to obtain further insights into the biological function of these modifications. This tool will need to be employed to investigate the spatiotemporal properties of mRNA methylations, for example in different tissues and in response to environmental stimuli. It can be predicted that higher spatial resolution and dissection of the temporal dynamics of methylation will greatly enhance our understanding of the fine tuning of gene expression.

**Acknowledgement** We apologize to those colleagues whose work could not be cited due to space limitations. This work was funded by the Deutsche Forschungsgemeinschaft (DFG, German Research foundation) grant 416961553 (CH 2292/1-1) to M.C. and the Leibniz University Hannover (Wege in die Forschung II to M.C.).

## References

Agarwala SD, Blitzblau HG, Hochwagen A et al (2012) RNA methylation by the MIS complex regulates a cell fate decision in yeast. PLoS Genet 8:e1002732

Akichika S, Hirano S, Shichino Y et al (2018) Cap-specific terminal N6-methylation of RNA by an RNA polymerase II-associated methyltransferase. Science 363(6423):eaav0080. https://doi.org/10.1126/science.aav0080

Alarcón CR, Goodarzi H, Lee H et al (2015) HNRNPA2B1 Is a mediator of $m^6A$-dependent nuclear RNA processing events. Cell 162:1299–1308

Arango D, Sturgill D, Alhusaini N et al (2018) Acetylation of cytidine in mRNA promotes translation efficiency. Cell 175:1–15

Arribas-Hernández L, Bressendorff S, Hansen MH et al (2018) An $m^6A$-YTH module controls developmental timing and morphogenesis in Arabidopsis. Plant Cell 30:952–967

Belanger F, Stepinski J, Darzynkiewicz E et al (2010) Characterization of hMTr1, a human Cap1 2′-*O*-ribose methyltransferase. J Biol Chem 285:33037–33044

Bodi Z, Button JD, Grierson D et al (2010) Yeast targets for mRNA methylation. Nucleic Acids Res 38:5327–5335

Bokar JA, Rath-Shambaugh ME, Ludwiczak R et al (1994) Characterization and partial purification of mRNA $N^6$-adenosine methyltransferase from HeLa cell nuclei. Internal mRNA methylation requires a multisubunit complex. J Biol Chem 269:17697–17704

Bokar JA, Shambaugh ME, Polayes D et al (1997) Purification and cDNA cloning of the AdoMet-binding subunit of the human mRNA (N-6-adenosine)-methyltransferase. RNA 3:1233–1247

Chen M, Urs MJ, Sánchez-González I et al (2018) $m^6A$ RNA degradation products are catabolized by an evolutionarily conserved $N^6$-methyl-AMP deaminase in plant and mammalian cells. Plant Cell 30:1511–1522

Chu J-M, Ye T-T, Ma C-J et al (2018) Existence of internal $N^7$-methylguanosine modification in mRNA determined by differential enzyme treatment coupled with mass spectrometry analysis. ACS Chem Biol 13(12):3243–3250. https://doi.org/10.1021/acschembio.7b00906

Clancy MJ, Shambaugh ME, Timpte CS et al (2002) Induction of sporulation in *Saccharomyces cerevisiae* leads to the formation of $N^6$-methyladenosine in mRNA: a potential mechanism for the activity of the IME4 gene. Nucleic Acids Res 30:4509–4518

Cui X, Liang Z, Shen L et al (2017) 5-methylcytosine RNA methylation in *Arabidopsis thaliana*. Mol Plant 10:1387–1399

Daffis S, Szretter KJ, Schriewer J et al (2010) 2′-O methylation of the viral mRNA cap evades host restriction by IFIT family members. Nature 468:452–456

Dai X, Wang T, Gonzalez G et al (2018) Identification of YTH domain-containing proteins as the readers for $N^1$-Methyladenosine in RNA. Anal Chem 90:6380–6384

David R, Burgess A, Parker B et al (2017) Transcriptome-wide mapping of RNA 5-methylcytosine in Arabidopsis mRNAs and noncoding RNAs. Plant Cell 29:445–460

Delatte B, Wang F, Ngoc LV et al (2016) Transcriptome-wide distribution and function of RNA hydroxymethylcytosine. Science 351:282–285

Desrosiers R, Friderici K, Rottman F (1974) Identification of methylated nucleosides in messenger RNA from Novikoff Hepatoma Cells. Proc Natl Acad Sci USA 71:3971–3975

Dominissini D, Moshitch-Moshkovitz S, Schwartz S et al (2012) Topology of the human and mouse $m^6A$ RNA methylomes revealed by $m^6A$-seq. Nature 485:201–206

Dominissini D, Nachtergaele S, Moshitch-Moshkovitz S et al (2016) The dynamic $N^1$-methyladenosine methylome in eukaryotic messenger RNA. Nature 530:441–446

Du H, Zhao Y, He J et al (2016) YTHDF2 destabilizes $m^6A$-containing RNA through direct recruitment of the CCR4-NOT deadenylase complex. Nat Commun 7:12626

Duan H-C, Wei L-H, Zhang C et al (2017) ALKBH10B Is an RNA $N^6$-methyladenosine demethylase affecting Arabidopsis floral transition. Plant Cell 29:2995–3011

Fu L, Guerrero CR, Zhong N et al (2014) Tet-mediated formation of 5-hydroxymethylcytosine in RNA. J Am Chem Soc 136:11582–11585

Gerken T, Girard CA, Tung Y-CL et al (2007) The obesity-associated FTO gene encodes a 2-oxoglutarate-dependent nucleic acid demethylase. Science 318:1469–1472

Geula S, Moshitch-Moshkovitz S, Dominissini D et al (2015) Stem cells. $m^6A$ mRNA methylation facilitates resolution of naïve pluripotency toward differentiation. Science 347:1002–1006

Hartmann AM, Nayler O, Schwaiger FW et al (1999) The interaction and colocalization of Sam68 with the splicing-associated factor YT521-B in nuclear dots is regulated by the Src family kinase p59(fyn). Mol Biol Cell 10:3909–3926

Haussmann IU, Bodi Z, Sanchez-Moran E et al (2016) $m^6A$ potentiates Sxl alternative pre-mRNA splicing for robust Drosophila sex determination. Nature 540:301–304

Hongay CF, Orr-Weaver TL (2011) Drosophila Inducer of MEiosis 4 (IME4) is required for Notch signaling during oogenesis. Proc Natl Acad Sci USA 108:14855–14860

Horiuchi K, Kawamura T, Iwanari H et al (2013) Identification of Wilms' tumor 1-associating protein complex and its role in alternative splicing and the cell cycle. J Biol Chem 288:33292–33302
Horowitz S, Horowitz A, Nilsen TW et al (1984) Mapping of $N^6$-methyladenosine residues in bovine prolactin mRNA. Proc Natl Acad Sci USA 81:5667–5671
Hsu PJ, Zhu Y, Ma H et al (2017) Ythdc2 is an $N^6$-methyladenosine binding protein that regulates mammalian spermatogenesis. Cell Res 27:1115–1127
Huang H, Weng H, Sun W et al (2018) Recognition of RNA $N^6$-methyladenosine by IGF2BP proteins enhances mRNA stability and translation. Nat Cell Biol 20:285–295
Huber SM, van Delft P, Mendil L et al (2015) Formation and abundance of 5-hydroxymethylcytosine in RNA. Chembiochem 16:752–755
Jackson RJ, Hellen CUT, Pestova TV (2010) The mechanism of eukaryotic translation initiation and principles of its regulation. Nat Rev Mol Cell Biol 11:113–127
Jia G, Yang C-G, Yang S et al (2008) Oxidative demethylation of 3-methylthymine and 3-methyluracil in single-stranded DNA and RNA by mouse and human FTO. FEBS Lett 582:3313–3319
Jia G, Fu Y, Zhao X et al (2011) $N^6$-Methyladenosine in nuclear RNA is a major substrate of the obesity-associated FTO. Nat Chem Biol 7:885–887
Ke S, Pandya-Jones A, Saito Y et al (2017) $m^6A$ mRNA modifications are deposited in nascent pre-mRNA and are not required for splicing but do specify cytoplasmic turnover. Gene Dev 31:990–1006
Kretschmer J, Rao H, Hackert P et al (2018) The $m^6A$ reader protein YTHDC2 interacts with the small ribosomal subunit and the 5′–3′ exoribonuclease XRN1. RNA 24:1339–1350
Lee D-H, Jin S-G, Cai S et al (2005) Repair of methylation damage in DNA and RNA by mammalian AlkB homologues. J Biol Chem 280:39448–39459
Lence T, Akhtar J, Bayer M et al (2016) $m^6A$ modulates neuronal functions and sex determination in Drosophila. Nature 540:242–247
Li D, Zhang H, Hong Y et al (2014) Genome-wide identification, biochemical characterization, and expression analyses of the YTH domain-containing RNA-binding protein family in Arabidopsis and Rice. Plant Mol Biol Rep 32:1169–1186
Li X, Xiong X, Wang K et al (2016) Transcriptome-wide mapping reveals reversible and dynamic $N^1$-methyladenosine methylome. Nat Chem Biol 12:311–316
Li X, Xiong X, Zhang M et al (2017a) Base-resolution mapping reveals distinct $m^1A$ methylome in nuclear- and mitochondrial-encoded transcripts. Mol Cell 68:993–1005
Li A, Chen Y-S, Ping X-L et al (2017b) Cytoplasmic $m^5A$ reader YTHDF3 promotes mRNA translation. Cell Res 27:444–447
Linder B, Grozhik AV, Olarerin-George AO et al (2015) Single-nucleotide-resolution mapping of $m^6A$ and $m^6A_m$ throughout the transcriptome. Nat Methods 12:767–772
Liu J, Yue Y, Han D et al (2014) A METTL3-METTL14 complex mediates mammalian nuclear RNA N-6-adenosine methylation. Nat Chem Biol 10:93–95
Liu N, Dai Q, Zheng G et al (2015) $N^6$-methyladenosine-dependent RNA structural switches regulate RNA-protein interactions. Nature 518:560–564
Macon J, Wolfenden R (1968) 1-methyladenosine. Dimroth rearrangement and reversible reduction. Biochemistry 7:3453–3458
Martínez-Pérez M, Aparicio F, López-Gresa MP et al (2017) Arabidopsis $m^6A$ demethylase activity modulates viral infection of a plant virus and the $m^6A$ abundance in its genomic RNAs. Proc Natl Acad Sci USA 114:10755–10760
Mauer J, Luo X, Blanjoie A et al (2017) Reversible methylation of $m^6A_m$ in the 5′ cap controls mRNA stability. Nature 541:371–375
Meyer KD, Jaffrey SR (2017) Rethinking $m^6A$ readers, writers, and erasers. Annu Rev Cell Dev Biol 33:319–342
Meyer KD, Saletore Y, Zumbo P et al (2012) Comprehensive analysis of mRNA methylation reveals enrichment in 3′ UTRs and near stop codons. Cell 149:1635–1646

Meyer KD, Patil DP, Zhou J et al (2015) 5′ UTR $m^6A$ promotes cap-independent translation. Cell 163:999–1010
Motorin Y, Lyko F, Helm M (2009) 5-methylcytosine in RNA: detection, enzymatic formation and biological functions. Nucleic Acids Res 38:1415–1430
Ortega A, Niksic M, Bachi A et al (2003) Biochemical function of female-lethal (2)D/Wilms' tumor suppressor-1-associated proteins in alternative pre-mRNA splicing. J Biol Chem 278:3040–3047
Patil DP, Chen C-K, Pickering BF et al (2016) $m^6A$ RNA methylation promotes XIST-mediated transcriptional repression. Nature 537:369–373
Perry RP, Kelley DE (1974) Existence of methylated messenger RNA in mouse L cells. Cell 1:37–42
Pfaff C, Ehrnsberger HF, Flores-Tornero M et al (2018) ALY RNA-binding proteins are required for nucleocytosolic mRNA transport and modulate plant growth and development. Plant Physiol 177:226–240
Ping X-L, Sun B-F, Wang L et al (2014) Mammalian WTAP is a regulatory subunit of the RNA $N^6$-methyladenosine methyltransferase. Cell Res 24:177–189
Roundtree IA, Evans ME, Pan T et al (2017) Dynamic RNA modifications in gene expression regulation. Cell 169:1187–1200
Růžička K, Zhang M, Campilho A et al (2017) Identification of factors required for $m^6A$ mRNA methylation in Arabidopsis reveals a role for the conserved E3 ubiquitin ligase HAKAI. New Phytol 215:157–172
Safra M, Sas-Chen A, Nir R et al (2017) The $m^1A$ landscape on cytosolic and mitochondrial mRNA at single-base resolution. Nature 551:251–255
Schapira M (2016) Structural chemistry of human RNA methyltransferases. ACS Chem Biol 11:575–582
Schibler U, Perry RP (1977) The 5′-termini of heterogeneous nuclear RNA: a comparison among molecules of different sizes and ages. Nucleic Acids Res 4:4133–4149
Schibler U, Kelley DE, Perry RP (1977) Comparison of methylated sequences in messenger RNA and heterogeneous nuclear RNA from mouse L cells. J Mol Biol 115:695–714
Schwartz S, Mumbach MR, Jovanovic M et al (2014) Perturbation of $m^6A$ writers reveals two distinct classes of mRNA methylation at internal and 5′ sites. Cell Rep 8:284–296
Scutenaire J, Deragon J-M, Jean V et al (2018) The YTH domain protein ECT2 is an $m^6A$ reader required for normal trichome branching in Arabidopsis. Plant Cell 30:986–1005
Shi H, Wang X, Lu Z et al (2017) YTHDF3 facilitates translation and decay of N-6-methyladenosine-modified RNA. Cell Res 27:315–328
Śledź P, Jinek M (2016) Structural insights into the molecular mechanism of the $m^6A$ writer complex. eLife 5:e18434
Squires JE, Patel HR, Nousch M et al (2012) Widespread occurrence of 5-methylcytosine in human coding and non-coding RNA. Nucleic Acids Res 40:5023–5033
Stoilov P, Rafalska I, Stamm S (2002) YTH: a new domain in nuclear proteins. Trends Biochem Sci 27:495–497
Wang Y, Li Y, Toth JI et al (2014a) $N^6$-methyladenosine modification destabilizes developmental regulators in embryonic stem cells. Nat Cell Biol 16:191–198
Wang X, Lu Z, Gomez A et al (2014b) $N^6$-methyladenosine-dependent regulation of messenger RNA stability. Nature 505:117–120
Wang X, Zhao BS, Roundtree IA et al (2015) $N^6$-methyladenosine modulates messenger RNA translation efficiency. Cell 161:1388–1399
Wang P, Doxtader KA, Nam Y (2016a) Structural basis for cooperative function of mettl3 and mettl14 methyltransferases. Mol Cell 63:306–317
Wang X, Feng J, Xue Y et al (2016b) Structural basis of $N^6$-adenosine methylation by the METTL3-METTL14 complex. Nature 534:575–578
Wei J, Liu F, Lu Z et al (2018a) Differential $m^6A$, $m^6A_m$, and $m^1A$ demethylation mediated by FTO in the cell nucleus and cytoplasm. Mol Cell 71:973–985

Wei L-H, Song P, Wang Y et al (2018b) The $m^6A$ reader ECT2 controls trichome morphology by affecting mRNA stability in Arabidopsis. Plant Cell 30:968–985

Werner M, Purta E, Kaminska K et al (2011) 2′-*O*-ribose methylation of cap2 in human: function and evolution in a horizontally mobile family. Nucleic Acids Res 39:4756–4768

Wojtas MN, Pandey RR, Mendel M et al (2017) Regulation of m6A transcripts by the 3′→5′ RNA helicase YTHDC2 is essential for a successful meiotic program in the mammalian germline. Mol Cell 68:374–387

Wu B, Su S, Patil DP et al (2018) Molecular basis for the specific and multivariant recognitions of RNA substrates by human hnRNP A2/B1. Nat Commun 9:420

Xiao W, Adhikari S, Dahal U et al (2016) Nuclear $m^6A$ reader YTHDC1 regulates mRNA splicing. Mol Cell 61:507–519

Xu L, Liu X, Sheng N et al (2017) Three distinct 3-methylcytidine ($m^3C$) methyltransferases modify tRNA and mRNA in mice and humans. J Biol Chem 292:14695–14703

Yang X, Yang Y, Sun B-F et al (2017) 5-methylcytosine promotes mRNA export-NSUN2 as the methyltransferase and ALYREF as an $m^5C$ reader. Cell Res 27:606–625

Yoon K-J, Ringeling FR, Vissers C et al (2017) Temporal control of mammalian cortical neurogenesis by $m^6A$ methylation. Cell 171:877–889

Yu B, Yang Z, Li J et al (2005) Methylation as a crucial step in plant microRNA biogenesis. Science 307:932–935

Yue Y, Liu J, Cui X et al (2018) VIRMA mediates preferential $m^6A$ mRNA methylation in 3′UTR and near stop codon and associates with alternative polyadenylation. Cell Discov 4:10

Zhang H, Lang Z, Zhu J-K (2018) Dynamics and function of DNA methylation in plants. Nat Rev Mol Cell Biol 19:489–506

Zheng G, Dahl JA, Niu Y et al (2013) ALKBH5 is a mammalian RNA demethylase that impacts RNA metabolism and mouse fertility. Mol Cell 49:18–29

Zhong S, Li H, Bodi Z et al (2008) MTA is an Arabidopsis messenger RNA adenosine methylase and interacts with a homolog of a sex-specific splicing factor. Plant Cell 20:1278–1288

Zhou KI, Pan T (2018) An additional class of $m^6A$ readers. Nat Cell Biol 20:230–232

# The Role of mRNA m$^6$A in Regulation of Gene Expression

**Sicong Zhang**

## Contents



**Abstract** *N*$^6$-methyladenosine (m$^6$A) is the most prevalent internal methylation in messenger RNA (mRNA). This biochemically reversible modification is deposited by m$^6$A methyltransferases, removed by m$^6$A demethylases and recognized by different RNA-binding proteins. Depending on the localization of m$^6$A and its reader proteins, an array of cellular processes ranging from RNA maturation and export in nucleus, to degradation and translation in cytoplasm, can be affected and consequently lead to diverse cell fates. The essential role of m$^6$A in normal tissue development as well as tumor progression has been revealed in the past few years, emphasizing an additional layer of gene expression regulation.

**Keywords** *N*$^6$-methyladenosine · m$^6$A methylation · Methyltransferase · Demethylase

S. Zhang (✉)
Laboratory of Biochemistry and Molecular Biology, The Rockefeller University, New York, NY, USA
e-mail: szhang06@rockefeller.edu



S. Jurga, J. Barciszewski (eds.), *The DNA, RNA, and Histone Methylomes*, RNA Technologies, https://doi.org/10.1007/978-3-030-14792-1_14

# 1 Introduction

$N^6$-methyladenosine ($m^6A$) modification (Fig. 1), the most abundant internal methylation on mRNA, has emerged as a key regulatory mark on messenger RNA (mRNA). Despite the demonstration of methylation of ribosomal RNA (rRNA) and transfer RNA (tRNA) in mammalian cells by 1970, mRNA methylation was not ascertained until mRNA could be purified with polyA selection. With the pursuit of cap structure of mRNA (or heterogeneous nuclear RNA, hnRNA), Perry and Rottman laboratories found that the internal regions of mRNA and hnRNA are frequently methylated by $m^6A$ in mammalian cells, with an estimated occurrence of ~1–2 $m^6A$ per 1000 nucleotides (Desrosiers et al. 1974; Perry et al. 1975). Subsequent mutation and enzymatic footprinting assays by a number of groups revealed that $m^6A$ is deposited within a consensus sequence Pu(G>A)$m^6$AC(A/C/U) (where Pu represents purine) (Schibler et al. 1977; Wei and Moss 1977; Kane and Beemon 1987). In mammalian cells, the majority of $m^6A$ modifications are catalyzed by a multicomponent methylation complex identified by Rottman laboratory (Bokar et al. 1994), which contains the core catalytic component METTL3 (methyltransferase like 3, initially called MT-A70) (Bokar et al. 1997). $m^6A$ was thought to be irreversible for decades until the groundbreaking discovery by Chuan He laboratory with their exciting report of the first $m^6A$ demethylase FTO (fat mass and obesity-associated protein) in mammalian cells (Jia et al. 2011). Another member of AlkB subfamily of the Fe(II)/2-oxoglutarate (2OG) dioxygenase superfamily, ALKBH5 (alkB homologue 5), was soon discovered to have a similar $m^6A$ demethylase activity (Zheng et al. 2013). These remarkable discoveries revived the interest from RNA research community and led to the functional characterization of $m^6A$. Meanwhile, transcriptomic analysis afforded by high-throughput sequencing enables global mapping and measurement of $m^6A$ at levels of cells, tissues and species, with the use of an antibody that specifically recognizes $m^6A$ (Meyer et al. 2012; Dominissini et al. 2012; Schwartz et al. 2013; Batista et al. 2014; Wang et al. 2014b). These $m^6A$-seq studies, ranging from yeast to mammalian cells, confirmed that $m^6A$ residues within the consensus sequences are scattered in coding sequences (CDS), 3′ untranslated regions (UTRs) and 5′ UTRs, and are more enriched in the last exon around stop codon, suggesting a highly conserved $m^6A$ deposition

**Fig. 1** The structures of mRNA methylation

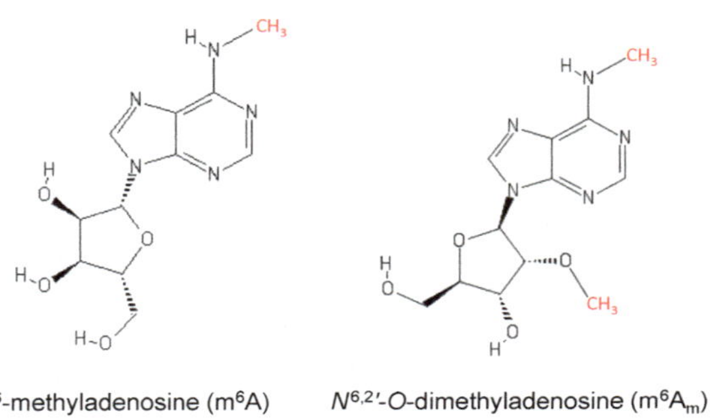

machinery. Pioneer studies by James E. Darnell and others (Sommer et al. 1978; Friderici et al. 1976) suggested a correlation between m$^6$A and mRNA instability, which has been elucidated by He laboratory and others with the demonstration of YTHDF2 (YTH domain-containing family protein 2)-dependent degradation of a subset of mRNAs bearing m$^6$A (Wang et al. 2014a; Du et al. 2016). Shortly after, m$^6$A has been shown to participate in many critical aspects of RNA regulation that include alternative splicing, nuclear export and translation, which are carried out by a variety of RNA-binding proteins and are ultimately essential for many major biological processes, such as embryonic stem cell (ESC) differentiation, circadian clock, and spermatogenesis (Geula et al. 2015; Batista et al. 2014; Wang et al. 2014b; Fustin et al. 2013; Zheng et al. 2013).

## 2 m$^6$A Sequencing Technologies Enable "Epitranscriptomic" Studies

The transcriptome-wide profile of m$^6$A in eukaryotic cells was first accomplished by two m$^6$A-seq (or MeRIP-seq) studies based on the immunoprecipitation with a specific anti-m$^6$A antibody and high-throughput sequencing (Dominissini et al. 2012; Meyer et al. 2012). m$^6$A-seq starts with purified RNAs that are subjected to polyA-selection or ribosomal RNA depletion, fragmentation, and immunoprecipitation followed by high-throughput sequencing. This straightforward method led to the identification of m$^6$A-marked transcripts corresponding to more than 7000 human genes. When coupled with methyltransferase depletion to reduce the detection background, this approach can achieve a relatively high resolution (Schwartz et al. 2013). Given the limitation of resolution due to the cumulative signals from multiple m$^6$A residues residing within the same ~100–200 nucleotide fragments, the CLIP-based m$^6$A-CLIP (Cross-Linking and Immunoprecipitation) and PA-m$^6$A-seq (photocrosslinking-assisted m$^6$A sequencing) were introduced to facilitate higher resolution. Both methods use UV irradiation to capture RNA-antibody interactions by creating a covalent crosslink between the antibody and its bound RNA. However, unlike the m$^6$A-CLIP that allows the antibody to be crosslinked directly at m$^6$A sites (Ke et al. 2015; Linder et al. 2015), PA-m$^6$A-seq UV irradiation induces T-to-C transitions at 4-thiouridine residues in m$^6$A nearby regions, which therefore requires the search of consensus sequence to infer the precise m$^6$A sites (Chen et al. 2015). Consistent with the early observations regarding a biased m$^6$A distribution in 3′ end of mRNA (Perry et al. 1975), nearly all m$^6$A-seq studies have m$^6$A peaks mapped enriched in 3′ UTRs and near stop codons. It should be noted that this antibody also binds to m$^6$A$_m$, a frequent modification adjacent to the cap of mRNA, which yields "false" m$^6$A peaks within 500 nt from the transcription start site (TSS). Nonetheless, these peaks can be extracted from m$^6$A-CLIP mapped sites by in silico analysis of their overlap with TSSs and their localization in a sequence context matching the core initiator motif (Mauer et al. 2017).

## 3 m$^6$A Is Primarily Deposited by the METTL3–METTL14 Methyltransferase

The first cloned mRNA m$^6$A methyltransferase METTL3 was identified from a partially purified Hela cell nuclear fraction referred to as MT-A, which contains the subunit that binds to S-adenosylmethionine (SAM), the catalytic co-factor for methyltransferase (Bokar et al. 1997; Bokar et al. 1994). METTL3 has a co-localization with nuclear speckles and a widespread expression pattern in human tissues (Bokar et al. 1997). A phylogenomic analysis revealed over 40 bacterial and eukaryotic proteins that share sequence similarity with the methyltransferase domain of human METTL3 (Bujnicki et al. 2002). METTL3 orthologs are conserved throughout all eukaryotes that include *Saccharomyces cerevisiae* (IME4), *Drosophila melanogaster* (IME4) and *Arabidopsis Thaliana* (MTA). In *S. cerevisiae*, m$^6$A can be detected in sporulating cells at low levels that requires the catalytic activity of Ime4 (Inducer of Meiosis 4) (Clancy et al. 2002). In *D. melanogaster*, *Ime4* mutants are viable but develop behavioural defects with a sex bias towards maleness due to reduced female-specific splicing of the Sex lethal (Sxl), a master switch gene that controls sex determination and dosage compensation (Haussmann et al. 2016; Lence et al. 2016). The METTL3 orthologs are essential for developing embryo to progress past the globular stage in *A. thaliana* (Zhong et al. 2008) and required for embryonic development in mammalian cells (Geula et al. 2015). During the course of evolution, METTL3-mediated m$^6$A RNA modification turns to be more actively involved in critical biological processes. Noteworthy, m$^6$A is also present in bacterial mRNAs that carry a unique consensus motif of GCCAU. With a distinct modification pattern from those in eukaryotes, these m$^6$A-modified genes are associated with respiration, amino acids metabolism, stress response and small RNAs. Homologs of mammalian m$^6$A methyltransferases have not been reported in bacteria, suggesting a distinct bacterial pathway of m$^6$A modification (Deng et al. 2015).

The MT-A fraction most likely contains METTL14 (methyltransferase-like 14) as well, which partners with METTL3 to form a heterodimer (Liu et al. 2014; Wang et al. 2014b), since monomeric METTL3 is nearly inactive (Wang et al. 2016a; Scholler et al. 2018). Beyond the mammalian cells, METTL14 orthologs are present in *A. thaliana* (MTB/EMB1691) and *D. melanogaster* (CG7818). Structural analyses revealed that METTL14 has no catalytic ability, but instead, it provides an RNA-binding platform and stabilizes METTL3 conformation in the tight asymmetric heterodimer (Wang et al. 2016a, b; Sledz and Jinek 2016). A narrow groove lined with conserved positively charged residues that are contributed by the two methyltransferase domains (MTDs) in the heterodimeric METTL3-METTL14 complex, allows single-stranded RNA to fit in, so that the METTL3-METTL14 complex can methylate target adenosine in the consensus sequence of single-stranded RNAs, rather than long stable duplex structures (Wang et al. 2016a, b; Sledz and Jinek 2016; Narayan et al. 1994). In addition to the MTDs, the methyltransferase activity requires two CCCH-type zinc fingers of METTL3 and a domain of Arginine-Glycine-Glycine (RGG) repeats on the C-terminus of METTL14 (Scholler et al.

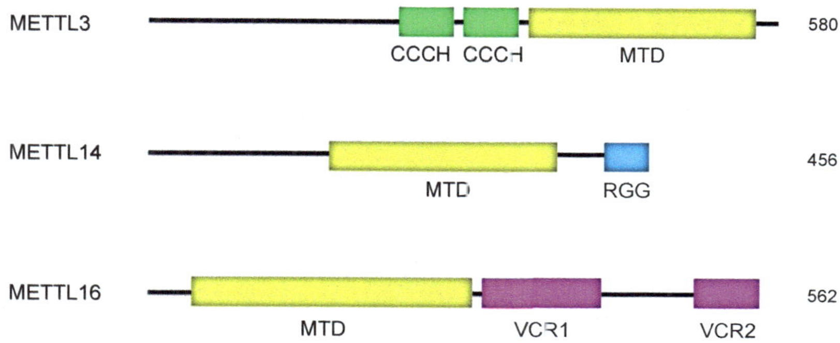


**Fig. 2** Schematic domain structures of human mRNA m$^6$A methyltransferases

2018; Wang et al. 2016a) (Fig. 2). Moreover, METTL3-METTL14 can form heterotetramers through the METTL14 homodimer (Ruzicka et al. 2017; Liu et al. 2014), which may affect substrate recognition or catalytic activity.

Interestingly, whereas the METTL3-containing fraction (MT-A) from Hela cell nuclear extract displayed low methyltransferase activity, the addition of an METTL3-absent, methylation-deficient fraction (MT-B) fully restored the activity (Bokar et al. 1994, 1997). There are two possibilities that may be not mutually exclusive: MT-B releases METTL3-METTL14 from an inhibitory state of MT-A; or METTL3-METTL14 interacts with the cofactors in MT-B to create a highly active "Super Methylation Complex" (SMC). Although the exact composition of MT-B fraction (875 kDa) has yet to be defined, some lines of evidence suggest that METTL3-METTL14 may form a SMC through direct or indirect interactions with the cofactors. The evolutionally conserved SMC is consisted of METTL3-METTL14 core methylation subunits and some cofactors that at least include WTAP (Wilms tumour 1-associated protein), KIAA1429 (VIRMA, vir like m$^6$A methyltransferase associated), RBM15/RBM15B (RNA binding motif protein 15/15B), HAKAI (E3 ubiquitin-protein ligase Hakai) and ZC3H13 (zinc finger CCCH domain-containing protein 13) (Schwartz et al. 2014; Horiuchi et al. 2013; Knuckles et al. 2018; Yue et al. 2018; Zhong et al. 2008). Whereas the core methylation subunits METTL3 and METTL14 tend to have comparable stoichiometry and remain a biochemically stable complex, the cofactors are less associated with the core subunits (Knuckles et al. 2018), which suggests that a highly active SMC may be short-lived to prevent hyperactivity of the methyltransferase, that methylation on different sites is tightly controlled by cofactors, and that heterogenous SMCs may exist in cells.

HAKAI, an E3-ubiquitin ligase that interacts with MTB in *A. thaliana* and WTAP complex through the RING finger domain, has been shown to affect m$^6$A levels to some extent (Ruzicka et al. 2017; Yue et al. 2018; Horiuchi et al. 2013). WTAP has no methyltransferase capacity, but it directly binds to METTL3 and is required for efficient methylation in vivo (Ping et al. 2014; Liu et al. 2014; Zhong et al. 2008). WTAP orthologs that include FIP37 in *A. thaliana,* Fl(2)d in *D. melanogaster* and Mum2 in *S. cerevisiae,* are all METTL3 partners. In addition, GST-WTAP has been shown to bind to the 3′ UTR of Cyclin A2 mRNA directly in vitro (Horiuchi et al. 2006), suggesting its potential role in selectivity of m$^6$A methylation.

On the other hand, the association between RBM15/RBM15B proteins and METTL3-METTL14 is WTAP-dependent. An RBM15/RBM15B-containing SMC is likely to be recruited to RBM15/RBM15B binding sites to methylate adjacent consensus motifs (Patil et al. 2016; Kan et al. 2017). Depletion of VIRMA, another WTAP-associated protein, reduces 3′ UTR methylation and seems to increase some 5′ UTR methylation as well (Yue et al. 2018). Likewise, ZC3H13 mainly affects 3′ UTR methylation and appears to bridge RBM15/RBM15B and WTAP through its C-terminal region (Wen et al. 2018; Knuckles et al. 2018). Interestingly, loss of Zc3h13 induces translocation of a great majority of WTAP, Virma, Hakai, Mettl3 and Mettl4 from nucleus to cytoplasm, where WTAP remains to be associated with Virma, Hakai, Mettl3 and Mettl4 (Wen et al. 2018). Since m$^6$A modification in polyadenylated RNA has only been found to occur in the nucleus (Wen et al. 2018; Ke et al. 2017; Sommer et al. 1978), ZC3H13 with a related nuclear localization of the m$^6$A processing machinery is essential for m$^6$A methylation in vivo. Orthologs of VIRMA, RBM15/RBM15B, and ZC3H13 have also been found in *D. melanogaster*, suggesting an evolutionally conserved mechanism for m$^6$A regulation. Then, the question is if there is any SMC component that can stimulate the intrinsic METTL3-METTL14 activity. Thus far, it is not clear whether these proteins influence m$^6$A methylation directly or indirectly since these cofactors do not exist excusively in SMC—they exert cellular functions in other complexes as well. As revealed by genetic studies in *Drosophila, ime4*, *mettl14*, *YT521-B* mutants are homozygous viable, whereas *fl(2)d* (WTAP), *Virilizer* (VIRMA) and *Nito* (RBM15/RBM15B) mutants are homozygous lethal (Kan et al. 2017), which suggests that the participation of methyltransferase in a WTAP complex is not essentially required for the function of the latter, whose defects have broad effects on the gene expression beyond m$^6$A.

Since METTL3-METTL14 complexes prepared from *E. coli*, insect cells and mammalian cells display similar activities in vitro, post-translational modifications are not expected to have a substantial impact on the methylation activity (Wang et al. 2016a). Indeed, although multiple phosphorylation sites have been identified in METTL3 and METTL14, none of them seems to affect methylation (Scholler et al. 2018). However, unlike phosphorylation, SUMOylation of METTL3 in the N-terminal region near the two CCCH motifs is suggested to repress its methyltransferase activity with an unknown mechanism (Du et al. 2018).

## 4 METTL16 Is an m$^6$A Methyltransferase That Prefers Structured RNAs

METTL16, the mammalian ortholog of *Escherichia coli* methyltransferase for A1618 in the 23S ribosomal RNA, was reported to be an m$^6$A methyltransferase for U6 snRNA A43 methylation and for ~20% m$^6$As in either total RNA or polyadenylated RNA in human HEK293 cells (Pendleton et al. 2017; Warda et al. 2017). It is found that 82% of the METTL16-dependent $m^6A$ sites are in introns or spanned intron-exon boundaries (Pendleton et al. 2017), and 87% of the METTL16-bound pre-mRNA introns are constitutively spliced (Warda et al. 2017). Despite a far less methylation scope relative to the METTL3-METTL14 complex, homozygous Mettl16$^{-/-}$ knockout only allows mouse embryonic development until blastocyst stage (E3.5) and causes developmental arrest around the time of implantation (Mendel et al. 2018). Loss of Mettl16 leads to massive transcriptome dysregulation that includes reduced mRNA levels of the SAM synthetase Mat2a (Mendel et al. 2018), which encodes a catalytic subunit ($\alpha 2$) in the methionine adenosyltransferase isoenzyme MAT II. METTL16 introduces $m^6A$ at six vertebrate-conserved hairpin structures in the MAT2A 3′ UTR, wherein the binding of METTL16 at hp1 regulates the expression of a nuclear retained-intron isoform fated for nuclear decay. Different from hp1, $m^6As$ in the hp2–6 cluster promote degradation of MAT2A mRNA without affecting intron retention (Pendleton et al. 2017; Shima et al. 2017; Pendleton et al. 2018). When SAM level is low, METTL16 dwells longer on hp1, which promotes the splicing of an otherwise retained proximal upstream intron to facilitate MAT2A expression and to maintain SAM homeostasis. Of note, it is the binding of METTL16 rather than its methyltransferase activity that induces MAT2A splicing through the METTL16-VCR domains (Pendleton et al. 2017) (Fig. 2). Interestingly, MAT2A protein also associates with METTL16 in mammalian cells (Shima et al. 2017), suggesting an intimate involvement of METTL16 in the SAM-sensing feedback mechanism.

Apart from its interaction with U6 snRNA, METTL16 also associates with other non-coding RNAs, such as XIST (X inactive specific transcript) and MALAT1 (metastasis associated lung adenocarcinoma transcript 1) (Pendleton et al. 2017; Warda et al. 2017). MALAT1 contains a unique triple helix structure that is specifically recognized by METTL16 (Brown et al. 2016), suggesting that METTL16 may prefer structured RNAs. Structural analyses of METTL16 show that the RNA-binding groove comprising positively charged residues from the N-terminal module and the MTD allows RNA binding for subsequent methylation of the target adenosine, which is captured by a hydrophobic pocket (Ruszkowska et al. 2018; Doxtader et al. 2018; Mendel et al. 2018). The MTD contains a polypeptide loop that adopts an autoinhibitory conformation, characterized by a key lysine residue near the SAM binding site. Destabilizing mutations within this K-loop lead to a hyperactivity of METTL16 (Doxtader et al. 2018). Although a consensus sequence for METTL16 methylation is absent, the target adenosine must be unpaired and surrounded by stems. The transition region of the RNA stem-loop is a key region affecting catalysis, which possibly involves a disordered loop within the MTD that is not required for RNA binding (Doxtader et al. 2018; Mendel et al. 2018). Interestingly, the C-terminal region of METTL16 is required for an METTL16/MALAT1 RNA triple helix interaction with a stoichiometry of 2:1 (Ruszkowska et al. 2018), but it is unclear whether the two METTL16 molecules exist as monomers and homodimer in the complex.

# 5 m$^6$A Methylation Occurs Before Splicing and Depends on Transcription

The early radioactive labeling experiments by J. E. Darnell laboratory suggested that (1) m$^6$A deposition mainly occurs in exon regions of mRNA co-transcriptionally or shortly after transcription completion (Lavi et al. 1977; Sommer et al. 1978) and (2) methylation can occur before splicing (Chen-Kiang et al. 1979). These pioneer studies defined the m$^6$A-formation step in the course of mRNA biogenesis and processing in steady-state cells, although it is unknown when m$^6$A occurs in cells under stress conditions, such as heat shock and ultraviolet-induced DNA damage response (Zhou et al. 2015; Xiang et al. 2017). In agreement with these concepts, METTL3 can localize to transcription active sites in an RNA-dependent manner as marked by RNAP II (RNA polymerase II) co-localization (Haussmann et al. 2016). An association between METTL3 and RNAP II was observed when cells were treated with camptothecin, an topoisomerase I inhibitor (Slobodin et al. 2017). Moreover, upon transcription inhibition by actinomycin D, WTAP nulear localization is altered, which may partially explain the m$^6$A dependence on transcription (Little et al. 2000). On the other hand, the intronic existence of m$^6$A in eukaryotic cells (Carroll et al. 1990; Dominissini et al. 2012), albeit at a very low abundance, suggests that at least these mRNAs are methylated before spliced. The m$^6$A-CLIP study from Robert B. Darnell laboratory showed that a majority of m$^6$As are added to exons of nascent pre-mRNA (Ke et al. 2017). However, ~43–49% and ~29–34% of binding sites for METTL3/METTL14/WTAP identified by PAR-CLIP largely fall into intergenic regions and introns, respectively (Liu et al. 2014). Consistent with the PAR-CLIP study, ~57% of early m$^6$A peaks around consensus motif or motifs sharing an SAG core arose within introns by bromouridine (BrU) pulse-chase of nascent RNAs (Louloupi et al. 2018). The discrepancies could be caused by different factors that include microRNAs (Alarcon et al. 2015) in these regions, distinct kinetics of transcription elongation, RNA processing, methylation and demethylation, among others.

Despite the establishment of m$^6$A as a primarily co-transcriptional process, the mechanisms underlying the interplay of m$^6$A deposition and transcription are unclear. Several intriguing questions of m$^6$A formation are begging answers: (1) only selected consensus sequences are methylated; (2) methylation is nonstoichiometric, with most genes displaying less than 50% methylation levels; (3) methylation of a gene transcript may be changed in different conditions. While the selection of primary sequences may reflect a context effect of the RNA substrates which emphasizes the interaction with SMC components, the latter two facts are likely to be contributed by different transcription activators, co-factors or regulators. This conceptual model can be supported by several lines of evidence: (1) the recruitment or displacment of METTL3 at genomic regions can be induced by transcription activators in response to stimulus or development conditions; (2) an METTL3 accumulation in nuclear speckles where most proteins are not active in RNA processing suggests an inducible interaction between METTL3 and nascent RNAs that are synthesized in the chromatin-associated regions. This model also implies that histone modifications and DNA methylation may play indirect role in m$^6$A formation (Zhang 2018).

For the METTL3-METTL14 complex, it is tempting to hypothesize that the sequential recruitment of m$^6$A methylation cofactors to nascent transcripts following transcription activation nucleates a pre-methylation complex (PMC), which further assembles to an active SMC upon interaction with the METTL3-METTL14 core subunits (Zhang 2018). In this transcription → PMC (without core subunits) → SMC → m$^6$A methylation pathway, RNA-binding proteins (RBPs) with different target sequence preferences, such as RBM15 (and possibly WTAP), first bind to nascent transcripts and seed the PMC. If the methylation core subunits join, a SMC will be formed, stabilized and able to modify nearby consensus sequences around RBP binding sites. In this methylation pathway, the PMC formation on a near consensus sequence and its transition to a SMC would be rate-limiting after transcription activation. It would appear therefore that various associations between RBPs and SMC components, or potential heterogenous SMCs under different conditions, cooperatively define the precise temporal and spatial m$^6$A landscapes.

For METTL16 methyltransferase, the relationship between m$^6$A decoration and transcription is currently unknown. The METTL16 substrate U6 snRNA is transcribed by RNA polymerase III. An early study showed that U6 A43 can be methylated by HeLa whole cell S-100 extract with a high efficiency (Shimba et al. 1995), suggesting transcription might not be essential to this type of methylation. Whether m$^6$A methylation of RNA pol II transcribed genes by METTL16 depends on transcription warrants further study.

## 6 m$^6$A Demethylation Occurs on Nascent Transcripts in Mammalian Cells

The first identified m$^6$A demethylase FTO was initially shown to utilize α-ketoglutarate (α-KG)-and Fe(II) to demethylate 3-methylthymine (3-meT) and 3-methyluracil (3-meU) in single-stranded DNA and RNA, respectively, with low in vitro activities (Jia et al. 2008; Gerken et al. 2007). Later, FTO was found to demethylate ~8–20% cellular m$^6$As in mammalian cells (Wei et al. 2018a; Jia et al. 2011), with $N^6$-hydroxymethyladenosine (hm$^6$A) and $N^6$-formyladenosine (f$^6$A) as intermediate products during the demethylation (Fu et al. 2013). Soon after, the target spectrum of FTO in vivo was further extended to include the $N^6$, 2′-O-dimethyladenosine (m$^6$A$_m$) adjacent to the $N^7$-methylguanosine (m$^7$G) cap in mRNA (Mauer et al. 2017) (Fig. 1), $N^1$-methyladenosine (m$^1$A) in tRNA, m$^6$A in U6 RNA and m$^6$A$_m$ in several snRNAs (Wei et al. 2018a). Notably, whereas the demethylase activity of FTO on cap m$^6$A$_m$ is higher relative to internal m$^6$A, (Wei et al. 2018a; Mauer et al. 2017), the cellular levels of m$^6$A$_m$ in polyadenylated RNA are only ~1/10–1/25 of that of m$^6$A (Wei et al. 2018a). Upon demethylation, the m$^6$A$_m$ of m$^7$Gpppm$^6$A$_m$ is converted to 2′-$O$-methyladenosine (A$_m$) via an $N^6$-

hydroxymethyl intermediate ($m^7Gppphm^6A_m$). The consequence of demethylation was thought to reduce RNA stability (Mauer et al. 2017). However, a more careful estimation of demethylation effect on the transcripts containing $m^6A_m$ without any $m^6A$ suggests no statistically noticeable increase of RNA levels upon FTO knockdown in HEK293T cells (Wei et al. 2018a). Considering the lack of $m^6A_m$ in *Arabidopsis* and *Drosophila* polyadenylated RNA, whether the reversible modification near cap plays any important role in gene expression remains an open question. The other demethylase ALKBH5 has been reported to demethylate an estimated ~9–30% of total $m^6$As in human cells (Zheng et al. 2013; Zhang et al. 2017). Both FTO and ALKBH5 belong to the mammalian AlkB family of Fe(II)- and α-ketoglutarate-dependent dioxygenases that is consisted of nine AlkB homologs (ALKBH1-8, FTO). FTO and ALKBH5 are not essential for embryonic development, but each possesses non-redundant functions in specific tissues. Fto is widely expressed in adult and fetal tissues, with expression highest in the brain. $Fto^{-/-}$ mice show postnatal growth retardation with a significant reduction in adipose tissue and lean body mass (Fischer et al. 2009). Alkbh5 is also widely expressed, with most abundant expression in the testes where it is essential for spermatogenesis and male fertility (Zheng et al. 2013). The characterization of ALKBH10B and ALKBH9B, two *Arabidopsis* ALKBH5 orthologs with an $m^6A$ demethylase activity on single-stranded RNA, indicates that demethylation occurs not solely in mammalian cells. Inactivation of ALKBH10B demethylase activity destabilizes mRNA transcripts of *FT, SPL3*, and *SPL9*, delays flowering, and represses vegetative growth (Duan et al. 2017). ALKBH9B was identified as an interactive protein with coat protein (CP) of alfalfa mosaic virus (AMV) by an yeast two-hybrid screen. ALKBH9B positively regulates AMV infection and influences $m^6A$ levels in the AMV genome. In contrast, cucumber mosaic virus (CMV) CP fails to interact with ALKBH9B. Moreover, $m^6A$-modified viral RNAs of CMV differ from those of AMV in that neither the methylation levels nor viral infection is affected by ALKBH9B, suggesting a selectivity for substrate demethylation (Martinez-Perez et al. 2017). Interestingly, ALKBH9B exhibits a partial colocalization with siRNA bodies and P bodies in cytoplasm, thus raising the possibility of cytoplasmic demethylation in *A. thaliana*.

Because the mammalian demethylases were found to mainly localize in nucleus, demethylation in mammalian cells was thought to be a nuclear event in normal conditions (Jia et al. 2011; Zheng et al. 2013; Zhang et al. 2017). However, FTO is also found in the cytoplasm of some cell lines (Wei et al. 2018a). HEK293T cells expressing a cytoplasm-retained FTO mutant for lack of a nuclear localization signal at N-terminal region show reduced $m^6A_m$ but not $m^6A$ in polyadenylated RNA. A more detailed analysis on nuclear and cytoplasmic fractions suggested that the FTO-mediated demethylation of $m^6A$ is prominent in the nucleus, whereas $m^6A_m$ is more affected in the cytoplasm of HEK293T cells. This observation contrasts with the finding that FTO knockdown in some acute myeloid leukemia (AML) cell lines leads to increased $m^6A$ levels in both nucleus and cytoplasm, although the $m^6A_m$ levels only increase in the cytoplasm (Wei et al. 2018a). These findings unambiguously favor nucleus as the primary site for FTO-mediated $m^6A$ demethylation and also point out that FTO-mediated demethylation of $m^6A_m$ mainly occurs in cytoplasm. The concomitant $m^6A$ change in cytoplasm could be caused by RNA export and turnover of the $m^6A$-methylated fraction.

Since $m^6A$ of mammalian cell mRNA is added and removed in nucleus, an interesting question is whether demethylation takes place on mature mRNA or nascent transcripts. Based on the subcellular fractionation experiments, Zhang et al. (2017) measured the levels of ALKBH5 protein and FOXM1 (Forkhead box M1) transcripts in three subcellular fractions: a cytoplasmic fraction, a soluble nuclear fraction (nucleoplasmic) and an insoluble fraction that contains chromatin and associated ternary transcription complexes (RNAPII with attached nascent transcripts). It is found that, in glioblastoma cells, FOXM1 nascent transcripts are exclusively present in the insoluble fraction, where there is also a large amount of ALKBH5 proteins. The depletion of ALKBH5 increases $m^6A$ levels on FOXM1 nascent transcripts and mature mRNAs equally. The native RNA immunoprecipitation with nuclear ALKBH5 confirmed its interaction with FOXM1 nascent rather than mature transcripts even though the latter is readily detected in nucleus, indicating that nascent transcripts attached to the chromatin are the nuclear substrates for ALKBH5 (Zhang et al. 2017). Similar subcellular fractionation experiments carried out by R.B. Darnell laboratory were further coupled with $m^6A$-CLIP to map and quantify each $m^6A$ in RNAs from the three fractions (Ke et al. 2017). In HeLa cells, about 10% of the $m^6As$ in pre-mRNA exons disappear when these RNAs are released from chromatin, whereas >99% of nucleoplasmic $m^6As$ remain unchanged when they become cytoplasmic (Ke et al. 2017). These two studies pointed out that, in steady state, $m^6A$ demethylation takes place in the nascent transcripts associated with chromatin. However, where demethylation occurs in response to stress is not known (Xiang et al. 2017; Zhou et al. 2018). Since the demethylase FTO is active in cytoplasm where $m^6A$ demethylation rarely happens, it is tempting to speculate an unidentified factor that directs demethylases to nascent transcripts and perhaps stimulates their activities as well. The possible candidates include protein modifications, long non-coding RNAs and other trans and cis factors, depending on what is achieved from certain contexts. A relevant example that has been illustrated above, is the specific interaction of ALKBH9B and AMV CP. Similarly, viral infection induces a critical acetylation of the nuclear DEAD-box helicase member DDX46 at Lys470 that allows ALKBH5 recruitment via DDX46's DEAD helicase domain to demethylate $m^6A$-modified antiviral transcripts, such as MAVS, TRAF3 and TRAF6, and therefore to enforce their retention in the macrophage nucleus and to reduce their translation (Zheng et al. 2017) (the molecular mechanism of nuclear retention will be discussed later). Long non-coding RNAs with sequence complementary to pre-mRNA have also been suggested to enhance ALKBH5-mediated demethylation on specific nascent transcripts (Zhang et al. 2017).

In contrast to the preferential exonic deposition of $m^6A$ as reported by R. B. Darnell laboratory, two studies addressing FLAG-FTO targets by CLIP showed the preferential binding of FTO in introns relative to exons (Bartosovic et al. 2017; Wei et al. 2018a). It should be mentioned that the $m^6A$ consensus motif is not enriched in FTO binding sequences and that the FTO binding sites appear not to overlap with $m^6A$ sites very well (Bartosovic et al. 2017). Nevertheless, it agrees with the reports that show a large fraction of early $m^6$As and METTL3-METTL14 binding sites in introns (Liu et al. 2014; Louloupi et al. 2018). An explanation would be these intronic $m^6$As are extremely short-lived due to frequent demethylation. In fact, this is not entirely impossible because upon UV stress, $m^6$A can be efficiently added by METTL3 and removed by FTO both within a few minutes (Xiang et al. 2017). Future studies should provide direct evidence to prove the emergence of a large group of intronic $m^6$As in cells depleted of any $m^6$A demethylase activity, if the undetectable intronic $m^6$As are indeed because of rapid demethylation.

## 7 Nuclear $m^6$A Regulates Alternative Splicing and Export

Since $m^6$A methylation occurs before splicing completion, $m^6$A was conceived to participate in the regulation of RNA processing a few decades ago (Stoltzfus and Dane 1982; Camper et al. 1984). However, embryonic stem cells depleted of Mettl3 have all $m^6$A-containing constitutive exons spliced normally, with only a minor change of alternative splicing events, which suggests no common obligatory role of $m^6$A in splicing (Ke et al. 2017). It is of importance to be noted that any change of alternative splicing by depletion of methyltransferase or demethylase cannot be simply interpreted as a direct effect of $m^6$A, because some small non-coding RNAs involved in splicing are possibly regulated by $m^6$A or $m^6A_m$ and, as exemplified by METTL16, methylation activity is not required for its regulation of MAT2A intron retention. A direct contribution of $m^6$A to alternative splicing is that $m^6$A destabilizes the hairpin structure to increase the accessibility of its surrounding single-stranded RNA binding motifs to RNA binding proteins, such as heterogeneous nuclear ribonucleoprotein C (HNRNPC) and HNRNPG, which modulate the alternative splicing of some gene transcripts (Liu et al. 2015, 2017). In the case of HNRNPC, the presence of methylation in an $m^6$A-containing stem loop enhances the HNRNPC interaction with the opposing poly-U strand that is otherwise buried within their local RNA structures. The majority of these $m^6$A-switches are found in introns and depend on METTL3-METTL14 mediated methylation. Knockdown of HNRNPC or METTL3-METTL14 resulted in differentially expressed splice variants associated with 221 $m^6$A-switch-containing genes in HEK293T cells (Liu et al. 2015). Similarly, $m^6$A-switches also modulate the binding affinity of HNRNPG to RNA (Liu et al. 2017). On the other hand, the cofactors of methyltransferase with additional roles in splicing may provide a connection with $m^6$A, such as WTAP, which facilitates alternative splicing (Ortega et al. 2003). Likewise, YT521-B, the ortholog of human $m^6$A binding protein YTHDC1, regulates alternative splicing of Sxl in *D. melanogaster* (Haussmann et al. 2016; Lence et al. 2016). Moreover, *Alkbh5* knockout in the mouse testes leads to several hundred of shorter 3′ UTR transcripts during spermatogenesis (Tang et al. 2018), likely due to alternative splicing or alternative polyA (APA) usage in the last exon (Ke et al. 2015). APA could be partially mediated by VIRMA, an SMC component that is associated with polyadenylation cleavage factors CPSF5 (cleavage and polyadenylation specificity factor subunit 5) and CPSF6 in an RNA-dependent manner (Yue et al. 2018). Depletion of VIRMA or METTL3 leads to 3′ UTR lengthening of a few hundred of genes in HeLa cells, indicating a potential role of m$^6$A in the selection of proximal polyadenylation sites (Yue et al. 2018). Again, despite the correlation, unresolved is if the m$^6$A within 3′ UTR *per se* regulates APA, or vice versa.

After splicing is completed, RNA export may also be influenced by m$^6$A. A delayed appearance of mRNA in the HeLa cell cytoplasm caused by methyltransferase inhibitors was reported by Rottman laboratory based on radioactive labeling (Camper et al. 1984). Subsequent studies showed that a delayed cytoplasmic appearance of bPRL mRNA was accompanied with nuclear accumulation of precursor RNA (Carroll et al. 1990). Similarly, the application of methylation inhibitors leads to a prolonged nuclear retention of two key clock genes, *PER2* (period circadian clock 2) and *ARNTL* (aryl hydrocarbon receptor nuclear translocator like), and thereby an altered schedule of gene expression that results in an elongated circadian period (Fustin et al. 2013). On the other hand, RNA export is enhanced by ALKBH5 knockdown (Zheng et al. 2013), which may be mediated by YTHDC1, the only nuclear exclusive YTH domain family member. Despite its role in *D. melanogaster*, YTHDC1 mainly binds to exons (Xu et al. 2014) and has very minor effect on splicing in Hela cells (Roundtree et al. 2017b). Depletion of YTHDC1 induces a nuclear accumulation of mature m$^6$A-marked mRNAs and a decrease in their cytoplasmic appearance. The nuclear export function of YTHDC1 is mediated by an RNA-independent interaction between the C-terminal YTHDC1 and SRSF3 (serine/arginine-rich splicing factor 3), which interacts with the canonical mRNA export receptor NXF1 (nuclear RNA export factor 1) (Roundtree et al. 2017b).

## 8 Cytoplasmic m$^6$A Regulates mRNA Stability and Translation Efficiency

Before the advance of the "omics" era, biochemical studies detected a positive correlation between m$^6$A decoration and mRNA decay (Friderici et al. 1976; Sommer et al. 1978). These radioactive labeling experiments provided the first evidence of the functional relevance of m$^6$A. The principle of m$^6$A-mediated post-transcriptional gene regulation is that RBPs recruited or repelled by m$^6$As modify the fate of mRNA, such as decay and translation, depending on the function of RBP *per se* or through the association with a traditional pathway. Among these RBPs, the proteins harboring a YTH domain are described as the m$^6$A "readers" that specifically recognize and regulate m$^6$A-decorated transcripts (Dominissini et al. 2012; Wang et al. 2014a). The mammalian YTH domain proteins that have been largely characterized by He laboratory include the cytoplasmic YTHDF1, YTHDF2, YTHDF3, YTHDC2 and the nuclear YTHDC1, all of which employ their conserved aromatic cage to recognize $m^6A$ (Xu et al. 2014; Zhu et al. 2014). With the list of "reader" proteins extended by other RBPs that are absent of a YTH domain, a critical regulatory network has emerged that impacts multiple aspects of mRNA homeostasis, which cannot be achieved by the primary sequence alone.

## 8.1 *m⁶A and mRNA Decay*

The first confirmed $m^6A$ reader-mediated effect on gene expression arose from the mechanistic studies of YTHDF2 (Wang et al. 2014a). YTHDF2 accelerates mRNA deadenylation and degradation, with the C-terminal YTH domain specifically recognizing $m^6A$-methylated RNA and the N-terminal region interacting with the CCR4–NOT deadenylase complex through CNOT1 (CCR4–NOT transcription complex subunit 1) (Wang et al. 2014a; Du et al. 2016). In early life of zebrafish embryos, the depletion of Ythdf2 stabilizes $m^6A$-modified maternal mRNAs and delays maternal-to-zygotic transition (Zhao et al. 2017). Likewise, Ythdf2 is maternally required for oocyte maturation (Ivanova et al. 2017). Moreover, Ythdf2 depletion leads to a disturbed $m^6A$-dependent degradation of neural development-related mRNA targets, and impaired proliferation and differentiation capabilities of neural stem/progenitor cells, consequently leading to defective neurogenesis. Homozygous *Ythdf2*$^{-/-}$ mice in C57BL/6 background are embryonic lethal, with the majority of *Ythdf2*$^{-/-}$ embryos lost between E14.5 and E18.5. Semi-lethality was observed for the *Ythdf2*$^{+/-}$ mice, with one third of the surviving mice having malfunctioning eyes (Li et al. 2018). It is not surprising that depletion of $m^6A$ methyltransferase, accompanied with a failure of $m^6A$-induced decay of methylated transcripts, would give rise to similar or more severe phenotypes. Indeed, a loss of $m^6A$ methyltransferase exhibits significant impacts on mouse ESC differentiation and preimplantation development (Batista et al. 2014; Wang et al. 2014b; Geula et al. 2015). *Mettl3* knockout mice display more severe early embryo lethality phenotypes than *Ythdf2* knockout, with E5.5-E7.5 *Mettl3*$^{-/-}$ embryos having abnormal characteristics (Geula et al. 2015), indicating additional mechanisms for $m^6A$-mediated effects as discussed later. Interestingly, depending on the dominance of already-expressed genes in different cell states to be stabilized, the loss of methyltransferase in murine ESCs can lead to opposite cell fates, i.e., an improved self-renewal in naïve state but a defective cell regeneration in primed state (Wang et al. 2014b; Geula et al. 2015; Batista et al. 2014). In general, mRNAs that encode key proteins with regulatory functions tend to have more $m^6A$s with faster turnover (Wang et al. 2014b; Ke et al. 2017). A proposed model describes that $m^6A$ induces rapid clearance of a group of mRNAs encoding key regulators, such as transcription factors, in a coordinated manner during the transition of cell states (Roundtree et al. 2017a). Additionally, $m^6A$ may destabilize some transcripts by repelling RBPs without a YTH domain, such as G3BP1 (G3BP stress granule assembly factor 1) and HuR (ELAV-like protein 1) (Edupuganti et al. 2017; Wang et al. 2014b).

Not only m$^6$A methyltransferase and its cofactors are evolutionally conserved, the "readers" coevolve from yeast to mammals as well (Schwartz et al. 2013). A phylogenetic analysis of 297 YTH-containing proteins from 32 representative species suggests that 57 of them belong to the DC group and 240 to the DF group (Scutenaire et al. 2018). Eleven of the 13 members of the *Arabidopsis* YTH protein family, namely ECT1-11 for Evolutionarily Conserved C Terminus, fall into the YTHDF clade. One of the two YTHDC-type proteins possesses three highly conserved N-terminal zinc fingers in addition to the YTH domain and is a member of the plant polyadenylation complex (CPSF30; AT1G30460). ECT2 is required for trichome branching and involved in the redundant control of timing of leaf formation with ECT3 (Arribas-Hernandez et al. 2018). Surprisingly, by formaldehyde-assisted crosslinking of ECT2 and associated RNA in *A. thaliana*, G. Jia laboratory found that the majority of ECT2-binding sites are located within the 3′ UTR of mRNA containing a unique URUAY motif (R=G>A, Y=U>A, where the majority [over 90%] UGUAY)(Wei et al. 2018b). ECT2 appears to bind methylated UGUAA and methylated GGACU with similar affinities (Wei et al. 2018b), suggesting the sequence bias may reflect the relative abundance of methylated motifs in the cells, and in line with this, *Arabidopsis* nuclear extracts exhibit higher methylation activity for UGUA relative to GGACU. However, the RRACH is more significantly enriched than UGUAY in overall m$^6$A peaks, and interestingly, UGUAY is identical to a specific UGUAACA methylation signature near the start codon (Luo et al. 2014). The ECT2 binding motifs also share sequence similarity with far upstream elements that regulate polyadenylation, likely lending evidence of a potential regulatory role in RNA processing (Wei et al. 2018b). However, ECT2 mostly accumulates in the cytoplasm with a diffuse pattern or aggregates in stress granules upon heat stress (Scutenaire et al. 2018; Arribas-Hernandez et al. 2018), which suggests its function on cytoplasmic RNAs. This is partially evidenced by ECT2-dependent stabilization of some RNA transcripts associated with trichome morphogenesis (Wei et al. 2018b).

ECT2 is not the first reader that was found to stabilize m$^6$A-marked transcripts. In mammalian cells, insulin-like growth factor 2 mRNA-binding proteins (IGF2BPs; including IGF2BP1/2/3) were reported to enhance RNA stability and translation under normal and stress conditions through the K homology domains that recognize m$^6$A residues (Huang et al. 2018). Moreover, depletion of FTO appears to increase the levels of some m$^6$A-methylated transcripts (Wei et al. 2018a). Therefore, despite the m$^6$A-mediated destabilization of a majority of transcripts, a subgroup of transcripts become more stable, which diverts the fates of mRNAs for dynamic physiological roles.

## 8.2 m$^6$A and mRNA Translation

In bacterial translation systems, m$^6$A within coding regions acts as a barrier to tRNA accommodation during translation elongation (Choi et al. 2016). However, m$^6$A was found to associate with enhanced translation efficiency of dihydrofolate reductase transcripts (Tuck et al. 1999). Recent studies support the positive role of $m^6A$ in translation through several mechanisms. First, the cap-dependent translation can be enhanced by the $m^6A$-bound YTHDF1 through recruitment of eIF3 (translation initiation factor complex 3) (Wang et al. 2015). Interestingly, YTHDF1 was found to associate with G3BP1 in an RNA-dependent manner (Wang et al. 2015). Although genetic deletion of YTHDF1 allows mice to develop normally at least to four months of age, these mice show learning and memory defects, impaired hippocampal synaptic transmission and long-term potentiation due to the loss of YTHDF1-dependent expedited translation of $m^6A$-methylated neuronal mRNAs in response to neuronal stimulation in the hippocampus (Shi et al. 2018). Moreover, YTHDF3 also promotes translation through interaction with YTHDF1 (Shi et al. 2017). Surprisingly, cytoplasmic METTL3 enhances translation of mRNAs bearing $m^6As$ that are close to stop codon independently of its methyltransferase activity or YTHDF1. In cytoplasm, in addition to a direct METTL3-$m^6A$ interaction, the N-terminal region of METTL3 binds eIF3h (translation initiation factor 3 subunit h) to facilitate mRNA looping, thus promoting translation of a large subset of oncogenic mRNAs, such as BRD4 (bromodomain-containing protein 4) in human primary lung tumors (Lin et al. 2016; Choe et al. 2018). Second, $m^6As$ within 5′ UTR can be recognized by eIF3, which recruits the 43S complex and initiates translation in a cap- and internal ribosome entry site (IRES)-independent manner (Meyer et al. 2015; Zhou et al. 2015). ABCF1 (ATP-binding cassette sub-family F member 1) and YTHDF3 appear to contribute to this type of $m^6A$-facilitated mRNA translation to some extent (Coots et al. 2017). Other RBPs implicated in translation include YTHDC2 that increases translation efficiency (Hsu et al. 2017), and FMR1 (fragile X mental retardation 1) (Edupuganti et al. 2017), a polyribosome-associated neuronal RNA-binding protein associated with Fragile X syndrome that represses translation by stalling ribosome translocation.

Elucidating $m^6A$ function at different conditions is central for unraveling the role of $m^6A$ in more complex processes. Upon heat shock, YTHDF2 translocates into nucleus and binds to the $m^6A$ within 5′ UTR, which blocks FTO-mediated demethylation and hence enhances translation of stress-induced transcripts (Zhou et al. 2015). During integrated stress response, the 5′ UTR-located $m^6A$ controls ribosome scanning and start codon selection, while demethylation promotes the reinitiation of ATF4 (activating transcription factor 4) translation (Zhou et al. 2018). These complex $m^6A$-mediated pathways of translational regulation undoubtedly add new dimensions to the dynamic effects of $m^6A$ in relation to gene expression.

## 9 $m^6A$ Methyltransferases and Demethylases Play Important Roles in Cancer

The comprehensive functional analyses for $m^6A$ regulators in cancer were initiated by an FTO study in leukemia and an ALKBH5 study in glioblastoma (Li et al. 2017; Zhang et al. 2017), and have been extended by a plethora of studies focused on the dysregulation of m$^6$A methylation and associated proteins. FTO is highly expressed in several subtypes of AMLs, including those that carry t(11q23)/*MLL*-rearrangements, t(15;17)/*PML-RARA* translocation, *FLT3* gene internal tandem duplication (ITD) mutations and/or *NPM1* mutation. FTO demethylates 3′ or/and 5′ UTR-located m$^6$As on the transcripts of *ASB2* (ankyrin repeat and SOCS box containing 2) and *RARA* (retinoic acid receptor alpha), which encode key regulators in all-trans-retinoic acid (ATRA)-induced differentiation of leukemia cells. It therefore enhances ASB2 and RARA mRNA decay and induces downregulation of these two genes. The oncogenic role of FTO in leukemic cell transformation and leukemogenes was confirmed by a series of in vitro and in vivo assays (Li et al. 2017). Interestingly, the oncogenic activity of FTO is inhibited by 2-hydroxyglutarate (R-2HG), an oncometabolite competitive inhibitor of α-KG-dependent dioxygenases that is produced by the cancer-associated mutant *IDH1/2* (isocitrate dehydrogenase 1/2) (Su et al. 2018). In leukemic cells that highly express FTO with a "medium" level of MYC, FTO affects the stability of MYC transcripts by changing m$^6$A levels at MYC 5′ UTR and CDS regions, which confers R-2HG sensitivity in these cells. However, hyperactivation of MYC-associated pathways renders leukemic cells resistant to R-2HG. On the other hand, in glioblastoma multiforme (GBM), ALKBH5 is highly expressed in the patient-derived glioma stem cells (GSCs) and informs poor survival for GBM patients (Zhang et al. 2017). ALKBH5 depletion impaires GSC growth in vitro and in vivo due to the loss expression of FOXM1, an essential transcription factor for GSC self-renewal and proliferation. ALKBH5 binds to and demethylates the 3′ UTR of *FOXM1* nascent transcripts, thereby enhancing FOXM1 mRNA expression through interaction with HuR, an abundant nuclear RBP that promotes pre-mRNA expression. The demethylation seems to be facilitated by a long noncoding RNA transcribed antisense to the *FOXM1*, which harbors an overlapping region with FOXM1 3′ UTR.

Interestingly, although these demethylases behave as oncogenes in these cancers, the methyltransferase complex does not simply play the opposite role, i.e., METTL3-METTL14 being oncogenic in AMLs (Barbieri et al. 2017; Vu et al. 2017; Weng et al. 2018). Studies from different laboratories suggest that METTL3-METTL14 complex may employ multiple mechanisms to activate oncogenic pathways that at least include some key transcription factors, such as MYC (Barbieri et al. 2017; Vu et al. 2017; Weng et al. 2018). Starting with a catalytic activity domain-focused Cas9-induced indel mutation screen in mouse primary leukemia cells, Barbieri *et al.* (Barbieri et al. 2017) confirmed the requirement of *METTL3*, *METTL14* and *METTL16* for cell growth in ten human AML cell lines. They also found that METTL3 and METTL14 can be crosslinked to TSSs of distinct groups of coding genes that are decorated with bimodal trimethylation of lysine 4 on histone H3 (H3K4me3) marks by chromatin immunoprecipitation assays. The promoter-bound METTL3, likely recruited by CEBPZ (CCAAT enhancer binding protein zeta), enhances translation of target genes without affecting mRNA levels, such as the transcription factors *SP1* and *SP2*, which activate the transcription of *MYC* oncogene. Similarly, Vu et al. (2017) reported METTL3 as an essential gene to maintain the undifferentiated state and survival of hematopoietic stem/progenitor cells (HSPCs) and myeloid leukemia cell lines. METTL3 depletion reduces translational efficiency and causes decreased levels of MYC, BCL2 and PTEN (phosphatase and tensin homolog) protein expression in AML cells. Weng et al. (2017) showed that METTL14 is expressed at high levels in HSPCs and AML cells carrying t(11q23), t(15;17), or t(8;21) and is downregulated during myeloid differentiation. METTL14 sustains the development and maintenance of AMLs and self-renewal of leukemia stem/initiation cells through promoting stability and translation of MYB and MYC mRNAs. METTL14 itself is transcriptionally suppressed by the transcription factor SPI1. In addition, mutations of methyltransferases have also been implicated in tumor development. The hotspot R298P mutation of METTL14 in endometrial tumors is correlated with reductions in $m^6A$ methylation, and promotes proliferation and tumorigenicity of endometrial cancer cells through activation of the AKT pathway (Liu et al. 2018).

The connections of $m^6A$ and tumor development are not limited to aforementioned types of tumor. A growing body of studies have indicated that multiple types of cancer take advantage of the tunable regulatory mechanisms to survive and propagate, through mutations or aberrant expression of components of the $m^6A$ regulatory network (e.g., writers, erasers and readers).

## 10 Conclusion and Future Perspective

mRNA $m^6A$ methylation was discovered over 40 years ago. However, the attempts from RNA biologists to understand the function and regulation of this abundant modification at molecular, cellular and tissue levels have just started in very recent years, because of the important biological functions of reversible RNA methylation revealed by a few groups (Zheng et al. 2013; Li et al. 2017; Zhang et al. 2017) and the advance of new technologies afforded by high-throughput sequencing. A current view of $m^6A$ on mRNA is that it is deposited, removed, sensed and perhaps regulated by several conserved machineries from fly to human, and that $m^6A$ plays distinct roles in life when incorporated into diverse biological processes. In Fig. 3, a simplified model of METTL3-METTL14-mediated $m^6A$ deposition is hypothesized—the recruitment of a Pre-Methylation Complex (PMC) with subsequent assembly to a Super Methylation Complex (SMC) allows efficient $m^6A$ methylation at consensus sequence adjacent to the RBP binding sites. The intracellular and extracellular signals that trigger transcription allow $m^6A$ methylation for specific genes, while the selection of consensus motif to be modified is decided by the PMC (Fig. 3a). The signals may also regulate PMC component expression, modification, or localization, etc., and as a result, $m^6A$ distribution. On the other hand, the "free" methyltransferase core subunits have intrinsic ability to bind and catalyze single-stranded RNA with low activity, likely giving rise to abortive methylation in the absence of PMC or low methylation "noise" in cells (Fig. 3b). Therefore, the sequential recruitment of SMC components ensures accurate $m^6A$ deposition in a large number of transcripts. After methylation, $m^6A$

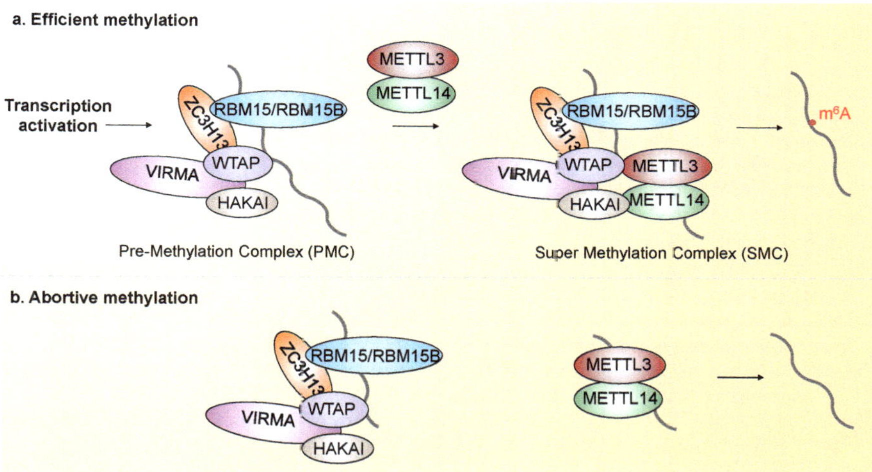


**Fig. 3** A hypothetical model of METTL3-METTL14-dependent m$^6$A deposition. The overlaps of proteins indicate confirmed or proposed interactions

demethylation occurs on a subset of nascent transcripts in mammalian cells, which is regulated by a poorly understood mechanism. Depending on the functions of the RBPs that display higher or lower binding affinities to m$^6$A-marked transcripts, a wide range of events throughout RNA life are influenced by m$^6$As that include RNA processing, localization, decay and translation, which altogether modulate gene expression for dynamic physiological functions. Elucidating how m$^6$A affects gene expression not only aids our understanding of essential physiology but also enables novel therapeutic approaches for human diseases.

## References


Alarcon CR, Lee H, Goodarzi H et al (2015) N6-methyladenosine marks primary microRNAs for processing. Nature 519:482–485

Arribas-Hernandez L, Bressendorff S, Hansen MH et al (2018) An m(6)A-YTH module controls developmental timing and morphogenesis in arabidopsis. Plant Cell 30:952–967

Barbieri I, Tzelepis K, Pandolfini L et al (2017) Promoter-bound METTL3 maintains myeloid leukaemia by m(6)A-dependent translation control. Nature 552:126–131

Bartosovic M, Molares HC, Gregorova P et al (2017) N6-methyladenosine demethylase FTO targets pre-mRNAs and regulates alternative splicing and 3′-end processing. Nucleic Acids Res 45:11356–11370

Batista PJ, Molinie B, Wang J et al (2014) m(6)A RNA modification controls cell fate transition in mammalian embryonic stem cells. Cell Stem Cell 15:707–719

Bokar JA, Rath-Shambaugh ME, Ludwiczak R et al (1994) Characterization and partial purification of mRNA N6-adenosine methyltransferase from HeLa cell nuclei. Internal mRNA methylation requires a multisubunit complex. J Biol Chem 269:17697–17704

Bokar JA, Shambaugh ME, Polayes D et al (1997) Purification and cDNA cloning of the AdoMet-binding subunit of the human mRNA (N6-adenosine)-methyltransferase. RNA 3:1233–1247

Brown JA, Kinzig CG, DeGregorio SJ et al (2016) Methyltransferase-like protein 16 binds the 3′-terminal triple helix of MALAT1 long noncoding RNA. Proc Natl Acad Sci USA 113:14013–14018

Bujnicki JM, Feder M, Radlinska M et al (2002) Structure prediction and phylogenetic analysis of a functionally diverse family of proteins homologous to the MT-A70 subunit of the human mRNA:m(6)A methyltransferase. J Mol Evol 55:431–444

Camper SA, Albers RJ, Coward JK et al (1984) Effect of undermethylation on mRNA cytoplasmic appearance and half-life. Mol Cell Biol 4:538–543

Carroll SM, Narayan P, Rottman FM (1990) N6-methyladenosine residues in an intron-specific region of prolactin pre-mRNA. Mol Cell Biol 10:4456–4465

Chen K, Lu Z, Wang X et al (2015) High-resolution N(6)-methyladenosine (m(6)A) map using photo-crosslinking-assisted m(6)A sequencing. Angew Chem Int Ed Engl 54:1587–1590

Chen-Kiang S, Nevins JR, Darnell JE Jr (1979) N-6-methyl-adenosine in adenovirus type 2 nuclear RNA is conserved in the formation of messenger RNA. J Mol Biol 135:733–752

Choe J, Lin S, Zhang W et al (2018) mRNA circularization by METTL3-eIF3h enhances translation and promotes oncogenesis. Nature 561:556–560

Choi J, Ieong KW, Demirci H et al (2016) N(6)-methyladenosine in mRNA disrupts tRNA selection and translation-elongation dynamics. Nat Struct Mol Biol 23:110–115

Clancy MJ, Shambaugh ME, Timpte CS et al (2002) Induction of sporulation in *Saccharomyces cerevisiae* leads to the formation of N6-methyladenosine in mRNA: a potential mechanism for the activity of the IME4 gene. Nucleic Acids Res 30:4509–4518

Coots RA, Liu XM, Mao Y et al (2017) m(6)A facilitates eIF4F-independent mRNA translation. Mol Cell 68:504–514 e7

Deng X, Chen K, Luo GZ et al (2015) Widespread occurrence of N6-methyladenosine in bacterial mRNA. Nucleic Acids Res 43:6557–6567

Desrosiers R, Friderici K, Rottman F (1974) Identification of methylated nucleosides in messenger RNA from Novikoff hepatoma cells. Proc Natl Acad Sci USA 71:3971–3975

Dominissini D, Moshitch-Moshkovitz S, Schwartz S et al (2012) Topology of the human and mouse m6A RNA methylomes revealed by m6A-seq. Nature 485:201–206

Doxtader KA, Wang P, Scarborough AM et al (2018) Structural basis for regulation of METTL16, an S-adenosylmethionine homeostasis factor. Mol Cell 71:1001–1011.e4

Du H, Zhao Y, He J et al (2016) YTHDF2 destabilizes m(6)A-containing RNA through direct recruitment of the CCR4-NOT deadenylase complex. Nat Commun 7:12626

Du Y, Hou G, Zhang H et al (2018) SUMOylation of the m6A-RNA methyltransferase METTL3 modulates its function. Nucleic Acids Res 46:5195–5208

Duan HC, Wei LH, Zhang C et al (2017) ALKBH10B is an RNA N(6)-methyladenosine demethylase affecting Arabidopsis floral transition. Plant Cell 29:2995–3011

Edupuganti RR, Geiger S, Lindeboom RGH et al (2017) N(6)-methyladenosine (m(6)A) recruits and repels proteins to regulate mRNA homeostasis. Nat Struct Mol Biol 24:870–878

Fischer J, Koch L, Emmerling C et al (2009) Inactivation of the Fto gene protects from obesity. Nature 458:894–898

Friderici K, Kaehler M, Rottman F (1976) Kinetics of Novikoff cytoplasmic messenger RNA methylation. Biochemistry 15:5234–5241

Fu Y, Jia G, Pang X et al (2013) FTO-mediated formation of N6-hydroxymethyladenosine and N6-formyladenosine in mammalian RNA. Nat Commun 4:1798

Fustin JM, Doi M, Yamaguchi Y et al (2013) RNA-methylation-dependent RNA processing controls the speed of the circadian clock. Cell 155:793–806

Gerken T, Girard CA, Tung YC et al (2007) The obesity-associated FTO gene encodes a 2-oxoglutarate-dependent nucleic acid demethylase. Science 318:1469–1472

Geula S, Moshitch-Moshkovitz S, Dominissini D et al (2015) Stem cells. m6A mRNA methylation facilitates resolution of naive pluripotency toward differentiation. Science 347:1002–1006

Haussmann IU, Bodi Z, Sanchez-Moran E et al (2016) m(6)A potentiates Sxl alternative pre-mRNA splicing for robust Drosophila sex determination. Nature 540:301–304
Horiuchi K, Umetani M, Minami T et al (2006) Wilms' tumor 1-associating protein regulates G2/M transition through stabilization of cyclin A2 mRNA. Proc Natl Acad Sci USA 103:17278–17283
Horiuchi K, Kawamura T, Iwanari H et al (2013) Identification of Wilms' tumor 1-associating protein complex and its role in alternative splicing and the cell cycle. J Biol Chem 288:33292–33302
Hsu PJ, Zhu Y, Ma H et al (2017) Ythdc2 is an N(6)-methyladenosine binding protein that regulates mammalian spermatogenesis. Cell Res 27:1115–1127
Huang H, Weng H, Sun W et al (2018) Recognition of RNA N(6)-methyladenosine by IGF2BP proteins enhances mRNA stability and translation. Nat Cell Biol 20:285–295
Ivanova I, Much C, Di Giacomo M et al (2017) The RNA m(6)A Reader YTHDF2 is essential for the post-transcriptional regulation of the maternal transcriptome and oocyte competence. Mol Cell 67:1059–1067 e4
Jia G, Yang CG, Yang S et al (2008) Oxidative demethylation of 3-methylthymine and 3-methyluracil in single-stranded DNA and RNA by mouse and human FTO. FEBS Lett 582:3313–3319
Jia G, Fu Y, Zhao X et al (2011) N6-methyladenosine in nuclear RNA is a major substrate of the obesity-associated FTO. Nat Chem Biol 7:885–887
Kan L, Grozhik AV, Vedanayagam J et al (2017) The m(6)A pathway facilitates sex determination in Drosophila. Nat Commun 8:15737
Kane SE, Beemon K (1987) Inhibition of methylation at two internal N6-methyladenosine sites caused by GAC to GAU mutations. J Biol Chem 262:3422–3427
Ke S, Alemu EA, Mertens C et al (2015) A majority of m6A residues are in the last exons, allowing the potential for 3′ UTR regulation. Genes Dev 29:2037–2053
Ke S, Pandya-Jones A, Saito Y et al (2017) m(6)A mRNA modifications are deposited in nascent pre-mRNA and are not required for splicing but do specify cytoplasmic turnover. Genes Dev 31:990–1006
Knuckles P, Lence T, Haussmann IU et al (2018) Zc3h13/Flacc is required for adenosine methylation by bridging the mRNA-binding factor Rbm15/Spenito to the m(6)A machinery component Wtap/Fl(2)d. Genes Dev 32:415–429
Lavi U, Fernandez-Munoz R, Darnell JE Jr (1977) Content of N-6 methyl adenylic acid in heterogeneous nuclear and messenger RNA of HeLa cells. Nucleic Acids Res 4:63–69
Lence T, Akhtar J, Bayer M et al (2016) m(6)A modulates neuronal functions and sex determination in Drosophila. Nature 540:242–247
Li Z, Weng H, Su R et al (2017) FTO plays an oncogenic role in acute myeloid leukemia as a N(6)-methyladenosine RNA demethylase. Cancer Cell 31:127–141
Li M, Zhao X, Wang W et al (2018) Ythdf2-mediated m(6)A mRNA clearance modulates neural development in mice. Genome Biol 19:69
Lin S, Choe J, Du P et al (2016) The m(6)A methyltransferase METTL3 promotes translation in human cancer cells. Mol Cell 62:335–345
Linder B, Grozhik AV, Olarerin-George AO et al (2015) Single-nucleotide-resolution mapping of m6A and m6Am throughout the transcriptome. Nat Methods 12:767–772
Little NA, Hastie ND, Davies RC (2000) Identification of WTAP, a novel Wilms' tumour 1-associating protein. Hum Mol Genet 9:2231–2239
Liu J, Yue Y, Han D et al (2014) A METTL3-METTL14 complex mediates mammalian nuclear RNA N6-adenosine methylation. Nat Chem Biol 10:93–95
Liu N, Dai Q, Zheng G et al (2015) N(6)-methyladenosine-dependent RNA structural switches regulate RNA-protein interactions. Nature 518:560–564
Liu N, Zhou KI, Parisien M et al (2017) N6-methyladenosine alters RNA structure to regulate binding of a low-complexity protein. Nucleic Acids Res 45:6051–6063
Liu J, Eckert MA, Harada BT et al (2018) m(6)A mRNA methylation regulates AKT activity to promote the proliferation and tumorigenicity of endometrial cancer. Nat Cell Biol 20:1074–1083

Louloupi A, Ntini E, Conrad T et al (2018) Transient N-6-methyladenosine transcriptome sequencing reveals a regulatory role of m6A in splicing efficiency. Cell Rep 23:3429–3437
Luo GZ, MacQueen A, Zheng G et al (2014) Unique features of the m6A methylome in Arabidopsis thaliana. Nat Commun 5:5630
Martinez-Perez M, Aparicio F, Lopez-Gresa MP et al (2017) Arabidopsis m(6)A demethylase activity modulates viral infection of a plant virus and the m(6)A abundance in its genomic RNAs. Proc Natl Acad Sci USA 114:10755–10760
Mauer J, Luo X, Blanjoie A et al (2017) Reversible methylation of m(6)Am in the 5′ cap controls mRNA stability. Nature 541:371–375
Mendel M, Chen KM, Homolka D et al (2018) Methylation of structured RNA by the m(6)A writer METTL16 is essential for mouse embryonic development. Mol Cell 71:986–1000. e11
Meyer KD, Saletore Y, Zumbo P et al (2012) Comprehensive analysis of mRNA methylation reveals enrichment in 3′ UTRs and near stop codons. Cell 149:1635–1646
Meyer KD, Patil DP, Zhou J et al (2015) 5′ UTR m(6)A promotes cap-independent translation. Cell 163:999–1010
Narayan P, Ludwiczak RL, Goodwin EC et al (1994) Context effects on N6-adenosine methylation sites in prolactin mRNA. Nucleic Acids Res 22:419–426
Ortega A, Niksic M, Bachi A et al (2003) Biochemical function of female-lethal (2)D/Wilms' tumor suppressor-1-associated proteins in alternative pre-mRNA splicing. J Biol Chem 278:3040–3047
Patil DP, Chen CK, Pickering BF et al (2016) m(6)A RNA methylation promotes XIST-mediated transcriptional repression. Nature 537:369–373
Pendleton KE, Chen B, Liu K et al (2017) The U6 snRNA m(6)A methyltransferase METTL16 regulates SAM synthetase intron retention. Cell 169:824–835.e14
Pendleton KE, Park SK, Hunter OV et al (2018) Balance between MAT2A intron detention and splicing is determined cotranscriptionally. RNA 24:778–786
Perry RP, Kelley DE, Friderici K et al (1975) The methylated constituents of L cell messenger RNA: evidence for an unusual cluster at the 5′ terminus. Cell 4:387–394
Ping XL, Sun BF, Wang L et al (2014) Mammalian WTAP is a regulatory subunit of the RNA N6-methyladenosine methyltransferase. Cell Res 24:177–189
Roundtree IA, Evans ME, Pan T et al (2017a) Dynamic RNA modifications in gene expression regulation. Cell 169:1187–1200
Roundtree IA, Luo GZ, Zhang Z et al (2017b) YTHDC1 mediates nuclear export of N(6)-methyladenosine methylated mRNAs. Elife 6:e31311
Ruszkowska A, Ruszkowski M, Dauter Z et al (2018) Structural insights into the RNA methyltransferase domain of METTL16. Sci Rep 8:5311
Ruzicka K, Zhang M, Campilho A et al (2017) Identification of factors required for m(6) A mRNA methylation in Arabidopsis reveals a role for the conserved E3 ubiquitin ligase HAKAI. New Phytol 215:157–172
Schibler U, Kelley DE, Perry RP (1977) Comparison of methylated sequences in messenger RNA and heterogeneous nuclear RNA from mouse L cells. J Mol Biol 115:695–714
Scholler E, Weichmann F, Treiber T et al (2018) Interactions, localization, and phosphorylation of the m(6)A generating METTL3-METTL14-WTAP complex. RNA 24:499–512
Schwartz S, Agarwala SD, Mumbach MR et al (2013) High-resolution mapping reveals a conserved, widespread, dynamic mRNA methylation program in yeast meiosis. Cell 155:1409–1421
Schwartz S, Mumbach MR, Jovanovic M et al (2014) Perturbation of m6A writers reveals two distinct classes of mRNA methylation at internal and 5′ sites. Cell Rep 8:284–296
Scutenaire J, Deragon JM, Jean V et al (2018) The YTH domain protein ECT2 is an m(6)A reader required for normal trichome branching in arabidopsis. Plant Cell 30:986–1005
Shi H, Wang X, Lu Z et al (2017) YTHDF3 facilitates translation and decay of N(6)-methyladenosine-modified RNA. Cell Res 27:315–328
Shi H, Zhang X, Weng YL et al (2018) m(6)A facilitates hippocampus-dependent learning and memory through YTHDF1. Nature 563:249–253
Shima H, Matsumoto M, Ishigami Y et al (2017) S-Adenosylmethionine synthesis is regulated by selective N(6)-adenosine methylation and mRNA degradation involving METTL16 and YTHDC1. Cell Rep 21:3354–3363

Shimba S, Bokar JA, Rottman F et al (1995) Accurate and efficient N-6-adenosine methylation in spliceosomal U6 small nuclear RNA by HeLa cell extract in vitro. Nucleic Acids Res 23:2421–2426

Sledz P, Jinek M (2016) Structural insights into the molecular mechanism of the m(6)A writer complex. Elife 5:e18434

Slobodin B, Han R, Calderone V et al (2017) Transcription impacts the efficiency of mRNA translation via co-transcriptional N6-adenosine methylation. Cell 169:326–337.e12

Sommer S, Lavi U, and Darnell J E, Jr. (1978) The absolute frequency of labeled N-6-methyladenosine in HeLa cell messenger RNA decreases with label time. J Mol Biol 124:487-499

Stoltzfus CM, Dane RW (1982) Accumulation of spliced avian retrovirus mRNA is inhibited in S-adenosylmethionine-depleted chicken embryo fibroblasts. J Virol 42:918–931

Su R, Dong L, Li C et al (2018) R-2HG exhibits anti-tumor activity by targeting FTO/m(6)A/MYC/CEBPA signaling. Cell 172:90–105 e23

Tang C, Klukovich R, Peng H et al (2018) ALKBH5-dependent m6A demethylation controls splicing and stability of long 3′-UTR mRNAs in male germ cells. Proc Natl Acad Sci USA 115: E325–EE33

Tuck MT, Wiehl PE, Pan T (1999) Inhibition of 6-methyladenine formation decreases the translation efficiency of dihydrofolate reductase transcripts. Int J Biochem Cell Biol 31:837–851

Vu LP, Pickering BF, Cheng Y et al (2017) The N(6)-methyladenosine (m(6)A)-forming enzyme METTL3 controls myeloid differentiation of normal hematopoietic and leukemia cells. Nat Med 23:1369–1376

Wang X, Lu Z, Gomez A et al (2014a) N6-methyladenosine-dependent regulation of messenger RNA stability. Nature 505:117–120

Wang Y, Li Y, Toth JI et al (2014b) N6-methyladenosine modification destabilizes developmental regulators in embryonic stem cells. Nat Cell Biol 16:191–198

Wang X, Zhao BS, Roundtree IA et al (2015) N(6)-methyladenosine modulates messenger RNA translation efficiency. Cell 161:1388–1399

Wang P, Doxtader KA, Nam Y (2016a) Structural basis for cooperative function of Mettl3 and Mettl14 methyltransferases. Mol Cell 63:306–317

Wang X, Feng J, Xue Y et al (2016b) Structural basis of N(6)-adenosine methylation by the METTL3-METTL14 complex. Nature 534:575–578

Warda AS, Kretschmer J, Hackert P et al (2017) Human METTL16 is a N(6)-methyladenosine (m (6)A) methyltransferase that targets pre-mRNAs and various non-coding RNAs. EMBO Rep 18:2004–2014

Wei CM, Moss B (1977) Nucleotide sequences at the N6-methyladenosine sites of HeLa cell messenger ribonucleic acid. Biochemistry 16:1672–1676

Wei J, Liu F, Lu Z et al (2018a) Differential m(6)A, m(6)Am, and m(1)A demethylation mediated by FTO in the cell nucleus and cytoplasm. Mol Cell 71(973–985):e5

Wei LH, Song P, Wang Y et al (2018b) The m(6)A Reader ECT2 controls trichome morphology by affecting mRNA stability in arabidopsis. Plant Cell 30:968–985

Wen J, Lv R, Ma H et al (2018) Zc3h13 regulates nuclear RNA m(6)A methylation and mouse embryonic stem cell self-renewal. Mol Cell 69:1028–1038.e6

Weng H, Huang H, Wu H et al (2017) METTL14 inhibits hematopoietic stem/progenitor differentiation and promotes leukemogenesis via mRNA m(6)A modification. Cell Stem Cell 22:191–205

Weng H, Huang H, Wu H et al (2018) METTL14 inhibits hematopoietic stem/progenitor differentiation and promotes leukemogenesis via mRNA m(6)A modification. Cell Stem Cell 22:191–205 e9

Xiang Y, Laurent B, Hsu CH et al (2017) RNA m(6)A methylation regulates the ultraviolet-induced DNA damage response. Nature 543:573–576

Xu C, Wang X, Liu K et al (2014) Structural basis for selective binding of m6A RNA by the YTHDC1 YTH domain. Nat Chem Biol 10:927–929

Yue Y, Liu J, Cui X et al (2018) VIRMA mediates preferential m(6)A mRNA methylation in 3'UTR and near stop codon and associates with alternative polyadenylation. Cell Discov 4:10

Zhang S (2018) Mechanism of N(6)-methyladenosine modification and its emerging role in cancer. Pharmacol Ther 189:173–183

Zhang S, Zhao BS, Zhou A et al (2017) m(6)A demethylase ALKBH5 maintains tumorigenicity of glioblastoma stem-like cells by sustaining FOXM1 expression and cell proliferation program. Cancer Cell 31:591–606.e6

Zhao BS, Wang X, Beadell AV et al (2017) m(6)A-dependent maternal mRNA clearance facilitates zebrafish maternal-to-zygotic transition. Nature 542:475–478

Zheng G, Dahl JA, Niu Y et al (2013) ALKBH5 is a mammalian RNA demethylase that impacts RNA metabolism and mouse fertility. Mol Cell 49:18–29

Zheng Q, Hou J, Zhou Y et al (2017) The RNA helicase DDX46 inhibits innate immunity by entrapping m(6)A-demethylated antiviral transcripts in the nucleus. Nat Immunol 18:1094–1103

Zhong S, Li H, Bodi Z et al (2008) MTA is an arabidopsis messenger RNA adenosine methylase and interacts with a homolog of a sex-specific splicing factor. Plant Cell 20:1278–1288

Zhou J, Wan J, Gao X et al (2015) Dynamic m(6)A mRNA methylation directs translational control of heat shock response. Nature 526:591–594

Zhou J, Wan J, Shu XE et al (2018) N(6)-methyladenosine guides mRNA alternative translation during integrated stress response. Mol Cell 69:636–647 e7

Zhu T, Roundtree IA, Wang P et al (2014) Crystal structure of the YTH domain of YTHDF2 reveals mechanism for recognition of N6-methyladenosine. Cell Res 24:1493–1496

# $m^6A$ mRNA Methylation in the Mammalian Brain: Distribution, Function and Implications for Brain Functions

**Mareen Engel and Alon Chen**

## Contents



**Abstract** RNA is abundantly modified by a range of covalent modifications, collectively termed the epitranscriptome. Of these modifications, $N^6$-methyladenosine ($m^6A$) is the most prevalent internal chemical tag in eukaryotic mRNA. Being cotranscriptionally deposited, it regulates almost all aspects of mRNA's lifetime

M. Engel
Ronin Institute, Montclair, NJ, USA

A. Chen (✉)
Department of Stress Neurobiology and Neurogenetics, Max Planck Institute of Psychiatry, Munich, Germany

Department of Neurobiology, Weizmann Institute of Science, Rehovot, Israel
e-mail: alon_chen@psych.mpg.de; alon.chen@weizmann.ac.il

 
S. Jurga, J. Barciszewski (eds.), *The DNA, RNA, and Histone Methylomes*, RNA Technologies, https://doi.org/10.1007/978-3-030-14792-1_15

including maturation into mRNA, stability, distribution and protein translation. While $m^6A$ is likely present in all developing and adult mammalian tissues, here we highlight its distribution and reported functions in the mammalian brain. Additionally, we describe its potential to act as an encoding mechanism for activity- and experience-dependent adaptation and memory-formation. Such alterations may be positive when adjusting to outer challenges or negative when involved in maladaptive processes of the brain such as in the development of psychopathologies.

Consequently, studying this layer of gene expression control in the brain, alongside posttranslational regulation of proteins and epigenetics may inform us as to the molecular mechanisms underlying normal and pathological behaviors. Unfortunately, measuring $m^6A$ levels, patterns and especially dynamics still poses a major technological challenge especially in such a complicated organ as the brain.

**Keywords** Epitranscriptome · $N^6$-methyladenosine · $m^6A$ · Methyltransferase · Demethylase · $m^6A$ reader · Brain functions · Psychiatric disorders

# 1 Introduction

Over 100 covalent base modifications have been found in all domains of life including prokaryotes and eukaryotes but also archaea and viruses. They appear on almost all types of RNA including mRNA, tRNA, rRNA and snRNA (Boccaletto et al. 2018). Although many of these modifications and their potential to post-transcriptionally regulate gene expression have been known since the 1960s and 1970s, the field—now known as epitranscriptomics—attracted little attention until recent technological developments. In mammals, the most diverse RNA species regarding modified nucleotides are tRNA and rRNA, while only a very limited set of modifications is present on mRNA. The most abundant internal modification in mammalian mRNA is $N^6$-methyladenosine, abbreviated to $m^6A$, but many others exist including $N^1$-methyladenosine, abbreviated to $m^1A$ (Dominissini et al. 2016; Li et al. 2016), pseudouridine $\Psi$ (Carlile et al. 2014; Schwartz et al. 2014a; Li et al. 2015), 5-methylcytosine $m^5C$ (Dubin and Taylor 1975; Squires et al. 2012), and A-to-I editing (Levanon et al. 2004; Li et al. 2009).

# 2 $m^6A$ mRNA Methylation

The biochemistry and cellular regulation of $m^6A$ has been described before in great detail here (Jia 2016) and elsewhere (e.g., Zhao et al. 2017). Internal mRNA $m^6A$ usually occurs in a fairly defined consensus motif DR$m^6$ACH (with D = A, U or G; R = G or C; and H = A, U, C) (Wei et al. 1975; Schibler and Perry 1977). $m^6A$ is preferentially localized to the 3′UTR near the stop codon and in the 5′UTR of mRNAs, and to some degree, in the coding sequence (exon), the transcription start

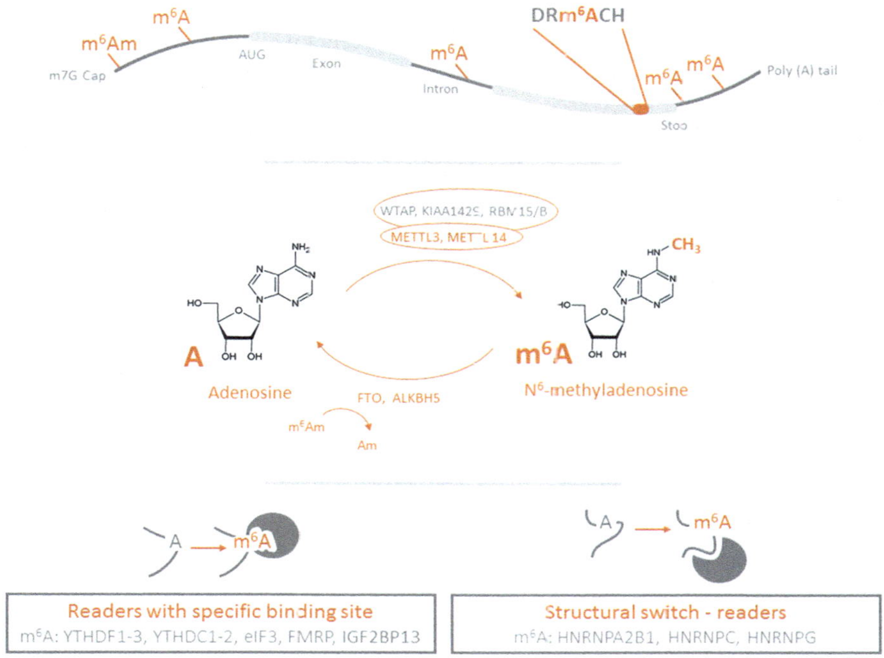


**Fig. 1** The m$^6$A regulatory system including writers, erasers and readers; distribution of m$^6$A on mRNA; and m$^6$A reader proteins (adapted from Engel and Chen 2018)

site (TSS) and in long internal exons (Dominissini et al. 2012; Meyer et al. 2012) (Fig. 1).

Adenosine methylation at the N$^6$-position, in contrast to e.g., N$^1$-methylation in m$^1$A, does not impair the Watson-Crick pairing with U but works via creating binding motifs increasing the accessibility for RNA-binding proteins (RBP; Liu et al. 2015, 2017), and modulating the mRNA secondary structure (Roost et al. 2015; Liu et al. 2015; Spitale et al. 2015).

The existence of both writer and eraser networks adding and removing m$^6$A (as described in the following) has been widely accepted to indicate that m$^6$A methylation is highly dynamic and a readily reversible system (Fig. 1).

## 2.1 *Writing m$^6$A: The Methyltransferase Complex*

A multiprotein methyltransferase complex transfers a methyl group from the donor substrate, S-adenosyl methionine, to the target RNA adenosines creating methylated adenosine (Bokar et al. 1994, 1997). This complex consists of two subunits with catalytic MT-A70 domains occurring in a heterodimer: METTL3 and METTL14 (Liu et al. 2014). Additional complex components include WTAP (Liu et al. 2014;

Ping et al. 2014), KIAA1429 (Schwartz et al. 2014b), RBM15/B (Patil et al. 2016), and others, which enable target tethering and specificity as well as establishing the distinct nuclear localization pattern of the complex. Deposition of $m^6A$ likely occurs co-transcriptionally, i.e., on nascent pre-mRNA that is still tethered to genomic DNA (Slobodin et al. 2017; Ke et al. 2017). These latter studies however argue for rather static levels of $m^6A$, i.e., conclude that once hnRNA has been released from the chromatin, $m^6A$ can only be removed by demethylation or mRNA decay. Posttranslational regulation of the methyltransferase proteins e.g., by phosphorylation has been described but may not necessarily regulate methylation activity per se (Schöller et al. 2018). Finally, conflicting roles for certain methyltransferase complex components aside from methylating nascent hnRNA or even those occurring in the cytoplasm have also been described but may be restricted to very special circumstances (Chen et al. 2015; Alarcón et al. 2015a; Lin et al. 2016).

## 2.2 Erasing $m^6A$: FTO and ALKBH5

There are two known $m^6A$ demethylating enzymes enabling a potentially reversible and thus fully dynamic regulation of $m^6A$: FTO (Jia et al. 2011) and ALKBH5 (Zheng et al. 2013). Interestingly, both were reported to have distinct subcellular and tissue distributions and thus potentially encode target- and tissue-specific regulation of $m^6A$ (Gerken et al. 2007; Vujovic et al. 2013; Zheng et al. 2013; Hess et al. 2013). Many reports even indicate that cellular regulation of e.g., FTO expression and activity is regulated thus enabling active regulation of $m^6A$ levels. However, recent results have dampened the original excitement over dynamic demethylation of $m^6A$ including the observation that FTO may preferentially demethylate a closely related and often co-detected modification, $N^6$,2′-O-dimethyladenosine $m^6Am$, in vitro and in vivo (Schwartz et al. 2014b; Linder et al. 2015; Mauer et al. 2017; Engel et al. 2018). These data suggest that reversibility of $m^6A$ may be less extensive than originally thought especially within physiological systems (Mauer et al. 2017; Mauer and Jaffrey 2018). Similar conclusions may be drawn from the fact that all known full mouse knockouts for FTO and ALKBH5 are, in contrast to all knockouts of the methyltransferases, viable after birth (Fischer et al. 2009; Zheng et al. 2013). In favor of active demethylation, localization of the enzymes and stoichiometry of $m^6A$ and $m^6Am$, may allow FTO to target $m^6A$ in significant amounts in vivo with more recent data showing demethylation of all three mRNA methylated adenosines, $m^6A$, $m^6Am$, $m^1A$, by FTO (Wei et al. 2018).

## 2.3 Readers of $m^6A$

Given the wide abundance of $m^6A$ and even more diverse cellular functions of $m^6A$ (detailed below), a large part of functional specificity has to be achieved by the diverse range of $m^6A$-interacting RBPs, the so called $m^6A$ readers. The most important family of m$^6$A-readers consists of the YTH-domain-containing proteins, which bind directly to m$^6$A. Currently known mammalian members of this family are YTHDF1-3 and YTHDC1-2. They have been assigned very diverse, often contradictory, yet sometimes cooperative cellular functions including promotion and inhibition of translation and decay (recently reviewed e.g., in Roundtree et al. 2017; Patil et al. 2018). The very diverse functions of the different YTH-family members may be regulated by several factors including cellular sub-localization, target-specificity and posttranslational regulation of the readers. Next to YTH-domain readers, m$^6$A has also been reported to directly recruit eIF3 leading to a promotion of cap-independent translation (Meyer et al. 2015). Other proteins lacking a YTH domain, e.g., the hnRNP proteins HNRNPA2B1, HNRNPC, and HNRNPG (Liu et al. 2015, 2017; Alarcón et al. 2015a), may bind m$^6$A instead via m$^6$A-specific structural features. Finally, additional classes of direct binder proteins without a YTH-domain or m$^6$A-specific structural features may exist including IGF2BP1–3 (Huang et al. 2018) and FMRP (Arguello et al. 2017; Edupuganti et al. 2017).

### 2.4 *Cellular Functions of m$^6$A*

m$^6$A cellular functions include the regulation of all stages of mRNA's lifetime and thus establish a layer of secondary gene expression regulation (recently reviewed e.g., in Roundtree et al. 2017; Patil et al. 2018). Starting at the very beginning of mRNA's life, m$^6$A has been described to regulate the maturation of pre-mRNA into mature mRNA including 5′ capping, 3′ polyadenylation, splicing, nuclear processing and nuclear export of mRNAs. Thereby, m$^6$A catalyzes differential splicing (Liu et al. 2015; Xiao et al. 2016; Ke et al. 2017) and differential polyA site usage (Ke et al. 2015; Molinie et al. 2016). m$^6$A further promotes and also inhibits mRNA translation depending on the respective mRNA-m$^6$A target and bound m$^6$A-reader (Wang et al. 2015; Zhou et al. 2015; Meyer et al. 2015; Li et al. 2017a; Shi et al. 2017). Indicating the end of mRNA lifetime, methylation generally appears to accelerate mRNA decay (Wang et al. 2014) yet even this relationship is likely more complex than seen on first sight. m$^6$A recognized by other effector proteins beyond the classical YTH-domain readers may have completely different effects on mRNA via third effector proteins. This includes the described interactions with ELAV-like RNA binding protein 1 (ELAV1/HuR) (Wang et al. 2014), intersections with miRNA biogenesis (Chen et al. 2015; Alarcón et al. 2015b), and interactions with the Toll-like receptor (TLR) family protein members TLR3 and TLR7 (Karikó et al. 2005).

## 3 m$^6$A Distribution and Function in the Brain

The brain is one of the most complexly structured and regulated mammalian organs both during development and in adulthood. The adult brain especially, is a unique organ regarding gene expression regulation as it not only consists largely of postmitotic cells with very limited regeneration- and repair-capacity, but also because of the huge diversity and specialization of those cells. Thus, a mechanism of secondary gene expression regulation such as RNA methylation may be especially crucial in such a system.

Several studies have begun to uncover the functional significance of $m^6A$ regulation in the central nervous system (CNS) and its role in normal brain physiology during all stages of life from development to adulthood and encoding cellular plasticity in the adult brain. Thereby, $m^6A$ is abundant in the brain during all developmental stages with increasing levels during development (Meyer et al. 2012). In the adult brain, $m^6A$ is likely found in all brain structures of the CNS but also in the peripheral nervous system (PNS) (Weng et al. 2018). Region-specific methylation levels and patterns have been reported (Chang et al. 2017; Engel et al. 2018).

## 3.1 *$m^6A$ in Brain Development*

Knockout of the $m^6A$ methyltransferases in embryonic stem cells leads to embryonic lethality in all known cases, usually together with severe nervous system malformation (Fukusumi et al. 2008; Geula et al. 2015). In line, $m^6A$ has been described to be essential for mammalian cortical neurogenesis (Yoon et al. 2017). Loss of $m^6A$ in murine neural progenitor cells by removal of either METTL3 or METTL14 leads to prolonged cell cycle progression and delayed neuronal differentiation via suppression of neuronal lineage markers, thus, extending the cortical neurogenesis widely into postnatal stages (Yoon et al. 2017). Similarly, conditional knockout of *Mettl3* using the prenatally expressing Nestin-Cre causes severe developmental defects both in cortical and cerebellar regions (Wang et al. 2018). Nestin-Cre *Mettl3* conditional knockout mice, i.e., those with a knockout in prenatal brain cells, display cerebellar hypoplasia caused by drastically enhanced apoptosis of newborn cerebellar granule cells in the external granular layer leading to severe motoric deficits and death within the first 3 weeks after birth (Wang et al. 2018). Loss of FTO or FTO function in both mice and human leads to postnatal growth retardation, including microcephaly and increased postnatal lethality (Boissel et al. 2009; Gao et al. 2010). Interestingly, full knockut of the reader genes *Ythdf1* or *Ythdf2* does not lead to any gross brain development abnormalities, lethality or motor deficits (Ivanova et al. 2017; Shi et al. 2018).

$m^6A$-profiling of human fetal forebrains and human brain organoids has revealed a conserved and unique $m^6A$ landscape similar to that in mouse embryonic forebrains (Yoon et al. 2017). In general, $m^6A$ has been described to be somewhat evolutionary conserved (Ma et al. 2017). Additional to stem cells in the developing brain, $m^6A$ is crucial for at least one of the two populations of neural stem cells (NSCs) remaining in the adult brain, the stem cells in the subgranular zone of the dentate gyrus (Li et al. 2017b). FTO loss in these cells reduces NSC proliferation and neuronal differentiation, reminiscent but not fully equal to the embryonic neurogenesis defect seen in *METTL3* and *METTL14* knockout mice. Finally, m$^6$A modified RNAs also play a key role in brain cancer (Zhang et al. 2017; Cui et al. 2017).

## 3.2 m$^6$A in the Adult Brain

Several detailed maps of m$^6$A in the adult mammalian brain are available (Meyer et al. 2012; Hess et al. 2013; Chang et al. 2017; Merkurjev et al. 2018; Engel et al. 2018), reporting a total of approximately 10,000–20,000 m$^6$A sites. Additionally, CNS RNA methylation has also been characterized in *Drosophila melanogaster* (Lence et al. 2016). In stark contrast to the deleterious effect of methyltransferases in the developing brain, loss of FTO, METTL3, or METTL14 in postnatal neurons only, e.g., via a conditional knockout using Camk2a-Cre driver lines, does not cause any major brain morphological changes or increase apoptosis of cells (Koranda et al. 2018; Engel et al. 2018; Zhang et al. 2018). The consequence of a loss of these enzymes in cells other than neurons, e.g., astrocytes, has not been investigated yet.

## 3.3 Sorting mRNAs in Complex Neurons by m$^6$A?

m$^6$A is involved in several mechanisms of regulating translocation of mRNA, including nuclear export (Zheng et al. 2013; Fustin et al. 2013) and sorting of mRNA into specific cytoplasmic aggregates like P-bodies and stress granules (Wang et al. 2015; Anders et al. 2018). Neurons are built more complexly than most mammalian cells, including higher polarization, higher fragmentation into specialized components like axons and dendrites, and a higher number of complex cell-to-cell connections. These cell-to-cell connections are highly regulated in the brain including changes of signal transmission efficacy and interactions of neurons with other cell types. Many such cellular changes are often realized via changes of gene expression control including compartmentalized regulation of protein translation e.g., at the synapse (Holt and Schuman 2013). Consequently, it has been speculated that m$^6$A-modification of mRNA may regulate spatial sorting and compartmentalized protein translation control within neurons into axons, pre-synaptic nerve terminals, dendrites, and dendritic spines (Fig. 2). While a final demonstration of such a mechanism is still lacking, enrichment of synaptic and neuronal projection gene ontology terms has been reported repeatedly during the recent efforts to map m$^6$A in the brain (Hess et al. 2013; Widagdo et al. 2016; Yoon et al. 2017; Merkurjev et al. 2018; Engel et al. 2018). Further, m$^6$A has been reported to be localized to axons and to regulate axonal growth via local translation via GAP-43 (Yu et al. 2018) and several components of the m$^6$A-machinery have been observed to be synaptically located including writers and erasers, classically considered to be nuclear proteins (Yu et al. 2018; Merkurjev

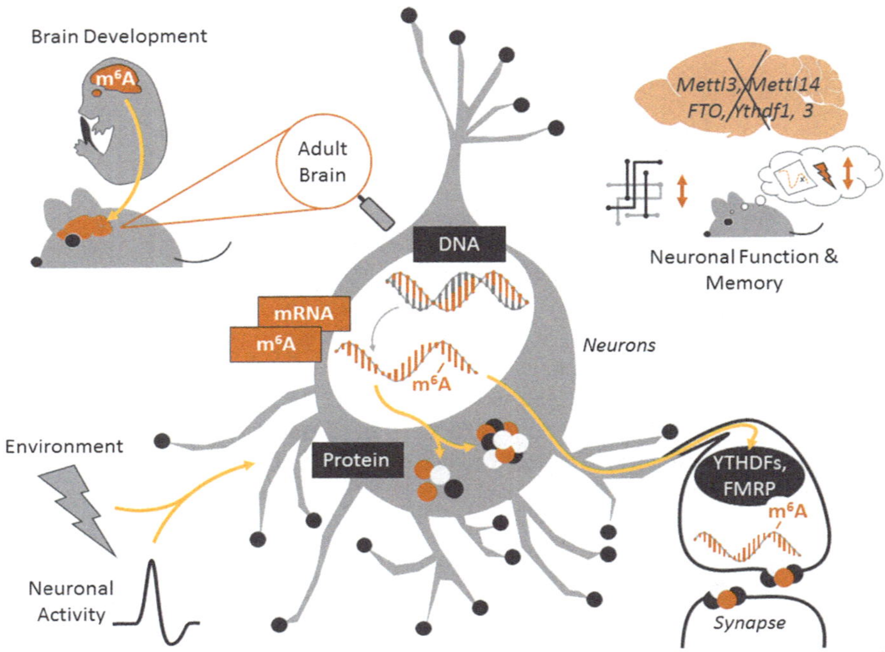


**Fig. 2** Functions of m$^6$A in the brain (adapted from Engel and Chen 2018)

et al. 2018). Consequently, synapses and neuronal somas have been reported to harbor their own specific epitranscriptome (Merkurjev et al. 2018).

## 3.4 *FMRP*

A potential key player involved in such a mechanism of localized translation regulation in neurons may be the fragile X mental retardation protein (FMRP). FMRP is a neuronal RNA-binding protein known for its role in metabotropic glutamate receptor (mGluR)-dependent signaling and synaptic plasticity (Waung and Huber 2009). It is found in neuronal RNA transport granules and regulates dendritic localization of RNAs. FMRP inhibits RNA local transcript translation including that occuring at the synapse (Holt and Schuman 2013). Overlap of the m$^6$A consensus motif with the one of FMRP and high co-occurrence of FMRP binding sites in m$^6$A modified target sites was observed early on (Anderson et al. 2016; Chang et al. 2017; Engel et al. 2018). Finally, actual binding of FMRP to m$^6$A was shown recently (Arguello et al. 2017; Edupuganti et al. 2017). Competitive binding of m$^6$A between FMRP and YTHDF1 and YTHDF2 was speculated upon as a potential mechanism of m$^6$A-FMRP action (Edupuganti et al. 2017; Zhang et al. 2018).

### 3.5 Activity Dependent Regulation

A special property of neurons is their constant regulation via electrical and chemical signaling leading to amongst others activity-dependent regulation of gene expression. This enables the critical adaptiveness of the brain to outer and inner stimuli via short- and long-term alterations in gene expression, neuronal morphology, connectivity and ultimately regulation and behavior.

Similarly to m$^6$A being involved in the basic cellular stress response (Dominissini et al. 2012; Zhou et al. 2015; Meyer et al. 2015; Xiang et al. 2017), it has also been described to be involved in cellular processes triggered upon neuronal signaling and activity-dependent regulation (Fig. 2). This includes both regulation in vitro, e.g., in primary neuronal cultures after KCl-induced neuronal depolarization (Widagdo et al. 2016), and in vivo in the adult brain after challenges in several brain regions, e.g., after fear conditioning (Widagdo et al. 2016; Walters et al. 2017; Zhang et al. 2018), stressful challenges (Engel et al. 2018), and also in the PNS after nerve injury (Weng et al. 2018). Levels of activity-dependent regulation include changes within the m$^6$A machinery, altered global and target-mRNA specific m$^6$A levels, and differential translation of downstream-effectors like immediate early genes (IEGs). In line with this, IEG function is widely impaired in *Mettl3* knockout primary cortical neurons after fear conditioning (Zhang et al. 2018).

Furthermore, while activity-dependent gene expression changes for m$^6$A enzymes and readers have been reported, the relation of dynamically regulated mRNA abundancy of e.g., *Mettl3* or *Fto* to the respective active protein levels and cellular consequences of regulated m$^6$A are still mostly unclear. This includes the very limited current knowledge on how the different m$^6$A-enzymes and readers are regulated at the protein level in response to activity, including their subcellular localization, activity, and target specificity. Investigating the posttranslational regulation of the m$^6$A-enzymes and readers, as for example shown or suggested for METTL3 and FTO via SUMOylation and ubiquitination (Tai et al. 2017; Zhu et al. 2018; Du et al. 2018) may provide valuable insight.

### 3.6 Brain Function, Electrophysiology and Behavior

Consistent with the concept of mRNA methylation being involved in the regulation of gene expression after neuronal activity, alterations of the m$^6$A system via m$^6$A enzyme knockouts were found to change neuronal electrophysiological properties. For example, long term potentiation (LTP) in the hippocampus was decreased after hippocampal knockout of *Mettl3* (Zhang et al. 2018) (but observe conflicting data Engel et al. 2018), *Fto* (Engel et al. 2018), and *Ythdf1* (Shi et al. 2018), while deletion of *Mettl14* in the striatum led to increased neuronal excitability and reduced spike frequency adaptation (Koranda et al. 2018).

While the lack of m$^6$A enzymes during development usually leads to severe developmental defects, enzyme deletion or depletion specifically in the adult brain causes only limited behavioral phenotypes, usually excluding effects on motor skills, movement properties, or anxiety-like behavior. Adult brain $m^6A$-manipulation, reported to date, leads fairly specifically to memory impairment: Conditional knockout of *Mettl3* in the hippocampus or forebrain excitatory neurons enhances cue-related memory consolidation after fear-conditioning and in the Morris Water Maze (MWM; Engel et al. 2018; Zhang et al. 2018), while full knockout of *Ythdf1* reduces memory consolidation in both of these tests (Shi et al. 2018). Additionally, $m^6A$ deficiency via *Mettl14* deletion in striatal neurons impairs learning and performance (Koranda et al. 2018). Conversely, knockout or knockdown of *Fto* in the prefrontal cortex or hippocampus enhances consolidation of cue- and or context-related fear memory (Wang et al. 2015; Widagdo et al. 2016; Walters et al. 2017) while impairing spatial learning and memory of mice in MWM and eight-arm maze test (Li et al. 2017b). Knockout of *Fto* also attenuates the response in cocaine-induced locomotion (Hess et al. 2013).

Together these studies show that although $m^6A$ seems to be much less crucial in the adult brain than during development, it is potentially important for specific brain functions. In line with $m^6A$ roles as a secondary mechanism of gene expression regulation, it may therefore be especially important for brain functions that require activity-dependent gene expression regulation, like memory formation.

## 4 $m^6A$: Possible Implications for Psychiatric Disorders

Obesity and type-2 diabetes have been repeatedly associated with genetic polymorphisms in the first intron of the human *FTO* gene (Dina et al. 2007; Scuteri et al. 2007; Frayling et al. 2007), although the variant reported on may actually not affect the *FTO* locus itself but rather neighboring genes (Smemo et al. 2014; Stratigopoulos et al. 2014; Claussnitzer et al. 2015). Consequently, the physiological roles of *FTO* in the context of energy metabolism and expenditure and food intake have been extensively investigated but led to mixed results and as such the mechanisms remain unknown (Hess and Brüning 2014). Beyond such metabolic functions, the *FTO* variant has also been associated with several psychiatric disorders including Major Depressive Disorder (MDD) (Samaan et al. 2013; Milaneschi et al. 2014), Alzheimer's Disease (AD) (Profenno et al. 2010; Keller et al. 2011; Reitz et al. 2012) and Attention Deficit Hyperactivity Disorder (ADHD) (Choudhry et al. 2013). It has further been indicated in memory processing capabilities in humans genome-wide association studies (Ho et al. 2010; Benedict et al. 2011; Keller et al. 2011). Additionally to *FTO*, a polymorphism related to *ALKBH5* has been reported to be associated to MDD in a relatively small candidate gene association study (Du et al. 2015). However, recent more powered genetic association studies and meta-analyses did not report any association of gene variants close to any of the $m^6A$-machinary genes to various psychiatric disorders, including MDD, Post-traumatic stress disorder, ADHD, and AD (Demontis et al. 2017*; Martin et al. 2017*; Purves et al. 2017*; Meier et al. 2018*; Duncan et al. 2018; Wray et al. 2018; Coleman et al. 2018*; Jansen et al. 2019, * not yet peer-reviewed preprints).

Beyond classic genetic association studies, increasing evidence suggests that dysregulation and maladaptation of transcriptional fine-tuning is central to the etiology of psychiatric disorders more so than monogenetic causes (Nestler et al. 2016). In agreement, our recent study comparing m$^6$A of MDD patients and healthy controls in blood and derived cells found hardly any differences between the two groups except for after stimulation of stress-response signaling pathways (Engel et al. 2018). While these measurements were performed only in blood or blood derived cells, peripheral DNA methylation signatures related to neurobiological phenotypes may have some, albeit limited, similarity to central signatures (Davies et al. 2012; Farré et al. 2015; Hannon et al. 2015). Taken together, this indicates that the m$^6$A-system may well be involved in the etiology of psychiatric disorders via changed regulation of gene expression, especially considering the encoding of activity-related gene expression regulation, and thus may lead to long-term changes contributing to psychiatric disorders.

# 5 Technological Challenges of Measuring m$^6$A in the Brain

Technical approaches to detect m$^6$A and challenges associated to them have been discussed in great detail here (Jia 2016) and elsewhere before (e.g., Schwartz and Motorin 2017; Helm and Motorin 2017; Schaefer et al. 2017). While the field of RNA modifications, especially m$^6$A, has seen an incredible increase in attention over the last 10 years, studying those modifications still poses a major challenge given the limited availability of appropriate molecular tools and methods. This is especially true when aiming to quantify m$^6$A dynamics. In the following section, we will summarize some of the current challenges related to the methods most used to detect m$^6$A in the brain in vivo.

## *5.1 Global Detection Techniques*

Two-dimensional thin layer chromatography (2-D TLC) and high-performance liquid chromatography coupled with triple-quadrupole tandem mass spectrometry (LC-MS/MS) were two of the earliest developed methods to detect and quantify modified nucleotides in RNA. Both require mRNA to be digested into single nucleotides first, which are then separated and detected based on their physico-chemical properties. This enables specific discrimination between different modifications, e.g., m$^1$A and m$^6$A. Stable isotope labelling approaches were developed to optimize these techniques but are mostly not applicable for analyzing intact mice or human organs (Popova and Williamson 2014; Kellner et al. 2014; Paulines and Limbach 2017). Both TLC and LC-MS/MS require comparably large amounts of input material and are usually performed on total RNA or mRNA preparations, and thus report an average global methylation signal derived across all nucleotides with a loss of sequence and target context. These techniques are thus greatly suited to detect major changes in levels of different modifications e.g., after knockout of enzymes or in some cases of experimental manipulations (Engel et al. 2018), but may fail to detect more subtle changes as often expected after stimulation of the adult brain.

Two additional techniques to measure global $m^6A$-levels are antibody-based enzyme-linked colorimetric methods like Dot Blots and ELISAs. As with TLC and LC-MS/MS, these methods are commonly performed on total RNA or mRNA preparations but without previous digestion of the RNA. In contrast with the earlier described methods, they allow for low time-consuming and high throughput screening of samples and thus may be an entry point for characterizing $m^6A$ regulation in vivo. However, beyond being only able to report large and global differences in methylation, these methods also potentially suffer from antibody-associated issues (discussed below).

In contrast, a related method not suffering from the problems arising from global averaging of $m^6A$ modification is the SCARLET method (site-specific cleavage and radioactive-labeling followed by ligation-assisted extraction and thin-layer chromatography, Liu et al. 2013) which quantifies nucleotides at one specific location with TLC. However, SCARLET also requires very high amounts of input material and radioactive labeling and thus is not readably applicable to study $m^6A$ dynamics in vivo.

## 5.2 *$m^6A$-Seq*

The development of mapping $m^6A$ in a transcriptome-wide approach revolutionized the field in 2012 ($m^6A$-Seq: Dominissini et al. 2012; meRIP-Seq: Meyer et al. 2012). Both, essentially equal, methods are based on affinity purification of fragmented mRNA with $m^6A$-specific antibodies followed by random primed cDNA library generation, adaptor ligation and high throughput small read sequencing. Antibody-based affinity purification is needed because $m^6A$, in contrast to other modifications as for example $m^1A$ (Dominissini et al. 2016), does not stop the most common reverse transcription enzymes or lead to base misincorporation. This may be overcome by using specific $m^6A$-sensitive polymerases (Harcourt et al. 2013) or engineered reverse transcriptase with increased misincorporation in the opposite strand (Aschenbrenner et al. 2018), although both methods are not yet commonly established. This feature of $m^6A$ also limits the current availability of PCR based techniques.

While $m^6A$-Seq enables a high throughput transcriptome-wide description of $m^6A$, it comes with many significant problems: First of all, it is dependent on the antibody used with potential differences between the used antibodies and even affinity-differences across different batches of the same antibody. Antibodies may also detect related modifications like $m^6Am$ (Linder et al. 2015). Click chemistry protocols may improve antibody specificity (Hartstock et al. 2018) but are not yet implemented or tested in the majority of protocols. So far, only a limited number of antibodies have been used, leading to somewhat different signatures (Zeng et al. 2018) but a comprehensive study comparing their detection patterns is not available yet. Furthermore, the binding properties of the antibodies may be influenced by the adjacent nucleotide sequence and RNA secondary structures. Antibodies also may suffer from substantial background signal with sequence-selective capture of certain unmodified fragments (Schwartz et al. 2013). Moreover, all of these biases of the sequence read distributions are mixed and amplified by the structural biases inherent to RNA-Seq itself. The resulting bias-mixture can only be partially remedied by the typical correlation to input RNA-Seq common in m$^5$A-Seq.

Secondly, the original m$^6$A-Seq protocols required substantial quantities of input material (several micrograms of purified mRNA). Recent low input protocols overcome this problem by using library kits optimized for ultra-low input RNA-Seq (Zeng et al. 2018). These low-input protocols may however introduce even more additional biases. Superficial comparisons of data received with such low-input protocols have already showed significant differences from the respective classical m$^6$A-Seq data set but this needs to be further investigated (Zeng et al. 2018).

Thirdly, classic m$^6$A-Seq protocols do not detect the modified nucleotide itself but a pile-up of fragments called m$^6$A-peaks which should, in principle, harbor the m$^6$A site in its center. Such peaks, by design, are around 200 nt long (when using 100 nt RNA fragments) but often experiments result in even bigger peaks likely due to several m$^6$A-sites close by. The m$^6$A-sites appearing in clusters (Ke et al. 2015; Linder et al. 2015) and their potential dynamics will thus be partially lost in m$^6$A-Seq. Further, due to commonly employed method of cDNA synthesis via random hexamer primers, m$^6$A-peaks will not include 5′ ends of the mRNA if methylated, although 5′ mRNA m$^6$A and m$^6$Am methylation may still be inferred from a peak located at the 5′ UTR.

Aiming to improve these protocols, several protocols have been recently developed to enable a nucleotide-specific transcriptome-wide mapping of m$^6$A (m$^6$A-CLIP: Ke et al. 2015; miCLIP: Linder et al. 2015). These methods use UV-crosslinking of an m$^6$A-antibody with mRNA-fragment leading to predictable mutation and truncation patterns in the cDNA strand during reverse-transcription that can be detected later in the sequencing data. Besides being prone to sequence and structural biases again, and the difficulty to map due to short fragments, the resulting data is often more noisy and less consistent than m$^6$A-Seq data with many more replicates needed for consistent mapping and unclear quantitative potential.

In the future, detection of m$^6$A via directly sequencing the RNA in its native form, e.g., while pulling through a nanopore, may overcome these challenges but just begins to be established for a wider audience (Liu et al. 2019).

## 5.3 *Quantification of m$^6$A-Seq*

Most bioinformatics pipelines to analyze m$^6$A-Seq data rely on comparisons of the enrichment of m$^6$A-immunoprecipitated fragments over standard RNA-Seq signal, employing cutoffs to defined peak ranges and minimum-occurrence across several replicates to call the presence or absence of a peak. As a result, analysis results are heavily dependent on the bioinformatics algorithm used and seemingly arbitrary cutoffs chosen by the respective scientist with no consensus yet of best practices. Increasing the number of replicates used may often help to filter out sites that are spuriously detected aiming to increase robustness and reproducibility of the achieved maps. More dedicated analytical approaches are continuously being developed to better identify modified sites, more effectively integrate background levels and better filter out noise.

While binary maps of peaks and non-peaks can then be compared between two conditions, most likely in a physiological regulation of $m^6A$, for example upon stimulation in the adult brain, quantitative regulation of $m^6A$ may be much more likely. While none of the $m^6A$-Seq protocols have been originally developed for quantitative comparison of methylation states, methods to compare $m^6A$-Seq signatures across conditions were proposed soon after the original methods description and are actively developed (Meng et al. 2014; Cui et al. 2015; Liu et al. 2016). For the special case of the mostly postmitotic adult brain, given enough replicates and a mild enough manipulation, underlying changes in mRNA expression may be assumed to be negligible, thus removing the need for extensive normalization to background levels.

Notably, a calibrated pulldown procedure called LAIC-Seq was establish recently as a tool to provide quantitative estimates of methylation stoichiometry (Molinie et al. 2016). However, this protocol uses immunoprecipitation of full-length mRNA rather than fragments with the $m^6A$-antibody, thus averages $m^6A$ content across the entire transcript and likely suffers from extensive noise due to structural biases. Similar methods have been developed using qPCR instead of RNA-Seq to measure the abundance of $m^6A$-containing (full-length) mRNA and $m^6A$- calibrator spike-ins to allow for immunoprecipitation efficiency correction (Engel et al. 2018).

The perhaps biggest challenge to quantifying $m^6A$ in the (adult) brain is the lack of techniques to profile $m^6A$ in different brain cell subpopulations, leaving all currently available $m^6A$-Seq data to be an average $m^6A$ signal from a widespread cell mixture. Employing knockout animals with conditional removal of $m^6A$-related genes from distinct cell types may provide some insight about cell-specific $m^6A$-signatures. Today, not even a comprehensive comparison of, for example, $m^6A$-signatures in neuron versus astrocytes, the two major brain cell types, is available. The use of neuronal cultures to detect neuron-specific signals was employed before (Widagdo et al. 2016), but may carry limited significance due to neurons normally being highly embedded and structured in vivo.

## 5.4 *Target Manipulation of $m^6A$-Sites*

Almost no reports are available describing the effects of manipulating $m^6A$ at a single target site (except Kane and Beemon 1987; Schwartz et al. 2013). To date, functional significance of RNA methylation has usually been shown by correlation, or broad manipulation of $m^6A$ levels by removal $m^6A$ enzymes. As a consequence, significance and cellular function of site-specific $m^6A$ is mostly unknown so far, although such manipulation should be easy to achieve using recent CRISPR/Cas9 techniques. One explanation may be that methyltransferases might set compensatory methylation at adjacent sites due to remaining $m^6A$-site context when only removing the target site (Narayan et al. 1994; observe contrary results in Schwartz et al. 2013).

Rather than genetic manipulation of a target site, using CRISPR/Cas9 technology to recruit epitranscriptomic modulators to mRNA, potentially even in a temporally- and cell-type controlled manner, may provide much more insight into the cellular functions of specific $m^6A$-sites (O'Connell et al. 2014).

## 6 Conclusions

The field of mRNA adenosine methylation has experienced a recent sudden take off, fueled by the transcriptome-wide mapping of $m^6A$ and many functional follow-up studies. While the core enzymes and reader protein have been well described, more and more details are added each year highlighting the function of $m^6A$ in adapting and fine-tuning gene expression especially after environmental stimulation and, in the brain, neuronal activity. Nevertheless, many questions about the regulatory mechanisms, complex interplay of RNA modifications, and the cellular consequences of RNA methylation remain unanswered. Brute force powered functional studies, based on removing enzymes from whole brain areas, have provided some valuable early insight into potential roles of $m^6A$ in the brain, but the actual functions of $m^6A$ in the brain and related pathology are still open. Methodological advancements to measure stimulated and activity-related $m^6A$-changes in a time- and cell-specific manner as well as functional assays focusing on specific $m^6A$ sites may in the future answer the question of the actual significance of $m^6A$ in the brain.

## References

Alarcón CR, Goodarzi H, Lee H et al (2015a) HNRNPA2B1 is a mediator of m6A-dependent nuclear RNA processing events. Cell 162:1299–1308. https://doi.org/10.1016/j.cell.2015.08.011

Alarcón CR, Lee H, Goodarzi H et al (2015b) N6-methyladenosine marks primary microRNAs for processing. Nature 519:482–485. https://doi.org/10.1038/nature14281

Anders M, Chelysheva I, Goebel I et al (2018) Dynamic m6A methylation facilitates mRNA triaging to stress granules. Life Sci Alliance 1:e201800113. https://doi.org/10.26508/lsa.201800113

Anderson BR, Chopra P, Suhl JA et al (2016) Identification of consensus binding sites clarifies FMRP binding determinants. Nucleic Acids Res 44:6649–6659. https://doi.org/10.1093/nar/gkw593

Arguello AE, DeLiberto AN, Kleiner RE (2017) RNA chemical proteomics reveals the N6-methyladenosine (m6A)-regulated protein-RNA interactome. J Am Chem Soc 139:17249–17252. https://doi.org/10.1021/jacs.7b09213

Aschenbrenner J, Werner S, Marchand V et al (2018) Engineering of a DNA polymerase for direct m6A sequencing. Angew Chem Int Ed 57:417–421. https://doi.org/10.1002/anie.201710209

Benedict C, Jacobsson JA, Rönnemaa E, Sällman-Almén M, Brooks S, Schultes B, Fredriksson R, Lannfelt L, Kilander L, Schiöth HB (2011) The fat mass and obesity gene is linked to reduced verbal fluency in overweight and obese elderly men. Neurobiol Aging 32(6):1159.e1–1159.e5. https://doi.org/10.1016/j.neurobiolaging.2011.02.006

Boccaletto P, Machnicka MA, Purta E et al (2018) MODOMICS: a database of RNA modification pathways. 2017 update. Nucleic Acids Res 46:D303–D307. https://doi.org/10.1093/nar/gkx1030

Boissel S, Reish O, Proulx K et al (2009) Loss-of-function mutation in the dioxygenase-encoding FTO gene causes severe growth retardation and multiple malformations. Am J Hum Genet 85:106–111. https://doi.org/10.1016/j.ajhg.2009.06.002

Bokar JA, Rath-Shambaugh ME, Ludwiczak R et al (1994) Characterization and partial purification of mRNA N6-adenosine methyltransferase from HeLa cell nuclei. Internal mRNA methylation requires a multisubunit complex. J Biol Chem 269:17697–17704

Bokar JA, Shambaugh ME, Polayes D et al (1997) Purification and cDNA cloning of the AdoMet-binding subunit of the human mRNA (N6-adenosine)-methyltransferase. RNA 3:1233–1247

Carlile TM, Rojas-Duran MF, Zinshteyn B et al (2014) Pseudouridine profiling reveals regulated mRNA pseudouridylation in yeast and human cells. Nature 515:143–146. https://doi.org/10.1038/nature13802

Chang M, Lv H, Zhang W et al (2017) Region-specific RNA m6A methylation represents a new layer of control in the gene regulatory network in the mouse brain. Open Biol 7:170166. https://doi.org/10.1098/rsob.170166

Chen T, Hao Y-J, Zhang Y et al (2015) m6A RNA methylation is regulated by MicroRNAs and promotes reprogramming to pluripotency. Cell Stem Cell 16:289–301. https://doi.org/10.1016/j.stem.2015.01.016

Choudhry Z, Sengupta SM, Grizenko N et al (2013) Association between obesity-related gene FTO and ADHD. Obesity (Silver Spring) 21:E738–E744. https://doi.org/10.1002/oby.20444

Claussnitzer M, Dankel SN, Kim K-H et al (2015) FTO obesity variant circuitry and adipocyte browning in humans. New Engl J Med 373:895–907. https://doi.org/10.1056/NEJMoa1502214

Coleman JRI, Peyrot WJ, Purves KL et al (2018) Genome-wide gene-environment analyses of major depressive disorder and reported lifetime traumatic experiences in UK Biobank. bioRxiv. https://doi.org/10.1101/247353

Cui X, Zhang L, Meng J et al (2015) MeTDiff: a novel differential RNA methylation analysis for MeRIP-Seq data. IEEE/ACM Trans Comput Biol Bioinform 15(2):526–534. https://doi.org/10.1109/TCBB.2015.2403355

Cui Q, Shi H, Ye P et al (2017) m6A RNA methylation regulates the self-renewal and tumorigenesis of glioblastoma stem cells. Cell Rep 18:2622–2634. https://doi.org/10.1016/j.celrep.2017.02.059

Davies MN, Volta M, Pidsley R et al (2012) Functional annotation of the human brain methylome identifies tissue-specific epigenetic variation across brain and blood. Genome Biol 13:R43. https://doi.org/10.1186/gb-2012-13-6-r43

Demontis D, Walters RK, Martin J et al (2017) Discovery of the first genome-wide significant risk loci for ADHD. bioRxiv. https://doi.org/10.1101/145581

Dina C, Meyre D, Gallina S et al (2007) Variation in FTO contributes to childhood obesity and severe adult obesity. Nat Genet 39:724–726. https://doi.org/10.1038/ng2048

Dominissini D, Moshitch-Moshkovitz S, Schwartz S et al (2012) Topology of the human and mouse m6A RNA methylomes revealed by m6A-seq. Nature 485:201–206. https://doi.org/10.1038/nature11112

Dominissini D, Nachtergaele S, Moshitch-Moshkovitz S et al (2016) The dynamic N1-methyladenosine methylome in eukaryotic messenger RNA. Nature 530:441–446. https://doi.org/10.1038/nature16998

Du T, Rao S, Wu L et al (2015) An association study of the m6A genes with major depressive disorder in Chinese Han population. J Affect Disord 183:279–286. https://doi.org/10.1016/j.jad.2015.05.025

Du Y, Hou G, Zhang H et al (2018) SUMOylation of the m6A-RNA methyltransferase METTL3 modulates its function. Nucleic Acids Res 46:5195–5208. https://doi.org/10.1093/nar/gky156

Dubin DT, Taylor RH (1975) The methylation state of poly A-containing messenger RNA from cultured hamster cells. Nucleic Acids Res 2:1653–1668

Duncan LE, Ratanatharathorn A, Aiello AE et al (2018) Largest GWAS of PTSD ($N$=20 070) yields genetic overlap with schizophrenia and sex differences in heritability. Mol Psychiatry 23:666–673. https://doi.org/10.1038/mp.2017.77

Edupuganti RR, Geiger S, Lindeboom RGH et al (2017) N6-methyladenosine (m6A) recruits and repels proteins to regulate mRNA homeostasis. Nat Struct Mol Biol 24:870–878. https://doi.org/10.1038/nsmb.3462

Engel M, Chen A (2018) The emerging role of mRNA methylation in normal and pathological behavior. Genes Brain Behav 17:e12428. https://doi.org/10.1111/gbb.12428

Engel M, Eggert C, Kaplick PM et al (2018) The role of m6A/m-RNA methylation in stress response regulation. Neuron 99:389–403.e9. https://doi.org/10.1016/j.neuron.2018.07.009

Farré P, Jones MJ, Meaney MJ et al (2015) Concordant and discordant DNA methylation signatures of aging in human blood and brain. Epigenetics Chromatin 8:19. https://doi.org/10.1186/s13072-015-0011-y

Fischer J, Koch L, Emmerling C et al (2009) Inactivation of the Fto gene protects from obesity. Nature 458:894–898. https://doi.org/10.1038/nature07848

Frayling TM, Timpson NJ, Weedon MN et al (2007) A common variant in the FTO gene is associated with body mass index and predisposes to childhood and adult obesity. Science 316:889–894. https://doi.org/10.1126/science.1141634

Fukusumi Y, Naruse C, Asano M (2008) Wtap is required for differentiation of endoderm and mesoderm in the mouse embryo. Dev Dyn 237:618–629. https://doi.org/10.1002/dvdy.21444

Fustin J-M, Doi M, Yamaguchi Y et al (2013) RNA-methylation-dependent RNA processing controls the speed of the circadian clock. Cell 155:793–806. https://doi.org/10.1016/j.cell.2013.10.026

Gao X, Shin Y-H, Li M et al (2010) The fat mass and obesity associated gene FTO functions in the brain to regulate postnatal growth in mice. PLoS One 5:e14005. https://doi.org/10.1371/journal.pone.0014005

Gerken T, Girard CA, Tung Y-CL et al (2007) The obesity-associated FTO gene encodes a 2-oxoglutarate-dependent nucleic acid demethylase. Science 318:1469–1472. https://doi.org/10.1126/science.1151710

Geula S, Moshitch-Moshkovitz S, Dominissini D et al (2015) m6A mRNA methylation facilitates resolution of naïve pluripotency toward differentiation. Science 347:1002–1006. https://doi.org/10.1126/science.1261417

Hannon E, Lunnon K, Schalkwyk L, Mill J (2015) Interindividual methylomic variation across blood, cortex, and cerebellum: implications for epigenetic studies of neurological and neuropsychiatric phenotypes. Epigenetics 10:1024–1032. https://doi.org/10.1080/15592294.2015.1100786

Harcourt EM, Ehrenschwender T, Batista PJ et al (2013) Identification of a selective polymerase enables detection of N6-methyladenosine in RNA. J Am Chem Soc 135:19079–19082. https://doi.org/10.1021/ja4105792

Hartstock K, Nilges BS, Ovcharenko A et al (2018) Enzymatic or in vivo installation of propargyl groups in combination with click chemistry for the enrichment and detection of methyltransferase target sites in RNA. Angew Chem Int Ed 57:6342–6346. https://doi.org/10.1002/anie.201800188

Helm M, Motorin Y (2017) Detecting RNA modifications in the epitranscriptome: predict and validate. Nat Rev Genet 18:275–291. https://doi.org/10.1038/nrg.2016.169

Hess ME, Brüning JC (2014) The fat mass and obesity-associated (FTO) gene: obesity and beyond? Biochim Biophys Acta 1842:2039–2047. https://doi.org/10.1016/j.bbadis.2014.01.017

Hess ME, Hess S, Meyer KD et al (2013) The fat mass and obesity associated gene (Fto) regulates activity of the dopaminergic midbrain circuitry. Nat Neurosci 16:1042–1048. https://doi.org/10.1038/nn.3449

Ho AJ, Stein JL, Hua X, Lee S, Hibar DP, Leow AD, Dinov ID, Toga AW, Saykin AJ, Shen L, Foroud T, Pankratz N, Huentelman MJ, Craig DW, Gerber JD, Allen AN, Corneveaux JJ, Stephan DA, DeCarli CS, DeChairo BM, Potkin SG, Jack CR Jr, Weiner MW, Raji CA, Lopez OL, Becker JT, Carmichael OT, Thompson PM, Alzheimer's Disease Neuroimaging Initiative (2010) A commonly carried allele of the obesity-related FTO gene is associated with reduced brain volume in the healthy elderly. Proc Natl Acad Sci U S A 107(18):8404–8409. https://doi.org/10.1073/pnas.0910878107

Holt CE, Schuman EM (2013) The central dogma decentralized: new perspectives on RNA function and local translation in neurons. Neuron 80:648–657. https://doi.org/10.1016/j.neuron.2013.10.036

Huang H, Weng H, Sun W et al (2018) Recognition of RNA N6-methyladenosine by IGF2BP proteins enhances mRNA stability and translation. Nat Cell Biol 20:285–295. https://doi.org/10.1038/s41556-018-0045-z

Ivanova I, Much C, Di Giacomo M et al (2017) The RNA m6A reader YTHDF2 is essential for the post-transcriptional regulation of the maternal transcriptome and oocyte competence. Mol Cell 67:1059–1067.e4. https://doi.org/10.1016/j.molcel.2017.08.003

Jansen IE, Savage JE, Watanabe K et al (2019) Genome-wide meta-analysis identifies new loci and functional pathways influencing Alzheimer's disease risk. Nat Genet. https://doi.org/10.1038/s41588-018-0311-9

Jia G (2016) RNA modification N6-methyladenosine in post-transcriptional regulation. In: Jurga S, Erdmann VA, Barciszewski J (eds) Modified nucleic acids in biology and medicine. Springer, Cham, pp 131–145

Jia G, Fu Y, Zhao X et al (2011) N6-methyladenosine in nuclear RNA is a major substrate of the obesity-associated FTO. Nat Chem Biol 7:885–887. https://doi.org/10.1038/nchembio.687

Kane SE, Beemon K (1987) Inhibition of methylation at two internal N6-methyladenosine sites caused by GAC to GAU mutations. J Biol Chem 262:3422–3427

Karikó K, Buckstein M, Ni H, Weissman D (2005) Suppression of RNA recognition by toll-like receptors: the impact of nucleoside modification and the evolutionary origin of RNA. Immunity 23:165–175. https://doi.org/10.1016/j.immuni.2005.06.008

Ke S, Alemu EA, Mertens C et al (2015) A majority of m6A residues are in the last exons, allowing the potential for 3′ UTR regulation. Genes Dev 29:2037–2053. https://doi.org/10.1101/gad.269415.115

Ke S, Pandya-Jones A, Saito Y et al (2017) m6A mRNA modifications are deposited in nascent pre-mRNA and are not required for splicing but do specify cytoplasmic turnover. Genes Dev 31:990–1006. https://doi.org/10.1101/gad.301036.117

Keller L, Xu W, Wang H-X et al (2011) The obesity related gene, FTO, interacts with APOE, and is associated with Alzheimer's disease risk: a prospective cohort study. J Alzheimers Dis 23:461–469. https://doi.org/10.3233/JAD-2010-101068

Kellner S, Ochel A, Thüring K et al (2014) Absolute and relative quantification of RNA modifications via biosynthetic isotopomers. Nucleic Acids Res 42:e142. https://doi.org/10.1093/nar/gku733

Koranda JL, Dore L, Shi H et al (2018) Mettl14 is essential for epitranscriptomic regulation of striatal function and learning. Neuron 99:283–292. https://doi.org/10.1016/j.neuron.2018.06.007

Lence T, Akhtar J, Bayer M et al (2016) m6A modulates neuronal functions and sex determination in Drosophila. Nature 540:242–247. https://doi.org/10.1038/nature20568

Levanon EY, Eisenberg E, Yelin R et al (2004) Systematic identification of abundant A-to-I editing sites in the human transcriptome. Nat Biotechnol 22:1001–1005. https://doi.org/10.1038/nbt996

Li JB, Levanon EY, Yoon J-K et al (2009) Genome-wide identification of human RNA editing sites by parallel DNA capturing and sequencing. Science 324:1210–1213. https://doi.org/10.1126/science.1170995
Li X, Zhu P, Ma S et al (2015) Chemical pulldown reveals dynamic pseudouridylation of the mammalian transcriptome. Nat Chem Biol 11:592–597. https://doi.org/10.1038/nchembio.1836
Li X, Xiong X, Wang K et al (2016) Transcriptome-wide mapping reveals reversible and dynamic N1-methyladenosine methylome. Nat Chem Biol 12:311–316. https://doi.org/10.1038/nchembio.2040
Li A, Chen Y-S, Ping X-L et al (2017a) Cytoplasmic m6A reader YTHDF3 promotes mRNA translation. Cell Res 27:444–447. https://doi.org/10.1038/cr.2017.10
Li L, Zang L, Zhang F et al (2017b) Fat mass and obesity-associated (FTO) protein regulates adult neurogenesis. Hum Mol Genet 26:2398–2411. https://doi.org/10.1093/hmg/ddx128
Lin S, Choe J, Du P et al (2016) The m6A methyltransferase METTL3 promotes translation in human cancer cells. Mol Cell 62:335–345. https://doi.org/10.1016/j.molcel.2016.03.021
Linder B, Grozhik AV, Olarerin-George AO et al (2015) Single-nucleotide-resolution mapping of m6A and m6Am throughout the transcriptome. Nat Methods 12:767–772. https://doi.org/10.1038/nmeth.3453
Liu N, Parisien M, Dai Q et al (2013) Probing N6-methyladenosine RNA modification status at single nucleotide resolution in mRNA and long noncoding. RNA 19:1848–1856. https://doi.org/10.1261/rna.041178.113
Liu J, Yue Y, Han D et al (2014) A METTL3-METTL14 complex mediates mammalian nuclear RNA N6-adenosine methylation. Nat Chem Biol 10:93–95. https://doi.org/10.1038/nchembio.1432
Liu N, Dai Q, Zheng G et al (2015) N6-methyladenosine-dependent RNA structural switches regulate RNA-protein interactions. Nature 518:560–564. https://doi.org/10.1038/nature14234
Liu L, Zhang S-W, Gao F et al (2016) DRME: count-based differential RNA methylation analysis at small sample size scenario. Anal Biochem 499:15–23. https://doi.org/10.1016/j.ab.2016.01.014
Liu N, Zhou KI, Parisien M et al (2017) N6-methyladenosine alters RNA structure to regulate binding of a low-complexity protein. Nucleic Acids Res 45:6051–6063. https://doi.org/10.1093/nar/gkx141
Liu H, Begik O, Lucas MC et al (2019) Accurate detection of m6A RNA modifications in native RNA sequences. bioRxiv 525741. https://doi.org/10.1101/525741
Ma L, Zhao B, Chen K et al (2017) Evolution of transcript modification by N6-methyladenosine in primates. Genome Res. https://doi.org/10.1101/gr.212563.116
Martin J, Walters RK, Demontis D et al (2017) A genetic investigation of sex bias in the prevalence of attention deficit hyperactivity disorder. bioRxiv 154088. https://doi.org/10.1101/154088
Mauer J, Jaffrey SR (2018) FTO, m6Am, and the hypothesis of reversible epitranscriptomic mRNA modifications. FEBS Lett 592:2012–2022. https://doi.org/10.1002/1873-3468.13092
Mauer J, Luo X, Blanjoie A et al (2017) Reversible methylation of m6Am in the 5′ cap controls mRNA stability. Nature 541:371–375. https://doi.org/10.1038/nature21022
Meier S, Trontti K, Als TD et al (2018) Genome-wide association study of anxiety and stress-related disorders in the iPSYCH Cohort. bioRxiv 263855. https://doi.org/10.1101/263855
Meng J, Lu Z, Liu H et al (2014) A protocol for RNA methylation differential analysis with MeRIP-Seq data and exomePeak R/bioconductor package. Methods 69:274–281. https://doi.org/10.1016/j.ymeth.2014.06.008
Merkurjev D, Hong W-T, Iida K et al (2018) Synaptic N 6-methyladenosine (m$^6$A) epitranscriptome reveals functional partitioning of localized transcripts. Nat Neurosci 1. https://doi.org/10.1038/s41593-018-0173-6
Meyer KD, Saletore Y, Zumbo P et al (2012) Comprehensive analysis of mRNA methylation reveals enrichment in 3′ UTRs and near stop codons. Cell 149:1635–1646. https://doi.org/10.1016/j.cell.2012.05.003
Meyer KD, Patil DP, Zhou J et al (2015) 5′ UTR m6A promotes cap-independent translation. Cell 163:999–1010. https://doi.org/10.1016/j.cell.2015.10.012

Milaneschi Y, Lamers F, Mbarek H et al (2014) The effect of FTO rs9939609 on major depression differs across MDD subtypes. Mol Psychiatry 19:960–962. https://doi.org/10.1038/mp.2014.4

Molinie B, Wang J, Lim KS et al (2016) m6A-LAIC-seq reveals the census and complexity of the $m^6A$ epitranscriptome. Nat Methods. https://doi.org/10.1038/nmeth.3898

Narayan P, Ludwiczak RL, Goodwin EC, Rottman FM (1994) Context effects on $N^6$-adenosine methylation sites in prolactin mRNA. Nucleic Acids Res 22:419–426. https://doi.org/10.1093/nar/22.3.419

Nestler EJ, Peña CJ, Kundakovic M et al (2016) Epigenetic basis of mental illness. Neuroscientist 22:447–463. https://doi.org/10.1177/1073858415608147

O'Connell MR, Oakes BL, Sternberg SH et al (2014) Programmable RNA recognition and cleavage by CRISPR/Cas9. Nature 516:263–266. https://doi.org/10.1038/nature13769

Patil DP, Chen C-K, Pickering BF et al (2016) m6A RNA methylation promotes XIST-mediated transcriptional repression. Nature 537:369–373. https://doi.org/10.1038/nature19342

Patil DP, Pickering BF, Jaffrey SR (2018) Reading m6A in the transcriptome: m6A-binding proteins. Trends Cell Biol 28:113–127. https://doi.org/10.1016/j.tcb.2017.10.001

Paulines M, Limbach P (2017) Comparative analysis of ribonucleic acid digests (CARD) by mass spectrometry. In: Lusser A (ed) RNA methylation. Springer, New York, pp 19–32

Ping X-L, Sun B-F, Wang L et al (2014) Mammalian WTAP is a regulatory subunit of the RNA N6-methyladenosine methyltransferase. Cell Res 24:177–189. https://doi.org/10.1038/cr.2014.3

Popova AM, Williamson JR (2014) Quantitative analysis of rRNA modifications using stable isotope labeling and mass spectrometry. J Am Chem Soc 136:2058–2069. https://doi.org/10.1021/ja412084b

Profenno LA, Porsteinsson AP, Faraone SV (2010) Meta-analysis of Alzheimer's disease risk with obesity, diabetes, and related disorders. Biol Psychiatry 67:505–512. https://doi.org/10.1016/j.biopsych.2009.02.013

Purves KL, Coleman JRI, Rayner C et al (2017) The common genetic architecture of anxiety disorders. bioRxiv 203844. https://doi.org/10.1101/203844

Reitz C, Tosto G, Mayeux R et al (2012) Genetic variants in the Fat and Obesity Associated (FTO) gene and risk of Alzheimer's disease. PLoS One 7:e50354. https://doi.org/10.1371/journal.pone.0050354

Roost C, Lynch SR, Batista PJ et al (2015) Structure and thermodynamics of N6-methyladenosine in RNA: a spring-loaded base modification. J Am Chem Soc. https://doi.org/10.1021/ja513080v

Roundtree IA, Evans ME, Pan T, He C (2017) Dynamic RNA modifications in gene expression regulation. Cell 169:1187–1200. https://doi.org/10.1016/j.cell.2017.05.045

Samaan Z, Anand SS, Anand S et al (2013) The protective effect of the obesity-associated rs9939609 A variant in fat mass- and obesity-associated gene on depression. Mol Psychiatry 18:1281–1286. https://doi.org/10.1038/mp.2012.160

Schaefer M, Kapoor U, Jantsch MF (2017) Understanding RNA modifications: the promises and technological bottlenecks of the 'epitranscriptome'. Open Biol 7. https://doi.org/10.1098/rsob.170077

Schibler U, Perry RP (1977) The 5′-termini of heterogeneous nuclear RNA: a comparison among molecules of different sizes and ages. Nucleic Acids Res 4:4133–4149

Schöller E, Weichmann F, Treiber T et al (2018) Interactions, localization, and phosphorylation of the m6A generating METTL3-METTL14-WTAP complex. RNA 24:499–512. https://doi.org/10.1261/rna.064063.117

Schwartz S, Motorin Y (2017) Next-generation sequencing technologies for detection of modified nucleotides in RNAs. RNA Biol 14:1124–1137. https://doi.org/10.1080/15476286.2016.1251543

Schwartz S, Agarwala SD, Mumbach MR et al (2013) High-resolution mapping reveals a conserved, widespread, dynamic mRNA methylation program in yeast meiosis. Cell 155:1409–1421. https://doi.org/10.1016/j.cell.2013.10.047

Schwartz S, Bernstein DA, Mumbach MR et al (2014a) Transcriptome-wide mapping reveals widespread dynamic-regulated pseudouridylation of ncRNA and mRNA. Cell 159:148–162. https://doi.org/10.1016/j.cell.2014.08.028

Schwartz S, Mumbach MR, Jovanovic M et al (2014b) Perturbation of m6A writers reveals two distinct classes of mRNA methylation at internal and 5′ sites. Cell Rep 8:284–296. https://doi.org/10.1016/j.celrep.2014.05.048

Scuteri A, Sanna S, Chen W-M et al (2007) Genome-wide association scan shows genetic variants in the FTO gene are associated with obesity-related traits. PLoS Genet 3:e115. https://doi.org/10.1371/journal.pgen.0030115

Shi H, Wang X, Lu Z et al (2017) YTHDF3 facilitates translation and decay of N6-methyladenosine-modified RNA. Cell Res 27:315–328. https://doi.org/10.1038/cr.2017.15

Shi H, Zhang X, Weng Y-L et al (2018) m6A facilitates hippocampus-dependent learning and memory through YTHDF1. Nature 563:249–253. https://doi.org/10.1038/s41586-018-0666-1

Slobodin B, Han R, Calderone V et al (2017) Transcription impacts the efficiency of mRNA translation via co-transcriptional N6-adenosine methylation. Cell 169:326–337.e12. https://doi.org/10.1016/j.cell.2017.03.031

Smemo S, Tena JJ, Kim K-H et al (2014) Obesity-associated variants within FTO form long-range functional connections with IRX3. Nature 507:371–375. https://doi.org/10.1038/nature13138

Spitale RC, Flynn RA, Zhang QC et al (2015) Structural imprints in vivo decode RNA regulatory mechanisms. Nature 519:486–490. https://doi.org/10.1038/nature14263

Squires JE, Patel HR, Nousch M et al (2012) Widespread occurrence of 5-methylcytosine in human coding and non-coding RNA. Nucleic Acids Res 40:5023–5033. https://doi.org/10.1093/nar/gks144

Stratigopoulos G, Martin Carli JF, O'Day DR et al (2014) Hypomorphism for RPGRIP1L, a ciliary gene vicinal to the FTO locus, causes increased adiposity in mice. Cell Metab 19:767–779. https://doi.org/10.1016/j.cmet.2014.04.009

Tai H, Wang X, Zhou J et al (2017) Protein kinase Cβ activates fat mass and obesity-associated protein by influencing its ubiquitin/proteasome degradation. FASEB J 31:4396–4406. https://doi.org/10.1096/fj.201601159RR

Vujovic P, Stamenkovic S, Jasnic N et al (2013) Fasting induced cytoplasmic Fto expression in some neurons of rat hypothalamus. PLoS One 8:e63694. https://doi.org/10.1371/journal.pone.0063694

Walters BJ, Mercaldo V, Gillon CJ et al (2017) The role of The RNA demethylase FTO (fat mass and obesity-associated) and mRNA methylation in hippocampal memory formation. Neuropsychopharmacology 42:1502–1510. https://doi.org/10.1038/npp.2017.31

Wang Y, Li Y, Toth JI et al (2014) N6-methyladenosine modification destabilizes developmental regulators in embryonic stem cells. Nat Cell Biol 16:191–198. https://doi.org/10.1038/ncb2902

Wang X, Zhao BS, Roundtree IA et al (2015) N6-methyladenosine modulates messenger RNA translation efficiency. Cell 161:1388–1399. https://doi.org/10.1016/j.cell.2015.05.014

Wang C-X, Cui G-S, Liu X et al (2018) METTL3-mediated m6A modification is required for cerebellar development. PLoS Biol 16:e2004880. https://doi.org/10.1371/journal.pbio.2004880

Waung MW, Huber KM (2009) Protein translation in synaptic plasticity: mGluR-LTD, Fragile X. Curr Opin Neurobiol 19:319–326. https://doi.org/10.1016/j.conb.2009.03.011

Wei C-M, Gershowitz A, Moss B (1975) N6, O2′-dimethyladenosine a novel methylated ribonucleoside next to the 5′ terminal of animal cell and virus mRNAs. Nature 257:251–253. https://doi.org/10.1038/257251a0

Wei J, Liu F, Lu Z et al (2018) Differential m6A, m6Am, and m1A demethylation mediated by FTO in the cell nucleus and cytoplasm. Mol Cell 71:973–985.e5. https://doi.org/10.1016/j.molcel.2018.08.011

Weng Y-L, Wang X, An R et al (2018) Epitranscriptomic m6A regulation of axon regeneration in the adult mammalian nervous system. Neuron 97:313–325.e6. https://doi.org/10.1016/j.neuron.2017.12.036

Widagdo J, Zhao Q-Y, Kempen M-J et al (2016) Experience-dependent accumulation of N6-methyladenosine in the prefrontal cortex is associated with memory processes in mice. J Neurosci 36:6771–6777. https://doi.org/10.1523/JNEUROSCI.4053-15.2016

Wray NR, Ripke S, Mattheisen M et al (2018) Genome-wide association analyses identify 44 risk variants and refine the genetic architecture of major depression. Nat Genet 50:668. https://doi.org/10.1038/s41588-018-0090-3

Xiang Y, Laurent B, Hsu C-H et al (2017) RNA m6A methylation regulates the ultraviolet-induced DNA damage response. Nature 543:573–576. https://doi.org/10.1038/nature21671

Xiao W, Adhikari S, Dahal U et al (2016) Nuclear m6A reader YTHDC1 regulates mRNA splicing. Mol Cell 61:507–519. https://doi.org/10.1016/j.molcel.2016.01.012

Yoon K-J, Ringeling FR, Vissers C, et al (2017) Temporal control of mammalian cortical neurogenesis by m6A methylation. Cell 0: doi: https://doi.org/10.1016/j.cell.2017.09.003

Yu J, Chen M, Huang H et al (2018) Dynamic m6A modification regulates local translation of mRNA in axons. Nucleic Acids Res 46:1412–1423. https://doi.org/10.1093/nar/gkx1182

Zeng Y, Wang S, Gao S et al (2018) Refined RIP-seq protocol for epitranscriptome analysis with low input materials. PLoS Biol 16:e2006092. https://doi.org/10.1371/journal.pbio.2006092

Zhang S, Zhao BS, Zhou A et al (2017) m6A demethylase ALKBH5 maintains tumorigenicity of glioblastoma stem-like cells by sustaining FOXM1 expression and cell proliferation program. Cancer Cell 31:591–606.e6. https://doi.org/10.1016/j.ccell.2017.02.013

Zhang Z, Wang M, Xie D et al (2018) METTL3-mediated N 6-methyladenosine mRNA modification enhances long-term memory consolidation. Cell Res 28:1050. https://doi.org/10.1038/s41422-018-0092-9

Zhao BS, Roundtree IA, He C (2017) Post-transcriptional gene regulation by mRNA modifications. Nat Rev Mol Cell Biol 18:31–42. https://doi.org/10.1038/nrm.2016.132

Zheng G, Dahl JA, Niu Y et al (2013) ALKBH5 is a mammalian RNA demethylase that impacts RNA metabolism and mouse fertility. Mol Cell 49:18–29. https://doi.org/10.1016/j.molcel.2012.10.015

Zhou J, Wan J, Gao X et al (2015) Dynamic m6A mRNA methylation directs translational control of heat shock response. Nature 526:591–594. https://doi.org/10.1038/nature15377

Zhu T, Yong XLH, Xia D et al (2018) Ubiquitination regulates the proteasomal degradation and nuclear translocation of the fat mass and obesity-associated (FTO) protein. J Mol Biol 430:363–371. https://doi.org/10.1016/j.jmb.2017.12.003

# G9a and G9a-Like Histone Methyltransferases and Their Effect on Cell Phenotype, Embryonic Development, and Human Disease

**Carol A. Eisenberg and Leonard M. Eisenberg**

## Contents



**Abstract** Post-translational modifications of histone proteins alter the topology of the chromatin, which affect genome accessibility and thus, impact gene regulation. One type of histone modification that has a profound effect on the chromatin is methylation, where specific marking of lysine or arginine residues within the N-terminal tails of the histone core proteins control chromatin packaging and transcriptional dynamics throughout the genome. Two important histone methyltransferases are G9a and G9a-like protein-1 (GLP), whose enzymatic activities have a major influence on cell phenotype, embryonic development, cancer progression, and other human diseases. This review will provide an overview of the structural and enzymatic properties of G9a and GLP, the development of pharmacological reagents that target their catalytic activities, the impact of these enzymes on cell biology, embryogenesis, cancer progression, neurological and other human diseases. In addition, we will examine the known non-histone targets of the enzymes, to understand how G9a and GLP influence biological responses beyond their effect on histone proteins. Emphasis will be placed on highlighting the latest discoveries and examining the experimental details of several

---

C. A. Eisenberg · L. M. Eisenberg (✉)
Departments of Physiology and Medicine, New York Medical College, Valhalla, NY, USA
e-mail: carol_eisenberg@nymc.edu; leonard_eisenberg@nymc.edu



S. Jurga, J. Barciszewski (eds.), *The DNA, RNA, and Histone Methylomes*, RNA Technologies, https://doi.org/10.1007/978-3-030-14792-1_16

key studies that have contributed to our current understanding of the biological significance of G9a and GLP.

**Keywords** Histone methyltransferases · Epigenetics · G9a · GLP · EHMT1 · EHMT2 · Small molecule inhibitors · Cell proliferation · Embryonic development · Cancer · Kleefstra syndrome

## 1 Introduction

The genetic material of eukaryotic organisms is contained within elaborate three-dimensional structures called chromosomes, which comprise not only DNA, but also a protein edifice that serves to both package and regulate access to the genes (Venkatesh and Workman 2015). The overall structure of the chromosome is built on the repetitive framework of the nucleosome, which consists of ~147 base pair stretches of DNA wrapped around a core protein complex that consists of pairs of four histone proteins H2A, H2B, H3, and H4 (Fig. 1). Each of these proteins is composed of a central globular carboxy-terminal core, from which extends a lysine-rich N-terminal tail. This nucleosome protein core is anchored to DNA via Histone H1 protein. There are several types of post-translational modifications of histone core proteins that alter chromatin topology, which affects genome accessibility and thus, impacts gene regulation. Post-translational modifications of histones include, the conversion of arginine residues to citrulline, the addition of ubiquitin or SUMOs (small ubiquitin-like modifiers) to the protein, phosphorylation, and glycosylation of amino acid residues (Bannister and Kouzarides 2011; Casciello et al. 2015). The histone modifications that have received the most research attention have been acetylation and methylation, which typically occur on lysine residues of histone H3—although arginine residues also serve as substrates for methylation. There are many demethylating and methylating enzymes that affect the methylation pattern of histones, and greatly impact cell biology. This chapter will focus on two of these enzymes G9a and G9a-like protein-1 (GLP), whose histone methylation activities have profound influence on cell phenotype, embryonic development, cancer progression, and other human diseases.

## 2 An Overview of the Molecular, Structural, and Enzymatic Properties of the Histone Methyltransferases G9a and GLP

The histone methyltransferase G9a is a methylating enzyme that acts on specific lysine residues on histone H3 (Tachibana et al. 2001, 2005; Shinkai and Tachibana 2011). Although G9a is the most commonly used designation for this histone methyltransferase, it is also referred by other names (Allis et al. 2007; Chaturvedi

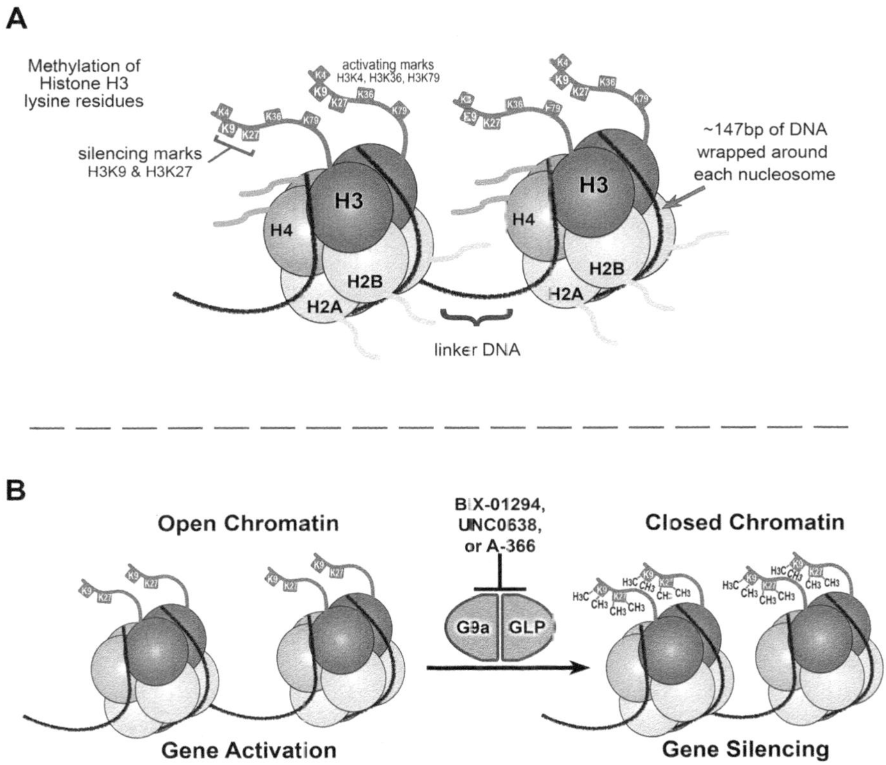


**Fig. 1** (**a**) Schematic representation of the structure of the nucleosome, which consists of pairs of four histone proteins: H2A, H2B, H3 H4. Each of these histones is composed of a central globular carboxy-terminal core, from which extends a lysine-rich N-terminal tail. The location of key lysine residues are indicated on the histone H3 N-terminal tail, with methylation of H3K4, K36, and K79 associated with gene activation. In contrast, methylated H3K9 and H3K27, which are targets of G9a/GLP catalytic activity, are linked to gene silencing. (**b**) Dimethylation of H3K9 and H3K27 by G9a and GLP promotes a closed chromatin that is inaccessible for gene transcription. In the absence of G9a and GLP, or when their activities are either inhibited or counteracted by histone demethylases, the resulting unmethylated state of H3K9 and H3K27 promotes an open chromatin configuration that allows active gene transcription. Indicated on this drawing are several pharmacological inhibitors that are specific for G9a and GLP

and Brand 2014). These include lysine methyltransferase-1C (KMT1C), euchromatic histone-lysine N-methyltransferase 2 (EHMT2), and HLA-B associated transcript 8 (BAT8), the name of which alludes to the location of the encoding gene within the human major histocompatibility complex class III region (Milner and Campbell 1993). The genetically related sister protein GLP, which is also known as EHMT1 and KMT1D, possesses similar activities to G9a (Allis et al. 2007; Shinkai and Tachibana 2011). These twin proteins often, although not always, act in concert as heterodimeric subunits that are part of larger heteromeric protein complexes (Tachibana et al. 2001, 2002, 2005; Shinkai and Tachibana 2011). However, G9a and GLP are able to exert their methyltransferase activities independently of one another by forming homodimers, although the prevalence of homodimer formation outside of experimental in vitro environments has not been determined. Historically, G9a has received far greater research emphasis than its highly similar twin protein GLP. Accordingly, the majority of studies cited in this review have a primary emphasis on the activities of G9a, even though GLP may have had an equally important role in the biological phenomena described in those investigations. Despite the discrepancy in the attention these two enzymes have received, this review will try to balance known information of G9a and GLP to get a full picture of their cooperative functions in regulating the chromatin, while pointing out any distinctions the twin proteins have with each other.

G9a and GLP are members of members of the Su(var)3-9 gene family first identified in Drosophila melanogaster (Schotta et al. 2002). The Drosophila gene Su(var)3-9 is so-named because it is a suppressor of variegation (i.e. color variations), whose activity results from its methylation of histone H3 on lysine 9 (H3K9). A major characteristic of Su(var)3-9 proteins are the presence of the highly conserved SET domain, which was first identified by a shared sequence in the three Drosophila chromatin regulatory proteins Su(var)3-9, Enhancer of zeste, and trithorax. SET, which is an acronym of these three molecules, consists of a 130 to 140 amino acid long module that is exhibited in most mammalian histone methyltransferases (Dillon et al. 2005). The SET domain of G9a and GLP (Fig. 2) is the methylating component of these enzymes that utilizes S-adenosyl-L-methionine (SAM) as a methyl donor.

Another structural characteristic of G9a and GLP is the presence of seven ankyrin repeats (Fig. 2), which are so-named because they were first described within the ankyrin family of proteins (Collins et al. 2008). The ankyrin repeat domain is a protein-protein interaction module of ~33 amino acids, which serves as the methyl-lysine binding component of G9a and GLP for histone H3. Other structural attributes of G9a include an N-terminal activation domain that contains nuclear localization sequences (Estève et al. 2005), E-rich domain consisting of 24 contiguous glutamic acid residues, and a 12 cysteine (Cys) containing domain (Tachibana et al. 2005; Shinkai and Tachibana 2011) (Fig. 2). GLP has a similar architectural layout, except that the region analogous to the E-rich domain of G9a consists of a sequence of repeated glutamic and aspartic acid residues (Tachibana et al. 2005; Shinkai and Tachibana 2011). In humans, G9a histone methyltransferase is exhibited primarily by two isoforms: a full-length 1210 amino acid protein (isoform a) derived from 24 exons of the G9a gene, and a splice variant of 1176 amino acid residues (isoform b) with a short region (exon 10) deleted between the E-rich and Cys-rich domains (Brown et al. 2001). In mouse, full-length G9a protein is slightly larger than its human ortholog with 1263 amino acids (Estève et al. 2005; Shinkai and Tachibana 2011). GLP in its full-length form contains 1298 or 1296 amino acids for human and mouse, respectively. G9a and GLP share >45% sequence identity and ~70% sequence similarity, with the SET domain being the region with the greatest conservation between the two molecules with >80% sequence identity (Chang et al. 2009; Shankar et al. 2013).

The principal targets of G9a/GLP are lysine 9 and lysine 27 of histone H3 (H3K9 and H3K27) (Tachibana et al. 2001; Wu et al. 2011; Yang et al. 2017). Both of these lysine residues are present on the tail region of histone H3 (Fig. 1a). G9a and GLP

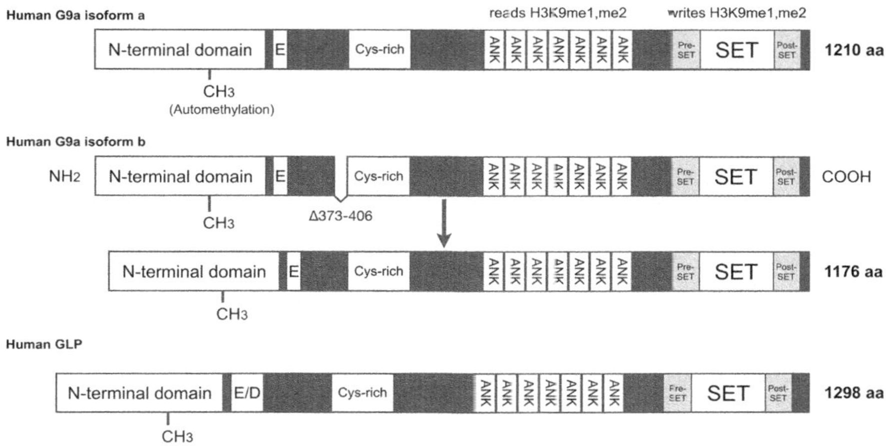


**Fig. 2** Schematic drawing showing the domain structure of human G9a and GLP. There are two principal G9a isoforms exhibited in human cells, a 1210 amino acid full-length isoform, and an 1176 amino acid splice variant that arises by the excision of exon 10. Full length GLP is a slightly larger protein at 1298 amino acids. G9a and GLP exhibit several major domains that are conserved between the proteins. These include the catalytic SET domain, flanked by the pre-SET and post-SET regions that are important for sequestering zinc, which is a necessary cofactor for the proper folding and enzymatic activity-of G9a and GLP (Dillon et al. 2005). The ankyrin domain consists of seven ankyrin repeats that are required for the binding of the enzymes to mono- and dimethylated H3K9. Other notable structural components of G9a and GLP include the cysteine-rich domain, and the N-terminal domain containing a lysine residue that is automethylated by these enzymes. One noted difference between the two proteins is that G9a has an E-rich domain that consists of 24 contiguous glutamic acids, while the corresponding E/D-rich region of GLP contains multiple aspartic acid residues interspersed among the repeated glutamic acids

are associated with heterochromatin-and transcriptionally silent regions of euchromatin, where their methylation of histone H3 promotes a closed chromatin configuration and gene inactivation (Fig. 1b). Both G9a and GLP catalyze the mono- and dimethylation of H3K9 and H3K27 via the SET domain (Chang et al. 2009). The ankyrin repeats of these enzymes bind to the resultant methylated histone H3 tail region, allowing G9a and GLP to serve as a platform for the formation of multimeric protein complexes with other histone modification proteins. For example, G9a and GLP form a complex with the lysine histone methyltransferase enzymes Suv39h and SETDB1 in establishing areas of heterochromatin and maintaining chromosomal stability (Fritsch et al. 2010). Within the euchromatin, cooperative interactions of G9a and GLP with the Polycomb Repressive Complex 2-proteins, including the histone-lysine methyltransferase enzyme EZH2, may help maintain transcriptionally silent regions within the genome (Mozzetta et al. 2014). G9a and GLP localization to the chromatin may itself be dependent on their interactions with two zinc finger proteins ZNF644 and WIZ (widely interspaced zinc finger motifs protein) (Bian et al. 2015; Olsen et al. 2016). G9a and GLP, which dimerize by attachment at their SET-containing C-terminal catalytic domains, form a chromatin binding complex with ZNF644 and WIZ that independently bind to the respective N-terminal domains of the two methyltransferases (Bian et al. 2015).

In vitro studies have demonstrated that G9a and GLP can exert their histone methyltransferase activity independently of one another. Yet in vivo, the methylating activity of these two proteins largely occurs in tandem. Not only do G9a and GLP preferentially heterodimerize in situ, but ablating either the G9a or GLP gene generated near identical phenotypes in the developing embryo and equivalent high level reductions of global H3K9 methylation in both the embryo and embryonic stem cell (ESC) lines (Tachibana et al. 2002, 2005). Moreover, no further reductions in H3K9 methylation were observed in ESCs that were G9a/GLP dual-deficient (Tachibana et al. 2002). However, these data do not mean that G9a and GLP play identical roles in promoting histone methylation in situ. The SET domain of G9a may play a dominant role in the methyltransferase activities within the G9a/GLP heterodimers, as mutation of the GLP SET domain to a catalytic inactivate subunit did not reduce embryonic viability (Inagawa et al. 2013; Kramer 2016). In addition, the ankyrin domains of G9a and GLP display distinct binding affinities, as they preferentially bind to mono- and dimethylated H3K9, respectively (Collins et al. 2008; Liu et al. 2015a; Kramer 2016). The ability of G9a and GLP to both change the methylation marking of histone H3 and differentially bind to the modified residues may serve as the means to allocate functional responsibilities of the two proteins, where one acts on marking histone H3, while the binding of the other to the histone H3 tail serves as a template for recruiting other accessory proteins.

Cross-talk between various epigenetic regulatory mechanisms has a major impact on gene expression (Collins and Cheng 2010; Suganuma and Workman 2011; Zentner and Henikoff 2013). There is a complex relationship between G9a/GLP methylation of H3K9 and H3K27 and the removal of those methylation marks by histone demethylation enzymes (Huang et al. 2010a; Suganuma and Workman 2011; Upadhyay and Cheng 2011; Choi et al. 2017). Other enzymes modify the histone H3 residues H3K4, H3K36, and H3K79, whose methylation confers gene activation (Zentner and Henikoff 2013). There is also a competitive balance between activating acetylation and silencing methylation markings of H3K9 and H3K27 (Wang et al. 2008; Yang et al. 2017). These types of histone modifications are among the interconnected regulatory streams that impact overall cell, tissue, and organ biology. It is beyond the scope of this chapter to explore these symbiotic interactions that determine and maintain the phenotypic functioning of individual cell types. Instead, this review will focus on the role of G9a and GLP in the overall epigenetic regulation of cell and tissue genotype and phenotype, and how that regulation impacts the biology of the organism, including the human organism.

## 3 Identification of Pharmacological Reagents as Tools for Understanding the Biological Significance of G9a and GLP Activity

All cells and tissues require a fully operative epigenetic machinery to properly function. The major advance in dissecting the role of G9a in modulating the overall epigenetic environment was the development of drugs that specifically target this enzyme. The first discovered and still the most highly used G9a inhibitor is BIX-01294, which was identified from a screen of ~125,000 compounds from a *B*oehringer *I*ngelheim chemical library that initially selected seven compounds able to inhibit G9a methylation of H3K9. Most of those compounds showed a broader spectrum of inhibitory activity against other SET-containing lysine-specific methyltransferases, with the exception of BIX-01294 that specifically inhibited G9a (Kubicek et al. 2007). Subsequently, it was shown that BIX-01294 was equally effective in suppressing the catalytic activities of GLP (Chang et al. 2009). Despite that correction, most studies that used BIX-01294 referred to it as a G9a selective inhibitor. In reality, BIX-01294 inhibition probably should be interpreted as a specific for the joint activities of G9a and GLP.

The identification of BIX-01294 as an inhibitor of G9a and GLP was a major advance in understanding the functional importance of these enzymes. However, it became apparent that there may be limitations on the utility of BIX-01294 due to the constricted range of its effective concentration, as doses of this drug only a few-fold higher than required to generate a robust inhibition of G9a and GLP activities are toxic to cultured tissues and cells. For example, in studies with mouse mesenchymal stem cells (MSCs), the optimal concentrations of BIX-01294 that inhibited histone H3 methylation and promoted changes in gene expression and cell phenotype, ranged from 2 to 8 μM without a noticeable decrease in cellular health (Fig. 3). However, MSC cultures exposed to a BIX-01294 dosage of 12 μM or greater displayed decreased cell viability (Mezentseva et al. 2013; Yang et al. 2015). Thus, the indication that BIX-01294 exhibits cytotoxic effects at concentrations just above the effective dosage has led to efforts to develop alternative G9a and GLP inhibitors (Fig. 4).

Initial alternatives to BIX-01294 were based on that drug's chemical structure (Liu et al. 2009; Liu et al. 2011). BIX-01294 is a dimethoxy diazepine-quinazoline-amine derivative (Fig. 4) that inhibits G9a and GLP activity by binding to the histone peptide binding site of the SET domain (Chang et al. 2009). Among the first alternative G9a/GLP inhibitors reported were UNC0224 and UNC0638, which were identified by screening molecules that shared a dimethoxy quinazoline moiety with BIX-01294 (Liu et al. 2009, 2010; Vedadi et al. 2011). Like BIX-01294, both UNC0224 and UNC0638 are able to inhibit G9a- and GLP- mediated H3K9 and H3K27 methylation without suppressing the activities of other lysine methyltransferases. Based on published reports, UNC0638 appears to exhibit a broader effective concentration range than BIX-01294. Additional G9a/GLP inhibitors that share the quinazoline core with BIX-01294 were also described. These include UNC0642, which exhibits similar low toxicity and high potency as UNC0638, but with improved pharmacokinetic properties that allow this drug to be utilized for in vivo studies (Liu et al. 2013).

Concurrent with efforts to generate improved BIX-01294-like inhibitors, were attempts to generate chemically distinct molecules that would specifically inhibit G9a and GLP. A-366 was identified using peptide-based AlphaLISA assays to screen chemically diverse compounds for their inhibition of G9a dimethylation of H3K9 (Sweis et al. 2014). A molecule that stood out for its inhibition of H3K9 dimethylation in the primary screen was spiro[cyclobutane-1,3′- indol]-2′-amine. From this initially identified structure, specificity toward G9a was optimized further

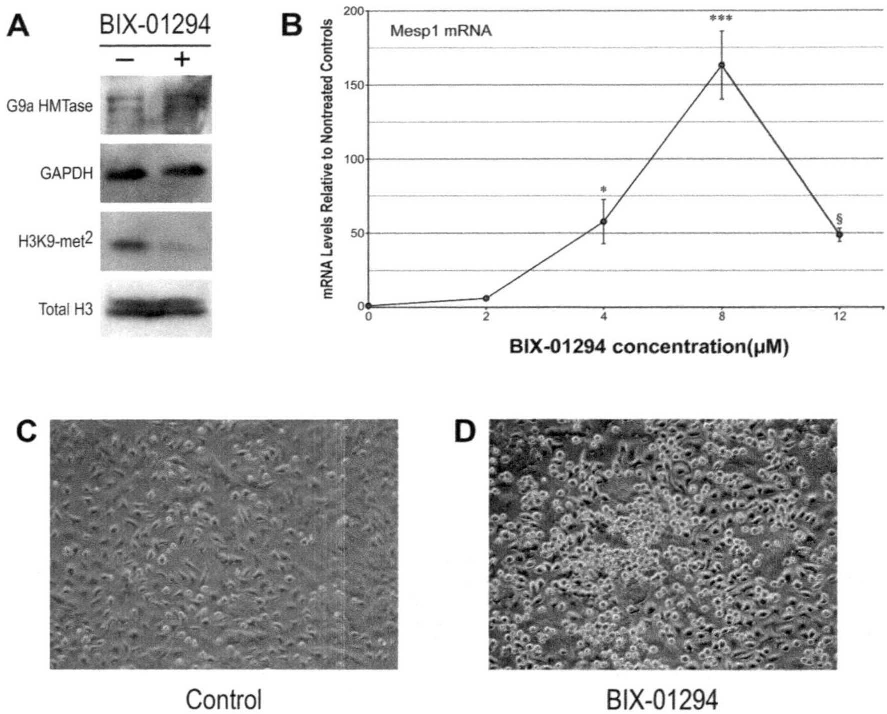


**Fig. 3** Effect of BIX-01294 inhibition of G9a/GLP on bone marrow stem cells. (**a**) Bone marrow stem cells were cultured in the absence or presence of 8 μM BIX01294 for 48 h, prior to the isolation and immunoblotting of the protein. While G9a levels were unaffected by BIX-01294, methylation activity of G9a was dramatically reduced, as indicated by diminished immunoreactivity against the dimethylated form of H3K9. (**b**) BIX-01294 induced MSC expression of Mesp1. Cells were cultured in the absence or presence of various doses of BIX01294 for 48 h and assayed for Mesp1 expression by real time qualitative (q)PCR. MSCs exhibited the highest Mesp1 levels with 8 μM BIX01294. However, this response was greatly diminished when the concentration was increased to 12 μM due to a noticeable decrease in cellular health at concentrations just above the effective dose. (**c**, **d**) MSCs were incubated for 48 h without or with BIX-01294, respectively, and then cultured for 4 additional days in high serum. In response to BIX-01294 there was a dramatic expansion in the amount of blast like cells within the cultures. Data from panels **a** and **b** was previously published in Yang et al. (2015, 2017)

by alteration in the side chains, with the highest inhibitory activity observed with modifications that displayed mimetics of lysine. This process led to the derivation of A-366, which suppressed G9a and GLP activity without affecting other methyltransferases. Like BIX-01294, A-366 prevents H3K9 methylation by its competitive inhibition of the substrate binding site of the SET domain on G9a and GLP (Sweis et al. 2014). Moreover, A-366 not only appears to be less cytotoxic than BIX-01294, but also UNC0638 (Pappano et al. 2015).

An alternative strategy for generating new structurally divergent G9a/GLP inhibitors was to screen for small molecules that were able to adhere to the G9a substrate

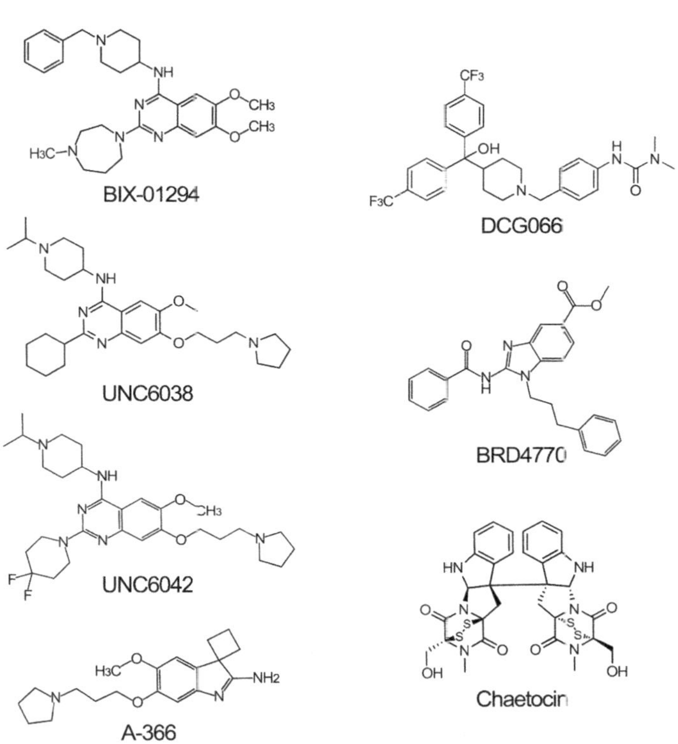


**Fig. 4** Chemical structures of various small-molecule inhibitors that suppress the activities of G9a and GLP

binding site at high affinity, while possessing a different chemical core than BIX-01294 (Kondengaden et al. 2016). DCG066 was discovered by structure-based virtual screening analysis (SBVS) of the SPECS public database (http://www.specs.net/) small molecule library. The computer-based screening algorithm, based on the crystal structure of G9a in complex with UNC0638, identified 125 candidate molecules possessing geometric and chemical properties with a potential to dock into the G9a binding pocket in a similar fashion as UNC0638. Those molecular candidates were then tested directly for binding to G9a using surface plasmon resonance (SPR) to identify molecules that exhibited high affinity for G9a. Molecules characterized for their G9a binding were subsequently tested for their ability to inhibit H3K9 methylation and provoke leukemia cells to exhibit biological responses akin to those previously described for BIX-01294 and UNC0638. Of these molecules, DCG066 was able to show levels of G9a/GLP inhibition that was comparable to BIX-01294 (Kondengaden et al. 2016).

Other avenues for generating G9a/GLP inhibitors were also pursued that did not target the substrate binding site of the SET domain. BIX-01338 was identified in the original Boehringer Ingelheim chemical compound library screen that singled out BIX-01294 as a G9a and GLP-specific inhibitor (Kubicek et al. 2007). BIX-01338 also significantly inhibited the activity of these two enzymes, however, unlike BIX-01294, it also suppressed a wide range of other methyltransferases. What was distinctive about BIX-01338 was that it did not interfere with the SET domain, but functioned as a SAM-competitive inhibitor. The problem with targeting the interactions of this methyl donor with G9a and GLP is that their SAM binding regions exhibit high sequence similarity with many other methyltransferases (Kubicek et al. 2007). BRD4770 is a SAM mimetic inhibitor that is a modified analogue of the aminobenzimidazole structure of BIX-01338, and which has become one of the more highly used alternatives to BIX-01294 (Yuan et al. 2012). Although marketed as a G9a/GLP-specific inhibitor, questions remain about the specificity of BRD4770. While the inhibitory effects of this drug toward methyltransferases may not be as undiscriminating as BIX-01338, available evidence indicates that BRD4770 may be similarly suppressive to Enhancer of zeste homolog 2 (EZH2), which is the enzymatic component of the Polycomb Repressive Complex 2, as it is toward G9a and GLP. Moreover, BRD4770 shows significant inhibition against NSD2 (Nuclear Receptor Binding SET Domain Protein 2) activities, albeit to a lesser extent than G9a, GLP, and EZH2. Thus, despite its effect on G9a and GLP-function, BRD4770 has yet to be vetted thoroughly enough against other proteins to fully understand the range of enzymatic targets that are affected by this reagent.

Another class of molecules that inhibits G9a and GLP activity, although with a cross-specificity for additional methyltransferases, is chaetocin and its analogues (Iwasa et al. 2010). Chaetocin is a complex epidithiodiketopiperazine produced by the fungi species Chaetomium, and has a broad specificity for several members of the Su(var)3-9 class of methyltransferases. Its inhibition of these enzymes, like BIX-01338 and BRD4770, is due to its interference with the SAM binding site. The specificity of chaetocin for Su(var)3-9 methyltransferases results from its interference with the C-terminal cysteine-rich domain that is present in this class of enzymes (Chaib et al. 2012), which causes the ejection of a structurally important zinc ion that affects both SAM and substrate binding by the enzyme. Despite its broad specificity, chaetocin, as well as other epidithiodiketopiperazines, have shown therapeutic utility for their antibacterial and anticancer activity (Isham et al. 2007; Kowolik et al. 2016). Furthermore, the targeting of the structural zinc by chaetocin in interfering with methyltransferases has provided another route for developing inhibitors that are specific for G9a and GLP (Lenstra et al. 2018).

An interesting development has been the identification of drug inhibitors that are selective for GLP alone. The small molecule inhibitors described above have been shown to be either dual selective for G9a and GLP or have a broader selectivity for additional methyltransferases. For the identification of GLP-selective reagents, molecules containing side chain modifications off the BIX-01294 and UNC0638 dimethoxy quinazoline core were examined. These synthetic molecules were evaluated for their ability to inhibit either G9a or GLP in biochemical assays that measured the transfer of the methyl group from tritiated SAM to lysine 9 of a 25-mer H3 peptide. From these studies, three molecules were identified—MS0124, MS3748, and MS3745—that possessed 34 to 65-fold higher potency for inhibiting GLP over G9a (Xiong et al. 2017a, b).

A number of issues need to be addressed in evaluating the specificity and utility of the various G9a and GLP inhibitors that have been reported to date. Not all the inhibitors that have been reported to target these two enzymes have been fully vetted. It is not clear how discriminating some of the inhibitors are for G9a and/or GLP. Nor has it yet been determined how equivalent are the multiple G9a and GLP inhibitors, as comprehensive comparative biochemical and biological functional analyses of these various small molecules have yet to be undertaken. In a similar vein, it is unclear how well the biochemical assays used for analyzing G9a and GLP activities correspond to in vivo function. An analogy can be made between the usage of small molecule inhibitors in different experimental contexts and similar issues concerning antibody validation, as it is well known that not all antibodies that specifically recognize a protein-specific peptide in a biochemical assay will show the same discrimination for the native protein within the cell, tissue, and organ (Uhlen et al. 2016).

An additional group of questions concerns the development of drugs that can discriminate between G9a and GLP. While it is evident that these drugs would be able to fully discriminate between the two enzymes in biochemical assays where the enzymes tested are either solely G9a or GLP, that may not be the case in vivo where the enzymes often act in tandem as part of a heteromeric complex. Thus, the utility of these new drugs would need to be tested in cells that have an aberrant phenotype due to abnormally high levels of G9a or GLP.

## 4 G9a and GLP Activity and Their Impact on Cell Phenotype, Differentiation, and Proliferation

It was about 30 years ago when G9a was first described and its importance as an epigenetic regulator began to be uncovered (Spies et al. 1989; Dunham et al. 1990). Yet what brought G9a to the attention of many researchers were reports that BIX-01294 had utility for generating induced pluripotent stem cells (iPSCs). The development of iPSCs, which involves the conversion of adult somatic cells (usually fibroblasts) to embryonic stem cell (ESC)-like pluripotent cells, is arguably the most important research breakthrough in the biological sciences during the past dozen years. Because the cells are pluripotent, iPSCs can be induced to differentiate into all of the various specialized cell types of the body. Initially, iPSCs were generated by the transgenic expression of four transcription factors: Oct3/4, Sox2, Klf4, and c-Myc (Takahashi and Yamanaka 2006). Shortly thereafter, efforts were made to find alternative non-genetic, pharmacological methods for iPSC generation. The molecule that showed the greatest efficacy in the initial attempts to establish non-genetic means for producing iPSCs, and the earliest evidence that a pharmacological approach to iPSC generation was feasible, was BIX-01294 (Shi et al. 2008a, b). In the protocol for producing iPSCs, BIX-01294 substituted for Sox2 and c-Myc, and thus reduced the required transgenes to Oct3/4 and Klf4. In recent years, efforts to generate iPSCs solely using pharmacological cocktails have been successful, with many, although not all, the combinations containing BIX-01294 (Wang et al. 2016).

Despite the success in using BIX-01294 in producing iPSCs, it is has yet to be determined why inhibiting G9a (and presumably GLP) activity is useful for establishing a pluripotent phenotype from somatic cells. However, since BIX-01294 can replace both Sox2 and c-Myc for generating iPSCs, it can be surmised that BIX-01294 exposure provides a functional substitute for these two genes. In accordance with the function of these genes in a variety of contexts, such as stem cell production, injury response, and pathogenesis, BIX-01294 exposure may replicate the requirements of Sox2 in initiating dedifferentiation and c-Myc as a cell cycle regulator (Herreros-Villanueva et al. 2013; Bretones et al. 2015).

The responses of adult mouse bone marrow MSC cultures to BIX-01294 are consistent with the potential role this reagent may play in provoking cellular dedifferentiation (Mezentseva et al. 2013; Yang et al. 2015, 2017). The bone marrow is a mesodermal-derived tissue, whose constituent MSCs have an intrinsic capacity to generate multiple tissues and cell types such as cartilage, fat, bone, connective tissue, and skeletal muscle. Because of the accessibility and abundance of bone marrow stem cells, this tissue has long been studied as a potential resource for generating new heart tissue, even though bone marrow MSCs display only a limited innate cardiac potential (Dai and Kloner 2011; Wysoczynski et al. 2014). However, a two day exposure to BIX-01294 appeared to dedifferentiate bone marrow MSCs, as this treatment induced expression of the transcriptional factors Mesp1 and brachyury (Mezentseva et al. 2013; Yang et al. 2015, 2017) (Fig. 5a), which are markers of the early embryonic mesoderm and a pan-mesodermal progenitor phenotype. As further evidence that BIX-01294 promoted MSC dedifferentiation, and thus converted these cells to a more multipotent stem cell phenotype, was the newly acquired competency of the MSC-derived cells to differentiate into a broader range of mesoderm-derived phenotypes. Evidence for a more potent cardiac capacity of MSC-derived cells was indicated by the display of multiple cardiac genes and proteins (Fig. 5b–d), such as GATA4, Nkx2.5, myocardin, sarcomeric α-actinin, and titin, when sequentially treated with BIX-01294 and the cardiac inducer protein Wnt11 (Mezentseva et al. 2013; Yang et al. 2015, 2017). An important detail in these experiments is that MSCs will not generate differentiated phenotypes in the continued presence of BIX-01294. That is, the G9a/GLP inhibition must be transient in order for the cells to subsequently respond to the differentiation signals.

While the iPSC and MSC studies suggest that G9a/GLP inhibition by BIX-01294 can initiate dedifferentiation, other evidence has shown the activities of these two methyltransferases are important components of cell differentiation. For example, neuronal, skeletal myogenic, tendon, red blood, and immune cell differentiation are all highly dependent on G9a function (Chaturvedi et al. 2009; Ling et al. 2012; Wada et al. 2015; Fiszbein et al. 2016; Scheer and Zaph 2017; Verbaro et al. 2018). Data from these and other studies indicate that G9a and/or GLP expression plays a supportive role in promoting and/or regulating cell differentiation in adult tissues.

The mouse neuronal progenitor cell line Neuro-2a provides a cell culture model for neural differentiation based on their ability to form neurite outgrowths in response to treatment with retinoic acid (Wu et al. 1998). In a recent study, both G9a knockdown with small-interfering (si)-RNA and G9a/GLP inhibition with

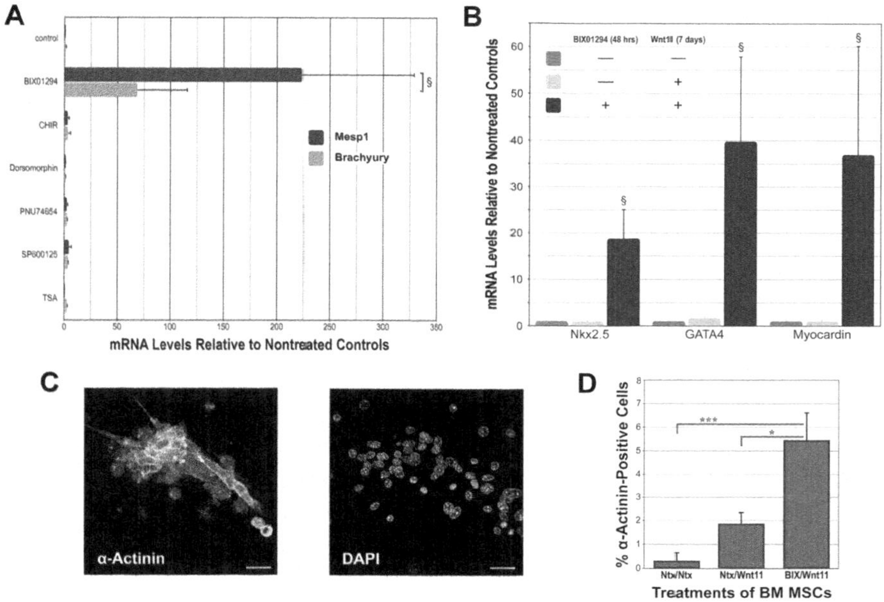


**Fig. 5** BIX01294 inhibition of G9a can provoke cardiac differentiation by MSCs. (**a**) MSCs were exposed to multiple compounds, with RNA harvested and analyzed by qPCR. Chart summarizes experimental results with inhibitors specific for G9a (BIX01294), GSK3β (CHIR99021), BMP (dorsomorphin), β-catenin (PNU74654), c-Jun N-terminal kinase (SP600125), and histone deacetylase (trichostatin A; TSA). (**b**) BIX01294 promotes responsiveness to the cardiogenic stimulating protein Wnt11. MSCs were cultured plus or minus 8 μM BIX01294 for 48 h prior to seven day culture with or without Wnt11. Analysis by qPCR indicated that MSCs upregulated the cardiac transcription factors Nkx2.5, GATA4, and myocardin when treated sequentially with BIX01294 and WNT11. Statistical significance is indicated by § $p < 0.05$. (**c**) MSCs treated sequentially with BIX-01294 and Wnt11 formed cellular aggregates that displayed myocardial structural proteins, as shown by immunoreactivity for sarcomeric $\alpha$-actinin (left), along with corresponding DAPI nuclear counterstain (right). Scale bar = 25 μm. (**d**) Tabulation of sarcomeric $\alpha$-actinin positive cells that arose from the MSCs significantly increased the number of cardiac protein expressing cells when treated sequentially with BIX01294 and WNT11, as compared to control conditions. Statistical significance is indicated by * $p < 0.05$; *** $p < 0.001$. Data from this figure is from Yang et al. (2015, 2017)

BIX-01294 prevented the formation of differentiated neuronal phenotypes (Fiszbein et al. 2016). An interesting aspect of that study was the demonstration that alternative splicing of G9a provides a regulatory mechanism for its activity, as there was preferential nuclear localization of the full-length version of this molecule (isoform a) in comparison with the exon 10-minus splice variant (isoform b). Consistent with that observation was data indicating that specific inhibition of G9a activity by knockdown of full-length G9a using siRNAs targeting exon 10 sequences, suppressed the neuronal differentiation of the Neuro-2a cells. A further finding was that differentiated and progenitor cells showed distinct G9a isoform distributions, with differentiated phenotypes displaying primarily the full-length isoform and non-differentiated phenotypes showing a preference for the exon 10-minus splice variant. Since exon E10 does not contain its own nuclear localization signal, it was speculated, although not proven, that its inclusion within full-length G9a causes a conformational change that provides access to the nuclear localization sequence within the N-terminal domain of the protein (Fiszbein et al. 2016).

Skeletal muscle is another tissue whose differentiation in vitro appears to be G9a dependent, although expression of this enzyme becomes reduced significantly when the cells become terminally differentiated (Ling et al. 2012; Battisti et al. 2016). During the differentiation of primary myoblast cultures, G9a appears to undergo a similar isoform shift to the full-length molecule (Battisti et al. 2016), as was observed in the neural cells (Fiszbein et al. 2016). GLP may play a similar regulatory role in skeletal myogenesis as G9a, although gain and loss of function experiments for these two enzymes indicated differences in their influence on specific myogenic genes—suggesting that the regulatory activities of these two sister proteins are not always in tandem (Battisti et al. 2016).

One of the most studied biological tissues in regards to G9a regulation is the immune system. In the mouse, targeted ablation of G9a in lymphocyte lineages produced a number of defects in the diversification of lymphocyte populations. For example, G9a deficiency resulted in the blurring of T cell phenotypes, as was indicated by experiments where a floxed G9a gene was ablated using a CD4-positive Cre-recombinase-expressing (Cre)-driver. This resulted in the specific removal of the G9a gene when CD4 is first exhibited in CD4/CD8 double-positive thymocytes, which upon subsequent lineage diversification produced cytolytic CD8-positive T cells that aberrantly exhibited helper T cell lineage genes (Verbaro et al. 2018). Helper T cell differentiation and function were also shown to be impaired by the lack of G9a when deleted using the broader hematopoietic lineage vav1 gene as Cre-driver (Lehnertz et al. 2010). In wild-type animals, G9a expression levels and a corresponding H3K9 dimethylation in T lineage cells are highest among less-differentiated naive phenotypic states and decrease upon T cell activation and differentiation into various T cell subclasses (Antignano et al. 2014; Scheer and Zaph 2017). Because types of T cells that are activated are dependent on the pathogenic environment, disregulation of G9a/GLP can disrupt immune responses. Manipulation of G9a/GLP activity may thus have utility if immune responses themselves become pathogenic. In one study, suppression of G9a activity, either by targeted gene deletion or pharmacological inhibition, increased yields of regulatory T cells (Tregs) (Antignano et al. 2014). Since Tregs are capable of damping down the activities of pathogenic T cells, opportune inhibition of G9a may have potential as a tool to alleviate the consequences of an overactive immune system.

In regards to the role of G9a and/or GLP in cell proliferation, the ability of BIX-01294 to substitute for c-Myc might suggest that inhibition of these two histone methyltransferases might enhance cell proliferation. To date, most studies on G9a/GLP and cell proliferation have been with cancer cell lines. As will be discussed below in Sect. 6 on cancer and disease, many cancers exhibit aberrantly high levels of G9a and/or GLP. Since cancer is characterized by uncontrolled growth rates of cells and tissues, the hypothesis that has emerged is that G9a/GLP activity is a direct stimulus for increased rates of proliferation. This hypothesis became the consensus based on many investigations using cancer cell lines and/or transformed cells, which exhibit G9a and GLP at higher than normal levels.

The proposed relationship between higher G9a and/or GLP levels and higher rates of proliferation was not what we observed when examining the effect of G9a/GLP on primary cell cultures (Yang et al. 2015; Kaur et al. 2016). In fact, we saw the opposite effect. Our studies showed G9a/GLP inhibition promotes proliferation (Fig. 3c, d)—at least in regards to stem and/or progenitor cells. In addition to the broadening of the differentiation potential of MSCs, as described above, BIX-01294 also promoted an expansion of progenitor cells within the cultures (Yang et al. 2015). This expansion effect was also observed with primary cultures of adult cardiac tissue, where BIX-01294 significantly enhanced cardiac progenitor cell (CPC) proliferation (Kaur et al. 2016).

Several studies have provided evidence that G9a promotes cell cycle exit. In the zebrafish embryo, G9a knockdown by morpholino antisense oligos increased the number of proliferating cells in the developing retina (Olsen et al. 2016). Similar results were obtained in the mouse embryo, where a conditional G9a knockout in retinal progenitor cells perpetuated their continual expansion at the expense of their terminal differentiation (Katoh et al. 2012). Gene specific short hairpin RNA (shRNA) used to suppress G9a expression in the human cell line HEK293T significantly increased cell proliferation, but only when subjected to etoposide-induced cellular stress. Under these conditions of low G9a expression, stress-induced apoptosis was reduced. Consistent with these data, overexpression of G9a by plasmid transfection reduced both the proliferation and survival rates of etoposide-treated HEK293T cells (Oh et al. 2014).

An understanding of the role G9a and GLP play in regulating cell phenotype, differentiation, and proliferation has yet to come into focus. Data reported to date provokes more questions than answers. Do G9a and GLP have universal roles in controlling cellular biology, or does their functional importance differ greatly among distinct cell lineages, tissue types, and differentiation stages? What is the function of G9a/GLP inhibition in promoting the formation of a pluripotent phenotype and how does that relate to their overall influence in maintaining non-differentiated and differentiated states? Is there a difference in G9a's effects on cell proliferation in stem/progenitor cells vs. differentiated cells? Does G9a serve as a simple on/off switch for cell cycle or does it possess a more complicated regulatory mechanism for managing cellular proliferation? Does expressing abnormally high levels of G9a and GLP exert an abnormal influence on the expression of other molecules, such as tumor repressor genes, that might affect cell cycle exit?

## 5 The Importance of G9a and GLP Function for Embryonic Development

The ability of BIX-01294 to facilitate the formation of a pluripotent phenotype and contribute to the formation of iPSCs would suggest that G9a and/or GLP play important roles in embryonic development. Moreover, it is highly likely that most histone regulatory proteins, not only G9a and GLP, are essential to the formation of a healthy embryo since these molecules control chromatin packaging and transcriptional accessibility throughout the genome. The essential global importance of epigenetic regulation of the genome makes it difficult to fully assess the embryonic function of individual molecules, such as G9a and GLP, by genetic knockout (KO), since gene ablation will likely impact most developing tissues, whose maldevelopment negatively feedbacks on the development of adjacent tissues in the growing organism. Nonetheless, genetic ablation and mutational studies in the animal model have not only provided significant insights into the importance of G9a and GLP for embryonic development, but also novel information on the general functions of these two enzymes.

It was not surprising that ablation of G9a and GLP would have great effect on embryonic development. When mice were genetically engineered to be G9a-deficient, the embryos displayed delayed development, growth arrest by the earliest stages monitored, and were no longer viable by embryonic day (ED) 9.5 (Tachibana et al. 2002). Similarly, retarded growth and gross morphological abnormalities at early embryonic stages were observed in GLP-deficient mouse embryos, with significant reduction in embryonic viability by ED9.5 (Tachibana et al. 2005). Histones extracted from still viable, early stage G9a- and GLP-deficient embryos showed greatly reduced levels of H3K9 dimethylation (Tachibana et al. 2002, 2005).

The importance G9a or GLP for embryonic development has been further explored by targeted ablation of these enzymes within specific developing tissues and organs. It was reported (Inagawa et al. 2013) that specifically ablating either G9a or GLP in the cardiomyocyte lineage did not produce an embryonic lethal defect, although almost all GLP-deficient mice died shortly after birth. The cardiomyocyte-specific G9a knockout mice were described as being normal. The myocardial-lineage GLP-deficient mice suffered from ventricular and atrioventricular septal defects, as well as malformed tricuspid and mitral valves. These are relatively common defects observed when key cardiac genes are either ablated or mutated in mice (Nakanishi et al. 2016), and which are often caused by disrupted morphogenesis of the developing heart as it undergoes a complex series of remodeling events during the transition from a one-to-four chambered structure. Yet looking over the data for G9a and GLP removal within the embryonic heart, the impression is that the influence of these molecules in heart development is probably understated. The comparison of the impact between G9a and GLP deficiency was hampered by the different Cre-drivers that were utilized, as G9a was removed using a myosin heavy chain 6:cre that would be activated in differentiated cardiomyocytes, while GLP was removed using Nkx2.5:cre that would be activated in myocardial lineage cells from the progenitor to differentiated state. As discussed above, there may be differences in the levels and/or activities of G9a and GLP among cells at various states of differentiation. Additionally, the reduction of H3K9 methylation among cells that were targeted for G9a and GLP reduction is not as low as would have been expected in comparison to other investigations that ablated or mutated G9a or GLP in the mouse genome (Tachibana et al. 2002, 2005). This suggests that the cardiac-specific reduction of G9a and GLP may not have been fully penetrant and that the role of these enzymes in heart development needs to be revisited.

Targeted disruption of G9a in the developing mouse retina generated a tissue that was highly disorganized (Katoh et al. 2012). This was due in large part to the effect of the G9a deficit on retinal progenitor cells, which in the absence of this methyltransferase remained highly proliferative, while displaying an impaired ability to mature into the specialized components of the retina. Conversely, the abnormal development of the conditional G9a knockout also caused an increase in apoptotic cells within the developing retina (Katoh et al. 2012). Similar results were obtained in zebrafish embryos when G9a expression was knocked down by morpholino antisense oligos (Olsen et al. 2016).

Neural disorders were also detected in a GLP haploinsufficiency ($GLP^{+}/GLP^{-}$) mouse model. The rationale for examining GLP haploinsufficient mice is that there is a human neurological disorder, referred to as Kleefstra syndrome, associated with low levels of functional GLP. Mice engineered with only one functional GLP allele appear to phenocopy many of the traits of Kleefstra syndrome (Balemans et al. 2010, 2014; Benevento et al. 2017), including developmental delay, cognitive disabilities, and facial dysmorphism. Postnatal developmental delay in the $GLP^{+}/GLP^{-}$ mice was indicated by a lag in the timing when the upper incisors erupt, and the ears and eyes open. $GLP^{+}/GLP^{-}$ mouse pups displayed a marked instability in their walking as compared to wild-type littermates. Mice that were haploinsufficient for GLP often exhibited cranial abnormalities such as a bent nose and incompletely fused left and right frontal bones (Balemans et al. 2014). Another phenotypic manifestation of reduced GLP expression in the adult mouse was increased cell proliferation in the hippocampus (Benevento et al. 2017). GLP-deficient mice displayed reduced activity and exploration when placed in a new environment, and increased anxiety when subjected to the light-dark box test (Balemans et al. 2014). However, GLP insufficiency did not lead to any distinguishable general learning deficits (Benevento et al. 2017); although a more recent study that used a forebrain-specific enhancer from the Dach1 gene as a Cre-driver to promote a more targeted GLP haploinsufficiency, reported mice that exhibited deficits in information processing and memory (Davis et al. 2018). G9a has not received as much attention as GLP for its importance in cognitive and behavioral development, although select studies suggest that altered G9a signaling may also result in neural dysfunction (Maze et al. 2010; Gupta-Agarwal et al. 2012; Benevento et al. 2015).

Several other tissues and cell lineages have been shown to be dependent on G9a function. G9a is required for establishing the adult erythroid cell phenotype. Suppression of G9a activity, either by gene knockdown or pharmacological inhibition, promotes a re-emergence of a fetal gene program as exhibited by the switch in expression from adult to fetal β-globin isoforms (Chaturvedi et al. 2009; Chaturvedi et al. 2012; Krivega et al. 2015). Interestingly, G9a appears to play both a negative and positive role in the shift from a fetal to adult gene program. G9a and GLP form a repressor complex with lysine demethylase 5A (KDM5A) on embryonic β-like globin gene Ey, where the combination of G9a and GLP dimethylation of H3K9 and H3K27, and KDM5A removal of methyl groups from tri- and dimethylated H3K4, results in gene silencing. Concurrently, at the $\beta^{maj}$-globin gene, the RNA polymerase II-Mediator core initiation complex is stabilized by the association of

G9a, which allows for normal activation of this gene in the adult (Chaturvedi et al. 2009; Chaturvedi et al. 2012).

GLP may also regulate switching among phenotypic fates, as was observed in the developing somites. Targeted deletion of the GLP gene among dermatomal precursors resulted in a sharp decline in brown fat and increased skeletal muscle differentiation. This phenotypic switch occurred due to the reduced methylation of H3K9 that activated myogenic gene expression, while having the opposite effect on the PRDM16 transcription factor that controls brown adipocyte differentiation (Ohno et al. 2013; Inagaki et al. 2016).

An interesting corollary to the brown fat studies is that conditional knockout of G9a in the skeletal muscle lineage may have little effect on skeletal myogenesis (Zhang et al. 2016a). In an initial set of experiments, the G9a gene was removed from the skeletal myogenic lineage using a MyoD Cre-driver, without any perceived deficit in muscle development, or whole muscle weight and myofiber size in adult mice. Muscle satellite cells isolated from these MyoD-Cre G9a knockout mice displayed both normal rates of proliferation and capacity to differentiate into myosin-positive myotubes in culture. In a follow-up set of experiments, a tamoxifen-activated Pax7 Cre-driver was used to specifically remove the G9a gene from adult muscle satellite cells. The genetically altered mice were then injected in the tibialis anterior muscle with the myotonic peptide notexin, either with or without prior intraperitoneal administration of tamoxifen. Even though the tamoxifen injection induced Cre-driven G9a removal in up to 70% of the satellite cell population, this genetic alteration did not lower the capacity of the muscle to heal following notexin injection (Zhang et al. 2016a).

The reported negative impact of G9a on skeletal myogenesis using a genetically manipulated mouse model differs from what was described using in vitro experimentation (Ling et al. 2012; Battisti et al. 2016). One interpretation of this discrepancy could be that whole animal models have greater physiological relevance than the cell culture environment. However, for any experimental results analyzing G9a and GLP activities, it is important that the finding be verified with alternative methods for inhibiting these molecules, whether by using alternative Cre-drivers, gene knockdown, or pharmacological methods. For genetic manipulation in a whole animal model, is important to demonstrate that results from a conditional gene knockout or knock-in is not due to the peculiarity and/or efficiency of the Cre-driver. In regards to G9a ablation, assaying for global reduction of dimethylated H3K9 through the targeted tissue may not be sufficient for assessing this enzyme's importance in the development of a cell lineage—especially when that reduction in H3K9 methylation is far from complete. The incomplete reduction in overall enzymatic function could reflect that the entirety of the genome is not equally affected by the overall reduction of dimethylated H3K9, and thus the lineage-specific genes that are the focus of a study may not be adversely regulated. Furthermore, the limited reduction of dimethylated H3K9 could potentially be due to GLP compensation, where homodimers of this methyltransferase could substitute as histone modifiers in the absence of G9a protein. This last point brings up issues that relate to how overlapping and/or independent are G9a and GLP in their activities in situ. An interesting line of inquiry would be to determine if G9a haploinsufficiency would mimic the neural dysfunction observed in $GLP^{+}/GLP^{-}$ mice. The extent that independent reduction of G9a and GLP expression phenocopies one another may be informative in understanding the division of labor between these two sister proteins in regulating embryonic and adult biology.

## 6 Aberrant G9a and GLP Expression in Cancer and Disease

Cancers are diseases that arise from abnormal gene expression and/or function. Many cancers are characterized by aberrant levels of expression and functional abnormalities of histone modifying proteins, and atypical patterns of histone modifications (Wan et al. 2018). For example, overexpression of G9a or GLP have been associated with many types of cancers including those of the biliary tract, bladder, blood, breast, colon, esophagus, stomach, liver, lung, neural, ovary, uterus, and squamous cells of the head and neck (Lu et al. 2013; Shankar et al. 2013; Casciello et al. 2015; Liu et al. 2015b; Alves-Silva et al. 2018; Hu et al. 2018; Mayr et al. 2018; Yang et al. 2018). Abnormally high levels of G9a and/or GLP expression in these tissues are associated with poor cancer survival (Hua et al. 2014; Zhong et al. 2015; Alves-Silva et al. 2018). Linkage has also been shown between G9a and GLP genetic variations and malignancy (Cheung et al. 2012; Spinella et al. 2016).

A majority of reported investigations analyzing how G9a and GLP facilitate a cancerous phenotype have employed various cancer cell lines as in vitro models of cancer. Many of these studies used transfection of siRNA or lentivirus infection of shRNA constructs to knockdown G9a or GLP expression, or pharmacological reagents (e.g., BIX-01294, UNC6038) to inhibit the activities of these enzymes. For example, there was a marked decrease reported in the metastatic phenotype of the human breast cancer cell line MDA-MB-231, with cell migration and invasion being reduced over a 48 hr assay period by 42% and 57%, or 30% and 31%, respectively, by transfection with G9a siRNAs or treatment with 2 μM BIX-01294 (Kim et al. 2018). Knockdown of G9a with individual gene-specific shRNAs reduced cell migration and invasion by 35 to 79% and 41 to 88%, respectively, in CL1-5 lung adenocarcinoma cells, and by 55 to 72% and 47 to 83%, respectively, in H1299 lung carcinoma cells (Chen et al. 2010). The targeting of GLP expression in the human gastric cancer cell line BGC-803 by siRNA knockdowns lessened cell migration and invasion by an average of 56% and 42%, respectively (Yang et al. 2018). Similar results have been reported using cell lines characteristic of other cancer types (Hua et al. 2014; Liu et al. 2015b; Huang et al. 2017; Hu et al. 2018).

The relevance of the culture data with cancer cell lines has been supported by various in vivo studies showing inhibition of G9a and/or GLP can decrease tumorigenesis. In one set of experiments, variants of the ovarian cancer cell line SKOV-3 that were modified by lentiviral infection of-non-target control or G9a-specific shRNA, were injected intraperitoneally into severe combined immunodeficient (SCID) mice. The mice provided with the control cells displayed widespread tumor formation, with cancerous growths spread throughout the peritoneum, mesentery, diaphragm, kidney, and liver. In contrast, injection of cells with knocked down G9a developed much smaller tumors with far lesser peritoneal carcinomatosis (Hua et al. 2014). Comparable outcomes were observed with various cancer cell lines modified with control and either G9a- or GLP-specific shRNAs and injected subcutaneously or intraperitoneally, respectively, into BALB/c nude mice (Wei et al. 2017; Yang et al. 2018). In both instances, inhibited expression of the respective histone methyltransferase significantly decreased tumorigenicity and metastasis in vivo. Another study with BALB/c nude mice utilized two gastric cancer cell lines, BGC-823 and MKN-28, which displayed high and low expression levels of G9a, respectively. Gene-specific knockdown with G9a shRNA lentivirus reduced by 65% the number of tumor nodules that would develop from an intraperitoneal injection with BGC-823 cells. Accordingly, enhancing G9a expression in MKN-28 cells with an enzyme-encoding lentivirus increased the number of peritoneal nodules by 2.2-fold (Hu et al. 2018). Further evidence that G9a/GLP play a major role in tumor formation was shown by in vivo inhibition of these enzymes with pharmacological reagents. Mammary tumor cells were injected subcutaneously into syngeneic recipient mice, and after allowing tumors to develop for 2 weeks, the animals were then injected intraperitoneally with UNC0642 or DMSO vehicle every two days, with the drug reducing tumor size by about 50% after 5 additional weeks (Casciello et al. 2017).

The investigations described above established that high levels of G9a and/or GLP promote the metastatic spread, invasiveness, and tumorigenesis of cancer cells. An additional property of cancer cells that was examined for G9a and GLP regulation was their high proliferation rate. Growth rates of multiple cancer cell lines were shown to be significantly decreased by siRNA or shRNA knockdown of G9a or GLP (Cho et al. 2011; Ding et al. 2013; Huang et al. 2017; Wei et al. 2017). In one study, growth of the human lung adenocarcinoma cell lines PC9 and A549 were reduced by 23% and 33% over a 6 day period in response to G9a shRNA knockdown (Huang et al. 2017). Pharmacological inhibition of G9a and GLP, either with BIX-01294 or UNC0638, was also reported to suppress cancer cell proliferation (Cho et al. 2011; Ding et al. 2013; Ke et al. 2014; Lehnertz et al. 2014; Huang et al. 2017). Interestingly, a comparative analysis with UNC0638 and A-366 indicated that these G9a/GLP inhibitors had disparate effects on the proliferation of multiple cancer cell lines. Although UNC0638 greatly diminished the growth rate of these cells over a five day period, A-366 showed more limited effects on cell proliferation, even though the concentrations used for both reagents had an equally suppressive effect on H3K9 dimethylation (Pappano et al. 2015).

As described in the preceding sections of this review, data from our laboratory and other investigators indicated that inhibition of G9a and GLP can enhance cellular proliferation—at least among progenitor and stem cells (Katoh et al. 2012; Yang et al. 2015; Kaur et al. 2016; Olsen et al. 2016). Yet many of the reports with cancer lines, as described in the preceding paragraphs, reached the opposite conclusion that higher G9a and/or GLP levels correlate with higher rates of proliferation. The question is why is there a reported divergence in the effect that G9a and GLP have on cell proliferation? Part of that conundrum may be due to technical issues, and part may deal with a broader consideration on how G9a and GLP regulate cell proliferation and biology.

A few observations on technical issues that complicate the interpretation of the function of G9a and GLP in promoting cell proliferation. Without pointing out individual studies, it should be noted that for some of the proliferation studies using cancer cells, the cultures treated with BIX-01294 or UNC0638 did not really proliferate over a multi-day treatment period. Only a couple of the reports showed images of the cultures, with those cells that had been treated with these drugs exhibiting an unhealthy, vacuolated phenotype. When growth rates were reported to be greatly reduced, the growth curves that were shown were flat or trending downward. Alternatively, if only the final yield of cells assayed at the end point were provided, some of these studies reported an 80–90% reduction in total cells as compared to the control group. When that is the experimental result, then one must call into question the health of the cells in response to the drug. What an investigator is assaying in those experiments is not necessarily the regulation of cell cycle and cellular proliferation, but cellular health. Yet it is clear that for some of the G9a knockdown experiments reported that a reduction in cell proliferation was observed that was not due to decreased cell viability (Huang et al. 2017). Thus, if abnormally high levels of G9a (and maybe GLP) enhances proliferation, the question is why the proliferation of cancer cell lines may be affected in way that differs from primary cultures of stem cells whose proliferation is enhanced when G9a/GLP activity is inhibited? This may be attributed to a 'too little-too much' phenomena, where abnormally high and low levels of G9a and/or GLP effect cell proliferation in analogous ways, with extreme levels of expression at both the high and low end causing a general epigenetic disregulation of the cell cycle machinery. Alternatively, abnormally high levels of G9a or GLP expression, as occurs in cancers, may provoke a unique set of regulatory interactions that does not occur in non-cancerous cells, such as blockage of tumor suppressor gene expression.

Evidence has been reported that high G9a and GLP levels suppress the activities of several tumor suppressor molecules. As discussed in the following section, the tumor repressor protein p53 is directly regulated by G9a and GLP. The p53 protein is a transcription factor whose activity plays an important role in preventing cellular carcinogenesis (Duffy et al. 2014; Vieler and Sanyal 2018). Although many cancers are characterized by p53 mutations, there are many cancers that exhibit a normal, wild-type gene. Overexpression of G9a and/or GLP can push the kinetics to an increased interaction with p53, which leads to the inactivation of this tumor suppressor and contributes to cancer formation (Huang et al. 2010b; Zhang et al. 2018). Another tumor repressor is RARRES3 (Retinoic acid receptor responder protein 3), which regulates cell proliferation, differentiation, and cell adhesion, and whose down-regulation is associated with several different types of cancers (Morales et al. 2014; Anderson et al. 2017). In a recent study, microarray analysis identified RARRES3 as a potential downstream target of G9a in liver cancer (Wei et al. 2017).

In hepatocellular carcinoma cell lines, high G9a levels correlated with the silencing of the RARRES3 gene, whose expression could be significantly enhanced by treatment with BIX-01294 or UNC0638. The ablation of G9a in these carcinoma cell lines using CRISPR-Cas9 technology reduced the cells' proliferative capacity, while enhancing their RARRES3 expression. However, the highly proliferative phenotype of these cancer cell lines could be restored from the G9a-minus variant lines by siRNA inhibition of RARRES3, showing a linkage between G9a silencing of RARRES3 expression and the formation of a cancer phenotype (Wei et al. 2017). Overexpression of G9a may play a similar role in initiating carcinogenesis of the stomach, as indicated by experimentation with cell line models of gastric cancer, where G9a dependent H3K9 methylation turns off expression of the tumor suppressor gene RUNX3 (Lee et al. 2009).

Other mechanisms have been implicated for explaining how high levels of G9a and GLP promote cancer formation. Experiments with lung cancer cells indicated that direct silencing of the Casp1 gene by G9a suppresses apoptosis and thereby leads to uncontrolled cell growth (Huang et al. 2017). A hypothesis on how G9a influences the development of breast cancers has been proposed based on its negative regulation of hephaestin expression (Wang et al. 2017). Hephaestin plays an important role in cellular iron metabolism (Jiang et al. 2015), which is essential for cell proliferation and growth (Oliveira et al. 2014). Excessive iron accumulation can also contribute to cancer initiation and progression (Torti and Torti 2013; Raza et al. 2014). Sectioned breast cancer tissue showed an inverse relationship between G9a and hephaestin antibody staining. High G9a levels suppress hephaestin expression, leading to iron accumulation and cancer enhancement (Wang et al. 2017).

Another example of how excessive G9a expression can promote cancer formation via altered cellular metabolism has been indicated for serine-glycine biosynthesis. High serine production can assist cancer cell survival and proliferation. G9a via its methylation of H3K9 on key target genes promotes the expression of several component enzymes of the serine biosynthetic pathway, which in turn enhances cellular metabolism required for supporting cancer growth (Ding et al. 2013).

Besides cancer, the greatest linkage between aberrant G9a and GLP activity and disease has been with neural disorders. Lysine methylation of histone H3 plays an important role in learning and memory. Two critical chromatin markings for these intellectual processes are trimethylated H3K4, which is governed by the activity of histone-lysine N-methyltransferase 2A (KMT2A) and dimethylated H3K9, which is regulated by the G9a/GLP gene repressor complex (Benevento et al. 2015; Karpova et al. 2017). In vivo experimentation with rats showed that administration of BIX-01294 or UNC0224 disrupted H3K9 methylation patterns in selected areas of the brain and interfered with long-term memory (Gupta-Agarwal et al. 2012). Studies in the mouse indicated G9a is a determinate factor in the brain's response to cocaine, with repeated exposure to this drug causing levels of dimethylated H3K9 to decrease within the nucleus accumbens. Targeted ablation of G9a in this region of the forebrain was able to mimic the effects of repeated cocaine administration in increasing dendritic spine density. Moreover, ectopic overexpression of this enzyme by injection of a G9a viral construct into the nucleus accumbens counteracted the cocaine-induced changes in dendritic morphology, suggesting that G9a histone methyltransferase activity may be centrally involved in the biology of drug addiction (Maze et al. 2010).

The best established linkage between neural disease and the histone methyltransferases that are the subject of this review, is Kleefstra syndrome. This genetic disorder, initially referred to as 9q34.3 subtelomeric deletion syndrome, is characterized by moderate to severe intellectual disability, autistic behavior, childhood hypotonia (decreased muscle tone), developmental speech delay, distinct facial features, and various other additional clinical features, including cardiovascular anomalies (Harada et al. 2004; Kleefstra et al. 2006; Iwase et al. 2017). The genetic basis of Kleefstra syndrome was demonstrated to be due to loss-of-function mutations within one of the alleles of GLP, which in most clinical studies on this disease is referred by its alternative name of EHMT1 (Kleefstra et al. 2006; Kleefstra et al. 2009; Bock et al. 2016; He et al. 2016; Blackburn et al. 2017; Yamada et al. 2018).

There have been a wide variety of GLP mutations that have been identified in Kleefstra syndrome patients. Almost all of the GLP mutations associated with Kleefstra syndrome arise de novo in the germ line. Most of the Kleefstra syndrome-associated GLP variants described to date are due to deletions, frameshift mutations, or nonsense mutations that generate a severely truncated protein (Kleefstra et al. 2006, 2009; He et al. 2016). There have been a few missense mutations have been described (Yamada et al. 2018), such as a GLP variant with an alteration in the ankyrin repeat region that prevents the proper folding of the protein and ability to bind to methylated H3K9 substrate (Blackburn et al. 2017). Another GLP abnormality exhibited by a Kleefstra syndrome patient was a nonsense mutation that leads to the generation of a molecule lacking a functional SET domain. Interestingly, while this patient exhibited features of autism, they displayed normal intellectual performance (Bock et al. 2016). As discussed in the preceding section, deletion of the SET domain in GLP genetically altered mice did not result in embryonic lethality, as long as a fully functional G9a gene was present (Inagawa et al. 2013; Kramer 2016). In the case with the Kleefstra syndrome patient, having one GLP allele without a functional SET domain produced neural disabilities that were relatively mild.

Cases of familial GLP variants in Kleefstra syndrome patients have been rare (Willemsen et al. 2011; Rump et al. 2013). One interesting instance is where a mother, who exhibited no characteristics of Kleefstra syndrome, passed on to her son a GLP mutant allele that generated abnormal splice variants (Rump et al. 2013). This resulted in the generation of GLP variants containing either a four nucleotide frameshift mutation that produced a truncated and probably nonfunctional protein, or a 35 amino acid deletion in frame that gave rise to an enzyme lacking one of the ankyrin repeats. An analysis of variant mRNAs produced by the mother and son indicated that 10% and 40%, respectively, of their GLP transcripts were abnormal. Although the difference in representation of variant GLP transcripts may explain why only the son had Kleefstra syndrome, the reason why the mother's overall genetic background led to a more normal splicing of GLP has not been determined.

Because Kleefstra syndrome is a disorder that arises from a genetic haploinsufficiency, a mouse model of this disease could be generated that faithfully replicated the human syndrome. As described in detail in the previous section, mice engineered with only one functional GLP allele appeared to mimic many of the traits exhibited by Kleefstra syndrome patients, and therefore should provide utility for understanding the etiology of the disease and generating potential therapies for alleviating the development of the illness after birth (Balemans et al. 2010, 2014; Benevento et al. 2017; Davis et al. 2018).

An intriguing question about Kleefstra syndrome concerns whether the involvement of G9a is important for the etiology of this disease. One expects G9a would be, yet why haven't mutations of this gene been identified with Kleefstra syndrome patients? There are a few possible explanations why Kleefstra associated G9a variants have yet to be found, even though a large variety of GLP variants have been discovered for this disease. One issue may relate to the finding that truncations or mutations that affect the SET domain of G9a are embryonic lethal, while similar genetic variations in GLP proteins may manifest itself in congenital disease but are survivable (Kramer 2016). Thus, the corresponding type of alterations in the GLP SET domain that have been described for Kleefstra syndrome would not be found in G9a, because those mutations would be selected against because of embryonic lethality. Another issue may be due the ankyrin domains, which provide G9a and GLP with distinct binding affinities for methylated forms of H3K9 (Collins et al. 2008; Liu et al. 2015a; Kramer 2016). Since the attachment of these methyltranferases to histone H3 via their ankyrin domain serves as a platform for the formation of multimeric protein complexes, alterations of GLP in this region may affect the interactions of G9a/GLP heterodimers with specific binding partners that are uniquely important in the development of Kleefstra syndrome. However, the absence of Kleefstra-associated G9a mutations should not be interpreted as an indication that this enzyme is an unimportant component of the disease, since in most instances, G9a and GLP work hand in hand to exert their function.

The involvement of epigenetic mechanisms in cardiovascular diseases has been a topic of intense study. Most of the investigative focus has been on histone acetylation, DNA methylation, and micro RNAs, and until recently, less on histone methylation (Martinez et al. 2015; Kim et al. 2016b). One obvious group who suffer from cardiac abnormalities due to aberrant expression of histone methylation enzymes are Kleefstra syndrome patients. Approximately 40 to 45% of these patients suffer from congenital malformations of the heart including patent foramen ovale and ductus arteriosus, atrial and ventricular septal defects, and maldeveloped valves such as bicuspid aortic valve and pulmonary stenosis (Kleefstra et al. 2006; Willemsen et al. 2012; Vargiami et al. 2016). A recent study has reported a potential association between pulmonary hypertension and Kleefstra syndrome (Okur et al. 2018).

To date, the linkage between GLP insufficiency and heart disease has provided the strongest evidence that G9a and/or GLP are necessary for both proper development of the heart and maintaining a healthy cardiovascular system. It was recently reported that G9a and GLP activity may also influence adult onset cardiac disease. Studies on the rat and mouse heart, showed that G9a, GLP, and dimethylated H3K9 were highly expressed in most adult cardiomyocytes, but were significantly downregulated following aortic banding-induced cardiac hypertrophy. Inhibition of G9a and GLP activity by A-366 administration, or conditional cardiomyocyte-specific G9a knockout using a tamoxifen-inducible myosin heavy chain 6-Cre driver, reduced dimethylated H3K9 markings in cardiomyocytes, and promoted cardiac hypertrophy with reduced ejection fractions. Preliminary analysis of hypertrophied and normal hearts from deceased human males indicated human cardiac hypertrophy is also accompanied by decreased G9a and GLP expression, and H3K9 dimethylation (Thienpont et al. 2017), which suggests that suppression of G9a and GLP may be part of a pathway of cardiac pathogenesis.

## 7 Non-histone Targets of G9a and GLP Histone Methyltransferases

Most if not all chromatin modifying proteins act on non-histone targets (Zhang et al. 2015). G9A and GLP are thus not unusual in their ability to modify and/or regulate molecules other than histones. There are a number of molecules, some of which have been discussed above (e.g., RARRES3, Casp1), whose activities are suppressed by gene silencing due to G9a and GLP H3K9 and H3K27 methylation marking at their genetic loci (Huang et al. 2017; Wei et al. 2017). However, there are multiple non-histone proteins whose activities are regulated by direct interaction with G9a and/or GLP.

G9a mediated lysine methylation alters the function of CCAAT enhancer-binding protein β (C/EBPβ), which is a transcription factor that is essential for development and/or function of the liver, ovaries, mammary gland, blood and immune system (Huber et al. 2012). G9a methylation of C/EBPβ at lysine 39 suppresses the transcriptional activity of this protein, which may be part of the normal regulatory process that determines its tissue-specific and temporal pattern of expression (Pless et al. 2008). Sirtuin 1 (SIRT1) is a NAD-dependent lysine deacetylase, which is involved in regulating a diverse set of biological processes and is also a target of G9a methylation (Moore et al. 2013). Interestingly, high SIRT1 signaling in the nucleus accumbens facilitates the addictive response to cocaine and morphine (Ferguson et al. 2013). Since cocaine-induced changes in this area of the brain are related to decreased G9a activity, with high levels of this methyltransferase counteracting the effects of cocaine (Maze et al. 2010), these results suggest that the direct regulation of G9a on SIRT1 activity plays a role in the biochemistry of drug addiction.

The skeletal myogenic factor MyoD is also a G9a substrate, with methylation at K104 decreasing the activity of this transcriptional protein, and restricting cell differentiation (Ling et al. 2012). As discussed in Sect. 4, G9a levels are reduced in differentiated cells, which perhaps provides a mechanism by which lowered G9a levels allows MyoD to stimulate skeletal muscle differentiation. G9a also methylates WIZ (Rathert et al. 2008), which is the GLP binding protein that assists the attachment of G9a/GLP heterodimers to the chromatin-binding complex (Bian et al. 2015). An additional target of G9a is ACINUS (apoptotic chromatin condensation inducer in the nucleus), which is a caspase-3-activated protein that facilitates chromatin condensation during apoptosis. Also directly regulated by G9a is the chromodomain Y-like protein (CDYL1), which is exhibited in various multimeric protein complexes associated with the chromatin, and is additionally involved in DNA damage repair (Sahara et al. 1999; Rathert et al. 2008; Abu-Zhayia et al. 2018).

The activity of the tumor suppressor protein p53 is also directly controlled by G9a and GLP. Methylation of p53 at lysine 373 by these methyltransferases inactivates this tumor suppressor, providing a mechanism for how high level G9a and/or GLP contributes to a cancerous phenotype (Huang et al. 2010b; Zhang et al. 2018). Pontin and Reptin are chromatin-remodeling factors that are important for normal cellular physiology, whose overexpression is also characteristic of many cancers. The activities of both proteins are regulated by G9a methylation under hypoxic conditions, where Reptin suppresses and Pontin enhances subsets of hypoxia-inducible factor 1 (HIF-1) responsive genes (Lee et al. 2010, 2011).

G9a and GLP additionally possess regulatory properties that are not related to their histone methyltransferase activity. During replication, G9a and DNA methyltransferase 1 (DNMT1) physically interact to coordinate H3K9 and DNA methylation during cell division (Estève et al. 2006). Cooperativity of G9a/GLP with other DNA methyltransferases, DNMT3a and DNMT3b, plays a role in gene inactivation during cell differentiation, where local methylation of H3K9 is followed by local DNA methylation, as exemplified in the silencing of the Oct3/4 during early embryogenesis (Feldman et al. 2006).

The principal mechanism employed by G9a and GLP in promoting changes in cell biology involves gene silencing. However, a growing body of literature describes an additional function of these proteins as transcriptional coactivators of glucocorticoid, androgen, and estrogen signaling (Lee et al. 2006; Purcell et al. 2011; Bittencourt et al. 2012; Poulard et al. 2017). A model has emerged where G9a/GLP heterodimers and the cognate receptors for these steroids form a multimeric complex with glutamate receptor interacting protein 1 (GRIP1), coactivator-associated arginine methyltransferase 1 (CARM1), p300 histone acetyltransferase, and heterochromatin protein 1 gamma (HP1γ). This assembly of proteins allows CARM1 and p300 to place activation marks on histones H3 and H4, and enables HP1γ to interact with RNA polymerase II, which links the recognition of hormone response elements to transcriptional initiation. The assistance provided by G9a and GLP for upregulation of steroid receptor-mediated gene expression does not require their histone methyltransferase activity, as removal of the SET domain of these enzymes did not decrease cooperative transcriptional activation in vitro, although the N-terminal domain is required for the coactivation of steroid-driven transcription (Purcell et al. 2011; Bittencourt et al. 2012; Poulard et al. 2017).

In contrast to the above model of G9a/GLP cooperative regulation of steroid signaling, a recent report suggested that G9a methyltransferase activity can play a role in coactivating estrogen-mediated gene expression via a multimeric complex containing a different set of protein partners (Zhang et al. 2016b). The binding of nuclear receptor coactivator 2 (NCOA2) to G9a and the ligand bound estrogen receptor permits G9a to dimethylate lysine 235 of the estrogen receptor. This in turn allows the steroid receptor to be recognized by plant homeodomain finger protein 20 (PHF20), which then recruits the lysine acetyltransferase MOF (males absent on the first; also referred as MYST1 or KAT8), whose acetylation of H4K16 activates transcription. Whether the two co-activation pathways described for estrogen signaling operate simultaneously or are independent processes that may be cell, temporally, and/or gene specific, has not been determined. It is also unclear whether the co-regulation of gene expression involving G9a, PHF20, and MOF is utilized for other steroid hormones beyond estrogen. Neither is it clear, nor has it yet been fully addressed, whether GLP is necessary for the co-activation of steroid responsive gene expression.

The most surprising non-histone target of G9a and GLP are themselves. G9a and GLP are methylated at lysine 185 or 206, respectively, either intramolecularly or in trans by the sister protein (Chin et al. 2007; Sampath et al. 2007; Poulard et al. 2017). Substitutions of these lysine residues in G9a and GLP prevent its ability to both bind with HP1γ (Chin et al. 2007) and enhance transcription of glucocorticoid responsive genes (Poulard et al. 2017). Other binding partners of HP1γ are DNMT1 (Smallwood et al. 2007), which is known to associate with G9a in regulating the epigenetic markings during DNA replication (Estève et al. 2006), and various protein components of the DNA damage response pathway (Kim et al. 2016a). High levels of G9a, as seen in cancers, inhibit the DNA damage response. Inhibition of G9a in colorectal cancer cells inhibits their growth, by activating the DNA damage response, which leads to cellular senescence (Zhang et al. 2016a). These data suggest a pathway involving the overexpression and automethylation of G9a, its binding to HP1γ, and the prevention of cell cycle arrest in cancer cells, as well as providing another example of the widespread influence G9a has in affecting normal cell and tissue biology, and the pathogenesis of human disease.

## 8 Concluding Remarks

Almost three decades have passed since G9a was discovered, and although a lot is now known about how this enzyme and its sister protein GLP both write and read histone markings, there is much about these molecules that is poorly understood. Reviewing what is known about G9a and GLP seems like pooling disparate parts of information together. As complex as the molecular biology that underlies their catalytic activities, which involve a myriad of different binding partners, the involvement of G9a and GLP in regulating the genome is just a small part of the overall epigenetic regulation of chromosomal stability, gene expression, cell biology, and function. Yet, everywhere one looks, G9a and GLP seem to have major impact, whether it is Kleefstra syndrome or other neurological disorders, cancer progression, immune cell diversification, and perhaps cardiac pathogenesis. As research into G9a and GLP has expanded to move outside the realm of histones, we now learn that G9a and GLP may be major regulators of steroid hormone activities.

Much of the future emphasis of G9a and GLP research will be on drug development and testing. Drugs are needed that have low toxicity and can be administered in vivo.

Can G9a and GLP inhibitors be used to treat cancers without adverse effect on normal, healthy cells? Can low dose of G9a/GLP inhibitors convert cancer cells to a normal phenotype? Can G9a/GLP inhibition both be used to reduce cancer cell proliferation and yet have utility as a tool for stem cell expansion? Can G9a and GLP inhibitors selectively suppress histone and non-histone targets independently? Will G9a and GLP-selective drugs be effective treatments by themselves or will they work more effectively in combination with drugs that target their regulatory partners, such as HP1γ, or EZH2? In that regard, a recent study has shown that growth of breast cancer cells can be more efficiently inhibited by joint, rather than independent pharmacological targeting of EZH2 and G9a (Curry et al. 2015). Future development of G9a and GLP-specific drugs will not only impact medicine, but also basic science, as discovery of these pharmacological tools have already greatly accelerated research into the biological impact of these methyltransferases. Although there is lot more to discover about G9a and GLP, it is already clear that treatments that regulate their activities provide a most promising therapeutic avenue for remedying human diseases.

## References

Abu-Zhayia ER, Awwad SW, Ben-Oz BM et al (2018) CDYL1 fosters double-strand break-induced transcription silencing and promotes homology-directed repair. J Mol Cell Biol 10:341–357

Allis CD, Berger SL, Cote J et al (2007) New nomenclature for chromatin-modifying enzymes. Cell 131:633–636

Alves-Silva JC, de Carvalho JL, Rabello DA et al (2018) GLP overexpression is associated with poor prognosis in chronic lymphocytic leukemia and its inhibition induces leukemic cell death. Investig New Drugs 36:955–960

Anderson AM, Kalimutho M, Harten S et al (2017) The metastasis suppressor RARRES3 as an endogenous inhibitor of the immunoproteasome expression in breast cancer cells. Sci Rep 7:39873

Antignano F, Burrows K, Hughes MR et al (2014) Methyltransferase G9A regulates T cell differentiation during murine intestinal inflammation. J Clin Invest 124:1945–1955

Balemans MC, Huibers MM, Eikelenboom NW et al (2010) Reduced exploration, increased anxiety, and altered social behavior: autistic-like features of euchromatin histone methyltransferase 1 heterozygous knockout mice. Behav Brain Res 208:47–55

Balemans MC, Ansar M, Oudakker AR et al (2014) Reduced euchromatin histone methyltransferase 1 causes developmental delay, hypotonia, and cranial abnormalities associated with increased bone gene expression in Kleefstra syndrome mice. Dev Biol 386:395–407

Bannister AJ, Kouzarides T (2011) Regulation of chromatin by histone modifications. Cell Res 21:381–395

Battisti V, Pontis J, Boyarchuk E et al (2016) Unexpected distinct roles of the related histone H3 lysine 9 methyltransferases G9a and G9a-like protein in myoblasts. J Mol Biol 428:2329–2343

Benevento M, van de Molengraft M, van Westen R et al (2015) The role of chromatin repressive marks in cognition and disease: a focus on the repressive complex GLP/G9a. Neurobiol Learn Mem 124:88–96

Benevento M, Oomen CA, Horner AE et al (2017) Haploinsufficiency of EHMT1 improves pattern separation and increases hippocampal cell proliferation. Sci Rep 7:40284

Bian C, Chen Q, Yu X (2015) The zinc finger proteins ZNF644 and WIZ regulate the G9a/GLP complex for gene repression. elife 4

Bittencourt D, Wu DY, Jeong KW et al (2012) G9a functions as a molecular scaffold for assembly of transcriptional coactivators on a subset of glucocorticoid receptor target genes. Proc Natl Acad Sci U S A 109:19673–19678

Blackburn PR, Tischer A, Zimmermann MT et al (2017) A novel Kleefstra syndrome-associated variant that affects the conserved TPLX motif within the ankyrin repeat of EHMT1 leads to abnormal protein folding. J Biol Chem 292:3866–3876

Bock I, Németh K, Pentelényi K et al (2016) Targeted next generation sequencing of a panel of autism-related genes identifies an EHMT1 mutation in a Kleefstra syndrome patient with autism and normal intellectual performance. Gene 595:131–141

Bretones G, Delgado MD, León J (2015) Myc and cell cycle control. Biochim Biophys Acta 1849:506–516

Brown SE, Campbell RD, Sanderson CM (2001) Novel NG36/G9a gene products encoded within the human and mouse MHC class III regions. Mamm Genome 12:916–924

Casciello F, Windloch K, Gannon F et al (2015) Functional role of G9a histone methyltransferase in cancer. Front Immunol 6:487

Casciello F, Al-Ejeh F, Kelly G et al (2017) G9a drives hypoxia-mediated gene repression for breast cancer cell survival and tumorigenesis. Proc Natl Acad Sci U S A 114:7077–7082

Chaib H, Nebbioso A, Prebet T et al (2012) Anti-leukemia activity of chaetocin via death receptor-dependent apoptosis and dual modulation of the histone methyl-transferase SUV39H1. Leukemia 26:662–674

Chang Y, Zhang X, Horton JR et al (2009) Structural basis for G9a-like protein lysine methyltransferase inhibition by BIX-01294. Nat Struct Mol Biol 16:312–317

Chaturvedi CP, Brand M (2014) EHMT2 (euchromatic histone-lysine N-methyltransferase 2). Atlas Genet Cytogenet Oncol Haematol 18:38–45

Chaturvedi CP, Hosey AM, Palii C et al (2009) Dual role for the methyltransferase G9a in the maintenance of beta-globin gene transcription in adult erythroid cells. Proc Natl Acad Sci U S A 106(43):18303–18308

Chaturvedi CP, Somasundaram B, Singh K et al (2012) Maintenance of gene silencing by the coordinate action of the H3K9 methyltransferase G9a/KMT1C and the H3K4 demethylase Jarid1a/KDM5A. Proc Natl Acad Sci U S A 109:18845–18850

Chen MW, Hua KT, Kao HJ et al (2010) H3K9 histone methyltransferase G9a promotes lung cancer invasion and metastasis by silencing the cell adhesion molecule Ep-CAM. Cancer Res 70:7830–7840

Cheung HC, Yatsenko SA, Kadapakkam M et al (2012) Constitutional tandem duplication of 9q34 that truncates EHMT1 in a child with ganglioglioma. Pediatr Blood Cancer 58:801–805

Chin HG, Estève PO, Pradhan M et al (2007) Automethylation of G9a and its implication in wider substrate specificity and HP1 binding. Nucleic Acids Res 35:7313–7323

Cho HS, Kelly JD, Hayami S et al (2011) Enhanced expression of EHMT2 is involved in the proliferation of cancer cells through negative regulation of SIAH1. Neoplasia 13:676–684

Choi JY, Yoon SS, Kim SE et al (2017) KDM4B histone demethylase and G9a regulate expression of vascular adhesion proteins in cerebral microvessels. Sci Rep 7:45005

Collins R, Cheng X (2010) A case study in cross-talk: the histone lysine methyltransferases G9a and GLP. Nucleic Acids Res 38:3503–3511

Collins RE, Northrop JP, Horton JR et al (2008) The ankyrin repeats of G9a and GLP histone methyltransferases are mono- and dimethyllysine binding modules. Nat Struct Mol Biol 15:245–250

Curry E, Green I, Chapman-Rothe N et al (2015) Dual EZH2 and EHMT2 histone methyltransferase inhibition increases biological efficacy in breast cancer cells. Clin Epigenetics 7:84

Dai W, Kloner RA (2011) Bone marrow-derived cell transplantation therapy for myocardial infarction: lessons learned and future questions. Am J Transplant 11:2297–2301

Davis BA, David F, O'Regan C et al (2018) Ehmt1 forebrain haploinsufficiency leads to impaired memory, sensory gating and information processing. bioRxiv:257626

Dillon SC, Zhang X, Trievel RC et al (2005) The SET-domain protein superfamily: protein lysine methyltransferases. Genome Biol 6:227

Ding J, Li T, Wang X et al (2013) The histone H3 methyltransferase G9A epigenetically activates the serine-glycine synthesis pathway to sustain cancer cell survival and proliferation. Cell Metab 18:896–907
Duffy MJ, Synnott NC, McGowan PM et al (2014) p53 as a target for the treatment of cancer. Cancer Treat Rev 40:1153–1160
Dunham I, Sargent CA, Kendall E et al (1990) Characterization of the class III region in different MHC haplotypes by pulsed-field gel electrophoresis. Immunogenetics 32:175–182
Estève PO, Patnaik D, Chin HG et al (2005) Functional analysis of the N- and C-terminus of mammalian G9a histone H3 methyltransferase. Nucleic Acids Res 33:3211–3223
Estève PO, Chin HG, Smallwood A et al (2006) Direct interaction between DNMT1 and G9a coordinates DNA and histone methylation during replication. Genes Dev 20:3089–3103
Feldman N, Gerson A, Fang J et al (2006) G9a-mediated irreversible epigenetic inactivation of Oct-3/4 during early embryogenesis. Nat Cell Biol 8:188–194
Ferguson D, Koo JW, Feng J et al (2013) Essential role of SIRT1 signaling in the nucleus accumbens in cocaine and morphine action. J Neurosci 33:16088–16098
Fiszbein A, Giono LE, Quaglino A et al (2016) Alternative splicing of G9a regulates neuronal differentiation. Cell Rep 14:2797–2808
Fritsch L, Robin P, Mathieu JR et al (2010) A subset of the histone H3 lysine 9 methyltransferases Suv39h1, G9a, GLP, and SETDB1 participate in a multimeric complex. Mol Cell 37:46–56
Gupta-Agarwal S, Franklin AV, Deramus T et al (2012) G9a/GLP histone lysine dimethyltransferase complex activity in the hippocampus and the entorhinal cortex is required for gene activation and silencing during memory consolidation. J Neurosci 32:5440–5453
Harada N, Visser R, Dawson A et al (2004) A 1-Mb critical region in six patients with 9q34.3 terminal deletion syndrome. J Hum Genet 49:440–444
He X, Caluseriu O, Srivastava R et al (2016) Reversible white matter lesions associated with mutant EHMT1 and Kleefstra syndrome. Neurol Genet 2:e58
Herreros-Villanueva M, Zhang JS, Koenig A et al (2013) SOX2 promotes dedifferentiation and imparts stem cell-like features to pancreatic cancer cells. Oncogene 2:e61
Hu L, Zang MD, Wang HX et al (2018) G9A promotes gastric cancer metastasis by upregulating ITGB3 in a SET domain-independent manner. Cell Death Dis 9:278
Hua KT, Wang MY, Chen MW et al (2014) The H3K9 methyltransferase G9a is a marker of aggressive ovarian cancer that promotes peritoneal metastasis. Mol Cancer 13:189
Huang C, Xiang Y, Wang Y et al (2010a) Dual-specificity histone demethylase KIAA1718 (KDM7A) regulates neural differentiation through FGF4. Cell Res 20:154–165
Huang J, Dorsey J, Chuikov S et al (2010b) G9a and Glp methylate lysine 373 in the tumor suppressor p53. J Biol Chem 285:9636–9641
Huang T, Zhang P, Li W et al (2017) G9A promotes tumor cell growth and invasion by silencing CASP1 in non-small-cell lung cancer cells. Cell Death Dis 8:e2726
Huber R, Pietsch D, Panterodt T et al (2012) Regulation of C/EBPβ and resulting functions in cells of the monocytic lineage. Cell Signal 24:1287–1296
Inagaki T, Sakai J, Kajimura S (2016) Transcriptional and epigenetic control of brown and beige adipose cell fate and function. Nat Rev Mol Cell Biol 17:480–495
Inagawa M, Nakajima K, Makino T et al (2013) Histone H3 lysine 9 methyltransferases, G9a and GLP are essential for cardiac morphogenesis. Mech Dev 130:519–531
Isham CR, Tibodeau JD, Jin W et al (2007) Chaetocin: a promising new antimyeloma agent with in vitro and in vivo activity mediated via imposition of oxidative stress. Blood 109:2579–2588
Iwasa E, Hamashima Y, Fujishiro S et al (2010) Total synthesis of (+)-chaetocin and its analogues: their histone methyltransferase G9a inhibitory activity. J Am Chem Soc 132:4078–4079
Iwase S, Bérubé NG, Zhou Z et al (2017) Epigenetic etiology of intellectual disability. J Neurosci 37:10773–10782
Jiang R, Hua C, Wan Y et al (2015) Hephaestin and ceruloplasmin play distinct but interrelated roles in iron homeostasis in mouse brain. J Nutr 145:1003–1009
Karpova NN, Sales AJ, Joca SR (2017) Epigenetic basis of neuronal and synaptic plasticity. Curr Top Med Chem 17:771–793

Katoh K, Yamazaki R, Onishi A et al (2012) G9a histone methyltransferase activity in retinal progenitors is essential for proper differentiation and survival of mouse retinal cells. J Neurosci 32:17658–17670

Kaur K, Yang J, Edwards JG et al (2016) G9a histone methyltransferase inhibitor BIX01294 promotes expansion of adult cardiac progenitor cells without changing their phenotype or differentiation potential. Cell Prolif 49:373–385

Ke XX, Zhang D, Zhu S et al (2014) Inhibition of H3K9 methyltransferase G9a repressed cell proliferation and induced autophagy in neuroblastoma cells. PLoS One 9:e106962

Kim H, Choi JD, Kim BG et al (2016a) Interactome analysis reveals that heterochromatin protein 1γ (HP1γ) Is associated with the DNA damage response pathway. Cancer Res Treat 48:322–333

Kim SY, Morales CR, Gillette TG et al (2016b) Epigenetic regulation in heart failure. Curr Opin Cardiol 31:255–265

Kim K, Son MY, Jung CR et al (2018) EHMT2 is a metastasis regulator in breast cancer. Biochem Biophys Res Commun 496:758–762

Kleefstra T, Brunner HG, Amiel J et al (2006) Loss-of-function mutations in euchromatin histone methyl transferase 1 (EHMT1) cause the 9q34 subtelomeric deletion syndrome. Am J Hum Genet 79:370–377

Kleefstra T, van Zelst-Stams WA, Nillesen WM et al (2009) Further clinical and molecular delineation of the 9q subtelomeric deletion syndrome supports a major contribution of EHMT1 haploinsufficiency to the core phenotype. J Med Genet 46:598–606

Kondengaden SM, Luo LF, Huang K et al (2016) Discovery of novel small molecule inhibitors of lysine methyltransferase G9a and their mechanism in leukemia cell lines. Eur J Med Chem 122:382–393

Kowolik CM, Lin M, Xie J et al (2016) NT1721, a novel epidithiodiketopiperazine, exhibits potent in vitro and in vivo efficacy against acute myeloid leukemia. Oncotarget 7:86186–86197

Kramer JM (2016) Regulation of cell differentiation and function by the euchromatin histone methyltranserfases G9a and GLP. Biochem Cell Biol 94:26–32

Krivega I, Byrnes C, de Vasconcellos JF et al (2015) Inhibition of G9a methyltransferase stimulates fetal hemoglobin production by facilitating LCR/γ-globin looping. Blood 126:665–672

Kubicek S, O'Sullivan RJ, August EM et al (2007) Reversal of H3K9me2 by a small-molecule inhibitor for the G9a histone methyltransferase. Mol Cell 25:473–481

Lee DY, Northrop JP, Kuo MH et al (2006) Histone H3 lysine 9 methyltransferase G9a is a transcriptional coactivator for nuclear receptors. J Biol Chem 281:8476–8485

Lee SH, Kim J, Kim WH et al (2009) Hypoxic silencing of tumor suppressor RUNX3 by histone modification in gastric cancer cells. Oncogene 28:184–194

Lee JS, Kim Y, Kim IS et al (2010) Negative regulation of hypoxic responses via induced Reptin methylation. Mol Cell 39:71–85

Lee JS, Kim Y, Bhin J et al (2011) Hypoxia-induced methylation of a pontin chromatin remodeling factor. Proc Natl Acad Sci U S A 108:13510–13515

Lehnertz B, Northrop JP, Antignano F et al (2010) Activating and inhibitory functions for the histone lysine methyltransferase G9a in T helper cell differentiation and function. J Exp Med 207:915–922

Lehnertz B, Pabst C, Su L et al (2014) The methyltransferase G9a regulates HoxA9-dependent transcription in AML. Genes Dev 28:317–327

Lenstra DC, Al Temimi AHK, Mecinović J (2018) Inhibition of histone lysine methyltransferases G9a and GLP by ejection of structural Zn(II). Bioorg Med Chem Lett 28(7):1234–1238

Ling BM, Bharathy N, Chung TK et al (2012) Lysine methyltransferase G9a methylates the transcription factor MyoD and regulates skeletal muscle differentiation. Proc Natl Acad Sci U S A 109:841–846

Liu F, Chen X, Allali-Hassani A et al (2009) Discovery of a 2,4-diamino-7-aminoalkoxyquinazoline as a potent and selective inhibitor of histone lysine methyltransferase G9a. J Med Chem 52:7950–7953

Liu F, Chen X, Allali-Hassani A et al (2010) Protein lysine methyltransferase G9a inhibitors: design, synthesis, and structure activity relationships of 2,4-diamino-7-aminoalkoxy-quinazolines. J Med Chem 53:5844–5857
Liu F, Barsyte-Lovejoy D, Allali-Hassani A et al (2011) Optimization of cellular activity of G9a inhibitors 7-aminoalkoxy-quinazolines. J Med Chem 54:6139–6150
Liu F, Barsyte-Lovejoy D, Li F et al (2013) Discovery of an in vivo chemical probe of the lysine methyltransferases G9a and GLP. J Med Chem 56:8931–8942
Liu N, Zhang Z, Wu H et al (2015a) Recognition of H3K9 methylation by GLP is required for efficient establishment of H3K9 methylation, rapid target gene repression, and mouse viability. Genes Dev 29:379–393
Liu S, Ye D, Guo W et al (2015b) G9a is essential for EMT-mediated metastasis and maintenance of cancer stem cell-like characters in head and neck squamous cell carcinoma. Oncotarget 6:6887–6901
Lu Z, Tian Y, Salwen HR et al (2013) Histone-lysine methyltransferase EHMT2 is involved in proliferation, apoptosis, cell invasion, and DNA methylation of human neuroblastoma cells. Anti-Cancer Drugs 24:484–493
Martinez SR, Gay MS, Zhang L (2015) Epigenetic mechanisms in heart development and disease. Drug Discov Today 20:799–811
Mayr C, Helm K, Jakab M et al (2018) The histone methyltransferase G9a: a new therapeutic target in biliary tract cancer. Hum Pathol 72:117–126
Maze I, Covington HE, Dietz DM et al (2010) Essential role of the histone methyltransferase G9a in cocaine-induced plasticity. Science 327:213–216
Mezentseva NV, Yang J, Kaur K et al (2013) The histone methyltransferase inhibitor BIX01294 enhances the cardiac potential of bone marrow cells. Stem Cells Dev 22:654–667
Milner CM, Campbell RD (1993) The G9a gene in the human major histocompatibility complex encodes a novel protein containing ankyrin-like repeats. Biochem J 290:811–818
Moore KE, Carlson SM, Camp ND et al (2013) A general molecular affinity strategy for global detection and proteomic analysis of lysine methylation. Mol Cell 50:444–456
Morales M, Arenas EJ, Urosevic J et al (2014) RARRES3 suppresses breast cancer lung metastasis by regulating adhesion and differentiation. EMBO Mol Med 6:865–881
Mozzetta C, Pontis J, Fritsch L et al (2014) The histone H3 lysine 9 methyltransferases G9a and GLP regulate polycomb repressive complex 2-mediated gene silencing. Mol Cell 53:277–289
Nakanishi T, Markwald RR, Baldwin HS et al (2016) Etiology and morphogenesis of congenital heart disease: from gene function and cellular interaction to morphology. Springer, Tokyo
Oh ST, Kim KB, Chae YC et al (2014) H3K9 histone methyltransferase G9a-mediated transcriptional activation of p21. FEBS Lett 588:685–691
Ohno H, Shinoda K, Ohyama K et al (2013) EHMT1 controls brown adipose cell fate and thermogenesis through the PRDM16 complex. Nature 504:163–167
Okur V, Nees S, Chung WK et al (2018) Pulmonary hypertension in patients with 9q34.3 microdeletion-associated Kleefstra syndrome. Am J Med Genet A 176:1773–1777
Oliveira F, Rocha S, Fernandes R (2014) Iron metabolism: from health to disease. J Clin Lab Anal 28:210–218
Olsen JB, Wong L, Deimling S et al (2016) G9a and ZNF644 physically associate to suppress progenitor gene expression during neurogenesis. Stem Cell Reports 7:454–470
Pappano WN, Guo J, He Y et al (2015) The histone methyltransferase inhibitor A-366 uncovers a role for G9a/GLP in the epigenetics of leukemia. PLoS One 10:e0131716
Pless O, Kowenz-Leutz E, Knoblich M et al (2008) G9a-mediated lysine methylation alters the function of CCAAT/enhancer-binding protein-beta. J Biol Chem 283:26357–26363
Poulard C, Bittencourt D, Wu DY et al (2017) A post-translational modification switch controls coactivator function of histone methyltransferases G9a and GLP. EMBO Rep 18:1442–1459
Purcell DJ, Jeong KW, Bittencourt D et al (2011) A distinct mechanism for coactivator versus corepressor function by histone methyltransferase G9a in transcriptional regulation. J Biol Chem 286:41963–41971

Rathert P, Dhayalan A, Murakami M et al (2008) Protein lysine methyltransferase G9a acts on non-histone targets. Nat Chem Biol 4:344–346
Raza M, Chakraborty S, Choudhury M et al (2014) Cellular iron homeostasis and therapeutic implications of iron chelators in cancer. Curr Pharm Biotechnol 15:1125–1140
Rump A, Hildebrand L, Tzschach A et al (2013) A mosaic maternal splice donor mutation in the EHMT1 gene leads to aberrant transcripts and to Kleefstra syndrome in the offspring. Eur J Hum Genet 21:887–890
Sahara S, Aoto M, Eguchi Y et al (1999) Acinus is a caspase-3-activated protein required for apoptotic chromatin condensation. Nature 401:168–173
Sampath SC, Marazzi I, Yap KL et al (2007) Methylation of a histone mimic within the histone methyltransferase G9a regulates protein complex assembly. Mol Cell 27:596–608
Scheer S, Zaph C (2017) The lysine methyltransferase G9a in immune cell differentiation and function. Front Immunol 8:429
Schotta G, Ebert A, Krauss V et al (2002) Central role of Drosophila SU(VAR)3-9 in histone H3-K9 methylation and heterochromatic gene silencing. EMBO J 21:1121–1131
Shankar SR, Bahirvani AG, Rao VK et al (2013) G9a, a multipotent regulator of gene expression. Epigenetics 8:16–22
Shi Y, Desponts C, Do JT et al (2008a) Induction of pluripotent stem cells from mouse embryonic fibroblasts by Oct4 and Klf4 with small-molecule compounds. Cell Stem Cell 3:568–574
Shi Y, Do JT, Desponts C et al (2008b) A combined chemical and genetic approach for the generation of induced pluripotent stem cells. Cell Stem Cell 2:525–528
Shinkai Y, Tachibana M (2011) H3K9 methyltransferase G9a and the related molecule GLP. Genes Dev 25:781–788
Smallwood A, Estève PO, Pradhan S et al (2007) Functional cooperation between HP1 and DNMT1 mediates gene silencing. Genes Dev 21:1169–1178
Spies T, Bresnahan M, Strominger JL (1989) Human major histocompatibility complex contains a minimum of 19 genes between the complement cluster and HLA-B. Proc Natl Acad Sci U S A 86:8955–8958
Spinella JF, Cassart P, Richer C et al (2016) Genomic characterization of pediatric T-cell acute lymphoblastic leukemia reveals novel recurrent driver mutations. Oncotarget 7:65485–65503
Suganuma T, Workman JL (2011) Signals and combinatorial functions of histone modifications. Annu Rev Biochem 80:473–499
Sweis RF, Pliushchev M, Brown PJ et al (2014) Discovery and development of potent and selective inhibitors of histone methyltransferase G9a. ACS Med Chem Lett 5:205–209
Tachibana M, Sugimoto K, Fukushima T et al (2001) Set domain-containing protein, G9a, is a novel lysine-preferring mammalian histone methyltransferase with hyperactivity and specific selectivity to lysines 9 and 27 of histone H3. J Biol Chem 276:25309–25317
Tachibana M, Sugimoto K, Nozaki M et al (2002) G9a histone methyltransferase plays a dominant role in euchromatic histone H3 lysine 9 methylation and is essential for early embryogenesis. Genes Dev 16:1779–1791
Tachibana M, Ueda J, Fukuda M et al (2005) Histone methyltransferases G9a and GLP form heteromeric complexes and are both crucial for methylation of euchromatin at H3-K9. Genes Dev 19:815–826
Takahashi K, Yamanaka S (2006) Induction of pluripotent stem cells from mouse embryonic and adult fibroblast cultures by defined factors. Cell 126:663–676
Thienpont B, Aronsen JM, Robinson EL et al (2017) The H3K9 dimethyltransferases EHMT1/2 protect against pathological cardiac hypertrophy. J Clin Invest 127:335–348
Torti SV, Torti FM (2013) Iron and cancer: more ore to be mined. Nat Rev Cancer 13:342–355
Uhlen M, Bandrowski A, Carr S et al (2016) A proposal for validation of antibodies. Nat Methods 13:823–827
Upadhyay AK, Cheng X (2011) Dynamics of histone lysine methylation: structures of methyl writers and erasers. Prog Drug Res 67:107–124
Vargiami E, Ververi A, Al-Mutawa H et al (2016) Multiple coronary artery microfistulas in a girl with Kleefstra syndrome. Case Rep Genet 2016:3056053

Vedadi M, Barsyte-Lovejoy D, Liu F et al (2011) A chemical probe selectively inhibits G9a and GLP methyltransferase activity in cells. Nat Chem Biol 7:566–574
Venkatesh S, Workman JL (2015) Histone exchange, chromatin structure and the regulation of transcription. Nat Rev Mol Cell Biol 16:178–189
Verbaro DJ, Sakurai N, Kim B et al (2018) Cutting edge: the histone methyltransferase G9a Is required for silencing of helper T lineage-associated genes in proliferating CD8 T cells. J Immunol 200:3891–3896
Vieler M, Sanyal S (2018) p53 isoforms and their implications in cancer. Cancers (Basel) 10(9):288
Wada S, Ideno H, Shimada A et al (2015) H3K9MTase G9a is essential for the differentiation and growth of tenocytes in vitro. Histochem Cell Biol 144:13–20
Wan YCE, Liu J, Chan KM (2018) Histone H3 mutations in cancer. Curr Pharmacol Rep 4:292–300
Wang Z, Zang C, Rosenfeld JA et al (2008) Combinatorial patterns of histone acetylations and methylations in the human genome. Nat Genet 40:897–903
Wang Y, Qin J, Wang S et al (2016) Conversion of human gastric epithelial cells to multipotent endodermal progenitors using defined small molecules. Cell Stem Cell 19:449–461
Wang YF, Zhang J, Su Y et al (2017) G9a regulates breast cancer growth by modulating iron homeostasis through the repression of ferroxidase hephaestin. Nat Commun 8:274
Wei L, Chiu DK, Tsang FH et al (2017) Histone methyltransferase G9a promotes liver cancer development by epigenetic silencing of tumor suppressor gene RARRES3. J Hepatol 67:758–769
Willemsen MH, Beunders G, Callaghan M et al (2011) Familial Kleefstra syndrome due to maternal somatic mosaicism for interstitial 9q34.3 microdeletions. Clin Genet 80:31–38
Willemsen MH, Vulto-van Silfhout AT, Nillesen WM et al (2012) Update on Kleefstra syndrome. Mol Syndromol 2:202–212
Wu G, Fang Y, Lu ZH et al (1998) Induction of axon-like and dendrite-like processes in neuroblastoma cells. J Neurocytol 27:1–14
Wu H, Chen X, Xiong J et al (2011) Histone methyltransferase G9a contributes to H3K27 methylation in vivo. Cell Res 21:365–367
Wysoczynski M, Hong KU, Moore JB (2014) Bone marrow cell therapies in ischemic cardiomyopathy. Expert Opin Biol Ther 14:1229–1232
Xiong Y, Li F, Babault N et al (2017a) Discovery of potent and selective inhibitors for G9a-like protein (GLP) lysine methyltransferase. J Med Chem 60:1876–1891
Xiong Y, Li F, Babault N et al (2017b) Structure-activity relationship studies of G9a-like protein (GLP) inhibitors. Bioorg Med Chem 25:4414–4423
Yamada A, Shimura C, Shinkai Y (2018) Biochemical validation of EHMT1 missense mutations in Kleefstra syndrome. J Hum Genet 63:555–562
Yang J, Kaur K, Ong LL et al (2015) Inhibition of G9a histone methyltransferase converts bone marrow mesenchymal stem cells to cardiac competent progenitors. Stem Cells Int 2015:270428
Yang J, Kaur K, Edwards JG et al (2017) Inhibition of histone methyltransferase, histone deacetylase, and β-catenin synergistically enhance the cardiac potential of bone marrow cells. Stem Cells Int 2017:3464953
Yang Y, Shen J, Yan D et al (2018) Euchromatic histone lysine methyltransferase 1 regulates cancer development in human gastric cancer by regulating E-cadherin. Oncol Lett 15:9480–9486
Yuan Y, Wang Q, Paulk J et al (2012) A small-molecule probe of the histone methyltransferase G9a induces cellular senescence in pancreatic adenocarcinoma. ACS Chem Biol 7:1152–1157
Zentner GE, Henikoff S (2013) Regulation of nucleosome dynamics by histone modifications. Nat Struct Mol Biol 20:259–266
Zhang X, Huang Y, Shi X (2015) Emerging roles of lysine methylation on non-histone proteins. Cell Mol Life Sci 72:4257–4272
Zhang RH, Judson RN, Liu DY et al (2016a) The lysine methyltransferase Ehmt2/G9a is dispensable for skeletal muscle development and regeneration. Skelet Muscle 6:22

Zhang X, Peng D, Xi Y et al (2016b) G9a-mediated methylation of ERα links the PHF20/MOF histone acetyltransferase complex to hormonal gene expression. Nat Commun 7:10810
Zhang J, Wang Y, Shen Y et al (2018) G9a stimulates CRC growth by inducing p53 Lys373 dimethylation-dependent activation of Plk1. Theranostics 8:2884–2895
Zhong X, Chen X, Guan X et al (2015) Overexpression of G9a and MCM7 in oesophageal squamous cell carcinoma is associated with poor prognosis. Histopathology 66:192–200

# Biomolecular Recognition of Methylated Histones

**Miriam R. B. Porzberg, Bas J. G. E. Pieters, and Jasmin Mecinović**

## Contents



**Abstract** Lysine and arginine methylations are among the most abundant posttranslational modifications found on histone proteins. The recognition of methylated lysine and arginine residues by epigenetic reader proteins provides an important molecular requirement for regulation of human genes. Recent structural and mechanistic studies importantly advanced our basic understanding of biomolecular recognition of methylated histones by diverse classes of epigenetic readers. In this chapter, we describe chemical biological studies on the recognition of methylated histones by the aromatic cage-containing reader proteins.

**Keywords** Biomolecular recognition · Histone · Epigenetics · Methylation · Reader proteins

---

M. R. B. Porzberg
Department of Physics, Chemistry and Pharmacy, University of Southern Denmark, Odense, Denmark

B. J. G. E. Pieters
Institute for Molecules and Materials, Radboud University, Nijmegen, The Netherlands

J. Mecinović (✉)
Department of Physics, Chemistry and Pharmacy, University of Southern Denmark, Odense, Denmark

Institute for Molecules and Materials, Radboud University, Nijmegen, The Netherlands
e-mail: mecinovic@sdu.dk

 
S. Jurga, J. Barciszewski (eds.), *The DNA, RNA, and Histone Methylomes*, RNA Technologies, https://doi.org/10.1007/978-3-030-14792-1_17

# 1 Introduction

Eukaryotic DNA is wrapped around four histone proteins (H2A, H2B, H3 and H4 each in a dimeric form) to form the nucleosome, the basic repeating unit of chromosomes, which provides the first level of compaction of the abundant genetic material in the cell nucleus (Fig. 1a). Histone proteins undergo extensive posttranslational modifications (PTMs), including lysine and arginine methylation (Smith and Denu 2009). Lysine residues are found to be monomethylated (Kme), dimethylated (Kme2) or trimethylated (Kme3), whereas arginine residues can be monomethylated (Rme), symmetrically dimethylated (Rme2s) or asymmetrically dimethylated (Rme2a) (Fig. 1b). Methylation of lysine and arginine residues increases the hydrophobicity of the ammonium/guanidinium groups, but keeps the positive charge unaltered under physiological conditions. Methylation of both residues, however, decreases their ability to form hydrogen bonds with water and interacting biomolecules. The installation of methyl group(s) on lysine and arginine is catalyzed by histone lysine methyltransferases (KMTs) and protein arginine methyltransferases (PRMTs), respectively, members of a superfamily of S-adenosylmethionine (SAM) dependent enzymes (Greer and Shi 2012). The opposite reaction, i.e. the removal of methylated lysine and arginine residues, is predominantly catalyzed by nonheme Fe(II) and 2-oxoglutarate dependent histone demethylases, although flavin dependent amine oxidases also remove Kme2 and Kme marks (Mosammaparast and Shi 2010).

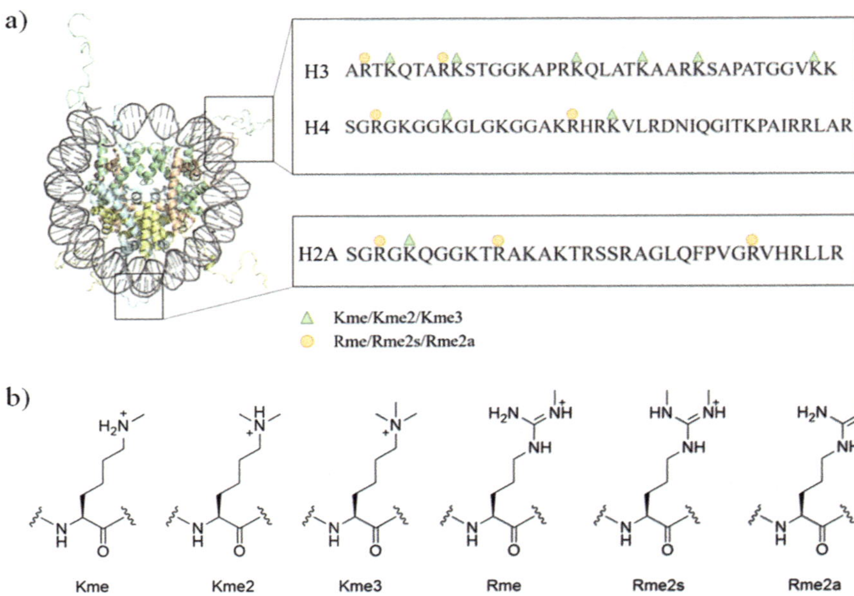


**Fig. 1** Lysine and arginine methylation of histone tails. (**a**) Methylation sites on histone tails. (**b**) Structures of methylated lysine and arginine residues

Methylated histones can function as recognition and docking sites for reader domains of numerous proteins or protein complexes that control the epigenetic landscape of human genome (Taverna et al. 2007; Patel and Wang 2013). Although PTMs can have a direct effect on the nucleosome/chromatin structure, specific recognition of PTMs by epigenetic reader proteins can also contribute to transcriptional activation or repression (known as the histone code) (Strahl and Allis 2000; Jenuwein and Allis 2001). Depending on the histone position and the methylation mark, methylated lysine residues have been observed in transcriptionally active (e.g. H3K4me3, H3K36me3) and silent (e.g. H3K9me3, H3K27me3, H4K20me3) regions of human genome.

Reader domains can recognize specific PTMs within sequence context with high degree of affinity and specificity. The recognition domains of epigenetic readers are largely involved in generating ligand specificity: location and size of the binding pocket may induce steric hindrance, whereas specific amino acid residues (e.g. charged, hydrophilic or hydrophobic) may attract or repulse certain PTMs. Due to a broad diversity of PTMs found on histones, a wide variety of reader domains has evolved. The number of readers that recognize histone modifications is already quite extensive, and new readers are still being discovered and investigated (Andrews et al. 2016). Here we describe recent chemical biological studies on recognition of methylated histones by epigenetic reader proteins.

## 2 Recognition of Methylated Lysines

Methylation on histone lysine residues is a well-studied posttranslational modification that can be found on unstructured histone tails of H2A, H3 and H4 (Fig. 1a), and on the histone core (e.g. H3K64 and H3K79). Two major classes of methyllysine recognition domains exist, namely the Royal family, which includes the chromodomain, tudor domain, MBT domain and PWWP domain, and the PHD (Plant HomeoDomain) zinc fingers (Maurer-Stroh et al. 2003; Taverna et al. 2007). Below we will describe separately epigenetic readers of the lower methylation marks (Kme and Kme2) and the higher methylation state (Kme3).

### *2.1 Monomethyllysine and Dimethyllysine*

Malignant brain tumor (MBT) domain proteins recognize monomethylated (Kme) and dimethylated lysine (Kme2) residues on H3 and H4 tails (Bonasio et al. 2010). The MBT domain is found in Polycomb (Pc) group proteins and the L(3)MBT tumor suppressor family consisting of L3MBTL1, L3MBTL2, and L3MBTL3, and it has a high sequence similarity with both, tudor and chromo domains (Maurer-Stroh et al. 2003; Bonasio et al. 2010). Generally, the methyllysine binding pocket of these proteins is present in one of the 2–4 MBT domain repeats, which contain both a

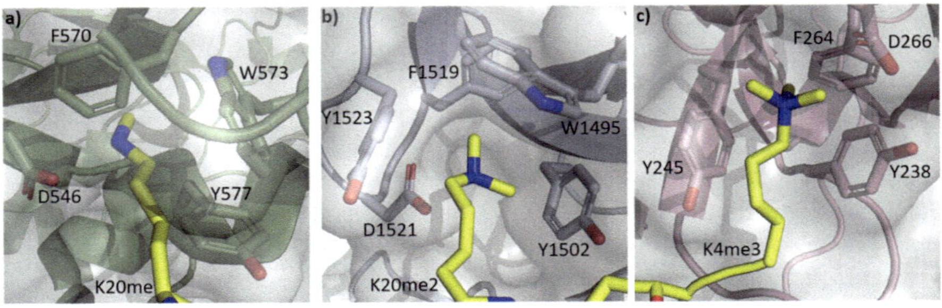


**Fig. 2** Views on crystal structures of the methyllysine binding pockets of (**a**) MBT domain protein L3MBTL2 in complex with H4K20me (PDB ID: 3F70); (**b**) tandem tudor domain protein 53BP1 in complex with H4K20me2 (PDB ID: 2IG0); and (**c**) tandem tudor domain SGF29 in complex with H3K4me3 (PDB ID: 3MEA)

30–50 residue N-terminal arm motif followed by a five-stranded antiparallel β-barrel fold (Li et al. 2007; Zhou 2015). The binding pocket of L3MBTL1, which binds Kme and Kme2, is situated in the second out of three MBTs and contains a conserved triad of Phe-Trp-Tyr residues, complemented with a negatively charged aspartate residue that forms a hydrogen bond or salt-bridge with the Kme and Kme2 methylammonium group, thus inducing Kme and Kme2 binding specificity via the cavity inserting binding mode (Fig. 2a) (Li et al. 2007; Min et al. 2007). Further specificity over Kme3 and K is generated by steric exclusion of the bulky Kme3 moiety, as the aromatic cage is situated in a pocket lined with hydrophobic residues, which preferentially interact with the more hydrophobic Kme and Kme2 residues relative to unmethylated lysine. For L3MBTL1, the strongest binding affinities were observed for the methyllysine marks H4K20me ($K_d = 5$ μM) and H4K20me2 ($K_d = 6$ μM), whereas bindings of H4K20 ($K_d = 410$ μM) and H4K20me3 ($K_d = 190$ μM) are significantly weaker (Li et al. 2007).

Along with MBT domain proteins, tudor domain containing proteins, e.g. the tandem tudor protein 53BP1, are readers of Kme and Kme2 (Botuyan et al. 2006). 53BP1 binds highly specifically to H4K20me2 ($K_d = 20$ μM), whereas weaker binding to H3K79me2 was also detected, even though these histone residues have no sequence similarities. The binding cage of 53BP1 consists of four aromatic residues, which stabilize binding of Kme2 by van der Waals and cation-π interactions, and an aspartic acid residue, which through a formation of salt-bridge with the N-$H^+$ group of the dimethylammonium cation contributes to the selectivity of Kme2 over the other methylation states; whereas Kme3 is too bulky, K and Kme have less optimal van der Waals and cation-π interactions when binding to 53BP1 (Fig. 2b) (Botuyan et al. 2006).

Furthermore, Kme2 is recognized by the BAH domain (bromo adjacent homology), which is present in the effector protein ORC1 (Kuo et al. 2012). The BAH domain binds strongly to H4K20me2 ($K_d = 5$ μM) and is highly selective for H4K20me2 over H4K20me, H4K20me3 and other peptides containing Kme2. The aromatic cage consists of two tyrosine, two tryptophan residues, and negatively charged Glu93 that contributes to stabilization of H4K20me2 by hydrogen bonding, thereby inducing selectivity for Kme2 over other marks (Kuo et al. 2012).

The ankyrin repeat domain, another reader domain of Kme and Kme2, is found six times at the N-terminal regions of the SET-domain containing methyltransferases G9a and GLP (G9a-like protein) (Collins et al. 2008). In contrast to G9a, which prefers binding of H3K9me2 ($K_d = 6\ \mu M$) over H3K9me ($K_d = 14\ \mu M$), GLP ankyrin repeats bind H3K9me ($K_d = 5\ \mu M$) more efficiently than H3K9me2 ($K_d = 7\ \mu M$), which is probably due to differences in the size of the binding cage. G9a's and GLP's ankyrin repeats adopt a helix-turn-helix-β-turn structure, which encloses H3 between repeats 4 and 5 and binds H3K9me2 in an aromatic pocket consisting of three tryptophans and a glutamic acid (Collins et al. 2008).

Methylated lysine residues can easily be incorporated into recombinant histones via site-specific installation and alkylation of cysteine residues (Simon et al. 2007). The cysteine-derived methyllysine and dimethyllysine analogues incorporated at position H3K9 are strongly recognized by the corresponding antibodies and cannot be distinguished from natural methyllysine and dimethyllysine residues in binding assays. Binding studies on H3K9me2 and its cysteine-derived analogue by the HP1 chromodomain indicates that both peptides bind strongly with only minor differences, thereby demonstrating the functionality of methyllysine analogues in intact histones. The nucleosome that contains the cysteine-derived Kme2 at position K9 does not only interact with the recombinant HP1, but also with HP1 from nuclear extracts (Simon et al. 2007).

## 2.2 *Trimethyllysine*

Most, if not all, readers of Kme3 contain the aromatic cage that is comprised of side chains of several aromatic residues flanking the trimethylammonium moiety (Taverna et al. 2007). A cage with four aromatic residues is often referred to as a full cage, whereas a cage flanking trimethylammonium on one or three sides is referred to as a half-cage structure (Yun et al. 2011). In the case of half-cages, the aromatic cage is usually complemented by a negatively charged residue that can form electrostatic interactions with the positively charged trimethylammonium moiety (Taverna et al. 2007). Binding trends for readers of Kme3 in general follow: Kme3 > Kme2 > Kme3 >> K (Sims and Reinberg 2006). Recent physical-organic chemistry studies revealed that biomolecular recognition of trimethyllysine by aromatic cages of readers is driven by energetically favorable cation-π interactions and Kme3-mediated release of high energy water molecules that occupy aromatic cages in the free state (Kamps et al. 2015; Hughes et al. 2007). Thermodynamic results showed that the association between human reader proteins and positively charged trimethyllysine was enthalpy-driven, whereas binding of the analogous neutral analogue of trimethyllysine was driven by more favorable entropy, thus implying a presence of strong cation-π interactions in the readout of Kme3 (Kamps et al. 2015). Advanced quantum chemical analysis on the aromatic cage-Kme3 interactions further verified the contribution of Pauli repulsion, electrostatic interactions, orbital interactions and dispersion interactions, along with the essential role of desolvation. In addition to favorable cation-π interactions, desolvation of the aromatic cage of readers presents an important contribution to the overall molecular readout of Kme3. WaterMap calculations revealed that water molecules that occupy the aromatic cages of a panel of human reader proteins are energetically unfavorable and that the release of such water by the Kme3 side chain contributes favorably to overall binding (Kamps et al. 2015). To probe the strength of cation-π interactions in the readout of trimethyllysine by reader proteins, several recent studies focused on subtle alterations of the aromatic cage residues (Lee et al. 2016; Baril et al. 2017) and the trimethyllysine side chain (Belle et al. 2017; Al Temimi et al. 2018). An incorporation of electron-deficient fluorinated phenylalanines in Mpp8 chromodomain led to a significantly weaker binding affinity for H3K9me3; binding affinities decreased with an increased number of fluorine substituents (Lee et al. 2016). Similarly, a replacement of tyrosine residues that constitute the aromatic cage of HP1 chromodomain by electron rich analogues led to strong interactions, whereas a substitution by electron poorer tyrosine analogues led to weaker interactions with Kme3, demonstrating the essential role of the π aromatic system on the strength of cation-π interactions. Thermodynamic and computational work revealed that reader proteins also recognize histones that possess trimethylornithine and trimethylhomolysine, the simplest trimethyllysine analogues that differ in the length of the side chain (Al Temimi et al. 2018). Both Kme3 analogues typically display weaker binding affinity than Kme3, as a result of somewhat weaker enthalpy of binding, suggesting that the positioning of the Kme3 side chain in the aromatic cage is optimal, thus leading to stronger cation-π interactions. Surprisingly, readers of trimethyllysine not only recognize histones that bear trimethyllysine with L-stereochemistry (i.e. L-Kme3), but also associate relatively well with configurationally distinct D-Kme3 counterparts (Belle et al. 2017). Molecular dynamic simulations revealed that the histone backbone can reorient to associate the D-Kme3 side chain in the aromatic cage, thus prioritizing favorable cation-π interactions. Readers of Kme3 also recognize the cysteine-derived Kme3 analogue, both on histone peptides and intact histones (Simon et al. 2007; Seeliger et al. 2012; Chen et al. 2018) In general, comparable binding affinities for Kme3 and the cysteine-derived Kme3 with readers were observed with histone peptides, suggesting that the cysteine-derived Kme3 could be used as a good Kme3 mimic for studies on intact histones and more complex nucleosome (Seeliger et al. 2012). Due to the fact that out of four histones, that constitute the nucleosome, only histone H3 possesses a cysteine residue (C110), the site-specific incorporation of Kme3 analogues can be easily achieved on intact histones by an alkylation of the cysteine residue introduced via point mutagenesis (Simon et al. 2007; Chen et al. 2018). An incorporation of the cysteine-derived trimethyllysine at position K9 of intact H3 further enabled the constitution of the nucleosome that bears specific posttranslational modifications (Simon et al. 2007). Unlike histone peptides, intact histones appeared to exhibit a weaker (~ 13-fold) binding affinity for the cysteine-derived Kme3 relative to Kme3 (Chen et al. 2018). As described below, recent structural and mechanistic work revealed structurally diverse classes of readers of trimethyllysine.

Chromatin organization modifier (Chromo) domain proteins contain a 40–60 residue domain that has first been described in heterochromatin 1 (HP1) and polycomb (Pc) proteins (Blus et al. 2011; Teske and Hadden 2017). The majority of chromodomains are associated with trimethyllysines in H3 and H4, e.g. HP1 primarily recognizes H3K9me3 (Blus et al. 2011). The common mechanism of Kme3 recognition by chromodomains is achieved by a conserved Tyr-Trp-Tyr motif, which constitutes a half-cage structure capable of binding Kme3 and Kme2. Chromodomain proteins contain a small hydrophobic pocket next to the aromatic binding cage that binds alanine residues that are located at a distance of two amino acids from H3K9me3 and H3K27me3, thereby additionally stabilizing binding of the histone peptides (Teske and Hadden 2017). The residues surrounding this binding groove are involved in generating substrate specificity, as they form non-covalent interactions with the interacting histone peptide. By subtly varying the amino acids that interact with the histone peptide, sequence specificity can be generated for Kme3 marks (Taverna et al. 2007).

The Tudor domain, named after the *Drosophila* Tudor protein, is another member of the Royal Family of readers (Boswell and Mahowald 1985; Lu and Wang 2013). Tudor domain proteins can recognize various PTMs, as some interact with Kme2 or Kme3 residues, whereas others recognize methylated arginines (see below). Common histone targets of Tudor domain proteins include H4K20me3 (e.g. JMJD2A, 53BP1), H3K4me3 (e.g. JMJD2A, SGF29, Spindlin1), and H3K36me3 (e.g. PHF1, PHF19). The domain itself is comprised of approximately 60 amino acid residues that form a barrel-like structure composed out of 4–5 antiparallel β-strands (Lu and Wang 2013). As common for other domains, Kme3 associates with electron-rich aromatic cages of Tudor domains (Fig. 2c). In addition to the Kme3 association with the aromatic cage, adjacent residues define the binding specificity, e.g. the H3A1 binding pocket contributes to specific recognition of H3K4me3 over H3K9me3 and other sequences (Pieters et al. 2013). Tudor domain proteins often accommodate another reader domain in near proximity, resulting in a combinatorial readout of specific histone tails. In UHRF1, binding to H3 is predominantly achieved by high affinity of the N-terminus, which binds to the PHD finger motif, whereas R8 and K9me3 are weakly bound to the tandem tudor domain, thereby enhancing selectivity (Lu and Wang 2013).

The PWWP domain is named after its semi-conserved Pro-Trp-Trp-Pro motif (Qin and Min 2014). The PWWP aromatic cage is comprised of three aromatic residues: a F/Y/W residue immediately preceding the PWWP motif, a W/Y residue that stems from the third residue of the PWWP motif and a F/Y/W residue which stems from the third β-strand of the PWWP domain. PWWP domains have been shown to specifically recognize H3K36me3 and H4K20me3 sequences, which have been associated with active and inactive genes, respectively (Qin and Min 2014). Along with histone peptides, PWWP domain proteins also bind to DNA, however, binding occurs via electrostatic interactions with the phosphate backbone of DNA and not via sequence specificity. Therefore, binding of the PWWP protein PSIP1 to the nucleosome containing H3K36me3 is stronger ($K_d$ = 1.5 μM) than separate binding to either the H3K36 peptide ($K_d$ = 17 mM) or the DNA ($K_d$ = 150 μM) (Qin and Min 2014).

The plant homeodomain (PHD) zinc fingers can be divided in several subsets, including H3K4me3 binders and H3K4 binders (Champagne and Kutateladze 2009; Sanchez and Zhou 2011). The general PHD finger structure comprises approximately 40–70 amino acid residues with low sequence conservation, although several highly conserved cysteine residues and a conserved histidine residue are present in all PHD fingers. These residues chelate two zinc ions in a cross-braced manner generally using the $Cys_4$-His-$Cys_3$ PHD finger motif. Upon binding, the histone forms a third antiparallel β-strand, which pairs with the double stranded β-sheet of the PHD domain, thereby inserting Kme3 in the aromatic cage (Champagne and Kutateladze 2009; Sanchez and Zhou 2011).

Unlike the histone reader domains discussed above, recognition by the ADD domain is obtained by a combinatorial readout of both, unmethylated H3K4 and H3K9me2/3 (Eustermann et al. 2011; Iwase et al. 2011). A peptide containing unmodified H3K4 and H3K9me3 ($K_d$ = 0.3 μM) or H3K9me2 ($K_d$ = 0.4 μM) binds strongly to the ADD domain of the gene encoding protein ATRX. Both residues are recognized simultaneously with K4 being bound by a PHD zinc finger containing acid residues, while K9me3 binding takes place in a non-aromatic binding pocket of a second zinc finger called GATA (Eustermann et al. 2011; Iwase et al. 2011).

Mutational analyses on aromatic cages that possess a negatively charged residue revealed the role of electrostatic interactions in the readout of Kme3/Kme2, and that a further selectivity for binding of Kme3 over Kme2 can be achieved (Eisert and Waters 2011; Pieters et al. 2015). Substituting the negatively charged E52 in HP1 chromodomain by neutral amino acids increased the selectivity from 1.2 (for WT HP1) to 3.4 (for E52Q variant), implying that electrostatic interactions and hydrogen bonding play an important role in the readout of Kme2. Replacing D266 in SGF29 tudor domain by a series of amino acids led to the same trend in lower binding affinities for both Kme3 and Kme2, suggesting that the same type of electrostatic interactions contribute to the molecular readout of Kme3 and Kme2 by SGF29.

## 3 Recognition of Methylated Arginines

Methylation on arginine residues is an abundant posttranslational modification that has been found on several sites on histones, including H2AR29, H3R2, H3R8, H3R17, H3R26, H3R42, and H4R3 (Fig. 1a) (Jahan and Davie 2015). Installation of methyl groups is catalyzed by Protein Arginine Methyltransferases (PRMTs) of which nine are currently known in mammals. Methylation of arginines can lead to three different products: monomethylated arginine (Rme, also known as MMA), symmetrically dimethylated arginine (Rme2s or SDMA) and asymmetrically dimethylated arginine (Rme2a or ADMA) (Fig. 1b) (Gayatri and Bedford 2014).

A few reader proteins that specifically recognize Rme2a or Rme2s marks in histones and non-histone proteins have been recently characterized, and more reader proteins and methylation sites are expected to be discovered.

Recent biological work showed that TDRD3, Spindlin1, SMN and SPF30 recognize Rme2a in histones and other proteins, where methylation predominantly takes place on Glycine/Arginine-Rich (GAR) motifs (Gayatri and Bedford 2014). All readers of methylarginine possess an aromatic cage that predominantly binds methylarginine by cation-π interactions and π-π stacking of the guanidinium group with two aromatic residues (James et al. 2013; Gayatri and Bedford 2014). Methylarginine binding pockets are generally narrower when compared to the binding pocket of methyllysine readers (see above) (Gayatri and Bedford 2014). In histones, the Tudor domain protein TDRD3 specifically binds H3R17me2a and H4R3me2a, which are associated with transcriptional activation, whereas binding of H3R2me2a appears to be weaker (Yang et al. 2010). Specificity for Rme2a over Rme2s is due to presence of Tyr566, one out of four aromatic residues in the binding cage, which upon mutation weakens binding to Rme2a and strengthens binding to Rme2s (Sikorsky et al. 2012). Recombinant intact histones that bear a cysteine-derived dimethylarginine analogue incorporated by site-specific conjugation via cysteine alkylation are also recognized by TDRD3 (Le et al. 2013). Along with histones, TDRD3 also associates with other proteins, including the C-terminal domain of RNA Polymerase II (Fig. 3a), and Sm proteins that arise in the cytoplasm (Côté and Richard 2005; Sikorsky et al. 2012). The recognition of Rme2a by the aromatic cage of TDRD3 is stabilized by NH-O hydrogen bond and cation-π interactions; the tudor domain itself without C- and N-terminal extensions is also capable of ligand binding (Liu et al. 2012; Sikorsky et al. 2012).

Unlike TDRD3, Spindlin1 is not a Tudor domain protein, but it consists of three Spin/Ssty domains that fold in a similar way as Tudor domain proteins (Su et al. 2014). In addition to the methylarginine mark H3R8me2a, Spindlin1 also recognizes the adjacent trimethyllysine mark H3K4me3. In 10-mer peptides, the binding affinity of H3R8me2a to Spindlin1 is $K_d = 22\ \mu M$, whereas H3K4me3 binds even stronger with $K_d = 147$ nM. When both modifications are combined in the same

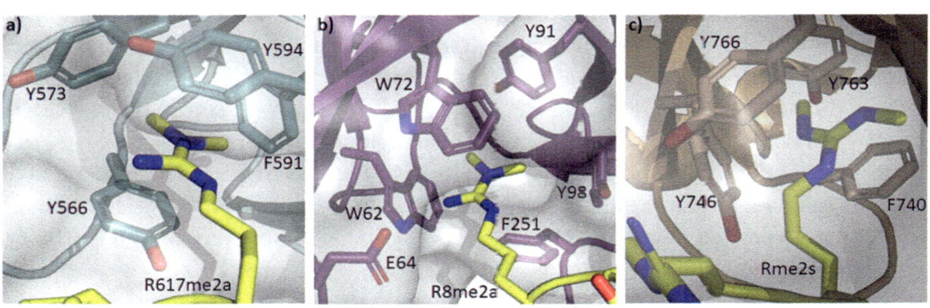


**Fig. 3** Views on crystal structures of the methylarginine binding pockets of (**a**) TDRD3 with a DNA-directed RNA polymerase II subunit (PDB ID: 2LTO); (**b**) Spindlin1 in complex with the H3K4me3R8me2a peptide (PDB ID: 4MZF); (**c**) SND1 with a PIWIL1 peptide (PDB ID: 3OMG)

peptide (H3K4me3R8me2a), the binding affinity with Spindlin1 was found to be significantly higher ($K_d = 45$ nM), suggesting a positive cooperative effect (Su et al. 2014). The crystal structure of Spindlin1-H3K4me3R8me2a revealed that H3K4me3 and H3R8me2a bind to two distinct binding pockets consisting of aromatic residues that lie close to each other (Fig. 3b). A combination of cation-π interactions and CH-π interactions with the aromatic residues, a salt-bridge formation with E64 and a hydrogen bonding with Y98 stabilizes R8me2a binding to Spindlin1 (Fig. 3b); point mutation of one of the aromatic residues into alanine leads to weaker binding affinity (Su et al. 2014). Binding of R8me2a takes place by a "cavity insertion" mode, meaning that the binding pocket is deep and narrow, in contrast to a "surface groove" mode. On the other hand, R8me2s typically does not bind to the same narrow binding pocket, presumably due to its large size and geometry. In other proteins, Rme2s is often recognized by a "surface groove" mode (Su et al. 2014).

In Rme2s, the methyl groups can adapt two distinct conformations: *anti-anti* (two methyl groups point in the same direction) and *anti-syn* (two methyl groups point in opposite directions) (Supekar et al. 2018). Interestingly, readers of Rme2s have different preferences over one of the two conformations. The Tudor domain protein Survival Motor Neuron (SMN), for example, preferentially binds to H3R2me2s in the *anti-anti* conformation and to H3R2me2a (Supekar et al. 2018). Like in TDRD3, the tudor domain of SMN adopts a five-stranded β-barrel fold, which has a sequence identity of 37% with the TDRD3 tudor domain. SMN's binding groove near the binding pocket is more spacious, thereby enabling binding to additional motifs like the glycine- and methionine-rich (PGM) motif, whereas TDRD3 only recognizes proteins containing glycine and arginine-rich regions (GAR motifs) and isolated Rme2a histone marks (Liu et al. 2012). In SMN, the *anti-anti* conformation is higher in free energy than the *anti-syn* conformation, and forms stronger cation-π interactions with the aromatic cage. However, upon single point mutation of Trp102 that lies in the binding pocket into a smaller Phe, the strength of cation-π interactions changed, resulting in a preference for *anti-syn* binding (Supekar et al. 2018).

SPF30 is another reader of Rme2s, possessing a Tudor domain that has a sequence identity of 45% with the Tudor domain of SMN (Tripsianes et al. 2011; Liu et al. 2012). Known substrates of both SMN and SPF30 include the GAR rich C-terminal tails of Sm proteins and PIWI proteins that contain a GAR motif (Tripsianes et al. 2011; Liu et al. 2012). However, no histones that associate with SPF30 have been found yet (Liu et al. 2012). Overall, SMN has a higher binding affinity than SPF30 for Rme2s. In the aromatic binding pocket, the main difference between SMN and SPF30 lies in the backbone: SMN contains a tyrosine residue (Tyr127) that is involved in a hydrogen bond triangle with Glu134 and Gln136, whereas SPF30 contains a phenylalanine residue instead (Phe108). Upon replacement of Phe108 by Tyr108, an additional hydrogen bond is formed, resulting in an improved binding affinity, while double mutation of F108Y and T115Q is required for SPF30 to ensue equally strong binding interactions through a complete hydrogen bond triangle, like SMN. The crystal structures of SMN and SPF30 with both Rme2a and Rme2s reveal that the guanidinium group is stacked parallel to the aromatic tryptophan and tyrosine rings to maximize cation-π interactions. Another tyrosine and the fourth aromatic residue are oriented orthogonally to it, thereby favoring T-shaped interactions (Tripsianes et al. 2011).

SND1, another Tudor domain protein, recognizes H3R2me2s ($K_d$ = 99 μM), H4R3me2s ($K_d$ = 63 μM) and H3R2me2a with a lower binding affinity ($K_d$ > 300 μM). In addition to these histone marks, SND1 also recognizes Rme2s-containing Piwi peptides with a strong binding affinity in the lower micromolar range, which is probably due to several RA/RG repeats. Unlike TDRD3, SND1 itself is not capable of ligand binding and requires folding of C- and N-terminal extensions to be active (Liu et al. 2010; Liu et al. 2012). Contrary to SMN and SPF30, SND1 appears to prefer binding of Rme2s substrates in the *anti-syn* conformation, even though some substrates are the same (Fig. 3c).

Recently, another reader protein for H4R3me2s, called PHD finger protein 1 (PHF1), was discovered (Liu et al. 2018b). PHF1 contains a tudor domain that associates with H3K36me3, two PHD fingers, a domain of extended homology and a chromo-like domain at the C-terminal. PHF1 is physically associated with PRMT5-WDR77, a methyltransferase complex that symmetrically dimethylates H4R3, H3R8 and H3R2. However, the N-terminal PHD finger domain (PHD1) of PHF1 only associates with H4R3me2s ($K_d$ = 13.4 μM) and not with other Rme2s or Rme2a histone marks. In addition to the N-terminal PHD finger domain (PHD1), PHF1 also harbors a tudor domain that specifically recognizes H3K36me3. Like other reader proteins, PHF1 contains an aromatic cage, which upon single point mutation of one of aromatic residues into an smaller nonaromatic Ala completely loses its ability to recognize Rme2s (Liu et al. 2018b). Expression of PHF1 serves as a potential biomarker, since it might promote tumorigenesis and metastasis in breast cancer cells.

Unlike other methylarginine reader proteins, WDR5 is a WD40 domain protein and not a tudor domain protein (Migliori et al. 2012). Upon binding, this domain folds into a seven-blade β-propeller that surrounds the guanidinium group of methylated arginine. WDR5 is known to be part of several coactivator complexes, and its WD40 domain strongly binds H3R2me2s with a binding affinity of $K_d$ = 0.1 μM, while a weaker binding affinity for the unmodified H3 peptide occurs ($K_d$ = 5.6 μM). Binding of the unmodified H3 is achieved by two water-mediated hydrogen bonds with Ser175 and Ser218, while in the case of H3R2me2s, one of the water molecules is replaced and hydrophobic interactions with Phe219 contribute to stabilization of the complex (Migliori et al. 2012).

## 4 Recognition of Methylated Histones by Small Molecules

Following important biomolecular recognition processes of posttranslationally modified histones, histone posttranslational modifications have also been studied by supramolecular hosts, of which some are known to bind methyllysine and methylarginine marks in histone peptides (Hof 2016). Applying host-guest chemistry to epigenetic methylation marks is challenging, since the synthetic host molecule that mimics the aromatic cage needs to be functional in an aqueous environment,

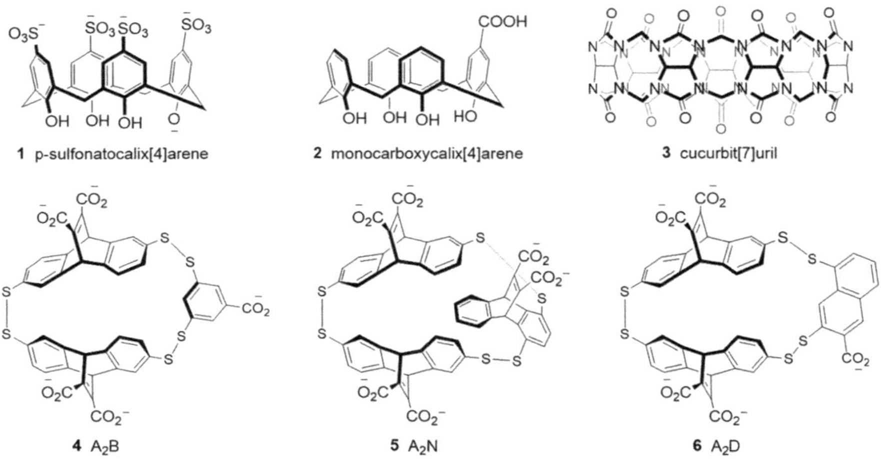


**Fig. 4** Supramolecular hosts for recognition of methylated lysine and arginine residues

which often contains additional salts. The calixarene p-sulfonatocalix[4]arene (**1**, CX4) fits these criteria, and it has been reported as one of the first methyllysine hosts with a strong affinity for H3K9me3 ($K_d$ = 7.2 μM) over other amino acids with a general trend of Kme3 > Kme2 > Kme > K (Fig. 4). Upon binding, the trimethylammonium group is encapsulated in the aromatic cavity, while sulfonate groups attached on the host interact with the neighboring amino acids of the histone peptide, resulting in a stronger binding affinity when compared to the free Kme3. In addition to H3K9me3, CX4 also binds H3K27me3, which is recognized by the native reader protein Chromobox Homolog 7 (CBX7) with a relatively low binding affinity of $K_d$ = 10–100 μM (Hof 2016). In competition experiments with CBX7, CX4 disrupted the natural CBX7-H3K27me3 interaction by targeting H3K27me3, thereby serving as an inhibitor (Daze et al. 2012). Along with methyllysine in histones, CX4 also binds to lysozyme, which possesses multiple lysine residues. Binding occurs selectively towards Kme2, especially K116me2, which is the most accessible lysine residue on the protein surface, whereas accessible arginine residues are recognized with a 50-fold decreased binding affinity. CX4 also binds methylarginine residues, however, there is no selectivity for one dimethylarginine state over the other (McGovern et al. 2015). Monocarboxycalix[4]arene is another calixarene that predominantly binds di- and trimethylated lysine residues (**2**) (Fig. 4). In unmodified lysine and monomethyllysine, the positively charged ammonium group does not occupy the aromatic pocket and points out, whereas di- and trimethylated lysine residues, both in histone peptides and as free amino acids, bind to the calixarene **2** with stronger binding affinities ($K_d$ = 50 μM for H3K9me2 and H3K9me3; $K_d$ > 500 μM for H3K9me) (Hanauer et al. 2017).

Cucurbit[7]uril (**3**, CB[7]) is a host that binds methyllysine residues with a general trend of Kme3 > Kme2 > Kme > K (Fig. 4). Host **3** also binds Rme2s with a threefold increased selectivity over Rme2a, which is due to a distinct binding mode; Rme2a is almost completely encapsulated by CB[7], whereas in the case of Rme2s, only the dimethylguanidinium group is enclosed (Gamal-Eldin and Macartney 2013).

$A_2B$ (**4**) is a host that binds H3K9me2 and H3K9me3 with similar affinity to HP1 (Fig. 4). Using the dynamic combinatorial chemistry (DCC) approach, an even more hydrophobic host than $A_2B$ has been discovered: $A_2N$ (**5**) contains monomer N, which enables stronger cation-$\pi$ interaction and enhances the binding (Fig. 4). Compound **5** binds H3K9me3 with $K_d = 300$ nM, whereas the binding affinity of $A_2B$ with H3K9me3 is $K_d = 2.6$ μM. Besides binding affinity, selectivity towards Kme3 over the other methylation states is increased in $A_2N$ when compared to $A_2B$ (Pinkin and Waters 2014).

$A_2D$ (**6**) is another analogue of $A_2B$ that has been discovered by DCC (Fig. 4). Inhibitor **6** strongly interacts with H3R8me2a ($K_d = 5$ μM) with a sevenfold selectivity over H3R8me2s and tenfold selectivity over nonmethylated H3. Selectivity towards Rme2a over Rme2s might be due to the bulkiness of Rme2s, resulting in only partial association of Rme2s with $A_2D$ (James et al. 2013).

## 5 Inhibition of Readers of Methylated Histones

Since some epigenetic reader proteins are associated with human diseases, including cancer, studying small molecule inhibitors that bind in the aromatic cages gains an importance in rational drug design (Dawson and Kouzarides 2012). For MBT domain readers, a few small molecule inhibitors that target the binding domain are known, especially for L3MBTL1 and L3MBTL3 (Teske and Hadden 2017). However, designing inhibitors that specifically bind to only one MBT domain is challenging due to their sequence similarities, and therefore, most inhibitors known to date show similar affinity for L3MBTL1 and L3MBTL3. Compound **7** is an example of an inhibitor that specifically targets L3MBTL3 with a high binding affinity ($IC_{50} = 0.048$ μM), whereas a lower binding affinity towards L3MBTL1 has been obtained ($IC_{50} = 86$ μM) (Fig. 5). Small molecules targeting 53BP1 are also known, with **8** (UNC2170) being one of the most promising candidates ($IC_{50} = 22$ μM) with high selectivity towards 53BP1 over other methyllysine readers (Fig. 5). For the PHD3 finger domain of JARID1A, several small molecule inhibitors have been recently identified, including **9** (di-*N*-desethylamiodarone) with $IC_{50} = 26$ μM (Fig. 5) (Wagner et al. 2012; Bhushan et al. 2018) Small molecule inhibitors of chromodomain CBX7 are mainly based on peptides (e.g. **10**, UNC3866, Fig. 5). Compound **10** shows a significantly higher binding affinity towards CBX7 ($K_d = 97$ nM) when compared to H3K27me3 ($K_d = 110$ μM) (Teske and Hadden 2017). For PWWP proteins, no small molecule inhibitors that target the methyllysine binding domain are currently known (Teske and Hadden 2017).

In contrast to inhibitors of methyllysine readers, small molecule inhibitors that bind to readers of methylarginine remain to be developed. Only recently, out of a library containing 890 molecules, 14 fragments that bind to TDRD3 were identified

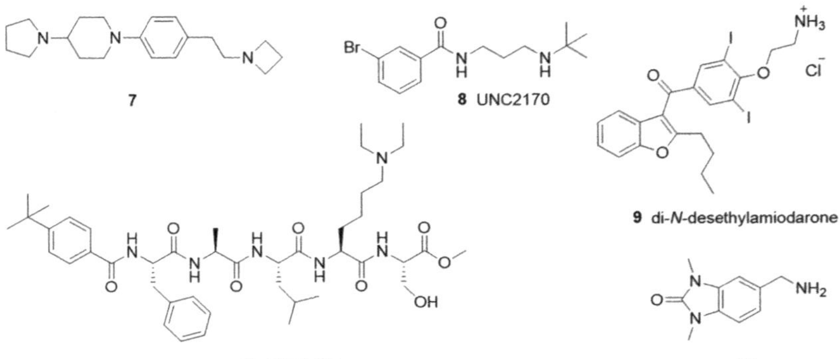


**Fig. 5** Small molecule inhibitors of methyllysine reader domains (**7–10**) and methylarginine reader TDRD3 (**11**)

by NMR (Liu et al. 2018a). Compound **11** was found to be a promising candidate that binds selectively to TDRD3 with $K_d = 48\ \mu M$, while its binding affinity for SMN is somewhat lower ($K_d = 170\ \mu M$) (Fig. 5). Interestingly, compound **11** differs in its binding mode from histone residues: it protrudes the aromatic binding pocket and forms an extra hydrogen bond with an Asn and π-π interactions with the two parallel Tyr residues in the binding pocket (Liu et al. 2018a). The discovery of **11** as ligand for TDRD3 is the first step towards the development of inhibitors that specifically target readers of methylarginine.

## 6 Conclusion and Prospects

The past two decades have witnesses significant advances in examinations of biomolecular recognition of posttranslationally modified histones. Numerous reader domain proteins, including those that recognize methylated lysine and arginine residues in histones, have been identified and characterized, and there are ongoing activities in discovering novel posttranslational modifications on histones and new reader proteins that specificially recognize such modifications. Structurally diverse families of reader proteins, including tudor domains, chromodomains and PHD zinc fingers, specifically recognize methylated lysine and arginine residues, most of them containing the aromatic cage where the recognition of posttranslationally modified residues takes place. Despite a large body of structural and biochemical data that examine structure-function relationships, we are currently still lacking in-depth molecular understanding of essential biomolecular recognition processes involved in epigenetic gene regulation. Although recent work showed that a combination of thermodynamic, structural and computational analyses importantly contributes to an improved basic understanding of biomolecular recognition in epigenetics, there are many binding processes that still need to be examined in great detail. It is envisioned that ongoing and future studies will elucidate the nature and strength of noncovalent interactions and the exact role of water in biomolecular recognition of epigenetic processes. We have just entered the phase in which the fundamental understanding of molecular readout of posttranslationally modified histones is associated with opportunities in drug discovery, and area of current biomedicinal interest.

**Acknowledgement** We thank the European Research Council for financial support (ERC Starting Grant to J.M., ChemEpigen-715691).

## References

Al Temimi AHK, Belle R, Kumar K et al (2018) Recognition of shorter and longer trimethyllysine analogues by epigenetic reader proteins. Chem Commun 54:2409–2412

Andrews FH, Strahl BD, Kutateladze TG (2016) Insights into newly discovered marks and readers of epigenetic information. Nat Chem Biol 12:662–668

Baril SA, Koenig AL, Krone MW et al (2017) Investigation of trimethyllysine binding by the HP1 chromodomain via unnatural amino acid mutagenesis. J Am Chem Soc 139:17253–17256

Belle R, Al Temimi AHK, Kumar K et al (2017) Investigating d-lysine stereochemistry for epigenetic methylation, demethylation and recognition. Chem Commun 53:13264–13267

Bhushan B, Erdmann A, Zhang Y et al (2018) Investigations on small molecule inhibitors targeting the histone H3K4 tri-methyllysine binding PHD-finger of JmjC histone demethylases. Bioorganic Med Chem 26:2984–2991

Blus BJ, Wiggins K, Khorasanizadeh S (2011) Epigenetic virtues of chromodomains. Crit Rev Biochem Mol Biol 46:507–526

Bonasio R, Lecona E, Reinberg D (2010) MBT domain proteins in development and disease. Semin Cell Dev Biol 21:221–230

Boswell RE, Mahowald AP (1985) tudor, a gene required for assembly of the germ plasm in *Drosophila melanogaster*. Cell 43:97–104

Botuyan MV, Lee J, Ward IM et al (2006) Structural basis for the methylation state-specific recognition of histone H4-K20 by 53BP1 and Crb2 in DNA repair. Cell 127:1361–1373

Champagne KS, Kutateladze TG (2009) Structural insight into histone recognition by the ING PHD fingers. Curr Drug Targets 10:432–441

Chen Z, Notti RQ, Ueberheide B et al (2018) Quantitative and structural assessment of histone methyllysine analogue engagement by cognate binding proteins reveals affinity decrements relative to those of native counterparts. Biochemistry 57:300–304

Collins RE, Northrop JP, Horton JR et al (2008) The ankyrin repeats of G9a and GLP histone methyltransferases are mono- and dimethyllysine binding modules. Nat Struct Mol Biol 15:245–250

Côté J, Richard S (2005) Tudor domains bind symmetrical dimethylated arginines. J Biol Chem 280:28476–28483

Dawson MA, Kouzarides T (2012) Cancer epigenetics: from mechanism to therapy. Cell 150:21–27

Daze KD, Pinter T, Beshara CS et al (2012) Supramolecular hosts that recognize methyllysines and disrupt the interaction between a modified histone tail and its epigenetic reader protein. Chem Sci 3:2695–2699

Eisert RJ, Waters ML (2011) Tuning HP1α chromodomain selectivity for di- and trimethyllysine. ChemBioChem 12:2786–2790

Eustermann S, Yang JC, Law MJ et al (2011) Combinatorial readout of histone H3 modifications specifies localization of ATRX to heterochromatin. Nat Struct Mol Biol 18:777–783

Gamal-Eldin MA, Macartney DH (2013) Selective molecular recognition of methylated lysines and arginines by cucurbit[6]uril and cucurbit[7]uril in aqueous solution. Org Biomol Chem 11:488–495
Gayatri S, Bedford MT (2014) Readers of histone methlyarginine marks. Biochim Biophys Acta 1839:702–710
Greer EL, Shi Y (2012) Histone methylation: a dynamic mark in health, disease and inheritance. Nat Rev Genet 13:343–357
Hanauer T, Hopkinson RJ, Patel K et al (2017) Selective recognition of the di/trimethylammonium motif by an artificial carboxycalixarene receptor. Org Biomol Chem 15:1100–1105
Hof F (2016) Host-guest chemistry that directly targets lysine methylation: synthetic host molecules as alternatives to bio-reagents. Chem Commun 52:10093–10108
Hughes RM, Wiggins KR, Khorasanizadeh S et al (2007) Recognition of trimethyllysine by a chromodomain is not driven by the hydrophobic effect. Proc Natl Acad Sci U S A 104:11184–11188
Iwase S, Xiang B, Ghosh S et al (2011) ATRX ADD domain links an atypical histone methylation recognition mechanism to human mental-retardation syndrome. Nat Struct Mol Biol 18:769–777
Jahan S, Davie JR (2015) Protein arginine methyltransferases (PRMTs): role in chromatin organization. Adv Biol Regul 57:173–184
James LI, Beaver JE, Rice NW et al (2013) A synthetic receptor for asymmetric dimethyl arginine. J Am Chem Soc 135:6450–6455
Jenuwein T, Allis CD (2001) Translating the histone code. Science 293:1074–1080
Kamps JJAG, Huang J, Poater J et al (2015) Chemical basis for the recognition of trimethyllysine by epigenetic reader proteins. Nat Commun 6:8911
Kuo AJ, Song J, Cheung P et al (2012) The BAH domain of ORC1 links H4K20me2 to DNA replication licensing and Meier-Gorlin syndrome. Nature 484:115–121
Le DD, Cortesi AT, Myers SA et al (2013) Site-specific and regiospecific installation of methylarginine analogues into recombinant histones and insights into effector protein binding. J Am Chem Soc 135:2879–2882
Lee YJ, Schmidt MJ, Tharp JM et al (2016) Genetically encoded fluorophenylalanines enable insights into the recognition of lysine trimethylation by an epigenetic reader. Chem Commun 52:12606–12609
Li H, Fischle W, Wang W et al (2007) Structural basis for lower lysine methylation state-specific readout by MBT repeats of L3MBTL1 and an engineered PHD finger. Mol Cell 28:677–691
Liu K, Chen C, Guo Y et al (2010) Structural basis for recognition of arginine methylated Piwi proteins by the extended tudor domain. Proc Natl Acad Sci U S A 107:18398–18403
Liu K, Guo Y, Liu H et al (2012) Crystal structure of TDRD3 and methyl-arginine binding characterization of TDRD3, SMN and SPF30. PLoS One 7:e30375
Liu J, Zhang S, Liu M et al (2018a) Structural plasticity of the TDRD3 tudor domain probed by a fragment screening hit. FEBS J 285:2091–2103
Liu R, Gao J, Yang Y et al (2018b) PHD finger protein 1 (PHF1) is a novel reader for histone H4R3 symmetric dimethylation and coordinates with PRMT5–WDR77/CRL4B complex to promote tumorigenesis. Nucleic Acids Res 46:6608–6626
Lu R, Wang GG (2013) Tudor: a versatile family of histone methylation “readers”. Trends Biochem Sci 38:546–555
Maurer-Stroh S, Dickens NJ, Hughes-Davies L et al (2003) The tudor domain “Royal Family”: tudor, plant Agenet, Chromo, PWWP and MBT domains. Trends Biochem Sci 28:69–74
McGovern RE, Snarr BD, Lyons JA et al (2015) Structural study of a small molecule receptor bound to dimethyllysine in lysozyme. Chem Sci 6:442–449
Migliori V, Müller J, Phalke S et al (2012) Symmetric dimethylation of H3R2 is a newly identified histone mark that supports euchromatin maintenance. Nat Struct Mol Biol 19:136–145
Min J, Allali-Hassani A, Nady N et al (2007) L3MBTL1 recognition of mono- and dimethylated histones. Nat Struct Mol Biol 14:1229–1230
Mosammaparast N, Shi Y (2010) Reversal of histone methylation: biochemical and molecular mechanisms of histone demethylases. Annu Rev Biochem 79:155–179

Patel DJ, Wang Z (2013) Readout of epigenetic modifications. Annu Rev Biochem 82:81–118
Pieters B, Belle R, Mecinovic J (2013) The effect of the length of histone H3K4me3 on recognition by reader proteins. ChemBioChem 14:2408–2412
Pieters BJGE, Meulenbroeks E, Belle R et al (2015) The role of electrostatic interactions in binding of histone H3K4me2/3 to the Sgf29 tandem tudor domain. PLoS One 10:e0139205
Pinkin NK, Waters ML (2014) Development and mechanistic studies of an optimized receptor for trimethyllysine using iterative redesign by dynamic combinatorial chemistry. Org Biomol Chem 12:7059–7067
Qin S, Min J (2014) Structure and function of the nucleosome-binding PWWP domain. Trends Biochem Sci 39:536–547
Sanchez R, Zhou MM (2011) The PHD finger: a versatile epigenome reader. Trends Biochem Sci 36:364–372
Seeliger D, Soeroes S, Klingberg R et al (2012) Quantitative assessment of protein interaction with methyl-lysine analogues by hybrid computational and experimental approaches. ACS Chem Biol 7:150–154
Sikorsky T, Hobor F, Krizanova E et al (2012) Recognition of asymmetrically dimethylated arginine by TDRD3. Nucleic Acids Res 40:11748–11755
Simon MD, Chu F, Racki LR et al (2007) The site-specific installation of methyl-lysine analogs into recombinant histones. Cell 128:1003–1012
Sims RJ, Reinberg D (2006) Histone H3 Lys 4 methylation: caught in a bind? Genes Dev 20:2779–2786
Smith BC, Denu JM (2009) Chemical mechanisms of histone lysine and arginine modifications. Biochim Biophys Acta 1789:45–57
Strahl BD, Allis CD (2000) The language of covalent histone modifications. Nature 403:41–45
Su X, Zhu G, Ding X et al (2014) Molecular basis underlying histone H3 lysine – arginine methylation pattern readout by Spin/Ssty repeats of Spindlin1. Genes Dev 28:622–636
Supekar S, Papageorgiou AC, Gemmecker G et al (2018) Conformational selection of dimethylarginine recognition by the survival motor neuron tudor domain. Angew Chemie Int Ed 57:486–490
Taverna SD, Li H, Ruthenburg AJ et al (2007) How chromatin-binding modules interpret histone modifications: lessons from professional pocket pickers. Nat Struct Mol Biol 14:1025–1040
Teske KA, Hadden MK (2017) Methyllysine binding domains: structural insight and small molecule probe development. Eur J Med Chem 136:14–35
Tripsianes K, Madl T, MacHyna M et al (2011) Structural basis for dimethylarginine recognition by the tudor domains of human SMN and SPF30 proteins. Nat Struct Mol Biol 18:1414–1420
Wagner EK, Nath N, Flemming R et al (2012) Identification and characterization of small molecule inhibitors of a plant homeodomain finger. Biochemistry 51:8293–8306
Yang Y, Lu Y, Espejo A et al (2010) TDRD3 is an effector molecule for arginine methylated histone marks. Mol Cell 40:1016–1023
Yun M, Wu J, Workman JL et al (2011) Readers of histone modifications. Cell Res 21:564–578
Zhou MM (2015) Histone recognition. Springer, Cham

# The Role of Protein Lysine Methylation in the Regulation of Protein Function: Looking Beyond the Histone Code

**Hemanta Adhikary, Orneala Bakos, and Kyle K. Biggar**

## Contents



**Abstract** Histone proteins and their diverse array of post-translational modifications have been subject to exquisite evolutionary conservation in eukaryotes. Accordingly, the factors that control the deposition, removal, and interpretation of histone modifications are themselves deeply conserved, with many strongly impacting development and disease in humans. Of these modifications, lysine methylation has in recent years emerged as a prevalent modification occurring on histone proteins. However, although numerous lysine methyltransferase and demethylase enzymes have been extensively characterized with respect to their ability to control methylation at

H. Adhikary · O. Bakos · K. K. Biggar (✉)
Institute of Biochemistry, Carleton University, Ottawa, ON, Canada
e-mail: kyle_biggar@carleton.ca



S. Jurga, J. Barciszewski (eds.), *The DNA, RNA, and Histone Methylomes*, RNA Technologies, https://doi.org/10.1007/978-3-030-14792-1_18

specific histone residues, their known targets have been rapidly expanding to include the methylation of non-histone proteins as well. These findings extend the role of lysine methylation well-beyond the established histone code and its role in epigenetic regulation. To date, this lysine methylation has been found to directly regulate protein sub-cellular localization, protein-protein interactions, and has also been found to interplay with other post-translational modifications. As a result, lysine methylation is now known to coordinate protein function and be a key driving of a growing list of cellular signaling events, including apoptosis, DNA damage repair, protein translation, cell growth, and signal transduction among others. This chapter will provide insight into the role of protein lysine methylation and its role in regulating protein function and its impact on human development and disease.

**Keywords** Lysine methylation · Non-histone methylation · Methyllysine proteomics

# 1 Preface

There is a kink (shoulder) on [the] Lys peak... Richard P. Ambler (1959)

These words marked the initial discovery of lysine methylation and introduced a segue into a brand new field of scientific research. At the time, Ambler was a graduate student working in the laboratory of Dr. Maurice W. Reese at the University of Cambridge, working on the amino acid composition of bacterial flagellin (Ambler and Rees 1959). Through ion-exchange and 2D-chromatography experiments, a unique "kink" in a chromatograph was interpreted as a new amino acid, the ε-N-methyl-lysine. This new amino acid was discovered from the hydrolysates (proteins digested into smaller fragments, peptides, and amino acids) of *Salmonella typhimurium* flagellin, and provided the first insight that protein lysine methylation occurs amongst living cells. Although initially sparking a surge of research interest for a number of years, focus on lysine methylation quickly faded as a result of the inherent difficulty and lack of suitable technologies to study this very small, uncharged protein modification. Consequently, the functional implications of lysine methylation have only now begun to be established.

By the time protein methylation emerged as a field of interest, research into other post-translational modifications (PTMs; a chemical modification made to proteins that alter the host protein fate or function) more recently discovered was already firmly underway. For example, the discovery of lysine methylation (Kme) predates tyrosine phosphorylation by two decades following its discovery on v-Src-associated kinase (Anderson et al. 1990).

This chapter will discuss the expanding field of lysine methylation, along with its historical context and some of the key discoveries that have set the stage for a greater understanding of this intriguing post-translational modification. This chapter will also introduce several key examples of how lysine methylation is currently known to regulate protein function, drive in disease pathologies, and finally, new technologies utilized for its discovery.

## 2 Lysine Methylation: A Brief History in Its Discovery

The initial discovery hallmarked by Ambler and Rees' observation of methyllysine in the flagellin of *Salmonella typhimurium*, provided the scientific community with its first evidence of protein methylation in living cells (Ambler and Rees 1959). Additional to this pivotal discovery, subsequent findings also led to the identification of a separate gene that influenced the presence, or absence, of the methyllysine modification—this demonstrated that methylation was a modification that occurred post-translationally (Stocker and McDonough 1961). It was then further reasoned that a specific enzyme must act to add the methylation modification directly to protein lysine residues. Impressively, these early theories posited the fundamental principles of which future revelations have been realized within the field.

Indeed, the lysine methylation of proteins have since been established to regulate many cellular processes, including protein interactions and cellular signaling transduction (Biggar and Li 2015; Wu et al. 2017). However, although the first lysine methylation event was found to occur in a non-histone protein, the methylation of histone proteins and its role in regulating chromatin structure became the impetus in driving the lysine methylation research for the following decades. It has now been established that hundreds of proteins are methylated at lysine residues and that this PTM is involved in regulating a growing number of cellular events, including growth signaling and DNA damage response (Carlson and Gozani 2016; Cao and Garcia 2016).

Although the physiological and regulatory roles of other PTMs, such as phosphorylation, were already being established, the 1960s brought important contributions to the most basic understanding of methylation. For example, in 1964 Kenneth Murray discovered the presence of methyllysine modified histone proteins (Murray 1964). Others have demonstrated that methyllysine could not be conjugated to tRNAs, thus resolving a persisting question on when the methylation of lysine occurred (Kim and Paik 1965). This discovery confirmed that histones were methylated after translation and not through the tRNA-mediated incorporation of a modified lysine residue. Building on these insights, Vincent Allfrey, and fellow researchers posited what, at the time, would have been a truly insightful hypothesis: that methylation of histones could regulate gene transcription (Allfrey et al. 1964).

Following these initial discoveries, there was a precipitous drop in research in subsequent decades. Throughout the 1960s and 1970s, Kim and Paik had diverted their focus towards the identification of the enzymes proposed to be involved in methylation. This was a fortunate detour, as they were able to establish the first methyltransferase activity, which involved the transfer of a methyl group from S-adenosylmethionine (SAM) to lysine, arginine, aspartic acid or glutamic acid

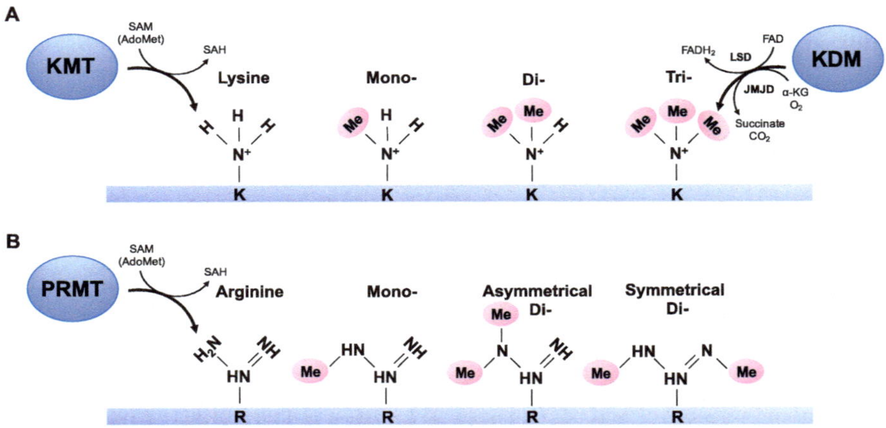


**Fig. 1** Mechanism of lysine (K) and arginine (R) methylation. Lysine methyltransferase enzymes (KMTs) facilitate methylation through the use of S-adenosyl-L-methionine (SAM/AdoMet) as a methyl donor, yielding a methylated lysine residue and S-Adenosyl homocysteine (SAH). Specific to lysine, up to three methylation groups can be added a single lysine residue resulting in the formation of mono-, di- or tri-methyllysine. Lysine demethylases (KDMs) facilitate the removal of these methyl groups. Lysine specific demethylases (LSDs) target mono- and di-methylated lysines, reducing FAD to $FADH_2$ in the process. Jumonji domain containing demethylases (JMJDs) target mono-, di- and tri-methylated groups, carrying out oxidative decarboxylation and hydroxylation reaction with their associated co-factors, α-KG and $Fe^{2+}$. Arginine methyltransferases (PRMTs) facilitate arginine methylation through the use of SAM/AdoMet as a methyl donor, yielding mono-, asymmetrical di-, and symmetrical di-methylation

residues (Kim and Paik 1965). In the case of lysine methyltransferase (KMT) enzymes, it was determined that these enzymes were able to add a maximum of 3 methyl groups to the $\varepsilon$-nitrogen of the lysine residue (Fig. 1). It was not until several decades later that hints of a functional role for lysine methylation were finally beginning to be resolved, driven through advancements in genetics and molecular biology; notably through the study of gene expression and chromatin biology.

Methylation is the smallest PMT with little steric bulk and not contributing charge. This modification can occur on the side chains of at least 9 out of 20 amino acids, with lysine and arginine the most commonly methylated residues. To help direct the function of methylated protein, methylated lysine/arginine residues can also be recognized by proteins which "read" the adjacent amino acid sequence and the aromatic cage pockets of the methylated residues (Gayatri and Bedford 2014; Lachner et al. 2001); these modular protein domains are collectively referred to as methyl-binding domains (MBDs) and will be discussed periodically throughout the following sections of this chapter. These methyl-dependent interactions are stabilized through the strong attractive forces of the cation and the negative $\pi$-surface of the aromatic ring. Conversely, a non-methylated lysine residue displays acidic residues thus allowing for readers to be selective based on the ratio of aromatic to acidic residues.

Although it was first discovered in 1959, only in recent decades has our knowledge of protein methylation as a PTM has become a more prolific area of discovery. What we know of its properties and significance in biological functions leaves many unanswered questions, which makes it all the more intriguing for researchers to explore.

## 3 Protein Lysine Methylation: A Dynamic Post-Translational Modification

Estimated at over 21,000 different genes, the human genome provides greater proteome diversity through alternative mRNA splicing, giving rise to a number of proteins from a single gene. However, due to the myriad biochemical reactions present within a cell, even more protein diversity is required. Provided through the covalent addition of small moieties to specific amino acids, PTMs provide variations to protein function through modifications in electrostatic and structural properties, in addition to affecting the protein-protein interaction (PPI) that may be associated with the particular protein. As a result, this provides a diverse number of functions and interactions for a single protein, affecting a series of biochemical pathways and reactions within the cell (Duan and Walther 2015).

The nucleosome (i.e., the fundamental subunit of chromatin) is subjected to various PTMs (including phosphorylation, methylation, ubiquitylation, sumoylation, and acetylation) that work together to comprise what is known as the "histone code" for regulation of gene expression. Through various dynamic combinations of these PTMs, each cell can differentiate with unique morphology and biochemistry associated with its function. Among the most abundant of these PTMs, histone methylation has been established to play a critical role in transcriptional activation or repression—with the methylation of histone H3 at lysine 4 (H3K4me) known as a marker of gene activation, and both H3K9me and H3K27me as markers of gene repression (Arrowsmith et al. 2012). The dynamics of histone methylation and its control over gene expression can be reviewed in Hyun et al. (2017). Expanding beyond this 'histone code', sequence similarities between histone and other non-histone substrates have allowed for the novel identification of many other dynamically methylated substrates; in recent years, this has resulted in the methyllysine proteome expanding beyond histone methylation and chromatin regulation (Biggar and Li 2015). This expanded role of methylation has now been shown to include neoplastic growth and development, shedding light on the effects of methylation with regards to apoptosis, hypoxia, cell cycle arrest, and various other stress stimuli.

Predominantly favoring lysine and arginine residues in eukaryotic organisms (Clarke 2013), the addition of a methyl group to an amino acid requires the presence of a methyl-donor. The metabolite SAM (AdoMet), acts in this capacity and donates a methyl ($CH_3$) group to the recipient amino acid in a reaction that is facilitated by a methyltransferase enzyme (Fig. 1). Dependent on the substrate, the methylation reaction may occur in a sequential fashion—adding one, two, or three methylation groups. In the presence of SAM and KMTs, methylation of the ε-amino group on lysine residues is open to mono-, di- or tri-methylation modification. While a similar process occurs with protein arginine methyltransferases (PRMTs) and the guanidine nitrogen of arginine residues, modification level is limited to mono- or di-methylation, with di-methylation expressed as either an asymmetrical or symmetrical modification. The ability to interact and recognize with specific lysine and arginine methylation events is separated within the methyltransferase families, with each class expressing specificity towards particular amino acids—thereby aiding in substrate specificity. As a result of the chemical nature of methylation, no effect has been observed towards protein integrity, as the addition of a methyl group itself provides minor size change and no direct charge difference. However, the modification leads to an increase in lysine basic nature, leading to an increased hydrogen bonding potential and thus increase recognition by other proteins (Hamamoto et al. 2015).

Following the discovery of KDM1A (LSD1), a histone-specific demethylase enzyme, the process of lysine methylation began to be understood as a dynamic modification—a modification that could be readily written (by KMTs) and removed (by KDMs) to regulate function (Shi et al. 2004). Similar to their methyltransferase counterparts, the demethylase family is subdivided into two main classes based on its catalytic domain, mechanism of demethylation, and interacting partners. Discovered as the first group of active demethylases, lysine-specific demethylases (LSDs) mainly target mono- and di-methyl substrates. In contrast, greater substrate diversity is observed with jumonji-domain containing demethylases (JmjCs), further subdividing the family of lysine demethylases (Accari and Fisher 2015). Utilizing α-ketoglutarate (αKG) and $Fe^{2+}$ cofactors, methyllysine binding of JmjC-domain-containing enzymes follows a distinct mechanism involving the formation of a hydrogen bond network between the oxygen atoms of the catalytic residues and the methyl groups of the substrate. This non-classical methyl-binding mechanism allows correct positioning of the tri-methylated substrates to the $Fe^{2+}$ cofactor, allowing ideal reaction conditions and the demethylation reaction to occur. In contrast, LSD enzymes require the presences of a lone electron pair at the methylated amine, opting out the possibility of LSD-catalyzed demethylation of tri-methyl substrates (Hou and Yu 2010).

## *3.1 SET Domain (Class V) Methyltransferases*

Perhaps the most well-studied KMTs, lysine methylation is carried out by a class V methyltransferase that each contain a conserved catalytic SET domain (Fig. 2), consisting of four conserved active motifs GXG, YXG, NHXCXPN and ELXFDY that are composed of eight, curved β-sheet pseudo-knot-like structures (Fig. 2a). During the methyl-transfer reaction, the GXG motif aids in the correct positioning of the methyl donor SAM, while the hydrophobic pocket formed by the NHXCXPN

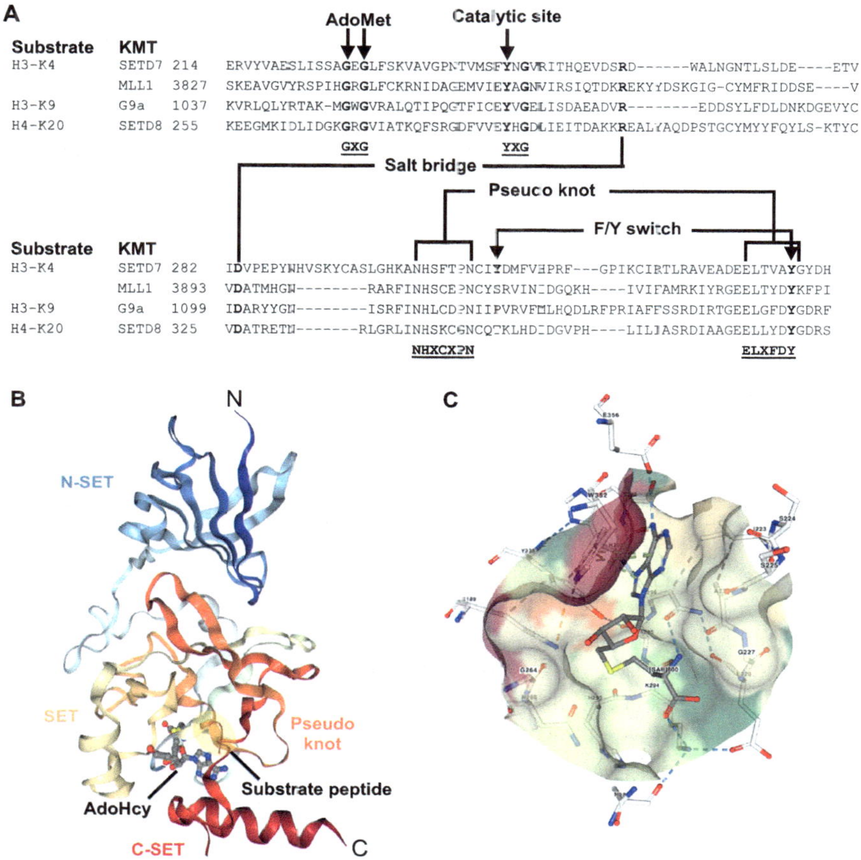


**Fig. 2** SET domain lysine methyltransferases. (**a**) A protein sequence alignment of SET domains from several SET-domain containing lysine methyltransferase enzymes (KMTs). The involvement of residues in binding to AdoMet, catalysis, the structural pseudo knot, an intra-molecular interacting salt bridge, and an F/Y switch controlling whether the product is a mono-, di-, or tri-methylated lysine are indicated. (**b**) Representative structure of SETD7 KMT (3M53.pdb). The N-SET, SET, C-SET, pseudo knot, AdoHcy and substrate binding pockets in SET7/9 are indicated. (**c**) Structure of the co-factor AdoMet/AdoHcy binding site of SETD7 KMT

and ELXFDY motif aid in the recruitment and positioning of the methyl substrate on each side of its methyl transfer channel (Fig. 2b, c). This correct orientation allows for the $SN_2$ reaction (Helin and Dhanak 2013) carried out by the catalytic tyrosine present at the YXG motif (Petrossian and Clarke 2009a), transferring a methyl group from SAM to the ε-amine group of the lysine (Petrossian and Clarke 2009b). While sequence similarity is shared by all SET proteins at both N and C-terminal ends, it is the knot-like structure located at the C-terminal that is hypothesized to determine substrate specificity (Fig. 2b), in addition to the type of methylation carried out by the particular methyltransferase (Petrossian and Clarke 2009a).

## 3.2 Seven β-Strand (Class I) Methyltransferases

Found within all three domains of life (Lanouette et al. 2014), the class I methyltransferases, or seven β-strand (7BS) methyltransferases, comprise a larger superfamily of methyltransferases known to methylate a large variety of substrates such as DNA and RNA, in addition to a variety of amino acids such as arginine, glutamine, aspartate, histidine and lysine (Clarke 2013; Lanouette et al. 2014). The enzymes possess the conserved Rossmann fold characterized as several twisted beta sheets sandwiched between a series of alpha-helices with a C-terminal beta-hairpin (Petrossian and Clarke 2009a). Separated into four motifs (I, Post I, II and III), the first two motifs contain a conserved aspartate amino acid for charge stabilization and proper orientation, while the last two take part in methyl-substrate recruitment and binding (Zhang et al. 2000).

Forming a subdivision within the 7BS methyltransferase family, PRMTs catalyze the methylation of arginine residues resulting in either mono- or di-methylation. Unlike KMTs, arginine di-methylation through PRMTs can result in either symmetric or asymmetric methylation conformations (Smith and Denu 2009). Dependent on the type of PRMT catalyzing the reaction, further division can be made based on the type of di-methylation form that is facilitated by the enzyme. The most common, type I PRMTs, recognize terminal nitrogen atoms facilitating asymmetric di-methylation through the addition of two methyl groups (Kim et al. 2016a, b), or mono-methylation (Debler et al. 2016). In contrast, type II PRMTs carry out symmetric di-methylation though the addition of a single methyl group to terminal nitrogen groups, in addition to mono-methylation (Debler et al. 2016). While type III PRMTs are able to facilitate the production of the mono-methylated arginine (Kim et al. 2016a, b). Whereas types I–III are found in all life lineages, type IV PRMTs are specific to yeast and plants (Debler et al. 2016), catalyzing mono-methylation of internal nitrogen atoms. Similar to lysine methylation, such modifications are often involved in signal transduction, DNA damage and repair, protein interaction, translocation, cellular proliferation, chromatin remodeling and RNA splicing (Kim et al. 2016a, b).

Until recently, the histone-specific methyltransferase, DOT1L, was the only identified eukaryotic 7BS KMT (Singer et al. 1998). However, a number of novel 7BS KMTs have now been discovered. For example, the methyltransferase-like (METTL) protein family, containing METTL21D, METTL22, and METTL21A KMTs, has been found to methylate a number of different non-histone substrates (Falnes et al. 2016). The type II ATPase VCP/p97 has been shown to be tri-methylated by METTL21D (also known as VCPKMT) at lysine K315, negatively regulating VCP/p97 function including ubiquitin-dependent protein degradation (Kernstock et al. 2012). METTL21A has been reported to tri-methylate the HSP70 family (including HSPA1, HSPA8, and HSPA5) of chaperone proteins at an unknown site (s). This methylation event is especially interesting, as has been shown to interfere with the interaction between HSPA8 and alpha-synuclein (Jakobsson et al. 2013)—the main protein aggregate found in Parkinson's disease (Spillantini et al. 1997). Additionally, the association of the DNA/RNA binding protein, KIN17, with chromatin is thought to be influenced through lysine K135 tri-methylation by METTL22 (Cloutier et al. 2014). Together these findings collectively showcase the ability of 7BS KMTs in the regulation of a broad range of non-histone protein targets and implication in diverse cellular functions.

## 3.3 *Lysine-Specific Demethylases*

Comprising the first group reported to function in histone lysine demethylation, lysine-specific demethylases (LSDs) comprise a sub-class of the amine oxidase superfamily (Smith and Denu 2009). Including only two members, LSD1 and LSD2, the pair share a conserved SWIRM domain located at the enzymes N-terminal. These domains form a globular core structure with the two amine oxidase domains (AOD) that contain the substrate and Flavin adenine dinucleotide (FAD) binding sites (Hou and Yu 2010; Liu et al. 2017). Specific to LSD1, a tower domain is formed between the two AODs by two antiparallel helices that function as a binding site for the binding partners CoREST, MTA2/NuRD, AR and AML (Marabelli et al. 2016; Yang et al. 2017). Utilizing a redox reaction, the mechanism results in the formation of an imine intermediate through FAD reduction and methyllysine oxidation. In order to produce the demethylated lysine, the imine intermediate is hydrolyzed to form a hemiaminal that breaks down to form an amine and formaldehyde. However, as the mechanism requires the presence of a methyllysine nitrogen lone electron pair, demethylation is limited to mono- and di-methylated substrates (Smith and Denu 2009). Nevertheless, recognition and binding to tri-methylated substrates persist with greater affinity than favored mono- and di-methyllysine (Hou and Yu 2010). While similar to LSD1 in catalytic mechanism and active structure, LSD2 expresses slight differences in function, structure, and kinetics. Lacking the tower domain, thus expressing no interaction with CoREST, LSD2 modifies its substrate binding core through interaction with the protein NPAC/GLYR1 (Fang et al. 2013; D'Oto et al. 2016). Binding in close proximity to the active site, the putative oxidoreductase allows tighter binding of substrate N-terminal residues through enlargement of the interaction surfaces (Marabelli et al. 2016; Fang et al. 2013). Additionally, LSD2 has been reported to feature a zinc-finger domain (Marabelli et al. 2016) and favors binding to transcribed coding regions, unlike its LSD1 counterpart which favors promoter regions (Chen et al. 2017).

## 3.4 *Jumonji Domain Demethylases*

Part of the 2-oxoglutarate (2OG)—and ferrous iron ($Fe^{2+}$) oxygenase superfamily, the JmjC-KMDs comprise the larger, second family of demethylases (Kooistra and Helin 2012). Sharing the characteristic JmjC domain, consisting of a jellyroll like β-fold homologous to the cupin metalloenzymatic superfamily, the enzymes maintain structural integrity and substrate specificity through a series of structural elements further surrounding the domain. Buried at its core, the domain carries the catalytic domain, in addition to the $Fe^{2+}$ and α-ketoglutarate (α-KG) binding sites and three essential residues, H188, E190, and H276 (as found in KDM4D), which in combination with α-KG, aid in the coordination of $Fe^{2+}$. Unlike their counterparts, JmjC-KDMs carry out an oxidative decarboxylation and hydroxylation reaction with their associated co-factors, α-KG, and $Fe^{2+}$ (Fig. 1; Klose et al. 2006; Chen et al. 2006a, b). This leads to the production of an unstable hemiaminal intermediate, which then breaks down to produce the demethylated substrate and formaldehyde. As the mechanism lacks the requirement of lone pair electrons, demethylation of tri-methylated substrates by the majority of JmjC enzymes is possible. During the reaction, binding to the methyllysine substrate occurs through a distinct mechanism that involves the formation of a hydrogen bond network between the oxygen atoms of the catalytic residues and the methyl groups of the substrate (Hou and Yu 2010). This non-classical methyl-binding domain allows correct positioning of the tri-methylated substrate to the $Fe^{2+}$ cofactor, allowing ideal conditions for reaction. Due to the reduced size of the mono- and di-methylated substrates, the formed hydrogen bonds separate the $Fe^{2+}$ from the methyl groups limiting catalytic reaction. However, through rotational movement of the di-methylated substrates, interaction with $Fe^{2+}$ becomes possible, allowing the demethylation reaction to occur. In the case of mono-methylated substrates, rotational movement produces no changes in orientation preventing their demethylation by some family members, such as KDM4A (Ng et al. 2007; Cloutier et al. 2014). However, due to steric hindrance from space limitation at the active core, other family members such as PHF8 and KDM7A express substrate specificity towards di-methylated lysine solely, while similar limitations as those associated with KDM4A govern their recognition of mono-methylated substrates. These slight differences in the JmjC core not only govern the substrate specificity of the enzymes, but also allow for their subdivision based on homology of the catalytic core (Horton et al. 2010; Yang et al. 2010). In addition to the characteristic JmjC domain, the majority of members possess other functional domains, including MBDs such as PHD, Tudor, as well as protein interaction F-box and TPR domains and DNA binding domains BRIGHT/ARID and $Zn^{2+}$ fingers that further aid in substrate specificity, family subdivision and recruitment of the enzymes to specific loci (Klose et al. 2006).

Although the discovery of the KDMs helped establish lysine methylation as a dynamic process (Biggar and Li 2015), reports of arginine demethylases (PRDMs) are limited and often controversial. While in recent years the JmjC family member JMJD6 has been reported to express PRDM activity (Poulard et al. 2014); such functions for the enzyme remain unconfirmed as equal reports express lack of PRDM activity (Walport et al. 2016).

## 4 Non-Histone Methylation: Functional Methylation and Regulation of Cellular Processes

Kenneth Murray for the first-time reported lysine methylation on bovine histone proteins of mammals (Murray 1964). Following this discovery, most research on lysine methylation had followed suit and focused on histone methylation due to its clear importance in chromatin biology and gene regulation. Until recently, research on non-histone lysine methylation was limited as there were no strategies to identify lysine methylation across the entire proteome. Starting in 2013, several research groups developed techniques to identify methylated proteins (Cao and Garcia 2016; Carlson and Gozani 2016; Liu et al. 2013). These proteomic studies have each revealed several hundreds of new methylated proteins and lysine residues. As a result, there is now an abundance of evidence demonstrating that, in addition to histones, lysine methylation also occurs on various non-histone proteins that are important for signal transduction events and epigenetic regulation of transcription and chromatin in eukaryotes.

Since its discovery in 1959, the role of histone and non-histone methylation has not only advanced our understanding of cellular regulation, but has also provided new target substrates for the development of therapeutics (Arrowsmith et al. 2012). The last decade has seen great advancement in substrate identification, enzyme characterization and functional characterization, such as those methylation events associated with the functional regulation of the p53 tumour suppressor protein (Scoumanne and Chen 2008). However, further research is still necessary to gauge the breadth of the methyllysine proteome and the cellular roles that it fulfills.

While the role of lysine methylation in histones was already being elucidated, in 1998 the discovery of methylation of RNA binding proteins (RBPs) began to expand the scope of protein methylation (Brahms et al. 2001). The methylation of RBPs was shown to have a regulatory role in ribonucleoprotein (RNP) assembly, pre-mRNA splicing, and mRNA stability. An important function of lysine methylation in the p53 tumor suppressor protein has been observed (Chuikov et al. 2004). In particular, SET domain-containing protein 7 (SETD7)-dependent methylation of lysine in p53 resulted in enhanced transcriptional activity, nuclear stability as well as apoptosis (Fig. 3). Subsequent studies revealed that p53 could function as an activator or repressor in response to the methylation of four other lysine and three arginine residues (Huang et al. 2006). In 2007, it has been found that p53 could also be demethylated and is a reversible protein modification. Specifically, LSD1 demethylated lysine K370 di-methylation thereby disrupting the methyl reader abilities of p53-binding protein 1 (53BP1) and decreasing its transcriptional activity (Huang et al. 2007). This discovery started a segue into the dynamic lysine methylation of non-histone proteins, a PTM with functional implications existing beyond epigenetics and chromatin organization.

Although the methylation of histone proteins have comprised the majority of lysine methylation research, the lysine methylation of non-histone proteins is also being realized to facilitate critical roles in the regulation of cellular stress, cell proliferation, and angiogenesis. A prototypical example includes the methyl-regulation of non-histone substrates by the SMYD3 KMT enzyme, which has been reported to have a significant role in oncogenic cell proliferation. The first insights into SMYD3 methylation of non-histone proteins were reported in 2007 (Kunizaki et al. 2007). They revealed that SMYD3 was able to methylate the vascular endothelial growth factor receptor 1 (VEGFR1) at lysine K831, a conserved residue located within the tyrosine kinase domain and proposed to regulate VEGFR1

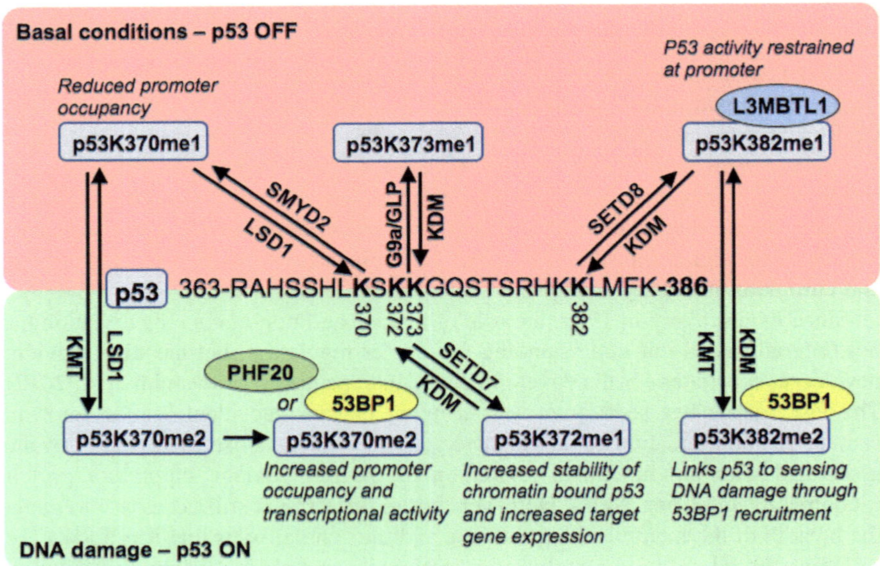


**Fig. 3** Control of p53 signaling network through dynamic lysine methylation. Lysine methyltransferase enzymes (KMTs) methylate p53 at several C-terminal locations, acting to differentially activate or inhibitor p53 transcriptional activity and/or signaling. Red shading indicates methylation events that are known to negatively influence p53 activity, whereas green indicated methylation events that are currently thought to promote p53 signaling in response to periods of DNA damage

kinase activity. Following this discovery, it was shown that SMYD3 mediated the methylation mitogen-activated protein-3 kinase2 (MAP 3K2) at lysine K260 to promote ERK1/2 signaling (Mazur et al. 2014). These events increased Ras signaling leading to increased cell proliferation and pancreatic tumorigenesis in an SMYD3-dependent manner.

## *4.1 Control Over p53 Transcriptional Activity by Combinatorial Methylation Signals*

The complexity of non-histone protein methylation in the regulation of protein function can be highlighted by the regulation of p53 by SET domain-containing KMTs (Fig. 3). The p53 tumor suppressor is currently known to be differentially regulated by a number of different KMT and KDM proteins (West and Gozani 2011). The function of p53 is controlled through at least four C-terminal lysine methylation sites, including K370, K372, K373, and K382. Collectively, these methylation events are controlled through the combined action of five KMTs, which include mono-methylation by SETD7 (K372me1) (Chuikov et al. 2004), SETD8 (K382me1) (Shi et al. 2007) and

SMYD2 (K370me1) (Huang et al. 2006), and di-methylation by G9a/GLP (K373me2) (Huang et al. 2010) KMT enzymes. Importantly, there still remains several methylation events with the direct implication in the regulation of p53 function with yet to be identified KMTs. For example, the di-methylation of p53 at K370 (i.e., p53K370me2) creates an interaction site for the tandem Tudor MBDs within 53BP1, increasing p53 promoter occupancy and increasing p53-dependent transcript of target genes. Although the KMT responsible for this di-methylation event at K370, it can be dynamically removed by the demethylase action of the LSD1 KDM, returning the K370 site back to mono-methylation status and resetting p53 activity by preventing the di-methylation-dependent 53BP1 association (Huang et al. 2007). This creates a simple 'switch'-like system that yields control of protein activity through the opposing action of KMT and KDM enzymes, however, this system fails to present the complexity of the overall methyllysine-regulatory system that acts to influence p53 transcriptional activity and cell fate.

Expanding upon the example of dynamic p53K370me2 methylation in the control of p53 transcriptional activity, several other methylation sites within p53 also exert regulatory influence over p53. For example, the mono-methylation of K370 by SMYD2 has been shown to be a methylation status correlated with reduced p53 promoter occupancy and lower p53 activity (Huang et al. 2006). Similarly, the mono-methylation modification imparted by SETD8 at K382 has been shown to restrain transcriptional activity at promoter sites through the mono-methylation-dependent interaction with the MBT MBD domains (3xMBT) of the L3MBTL1 protein (West et al. 2010). In contrast, nuclear mono-methylation at K372 by SETD7 has been shown to enhance activity through the stabilization of chromatin-bound p53 and has also been linked with the promotion of p53 acetylation in response to periods of cellular DNA damage (Chuikov et al. 2004). Lastly, the di-methylation of K373 by G9a/Glp has been classified as an inhibitory mark, reducing p53 activity in a methylation-dependent manner (Huang et al. 2010). Overall, p53 is an intriguing example of how dynamic lysine methylation events can exert regulatory control, how these methyl-modifications are sensed by MBD-containing proteins (such as 53BP1 and L3MBTL1), and how the act to modulate p53 function.

## 4.2 *HIF Regulation by Dynamic Lysine Methylation*

Research in recent years have begun to outline an important role for lysine methylation in tumorigenesis (Hamamoto et al. 2015), however much remains unknown regarding the mechanisms of which methylation mediates the initiation and progression of such diseases. As the microenvironment of malignant solid tumors is characterized by insufficient oxygen delivery, investigation of oxygen deprivation on the regulation of disease-relevant methylation events is essential for developing enhanced combination therapeutic strategies. Identified as a biomarker of a number of different carcinomas, interest in understanding how the LSD1 KDM contributes to cancer development has gained over the years as we continue to uncover its repertoire of non-histone substrates (Nagasawa et al. 2015). Following di-methylation at lysine K271, the receptor of activated protein C kinase 1 (RACK1) mediates ubiquitination-mediated degradation of the low oxygen sensor, hypoxia-inducible factor 1 alpha (HIF-1α), in an oxygen-independent manner through the RACK1-Hsp90 pathway (Yang et al. 2017). This di-methylation modification at K271 mediates the RACK1-HIF-1α interaction, an interaction that facilitates HIF-1α degradation. However, under hypoxic conditions LSD1 is reported to demethylate the K271 residue in RACK1, diminishing the methyl-dependent interaction between RACK1 and HIF-1α. In contrast, the activity of LSD1 deceases under chronic hypoxia as the biosynthesis of the FAD co-factor decreases. Characteristic in triple-negative breast cancer (TNBC), is a greater LSD1 gene expression, in addition to an altered FAD biosynthetic gene, has been shown to provide insight into patient prognosis. As a result, TNBC patients with increased LSD1 activity were found to correlate with a poor prognosis (Nagasawa et al. 2015; Marabelli et al. 2016).

Similar to the rather complex methyl-regulation that associated with p53 function, various lysine residues on HIF-1α have been found to be subject to dynamic methylation and demethylation, several with documented impact on HIF-1α cellular function. For example, both the SETD7 KMT and the LSD1 KDM have been found to work together to mediate the methylation of lysine K32 and K391 (Liu et al. 2015; Kim et al. 2016a, b; Lee et al. 2017). Occurring primarily within the nucleus, the K32 methylation is subjected to increase methylation under normoxia and prolonged hypoxia, while increased demethylation is observed during the early hypoxic transition (Liu et al. 2015). This methylation site has been proposed to regulate HIF-1 stability under normoxia and during late hypoxia, when activity is minimal. Speculated as a recruitment signal for an unknown E3 ligase, this methylation event is thought to function as a fine-tuning mechanism modulating "leaky pools" of remaining HIF-1 proteins under normoxia and late hypoxia. As such, it is theorized that remaining pools of HIF-1α that avoid cytosolic degradation undergo SETD7-mediation methylation once localized to the nucleus, leading to their ubiquitination induced proteasomal degradation (Kim et al. 2016a, b; Baek and Kim 2016).

## 4.3 Dynamic Lysine Methylation of FOXO Protein

The activity of the FOXO subfamily of transcriptional factors has been shown to be largely mediated through a number of different PTMs. In addition to the currently known regulatory phosphorylation, acetylation, ubiquitination, and arginine methylation PTMs that are associated to occur within the FOXO3a protein, methylation of lysine K270 by SETD7 has been shown to mediate oxidative stress-induced apoptosis (Xie et al. 2012). Interestingly, once methylated by SETD7, FOXO3a does not show any change in protein stability, localization, or other PTMs/interactions associated with its normal signaling pathways (i.e., PI3K/Akt) (Zhu 2012). Instead, methylation by SETD7 has been found decreasing the DNA-binding capability of FOXO3a, thus preventing expression of its target gene, *Bim*, a BH3-only protein. Due to the conserved nature of the K270 methylation site within the FOXO family, other family members, such as FOXO1, have also been shown to undergo SETD7 mediated methylation at their respective corresponding lysine residues; however, the functional outcome of this conserved methylation event has yet to be reported (Xie et al. 2012).

### *4.4 DNA Damage Repair Signaling Cascade*

As the most important bio-macromolecule in the cell, DNA is subject to damage induced by ionizing radiation, UV and other chemical environmental agents which induce double-strand breaks (DSBs). If this damage is not repaired in a timely fashion, this damage can signal cellular autophagy (controlled digestion of damaged organelles within a cell), apoptosis (programmed cell death), aging and can result in the progression of cancer. Therefore, upon the detection of DNA damage, it is necessary for the cell to immediately identify any DSB and initiate appropriate repair mechanisms. To accomplish this, eukaryotes have two major pathways to repair damaged DNA: (a) homologous recombination repair (HRR) and (b) non-homologous end joining (NHEJ) (Ciccia and Elledge 2010). The tumor suppressors 53BP1 and BRCA1 are the two factors that are enriched at sites of DSBs and are emerging as pivotal regulators of repair by either NHEJ and HRR, respectively. DSBs that occur within G1 phase of the cell cycle are repaired by NHEJ. Repair is initiated through the recruitment of the Ku70-Ku80 heterodimer, followed by ATM-related DNA-dependent protein kinase catalytic subunit (DNA-PKcs). Importantly, DNA-PKcs is responsible for maintaining the broken DNA ends within close proximity to each other, which is beneficial for recruiting end processing factors followed by re-ligation by DNA ligase complex. Previous studies have shown that DNA-PKcs undergoes active lysine methylation at K1150, K2746, and K3248 in response to DNA damage, and that loss of these methylation events impact repair capacity (Liu et al. 2013). Furthermore, the methyllysine interactions of the chromo MBD of heterochromatin protein (HP)1β are enriched with proteins involved in DNA damage repair (DDR), suggesting a central role for HP1β and methyl-dependent interactions in DDR. In this model, the HP1β chromo MBD interacts with DNA-PKcs in a methyllysine-dependent manner and regulates DNA-PKcs function in response to DNA damage.

On the other hand, HRR pathway is activated in response to DNA damage on S/G2 phase of the cell cycle. MRE11-Rad51-NBS1 (MRN) complex binds to the broken DNA ends followed by recruitment of CtIP (C-terminal binding protein interacting protein) and several nuclease machines to promote high throughput process of DNA end resection. Replication protein A (RPA) coat the generated 3′ ssDNA following resection. RAD51 displaces RPA to form a RAD51-ssDNA nucleofilament induced by BRCA1-PALB2-BRCA2 complex and RD51 paralogs. Finally, RAD51 nucleofilament searches for the complementary DNA template in the genome to synthesize and synapse to form a mature recombination product. HRR pathway is critically important for the cells to regulate normal cell behaviour. Mutations in the signature proteins of the pathway (BRCA1/2-PALB2-RAD51), fuel cancer and chemoresistance where cell loses the choice between the two pathways and continue repair with error prone NHEJ.

Emerging evidence indicate that the lysine methylation of histone and non-histone proteins can play important role in determining the repair pathway of choice, whether the cell should undergo HRR or NHEJ repair (Chen and Zhu 2016). Differentially methylated lysine on histone and non-histone proteins are currently thought to serve as the docking sites for HRR or NHEJ-related proteins, influencing the signaling of a particular repair pathway. For example, tri-methylated H3K36 is required for HR repair, while di-methylated H4K20 have been shown to recruit the 53BP1 for NHEJ repair (Ng et al. 2009; Freitag 2017). In recent years, it has become increasingly clear that methylation entails remodeling chromatin, and plays a major role in regulating DDR singling cascade which is quite obvious in disease like cancer.

## 4.5 *Lysine Methylation and Disease*

Lysine methylation on histone tails is a common PTM and is pivotal in the regulation of chromatin structure and gene transcription, spanning from growth and proliferation in physiological and pathological conditions such as cancer and neurodegenerative diseases (Esteller 2007; Greer and Shi 2012; Hamamoto et al. 2015). For example, an up-regulated expression of SMYD2 in oesophageal squamous cell carcinoma and bladder cancer cells has been observed (Cho et al. 2012), and further, an overexpressed SMYD3 in breast carcinoma has been shown to correlate with tumor proliferation (Hamamoto et al. 2006). Additionally, the KMT G9a is overexpressed in hepatocellular carcinoma and contributes to the invasiveness of lung and prostate cancer (Casciello et al. 2015). Correspondingly, lysine methylation has been reported to influence oncogenic pathways and hence provides a rationale for the involvement of KMTs in cancer.

SETD8 (also known as KMT5A), member of the SET domain family known to catalyze the mono-methylation of histone H4K20 (Nishioka et al. 2002). This methylation event is believed to be necessary in the methylation-dependent recruitment of signalling proteins like 53BP1 to site of double-strand DNA breaks (Dulev et al. 2014), or state of chromatin compaction (Lu et al. 2008; Jørgensen et al. 2007). SETD8 has also been reported to have implications in breast cancer through the dynamic methylation of Numb protein at lysine K158 and K163 (Dhami et al. 2013). Normally, the Numb protein exhibits tumor-suppressive ability through a direct with p53, stabilizing and promoting p53 transcriptional activity and cellular apoptosis. Interestingly, this stabilizing interaction with p53 is dynamically disrupted through the tandem methylation of Numb at K158 and K163 within its phosphotyrosine binding domain (PTB); the domain responsible for recruitment and p53 binding. Following the treatment of breast cancer cells with a chemotherapeutic agent (doxorubicin), the expression of SETD8 was found to be significantly reduced, decreasing Numb methylation and enhancing Numb-p53 mediated cellular apoptosis. Collectively, this work demonstrated SETD8-mediated Numb-p53 interaction as an important regulatory axis in breast cancer, and further highlighting one of the currently known roles that methyllysine facilitates in normal and disease cell biology.

Intriguingly, lysine methylation also has been proposed to play a role in bacterial pathogenicity. Vaccination efforts against typhus' agent *Rickettsia typhi* target the immunodominant antigen OmpB (Chao et al. 2004, 2008). The chemical methylation of lysine residues re-establishes serological reactivity of the OmpB fragment on a recombinant peptide (Chao et al. 2004). *Mycobacterium tuberculosis* adhesions (HBHA and LBP) important for adhesion to host cells are also heavily methylated (Biet et al. 2007; Delogu et al. 2011). Contemporary, methylation of *P. aeruginosa* Ef-Tu at K5 has been reported to mimic the ChoP epitope of human platelet-activating factor (PAF) further allowing association with PAF receptor and contributes to bacterial invasion and pneumonia onset (Barbier et al. 2013).

Taken together such findings demonstrate the infancy of the lysine methylation field on methyl-regulation function outside of epigenetics and chromatin biology. As a result, a number of questions still remain to be answered. For example, how many substrates do methyl-modifying enzymes regulate, and how expansive are the lysine methylation proteome and the cellular processes that it influences?

## 5 Methyllysine Proteomics: Methods to Discover Lysine Methylated Protein

Within the last decade of lysine methylation research, we have begun to define new and complex roles for this modification. Such cellular functions for this modification now include the facilitation of crosstalk between signaling cascades and connecting cellular signaling to nuclear effectors and chromatin regulation. Despite the rapid growth in our understanding of the function of lysine methylation, the field of lysine methylation has historically experienced limited growth as a result of a lack of suitable identification technology. Arginine has not experienced the same stunted growth as the identification of arginine methylation sites has been facilitated through the use of methylarginine-specific antibodies, enriching for arginine methylated proteins to be mapped and identified by mass spectrometry (Guo et al. 2014). In contrast, it has been difficult to develop suitable methyllysine-specific antibodies that are able to enrich for the lysine methylation modification without a high degree of non-specific interaction for unmodified protein. As a result, the identification and mapping of new lysine methylation sites have not undergone the same growth as that of arginine methylation.

### *5.1 Immunoaffinity-Based Annotation of Lysine Methylation Events*

Initially, efforts towards the global identification of lysine-methylated proteins utilized methyl-specific antibodies for the initial enrichment of methylated peptides prior to mass spectrometry-based detection and analysis. Although the first studies utilizing this approach were only able to identify several lysine methylation sites on histones H3 and H4, recent advancements in the development antibodies that display higher specificity towards methyllysine have begun to overcome the technical issues that previously plagued enrichment and identification. As methylation exists as a relatively small uncharged protein modification, it has been difficult to develop antibodies that do not suffer from low affinity and poor specificity, or that do not maintain specificity for the amino acid sequences surrounding the modified lysine. To overcome these technical issues, several labs have worked towards the development of methyllysine-specific antibodies with affinity and specificity appropriate to be used in methyllysine identification by immunoaffinity purification followed by tandem mass spectrometry (IP-MS/MS) (Fig. 4a). For example, one study utilized a panel of antibodies each specific against either mono-, di-, or tri-methylated lysine (Cao et al. 2013). They used these antibodies for immunoaffinity of trypsin-digested lysine-methylated peptides to be used for mass spectrometry, identifying 323 mono-methylation, 127 di-methylation, and 102 tri-methylation lysine modification sites within 413 proteins. Importantly, this study documented that it is possible to develop and utilize methyllysine-specific antibodies to be used in the IP-MS/MS identification of new lysine methylation sites.

## 5.2 *Methyl-Binding Domains for the Identification of Methyl-Directed Protein Interactions*

Although the use of antibody-based enrichment methods has begun to provide significant growth in the number of lysine-methylated sites that exist in the human proteome, a number of studies have begun to use MBDs for methyllysine enrichment. The use of these MBDs (such as the chromo, PHD, MBT, PWWP, WDR and Tudor domains) provide a means of natural methyl-specific affinity as a mechanism to enrich for lysine-methylated peptides prior to identification by mass spectrometry (Fig. 4b). This method has been successful in the mapping of the methyllysine proteome on a large scale by several labs (Liu et al. 2013; Carlson et al. 2014). As methyl-specific antibodies cannot provide information of direct physical interactions that may occur in the cell, this approach has been utilized for the mapping of methyl-depended complexes with MBDs, a collection of interactions referred to as the methyl-interactome (Liu et al. 2013). For example, Liu and colleagues use the chromo MBD from the HP1b protein to identify 29 methylated proteins. The associated HP1b methyl-interactome included a group of 14 proteins involved in the DNA damage response (including the aforementioned methylation of DNA-PKcs at lysine K1150 necessary for DNA-damage repair from Sect. 4.4), a cluster of 39 proteins involved in RNA splicing, and a group of eight ribosomal proteins (Liu et al. 2013). Another study successfully utilized the triple modular MBT domains (3xMBT) from the L3MBTL1 protein in an attempt to purify methyllysine modified proteins with

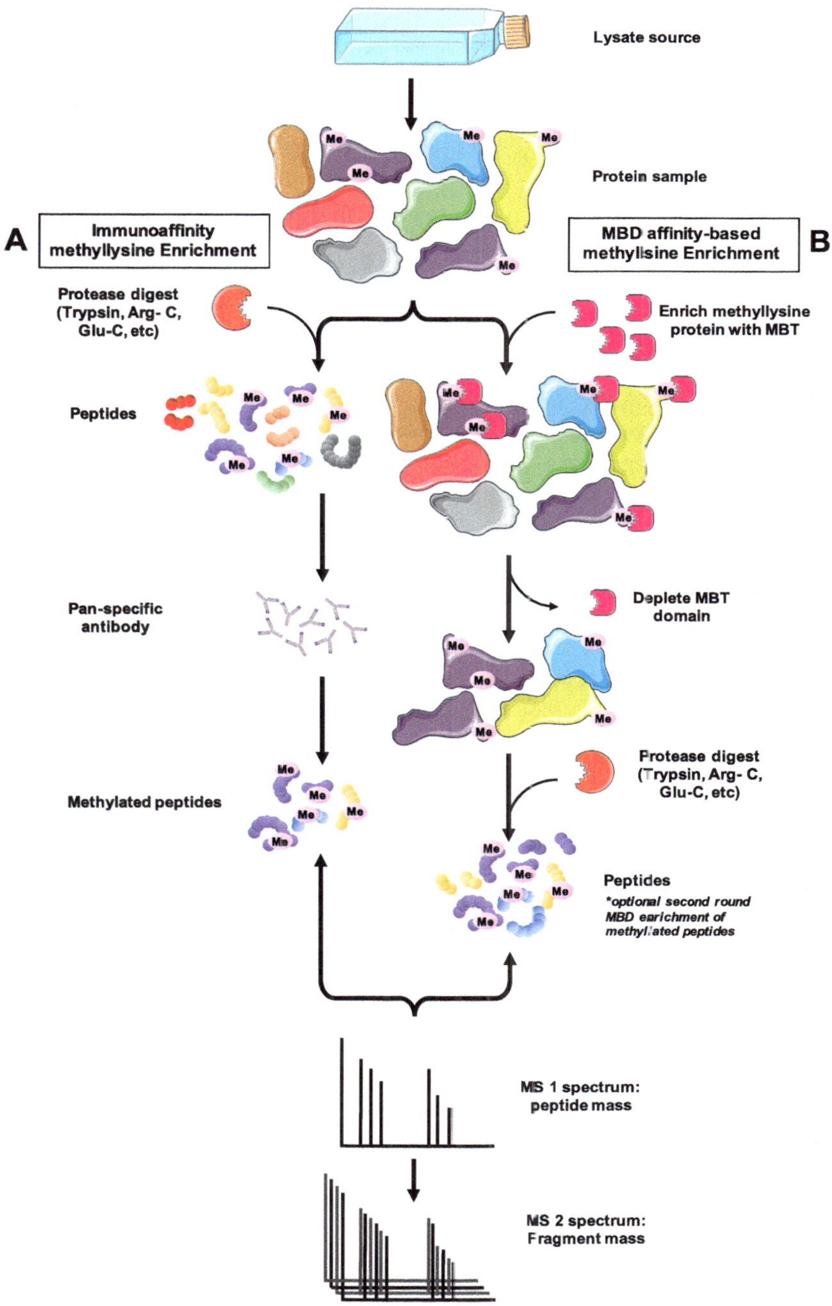


**Fig. 4** Identification of new methylated proteins. (**a**) Following protein isolation, peptide fragments are obtained through a digestion by specific proteases. Methylation-specific antibodies are then used to isolated methylated peptide fragments from their unmodified counterparts, and subjected to liquid chromatography-tandem mass spectrometry (LC-MS/MS) for identification.

little specificity for amino acid sequence neighboring the modified lysine residue (Carlson et al. 2014). Collectively, these two studies demonstrate the utility of using MBDs for the enrichment and annotation of lysine methylation sites, providing a deeper understanding of how methylation can integrate into broader biological processes through methyl-dependent protein interactions with MBD.

## *5.3 Computational Predictions the Methyllysine Proteome*

As previously mentioned, one of the largest challenges placed on the discovery of lysine methylated proteins, has been limitations in identification technology. However, the development of new *in silico* prediction resources hold the promise of aiding in the initial annotation of methyllysine on a proteome-wide scale. Although several affinity strategies that utilize commercial antibodies and natural MBDs (see above) have been remarkably successful in the identification of new lysine methylation events when coupled with mass spectrometry, these approaches are inherently biased towards the binding specificity of the protein used for enrichment. *In silico* prediction methods help to overcome this issue by predicting methylation events based on general underlying characteristics of known modified proteins.

During the past decade several attempts for developing methyllysine and methylarginine predictors have appeared in the scientific literature (Chen et al. 2006a, b; Hu et al. 2011; Qiu et al. 2014; Shao et al. 2009; Shi et al. 2012, 2015; Shien et al. 2009). These studies developed their models from the information of methylated sites extracted, mostly, from databases such as UniProtKB, PhosphoSite-Plus, and PubMed, gathering in total few hundreds of methylated sites. Regrettably, a certain number of deficiencies in the preparation of these datasets have been identified (Qiu et al. 2014; Shi et al. 2015), limiting the reliability in some of the currently available predictors. Unfortunately, in almost all cases these predictors omit the effect of the existing imbalance between known methylation sites and those that are assumed not to be subject to methylation during the evaluation of the models. Such an approach leads to optimistic estimations of the errors in the larger class (not-methylated sites) consequently increasing the precision of their outcomes in the validations. Such balanced datasets during evaluation (validation) do not match the challenging imbalanced scenario that these methods have to face when are used in real-life datasets like the entire Human proteome.

In addition to this issue, these predictors have an inherent limitation that undermine their applicability and trust and should be highlighted. Existing predictors have been to an average of only 200 non-redundant methyllysine sites for building and

**Fig. 4** (continued) (**b**) Methylated proteins can also be isolated by affinity purification through the use of specific, naturally occurring, modular methyl-binding domains (MBD). Methylated peptide fragments are obtained through protease digest and identified through LC-MS/MS

assessing their models, when the expected diversity of the sequence fragments carrying a methylated site can undoubtedly not be represented with such a few numbers of examples. For reference, as of 2018 the PhosphoSite database reports greater than 2000 Human methyllysine modification sites. The development of reliable *in silico* predictions of methylation does hold significant promise in its ability to annotate an initial enrichment dataset that could to be used to guide targeted mass spectrometry efforts. Future work will help determine how thoroughly MS identification experiments are able to probe the methyllysine proteome. It will also be critical to establish whether these identification technologies, either individually or used in conjunction with each other, will be able to provide a systems-level understanding of how lysine methylation impacts protein signaling, and how dynamic methylation acts to regulate protein, and cellular, function.

## 6 Summary

Although studies to date have already established that lysine methylation is a prevalent PTM occurring on non-histone substrates with diverse functional roles, it has become clear that we have only scratched the surface when it comes to delineating the complete breadth of the methyllysine proteome and the full spectrum of cellular and developmental processes that it regulates. Just how large is the methyllysine proteome? How is lysine modification dynamically controlled and coordinated in response to cell stimuli? These are critical questions that will be likely addressed in the near future, knowledge of which will provide a greater understanding of protein regulation and of the inner workings of cell biology.

**Acknowledgements** This work was supported by an NSERC discovery grant to K.K. Biggar.

## References

Accari SL, Fisher PR (2015) Emerging roles of JmjC domain-containing proteins. Int Rev Cell Mol Biol 319:165–220

Allfrey VG, Faulkner R, Mirsky AE (1964) Acetylation and methylation of histones and their possible role in the regulation of RNA synthesis. Proc Natl Acad Sci USA 51:786–794

Ambler R, Rees M (1959) Epsilon-N-methyl-lysine in bacterial flagellar protein. Nature 184:56–57

Anderson D, Koch CA, Grey L et al (1990) Binding of SH2 domains of phospholipase C gamma 1, GAP, and Src to activated growth factor receptors. Science 250:979–982

Arrowsmith CH, Bountra C, Fish PV et al (2012) Epigenetic protein families: a new frontier for drug discovery. Nat Rev Drug Discov 11:384–400

Baek SH, Kim K (2016) Regulation of HIF-1α stability by lysine methylation. BMB Rep 49:245–246

Barbier M, Owings JP, Martínez-Ramos I et al (2013) Lysine trimethylation of EF-Tu mimics platelet-activating factor to initiate *Pseudomonas aeruginosa* pneumonia. MBio 4 e00207–e00213

Biet F, de Melo Marques MA, Grayon M et al (2007) *Mycobacterium smegmatis* produces an HBHA homologue which is not involved in epithelial adherence. Microbes Infect 9:175–182

Biggar KK, Li SS (2015) Non-histone protein methylation as a regulator of cellular signalling and function. Nat Rev Mol Cell Biol 16:5–17
Brahms H, Meheus L, de Brabandere V et al (2001) Symmetrical dimethylation of arginine residues in spliceosomal Sm protein B/B' and the Sm-like protein LSm4, and their interaction with the SMN protein. RNA 7:1531–1542
Cao XJ, Garcia BA (2016) Global proteomics analysis of protein lysine methylation. Curr Protoc Protein Sci 86:24.8.1–24.8.19
Cao XJ, Arnaudo AM, Garcia BA (2013) Large-scale global identification of protein lysine methylation in vivo. Epigenetics 8:477–485
Carlson SM, Gozani O (2016) Nonhistone lysine methylation in the regulation of cancer pathways. Cold Spring Harb Perspect Med 6:11
Carlson SM, Moore KE, Green EM et al (2014) Proteome-wide enrichment of proteins modified by lysine methylation. Nat Protoc 9:37–50
Casciello F, Windloch K, Gannon F et al (2015) Functional role of G9a histone methyltransferase in cancer. Front Immunol 6:487
Chao CC, Wu SL, Ching WM (2004) Using LC-MS with de novo software to fully characterize the multiple methylations of lysine residues in a recombinant fragment of an outer membrane protein from a virulent strain of *Rickettsia prowazekii*. Biochim Biophys Acta 1702:145–152
Chao CC, Zhang Z, Wang H et al (2008) Serological reactivity and biochemical characterization of methylated and unmethylated forms of a recombinant protein fragment derived from outer membrane protein B of *Rickettsia typhi*. Clin Vaccine Immunol 15:684–690
Chen Y, Zhu WG (2016) Biological function and regulation of histone and non-histone lysine methylation in response to DNA damage. Acta Biochim Biophys Sin 48:603–616
Chen H, Xue Y, Huang N et al (2006a) MeMo: a web tool for prediction of protein methylation modifications. Nucleic Acids Res 34:W249–W253
Chen Z, Zang J, Whetstine J et al (2006b) Structural insights into histone demethylation by JMJD2 family members. Cell 125:691–702
Chen L, Vasilatos SN, Qin Y et al (2017) Functional characterization of lysine-specific demethylase 2 (LSD2/KDM1B) in breast cancer progression. Oncotarget 8:81737–81753
Cho HS, Hayami S, Toyokawa G et al (2012) RB1 methylation by SMYD2 enhances cell cycle progression through an increase of RB1 phosphorylation. Neoplasia 14:476–486
Chuikov S, Kurash JK, Wilson JR et al (2004) Regulation of p53 activity through lysine methylation. Nature 432:353–360
Ciccia A, Elledge SJ (2010) The DNA damage response: making it safe to play with knives. Mol Cell 40:179–204
Clarke SG (2013) Protein methylation at the surface and buried deep: thinking outside the histone box. Trends Biochem Sci 38:243–252
Cloutier P, Lavallée-Adam M, Faubert D et al (2014) Methylation of the DNA/RNA-binding protein Kin17 by METTL22 affects its association with chromatin. J Proteome 100:115–124
D'Oto A, Tian QW, Davidoff AM et al (2016) Histone demethylases and their roles in cancer epigenetics. J Med Oncol Ther 1:34
Debler EW, Jain K, Warmack RA et al (2016) A glutamate/aspartate switch controls product specificity in a protein arginine methyltransferase. Proc Natl Acad Sci USA 113:2068–2073
Delogu G, Chiacchio T, Vanini V et al (2011) Methylated HBHA produced in *M. smegmatis* discriminates between active and non-active TB disease among the QFT-IT-responders. Eur Respir J 38:1901
Dhami GK, Liu H, Galka M et al (2013) Dynamic methylation of Numb by Set8 regulates its binding to p53 and apoptosis. Mol Cell 50:565–576
Duan G, Walther D (2015) The roles of post-translational modifications in the context of protein interaction networks. PLoS Comput Biol 11:1–23
Dulev S, Tkach J, Lin S et al (2014) SET8 methyltransferase activity during the DNA double-strand break response is required for recruitment of 53BP1. EMBO Rep 15(11):1163–1174

Esteller M (2007) Cancer epigenomics: DNA methylomes and histone-modification maps. Nat Rev Genet 8:286–298
Falnes PØ, Jakobsson ME, Davydova E et al (2016) Protein lysine methylation by seven-β-strand methyltransferases. Biochem J 473:1995–2009
Fang R, Chen F, Dong Z et al (2013) LSD2/KDM1B and its cofactor NPAC/GLYR1 endow a structural and molecular model for regulation of H3K4 demethylation. Mol Cell 49:558–570
Freitag M (2017) Histone methylation by SET domain proteins in fungi. Annu Rev Microbiol 71:413–439
Gayatri S, Bedford MT (2014) Readers of histone methylarginine marks. Biochim Biophys Acta 1839:702–710
Greer EL, Shi Y (2012) Histone methylation: a dynamic mark in health, disease and inheritance. Nat Rev Genet 13:343–357
Guo A, Gu H, Zhou J et al (2014) Immunoaffinity enrichment and mass spectrometry analysis of protein methylation. Mol Cell Proteomics 13:372–387
Hamamoto R, Silva FP, Tsuge M et al (2006) Enhanced SMYD3 expression is essential for the growth of breast cancer cells. Cancer Sci 97:113–118
Hamamoto R, Saloura V, Nakamura Y (2015) Critical roles of non-histone protein lysine methylation in human tumorigenesis. Nat Rev Cancer 15:110–124
Helin K, Dhanak D (2013) Chromatin proteins and modifications as drug targets. Nature 502:480–488
Horton JR, Upadhyay AK, Qi HH et al (2010) Enzymatic and structural insights for substrate specificity of a family of jumonji histone lysine demethylases. Nat Struct Mol Biol 17:38–43
Hou H, Yu H (2010) Structural insights into histone lysine demethylation. Curr Opin Struct Biol 20:739–748
Hu LL, Li Z, Wang K et al (2011) Prediction and analysis of protein methylarginine and methyllysine based on Multisequence features. Biopolymers 95:763–771
Huang J, Perez-Burgos L, Placek BJ et al (2006) Repression of p53 activity by Smyd2-mediated methylation. Nature 444:629–632
Huang J, Sengupta R, Espejo AB et al (2007) p53 is regulated by the lysine demethylase LSD1. Nature 449:105–108
Huang J, Dorsey J, Chuikov S et al (2010) G9a and Glp methylate lysine 373 in the tumor suppressor p53. J Biol Chem 285:9636–9641
Hyun K, Jeon J, Park K et al (2017) Writing, erasing and reading histone lysine methylations. Exp Mol Med 49:e324
Jakobsson ME, Moen A, Bousset L et al (2013) Identification and characterization of a novel human methyltransferase modulating Hsp70 function through lysine methylation. J Biol Chem 288:27752–27763
Jørgensen S, Elvers I, Trelle MB et al (2007) The histone methyltransferase SET8 is required for S-phase progression. J Cell Biol 179:1337–1345
Kernstock S, Davydova E, Jakobsson M et al (2012) Lysine methylation of VCP by a member of a novel human protein methyltransferase family. Nat Commun 3:1038
Kim S, Paik WK (1965) Studies on the origin of epsilon-N-methyl-L-lysine in protein. J Biol Chem 240:4629–4634
Kim Y, Nam HJ, Lee J et al (2016a) Methylation-dependent regulation of HIF-1α stability restricts retinal and tumour angiogenesis. Nat Commun 7 10347
Kim JH, Yoo BC, Yang WS et al (2016b) The role of protein arginine methyltransferases in inflammatory responses. Mediat Inflamm 2016:4028353
Klose RJ, Kallin EM, Zhang Y (2006) JmjC-domain-containing proteins and histone demethylation. Nat Rev Genet 7:715–727
Kooistra SM, Helin K (2012) Molecular mechanisms and potential functions of histone demethylases. Nat Rev Mol Cell Biol 13:297–311
Kunizaki M, Hamamoto R, Silva FP et al (2007) The lysine 831 of vascular endothelial growth factor receptor 1 is a novel target of methylation by SMYD3. Cancer Res 67:10759–10765
Lachner M, O'Carroll D, Rea S et al (2001) Methylation of histone H3 lysine 9 creates a binding site for HP1 proteins. Nature 410:116–120

Lanouette S, Mongeon V, Figeys D et al (2014) The functional diversity of protein lysine methylation. Mol Syst Biol 10:724

Lee JY, Park JH, Choi HJ et al (2017) LSD1 demethylates HIF1α to inhibit hydroxylation and ubiquitin-mediated degradation in tumor angiogenesis. Oncogene 36:5512–5521

Liu H, Galka M, Mori E et al (2013) A method for systematic mapping of protein lysine methylation identifies functions for HP1β in DNA damage response. Mol Cell 50:723–735

Liu X, Chen Z, Xu C et al (2015) Repression of hypoxia-inducible factor α signaling by Set7-mediated methylation. Nucleic Acids Res 43:5081–5098

Liu Y, Wang Y, Chen C et al (2017) LSD1 binds to HPV16 E7 and promotes the epithelial-mesenchymal transition in cervical cancer by demethylating histones at the Vimentin promoter. Oncotarget 8:11329–11342

Lu X, Simon MD, Chodaparambil JV et al (2008) The effect of H3K79 dimethylation and H4K20 trimethylation on nucleosome and chromatin structure. Nat Struct Mol Biol 15:1122–1124

Marabelli C, Marrocco B, Mattevi A (2016) The growing structural and functional complexity of the LSD1/KDM1A histone demethylase. Curr Opin Struct Biol 41:135–144

Mazur PK, Reynoird N, Khatri P et al (2014) SMYD3 links lysine methylation of MAP 3K2 to Ras-driven cancer. Nature 510:283–287

Murray K (1964) The occurrence of epsilon-N-methyl lysine in histones. Biochemistry 3:10–15

Nagasawa S, Sedukhina AS, Nakagawa Y et al (2015) LSD1 overexpression is associated with poor prognosis in basal-like breast cancer, and sensitivity to PARP inhibition. PLoS One 10: e0118002

Ng SS, Kavanagh KL, McDonough M et al (2007) Crystal structures of histone demethylase JMJD2A reveal basis for substrate specificity. Nature 448:87–91

Ng SS, Yue WW, Oppermann U et al (2009) Dynamic protein methylation in chromatin biology. Cell Mol Life Sci 66:407–422

Nishioka K, Rice JC, Sarma K et al (2002) PR-Set7 is a nucleosome-specific methyltransferase that modifies lysine 20 of histone H4 and is associated with silent chromatin. Mol Cell 9:1201–1213

Petrossian TC, Clarke SG (2009a) Bioinformatic identification of novel methyltransferases. Epigenomics 1:163–175

Petrossian TC, Clarke SG (2009b) Multiple motif scanning to identify methyltransferases from the yeast proteome. Mol Cell Proteomics 8:1516–1526

Poulard C, Rambaud J, Hussein N et al (2014) JMJD6 regulates ERα methylation on arginine. PLoS One 9:e87982

Qiu WR, Xiao X, Lin WZ et al (2014) iMethyl-PseAAC: identification of protein methylation sites via a pseudo amino acid composition approach. Biomed Res Int 2014:947416

Scoumanne A, Chen X (2008) Protein methylation: a new regulator of the p53 tumor suppressor. Histol Histopathol 23:1143–1149

Shao J, Xu D, Tsai SN et al (2009) Computational identification of protein methylation sites through bi-profile Bayes feature extraction. PLoS One 4:e4920

Shi Y, Lan F, Matson C et al (2004) Histone demethylation mediated by the nuclear amine oxidase homolog LSD1. Cell 119:941–953

Shi X, Kachirskaia I, Yamaguchi H et al (2007) Modulation of p53 function by SET8-mediated methylation at lysine 382. Mol Cell 27:636–646

Shi SP, Qiu JD, Sun XY et al (2012) PMeS: prediction of methylation sites based on enhanced feature encoding scheme. PLoS One 7:e38772

Shi Y, Guo Y, Hu Y et al (2015) Position-specific prediction of methylation sites from sequence conservation based on information theory. Sci Rep 5:12403

Shien DM, Lee TY, Chang WC et al (2009) Incorporating structural characteristics for identification of protein methylation sites. J Comput Chem 30:1532–1543

Singer MS, Kahana A, Wolf AJ et al (1998) Identification of high-copy disruptors of telomeric silencing in *Saccharomyces cerevisiae*. Genetics 150:613–632

Smith BC, Denu JM (2009) Chemical mechanisms of histone lysine and arginine modifications. Biochim Biophys Acta 1789:45–57

Spillantini MG, Schmidt ML, Lee VM et al (1997) Alpha-synuclein in Lewy bodies. Nature 388: 839–840

Stocker B, McDonough MA (1961) Gene determining presence or absence of [varepsilon]-N-methyl-lysine in Salmonella flagellar protein. Nature 189:556–558

Walport LJ, Hopkinson RJ, Chowdhury R et al (2016) Arginine demethylation is catalysed by a subset of JmjC histone lysine demethylases. Nat Commun 7:11974

West LE, Gozani O (2011) Regulation of p53 function by lysine methylation. Epigenomics 3:363–369

West LE, Roy S, Lachmi-Weiner K et al (2010) The MBT repeats of L3MBTL1 link SET8-mediated p53 methylation at lysine 382 to target gene repression. J Biol Chem 285:37725–37732

Wu Z, Connolly J, Biggar KK (2017) Beyond histones: the expanding roles of lysine methylation. FEBS J 284:2732–2744

Xie Q, Hao Y, Tao L et al (2012) Lysine methylation of FOXO3 regulates oxidative stress-induced neuronal cell death. EMBO Rep 13:371–377

Yang Y, Hu L, Wang P et al (2010) Structural insights into a dual-specificity histone demethylase ceKDM7A from *Caenorhabditis elegans*. Cell Res 20:886

Yang SJ, Park YS, Cho JH et al (2017) Regulation of hypoxia responses by flavin adenine dinucleotide-dependent modulation of HIF-1α protein stability. EMBO J 36:1011–1028

Zhang X, Zhou L, Cheng X (2000) Crystal structure of the conserved core of protein arginine methyltransferase PRMT3. EMBO J 19:3509–3519

Zhu W (2012) Methylation of FoxO3 regulates neuronal cell death. Acta Pharmacol Sin 33:577

# Secondary Structures of Histone H3 Proteins with Unmethylated and Methylated Lysine-4 and -9 Residues: Characterization Using Circular Dichroism Spectroscopy

**Yudai Izumi**

## Contents




**Abstract** Circular dichroism (CD) spectroscopy, especially that using synchrotron radiation as a light source, is a powerful tool for analyzing secondary structures of proteins in solution. In particular, CD spectroscopy allows observations of structural changes following post-translational modifications, such as methylation. In this chapter, techniques and measurement protocols are introduced. Recent structural analyses of H3 proteins before and after methylation of lysine-4 and -9 residues are also shown. In these CD spectroscopy analyses, mono- and dimethylation of H3 increased the presence of α-helical structures and decreased β-strand contents, whereas trimethylation decreased α-helix and increased β-strand contents. These structural alterations occurred at adjacent and distant residues from the methylated site.



Y. Izumi (✉)
Hiroshima Synchrotron Radiation Center, Hiroshima University, Higashi-Hiroshima, Hiroshima, Japan
e-mail: izumi-yudai@hiroshima-u.ac.jp





S. Jurga, J. Barciszewski (eds.), *The DNA, RNA, and Histone Methylomes*, RNA Technologies, https://doi.org/10.1007/978-3-030-14792-1_19

**Keywords** Epigenetics · Post-translational modification · Synchrotron radiation · Chiroptical spectroscopy · Lysine-specific demethylase 1 (LSD1/KDM1)

# 1 Introduction

Nucleosomes are octamers of four core histone proteins (H2A, H2B, H3, and H4) around which 146–147 base pairs of DNA are wrapped, and these are the building blocks of chromatin in eukaryotic nuclei (Luger et al. 1997; Davey et al. 2002). Post-translational modifications, such as methylation, of histones have been shown to play substantial roles in cellular functions. For example, methylation of H3 on lysine-4 (K4) and lysine-9 (K9) residues has been associated with transcriptionally active chromatin and inactive chromatin, respectively, in higher eukaryotes (Lachner and Jenuwein 2002). On the other hand, since the absence of K4-methylated H3 is required for DNA methylation, defects of which are linked to serious human diseases, such as cancer (Gal-Yam et al. 2008), demethylation of methylated H3 is also essential. Such methylation states are controlled by specific histone methyltransferases and demethylases.

Lysine-specific demethylase 1 (LSD1), also known as KDM1, was the first histone demethylase to be identified (Shi et al. 2004). LSD1 belongs to the flavin adenine dinucleotide dependent amine oxidase family (Shi et al. 2004; Shi 2007; Forneris et al. 2008) and converts K4 monomethylated or dimethylated H3 (H3K4me1 and H3K4me2, respectively) to unmethylated H3. On the other hand, it does not catalyze K4 trimethylated H3 (H3K4me3) (Shi et al. 2004). LSD1 has been associated with various important cellular processes (Zheng et al. 2015), but is reportedly overexpressed in several cancer cells (Ota and Suzuki 2018). Accordingly, LSD1 inhibitors are potential anti-cancer drugs (Suzuki and Miyata 2011; Maes et al. 2015; McAllister et al. 2016; Niwa and Umehara 2017). Indeed, some inhibitors have been tested in clinical trials for treatment of acute leukemia (Harris et al. 2012; Mohammad et al. 2015), small cell lung cancers (Mohammad et al. 2015), and neurodegenerative disorders (Niwa and Umehara 2017). In addition, novel LSD1 inhibitors are under development (Ogasawara et al. 2013; Maiques-Diaz and Somervaille 2016; Amano et al. 2017; Ota and Suzuki 2018).

Most LSD1 inhibitors, such as *trans*-2-phenylcyclopropylamine (Schmidt and McCafferty 2007) and above inhibitors, are complex unnatural molecules. However, histone H3 proteins are worthy of consideration as LSD1 inhibitors. Because the 21-mer peptides of H3 act as inhibitors of LSD1 (Forneris et al. 2005, 2006) and the binding affinity between full-length of unmethylated H3 and LSD1 is nearly 100-fold higher than that of the peptides (Burg et al. 2016). Hence, full-length unmethylated H3 may act as a good inhibitor. Since Forneris et al. also showed peptides of methylated, and mutated H3 (21-mer) acted as inhibitors of LSD1 (Forneris et al. 2005, 2006), other methylated H3 proteins may show similar activities.

Conformations of (un)methylated H3 have important functional implications for use as inhibitors of LSD1, because these are closely related to protein–protein interactions (Tobi and Bahar 2005). However, unfortunately, yet no structural data for the full-length of non-nucleosomal H3 are available in the Protein Data Bank, although structural data for nucleosomes (Luger et al. 1997; Davey et al. 2002) and (un)methylated H3 peptides in complexes with various histone-binding proteins, including LSD1 (Yang et al. 2007), can be found.

Recently, our group analyzed secondary structures of full-length K4- or K9-methylated H3 using circular dichroism (CD) spectroscopy and compared these with unmethylated H3. These primary investigations of full-length (un)methylated H3 conformations showed that methylation at K4 or K9 induces structural changes in H3 at residues both adjacent and distant from methylated sites (Izumi et al. 2018a, b). In this chapter, these results are introduced to facilitate further studies of the interactions between full-length (un)methylated H3 and LSD1 and other histone binding proteins. Next section, experimental methods for analyzing protein structures using CD spectroscopy is introduced. In Sect. 3, the experimental results are shown. In the last section, brief summary is described.

## 2 CD Spectroscopy

Standard CD spectroscopy procedures for analyses of protein structures are reviewed thoroughly elsewhere (for example, Greenfield 2006; Miles and Wallace 2006). Thus, in this section, these techniques and measurement protocols are introduced, only briefly.

### 2.1 *Circularly Polarized Light*

Light is an electromagnetic wave. When the electric field of the light oscillates randomly in time, the light is called unpolarized light. Most light sources, such as sunlight, incandescent bulbs, and Xe lamps, emit unpolarized light. If the unpolarized light passes through appropriate prisms and filters, such as Pockels cells, the electric field oscillates sinusoidally in a single plane (Fig. 1a), and then the light is classified as linearly polarized light. When the electric field of the light consists of two perpendicular electromagnetic plane waves, which are equal in amplitude, but have a phase difference of 90° (quarter of the wavelength), the tip of the electric field vector rotates in a circle around the direction of propagation (Fig. 1b). Such polarized light is called circularly polarized light (CPL). More comprehensible animated graphics of the CPL can be found on the internet. CPL is often converted from linearly polarized light using quarter wave plates in laboratories. CPL is classified into two types depending on the rotation direction. Viewing from the receiver, if the vector of the CPL rotates clockwise, the CPL is called as right circularly polarized light (RCPL). In the opposite case, that is, the vector rotates counterclockwise, it is called as left circularly polarized light (LCPL).

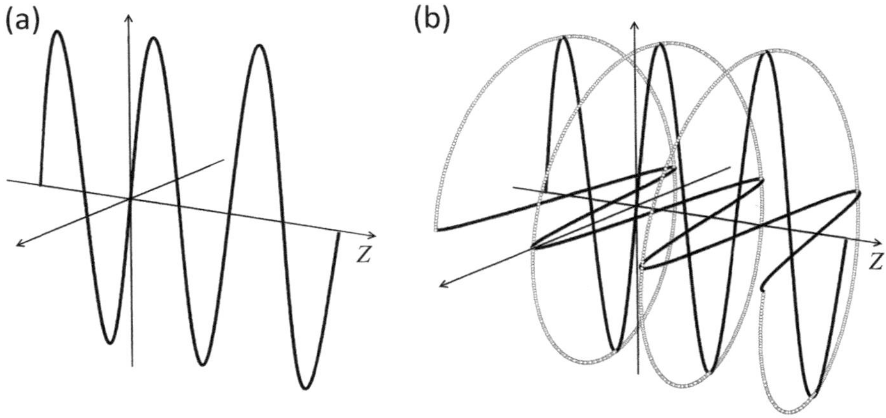


**Fig. 1** The electric fields of (**a**) linearly polarized light and (**b**) RCPL propagating along *Z*-axis

## 2.2 CD and Its Notations

CD is a phenomenon exhibited in absorption bands of optical active molecules, such as amino acids, riboses, and deoxyriboses, and defined as the difference between the molar absorption coefficient for LCPL ($\varepsilon_L$) and that for RCPL ($\varepsilon_R$) as in the following equation: $\Delta\varepsilon = \varepsilon_L - \varepsilon_R$ (in $M^{-1}$ $cm^{-1}$). As a consequence, CD spectroscopy is a variation of absorption spectroscopy. The CD value $\Delta\varepsilon$ is sometimes referred to as molar CD. However, for historical reasons, different notation, ellipticity, is also used to express the CD intensities. Indeed, commercial CD spectrophotometers often return CD spectra in terms of ellipticity.

Ellipticity $\theta'$ (in rad) is derived from the ratio of electric vector magnitudes of LCPL and RCPL ($E_L$ and $E_R$, respectively):

$$\tan\theta' = \frac{E_R - E_L}{E_R + E_L} \tag{1}$$

Molar CD and ellipticity are interconverted as follows. Using Beer-Lambert's law, the Eq. (1) is rewritten as

$$\tan\theta' = \frac{\exp\left(\frac{\ln 10}{2}\Delta\varepsilon\, C\, l\right) - 1}{\exp\left(\frac{\ln 10}{2}\Delta\varepsilon\, C\, l\right) + 1} \tag{2}$$

where $C$ and $l$ are the concentration of the sample (in M) and the path length of the cuvette (in cm). Since, in general, $\theta'$ and $\Delta\varepsilon$ are small, the above equation can be approximated to be

$$\theta' = \frac{\ln 10}{4}\Delta\varepsilon\, C\, l \tag{3}$$

In accord with tradition, converting the unit of ellipticity from radians to millidegrees and calculating the constant term, we can obtain

$$\Delta\varepsilon = \frac{\theta}{32980\ C\ l} \tag{4}$$

where θ is ellipticity (in millidegrees). Thus, the ellipticity provided as the data from CD spectrophotometers can be converted to molar CD using the Eq. (4). It is noted that, in the case of CD spectroscopy of proteins, mean residue molar concentration, namely molar concentration multiplied by number of amino acid residues in the protein, is used as *C*. Instead of molar CD, mean residues molar ellipticity [θ] (in degrees $cm^2$ $dmol^{-1}$) is also commonly used to describe CD intensities and is calculated as follows: $[\theta] = 3298\ \Delta\varepsilon$.

## 2.3 Advantages of CD Spectroscopy Using Synchrotron Radiation

CD spectroscopy can be performed using commercial CD spectrophotometers. However, the author would like to recommend the usage of synchrotron radiation (SR) CD beamlines (Miles and Wallace 2006). Since the photon flux from the Xe lamps, that are used as light sources for commercial CD spectrophotometers, severely decreases in the vacuum ultraviolet (VUV) region (wavelength < ~200 nm), it is hard to obtain meaningful data below ~190 nm. Alternatively, since photon flux of SR is higher than that for Xe lamps in the VUV region, the use of SR can extend CD spectra to the wavelength region below 190 nm and thereby can provide additional information which is unobtainable using commercial CD instruments. In particular, the use of SR is essential for CD measurements of chiral molecules which compose only from single bonds, such as saccharides, since CD peaks of assigned to $n \rightarrow \sigma^*$ and $\sigma \rightarrow \sigma^*$ transitions are detectable only below 190 nm (Arndt and Stevens 1993).

At the time of writing, SR-CD beamlines were available for use at the following eight SR facilities to my knowledge (Table 1): Hiroshima Synchrotron Radiation Center (HiSOR) of Hiroshima University in Japan (Matsuo and Gekko 2013), Institute for Storage Ring Facilities (ISA) of Aarhus University in Denmark (Miles et al. 2007), Beijing Synchrotron Radiation Facility (BSRF) in China (Tao et al. 2009), National Synchrotron Radiation Research Center (NSRRC) in Taiwan (Liu et al. 2010), Diamond Light Source (DLS) in UK (Hussain et al. 2012), Synchrotron SOLEIL in France (Réfrégiers et al. 2012), BESSY-II of the Helmholtz-Zentrum Berlin (Reichardt et al. 2001), and ANKA of Karlsruhe Institute of Technology (Bürck et al. 2015) in Germany. In general, these beamlines can be used after acceptance of proposal(s), which can be referred to websites of each SR facility (Table 1).

**Table 1** The list of the SR facilities equipped with SR-CD beamlines[a]

| Location | SR facility | URL |
|---|---|---|
| Japan | HiSOR | http://www.hsrc.hiroshima-u.ac.jp/english/index.html |
| Denmark | ISA | http://www.isa.au.dk/index.asp |
| China | BSRF | http://english.bsrf.ihep.cas.cn/ |
| Taiwan | NSRRC | http://www.nsrrc.org.tw/ |
| UK | DLS | https://www.diamond.ac.uk/Home.html |
| France | SOLEIL | https://www.synchrotron-soleil.fr/en |
| Germany | BESSY-II | https://www.helmholtz-berlin.de/quellen/bessy/index_en.html |
| Germany | ANKA | https://www.anka.kit.edu/ |

[a]At the time of writing

## 2.4 Structural Analyses of Proteins Using CD Spectroscopy

Structural information of proteins from CD spectra is limited compared with that from X-ray crystallography and nuclear magnetic resonance (NMR), both of which display three-dimensional structures with atomic-level resolutions. CD spectroscopy is, nonetheless, a powerful tool because it provides structural information, including structural dynamics, with greater ease than that using above techniques. In particular, (1) required sample quantities are only 1–10% of those required for X-ray crystallography and NMR (Kim et al. 2008) and (2) the samples can be prepared by simply dissolving the protein in a solvent. In addition, neither crystallization nor isotopic substitution is required, and sample losses and accidental denaturation during sample preparation are negligible in most cases.

Each protein forms individual conformation, but includes common structures called as secondary structures, for example, α-helices and β-strands, which compose β-sheets connecting laterally, turns and unordered structures, also known as random coil or disordered structures. Therefore, it is assumed that CD spectra of a protein are linear combinations of CD spectra from each of these secondary structures weighted by relative abundance, as indicated in the following equation (Greenfield 2006):

$$\mathrm{CD}(\lambda) = \sum_{s} f_s \Delta\varepsilon_s(\lambda) \tag{5}$$

where CD(λ) is the CD intensity of an unknown protein at wavelength λ, $f_s$ is the fraction of the secondary structure $s$ ($s = $ α-helix, β-strand, . . .), and $\Delta\varepsilon_s(\lambda)$ is the CD intensity of secondary structure $s$ at wavelength λ. Given a protein comprising some α-helices, β-strands, turns, and unordered structures at fractions of 25% ($f_s = 0.25$), CD spectra of the protein can be described as follows:

$$\mathrm{CD}(\lambda) = 0.25\left(\Delta\varepsilon_{\alpha\text{-helix}}(\lambda) + \Delta\varepsilon_{\beta\text{-strand}}(\lambda) + \Delta\varepsilon_{\text{turn}}(\lambda) + \Delta\varepsilon_{\text{Unordered}}(\lambda)\right) \tag{6}$$

Thus, if $\Delta\varepsilon_s(\lambda)$ is known, we can estimate secondary structure contents of unknown proteins $f_s$ by deconvoluting the CD spectrum. Values of $\Delta\varepsilon_s(\lambda)$ are often derived empirically from reference datasets of CD spectra for numerous proteins whose structures have been determined by X-ray crystallography. These secondary structures have characteristic CD peaks (for example, Matsuo et al. 2005) that are predominantly assigned to $n \rightarrow \pi^*$ and $\pi \rightarrow \pi^*$ transitions of peptide bonds in the UV–VUV region (Woody 1995). α-Helices exhibit a positive peak around 190 nm and two negative peaks at around 208 and 222 nm, and another positive peak appears as a shoulder around 175 nm. β-Strands are characterized by a negative peak and a positive peak at 218 and 195 nm, respectively, and unordered structures give negative peaks at around 200 nm and two positive peaks around 170 and 225 nm.

Whereas analytical procedures for CD spectroscopy may appear complex, various programs have been developed for analyzing CD spectra and these are publicly available (For example, Sreerama and Woody 2000; Whitmore and Wallace 2004; Micsonai et al. 2015). The analyses using these programs are easily performed only inputting CD data of proteins you measured and the calculations are completed in a short time using standard personal computers or the internet.

## 2.5 *Experimental Procedure for CD Spectroscopy*

### 2.5.1 Sample Preparation

CD spectra of proteins are usually measured in liquid solutions, which can be prepared easily by simply dissolving the protein in a solvent, as mentioned above. However, solvents should be carefully selected, especially for VUV-CD measurements (Miles and Wallace 2006). Chloride ions exhibit strong absorption bands in the VUV region and can hence interfere with VUV-CD measurements by severely decreasing the transmitted light intensity, even when using SR. Therefore, sodium chloride, which is often supplemented in protein solutions, had better be substituted with sodium fluoride or removed if possible. Similarly, Tris buffer should be acidified with phosphoric acid instead of HCl. Although phosphate buffers are preferable, concentrations should be kept as low as possible.

The path length of the sample-cells used in VUV-CD measurements is often below ~100 μm to reduce absorption of solvents, and it is shorter than that used in conventional CD measurements (1–10 mm). Observed CD intensities, such as ellipticity θ, are proportional to the sample concentration and the path length of the sample-cell. According to Beer-Lambert's law, when the path length decreases by 1/10, ten-fold sample concentrations are required to observe the same ellipticity. Thus, to obtain meaningful CD signals, concentrations of samples used in VUV-CD measurements need to be much higher than those used in conventional CD measurements. In general, optimal concentrations of proteins for use with a path length of 10 μm are 10 mg/mL for α-helical proteins and 15–20 mg/mL for β-sheet rich proteins (Miles and Wallace 2006).

### 2.5.2 Sample-Cell

Although various types of sample-cells are commercially available, custom-made sample-cells are often used at SR-CD beamlines (for example, Wien and Wallace 2005; Izumi and Matsuo 2018). Because short path length cells are preferable, demountable cells that are easier to clean are often used. In addition, sample cells for VUV-CD spectroscopy are often made of $CaF_2$ glasses (Wien and Wallace 2005) because transmittance of these in the VUV region is higher than that of the $SiO_2$ glasses used in conventional CD spectroscopy (cut-off wavelength, ~140 nm for $CaF_2$ vs. ~160 nm for $SiO_2$).

As an example, a sample cell recently developed is described (Izumi and Matsuo 2018). It is comprised of two circular glasses. One of the glasses has a counterbore hole and the other is flat, and the depth of the counterbore hole corresponds with the path length of the cell. After placing sample solution into the counterbore hole on the first glass, it is covered with another glass and the glasses are then fixed in the sample-cell holder and used for CD spectroscopy. Sample volumes are generally 2–3 μL to avoid foaming, although convenient volumes depend on the types of solutions and skills of users.

### 2.5.3 Measurements and Analyses of CD Spectra

The CD measurement systems used in SR-CD beamlines are similar to those of commercial CD spectrophotometers, although important details should be referred to in the publications and websites of each beamline (Reichardt et al. 2001; Miles et al. 2007; Tao et al. 2009; Liu et al. 2010; Hussain et al. 2012; Réfrégiers et al. 2012; Matsuo and Gekko 2013; Bürck et al. 2015). In general, SR emitted from a storage ring, which is linearly polarized light, is monochromated and converted to LCPL or RCPL using a phase shifter (photo-elastic modulator) (Fig. 2). Intensities of transmitted LCPL and RCPL passing through the sample are detected using a detector (photomultiplier tube) (Fig. 2) and CD spectra are then generated.

Before analyzing CD spectra, measured CD intensities, ellipticity θ, must be converted to molar CD (Δε) or mean residue molar ellipticity [θ] values, because magnitudes of θ depend on the concentration of the sample and the path length of the

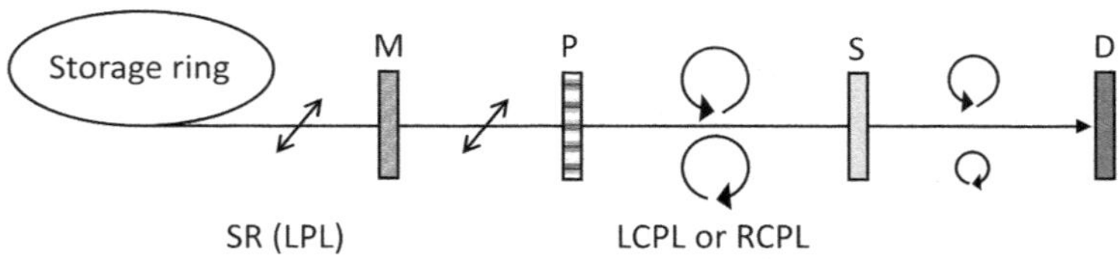


**Fig. 2** A schematic view of SR-CD beamlines. *SR* synchrotron radiation, *LPL* linearly polarized light, *M* monochromator, *P* phase shifter (photo-elastic modulator), *L(R)CPL* left (right) circularly polarized light, *S* sample, *D* detector (photomultiplier tube)

cuvette as shown in Sect. 2.2. Spectra can then be analyzed using empirical programs, as mentioned above (For example, Sreerama and Woody 2000; Whitmore and Wallace 2004; Micsonai et al. 2015).

# 3 Secondary Structures of K4- or K9-(Un)methylated H3

## *3.1 CD Spectra of K4- or K9-(Un)methylated H3*

Figure 3 shows the CD spectra of unmethylated H3, H3K4me1, H3K4me2, and H3K4me3 in 25-mM sodium phosphate buffer supplemented with 250-mM sodium fluoride. The CD spectrum for unmethylated H3 exhibited a positive peak at ~190 nm and two negative peaks at ~210 and ~220 nm. These peaks are characteristic of α-helix structures. H3K4me1 and H3K4me2 had similar CD spectra. It might be a reasonable result because both H3K4me1 and H3K4me2 are substrates of LSD1 (Shi et al. 2004), although these proteins are nucleosomal in cells. Positive peaks of H3K4me1 and H3K4me2 shifted toward the longer-wavelength region, whereas negative peaks shifted toward the shorter-wavelength region compared with those of unmethylated H3. CD peak intensities of H3K4me1 and H3K4me2 were higher than those of unmethylated H3. In contrast, negative CD peaks of H3K4me3 shifted further toward the shorter-wavelength region and their intensities were increased compared with those of H3K4me1 and H3K4me2. Although the positions of the positive peaks around 190 nm were similar for all three H3K4 methylation states, the intensities of those in H3K4me3 were decreased compared with H3K4me1 and H3K4me2 peaks.

Figure 4 shows the CD spectra of mono-, di-, and trimethylated H3 at residue K9 (H3K9me1, H3K9me2, and H3K9me3, respectively) in the same solvent. The CD spectrum of unmethylated H3 is also shown again for comparison. CD spectral shapes of H3K9me1 and H3K9me2 were similar, although the widths of positive peaks differed. Positive CD peak positions of H3K9me1 and H3K9me2 shifted toward the longer-wavelength region, and their intensities were the highest among

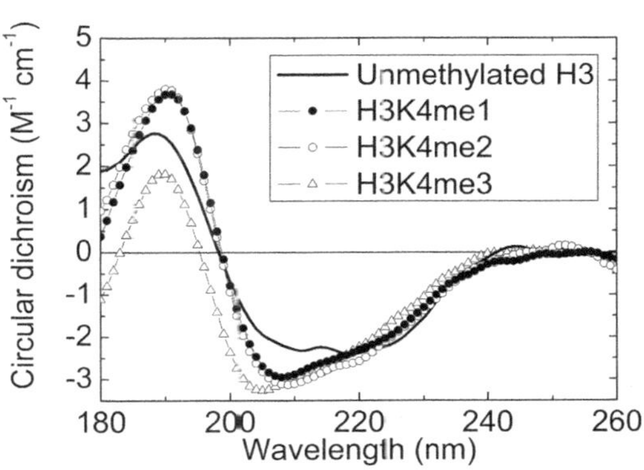


**Fig. 3** CD spectra of unmethylated H3 (solid line), H3K4me1 (closed circle), H3K4me2 (open circle), and H3K4me3 (open triangle)

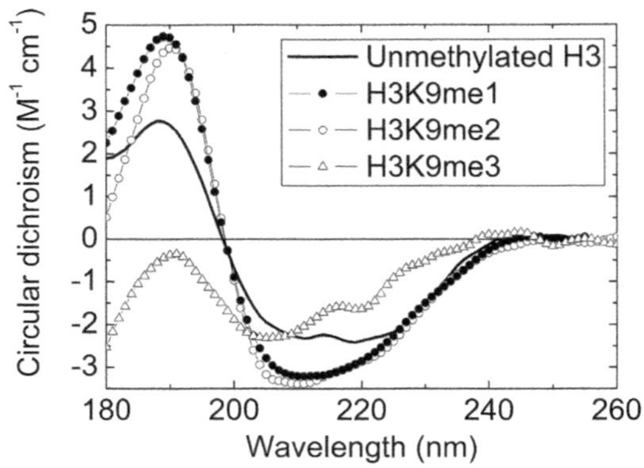


**Fig. 4** CD spectra of unmethylated H3 (solid line), H3K9me1 (closed circle), H3K9me2 (open circle), and H3K9me3 (open triangle)

the samples examined. Negative peak intensities of methylated H3K9 were also higher than those of unmethylated H3. Finally, H3K9me3 showed negative peaks at ~200–220 nm but no positive peak at ~190 nm, and the spectral shape differed substantially from those of other samples.

In comparisons of the CD spectra of K4- and K9-methylated H3 and unmethylated H3 (Figs. 3 and 4), spectral shapes depended on the positions and degrees of methylation. Because CD spectra reflect secondary structures of proteins, these results show that (1) methylation of K4 and K9 residues induces structural alterations of H3 and (2) that these methylated H3 form different structures from each other.

## *3.2 Secondary Structure Contents of K4- or K9-(Un) methylated H3*

Analyses of CD spectra were performed using the SELCON3 program (Sreerama et al. 1999; Sreerama and Woody 2000) with the reference dataset that was generated at HiSOR (Matsuo et al. 2004, 2005). The secondary structure contents of K4- or K9-(un)methylated H3 and their standard deviations are listed in Table 2, in which secondary structure contents are normalized to a total amount of 100%. Numbers of segments in secondary structures are also listed in Table 2. For ease of comparison, the results in Table 2 are shown in the proportional histograms of the secondary structure contents seen in Fig. 5. The contents of monomethylated H3 were within a standard deviation of those of dimethylated H3, and these are shown together in Table 2 and Fig. 5. H3K4me1 and H3K4me2 showed incremental differences in secondary structure contents and segment numbers of α-helix structures, compared with those of unmethylated H3. In contrast, the contents and numbers of β-strand structures were less in H3K4me1 and H3K4me2 than in unmethylated H3. Further methylation (trimethylation of K4 in H3; H3K4me3) decreased α-helix contents and incrementally increased β-strand contents, compared with those of H3K4me1 and H3K4me2. Decreased unordered structures were observed in H3K4me1 and

**Table 2** Secondary structure contents of (un)methylated H3

| Structure content (%) | H3 | H3K4me1<br>H3K4me2 | H3K4me3 | H3K9me1<br>H3K9me2 | H3K9me3 |
|---|---|---|---|---|---|
| α-Helix | 25.0 ± 1.2 | 30.7 ± 1.3 | 21.8 ± 0.8 | 36.5 ± 1.7 | 13.1 ± 0.8 |
| β-Strand | 21.3 ± 1.5 | 18.9 ± 2.0 | 25.1 ± 2.0 | 13.6 ± 2.6 | 29.6 ± 1.9 |
| Turn | 21.1 ± 1.0 | 23.5 ± 1.2 | 21.4 ± 0.7 | 23.1 ± 1.0 | 22.7 ± 1.2 |
| Unordered | 32.7 ± 1.7 | 27.0 ± 1.8 | 31.7 ± 1.6 | 27.8 ± 1.8 | 36.3 ± 2.4 |
| Numbers of α-Helices | 4 | 5 | 4 | 6 | 3 |
| Numbers of β-Strands | 6 | 5 | 7 | 4 | 8 |

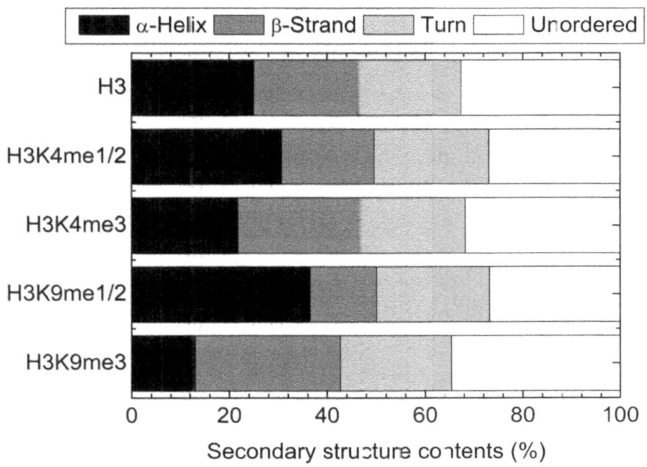


**Fig. 5** Comparison of secondary structure contents of (un)methylated H3 normalized to a total content of 100%

H3K4me2 compared with those in unmethylated H3, whereas those structures were almost equally prevalent in H3K4me3 and unmethylated H3.

The tendency of K9 methylation to induce structural alterations was similar to that of K4 methylation, except that more drastic structural changes were induced by the former. H3K9me1 (H3K9me2) showed the highest contents and numbers of α-helix structures and the smallest contents and segment numbers of β-strand structures. Conversely, contents and segment numbers of β-strand structures were greatest in H3K9me3, in which α-helix contents were the smallest.

## 3.3 Predicted Positions of α-Helices and β-Strands

Based on CD spectroscopy results, the positions of α-helices and β-strands in (un) methylated H3 were predicted using a neural network (NN) method, which is termed the VUVCD-NN combination method. This original method was developed at HiSOR, and the computational protocol is described elsewhere (Matsuo et al. 2008). Briefly, an NN algorithm was used to predict the positions of secondary structures using evolutionary sequence information based on the position-specific scoring matrices generated by the PSI-BLAST algorithm (Jones 1999). These computations were performed with reference to the numbers of α-helix and β-strand segments and the numbers of amino acid residues forming α-helix and β-strand structures, as determined in CD spectroscopy and SELCON3 analyses. The accuracy of the VUVCD-NN combination method is reportedly about 75% for 30 reference proteins (Matsuo et al. 2008).

Figure 6 shows predicted secondary structure sequences of samples. Turn and unordered structures were estimated using SELCON3 analysis and were classified as "others". In analyses of unmethylated H3, structures from the 1st to the 44th residues were assigned as others (turn or unordered structures), and were consistent with the crystal structure of nucleosomal H3 (Davey et al. 2002). In contrast, simulations of methylated H3 showed the formation of a β-strand structure from 5th to 8th residues. In H3K9me3, another β-strand structure was assigned at the 39th–42th residues. It was also predicted that structural alterations would not be limited to methylated regions. For example, α-helix formations were predicted at the 65th to 69th, 80th to 83rd, and 85th to 86th residues in H3K4me1 and H3K4me2, although these α-helices reverted to other structures in H3K4me3. Similar structural changes, such as increased α-helices with monomethylation and demethylation and decreased α-helices with trimethylation, were predicted for K9 methylation. However, the amount of structural changes induced by K4 methylation was less than those following K9 methylation.

In comparisons with predicted structures, the structure of H3K4me3 was almost the same as that of unmethylated H3, except for the β-strand at the N-terminal tail. Therefore, the full-length of H3K4me3 may act as an inhibitor of LSD1 with similar potency as that of full-length unmethylated H3. In agreement, the 21-mer peptide of H3K4me3 was characterized as an inhibitor of LSD1 previously (Forneris et al. 2006). However, the structures of K9-trimethylated H3 differed from those of K4-trimethylated and unmethylated H3, and therefore full-length K9-trimethylated H3 may have different effects on LSD1 or other proteins comparing to K4-trimethylated and unmethylated H3.

Although the causes of these differences have not yet been unidentified, structural alterations of H3 that depend on degrees and positions of methylation may be of interest for application as drugs. Future studies will examine binding affinities between full-length methylated H3 and various enzymes, including LSD1.

### *3.4 Structural Alterations Following K4 or K9 Methylation in Solution*

The VUVCD-NN combination method predicted that K4 and K9 methylation induces structural changes of H3 at adjacent and distal residues from the methylated sites. These data suggest that structural alterations are induced by interactions

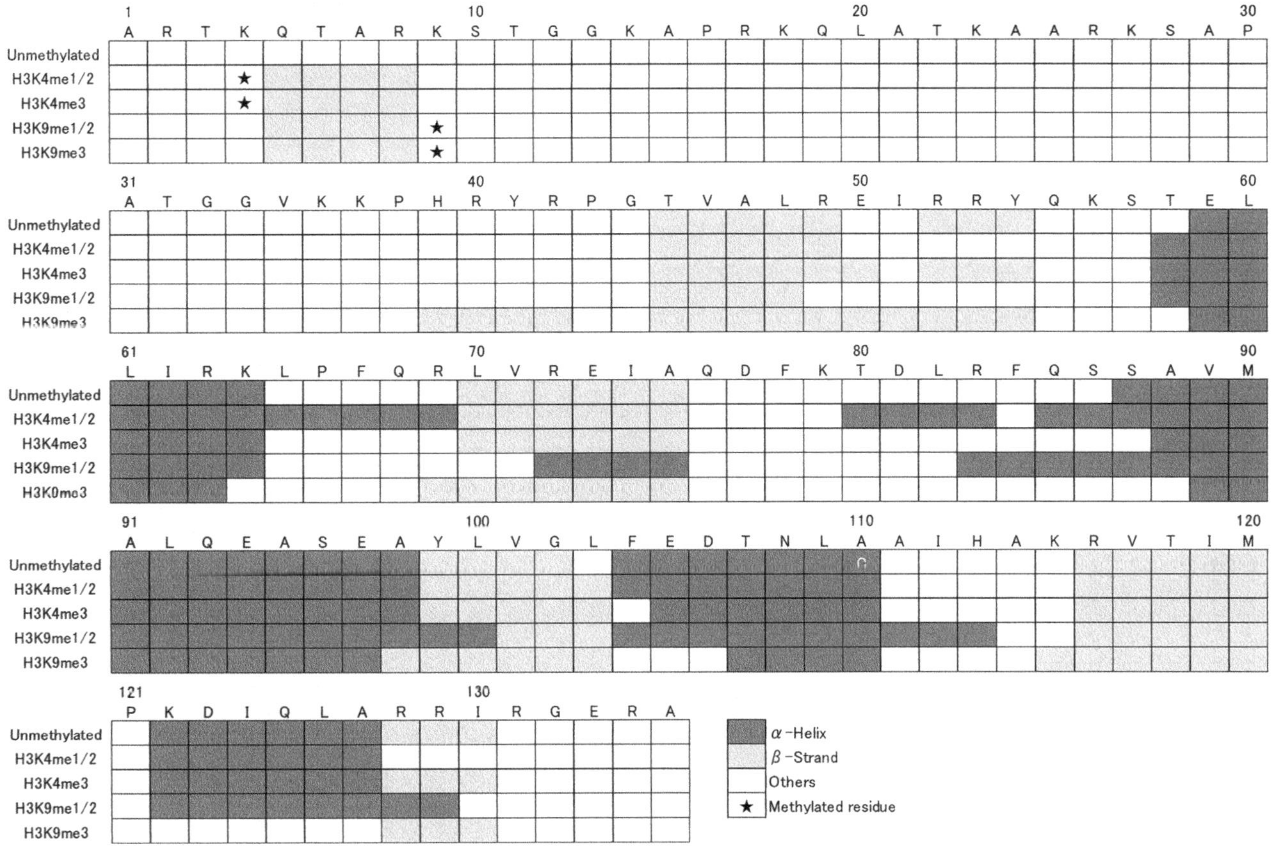


**Fig. 6** Sequence-based secondary structures of (un)methylated H3 as predicted using the VUVCD-NN method; α-helix, β-strand, and other structures are shown in dark gray, gray, and white rectangles, respectively. The star in the rectangle shows the methylated residue. The 110th residue of unmethylated H3 is cysteine (C) and differs from that of methylated H3 (alanine; A)

between residues that are distant from the methylation site, such as residues 65th to 86th in H3K4me1, and the methylated N-terminal tail. In the unmodified state, the N-terminal tail may not interact with other domains, as observed in nucleosomal H3 (Davey et al. 2002). Conversely, K4 or K9 mono- or dimethylation alters steric barriers and/or electrostatic interactions around the methylation site, and these likely drive the formation of secondary structures at the N-terminal tail. These methylation related structural changes could also promote interactions between the N-terminal tail and residues that are distant from the methylation site, leading to the formation of α-helix structure(s). Structural changes following trimethylation may be similar, except that the distant residues form β-strand and other structures through these interactions. Similar structural changes were previously observed following phosphorylation of OdHI (Barthe et al. 2009). Specifically, phosphorylation induced folding of the unordered region so that the phosphorylated residue bound to its own FHA domain, and an α-helix was formed at distal residues.

To describe mechanisms that lead to the structural changes induced by K4 or K9 methylation, more precise theoretical simulations, such as molecular dynamics simulations, are important and will be interesting subjects of future work.

## 4 Summary

SR-CD spectroscopy revealed that methylation of H3 at K4 or K9 residues induces structural alterations and the data suggest that these alterations occur in adjacent and distal residues from the methylated site. CD data also indicate that H3K4me3 forms a similar structure to that of unmethylated H3, which acts as an inhibitor of LSD1. Because these conformations are closely related to protein–protein interactions, H3K4me3 may also act as an inhibitor of LSD1. Future studies are warranted to investigate binding affinities between H3K4me3 and LSD1. Cyclopedic CD spectroscopy of other methylated histones will also form the basis of future studies into the properties of methylated histones as inhibitors of LSD1.

**Acknowledgments** CD spectroscopy was performed with the approval of the Hiroshima Synchrotron Radiation Center of Hiroshima University (proposal numbers: 15A48, 16AU007, and 16BG018). The author is grateful to Drs. Kentaro Fujii and Akinori Yokoya (National Institutes for Quantum and Radiological Science and Technology, QST), Profs. Koichi Matsuo, Hirofumi Namatame, and Masaki Taniguchi (Hiroshima University) for their kind support. This work was supported by JSPS KAKENHI grant numbers JP15K16130 and JP17K12825.

## References

Amano Y, Umezawa N, Sato S et al (2017) Activation of lysine-specific demethylase 1 inhibitor peptide by redox-controlled cleavage of a traceless linker. Bioorg Med Chem 25:1227–1234

Arndt ER, Stevens ES (1993) Vacuum ultraviolet circular dichroism studies of simple saccharides. J Am Chem Soc 115:7849–7853

Barthe P, Roumestand C, Canova MJ et al (2009) Dynamic and structural characterization of bacterial FHA protein reveals a new autoinhibition mechanism. Structure 17:568–578
Bürck J, Roth S, Windisch D et al (2015) UV-CD12: synchrotron radiation circular dichroism beamline at ANKA. J Synchrotron Radiat 22:844–852
Burg JM, Gonzalez JJ, Maksimchuk KR et al (2016) Lysine-specific demethylase 1A (KDM1A/LSD1): product recognition and kinetic analysis of full-length histones. Biochemist 55:1652–1662
Davey CA, Sargent DF, Luger K et al (2002) Solvent mediated interactions in the structure of the nucleosome core particle at 1.9 Å resolution. J Mol Biol 319:1097–1113
Forneris F, Binda C, Vanoni MA et al (2005) Human histone demethylase LSD1 reads the histone code. J Biol Chem 280:41360–41365
Forneris F, Binda C, Dall'Aglio A et al (2006) A highly specific mechanism of histone H3-K4 recognition by histone demethylase LSD1. J Biol Chem 281:35289–35295
Forneris F, Binda C, Battaglioli E et al (2008) LSD1: oxidative chemistry for multifaceted functions in chromatin regulation. Trends Biochem Sci 33:181–189
Gal-Yam EN, Saito Y, Egger G et al (2008) Cancer epigenetics: modifications, screening, and therapy. Annu Rev Med 59:267–280
Greenfield N (2006) Using circular dichroism spectra to estimate protein secondary structure. Nat Protoc 1:2876–2890
Harris WJ, Huang X, Lynch JT et al (2012) The histone demethylase KDM1A sustains the oncogenic potemtial of MLL-AF9 leukemia stem cells. Cancer Cell 21:473–487
Hussain R, Jávorfi T, Siligardi G (2012) Circular dichroism beamline B23 at the Diamond Light Source. J Synchrotron Radiat 19:132–135
Izumi Y, Matsuo K (2018) Sample-volume reduction using the Schwarzschild objective for a circular dichroism spectrophotometer and an application to the structural analysis of lysine-36 trimethylated histone H3 protein. Molecules 23:2865(1)–2865(12)
Izumi Y, Matsuo K, Fujii K et al (2018a) Circular dichroism spectroscopic study on structural alterations of histones induced by post-translational modifications in DNA damage responses: lysine-9 methylation of H3. J Radiat Res 59:108–115
Izumi Y, Matsuo K, Namatame H (2018b) Structural analysis of lysine-4 methylated histone H3 proteins using synchrotron radiation circular dichroism spectroscopy. Chirality 30:536–540
Jones DT (1999) Protein secondary structure prediction based on position-specific scoring matrices. J Mol Biol 292:195–202
Kim Y, Bigelow L, Borovilos M et al (2008) High-throughput protein purification for X-ray crystallography and NMR. Adv Protein Chem Struct Biol 75:85–105
Lachner M, Jenuwein T (2002) The many faces of histone lysine methylation. Curr Opin Cell Biol 14:286–298
Liu S-H, Lin Y-H, Huang L-J et al (2010) Design and construction of a compact end-station at NSRRC for circular-dichroism spectra in the vacuum-ultraviolet region. J Synchrotron Radiat 17:761–768
Luger K, Mäder AW, Richmond RK et al (1997) Crystal structure of the nucleosome core particle at 2.8 Å resolution. Nature 389:251–260
Maes T, Carceller E, Salas J et al (2015) Advances in the development of histone lysine demethylase innibitors. Curr Opin Pharmacol 23:52–60
Maiques-Diaz A, Somervaille TC (2016) LSD1: biologic roles and therapeutic targeting. Epigenomics 8:1103–1116
Matsuo K, Gekko K (2013) Construction of a synchrotron-radiation vacuum-ultraviolet circular-dichroism spectrophotometer and its application to the structural analysis of biomolecules. Bull Chem Soc Jpn 86:675–689
Matsuo K, Yonehara R, Gekko K (2004) Secondary-structure analysis of proteins by vacuum-ultraviolet circular dichroism spectroscopy. J Biochem 135:405–411
Matsuo K, Yonehara R, Gekko K (2005) Improved estimation of the secondary structures of proteins by vacuum-ultraviolet circular dichroism spectroscopy. J Biochem 138:79–88

Matsuo K, Watanabe H, Gekko K (2008) Improved sequence-based prediction of protein secondary structures by combining vacuum-ultraviolet circular dichroism spectroscopy with neural network. Proteins 73:104–112
McAllister TE, England KS, Hopkinson RJ et al (2016) Recent progress in histone demethylase inhibitors. J Med Chem 59:1308–1329
Micsonai A, Wien F, Kernya L et al (2015) Accurate secondary structure prediction and fold recognition for circular dichroism spectroscopy. Proc Natl Acad Sci USA 112:E3095–E3103. http://bestsel.elte.hu/index.php
Miles AJ, Wallace BA (2006) Synchrotron radiation circular dichroism spectroscopy of proteins and applications in structural and functional genomics. Chem Soc Rev 35:39–51
Miles AJ, Hoffmann SV, Tao Y et al (2007) Synchrotron radiation circular dichroism (SRCD) spectroscopy: new beamlines and new applications in biology. Spectroscopy 21:245–255
Mohammad HP, Smitheman KN, Kamat CD et al (2015) A DNA hypomethylation signature predicts antitumor activity of LSD1 inhibitors in SCLC. Cancer Cell 28:57–69
Niwa H, Umehara T (2017) Structural insight into inhibitors of flavin adenine dinucleotide-dependent lysine demethylases. Epigenetics 12:340–352
Ogasawara D, Itoh Y, Tsumoto H et al (2013) Lysine-specific demethylase 1-selective inactivators: Protein-targeted drug delivery mechanism. Angew Chem Int Ed 52:8620–8624
Ota Y, Suzuki T (2018) Drug design concepts for LSD1-selective inhibitors. Chem Rec 18:1–11
Réfrégiers M, Wien F, Ta H-P et al (2012) DISCO synchrotron-radiation circular-dichroism endstation at SOLEIL. J Synchrotron Radiat 19:831–835
Reichardt G, Bahrdt J, Schmidt J-S et al (2001) A 10 m-normal incidence monochromator at the quasi-periodic undulator U125-2 at BESSY II. Nucl Inst Methods Phys Res A 467-468:462–465
Schmidt DMZ, McCafferty DG (2007) *trans*-2-Phenylcyclopropylamine is a mechanism-based inactivator of the histone demethylase LSD1. Biochemist 45:4408–4416
Shi Y (2007) Histone lysine demethylases: emerging roles in development, physiology and disease. Nat Rev Genet 8:829–833
Shi Y, Lan F, Matson C et al (2004) Histone demethylation mediated by the nuclear amine oxidase homolog LSD1. Cell 119:941–953
Sreerama N, Woody RW (2000) Estimation of protein secondary structure from circular dichroism spectra: comparison of CONTIN, SELCON, and CDSSTR methods with an expanded reference set. Anal Biochem 287:252–260. https://sites.bmb.colostate.edu/sreeram/CDPro/
Sreerama N, Venyaminov SY, Woody RW (1999) Estimation of the number of α-helical and β-strand segments in proteins using circular dichroism spectroscopy. Protein Sci 8:370–380
Suzuki T, Miyata N (2011) Lysine demethylases inhibitors. J Med Chem 54:8236–8250
Tao Y, Huang Y, Gao Z et al (2009) Developing VUV spectroscopy for protein folding and material luminescence on beamline 4B8 at the Beijing Synchrotron Radiation Facility. J Synchrotron Radiat 16:857–863
Tobi D, Bahar I (2005) Structural changes involved in protein binding correlate with intrinsic motions of proteins in the unbound state. Proc Natl Acad Sci USA 102:18908–18913
Whitmore L, Wallace BA (2004) DICHROWEB, an online server for protein secondary structure analyses from circular dichroism spectroscopic data. Nucleic Acids Res 32:W688–W673. http://dichroweb.cryst.bbk.ac.uk/html/home.shtml
Wien F, Wallace BA (2005) Calcium fluoride micro cells for synchrotron radiation circular dichroism spectroscopy. Appl Spectrosc 59:1109–1113
Woody RW (1995) Circular dichroism. Methods Enzymol 246:34–71
Yang M, Culhane JC, Szewzuk LM et al (2007) Structural basis of histone demethylation by LSD1 revealed by suicide inactivation. Nat Struct Mol Biol 14:535–539
Zheng Y-C, Ma J, Wang Z et al (2015) A systematic review of histone lysine-specific demethylase 1 and its inhibitors. Med Res Rev 35:1032–1071

# Asymmetric Dimethylation on Arginine (ADMA) of Histones in Development, Differentiation and Disease

**Amit K. Behera and Tapas K. Kundu**

## Contents



**Abstract** Among myriads of histone modifications known today, asymmetric dimethylation of arginines (ADMA) have been found to have important implications in transcriptional regulation of gene expression. These modifications influence organismal development, regulate cellular differentiation of multiple lineages and modulate pathogenesis of various disease forms such as cancer, metabolic disorders and drug addiction. In this chapter, we discuss roles of ADMA of histones mediated

A. K. Behera
Transcription and Disease Laboratory, Molecular Biology and Genetics Unit, Jawaharlal Nehru Centre for Advanced Scientific Research, Bangalore, India

T. K. Kundu (✉)
Transcription and Disease Laboratory, Molecular Biology and Genetics Unit, Jawaharlal Nehru Centre for Advanced Scientific Research, Bangalore, India

CSIR-Central Drug Research Institute, Lucknow, India
e-mail: tapas@jncasr.ac.in

 
S. Jurga, J. Barciszewski (eds.), *The DNA, RNA, and Histone Methylomes*, RNA Technologies, https://doi.org/10.1007/978-3-030-14792-1_20

by different type I PRMTs in above mentioned physiological contexts and shed light on prospective therapeutic developments.

**Keywords** Asymmetric dimethylation on arginine (ADMA) · PRMTs · Histones · Transcription · Development · Cellular differentiation

## Abbreviations

| | |
|---|---|
| ADMA | Asymmetric dimethyl arginine |
| AMI | Arginine methyltransferase inhibitor |
| AML | Acute Myeloblastic Leukemia |
| AR | Androgen receptor |
| BRG1 | Brahma-related gene-1 |
| CaMKII | $Ca^{2+}$/calmodulin-dependent protein kinase II |
| CNC | Cranial neural crest |
| CVD | Cardiovascular diseases |
| DDAH | Dimethylarginine dimethylaminohydrolase |
| DM1 | Type I Diabetes mellitus |
| E2 | Estradiol |
| EGFR | Epidermal growth factor receptor |
| ER | Estrogen receptor |
| GST-P | Glutathione S-transferase placental form |
| HCC | Hepatocellular carcinoma |
| MEF | Mouse embryonic fibroblast |
| MEF2C | Myocyte enhancer factor-2C |
| MMA | Monomethyl arginine |
| MTA | Methylthioadenosine |
| NAc | Nucleus accumbens |
| NOS | Nitric oxide synthase |
| Nrf2 | Nuclear factor erythroid 2-related factor 2 |
| NS/PC | Neural stem/precursor cells |
| OIS | Oncogene induced senescence |
| PAD4 | Protein arginine deiminase 4 |
| PAF1c | Polymerase-Associated Factor 1 complex |
| PPAR-gamma | Peroxisome proliferator activated receptor-gamma |
| PRMT | Protein arginine methyltransferase |
| PSA | Prostate specific antigen |
| SAM | S-Adenosyl-L-Methionine |
| SDMA | Symmetric dimethylarginine |
| SRC | Steroid receptor coactivator |
| YY1 | Ying Yang 1 |

## 1 Introduction

Asymmetric dimethylation on arginine (ADMA) of histones has been shown to be mediated by type I PRMTs (Protein Arginine Methyltransferases) such as PRMT1, PRMT2, PRMT3, PRMT4/CARM1, PRMT6 and PRMT8 (Fig. 1). Methylation of H4R3 by PRMT1 and PRMT2, methylation of H3R17 and H3R26 mediated by PRMT4/CARM1, methylation of H2AR29 and H3R2 by PRMT6 have been found to influence gene expression in different cellular contexts having important implications in maintenance of pluripotency, cellular differentiation and tumorigenesis. Type I PRMTs play important roles as coregulators of transcription mediated by ER, p53, NF-kB, RUNX1, MEF2C, β-Catenin and PPAR-γ and contribute to normal development as well as patho-physiology of diseases such as cancer. In the last decade, development of therapeutics targeting type I PRMTs has gained momentum and small molecule inhibitors have been shown to successfully restrict growth of cancer cells. In this chapter, we will discuss the roles of ADMA of histones in regulating embryonic development, cellular differentiation and progression of diseases such as cancer, metabolic disorders, drug addiction and aging and highlight current efforts towards development of therapeutics to ameliorate above mentioned pathologies.

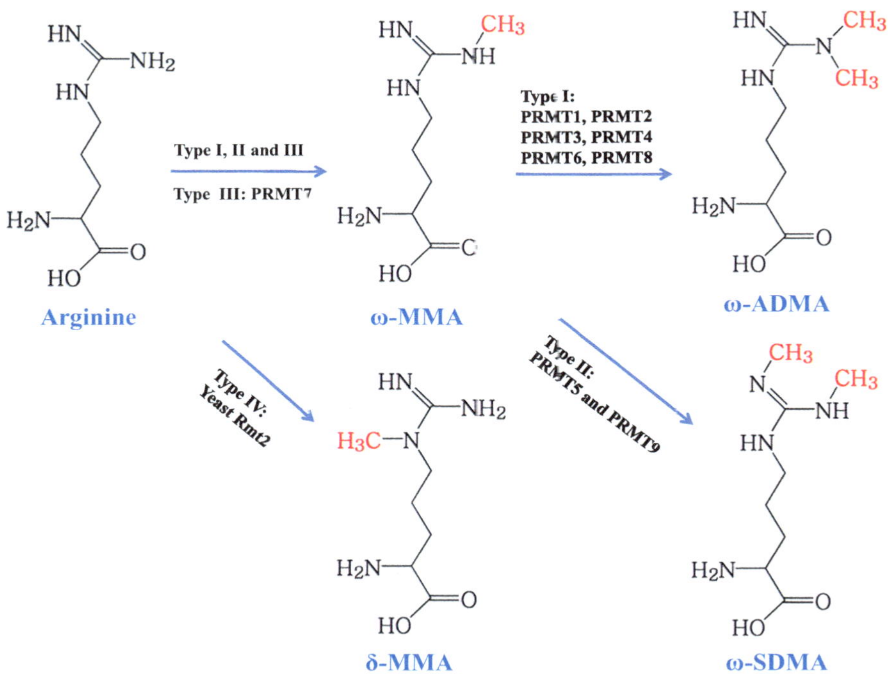


**Fig. 1** A schematic to show classification of PRMTs based on different modes of arginine methylation

## 2 Modes of Arginine Methylation Mediated by PRMTs

PRMTs are classified depending on the mode of methylation they bring about. Type I and type II enzymes both catalyze formation of ω-MMA (ω-monomethylarginine) intermediate, and type I PRMTs lead to the production of ADMA (ω-asymmetric dimethylarginine) whereas type II PRMTs catalyze formation of SDMA (ω-symmetric dimethylarginine). When two methyl groups are attached to single terminal N-atom of side chain of arginine it is designated as asymmetric dimethylation and when both the N-atoms of guanidine group of arginine get one methyl group each, it is designated as symmetric dimethylarginine (Fig. 1). PRMT7 is separated into type III PRMT, which catalyzes formation of monomethylation of arginine (ω-MMA) only. Another group of enzymes add methyl group on δ-N-atom of guanidine group of arginine, forming δ-MMA, a phenomenon so far has been documented only for yeast protein Rmt2 (Bedford and Clarke 2009).

## 3 Expression Status of Type I PRMTs and Their Splice Variants and Isoforms

The information on chromosomal localization of different type I PRMT genes have been enlisted in Table 1. The different splice variants of each type I PRMT members and cellular localization of corresponding protein isoforms have been summarized in Table 1. Except PRMT8, which is brain specific in expression, other type I PRMTs are ubiquitously/widely expressed in different tissue types. The type I PRMTs may achieve a degree of tissue specificity by alternative splicing (Wolf 2009). For example, in most human tissues CARM1ΔE15 (Isoform 1) seems to be the dominant form whereas CARM1FL (Isoform 3) appears to be the major isoform only in four human tissues e.g., heart, brain, skeletal muscle and testis (Wang et al. 2013).

## 4 Biochemical Properties of Type I PRMTs

All PRMTs have a common catalytic methyltransferase domain which consists of a highly conserved core region of around 310 amino acids and subdomains important for binding to the methyl donor S-adenosine-L-methionine (SAM) as well as to the substrate. Signature methyltransferase motifs: motif I, post I, II and III and conserved THW (threonine-histidine-tryptophan) loop are observed in all type I PRMTs. The individual PRMT family members differ in their N-terminal regions with variable length and distinct domain motifs which are believed to be involved in differential interactions with other proteins. PRMTs recognize mostly linear sequence motifs of the target substrates for arginine methylation. Recent studies show that recognition motifs for arginine methylation is truly diverse in nature and not limited to

**Table 1** Information on expression of various splice variants of type I PRMTs in human

| S. N. | Type I PRMTs | Cytogenetic band (locus) | Gene size (bp) (GRCh38/hg38) | Total exons | Splice variants | Amino acids | Cellular localization | Tissue specific protein expression |
|---|---|---|---|---|---|---|---|---|
| 1 | PRMT1 | 19q13.33 | 13,244 bp | 10 | Isoform 1 (V2) (C) | 371aa | Cytoplasmic | Ub |
| | | | | | Isoform 2 (V3) | 347aa | Unknown | Unknown |
| | | | | | Isoform 3 (V1) | 353aa | Nuclear | Ub |
| | | | | | Isoform 4 | 285aa | Unknown | Unknown |
| 2 | PRMT2 | 21q22.3 | 30,533 bp | 10 | Isoform I (C) | 433aa | Nuclear | WE |
| | | | | | Isoform II | 284aa | Unknown | Unknown |
| | | | | | Isoform III | 331aa | Unknown | Unknown |
| | | | | | Isoform PRMT2α | 289aa | Nuclear | Unknown |
| | | | | | Isoform PRMT2β | 301aa | Nuclear | Unknown |
| | | | | | Isoform PRMT2γ | 228aa | Nuclear | Unknown |
| | | | | | Isoform PRMT2L2 | 277aa | Nuclear | Unknown |
| 3 | PRMT3 | 11p15.1 | 121,804 bp | 16 | Isoform 1 (C) | 531aa | Nuclear | WE |
| | | | | | Isoform 2 | 469aa | Unknown | Unknown |
| 4 | PRMT4 | 19p13.2 | 51,558 bp | 16 | Isoform 1 | 585aa | Nuclear | Ub |
| | | | | | Isoform 2 | 384aa | Unknown | Unknown |
| | | | | | Isoform 3 (C) | 608aa | Nuclear | Ub |
| 5 | PRMT6 | 1p13.3 | 2650 bp | 1 | Isoform 1 (C) | 375aa | Nuclear | WE |
| | | | | | Isoform 2 | 292aa | Unknown | Unknown |
| 6 | PRMT8 | 12p13.32 | 212,625 bp | 10 | Isoform 1 (C) | 394aa | PM | Brain |
| | | | | | Isoform 2 | 385aa | Nuclear | Unknown |

*C* Canonical, *PM* Plasma membrane, *Ub* Ubiquitous, *WE* Widely expressed

traditionally believed "RGG" motif in the substrates (Wooderchak et al. 2008; Hamey et al. 2018).

The type I PRMTs share homology in the amino acid sequences thus exhibiting conservation in structural aspects as well, especially in the centrally located catalytic domain responsible for methyltransferase activity (Fig. 2). The N-terminal domains of the PRMTs seem to vary, being unique to individual enzymes and believed to regulate substrate specificity. The N terminal domains also harbor special motif regions which impart specific functional properties to different PRMTs (Fig. 2). For example, PRMT2 contains a SH3 motif, which modulates protein-protein interaction. PRMT3 contains a Zinc finger motif (ZnF) that helps in interaction and subsequent methylation of ribosomal protein rpS2 (Swiercz et al. 2007). PRMT8 contains a Myristoylation motif (Myr) in the N-terminus, which is important for plasma membrane localization. Among all the type I PRMTs, PRMT4 contains a unique C-terminal domain which is known to regulate coactivator properties of the protein (Teyssier et al. 2002; Troffer-Charlier et al. 2007) (Fig. 2).

PRMTs use S-Adenosyl-L-Methionine (SAM) as methyl donor for methylation of substrates. Experimental evidences from steady state kinetics and crystal structure analysis has revealed that the asymmetric dimethylation of arginine by type I PRMTs follow distributive mode of catalysis (Ordered sequential mechanism), where each event of catalysis is followed by release of the substrate (Kölbel et al. 2009; Lakowski and Frankel 2008, 2009). Therefore, both monomethylated and dimethylated end products will be formed by the enzymes; in contrast to processive dimethylation where only dimethylated end product would be expected to be formed. However, preference for monomethylated substrate over unmethylated substrate by the enzyme might make the distributive mode of catalysis seem processive, obscuring the interpretation of catalytic mechanism as observed with PRMT6 (Table 2).

Type I PRMTs such as PRMT1, PRMT2, PRMT4 and PRMT6 form homodimers and dimerization is known to be important for catalytic activity (Zhang and Cheng 2003; Lee et al. 2007). PRMT1 also forms higher order oligomers, relevance of which is not yet clearly understood. While PRMT3 forms monomers, PRMT8 seems to form either homo-tetramer (Lee et al. 2015) or homo-octamer (Toma-Fukai et al. 2016) for its physiological function (Table 2).

## 5 ADMA of Histones in Transcription

Asymmetric dimethylation of histones primarily regulate transcription from chromatin template via recruitment of transcriptional machinery. PRMT1 mediated H4R3 methylation has been shown to aid in nuclear receptor mediated transcription. In this context, a mutation in SAM binding site abrogated coactivation property of PRMT1 (Wang et al. 2001). PRMT2 and PRMT3 also methylate H4R3 in vitro and in vivo with putative transcriptional activation properties (Fig. 3; Table 3). PRMT2 mediated H3R8 methylation has been shown to be important in β-catenin mediated gene regulation (Blythe et al. 2010). Biochemical studies with peptide mapping

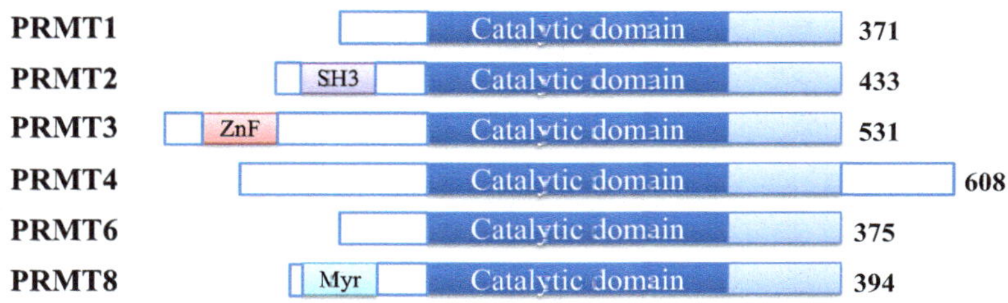


**Fig. 2** A schematic to show the domain organization of type I PRMTs. The canonical or the longest forms of PRMTs have been depicted with respective protein length (amino acids). Catalytic domain has been shown in blue (SAM binding domain in deep blue, substrate binding domain in light blue). Unique motifs present in N terminal domain in different members have been highlighted. *SH3* SRC homology 3 domain, *ZnF* Zinc finger motif, *Myr* Myristoylation motif

**Table 2** Information on catalytic domain span, oligomerization status of full length proteins and mechanism of catalysis of different type I PRMTs

| S. N. | Type I PRMTs | Catalytic domain (amino acid) | Functional Oligomerization | Mechanism of catalysis | Substrate preference (Unmethylated vs. monomethylated petptide) |
|---|---|---|---|---|---|
| 1 | PRMT1 | 34-322aa (rat) | Homo-dimer Homo-oligomer | Distributive | Similar |
| 2 | PRMT2 | 101-398aa (human) | Homo-dimer | Distributive | Similar |
| 3 | PRMT3 | 214-505aa (rat) | Monomer | Distributive | Similar |
| 4 | PRMT4 | 149-444aa (mouse) | Homo-dimer | Distributive | Similar |
| 5 | PRMT6 | 48-375aa (human) | Homo-dimer | Distributive | Preference for mono-methylated peptide |
| 6 | PRMT8 | 81-394aa (human) | Homo-tetramer Homo-octamer | Distributive | Unknown |

approach has revealed PRMT4/CARM1 mediated methylation at H3R2, H3R17, and H3R26 in the N-terminal region as well as on one or more of four arginines (128/129/131/134) at the C-terminal region of H3 in vitro (Schurter et al. 2001; Jacques et al. 2016). Arginine methylation of two of the N-terminal sites: H3R17 and H3R26 have been observed in vivo and map to the peptide sequence of KAXRK. Methylation of histone H3R17 by CARM1 recruits human RNA Polymerase-Associated Factor 1 complex (PAF1c) to the chromatin to activate transcription of estrogen responsive genes (Wu and Xu 2012; Bauer et al. 2002). CARM1 also methylates core region of histone H3 on residue H3R42 with subsequent activation of transcription (Casadio et al. 2013). PRMT6 is largely responsible for H3R2me2a deposition in cell with transcriptional repression attributes. While H2AR29me2a mediated by PRMT6 shown to aid in transcriptional repression (Waldmann et al. 2011), H3R42 methylation by PRMT6 might lead to transcriptional activation

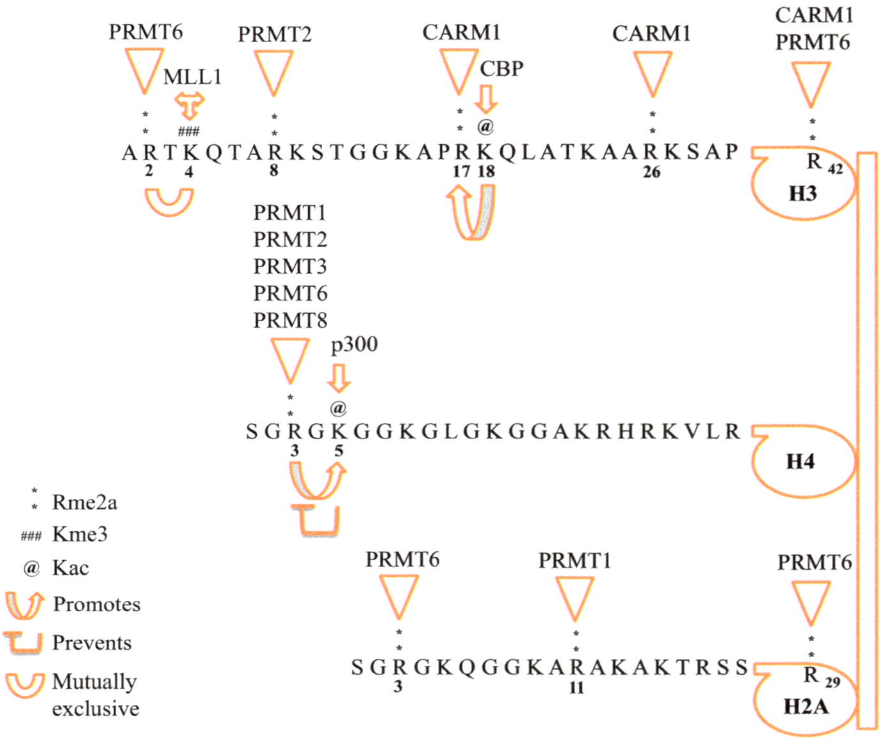


**Fig. 3** A schematic to show ADMA of histone H3 and H4 by different type I PRMTs and their cross-talk with different adjacent histone modifications

(Casadio et al. 2013). Relevance of methylation of H4R3 and H2AR3 by PRMT6 is yet to be elucidated.

Studies by different groups show that asymmetric dimethylation of arginines show definitive interplay with other histone modifications, which decides the transcriptional state of the chromatin template. Different cross-talks involving asymmetric dimethylation of arginines on histones have been summarized in Table 3 and depicted in Fig. 3. Mostly, ADMA of histones are prevented by acylation (acetylation, butyrylation, crotonylation etc.) of nearby lysine residues. For example, H4R3me2a by PRMT1, PRMT3 and PRMT8 are prevented by acylation of H4K5 (H4K5ac, H3K5bu, H3K5cr etc.) (Fulton et al. 2017; Feng et al. 2011). However, on the contrary, H3K18ac by CBP is known to facilitate H3R17me2a by CARM1 (Daujat et al. 2002). ADMA of histones seems to have a favorable effect on acetylation of nearby lysine residues in histones. For example, H4R3me2a by PRMT1 favors acetylation of H4 by p300 (Wang et al. 2001).

H3R2me2a by PRMT6 have been shown to prevent H3K4me3 at promoters of active genes rather than silent genes, thereby leading to suppression of gene expression (Guccione et al. 2007; Hyllus et al. 2007; Bouchard et al. 2018). H3R2me2a mediated impedance in H3K4me3 mediated transcription from promoters is believed

**Table 3** Various sites on core histones asymmetrically dimethylated by the type I PRMTs and their effect on transcription

| S. N. | Type I PRMTs | Histone modification | Involvement in cellular processes | Cross-talk with other histone modifications. |
|---|---|---|---|---|
| 1 | PRMT1 | H4R3me2a<br>H2AR11me2a | Transcriptional activation<br>Unknown | • H4K5ac prevents H4R3me2a by PRMT1<br>• H4R3me2a facilitates H4 acetylation by p300<br>• H3K9 and H3K27 methylation is indirectly impaired by H4R3me2a |
| 2 | PRMT2 | H4R3me2a<br>H3R8me2a | Transcriptional activation<br>Unknown | Unknown<br>Unknown |
| 3 | PRMT3 | H4R3me2a | Unknown | • H4K5me enhances H4R3me2a by PRMT3*<br>• H4K5ac prevents H4R3me2a by PRMT3* |
| 4 | PRMT4 | H3R17me2a<br>H3R26me2a<br>H3R42me2a | Transcriptional activation<br>Transcriptional activation<br>Transcriptional activation* | • H3K18ac by CBP promotes H3R17me2a |
| 5 | PRMT6 | H3R2me2a<br>H3R42me2a<br>H4R3me2a<br>H2AR3me2a<br>H2AR29me2a | Transcriptional repression<br>Transcriptional activation*<br>Unknown<br>Unknown<br>Transcriptional repression | • H3R2me2a prevents H3K4me3 by MLL complex at promoters and vice versa<br>• H3R2me2a facilitates H3K4me1 by KMT2D and H3K27ac by p300 at enhancers |
| 6 | PRMT8 | H4R3me2a*<br>H2AR3me2a* | Unknown<br>Unknown | • H4K5ac prevents H4R3me2a by PRMT8*<br>• H4K5me enhances H4R3me2a by PRMT8* |

Cross-talk with other histone marks also has been enlisted (*In vitro)

to fine tune expression of active genes and check against excessive unwanted transcriptional output. At the enhancers, however, H3R2me2a seem to promote deposition of H3K4me1 by KMT2D and H3K27ac by p300/CBP in NT2/D1 cells in neural differentiation dependent manner (Bouchard et al. 2018).

Type I PRMTs function as coactivators of different transcription factor mediated gene expression. PRMT1 can coactivate ERα, p53, YY1 and RUNX1. PRMT2 can coactivate AR and ERα. CARM1 can coactivate ERα, AR, SRC, β-catenin, p53, c-Fos, NF-kB, MEF2C and PPAR-γ (Di Lorenzo and Bedford 2011). PRMT6 can coactivate ERα, AR, GR and NF-kB (Sun et al. 2014; Di Lorenzo et al. 2014) and corepress RUNX1 (Lausen 2013). Moreover, PRMT1 and CARM1 show synergy with each other as well as with acetyltransferase p300/CBP as transcriptional coactivators (Stallcup et al. 2000; Xu et al. 2003; An et al. 2004). Coactivation of transcription from promoters largely seems to depend on catalytic activity of the enzymes involved. However, there are evidences to suggest catalytic activity independent coactivation by CARM1, where unique C-terminal domain might play an important role.

# 6 Role of ADMA of Histones in Development

The role of PRMT1 in embryonic development has been elucidated in zebra fish and mouse model with loss of function approach. Knock down of PRMT1 with morpholinos leads to impaired gastrulation (Tsai et al. 2011). Most of the morphants show defective epiboly at 10 hpf. PRMT1 plays an important role in craniofacial bone formation in mice. Cranial neural crest (CNC)-specific knock out of Prmt1 leads to defective craniofacial bone structure as analyzed by 3D-micro CT (micro computed tomography). Alterations observed in premaxilla, maxilla, frontal bone, palatine bone, mandible, incisor and alveolar bone seems to be phenotypically similar to that of Msx1-deficient mice. Moreover, depletion of PRMT1 led to downregulation of Msx1 expression in frontal bone and mandible primordium and calvaria-derived preosteoblast, suggesting upstream regulation of Msx1 expression by PRMT1 (Gou et al. 2018a). The same research group also demonstrated crucial role of PRMT1 in palate development. In absence of PRMT1 palatal mesenchymal cells showed reduced proliferation causing impaired palatogenesis. When PRMT1 function was disrupted in neural crest cells, it resulted in cleft palate phenotype in mice. At molecular level, loss of PRMT1 accompanied reduced levels of H4R3me2a mark and attenuated BMP signaling pathway. This study indicates the therapeutic potential of PRMT1 as a molecular target in prevention of cleft palate birth defects (Gou et al. 2018b).

The ability to regulate transcription of various genes has implicated an important role of CARM1 in the context of development. The blastomeres from four celled embryos with high arginine methylation on histone H3R17 and H3R26 have been shown to possess higher potential to support embryonic development compared to the blastomeres with low levels of arginine methylation on the above mentioned sites (Wu et al. 2009). Further investigation has unraveled an important role of CARM1 in maintenance of pluripotency of embryonic stem cells. With an epigenetic influence, CARM1 regulates expression of key pluripotent factors such as Oct4, Sox2 and Nanog which aid in self renewal as well as resistance to differentiation. Interestingly, CARM1 knock out did not result in embryonic lethality. However, CARM1−/− mice were smaller in size and died immediately after birth due to impaired respiratory functions (O'Brien et al. 2010). This observation strongly indicates an essential role of CARM1 in the process of normal development.

PRMT8 expression has been shown to be important for zebrafish embryonic development. While ubiquitous in expression in early embryonic stages, it becomes exclusively brain specific at 96 hpf. Defective developmental phenotypes were observed upon morpholino based knock down of PRMT8 (Lin et al. 2013). PRMT8 morphants possessed smaller brains, short trunks and curly tails indicating defects in neuronal development, epiboly and convergence/extension during embryonic development. Loss of HuC positive neurons due to apoptosis was also observed with knock-down of PRMT8. These defects could be rescued with introduction of

**Table 4** Knock-out phenotypes of mice deficient with different type I PRMTs as a reflection on the role in organismal development

| S. N. | Type I PRMTs | Knock-out phenotype in mice | Developmental defects |
|---|---|---|---|
| 1 | PRMT1 | Embryonic lethality around D6.5 to D7.5 | – |
| 2 | PRMT2 | KO mice are viable | No gross abnormality |
| 3 | PRMT3 | KO mice are viable | Smaller size during embryonic development and after birth, but achieve normal size when adult |
| 4 | PRMT4 | KO mice are smaller in size and die after birth due to impaired respiration. | Impaired lung development, impaired thymocyte development, impaired adipogenesis |
| 5 | PRMT6 | KO mice are viable | No gross abnormality |
| 6 | PRMT8 | KO mice are viable | Impaired brain development |

PRMT8 cRNAs, but not with catalytically inactive form of PRMT8, indicating importance of arginine methylation mediated by PRMT8 in embryonic development. PRMT8 has been shown to contribute to maintenance of pluripotency of hESC via PI3K/AKT/SOX2 axis (Jeong et al. 2017). Mice deficient with PRMT2, PRMT3 and PRMT6 are viable. PRMT3 knock-out mice are smaller during embryonic development; however, they achieve normal size during adulthood. PRMT2 and PRMT6 knock-out mice show no gross phenotypic developmental defects (Table 4).

# 7 Role of ADMA of Histones in Differentiation

The role of different type I PRMTs in regulation of cellular differentiation has been summarized in Table 5. PRMT1 seems to promote embryonic stem cell differentiation, neurogenesis, myogenesis, osteoclastogenesis (Choi et al. 2018). Functions of PRMT2 and PRMT3 in regulation of cellular differentiation have not yet been reported. CARM1 has been shown to promote neurogenesis, myogenesis, adipogenesis and chondrogenesis while repressing embryonic stem cell differentiation (Wu et al. 2009). PRMT6 has been shown to play a favorable role for embryonic stem cell differentiation (Lee et al. 2012). PRMT8 has been shown to suppress embryonic stem cell differentiation (Jeong et al. 2017) and promote neurogenesis (Lin et al. 2013). The role of ADMA of histones by different type I PRMTs in the context of neurogenesis, myogenesis and adipogenesis has been further discussed in greater details in subsequent sections.

## *7.1 Neurogenesis*

PRMT1 has been found to promote differentiation of neural stem/precursor cells (NS/PCs) towards astrocyte generation. In the context of astrocytic differentiation,

**Table 5** Regulatory role of different type I PRMTs in different types of cellular differentiation

| S. N. | Type I PRMTs | Promotes | Suppresses |
|---|---|---|---|
| 1 | PRMT1 | Embryonic stem cell differentiation<br>Neurogenesis<br>Myogenesis<br>Osteoclastogenesis | – |
| 2 | PRMT2 | Unknown | Unknown |
| 3 | PRMT3 | Unknown | Unknown |
| 4 | PRMT4 | Neurogenesis<br>Myogenesis<br>Adipogenesis<br>Chondrogenesis | Embryonic stem cell differentiation (Maintains pluripotency) |
| 5 | PRMT6 | Embryonic stem cell differentiation | – |
| 6 | PRMT8 | Neurogensis | Embryonic stem cell differentiation |

PRMT1 mediated methylation of STAT3 seems to play an important role (Honda et al. 2017). Histone modification by PRMT1 in the context of neurogenesis has not been addressed yet. CARM1 has been found to play an important role in maintaining glial cell population (Selvi et al. 2015). It was shown that targeted inhibition of H3R17 by ellagic acid and silencing of CARM1 could result in defective glial cell morphology in zebra fish model. The observed phenotypic sensory defects and cellular abnormality could be restored with exogenous overexpression of CARM1. In this context CARM1 was found to regulate neural development pathway through activation of Nanog expression via methylation of H3R17 on regulatory promoter elements. Nanog in turn was shown to regulate miRNAs (miR 17–92 network) that are associated with glial cell maintenance, which was negatively affected when CARM1 methyltransferase activity was inhibited with TBBD (Ellagic acid) (Fig. 4).

## 7.2 Myogenesis

Global ADMA level has been found to increase during the course of muscle differentiation (Shen et al. 2018). While expression of PRMT1 is induced during the course of differentiation, PRMT4 expression does not vary through myogenesis in a 7 day differentiation process in C2C12 cells. However, myonuclear localization of both PRMT1 and PRMT4 is increased with concomitant increase in levels of H4R3me2a and H3R17me2a in Day 7 myofibers compared to myoblast cells. Pharmacological inhibition of PRMT1 activity by TC-E led to reduced mitochondrial biogenesis, subsequently resulting in attenuated myogenesis; highlighting importance of ADMA of histones in the process of muscle differentiation (Shen et al. 2018). In the context of myogenesis it was found that CARM1 could act as a

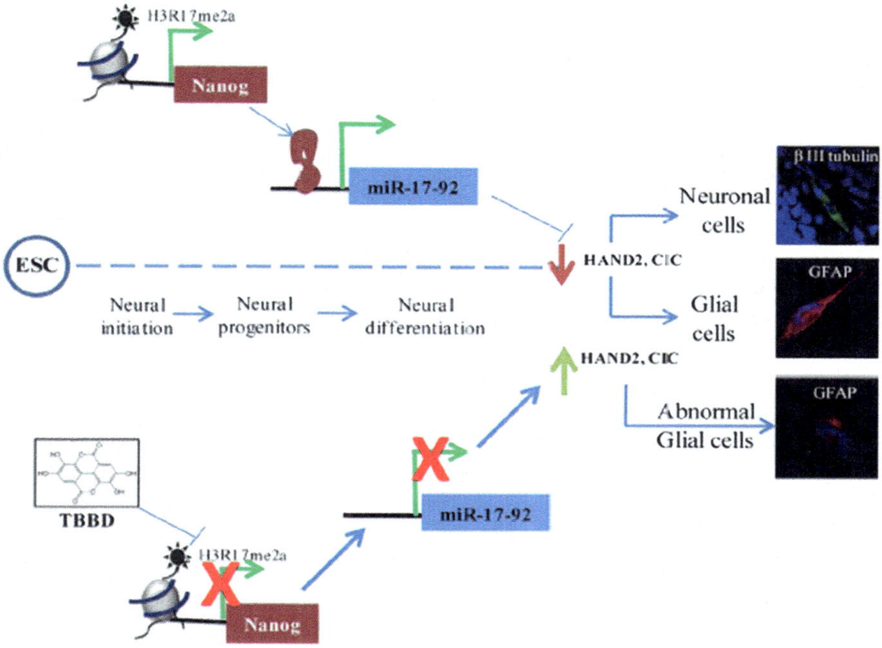


**Fig. 4** Schematic representation of the role of H3R17 methylation in specifying the astroglial lineage (Adapted from Selvi et al. 2015)

coactivaor for myocyte enhancer factor-2C (MEF2C) to support and potentiate muscle differentiation cascade by maintaining expression of myogenin and MEF2C and their target genes in myoblast cells (Chen et al. 2002). In a similar vein it has been demonstrated that CARM1 can positively regulate myogenin and MEF2C expression and thereby facilitate fast fiber formation and proper localization of slow muscle fiber in vivo in zebra fish model system (Batut et al. 2011). CARM1 was also shown to induce expression of myogenic miRNAs during the course of muscle differentiation, via H3R17me2a mediated recruitment of myogenin and BRG1 ATPase along with MyoD onto regulatory elements of miRNA genes for chromatin remodeling and subsequent gene expression regulation (Mallappa et al. 2011).

## 7.3 Adipogenesis

CARM1 has been convincingly shown to promote adipocyte differentiation as a bona-fide co-activator of PPAR-γ mediated gene regulation. A comprehensive analysis of gene expression unraveled a set of genes involved in lipid metabolism to be underrepresented in CARM1 knock-out mouse embryos compared to wild type control. In agreement to this finding, CARM1 knock-out embryos showed reduced brown fat mass compared to wild type counterparts. Additionally, CARM1 deficient cells exhibited lesser adipogenesis potential in terms of their conversion to mature adipocytes; further confirming its importance in regulating adipogenesis (Yadav et al. 2008). Pharmacological inhibition of CARM1 mediated H3R17 methylation by ellagic acid has been shown to lead to reduction in adipocyte differentiation in murine 3T3L1 cells (Behera et al. 2018). Similarly, ellagic acid treatment in human adipose tissue derived stem cells (hASCs) inhibited H3R17 methylation by CARM1 and suppressed subsequent adipogenesis (Kang et al. 2014).

# 8 Role of ADMA of Histones in Disease

Regulatory role of ADMA of histones by different type I PRMTs have been investigated from pathogenesis point of view in a few human diseases mainly in cancer and metabolic disorders. The type I PRMTs have been found to be overexpressed in multitude of cancer types and shown to be mostly oncogenic in nature, with a few exceptions where they seem to exhibit tumor suppressor function (Table 6). Role of ADMA of histones in a few major forms of cancers such as breast cancer, hepatocellular carcinoma, prostate cancer and colorectal cancer has been discussed with greater details in subsequent sections. Among metabolic disorders, cardiovascular diseases and diabetes have shown definitive association between ADMA of histones with pathophysiological outcome of the disease. The role of ADMA of histone in the context of above mentioned metabolic disorders have been discussed in greater detail in subsequent sections.

## *8.1 Cancer*

### 8.1.1 Breast Cancer

Two major functional isoforms of PRMT1 such as PRMT1v1 (nuclear) and PRMT1v2 (cytosolic) are overexpressed in breast cancer. PRMT1v2 variant has been shown to be important for breast cancer cell survival and invasiveness (Baldwin et al. 2012). Further studies have shown that PRMT1v1 and PRMT1v2 exhibit cellular localization specific interactome in breast cancer cells. An enriched nuclear protein interactome for PRMT1v1 suggested a role in gene expression regulation; while a predominantly cytosolic interactome for PRMT1v2 suggested involvement in regulation of cytoskeletal dynamics, thus justifying the favorable role in metastasis (Baldwin et al. 2015). According to another study, PRMT1 methylates H4R3 on ZEB1 promoter to activate its expression in the context of regulation of EMT (Epithelial to Mesenchymal Transition) in breast cancer (Gao et al. 2016).

**Table 6** Different cancer types where deregulated expression of type I PRMTs have been observed

| S. N. | Type I PRMTs | Up-regulated (in) | Down-regulated (in) | Promotes | Inhibits |
|---|---|---|---|---|---|
| 1 | PRMT1 | Head and neck cancer, glioma<br>Breast cancer<br>Lung cancer<br>HCC<br>Leukemia<br>Gastric cancer<br>Melanoma<br>Bladder cancer | Pancreatic cancer | Oral cancer, Glioma<br>Breast cancer<br>Lung cancer<br>HCC<br>Leukemia<br>Gastric cancer<br>Melanoma<br>Bladder cancer | Pancreatic cancer |
| 2 | PRMT2 | Breast cancer | – | – | – |
| 3 | PRMT3 | Breast cancer | AML | – | – |
| 4 | PRMT4 | Breast cancer<br>Lung cancer<br>HCC<br>Colon cancer<br>Prostate cancer<br>Osteosarcoma<br>Melanoma | – | Breast cancer<br>Lung cancer<br>HCC<br>Colon cancer<br>Prostate cancer | – |
| 5 | PRMT6 | Breast cancer<br>Lung cancer<br>Colon cancer<br>Prostate cancer<br>Bladder cancer<br>Cervical cancer | Melanoma<br>HCC | Breast cancer<br>Colon cancer<br>Prostate cancer<br>Bladder cancer | HCC |
| 6 | PRMT8 | Glioblastoma<br>Breast cancer<br>Prostate cancer<br>Ovarian cancer<br>Cervical cancer | AML | – | – |

Regulatory role of different type I PRMTs in different cancers, as evidenced from cellular studies, have been indicated (*AML* Acute Myeloblastic Leukemia, *HCC* Hepatocellular carcinoma)

CARM1 could act as a positive regulator of estrogen receptor (ER) mediated transcriptional activation in cell. This mechanism has been shown to play an essential role in estrogen induced cell cycle progression in MCF7 breast cancer cells. Estrogen stimulated expression of E2F1 was revealed to be CARM1 dependent, where E2F1 promoter chromatin was arginine methylated on H3R17 by CARM1 to facilitate recruitment of transcriptional regulators (Frietze et al. 2008). Another study on similar grounds claims an opposite function of CARM1 in ER positive breast cancer, where it can block cellular proliferation in MCF7 cells and induce cellular differentiation by modulating expression of genes regulated by ER-alpha. In this context CARM1 expression correlated to ER-alpha levels in ER positive tumors but was inversely correlated to tumor grade, indicating its putative tumor suppressive role. This was further confirmed in xenograft studies, in which knock down of CARM1 in MCF7 cells enhanced tumor growth. Authors of this study opine that consideration of coexpression of ER-alpha and CARM1 could serve as a better biomarker for well differentiated breast cancer (Al-Dhaheri et al. 2011). An investigation from another group suggest an oncogenic role of CARM1 in breast cancer, where CARM1 expression status positively correlated with large tumor size and high tumor grade along with biomarkers associated with luminal phenotype and poor prognosis such as HER2, EGFR, basal cytokeratins, CD71, Ki-67 and cyclin E. Outcome analyses has revealed its expression to be an independent predictor of shorter disease-free interval and patient survival in the whole series of invasive breast cancers and in the ER-positive subgroup (Habashy et al. 2013). In an attempt to understand the molecular pathways regulated by CARM1 in the context of breast cancer, a recent investigation uncovered tumorigenic function of chromatin remodeler BAF155 upon arginine methylation by CARM1. BAF155 shows differential recruitment on chromatin when modified by arginine methylation and subsequently activate c-myc pathway genes (Wang et al. 2014). PRMT6 has been shown to promote breast cancer growth and suppress senescence. Silencing of PRMT6 has been shown to lead to reduction in H3R2 methylation and derepression of p21 expression, resulting in cell cycle arrest and senescence in MCF7 cells (Phalke et al. 2012).

### 8.1.2 Hepatocellular Carcinoma

PRMT1 has been found to be upregulated in hepatocellular carcinoma (HCC) and shown to promote HCC growth and metastasis. PRMT1 expression shows association with poor prognosis in HCC patients (Zhang et al. 2018b; Ryu et al. 2017). Study with knock-down of PRMT1 by siRNA in HCC cells has been shown to lead to reduced proliferation, migration and invasion in vitro (Gou et al. 2017). According to a recent study PRMT1 might suppress hepatocyte proliferation via regulation of Hnf4a expression through arginine methylation of promoter (Zhao et al. 2018). In this study PRMT1 seemed to play a tumor suppressor role, where loss of PRMT1 led to higher proliferation of hepatocytes and resulted in 33% increase of liver size in hepatocyte specific knock-out mice. Moreover, mice fed with alcohol showed lower activity of PRMT1 in the liver. Authors demonstrate an interplay between PRMT1 and JMJD6 in regulating Hnf4a expression through arginine methylation and suggest a role for arginine methylation in alcohol induced liver cancer. Aberrant overexpression and oncogenic function of CARM1 in hepatocarcinoma has also been documented where it was found to activate promoter of GST-P (glutathione S-transferase placental form) through coactivation of transcription factor Nrf2 (Osada et al. 2013).

### 8.1.3 Prostate Cancer

Significant association between CARM1 overexpression and development of androgen-dependent prostate carcinoma as well as androgen-independent prostate carcinoma has been observed (Hong et al. 2004). CARM1 has been shown to play an important role in androgen receptor mediated gene regulation in prostate cancer, where androgen stimulation leads to CARM1 recruitment and methylation of H3R17 on AR responsive enhancers. In this context, silencing of CARM1 expression leads to reduction in androgen-dependent PSA (prostate specific antigen) and hK2 mRNA expression and subsequently resulting in reduced cellular proliferation (Majumder et al. 2006). PRMT6 has been found to be overexpressed and oncogenic in the context of prostate cancer. Knock-down of PRMT6 in PC3 cells reduced cell viability with induction of apoptosis. Decrease in migration and invasion was also observed with PRMT6 knock-down. Hypomethylation of H3R2 associated with PRMT6 silencing negatively affected expression of Akt, while enhancing expression of p27, PSA, AR in PC-3 cells (Almeida-Rios et al. 2016).

#### 8.1.4 Colorectal Cancer

Colorectal cancer has been shown to be associated with elevated levels of CARM1 expression (Kim et al. 2010). Further investigation has revealed coactivator role of CARM1 for β-catenin mediated gene expression in the context of colorectal cancer. CARM1 was found to be recruited to Wnt target genes through its interaction with β-catenin and mediate transcriptional activation via methylation of H3R17 on corresponding regulatory elements (Ou et al. 2011). PRMT6 has also been found to be upregulated in patient samples of colorectal cancer compared to adjacent normal tissue with strong association with clinicopathological features and patient survival (Lim et al. 2017). Knock-down of PRMT6 in HT29 colon cancer cells resulted in hypomethylation of H3R2 and subsequently derepression of p21 leading to cell cycle arrest as reflected by reduced cell proliferation. Moreover, PRMT6 silencing led to increase in levels of cleaved caspase 3 and cleaved PARP2 indicating higher levels of apoptosis.

## *8.2 Metabolic Disorders: Cardiovascular Diseases (CVDs) and Diabetes*

### 8.2.1 Cardiovascular Diseases

Canonical histones are usually incorporated to chromatin in replication dependent manner. Replication independent histone variant exchange is known to be employed by the cells to regulate gene expression. Turnover of both canonical and variant histones, which are arginine methylated by type I PRMTs would yield asymmetric dimethylarginine (ADMA) upon proteolysis. Turnover of arginine methylated non-histone proteins also contributes to the intracellular pool of ADMAs. ADMA is an endogenous inhibitor of nitric oxide synthase (NOS) and thereby linked to endothelial dysfunction in cardiovascular diseases (CVDs). While monomethylarginine

(MMA) exhibits similar inhibitory properties, symmetric dimethylarginine does not inhibit NOS and gets cleared by renal excretion without any toxic effects. ADMA and MMA are subjected to hepatic metabolism as part of detoxification process where they are degraded by dimethylarginine dimethylaminohydrolase (DDAH) to citrulline and mono or dimethylamine.

The L-Arg/ADMA ratio is suggested to regulate vascular homeostasis. Elevated plasma levels of ADMA has been observed in multiple types of cardiovascular disorders such as hypercholesterolemia, hypertriglyceridemia, hypertension, peripheral arterial disease, chronic heart failure, chronic renal failure etc. (Böger 2003). Hence, ADMA is considered as a biomarker or risk factor for cardiovascular diseases related to endothelial dysfunction. The physiological plasma concentration of L-Arg and ADMA are 100 μM and $0.42 \pm 0.06$ μM respectively. As ADMA levels need to reach approximately 10 μM in plasma (which is rarely achieved) to adversely affect NO generation by NOS in the cardiovascular system (Zakrzewicz and Eickelberg 2009), it most probably elicits its deleterious effects by inhibition of intracellular NOS in different cell types such as vascular endothelial cells. Therefore, the association studies with plasma levels of ADMA and cardiovascular complications need to be further investigated to identify the specific cell types responsible for disrupted vascular homeostasis for each type of disorder. Myocardial tissue of patients with atherosclerosis of coronary artery have been found to show high expression of PRMT1, PRMT3 and low expression of DDAH2, which seem to contribute to elevated levels of ADMA in the patients (Chen et al. 2006).

### 8.2.2 Diabetes

The levels of ADMA in plasma have been found to show negative correlationship with pathogenesis of diabetes, in contrast to CVD where it shows positive association. According to a study, pediatric patients with type I diabetes mellitus (DM1) contain significantly lower levels of ADMA compared to healthy young children (Huemer et al. 2011). In this context authors suggest that DM1 patients with lower levels of ADMA probably would be more vulnerable to NO induced oxidative stress. This could contribute to pathophysiology of DM1. A recent study further confirms this observation with the findings that ADMA concentration decreases with increase in duration of diabetes in young DM1 patients (Ersoy et al. 2018). Furthermore, carotid intima-media thickness showed negative association with ADMA levels in DM1 patients, indicating development of subclinical atherosclerosis as ADMA levels decreases with progression of DM1 in young patients. Doppler echocardiography also indicated similar negative relationship between ADMA levels and diastolic myocardial annular velocity. Therefore, decreased levels of ADMA in young DM1 patients seemed to correlate to higher risks of vasculopathy, which is in contradiction to the observation in many of the cardiovascular diseases as discussed above.

## 9 ADMA of Histones in Drug Addiction

A scientific study shows that cocaine administration enhances PRMT1 activity in Nucleus accumbens (NAc) of brain of C57BL/6 mice with concomitant upregulation in H4R3me2a levels. Cocaine seemed to induce expression of CaMKII and Cdk5 in NAc via PRMT1 mediated histone modification, which could be attenuated with genetic silencing of PRMT1 or treatment with small molecule inhibitors of PRMT1 (Li et al. 2015). On the other hand, cocaine administration has been found to lead to decreased levels of PRMT6 mediated H3R2me2a in NAc of mice and rats (Damez-Werno et al. 2016).

## 10 ADMA of Histones in Aging

CARM1 expression shows negative association with replicative senescence (Pang et al. 2013). Reduced expression of CARM1 has been correlated to reduced methylation, thereby attenuated function of HuR towards regulation of turnover of mRNAs. This evidence indicates that CARM1 function would suppress replicative senescence and slow aging of the cell. However, further investigations would be necessary to establish anti-aging role of CARM1. Another study finds reduced expression of PRMT1, CARM1 and PRMT6 during replicative senescence of diploid fibroblasts with concomitant reduction of ADMA of target proteins. Additionally, induction of premature senescence by treatment with sub-cytotoxic level of $H_2O_2$ showed similar results (Lim et al. 2008). Moreover, PRMT6−/− mouse embryonic fibroblasts have been found to exhibit early onset of senescence, which was not observed in PRMT6−/−, p53−/− MEFs. PRMT6 mediated H3R2me2a mark was found on the Trp53 promoter, indicating negative transcriptional regulation of p53 by PRMT6; where derepression of p53 and its target p21 expression was observed in absence of PRMT6 (Neault et al. 2012). In this context premature senescence observed is p53 dependent. Induction of senescence in breast cancer MCF7 cells, however, seemed to be p53 independent, where knock-down of PRMT6 led to reduction in H3R2me2a and resulted in derepression of p21 expression (Phalke et al. 2012). Similar relationship between PRMT6 and p21 was observed in models of oncogene induced senescence (OIS), where overexpression of PRMT6 can reduce OIS as a transcriptional repressor of p21 expression (Stein et al. 2012).

## 11 ADMA of Histones as Therapeutic Target

Abnormal arginine methylation of histones would alter transcriptional landscape and contribute to pathogenesis of different human disorders. In the last two decades small molecule inhibitors have been developed and investigated to assess PRMTs as therapeutic targets towards development of disease intervention. SAM analogs such as sinefungin and MTA (methylthioadenosine) have been found to inhibit PRMTs along with other methyltransferases which use SAM as methyl donor in cell, thus displaying very limited specificity. A class of inhibitors have been developed namely Arginine methyltransferase inhibitors (AMIs) which target only PRMTs, although with no selectivity among different members of PRMTs (Cheng et al. 2004). Therefore, AMIs are considered as pan-PRMT inhibitors.

In the last decade, individual PRMT specific inhibitors are being developed by several groups and at the moment only a few have been reported. Diamidine compounds (alkyl bis(oxy)dibenzimidamide derivatives) such as Decamidine, Furamidine, Stilbamidine etc. have been found to possess potent inhibitory activity against PRMT1 (Zhang et al. 2017a). Another group has demonstrated potent PRMT1 inhibitory activity of Hexamidine ($IC_{50} = 5.9 \pm 1.7$ μm) (Zhang et al. 2017b) further corroborating the utility of diamidine scaffold. NMR studies and molecular docking showed that hexamidine binds to substrate binding pocket of PRMT1 enzyme. Similarly, MS023, which binds to substrate binding pocket of PRMT6, have been shown to possess potent inhibitory activity against type I PRMTs, while being inactive towards type II and type III PRMTs, lysine methyltransferases and DNA methyltransferases (Eram et al. 2016). A potent inhibitor of CARM1, namely 1-benzyl-3,5-bis-(3-bromo-4-hydroxybenzylidene) piperidin-4-one ($IC_{50} = \sim 9.0$ μM) has been developed, which is specific against PRMT1 and PRMT6 (Cheng et al. 2011). Xenoestrogens have been found to possess arginine methyltransferase inhibitory activity. Among different xenoestrogens, tamoxifen and 4-hydroxytamoxifen showed greater inhibitory activity against CARM1 compared to other PRMTs (Cheng and Bedford 2011). Similarly, plant derived ellagic acid has been shown to inhibit methylation of H3R17 mediated by CARM1 in vitro and in vivo (Selvi et al. 2010). Structure activity relationship analysis and biochemical and biophysical characterization have led to identification of a few potent allosteric inhibitors of PRMT3, such as SGC707 ($IC_{50} = 31 \pm 2$ nM) (Kaniskan et al. 2015), (1-(benzo[d][1,2,3]thiadiazol-6-yl)-3-(2-cyclohexenylethyl) urea ($IC_{50} = 2.5$ μM) (Siarheyeva et al. 2012), 14u ($IC_{50} = 0.48$ μM) (Liu et al. 2013) with very good specificity. A potent and very specific inhibitor of CARM1 has recently been reported namely, TP-064 (N-methyl-N-((2-(1-(2-(methylamino)ethyl) piperidin-4-yl)pyridin-4-yl)methyl)-3-phenoxybenzamide), which shows anti-proliferative activity in multiple myeloma cell lines (Nakayama et al. 2018).

Inhibition of PRMT1 activity using pan-PRMT inhibitor AMI-1 has been found to enhance sensitivity of MDA-MB-468 cells (triple negative breast cancer cells) to cetuximab in the context of anti-EGFR therapy. PRMT1 mediated EGFR methylation seems to confer resistance to cetuximab (monoclonal antibody against EGFR), which could be reversed with pharmacological inhibition (Nakai et al. 2018). Inhibition of PRMT1 methyltransferase activity with Adenosine dialdehyde (AdOx) slowed proliferation and migration of oral cancer cell lines SAS and OECM with similar effects with genetic knock down of PRMT1 (Chuang et al. 2017). In a few preclinical disease models PRMT inhibitors have been found to be effective in ameliorating pathogenesis, especially in the context of cancer. However,

**Table 7** Pre-clinical studies with inhibitors of type I PRMTs in different cancer models

| S. N. | Inhibitors | Enzyme targets | Cancer model | Cell line used for xenograft/ transplantation in mice | Outcome of study | References |
|---|---|---|---|---|---|---|
| 1 | AMI-408 | PRMT1 | Acute myeloid leukemia (AML) | MLL-GAS7 | AMI-408 inhibited H4R3me2a; reduced disease penetrance and enhanced survival of mice | Cheung et al. (2016) |
| 2 | EZM2302 | CARM1 | Multiple Myeloma | RPMI-8226 | EZM2302 reduced tumor volume. | Drew et al. (2017) |

the therapeutic potential of small molecule modulators of type I PRMTs need rigorous pharmacological assessment before clinical studies could be attempted. The preclinical studies performed in different disease models targeting different type I PRMTs have been summarized in Table 7.

## 12 Future Perspective

Role of histone ADMA in regulation of gene expression in different biological pathways has begun to be understood only recently. Genome wide enrichment profile has been assessed only for H3R17me2a and H3R2me2a marks. With ChIP on ChIP analysis, H3R17me2a was found to be enriched on enhancer regions of ERα regulated genes in MCF7 cells, which are under the influence of E2 (Lupien et al. 2009). H3R2me2a marks have been found, from ChIP-Seq analysis, on non-bivalent active gene promoters as well as active enhancers (Bouchard et al. 2018). According to one study, TDRD3 functions as a reader of H3R17me2a and H4R3me2a marks on the transcriptionally active promoters and is enriched on the TSS region as revealed by ChIP-seq analysis (Yang et al. 2010). Another histone ADMA reader Spindlin1 (SPIN1) has been found to regulate rRNA gene expression (Zhang et al. 2018a) and wnt/β-catenin gene targets (Su et al. 2014) as a reader of H3R8me2a. Similar future studies will be necessary to correlate patterns of different ADMA marks on histones and occupancy of effector proteins to assign relevance and mode of gene regulation by different ADMA marks. Apart from transcription, ADMA on H3 (H3R17me2a) mediated by Mettl23 (Methyltransferase like family member 23) has recently been shown to play an important role in fertilization and zygotic development by reprogramming of paternal genome (Hatanaka et al. 2017). Future investigations would be necessary to unravel and distinguish physiological functions of different PRMT homologs as methyltransferases of histones in different cellular contexts.

Studies on reversal of arginine methylation are still in infancy. JMJD6 was shown to demethylate H3R2me2 and H4R3me2 (Chang et al. 2007). However, this observation has been questioned by other groups and JMJD6 is believed to be a lysyl hydroxylase rather than a true demethylase (Di Lorenzo and Bedford 2011). PAD4 mediated deimination converts arginine to citrulline, therefore it could antagonize ADMA/SDMA of histones and subsequently alter gene expression regulation. Identification of demethylases of ADMA marks in near feature would shed light into dynamicity of arginine methylation as a post-translational modification in cell.

ADMA of histones have been found to be deregulated in multitude of cancers and cardiovascular diseases. With the aim of development of therapeutics targeting arginine methylation, many research groups have developed various small molecules which show potent inhibitory activity against type I PRMTs in vitro and in vivo. Only a few studies have reported successful use of PRMT inhibitors to ameliorate pathophysiology of diseases including two pre-clinical studies mentioned in Table 7. Future investigations would be necessary to assess efficacy of different PRMT inhibitors in disease models, especially cancer, where type I PRMTs seem to be upregulated and exhibit oncogenic functions.

**Acknowledgement** TKK is recipient of JC Bose fellowship from Dept. of Science and technology, Govt. of India (SR/S2/JCB-28/2010). This work was supported by funding from Dept. of Biotechnology, Govt. of India (BT/01/CEIB/10/III/01) and JNCASR, India.

**Conflict of Interest** Authors declare that they have no conflict of interest.

## References

Al-Dhaheri M, Wu J, Skliris GP et al (2011) CARM1 is an important determinant of ER-α-dependent breast cancer cell differentiation and proliferation in breast cancer cells. Cancer Res 71:2118

Almeida-Rios D, Graça I, Vieira FQ et al (2016) Histone methyltransferase PRMT6 plays an oncogenic role of in prostate cancer. Oncotarget 7:53018–53028

An W, Kim J, Roeder RG (2004) Ordered cooperative functions of PRMT1, p300, and CARM1 in transcriptional activation by p53. Cell 117:735

Baldwin RM, Morettin A, Paris G et al (2012) Alternatively spliced protein arginine methyltransferase 1 isoform PRMT1v2 promotes the survival and invasiveness of breast cancer cells. Cell Cycle 11:4597

Baldwin RM, Bejide M, Trinkle-Mulcahy L et al (2015) Identification of the PRMT1v1 and PRMT1v2 specific interactomes by quantitative mass spectrometry in breast cancer cells. Proteomics 15(13):2187

Batut J, Duboé C, Vandel L (2011) The methyltransferases PRMT4/CARM1 and PRMT5 control differentially myogenesis in zebrafish. PLoS One 6:e25427

Bauer UM, Daujat S, Nielsen SJ et al (2002) Methylation at arginine 17 of histone H3 is linked to gene activation. EMBO Rep 3:39–44

Bedford MT, Clarke SG (2009) Protein arginine methylation in mammals: who, what and why. Mol Cell 33:1–13

Behera AK, Bhattacharya A, Vasudevan M et al (2018) p53 mediated regulation of coactivator associated arginine methyltransferase 1 (CARM1) expression is critical for suppression of adipogenesis. FEBS J 285:1730–1744

Blythe SA, Cha SW, Tadjuidje E et al (2010) beta-Catenin primes organizer gene expression by recruiting a histone H3 arginine 8 methyltransferase, Prmt2. Dev Cell 19:220

Böger RH (2003) The emerging role of asymmetric dimethylarginine as a novel cardiovascular risk factor. Cardiovasc Res 59:824
Bouchard C, Sahu P, Meixner M et al (2018) Genomic location of PRMT6-dependent H3R2 methylation is linked to the transcriptional outcome of associated genes. Cell Rep 24:3339–3352
Casadio F, Lu X, Pollock SB et al (2013) H3R42me2a is a histone modification with positive transcriptional effects. Proc Natl Acad Sci USA 110:14894
Chang B, Chen Y, Zhao Y et al (2007) JMJD6 is a histone arginine demethylase. Science 318:444
Chen SL, Loffler KA, Chen D et al (2002) The coactivator-associated arginine methyltransferase is necessary for muscle differentiation: CARM1 coactivates myocyte enhancer factor-2. J Biol Chem 277:4324
Chen X, Niroomand F, Liu Z et al (2006) Expression of nitric oxide related enzymes in coronary heart disease. Basic Res Cardiol 101:346
Cheng D, Bedford MT (2011) Xenoestrogens regulate the activity of arginine methyltransferases. Chembiochem 12:323
Cheng D, Yadav N, King RW, Swanson MS, Weinstein EJ, Bedford MT (2004) Small molecule regulators of protein arginine methyltransferases. J Biol Chem 279(23):23892–23899
Cheng D, Valente S, Castellano S et al (2011) Novel 3,5-bis(bromohydroxybenzylidene)piperidin-4-ones as coactivator-associated arginine methyltransferase 1 inhibitors: enzyme selectivity and cellular activity. J Med Chem 54:4928
Cheung N, Fung TK, Zeisig BB et al (2016) Targeting aberrant epigenetic networks mediated by PRMT1 and KDM4C in acute myeloid leukemia. Cancer Cell 29:32–48
Choi JH, Jang AR, Kim DI et al (2018) PRMT1 mediates RANKL-induced osteoclastogenesis and contributes to bone loss in ovariectomized mice. Exp Mol Med 50:111
Chuang CY, Chang CP, Lee YJ et al (2017) PRMT1 expression is elevated in head and neck cancer and inhibition of protein arginine methylation by adenosine dialdehyde or PRMT1 knockdown downregulates proliferation and migration of oral cancer cells. Oncol Rep 38:1115–1123
Damez-Werno DM, Sun H, Scobie KN et al (2016) Histone arginine methylation in cocaine action in the nucleus accumbens. Proc Natl Acad Sci USA 113:9623
Daujat S, Bauer UM, Shah V et al (2002) Crosstalk between CARM1 methylation and CBP acetylation on histone H3. Curr Biol 12:2090
Di Lorenzo A, Bedford MT (2011) Histone arginine methylation. FEBS Lett 585:2024
Di Lorenzo A, Yang Y, Macaluso M et al (2014) A gain-of-function mouse model identifies PRMT6 as a NF-κB coactivator. Nucleic Acids Res 42:8297
Drew AE, Moradei O, Jacques SL et al (2017) Identification of a CARM1 inhibitor with potent in vitro and in vivo activity in preclinical models of multiple myeloma. Sci Rep 7:17993
Eram MS, Shen Y, Szewczyk M et al (2016) A potent, selective, and cell-active inhibitor of human type I protein arginine methyltransferases. ACS Chem Biol 11:772–781
Ersoy B, Eroğlu N, Çetin M et al (2018) Asymmetric dimethylarginine levels and diabetes duration: relationship with measures of subclinical atherosclerosis and cardiac function in children and adolescents with type 1 diabetes. Diab Vasc Dis Res 15:196–203
Feng Y, Wang J, Asher S et al (2011) Histone H4 acetylation differentially modulates arginine methylation by an in Cis mechanism. J Biol Chem 286:20323
Frietze S, Lupien M, Silver PA et al (2008) CARM1 regulates estrogen-stimulated breast cancer growth through up-regulation of E2F1. Cancer Res 68:301
Fulton MD, Zhang J, He M et al (2017) Intricate effects of α-amino and lysine modifications on arginine methylation of the N-terminal tail of histone H4. Biochemistry 56:3539–3548
Gao Y, Zhao Y, Zhang J et al (2016) The dual function of PRMT1 in modulating epithelial-mesenchymal transition and cellular senescence in breast cancer cells through regulation of ZEB1. Sci Rep 6:19874
Gou Q, He S, Zhou Z (2017) Protein arginine N-methyltransferase 1 promotes the proliferation and metastasis of hepatocellular carcinoma cells. Tumour Biol 39:1010428317691419

Gou Y, Li J, Wu J et al (2018a) Prmt1 regulates craniofacial bone formation upstream of Msx1. Mech Dev 152:13–20
Gou Y, Li J, Jackson-Weaver O et al (2018b) Protein arginine methyltransferase PRMT1 is essential for palatogenesis. J Dent Res 97:1510–1518
Guccione E, Bassi C, Casadio F, Martinato F et al (2007) Methylation of histone H3R2 by PRMT6 and H3K4 by an MLL complex are mutually exclusive. Nature 449:933
Habashy HO, Rakha EA, Ellis IO et al (2013) The oestrogen receptor coactivator CARM1 has an oncogenic effect and is associated with poor prognosis in breast cancer. Breast Cancer Res Treat 140:307
Hamey JJ, Separovich RJ, Wilkins MR (2018) MT-MAMS: protein methyltransferase motif analysis by mass spectrometry. J Proteome Res 17:3485–3491
Hatanaka Y, Tsusaka T, Shimizu N et al (2017) Histone H3 methylated at arginine 17 is essential for reprogramming the paternal genome in zygotes. Cell Rep 20:2756–2765
Honda M, Nakashima K, Katada S (2017) PRMT1 regulates astrocytic differentiation of embryonic neural stem/precursor cells. J Neurochem 142(6):901–907. https://doi.org/10.1111/jnc.14123
Hong H, Kao C, Jeng MH et al (2004) Aberrant expression of CARM1, a transcriptional coactivator of androgen receptor, in the development of prostate carcinoma and androgen-independent status. Cancer 101:83
Huemer M, Simma B, Mayr D et al (2011) Low levels of asymmetric dimethylarginine in children with diabetes mellitus type I compared with healthy children. J Pediatr 158:602–606
Hyllus D, Stein C, Schnabel K et al (2007) PRMT6-mediated methylation of R2 in histone H3 antagonizes H3 K4 trimethylation. Genes Dev 21:3369
Jacques SL, Aquino KP, Gureasko J et al (2016) CARM1 preferentially methylates H3R17 over H3R26 through a random kinetic mechanism. Biochemistry 55:1635–1644
Jeong HC, Park SJ, Choi JJ et al (2017) PRMT8 controls the pluripotency and mesodermal fate of human embryonic stem cells by enhancing the PI3K/AKT/SOX2 axis. Stem Cells 35:2037–2049
Kang I, Okla M, Chung S (2014) Ellagic acid inhibits adipocyte differentiation through coactivator-associated arginine methyltransferase 1-mediated chromatin modification. J Nutr Biochem 25:946
Kaniskan HÜ, Szewczyk MM, Yu Z et al (2015) A potent, selective and cell-active allosteric inhibitor of protein arginine methyltransferase 3 (PRMT3). Angew Chem Int Ed Engl 54:5166
Kim YR, Lee BK, Park RY et al (2010) Differential CARM1 expression in prostate and colorectal cancers. BMC Cancer 10:197
Kölbel K, Ihling C, Bellmann-Sickert K et al (2009) Type I arginine methyltransferases PRMT1 and PRMT-3 act distributively. J Biol Chem 284:8274
Lakowski TM, Frankel A (2008) A kinetic study of human protein arginine N-methyltransferase 6 reveals a distributive mechanism. J Biol Chem 283:10015
Lakowski TM, Frankel A (2009) Kinetic analysis of human protein arginine N-methyltransferase 2: formation of monomethyl- and asymmetric dimethyl-arginine residues on histone H4. Biochem J 421:253
Lausen J (2013) Contributions of the histone arginine methyltransferase PRMT6 to the epigenetic function of RUNX1. Crit Rev Eukaryot Gene Expr 23:265
Lee DY, Ianculescu I, Purcell D et al (2007) Surface-scanning mutational analysis of protein arginine methyltransferase 1: roles of specific amino acids in methyltransferase substrate specificity, oligomerization, and coactivator function. Mol Endocrinol 21:1381
Lee YH, Ma H, Tan TZ et al (2012) Protein arginine methyltransferase 6 regulates embryonic stem cell identity. Stem Cells Dev 21:2613
Lee WC, Lin WL, Matsui T et al (2015) Protein arginine methyltransferase 8: tetrameric structure and protein substrate specificity. Biochemistry 54:7514
Li Y, Zhu R, Wang W et al (2015) Arginine methyltransferase 1 in the nucleus accumbens regulates behavioral effects of cocaine. J Neurosci 35:12890

Lim Y, Lee E, Lee J et al (2008) Down-regulation of asymmetric arginine methylation during replicative and H2O2-induced premature senescence in WI-38 human diploid fibroblasts. J Biochem 144:523
Lim Y, Yu S, Yun JA et al (2017) The prognostic significance of protein arginine methyltransferase 6 expression in colon cancer. Oncotarget 9:9010–9020
Lin YL, Tsai YJ, Liu YF et al (2013) The critical role of protein arginine methyltransferase prmt8 in zebrafish embryonic and neural development is non-redundant with its paralogue prmt1. PLoS One 8:e55221
Liu F, Li F, Ma A et al (2013) Exploiting an allosteric binding site of PRMT3 yields potent and selective inhibitors. J Med Chem 56:2110
Lupien M, Eeckhoute J, Meyer CA et al (2009) Coactivator function defines the active estrogen receptor alpha cistrome. Mol Cell Biol 29:3413
Majumder S, Liu Y, Ford OH et al (2006) Involvement of arginine methyltransferase CARM1 in androgen receptor function and prostate cancer cell viability. Prostate 66:1292
Mallappa C, Hu YJ, Shamulailatpam P et al (2011) The expression of myogenic microRNAs indirectly requires protein arginine methyltransferase (Prmt5) but directly requires Prmt4. Nucleic Acids Res 39:1243
Nakai K, Xia W, Liao HW et al (2018) The role of PRMT1 in EGFR methylation and signaling in MDA-MB-468 triple-negative breast cancer cells. Breast Cancer 25(1):74–80
Nakayama K, Szewczyk MM, Dela Sena C et al (2018) TP-064, a potent and selective small molecule inhibitor of PRMT4 for multiple myeloma. Oncotarget 9:18480–18493
Neault M, Mallette FA, Vogel G et al (2012) Ablation of PRMT6 reveals a role as a negative transcriptional regulator of the p53 tumor suppressor. Nucleic Acids Res 40:9513
O'Brien KB, Alberich-Jordà M, Yadav N et al (2010) CARM1 is required for proper control of proliferation and differentiation of pulmonary epithelial cells. Development 137:2147
Osada S, Suzuki S, Yoshimi C et al (2013) Elevated expression of coactivator-associated arginine methyltransferase 1 is associated with early hepatocarcinogenesis. Oncol Rep 30:1669
Ou CY, LaBonte MJ, Manegold PC et al (2011) A coactivator role of CARM1 in the dysregulation of β-catenin activity in colorectal cancer cell growth and gene expression. Mol Cancer Res 9:660
Pang L, Tian H, Chang N et al (2013) Loss of CARM1 is linked to reduced HuR function in replicative senescence. BMC Mol Biol 14:15
Phalke S, Mzoughi S, Bezzi M et al (2012) p53-Independent regulation of p21Waf1/Cip1 expression and senescence by PRMT6. Nucleic Acids Res 40:9534
Ryu JW, Kim SK, Son MY et al (2017) Novel prognostic marker PRMT1 regulates cell growth via downregulation of CDKN1A in HCC. Oncotarget 8:115444–115455
Schurter BT, Koh SS, Chen D et al (2001) Methylation of histone H3 by coactivator-associated arginine methyltransferase 1. Biochemistry 40:5747
Selvi BR, Batta K, Kishore AH et al (2010) Identification of a novel inhibitor of coactivator-associated arginine methyltransferase 1 (CARM1)-mediated methylation of histone H3 arg-17. J Biol Chem 285:7143–7152
Selvi BR, Swaminathan A, Maheshwari U, Nagabhushana A, Mishra RK, Kundu TK, Bronner M (2015) CARM1 regulates astroglial lineage through transcriptional regulation of Nanog and posttranscriptional regulation by miR92a. Mol Biol Cell 26(2):316–326
Shen NY, Ng SY, Toepp SL et al (2018) Protein arginine methyltransferase expression and activity during myogenesis. Biosci Rep 38. pii: BSR20171533
Siarheyeva A, Senisterra G, Allali-Hassani A et al (2012) An allosteric inhibitor of protein arginine methyltransferase 3. Structure 20:1425
Stallcup MR, Chen D, Koh SS et al (2000) Co-operation between protein-acetylating and protein-methylating co-activators in transcriptional activation. Biochem Soc Trans 28:415
Stein C, Riedl S, Rüthnick D et al (2012) The arginine methyltransferase PRMT6 regulates cell proliferation and senescence through transcriptional repression of tumor suppressor genes. Nucleic Acids Res 40:9522
Su X, Zhu G, Ding X et al (2014) Molecular basis underlying histone H3 lysine-arginine methylation pattern readout by Spin/Ssty repeats of Spindlin1. Genes Dev 28:622

Sun Y, Chung HH, Woo AR et al (2014) Protein arginine methyltransferase 6 enhances ligand-dependent and -independent activity of estrogen receptor α via distinct mechanisms. Biochim Biophys Acta 1843:2067

Swiercz R, Cheng D, Kim D et al (2007) Ribosomal protein rpS2 is hypomethylated in PRMT3-deficient mice. J Biol Chem 282:16917

Teyssier C, Chen D, Stallcup MR (2002) Requirement for multiple domains of the protein arginine methyltransferase CARM1 in its transcriptional coactivator function. J Biol Chem 277:46066

Toma-Fukai S, Kim JD, Park KE (2016) Novel helical assembly in arginine methyltransferase 8. J Mol Biol 428:1197–1208

Troffer-Charlier N, Cura V, Hassenboehler P et al (2007) Functional insights from structures of coactivator-associated arginine methyltransferase 1 domains. EMBO J 26:4391

Tsai YJ, Pan H, Hung CM et al (2011) The predominant protein arginine methyltransferase PRMT1 is critical for zebrafish convergence and extension during gastrulation. FEBS J 278:905

Waldmann T, Izzo A, Kamieniarz K et al (2011) Methylation of H2AR29 is a novel repressive PRMT6 target. Epigenetics Chromatin 4:11

Wang H, Huang ZQ, Xia L et al (2001) Methylation of histone H4 at arginine 3 facilitating transcriptional activation by nuclear hormone receptor. Science 293:853

Wang L, Charoensuksai P, Watson NJ et al (2013) CARM1 automethylation is controlled at the level of alternative splicing. Nucleic Acids Res 41:6870

Wang L, Zhao Z, Meyer MB, Saha S, Menggang Y, Guo A, Wisinski KB, Huang W, Cai W, Wesley Pike J, Yuan M, Ahlquist P, Wei X (2014) CARM1 methylates chromatin remodeling factor BAF155 to enhance tumor progression and metastasis. Cancer Cell 25(1):21–36

Wolf SS (2009) The protein arginine methyltransferase family: an update about function, new perspectives and the physiological role in humans. Cell Mol Life Sci 66:2109–2121

Wooderchak WL, Zang T, Zhou ZS et al (2008) Substrate profiling of PRMT1 reveals amino acid sequences that extend beyond the "RGG" paradigm. Biochemistry 47:9456

Wu J, Xu W (2012) Histone H3R17me2a mark recruits human RNA polymerase-associated factor 1 complex to activate transcription. Proc Natl Acad Sci USA 109:5675

Wu Q, Bruce AW, Jedrusik A et al (2009) CARM1 is required in embryonic stem cells to maintain pluripotency and resist differentiation. Stem Cells 27:2637

Xu W, Cho H, Evans RM (2003) Acetylation and methylation in nuclear receptor gene activation. Methods Enzymol 364:205

Yadav N, Cheng D, Richard S et al (2008) CARM1 promotes adipocyte differentiation by coactivating PPARgamma. EMBO Rep 9:193

Yang Y, Lu Y, Espejo A et al (2010) TDRD3 is an effector molecule for arginine-methylated histone marks. Mol Cell 40:1016

Zakrzewicz D, Eickelberg O (2009) From arginine methylation to ADMA: a novel mechanism with therapeutic potential in chronic lung diseases. BMC Pulm Med 9:5

Zhang X, Cheng X (2003) Structure of the predominant protein arginine methyltransferase PRMT1 and analysis of its binding to substrate peptides. Structure 11:509

Zhang J, Qian K, Yan C et al (2017a) Discovery of decamidine as a new and potent PRMT1 inhibitor. Med Chem Commun 8:440–444

Zhang WY, Lu WC, Jiang H et al (2017b) Discovery of alkyl bis(oxy)dibenzimidamide derivatives as novel protein arginine methyltransferase 1 (PRMT1) inhibitors. Chem Biol Drug Des 90:1260–1270

Zhang X, Zhu G, Su X et al (2018a) Nucleolar localization signal and histone methylation reader function is required for SPIN1 to promote rRNA gene expression. Biochem Biophys Res Commun 505:325–332

Zhang XP, Jiang YB, Zhong CQ et al (2018b) PRMT1 promoted HCC growth and metastasis in vitro and in vivo via activating the STAT3 signalling pathway. Cell Physiol Biochem 47:1643–1654

Zhao J, Adams A, Roberts B et al (2018) Protein arginine methyl transferase 1- and Jumonji C domain-containing protein 6-dependent arginine methylation regulate hepatocyte nuclear factor 4 alpha expression and hepatocyte proliferation in mice. Hepatology 67:1109–1126

# A Switch for Transcriptional Activation and Repression: Histone Arginine Methylation

**Tian-Shi Wang, Jin-Ke Cheng, Qun-Ying Lei, and Yi-Ping Wang**

## Contents



**Abstract** Histones can be methylated on both lysine (K) and arginine (R) residues. Histone arginine methylation is a prevalent post-translational modification catalyzed by protein arginine methyltransferases (PRMTs). As an epigenetic modification, histone arginine methylation is associated with signal transduction, cell differentiation, cellular metabolism, tissue homeostasis, immune and inflammatory responses etc. Methylation at arginine residues alters the properties of the nucleosome to regulate gene transcription and the interaction between nucleosome and other regulatory proteins. Histone arginine

T.-S. Wang · J.-K. Cheng
Department of Biochemistry and Molecular Cell Biology, Shanghai Key Laboratory for Tumor Microenvironment and Inflammation, Shanghai Jiao Tong University School of Medicine, Shanghai, P. R. China
e-mail: tianshi777@shsmu.edu.cn

Q.-Y. Lei · Y.-P. Wang (✉)
Fudan University Shanghai Cancer Center, Cancer Institute, Fudan University Cancer Hospital, and Cancer Metabolism Laboratory, Institutes of Biomedical Sciences, Fudan University, Shanghai, P. R. China
e-mail: yiping_wang@fudan.edu.cn

 
S. Jurga, J. Barciszewski (eds.), *The DNA, RNA, and Histone Methylomes*, RNA Technologies, https://doi.org/10.1007/978-3-030-14792-1_21

methylation results in either transcriptional repression or activation. This review focuses on the biochemistry, regulatory mechanism, and the functional significance of histone arginine methylation.

**Keywords** Arginine methylation · Protein arginine methyltransferases · Histone modification · Transcriptional regulation

## 1 Introduction

In the early 1960s, histones were found to be modified by post-translational methylation, which was correlated with gene expression level. Subsequently, methylation was revealed as a covalent modification that occurs not only on histone proteins but also on specific nuclear and cytoplasmic proteins. Histone is methylated at its lysine and arginine residues. The function of histone lysine methylation has been well-characterized in recent years. Similar to lysine methylation, arginine methylation alters the structural properties of the nucleosome and affects diverse cellular processes, including gene transcription, protein translation, DNA repair, and RNA processing. Studies in animal models indicate that methylation of histone arginine residues is associated with aging, development, stem cell homeostasis etc. Dysregulation of histone arginine methylation contributes to the onset and development of various diseases, including cancer, diabetes mellitus, cardiovascular disease, and immune deficiency (Cheung et al. 2007; Hashimoto et al. 2016; Neault et al. 2012).

The nucleosome is the fundamental packaging and regulatory unit of chromatin. Nucleosome comprises DNA and a histone octamer. This octamer contains two copies each of four histone proteins, H2A, H2B, H3, and H4. Every histone protein has an extended tail where the methyl groups and other epigenetic marks reside. At the tail end is the N-terminus of a histone protein. To a large extent, gene activation or repression is modulated by the methylation events at specific residue(s) in histone tails. Mass spectrometry studies revealed that H3 and H4 tails are hot spots for arginine methylation (Guccione et al. 2007; Hatanaka et al. 2017) (Fig. 1). For instance, H3R2 methylation (H3R2me) directed the chromatin-binding proteins to recognize histone H3. The methylation at different arginine residues in histone tails serves as distinctive epigenetic marks to recruit various proteins to interpret the epigenetic language, fine-tuning chromatin activation or repression (Swiercz et al. 2005). Of note, H4R55me, which was only detected by LC-MS/MS, has an unclear function in transcription.

Similar to other epigenetic modifications, histone arginine methylation has its unique writer, reader, and eraser. As writers, protein arginine methyltransferases (PRMT) deposit methyl groups on arginine residues in histone tails. The specificity of histone arginine methylation comes from PRMT (Table 1). Histone arginine methylation can be recognized by several conserved protein domains, called readers of arginine methylation (Table 2). Tudor domain-containing proteins serve as primary players in reading methylarginine marks. In addition, the domains previously

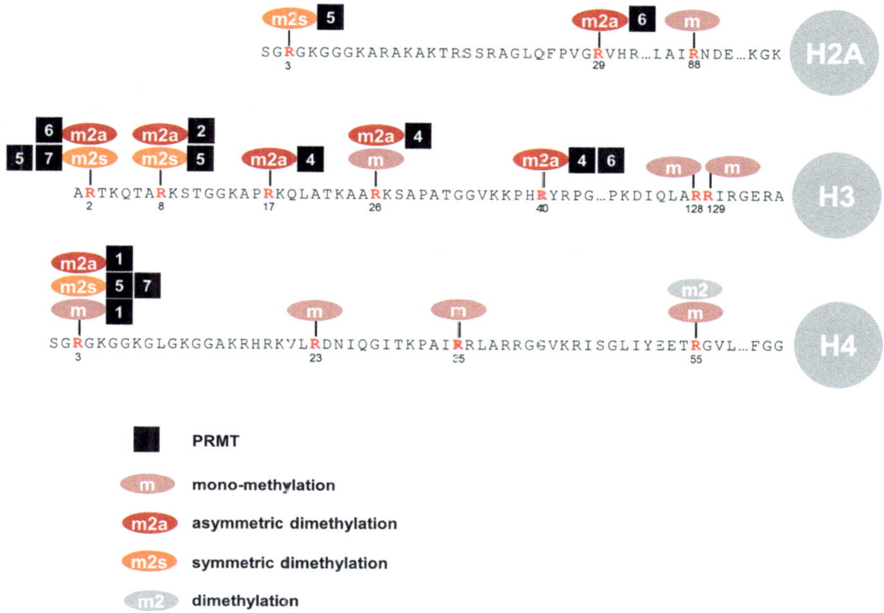


**Fig. 1** Modification of histone H2A, H3, and H4 by PRMT. The known arginine mono-methylation, asymmetric di-methylation and symmetric di-methylation sites of histone H2A, H3, and H4 are shown. Pink denotes mono-methylation, grey oval denotes di-methylation, red denotes asymmetric di-methylation and orange denotes symmetric di-methylation; Black Square represents PRMTs, and the number in the square indicates the specific PRMT responsible for methylation event(s)

characterized as methyllysine readers, such as Chromo, PHD, MBT, PWWP, Ank, BAH, and WD40 domains, also hold potential functions as methylarginine readers. Specifically, recombination activating gene RAG2 contains a PHD domain which coordinately bind to H3K4me3 and symmetrically dimethylated H3R2 (H3R2me2s) (Ramon-Maiques et al. 2007). DNA methyltransferase DNMT3A directly binds to H4R3me2s mark through its PHD motif (Zhao et al. 2009). The Tudor domain of TDRD3 recognizes asymmetric dimethylation at both H3R17 and H4R3 (H3R17me2a and H4R3me2a) to activate transcription (Yang et al. 2010). On the other hand, histone arginine methylation is postulated as a reversible modification. The activities of PRMTs can be offset by histone demethylase (Table 3). Although no bona fide arginine demethylase has been characterized in vivo, a number of α-KG-dependent histone demethylases exhibited arginine demethylase activity (Tsukada and Zhang 2006). Notably, when either histone or non-histone peptides were used as substrates in vitro, a handful of JmjC lysine demethylases showed arginine demethylase activity. Importantly, JMJD1B, known as an H3K9me2 demethylase, led to H4R3me2s demethylation in vitro and in hematopoietic stem/progenitor cells (HSPCs) (Li et al. 2018).

**Table 1** Site-specific modification of histone proteins by PRMT

| Arginine methyltransferase | Direct histone modifications | Subcellular location | Features of modification motifs | Functions |
|---|---|---|---|---|
| PRMT1 (Type I) | H4R3 | Nucleoplasm | Glycine and Arginine rich | Wide substrate specificity |
| PRMT2 (Type I) | H4 | Nucleoplasm and cytosol | Glycine and Arginine rich | Promote apoptosis and block NF-κb nuclear export |
| PRMT3 (Type I) | Not available | Cytosol | Glycine and Arginine rich | Balance of ribosomal subunits |
| PRMT4/CARM1 (Type I) | H3R17 and H3R26 | Nucleoplasm | XXPRX or XXRPX, where X is any amino acid | Regulate the coupling of transcription and splicing as a steroid receptor coactivator |
| PRMT5 (Type II) | H2A/H4R3 and H3R8 | Nucleus and cytosol | Glycine and Arginine rich | Transcriptional repressor |
| PRMT6 (Type I) | H3R2, H2A/H4R3 and H2AR29 | Nucleus and Nucleoli | Glycine and Arginine rich | Repress the transcription (H3R2) or activate transcription (H2AR3/H4R3)? |
| PRMT7 (Type III) | H4R3 and H3R2 | Nucleoplasm and Nucleoli fibrillar center | Glycine and Arginine rich | Stress response |
| PRMT8 (Type I) | Not available | Associated with plasma membrane | Glycine and Arginine rich | Functions in neurons |
| PRMT9 (Type II) | Not available | Microtubules, Nucleoplasm and Cytosol | Glycine and Arginine rich | Mediate protein-protein interactions |

## 2 Arginine Methyltransferases

The guanidino group of arginine residue contains a free amino group and a positively charged amino group; each of which is able to be methylated. Arginine dimethylation occurs in symmetric or asymmetric fashions (Fig. 2). PRMTs transfer a methyl group from S-adenosylmethionine (AdoMet or SAM) to the ω-guanidino nitrogen of arginine, resulting in the production of S-adenosylhomocysteine (AdoHcy) and methylarginine. PRMT harbors a catalytic pocket to bind to AdoMet, the methyl donor. The substrate specificity of PRMT is potentially determined by residues surrounding the catalytic pocket, which are responsible for substrate recognition.

### *2.1 Classification of PRMTs*

Human PRMT family has at least nine different members. Based on the catalytic property, these PRMTs are subdivided into three types. Type I PRMTs (PRMT1, 2, 3,

**Table 2** Readers of arginine methylation

| Reader | Domains | Functions | Writer |
|---|---|---|---|
| Tudor domains | SMN | Assembly of small nuclear ribonucleoproteins (snRNPs) | PRMT4/CARM1, PRMT5 |
| | SPF30 (SMNrp or SMNDC1) | Spliceosome maturation | PRMTs |
| | TDRD3 | DNA- and RNA-directed topoisomerase activity | PRMT1, PRMT4/CARM1 |
| | SND1 (TSN-p100 or TDRD11) | (1) Facilitate the acetylation of histone as a transcriptional coactivator<br>(2) Interact with the RISC in miRNA processing<br>(3) Pre-mRNA splicing<br>(4) Germ cells function | PRMTs |
| | TDRD1 | Interact with Piwi proteins through its Tudor domains | PRMTs |
| | TDRKH (TDRD2) | Repress transposition, regulate translation, and guide epigenetic programming in the germline | PRMT5 |
| | TDRD6 | Interact with Miwi and SDMA peptides | PRMT5, PRMT9 |
| | TDRD9 | Interact with Mili | PRMTs |
| RAG2 | PHD | Bind to H3R2me2s and H3K4me3 | PRMT5, PRMT7 |
| WDR5 | WD40 | Bind to H3R2me2s | PRMTs |
| DNMT3A | ADD | Bind to H4R3me2s | PRMT5 |
| MLL4 | PHD | Activate transcription by a trans-tail cross talk between H4R3me2s and H3K3me3 | PRMT5 |
| PAF complex | Not available | Interact with H3R17me2a | PRMT4/CARM1 |
| BRCA1 | BRCT | Interact with p300R754me2a | PRMT4/CARM1 |
| Polycomb 2 | Chromo | Bind to H4R3me2s in the presence IncRNA *TUG1* | PRMT5 |
| 7SK snRNA | Not available | Bind to H4R3me2a/s | PRMTs |
| PELP1 | Not available | Interact with asymmetric dimethylarginine | PRMT4/CARM1 |

**Table 3** Erasers of arginine methylation

| Eraser name | Domains | Substrates |
|---|---|---|
| JMJD6 | Jumonji | H3R2me2, H4R3me2 |
| PAPIs | Peptidylarginine deiminases | Deiminated proteins in H2A, H3 and H4 |
| KDM4E | JmjC | H3R2, H3R8, H3R26, H4R3me2a/s |
| KDM5C | | H3R2, H3R8me2a/s, H4R3me2a |
| JMJD1B | | H4R3me2s |

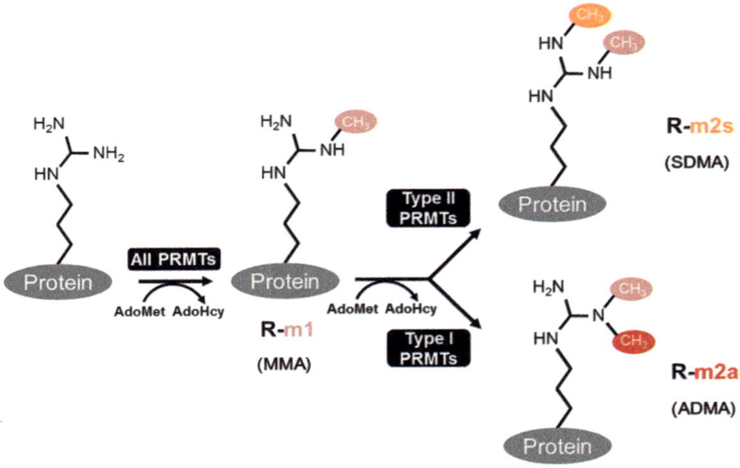


**Fig. 2** PRMTs transfer methyl group(s) onto arginine residues. Arginine residues in the tails of histones can be converted to monomethylarginine (MMA), asymmetric dimethylarginine (ADMA), and symmetric dimethylarginine (SDMA). The MMA form of arginine is an intermediate to the dimethylated state. All PRMTs can produce MMA, but the type I PRMTs transfer the second methyl group asymmetrically and type II PRMTs mediate symmetric di-methylation

4, 6, and 8) catalyze the production of asymmetric dimethylarginine (ADMA) at the arginine guanidino group. Type II PRMTs (PRMT5 and PRMT9) produce symmetric dimethylarginine (SDMA) (Table 1). Of note, both type I and type II PRMTs generate mono-methylarginine (MMA). However, type III enzyme (PRMT7) only catalyzes the formation of MMA (Blanc et al. 2017). In yeast, a type IV enzyme catalyzes the mono-methylation of the internal guanidine nitrogen atom, of which the function remains to be defined (Young et al. 2012).

In addition to the nine PRMTs (PRMT1-PRMT9), more methyltransferases have been identified to target arginine residues in recent years. NDUFAF7 is a member of SAM-dependent methyltransferase family. NDUFAF7 locates in mitochondria matrix and mediate symmetrical dimethylation of NDUFS2 to assist complex I assembly (Rhein et al. 2013). Moreover, the arginine methyltransferase activity of Mettl23 is critical for paternal genome reprogramming (Hatanaka et al. 2017).

## 2.2 *PRMTs Target Specific Arginine Residues in Histone*

Different PRMTs show unique selectivity on substrate peptide sequence and deposit methyl groups on specific arginine residues in histone tails. Bioinformatics studies indicate that certain motifs or domains in PRMTs are critical for their catalytic specificity, such as the SH3 domain of PRMT2, the zinc-finger domain of PRMT3, two putative AdoMet-binding motifs of both PRMT7 and PRMT9, a myristoylation motif of PRMT8, two tetratricopeptide repeats (TPR) of PRMT9 (Fig. 3).

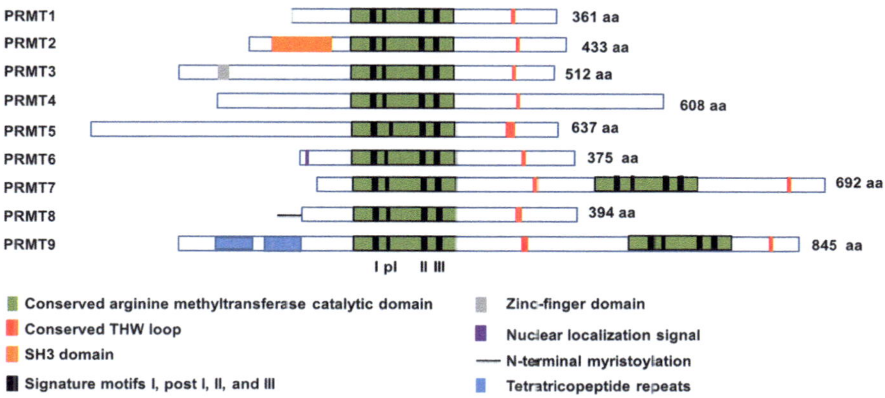


**Fig. 3** Domain organization of PRMT family. The domains of PRMTs are demonstrated by bioinformatics study to highlight certain motifs critical for their catalytic specificity, such as the SH3 domain of PRMT2, the zinc-finger domain of PRMT3, two putative AdoMet-binding motifs of both PRMT7 and PRMT9, a myristoylation motif of PRMT8, two tetratricopeptide repeats (TPR) of PRMT9

PRMT1 mediates H4R3me2a. PRMT1 is the first methylase identified to target arginine residues. Mass spectrometry studies suggest that PRMT1 accounts for approximately 85% of total protein arginine methylation in cells (Blanc et al. 2017). *PRMT1* knockdown decreases H4R3me2a, a transcription activation mark, by recruiting acetyltransferase p300 to histone tails in erythrocytes (Huang et al. 2005). Mixed lineage leukemia (MLL) fusion proteins interact with PRMT1 and promote H4R3 methylation and *HoxA9* expression (Cheung et al. 2007).

Similar to PRMT1, PRMT2 also mediates H4R3me2a. PRMT2 serves as a co-activator of gene expression. Its SH3 domain binds to the N-terminal domain of PRMT8 and modulates its methylase activity (Dong et al. 2018). PRMT2 upregulates both the androgen receptor and the estrogen receptor δ, and inhibits nuclear export of IκB-α to promote the apoptosis (Ganesh et al. 2006). Also, it stimulated glioblastoma multiforme (GBM) cell growth and even orthotopic tumor growth by upregulating H3R8me2a (Dong et al. 2018).

Evidence in PRMT3-mediated histone methylation is lacking. The zinc-finger domain at the amino-terminus of PRMT3 is critical for its methylase activity. PRMT3 modifies 40S ribosomal protein S2 (rpS2) and regulates ribosome biosynthesis (Swiercz et al. 2005). This observation suggests that PRMT3 potentially functions in nutrient sensing and protein synthesis.

PRMT4, also referred to CARM1, was identified as a transcription co-activator that generates H3R17me2 and H3R26me2 to modulate nutrient sensing and signaling (Liu et al. 2017). Nutrient starvation increase *CARM1* expression and H3R17me2 in an AMP-activated protein kinase (AMPK)-dependent manner. Additionally, CARM1 acts as the transcription co-activator of a wide spectrum of nuclear receptors and transcription factors to regulate downstream gene expression (Shin et al. 2016).

Besides, CARM1 methylates BAF155 to induce chromatin remodeling and upregulate target genes in the Myc pathway (Wang et al. 2014).

PRMT5, together with PRMT9, are the only enzymes capable of forming SDMA in mammalian cells. PRMT5 has been regarded as a transcriptional repressor, and it catalyzes the H2AR3me2s, H4R3me2s, and H3R8me2s (Cesaro et al. 2009). PRMT5 represses *aldolase A* gene by methylating H4R3 (Cesaro et al. 2009). Notably, PRMT5 is involved in the crosstalk of histone methylation and phosphorylation. Once kinase domain of TRPM6 is cleaved and released, it relocates into the nucleus and forms a complex with PRMT5 to modulate local histone arginine methylation (Krapivinsky et al. 2017). PRMT5 also regulates gene expression by modulating transcription- or splicing-related factors. PRMT5 transfers methyl groups to spliceosomal proteins to mediate the circadian regulation of pre-mRNA splicing (Sanchez et al. 2010). PRMT5 is also involved in transcriptional repression of *Cyclin E* and *CAD* (Pal et al. 2004).

PRMT6 produces H3R2me2a, H4R3me2a, and H2AR3me2a and plays a dual role in transcription regulation. PRMT6-mediated H3R2me2 antagonizes H3K4me, a transcriptional activation mark, to repel ASH2/WDR5/MLL methyltransferase complex (Guccione et al. 2007). However, PRMT6 methylates H4R3 and H2AR3 to coactivate transcription (Harrison et al. 2010).

PRMT7 harbors two putative AdoMet-binding motifs. An in vitro activity assay suggests that H2B is a highly preferred substrate for PRMT7 (Feng et al. 2013). Interestingly, PRMT7 exists in the BRG1-based hSWI/SNF chromatin remodeling complex. PRMT7 induces H2AR3me2 and H4R3me2 at the promoter of DNA repair genes to suppress their expression (Karkhanis et al. 2012). PRMT7 introduces H4R3me1 and H4R3me2s to *Bcl6* gene. As a result, *Bcl6* is suppressed to control germinal center formation (Ying et al. 2015).

PRMT8 shares high sequence similarity with PRMT1. The unique myristoylation motif of PRMT8 directs it to the plasma membrane. PRMT8 is able to methylate H2A/H2B dimer and the corresponding peptide in vitro (Lee et al. 2015). Whether PRMT8 modulates histone arginine methylation within the cell is yet to be determined. Known substrates of PRMT8 are non-histone proteins, including EWS and NIFK (Lee et al. 2015).

PRMT9 is featured by its two putative AdoMet-binding motifs and two tetratricopeptide repeats (TPR). TPR domain contributes to protein-protein interaction. The function of PRMT9 remains to be explored.

Of note, PRMT protein has no intrinsic activity to recognize specific DNA sequences. The recognition of target genes by PRMTs is potentially mediated by their binding partners, that is, sequence-specific DNA binding proteins. Indeed, transcription factor-directed recruitment of epigenetic enzymes serves as a prevalent regulatory mechanism for epigenetic enzymes to target specific genes (Wang et al. 2015; Wang and Lei 2018). In addition, PRMTs have a broad spectrum of non-histone substrates which are involved in RNA splicing, translation, signal transduction, and metabolism. PRMTs exert their biological activity by modulating protein-protein interaction, protein-DNA/RNA interaction, subcellular location, protein stability, and catalytic activity.

### 2.3 Mechanism of Arginine Methylation-Mediated Transcriptional Regulation

Gene transcription is tightly coupled with the open/close state of chromatin. Loose chromatin is necessary for active gene transcription. Histone methylation at different arginine sites regulates gene transcription by blocking or encouraging DNA access to transcription factors, thereby leading to either transcriptional repression or activation. CARM1-mediated H3R2me, H3R17me, and H3R26me are associated with transcriptional activation, and H3R17me activates the expression p53-responsive genes (Selvi et al. 2010). PRMT5 methylates H3R8 and H4R3, resulting in transcriptional repression of *CAD* and *Cyclin E* (Pal et al. 2004). Notably, arginine methylation marks are passed down to daughter cells during cell division.

Histone arginine methylation is also affected by environmental factors, such that nutrient starvation causes an increase in H3R17me2 (Shin et al. 2016). UV treatment leads to cytoplasmic accumulation PRMT1 and potentially affects histone arginine methylation (Suchankova et al. 2014).

### 2.4 The Physiological Function of PRMT: Clues from Mouse Models

PRMTs play a fundamental regulatory role in the development and tissue homeostasis. The direct evidence of PRMT's physiological function comes from genetic knockout mouse models. *PRMT1*-deficient mice are embryonic lethal at embryonic day 7.5 (Hashimoto et al. 2016). *PRMT1* disruption results in severe hypomyelination and developmental defects. In the brain of CNS-specific *PRMT1* knockout mice, myelin proteins were significantly decreased and oligodendrocyte maturation processes were dramatically suppressed (Hashimoto et al. 2016). *PRMT2*-deficient mice are lean and hypophagic, which is potentially linked to the absence of STAT3 methylation (Iwasaki et al. 2010). Compared to wildtype mice, *PRMT3*-deficient mice have a smaller size, possibly due to mRNA translation defects (Swiercz et al. 2007). *CARM1*-null mice die shortly after birth. CARM1 deletion results in defective adipogenesis, T cell differentiation, and hematopoiesis (Li et al. 2013; Yadav et al. 2008). PRMT5 may regulate neurogenesis by p53 pathway. *PRMT5* ablation results in embryonic lethality due to developmental defect, and PRMT5 has also been shown to maintain adult hematopoiesis (Liu et al. 2015). Embryonic fibroblasts of *PRMT6*-deficient mice show poor proliferation and cellular senescence (Neault et al. 2012). *PRMT7*-knockout mice die 5–10 days after birth with skeletal abnormalities, lower body mass and decreased red blood cell counts, and *PRMT7* depletion also results in disorders in germinal center formation (Ying et al. 2015) and satellite cell regeneration (Blanc et al. 2016). *PRMT8*-deficient mice have defects on motor coordination, which is potentially linked to its arginine methyltransferase and phospholipase activity in Purkinje cells (Kim et al. 2015). The in vivo function of PRMT9 remains to be defined.

# 3 Dysregulation of PRMT in Human Diseases

Aberrant histone arginine methylation has been linked to the onset of various human diseases (Fig. 4). Recent investigations have gained significant insight into how PRMT deregulation contributes to the development of cancer, degenerative disease, cardiovascular disease, and immune disorders.

## *3.1 Deregulation of PRMT in Cancer*

**Leukemia** Accumulating evidence has demonstrated that histone arginine methylation is an important regulator of tumor stem cells. PRMTs play important roles in the context of acute myeloid leukemia (AML) and acute lymphoblastic leukemia (ALL). PRMT5 regulates the self-renewal of imatinib-sensitive leukemia stem cells (Jin et al. 2016). Newly diagnosed patients with pediatric ALL show dramatic PRMT1 upregulation (Zou et al. 2012). Knockdown of *PRMT1* represses leukemia transformation. PRMT1 methylates the transcriptional factor RUNX1, which is critical for myeloid differentiation and lymphocyte development (Mizutani et al. 2015). Hematopoietic tissue-specific depletion of *CARM1* shows a minor effect on hematopoiesis. However, inhibition of CARM1 promotes myeloid leukemia cells to

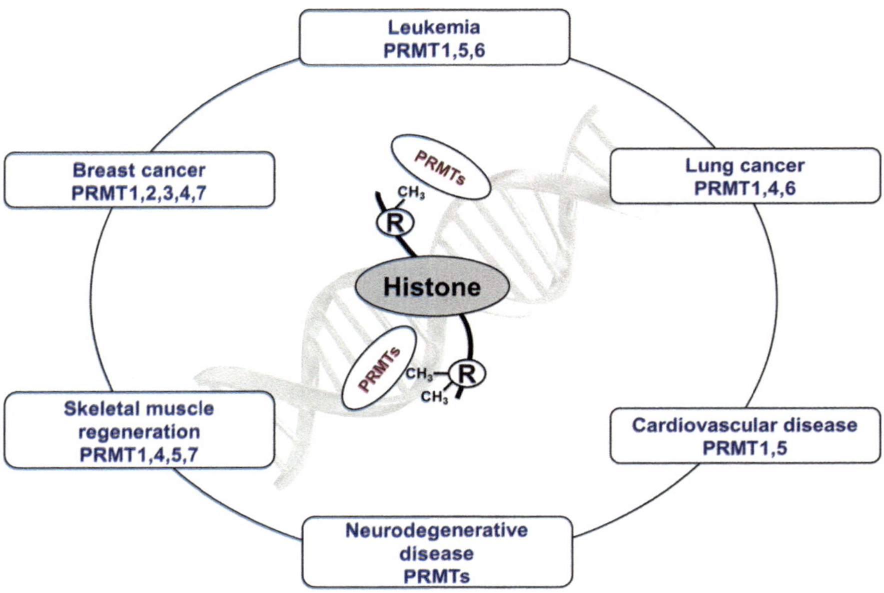


**Fig. 4** PRMT and human diseases. Dysregulation of PRMTs links to multiple human diseases

differentiate in vitro and disrupts the AML initiation and maintenance, supporting CARM1 as a promising therapeutic target for AML (Greenblatt et al. 2018).

**Lung Cancer** PRMT1, PRMT4, and PRMT6 show higher expression in lung cancer tissues, compared with normal tissue (Elakoum et al. 2014). PRMT1 represses E-cadherin and enhances epithelial-mesenchymal transition (EMT) in erlotinib-resistant non-small cell lung cancer (NSCLC) cells (Iderzorig et al. 2018). In *PRMT1*-silenced lung carcinoma, mitogenic factor neuromedin B receptor decreases and epithelial markers (cytokeratin 7 and 8) increase, indicating that PRMT1 reduces cell proliferation and promotes tumor differentiation (Elakoum et al. 2014).

**Breast Cancer** Breast cancer is classified into four subtypes: luminal, HER2$^+$, basal-like and normal breast-like. PRMT1, PRMT2, PRMT3, PRMT4, and PRMT7 are highly expressed in breast cancer tissues. Three splicing variants of *PRMT2* (*PRMT2α*, *PRMT2β*, and *PRMT2γ*) show increased expression in both breast cancer cell lines and human breast carcinoma samples, compared with normal ones. PRMT7 mediates H4R3me2a at *E-cadherin* promoter and induces EMT to increase migratory and invasive abilities of breast cancer cells (Dhar et al. 2012).

## 3.2 The Role of Histone Arginine Methylation in Regenerative Diseases

**Skeletal Muscle Regeneration** PRMTs are modulators of stem cell function and cell fate decision. *PRMT1* deficiency leads to muscle stem cell (MSC) expansion, increases self-renewal and impairs MSC differentiation. PRMT1-dependent methylation of Eya1 enhances its binding at *MyoD* promoter and increases its expression (Blanc et al. 2017). CARM1 is necessary for the asymmetric division of satellite stem cells in skeletal muscle, and upregulates *Myf5* to allow the entry of satellite stem cells into the myogenic process (Kawabe et al. 2012).

**Neurodegenerative Disease** PRMT5 is highly expressed in neuronal cells in mammals. Huntingtin (Htt) suppresses the activity of PRMT5 and reduces the symmetric dimethylation of H2A and H4 in neuron (Ratovitski et al. 2015). In addition, silencing *PRMT5* increases β-amyloid accumulation in primary neurons and human neuroblastoma cells (Quan et al. 2015).

## 3.3 PRMT Dysregulation and Cardiovascular Disease

Nitric oxide (NO) is a signal molecule to relieve the cardiovascular pressure and originates from nitrogen atoms of the arginine. ADMA generated by PRMTs inhibits NO synthesis (NOS) to increase the risk of endothelial dysfunction and cardiovascular disease (Bouras et al. 2013). In addition, SDMA, a structural isomer of ADMA, competitively inhibits NOS and it might be a clinical marker to determine the risk of coronary artery disease (Bode-Boger et al. 2006).

## 3.4 PRMT and Metabolic Disorders

Cellular metabolism is closely linked to epigenetic modifications (Wang and Lei 2017, 2018). Dietary nutrition modulates the cellular level of SAM, the methyl donor, and protein arginine methylation levels. B vitamins deficiency led to a significant decrease in protein arginine methylation in heart and brain. Specifically, H3R8 is hypomethylated in scarce of B vitamins (Esse et al. 2013). PRMT has been shown as a critical metabolic regulator by targeting both histone and non-histone proteins. Arginine methylation serves as a promising target in the therapeutic intervention of metabolic disorders. Therefore, we discuss arginine methylation to provide an inclusive understanding of the metabolic function.

**Adipogenesis** In *CARM1*-knockout embryo, multiple lipid metabolism genes are downregulated. Mechanistic studies revealed that CARM1 acts as a transcription co-activator of peroxisome proliferator-activated receptor γ (PPARγ) to regulate adipogenic genes and promotes the differentiation of adipocytes (Yadav et al. 2008). By mediating histone arginine dimethylation, PRMT5 also upregulates the adipogenic genes to induce chromatin remodeling and PPARγ2 recruitment (LeBlanc et al. 2012).

**Glucose Metabolism** Multiple glycolytic enzymes are modulated by PRMT. PRMT1 methylates PFKFB3 to inhibit F-2,6BP production. Hypomethylation of PFKFB3 results in enhancement of the pentose phosphate pathway to boost NADPH production (Yamamoto et al. 2014). CARM1 methylates glycolytic enzyme PKM2, but not PKM1, to activate glycolysis. PKM2 methylation promotes PKM2 tetramerization and the metabolic reprogramming of breast cancer cells (Liu et al. 2017). Interestingly, *Carm1* deletion increased the *de novo* serine synthesis in MEF cells, potentially by suppressing pyruvate kinase activity. CARM1 depletion provides a survival advantage with a limited supply of serine (Abeywardana et al. 2018). CARM1 also methylates another glycolytic enzyme GAPDH to suppress its interaction with the coenzyme $NAD^+$ and hence inhibit glycolysis. CARM1-mediated GAPDH methylation suppresses liver cancer growth (Zhong et al. 2018). In obese and insulin-resistant mouse models, PRMT6 methylates CRTC2 to enhance its binding with CREB, therefore promoting the expression of gluconeogenic enzymes (Han et al. 2014). In this scenario, PRMT5 also increases H3R2me and facilitates CRTC2-induced CREB phosphorylation to promote the chromatin accessibility of gluconeogenic genes and glucose production (Han et al. 2014; Tsai et al. 2013).

**Mitochondria Metabolism** Arginine methylation is involved in the regulation of mitochondria metabolism. PRMT1 was shown to be almost entirely responsible for mitochondria protein methylation in worms (Sha et al. 2017). Specifically, PRMT1 regulates mitochondrial ROS production and unfolded-protein response (Sha et al. 2017). MDH1, an enzyme in the malate-aspartate shuttle, is methylated by CARM1 in human cells. CARM1-dependent MDH1 methylation suppresses its dimerization and catalytic activity, and inhibits mitochondria respiration and glutamine utilization (Wang et al. 2016). The methylation of C-terminal of PGC1α by PRMT1 potentiates its function as a transcription co-activator to promote mitochondria biogenesis. Dysregulation of PGC1α methylation leads to defective fatty acid oxidation and aberrant energy metabolism of the myocardium, and PRMT1 is pathologically linked to perinatal cardiomyopathies (Garcia et al. 2011). In addition, PRMT1 regulates MICU1, which function as a shield to prevent mitochondria $Ca^{2+}$ overload (Madreiter-Sokolowski et al. 2016).

**Urea Cycle** The urea cycle is also under the control of arginine methylation. PRMT7 directly binds to ASS1, the rate-limiting enzyme of arginine synthesis in urea/citrulline-nitric oxide cycle. Arginine methylation of ASS1 is potentially linked to type I citrullinemia, which requires further investigation (Verma et al. 2017).

## 3.5 PRMTs Modulate Immune and Inflammatory Responses

PRMTs epigenetically modulate gene expression and signal transduction to control inflammatory and immune responses.

**PRMT1 and PRMT4 Modulates IL-4 Signaling** PRMT1 is highly expressed in a rat model of antigen-induced pulmonary inflammation (AIPI), compared with its expression in control animals. IL-4 was shown to be responsible for *PRMT1* overexpression by activating eotaxin-1 and STAT signaling. In the presence of IL-4, both PRMT1 and CARM1 regulate CBP/p300-interacting transactivator 2 (CITED2) to induce immune responses (Sun et al. 2015).

**PRMT1 and TGF-β Signaling** Transforming growth factor β (TGF-β) produced by IL-4-stimulated epithelial cells also leads to PRMT1 upregulation and pulmonary inflammation (Sun et al. 2015).

**PRMT5 Regulates TNF-α Signaling** TNF-α promotes PRMT1 to complex with p65 and poly (ADP-ribose) polymerase 1 (PARP1) (Hassa et al. 2008). In TNF-α-stimulated endothelial cells, PRMT5 methylates HOXA9. PRMT5 forms a complex with HOXA9 and binds to the promoter of *E-selectin* to induce its expression. Furthermore, PRMT5 promotes the expression of the pro-inflammatory protein and facilitates inflammation response of endothelial cells (Bandyopadhyay et al. 2012).

**PRMT1, CARM1, PRMT5, and PRMT6 Control NF-κB Activity** Multiple PRMTs have been shown to be involved in NF-κB signaling (An et al. 2004; Di Lorenzo et al. 2014; Hassa et al. 2008). PRMT1 forms a complex with p65 and PARP1 to activate NF-κB-target gene transcription (Hassa et al. 2008). CARM1 also functions as an NF-κB transcriptional co-activator and is involved in inflammatory diseases and autoimmune diseases (Covic et al. 2005). In addition, PRMT5 interacts with DR4 to suppress NF-κB signaling and the expression of inflammatory cytokine

(Tanaka et al. 2009). PRMT6 directly interacts with NF-κB and G-protein pathway suppressor 2 (GPS2) in response to inflammation (Di Lorenzo et al. 2014).

**PRMT1 Modulates Lymphocyte Activation and Macrophage Differentiation** CD28 signaling upregulates *PRMT1* expression and further links to T cell receptor (TCR) stimulation. In contrast, PRMT1 inhibition reduces the expression of TCR and IL-2 receptor. Also, PRMT1 deficiency impairs B cells proliferation and differentiation (Infantino et al. 2017). In addition, PRMT1 was essential for IL-4-induced macrophage M2 differentiation and the innate immune response was dependent on PPARγ (Tikhanovich et al. 2017).

## 4 Investigation of PRMTs Inhibitors

In order to discover compounds that are biologically active towards PRMT, researchers endeavor to identify the specific, potent, cell-permeable PRMTs inhibitors. In general, most pan-methyltransferase inhibitors are SAM analogs, such as methylthioadenosine (MTA), AzaAdoMet, Sinefungin, and AdoHcy. These inhibitors are used to determine the methyltransferase X-ray structure or alter the global methylation levels. However, these pan-methyltransferase inhibitors are disadvantageous in discriminating different PRMT paralogues. In this regard, PRMT-specific inhibitors have been developed based on biochemical screens.

**Targeting the Catalytic Center of PRMTs** The first panel of small molecules targeting PRMTs was defined in 2004, that is, AMI serial compounds 1 (AMI-1) and 2 (AMI-8). Of note, compound 6 from this panel dramatically reduces androgen-dependent transcription. These leading chemicals exert specific inhibitory effect against individual PRMT (Hu et al. 2016). Compound 17 serves as an effective inhibitor for PRMT1, PRMT6, and PRMT8 (Sun et al. 2015). The first reported specific inhibitor for PRMT3 is compound 18 (Hu et al. 2016). Compound 45, which is generated by conjugating propyl group to sinefungin, shows a strong increase in its potency for CARM1. Compound 45 exhibits an inhibition effect on CARM1 at low micromole doses (Hu et al. 2016). Compound 49 (EPZ015666), obtained from liver microsome, shows significant PRMT5-specific inhibition (Swiercz et al. 2007). Compounds 52 and 53 dose-dependently inhibit H3R2me and are highly selective for PRMT6 (Mitchell et al. 2015).

**Targeting Histone Recognition of PRMTs** An alternative strategy is to interrupt the interaction between PRMTs and histone. Multiple compounds specifically inhibit H3R2me by occupying the arginine-binding channel of PRMTs (Ferreira de Freitas et al. 2016). TBBD, a plant-derived ellagic acid, binds to histone preferentially at a “KAPRK” motif. Consequently, H3R17me is inhibited by TBBD, leading to decreased *p21* expression (Selvi et al. 2010).

**Targeting the Readers of Histone Arginine Methylation** Suppressing the reader of histone methylation is another promising way to develop biological active compounds. Efforts in targeting aromatic cages to block methylarginine readers are underway. Small chemical probes serve as receptors to recognize methylarginine, which allows the blockade of readers of arginine methylation. Among these chemicals, A2D shows higher affinity to methylarginine than most Tudor domains (James et al. 2013). However, UNC1215 binds to not only the MBT domains of L3MBTL3 and L3MBTL1 but also the Tudor domains of 53BP1 and PHF20 (James et al. 2013). Interestingly, installation of methylarginine analogs into specific sites in a recombinant protein can facilitate the functional analysis of arginine methylation. Improved knowledge of methylarginine recognition may help identify novel compounds for targeting cancer and degenerative diseases.

## 5 Conclusions and Future Prospects

Since the discovery of histone methylation, research progress in arginine methylation lagged far behind that in other histone modifications, including phosphorylation and acetylation. Technical advance in arginine methylation detection allowed a continuing expansion of arginine methylation field in last 20 years. To date, human protein arginine methylome has been shown to comprise thousands of proteins. Protein arginine methylation is an abundant modification that exists in almost every cellular compartment and protein family. Recent studies strongly suggest that arginine methylation is involved in numerous biological processes, including metabolism, signal transduction, cell cycle, and cell fate decision. However, only a handful of arginine methylase has been discovered in the human genome. Based on current knowledge, arginine methylation of such a broad spectrum of proteins seems to be under the precise control of a limited number of arginine methylases. The huge contrast between the size of arginine methylase family and arginine methylome promises a rapid growth in the appreciation of protein arginine methylation in the coming years.

**Define the Substrate Specificity of PRMTs** The physiological role of PRMT is coupled with its substrate specificity. For instance, PRMT2 methylates STAT3 to regulate leptin signaling and dietary-induced obesity (Iwasaki et al. 2010). PRMT2 also methylates Cobl, which is an actin nucleator, to promote neuronal morphogenesis (Hou et al. 2018). While PRMT5 modifies RNA splicing-related proteins to strengthen cytokine signaling (Inoue et al. 2018), it also induces H3R2me and regulates the transcription of gluconeogenic genes (Tsai et al. 2013). Moreover, CARM1 show either oncogenic or tumor-suppressive function by targeting different metabolic pathways in breast cancer and pancreatic cancer, respectively (Liu et al. 2017; Wang et al. 2016). These observations suggest that the function of PRMT is dependent on the biological context. Therefore, the substrate specificity of PRMTs across different cellular contexts, tissue types, and developmental stages is the basis to understand the biological function of PRMT. Cross-reference of PRMT interactome and quantitative arginine methylome will greatly advance our understanding of PRMT's substrate specificity. Importantly, ChIP-sequencing analysis

PRMTs and histone arginine methylation marks will help elucidate the role of PRMT in epigenetic regulation at a higher resolution.

**The Crosstalk of Arginine Methylation with Other Epigenetic Marks** The crosstalk of arginine methylation and other epigenetic marks remains poorly understood. Recent evidences strongly support that the interaction of histone arginine methylation and other epigenetic marks plays a major role in gene transcription. PRMT1-mediated H4R3me2a recruits p300/CBP-associated factor complex to induce hyperacetylation at histone H3K9 and H3K14, which further enhances the recruitment of transcriptional factors (Bedford and Clarke 2009). The crosstalk between histone methylation and acetylation modulates cell differentiation and cocaine response (Mitchell et al. 2015). Besides, PRMT6-induced H3R2me2a prevents the binding of MLL complex to the neighboring H3K4me3 mark (Neault et al. 2012). TRPM6-cleaved kinases (M6CKs) phosphorylate histones at serine and threonine residues and stops PRMT5 from deposit methylarginine marks (Krapivinsky et al. 2017). A deep understanding of how histone arginine methylation works in concert with other epigenetic marks will shed lights on the epigenetic function of PRMTs.

**Dual Role of Arginine Methylation in Gene Transcription** PRMTs play dual roles for epigenetic regulation in respect to gene expression. PRMT1 and CARM1 mostly function as transcription activators/co-activators, whereas PRMT5 and PRMT6 play the role of transcription repressors. Interestingly, H4R3 can be methylated by either PRMT1 or PRMT5. The consequent asymmetric or symmetric dimethylation has opposite effects on target gene transcription (Cheung et al. 2007; Huang et al. 2005). H4R3me2a recruits p300/CBP-associated complex to induce transcription activation (Bedford and Clarke 2009); while H4R3me2s is recognized by the DNA methyltransferase, leading to repression of the target gene (Zhao et al. 2009). Notably, PRMT6 not only acts as a repressor by generating H3R2me2a and H2AR29me2a (Waldmann et al. 2011) but also serves as a co-activator by producing H3R42me2a (Casadio et al. 2013). More comprehensive mapping of methylarginine marks is critical to elucidate the regulatory role of histone arginine methylation.

**Tissue-Specific Roles of Histone Arginine Methylation** Different PRMT may have different biological functions in the same tissue type. In the nervous system, PRMT1 is essential for myelination in CNS (Hashimoto et al. 2016); PRMT3 involves in the neuronal translation process and controls dendritic spine activity (Ikenaka et al. 2006); CARM1 methylates H3R17 to establish and maintain the astroglial lineage (Selvi et al. 2010); PRMT5 is required for oligodendrocyte differentiation (Huang et al. 2011); PRMT6 expression level decreases in response to chronic cocaine exposure (Damez-Werno et al. 2016), and PRMT8 acts as a rheostat alongside PRMT1 to drive late-stage differentiation (Simandi et al. 2015). In the muscular system, PRMT5 promotes chromatin remodeling during myogenesis (Dacwag et al. 2009); CARM1 regulates the expression of the glycogen-associated gene in muscle cells (Wang et al. 2012); PRMT1 and PRMT7 are essential for muscle homeostasis and regeneration (Blanc et al. 2016, 2017). Different arginine methylases work in a coordinated fashion to control cell differentiation and maintenance.

**Expanding the Toolkit of Arginine Methylation Studies** The increasing collection of PRMT inhibitors has significantly accelerated the progress in arginine methylation research. A valuable expansion to the toolkit will be compounds that target methylarginine readers and erasers. Besides, the detection and quantification of methylarginine rely heavily on mass spectrometry (Esse et al. 2013). Highly sensitive, high-throughput, and quantitative methods will remove the bottleneck of arginine methylation studies.

Asymmetric/symmetric dimethylarginine is the predominant product when PRMT activity is assayed in vitro. However, recent proteomic studies demonstrate that a large number of arginine residues are monomethylated in cells. This observation strongly suggests that mono-methylarginine not only serves as a catalytic intermediate in arginine methylation reaction but also functions as a specific post-translational modification itself. Therefore, the biochemistry of arginine methylation adds another layer of complexity to its biological function. Three different kinds of methylated arginine residues, monomethylarginine, asymmetric dimethylarginine, and symmetric dimethylarginine, potentially have different roles in regulating the histone and non-histone proteins: (1) arginine methylation possibly show dose-dependent effect on the target protein, because the addition of more methyl group results in stronger alteration in the steric hindrance and net charge of arginine side chain; (2) different types of methylation modification potentially have discrete impact on target protein, as each of them have unique spatial and biochemical properties.

To date, functional evaluation of arginine methylation at single-residue level remains a challenging task. Genetic code expansion has been successfully used to site-specifically incorporate non-canonical amino acids into recombinant protein (Chin 2017). Although this strategy has seen great success in lysine acetylation studies, an efficient site-specific coding system for inserting different forms of methylarginine into proteins is still lacking (Akahoshi et al. 2011). Further expansion of genetic code to enable site-specific installation of methylarginine will provide a powerful tool for arginine methylation studies.

## References

Abeywardana T, Oh M, Jiang L et al (2018) CARM1 suppresses de novo serine synthesis by promoting PKM2 activity. J Biol Chem 293:15290–15303

Akahoshi A, Suzue Y, Kitamatsu M et al (2011) Site-specific incorporation of arginine analogs into proteins using arginyl-tRNA synthetase. Biochem Biophys Res Commun 414:625–630

An W, Kim J, Roeder RG (2004) Ordered cooperative functions of PRMT1, p300, and CARM1 in transcriptional activation by p53. Cell 117:735–748

Bandyopadhyay S, Harris DP, Adams GN et al (2012) HOXA9 methylation by PRMT5 is essential for endothelial cell expression of leukocyte adhesion molecules. Mol Cell Biol 32:1202–1213

Bedford MT, Clarke SG (2009) Protein arginine methylation in mammals: who, what, and why. Mol Cell 33:1–13

Blanc RS, Vogel G, Chen T et al (2016) PRMT7 preserves satellite cell regenerative capacity. Cell Rep 14:1528–1539

Blanc RS, Vogel G, Li X et al (2017) Arginine methylation by PRMT1 regulates muscle stem cell fate. Mol Cell Biol 37:e00457-16

Bode-Boger SM, Scalera F, Kielstein JT et al (2006) Symmetrical dimethylarginine: a new combined parameter for renal function and extent of coronary artery disease. J Am Soc Nephrol 17:1128–1134

Bouras G, Deftereos S, Tousoulis D et al (2013) Asymmetric dimethylarginine (ADMA): a promising biomarker for cardiovascular disease? Curr Top Med Chem 13:180–200

Casadio F, Lu X, Pollock SB et al (2013) H3R42me2a is a histone modification with positive transcriptional effects. Proc Natl Acad Sci USA 110:14894–14899

Cesaro E, De Cegli R, Medugno L et al (2009) The Kruppel-like zinc finger protein ZNF224 recruits the arginine methyltransferase PRMT5 on the transcriptional repressor complex of the aldolase A gene. J Biol Chem 284:32321–32330

Cheung N, Chan LC, Thompson A et al (2007) Protein arginine-methyltransferase-dependent oncogenesis. Nat Cell Biol 9:1208–1215

Chin JW (2017) Expanding and reprogramming the genetic code. Nature 550:53–60

Covic M, Hassa PO, Saccani S et al (2005) Arginine methyltransferase CARM1 is a promoter-specific regulator of NF-kappaB-dependent gene expression. EMBO J 24:85–96

Dacwag CS, Bedford MT, Sif S et al (2009) Distinct protein arginine methyltransferases promote ATP-dependent chromatin remodeling function at different stages of skeletal muscle differentiation. Mol Cell Biol 29:1909–1921

Damez-Werno DM, Sun H, Scobie KN et al (2016) Histone arginine methylation in cocaine action in the nucleus accumbens. Proc Natl Acad Sci USA 113:9623–9628

Dhar SS, Lee SH, Kan PY et al (2012) Trans-tail regulation of MLL4-catalyzed H3K4 methylation by H4R3 symmetric dimethylation is mediated by a tandem PHD of MLL4. Genes Dev 26: 2749–2762

Di Lorenzo A, Yang Y, Macaluso M et al (2014) A gain-of-function mouse model identifies PRMT6 as a NF-kappaB coactivator. Nucleic Acids Res 42:8297–8309

Dong F, Li Q, Yang C et al (2018) PRMT2 links histone H3R8 asymmetric dimethylation to oncogenic activation and tumorigenesis of glioblastoma. Nat Commun 9:4552

Elakoum R, Gauchotte G, Oussalah A et al (2014) CARM1 and PRMT1 are dysregulated in lung cancer without hierarchical features. Biochimie 97:210–218

Esse R, Florindo C, Imbard A et al (2013) Global protein and histone arginine methylation are affected in a tissue-specific manner in a rat model of diet-induced hyperhomocysteinemia. Biochim Biophys Acta 1832:1708–1714

Feng Y, Maity R, Whitelegge JP et al (2013) Mammalian protein arginine methyltransferase 7 (PRMT7) specifically targets RXR sites in lysine- and arginine-rich regions. J Biol Chem 288:37010–37025

Ferreira de Freitas R, Eram MS, Szewczyk MM et al (2016) Discovery of a potent class I protein arginine methyltransferase fragment inhibitor. J Med Chem 59:1176–1183

Ganesh L, Yoshimoto T, Moorthy NC et al (2006) Protein methyltransferase 2 inhibits NF-kappaB function and promotes apoptosis. Mol Cell Biol 26:3864–3874

Garcia MM, Gueant-Rodriguez RM, Pooya S et al (2011) Methyl donor deficiency induces cardiomyopathy through altered methylation/acetylation of PGC-1alpha by PRMT1 and SIRT1. J Pathol 225:324–335

Greenblatt SM, Man N, Hamard PJ et al (2018) CARM1 is essential for myeloid leukemogenesis but dispensable for normal hematopoiesis. Cancer Cell 33:1111–1127.e5

Guccione E, Bassi C, Casadio F et al (2007) Methylation of histone H3R2 by PRMT6 and H3K4 by an MLL complex are mutually exclusive. Nature 449:933–937

Han HS, Jung CY, Yoon YS et al (2014) Arginine methylation of CRTC2 is critical in the transcriptional control of hepatic glucose metabolism. Sci Signal 7:ra19

Harrison MJ, Tang YH, Dowhan DH (2010) Protein arginine methyltransferase 6 regulates multiple aspects of gene expression. Nucleic Acids Res 38:2201–2216

Hashimoto M, Murata K, Ishida J et al (2016) Severe hypomyelination and developmental defects are caused in mice lacking protein arginine methyltransferase 1 (PRMT1) in the central nervous system. J Biol Chem 291:2237–2245
Hassa PO, Covic M, Bedford MT et al (2008) Protein arginine methyltransferase 1 coactivates NF-kappaB-dependent gene expression synergistically with CARM1 and PARP1. J Mol Biol 377: 668–678
Hatanaka Y, Tsusaka T, Shimizu N et al (2017) Histone H3 methylated at arginine 17 is essential for reprogramming the paternal genome in zygotes. Cell Rep 20:2756–2765
Hou W, Nemitz S, Schopper S et al (2018) Arginine methylation by PRMT2 controls the functions of the actin nucleator cobl. Dev Cell 45:262–275 e268
Hu H, Qian K, Ho MC et al (2016) Small molecule inhibitors of protein arginine methyltransferases. Expert Opin Investig Drugs 25:335–358
Huang S, Litt M, Felsenfeld G (2005) Methylation of histone H4 by arginine methyltransferase PRMT1 is essential *in vivo* for many subsequent histone modifications. Genes Dev 19: 1885–1893
Huang J, Vogel G, Yu Z et al (2011) Type II arginine methyltransferase PRMT5 regulates gene expression of inhibitors of differentiation/DNA binding Id2 and Id4 during glial cell differentiation. J Biol Chem 286:44424–44432
Iderzorig T, Kellen J, Osude C et al (2018) Comparison of EMT mediated tyrosine kinase inhibitor resistance in NSCLC. Biochem Biophys Res Commun 496:770–777
Ikenaka K, Miyata S, Mori Y et al (2006) Immunohistochemical and western analyses of protein arginine N-methyltransferase 3 in the mouse brain. Neuroscience 141:1971–1982
Infantino S, Light A, O'Donnell K et al (2017) Arginine methylation catalyzed by PRMT1 is required for B cell activation and differentiation. Nat Commun 8:891
Inoue M, Okamoto K, Terashima A et al (2018) Arginine methylation controls the strength of gammac-family cytokine signaling in T cell maintenance. Nat Immunol 19:1265–1276
Iwasaki H, Kovacic JC, Olive M et al (2010) Disruption of protein arginine N-methyltransferase 2 regulates leptin signaling and produces leanness *in vivo* through loss of STAT3 methylation. Circ Res 107:992–1001
James LI, Beaver JE, Rice NW et al (2013) A synthetic receptor for asymmetric dimethyl arginine. J Am Chem Soc 135:6450–6455
Jin Y, Zhou J, Xu F et al (2016) Targeting methyltransferase PRMT5 eliminates leukemia stem cells in chronic myelogenous leukemia. J Clin Invest 126:3961–3980
Karkhanis V, Wang L, Tae S et al (2012) Protein arginine methyltransferase 7 regulates cellular response to DNA damage by methylating promoter histones H2A and H4 of the polymerase delta catalytic subunit gene, POLD1. J Biol Chem 287:29801–29814
Kawabe Y, Wang YX, McKinnell IW et al (2012) Carm1 regulates Pax7 transcriptional activity through MLL1/2 recruitment during asymmetric satellite stem cell divisions. Cell Stem Cell 11:333–345
Kim JD, Park KE, Ishida J et al (2015) PRMT8 as a phospholipase regulates Purkinje cell dendritic arborization and motor coordination. Sci Adv 1:e1500615
Krapivinsky G, Krapivinsky L, Renthal NE et al (2017) Histone phosphorylation by TRPM6's cleaved kinase attenuates adjacent arginine methylation to regulate gene expression. Proc Natl Acad Sci USA 114:E7092–E7100
LeBlanc SE, Konda S, Wu Q et al (2012) Protein arginine methyltransferase 5 (Prmt5) promotes gene expression of peroxisome proliferator-activated receptor gamma2 (PPARgamma2) and its target genes during adipogenesis. Mol Endocrinol 26:583–597
Lee WC, Lin WL, Matsui T et al (2015) Protein arginine methyltransferase 8: tetrameric structure and protein substrate specificity. Biochemistry 54:7514–7523
Li J, Zhao Z, Carter C et al (2013) Coactivator-associated arginine methyltransferase 1 regulates fetal hematopoiesis and thymocyte development. J Immunol 190:597–604
Li S, Ali S, Duan X et al (2018) JMJD1B demethylates H4R3me2s and H3K9me2 to facilitate gene expression for development of hematopoietic stem and progenitor cells. Cell Rep 23:389–403
Liu F, Cheng G, Hamard PJ et al (2015) Arginine methyltransferase PRMT5 is essential for sustaining normal adult hematopoiesis. J Clin Invest 125:3532–3544

Liu F, Ma F, Wang Y et al (2017) PKM2 methylation by CARM1 activates aerobic glycolysis to promote tumorigenesis. Nat Cell Biol 19:1358–1370

Madreiter-Sokolowski CT, Klec C, Parichatikanond W et al (2016) PRMT1-mediated methylation of MICU1 determines the UCP2/3 dependency of mitochondrial Ca(2+) uptake in immortalized cells. Nat Commun 7:12897

Mitchell LH, Drew AE, Ribich SA et al (2015) Aryl pyrazoles as potent inhibitors of arginine methyltransferases: identification of the first PRMT6 tool compound. ACS Med Chem Lett 6:655–659

Mizutani S, Yoshida T, Zhao X et al (2015) Loss of RUNX1/AML1 arginine-methylation impairs peripheral T cell homeostasis. Br J Haematol 170:859–873

Neault M, Mallette FA, Vogel G et al (2012) Ablation of PRMT6 reveals a role as a negative transcriptional regulator of the p53 tumor suppressor. Nucleic Acids Res 40:9513–9521

Pal S, Vishwanath SN, Erdjument-Bromage H et al (2004) Human SWI/SNF-associated PRMT5 methylates histone H3 arginine 8 and negatively regulates expression of ST7 and NM23 tumor suppressor genes. Mol Cell Biol 24:9630–9645

Quan X, Yue W, Luo Y et al (2015) The protein arginine methyltransferase PRMT5 regulates Abeta-induced toxicity in human cells and *Caenorhabditis elegans* models of Alzheimer's disease. J Neurochem 134:969–977

Ramon-Maiques S, Kuo AJ, Carney D et al (2007) The plant homeodomain finger of RAG2 recognizes histone H3 methylated at both lysine-4 and arginine-2. Proc Natl Acad Sci USA 104:18993–18998

Ratovitski T, Arbez N, Stewart JC et al (2015) PRMT5-mediated symmetric arginine dimethylation is attenuated by mutant huntingtin and is impaired in Huntington's disease (HD). Cell Cycle 14:1716–1729

Rhein VF, Carroll J, Ding S et al (2013) NDUFAF7 methylates arginine 85 in the NDUFS2 subunit of human complex I. J Biol Chem 288:33016–33026

Sanchez SE, Petrillo E, Beckwith EJ et al (2010) A methyl transferase links the circadian clock to the regulation of alternative splicing. Nature 468:112–116

Selvi BR, Batta K, Kishore AH et al (2010) Identification of a novel inhibitor of coactivator-associated arginine methyltransferase 1 (CARM1)-mediated methylation of histone H3 Arg-17. J Biol Chem 285:7143–7152

Sha L, Daitoku H, Araoi S et al (2017) Asymmetric arginine dimethylation modulates mitochondrial energy metabolism and homeostasis in *Caenorhabditis elegans*. Mol Cell Biol 37:e00504-16

Shin HJ, Kim H, Oh S et al (2016) AMPK-SKP2-CARM1 signalling cascade in transcriptional regulation of autophagy. Nature 534:553–557

Simandi Z, Czipa E, Horvath A et al (2015) PRMT1 and PRMT8 regulate retinoic acid-dependent neuronal differentiation with implications to neuropathology. Stem Cells 33:726–741

Suchankova J, Legartova S, Sehnalova P et al (2014) PRMT1 arginine methyltransferase accumulates in cytoplasmic bodies that respond to selective inhibition and DNA damage. Eur J Histochem 58:2389

Sun Q, Liu L, Roth M et al (2015) PRMT1 upregulated by epithelial proinflammatory cytokines participates in COX2 expression in fibroblasts and chronic antigen-induced pulmonary inflammation. J Immunol 195:298–306

Swiercz R, Person MD, Bedford MT (2005) Ribosomal protein S2 is a substrate for mammalian PRMT3 (protein arginine methyltransferase 3). Biochem J 386:85–91

Swiercz R, Cheng D, Kim D et al (2007) Ribosomal protein rpS2 is hypomethylated in PRMT3-deficient mice. J Biol Chem 282:16917–16923

Tanaka H, Hoshikawa Y, Oh-hara T et al (2009) PRMT5, a novel TRAIL receptor-binding protein, inhibits TRAIL-induced apoptosis via nuclear factor-kappaB activation. Mol Cancer Res 7:557–569

Tikhanovich I, Zhao J, Olson J et al (2017) Protein arginine methyltransferase 1 modulates innate immune responses through regulation of peroxisome proliferator-activated receptor gamma-dependent macrophage differentiation. J Biol Chem 292:6882–6894

Tsai WW, Niessen S, Goebel N et al (2013) PRMT5 modulates the metabolic response to fasting signals. Proc Natl Acad Sci USA 110:8870–8875

Tsukada Y, Zhang Y (2006) Purification of histone demethylases from HeLa cells. Methods 40:318–326

Verma M, Charles RCM, Chakrapani B et al (2017) PRMT7 interacts with ASS1 and Citrullinemia mutations disrupt the interaction. J Mol Biol 429:2278–2289

Waldmann T, Izzo A, Kamieniarz K et al (2011) Methylation of H2AR29 is a novel repressive PRMT6 target. Epigenetics Chromatin 4:11

Wang YP, Lei QY (2017) Perspectives of reprogramming breast cancer metabolism. Adv Exp Med Biol 1026:217–232

Wang YP, Lei QY (2018) Metabolic recoding of epigenetics in cancer. Cancer Commun (Lond) 38:25

Wang SC, Dowhan DH, Eriksson NA et al (2012) CARM1/PRMT4 is necessary for the glycogen gene expression programme in skeletal muscle cells. Biochem J 444:323–331

Wang L, Zhao Z, Meyer MB et al (2014) CARM1 methylates chromatin remodeling factor BAF155 to enhance tumor progression and metastasis. Cancer Cell 25:21–36

Wang Y, Xiao M, Chen X et al (2015) WT1 recruits TET2 to regulate its target gene expression and suppress leukemia cell proliferation. Mol Cell 57:662–673

Wang YP, Zhou W, Wang J et al (2016) Arginine methylation of MDH1 by CARM1 inhibits glutamine metabolism and suppresses pancreatic cancer. Mol Cell 64:673–687

Yadav N, Cheng D, Richard S et al (2008) CARM1 promotes adipocyte differentiation by coactivating PPARgamma. EMBO Rep 9:193–198

Yamamoto T, Takano N, Ishiwata K et al (2014) Reduced methylation of PFKFB3 in cancer cells shunts glucose towards the pentose phosphate pathway. Nat Commun 5:3480

Yang Y, Lu Y, Espejo A et al (2010) TDRD3 is an effector molecule for arginine-methylated histone marks. Mol Cell 40:1016–1023

Ying Z, Mei M, Zhang P et al (2015) Histone arginine methylation by PRMT7 controls germinal center formation via regulating Bcl6 transcription. J Immunol 195:1538–1547

Young BD, Weiss DI, Zurita-Lopez CI et al (2012) Identification of methylated proteins in the yeast small ribosomal subunit: a role for SPOUT methyltransferases in protein arginine methylation. Biochemistry 51:5091–5104

Zhao Q, Rank G, Tan YT et al (2009) PRMT5-mediated methylation of histone H4R3 recruits DNMT3A, coupling histone and DNA methylation in gene silencing. Nat Struct Mol Biol 16: 304–311

Zhong X, Yuan XM, Xu YY et al (2018) CARM1 methylates GAPDH to regulate glucose metabolism and is suppressed in liver cancer. Cell Rep 24:3207–3223

Zou L, Zhang H, Du C et al (2012) Correlation of SRSF1 and PRMT1 expression with clinical status of pediatric acute lymphoblastic leukemia. J Hematol Oncol 5:42

# Aberrant Epigenomic Regulatory Networks in Multiple Myeloma and Strategies for Their Targeted Reversal

**Samrat Roy Choudhury and Brian A. Walker**

## Contents



**Abstract** Multiple Myeloma (MM), a subset of genetically complex paraproteinaemias, is characterized by abnormal clonal plasma cell expansion in the bone marrow, and accounts for about 13% of all patients with hematological malignancies. Primary genomic abnormalities include *IgH* translocations to *MMSET/FGFR3* (4p16), *CCND1* (11q13), *MAF* (16q23), or *MAFB* (20q12), as well as aneuploidy involving trisomies of several chromosomes, known as hyperdiploidy, and together are the hallmarks of the disease. Besides these structural abnormalities, recurrent mutations affecting key oncogenes and tumor suppressor genes are found, as well as aberrant modifications in epigenetic marks which deregulate key oncogenes in MM. Herein, we undertake to review the global epigenetic regulatory landscape of MM including DNA methylation, histone modifications, non-coding miRNA mechanisms or interactions from regulatory proteins such as CTCF and super-enhancers (SE), in

S. Roy Choudhury · B. A. Walker (✉)
Myeloma Center, University of Arkansas for Medical Sciences, Little Rock, AR, USA
e-mail: BWalker2@uams.edu



S. Jurga, J. Barciszewski (eds.), *The DNA, RNA, and Histone Methylomes*, RNA Technologies, https://doi.org/10.1007/978-3-030-14792-1_22

conjunction with gene expression and function in MM molecular subgroups at different stages of disease progression. Additionally, we discuss new perspectives in designing CRISPR/TAL-based synthetic proteins or novel small molecular drugs to target aberrant epigenetic marks with locus-specific precision, which may be an option for therapeutic intervention.

**Keywords** Multiple myeloma · DNA methylation · Histone modifications · miRNA mechanisms · Epigenetic therapeutics in myeloma

## 1 Introduction

Multiple Myeloma (MM) is characterized by an abnormal clonal plasma cell infiltration in the bone marrow, and accounts for about 13% of patients with hematological malignancies (Durie and Salmon 1975). MM begins with pathophysiologically distinct disease stages such as monoclonal gammopathy of undetermined significance (MGUS) or aymptomatic smoldering MM (SMM), which transform into symptomatic MM, extramedullary myeloma (EMM) or plasma cell leukaemias (PCL) at the progressive terminal stages (Palumbo and Anderson 2011). The clinicopathological properties, survival outcome and therapeutic options of MM exclusively depends on the diverse array of genomic aberrations, including chromosomal translocations, segmental deletion or duplications, copy number abnormalities, or global deregulation of key signaling pathways (Avet-Loiseau et al. 2009; Bergsagel and Kuehl 2001; Rajkumar et al. 2014; Tian et al. 2003). Besides genetic abnormalities, aberrant post-transcriptional modifications in DNA and DNA-adjacent histone proteins or recurrent mutations in epigenetic enzymes have been broadly identified as the key components of epigenetic dysregulation in MM (Galm et al. 2004). An increasing body of literature reports the pivotal role of multi-tier epigenetic modifications in the onset and progression of the disease (Dimopoulos et al. 2014), which were found to deregulate cell cycle, signal transduction pathways, cell adhesion and differentiation or transcription machinery. Herein, we review the different tiers of epigenetic dysregulation in the context of the pathobiology and therapeutic options in MM.

## 2 Clinicopathological Features of Multiple Myeloma

MM is a malignancy of terminally differentiated plasma cells which is more prevalent among African-Americans, compared to Caucasians, and more frequently reported in males than in females (Benjamin et al. 2003; Boyd et al. 2011). Moreover, individuals with an average age >50 years are more prone to an MGUS phase, which can progress to SMM and symptomatic MM at a rate of up to 1% per annum (Korde et al. 2011; Rasmussen et al. 2005).

At the molecular level, MM is defined by a genome-wide complex and heterogeneous genomic landscape. The primary genomic abnormalities include translocations between the *IgH locus* (14q32) and *MMSET/FGFR3* (4p16), *CCND1* (11q13), *MAF* (16q23), or *MAFB* (20q12), or chromosomal copy number aberrations resulting in hyperdiploidy, and are considered to be the etiological hallmarks of the disease (Fonseca et al. 2009; Walker et al. 2006; Zhan et al. 2006). In addition to the cytogenetic abnormalities, several other 'secondary drivers' of the disease have been identified. For instance, activating mutations (*KRAS* and *NRAS*) and deregulation in the expression of oncogenes (*MYC*), tumor suppressors (*p53* or *p18*), cell cycle regulators (*CCND1/2/3*), transcription factors (NF-κB) or amplifications at chromosome 1 (1q21) have been widely demonstrated as defining oncogenic events in MM (Chang et al. 2005; Dib et al. 2008; Kryukov et al. 2013; Walker et al. 2010a). Recently, an increasing body of studies started reporting the implications of aberrant posttranscriptional modifications or epigenetic alterations in the development of the disease (Dimopoulos et al. 2014). For instance, global DNA hypomethylation, clustered promoter-specific DNA hypermethylation of tumor suppressors, or overexpression of Multiple Myeloma SET Domain (MMSET) resulting in H3K36 di- and tri-methylation have already been established as prognostic epigenetic biomarkers of MM. However, unlike the substantial amount of effort made to determine the genetic aberrations, a relatively small number of studies have attempted to elucidate the epigenomic drivers of abnormal gene expression in MM. In the present review, we aim to create a compendium of epigenetic modifications combining DNA methylation, histone modifications, or non-coding miRNA mechanisms, both globally and in a gene-specific fashion, to illustrate the impact of epigenetic modifications at different stages of disease progression.

## 3 Epigenetics: A Brief Tale of Origin and Evolution

In 1956 Conrad Waddington coined the term 'epigenetics' (epi: above) to designate "the branch of biology, which studies the causal interactions between genes and their products, which bring the phenotype into being" and implied 'epigenetic landscape' to describe the trail of events during embryonic development (Goldberg et al. 2007). The explorative success of post 1960 epigenetic research have no doubt enhanced our understanding and knowledge about the diversity of epigenetic mechanisms and its correlation to different genetic disorders, with cancers at the forefront (Felsenfeld 2014). However, unlike the simplistic 'genetic code' of translating hereditary information from the DNA to functional proteins, the regulation and operation of the 'epigenetic code' are complex and yet to be deciphered (Turner 2007).

### *3.1 The Trail of DNA-Methylation Derivatives*

As suggested by Bryan M Turner in his seminal research, methylated CpGs (CpG) on a DNA strand can be well recognized by CpG-binding proteins, which in conjunction with the histone proteins constitute a regulatory network for stable and heritable epigenetic change in an organism (Turner 2007). DNA methylation is referred to as the addition of methyl (-$CH_3$) groups to the 5′-cytosine residue of DNA without alteration of the underlying sequence. Cytosine methylation (5mC) may occur in the CG dinucleotides (referred as CpG sites), or in the CHH or CHG (where H stands for A, T, or C residue) sites in plants or animals. However, the majority of functionally relevant 5mC modifications in humans have been reported in the context of GC rich CpG repeat sites, which are referred to as CpG islands (Gardiner-Garden and Frommer 1987). The addition of 5mCs are carried out by a group of enzymes, collectively referred to as DNA-methyltransferases (DNMTs). The three key genes encoding the DNMT enzyme family are *DNMT1*, *DNMT3A* and *DNMT3B*. *DNMT1* is responsible for the maintenance of methylation marks during developmental processes, while DNMT3A and DNMT3B are responsible for establishing de novo methylation marks (Okano et al. 1999). In addition, CpG binding proteins or methyl binding domain (MBD) proteins, such as MBD1, 2, or 3 have been reported to recruit histone-modifying protein complexes to the CpG sites, which cause heterochromatin formation and gene silencing (Parry and Clarke 2011) (Fig. 1).

The conversion of cytosines to 5mCs was considered to be a stably heritable trait until the recognition of acquired demethylation events during physiological processes such as embryonic development, somatic cell reprogramming or exclusion of gene imprinting in primordial germ cells (Morgan et al. 2005; Oswald et al. 2000; Sasaki and Matsui 2008). These demethylation events were subsequently correlated to the existence and catalytic activity of a family of enzymes, collectively referred to as 'Ten-Eleven Translocation (TET) dioxygenase'. TET enzymes including TET1, TET2 and TET3 enzymatically convert 5mCs to 5-hydroxymethylcytosines (5hmCs), which further oxidize to 5-formylcytosince (5fC) and 5-carboxylcytosine (5caC) (He et al. 2011; Ito et al. 2011; Kaas et al. 2013) (Fig. 1). TET modified oxidized 5mC derivatives are then frequently subjected to deamination or different DNA repair mechanisms, leading to their replacement with unmodified cytosines (Bhutani et al. 2011; Branco et al. 2012).

While around 1.5% of human genomic DNA is methylated and stably maintained, methylation patterns change dramatically during the onset and progression of genetic diseases including cancer (Kulis and Esteller 2010). In particular, global loss of DNA methylation, but acquired gene silencing through increased DNA-methylation or hypermethylation at gene promoters and clustered global hypomethylation at gene bodies and intergenic regions are considered to be the major drivers of epigenomic instabilities in cancer (Jones 2012). Furthermore, global DNA hypomethylation and the overreaching loss of genomic integrity, as mediated by chromosomal abnormalities with recurrent mutations in DNMTs or deregulated methylation patterns over DNA damage repair genes or retroposons also suggest that these events lead to maintenance of a severely chaotic intracellular state in cancers (Jin and Robertson 2013; Robertson 2005).

From the therapeutic perspective, 5-Azacytidine (Aza) and Decitabine (5-aza-2-′-deoxycytidine) are the two commercially available FDA approved drug options, which are used as DNA methylation inhibitors (Momparler 2005; Santi et al. 1984).

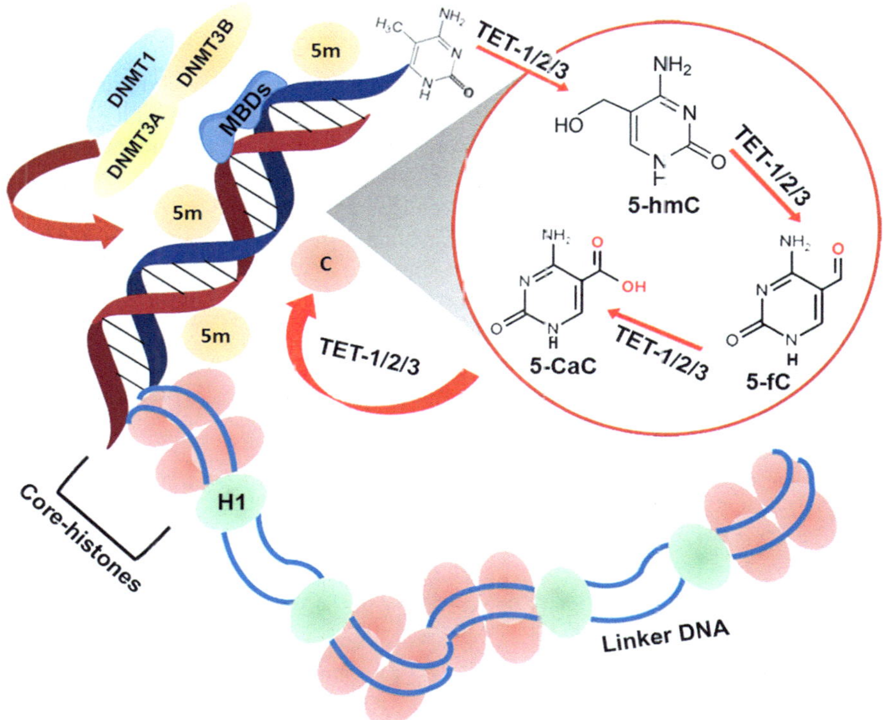


**Fig. 1** An illustration of the active DNA methylation and demethylation events. The covalent additions of methyl (5 m) groups to the cytosine (C) residue, generally in the CG rich repeat sequences are catalyzed de novo by DNMT3A and DNMT3B, while DNMT1 copies the methylated cytosines (5mC) to the daughter cells during cell division. Methyl binding domain (MBD) proteins aid DNMTs in the recognition of methylated sites on DNA. The demethylation events are catalyzed by the Ten-eleven translocation methylcytosine dioxygenases (TET) family of enzymes. TET1/2/3 catalyze successive oxidation of 5-mC to 5-hydroxymethyl- (5hmC), 5formyl- (5fC) and 5-carboxy- (5caC) cytosine. Methylation-demethylation cellular cycles are dynamic and the fate of the cycle changes in accordance to age or development of genetic disorders including cancers

Additionally, new generation therapeutics containing antisense oiligonucleotides or small methylation inhibiting drug molecules are also being used at the experimental level. For instance, MG98, a 20 bp antisense oligo has been successfully used in mouse bladder and colon cancer to prevent active transcription of DNMT1 (Davis et al. 2003; Winquist et al. 2006). Furthermore, RG108, a small molecular drug, has been demonstrated to inhibit DNMT1 activity and reactivate the tumor suppressor p16 in human cancer cells (Brueckner et al. 2005). In summary, aberrant DNA-methylation and hydroxymethylation modifications are associated with the majority of genetic disorders and are recognized as prime targets for prospective therapeutic options.

## 3.2 Histone Modification Codes

The nucleosome is the fundamental unit of chromatin architecture comprising of an octamer of core histones (H2A, H2B, H3, and H4), wrapped by DNA with an interval of 147 base pairs. The amino terminal exposed tail of all eight histones could be subjected to varying levels of enzymatic activities, including acetylation of lysines, methylation of lysines and arginines, and phosphorylation of serines and threonines (Bannister and Kouzarides 2011). These covalent histone modifications in combination with DNA methylation have been documented to impact the expression pattern and functional outcome of a gene (Cedar and Bergman 2009; Fuks 2005) (Fig. 2). Since the discovery of histone proteins (Allfrey et al. 1964), at least 16 different classes of histone modifications including methylation, acetylation, phosphorylation, ubiquitination or SUMOylation have been identified (Sawan and Herceg 2010; Tan

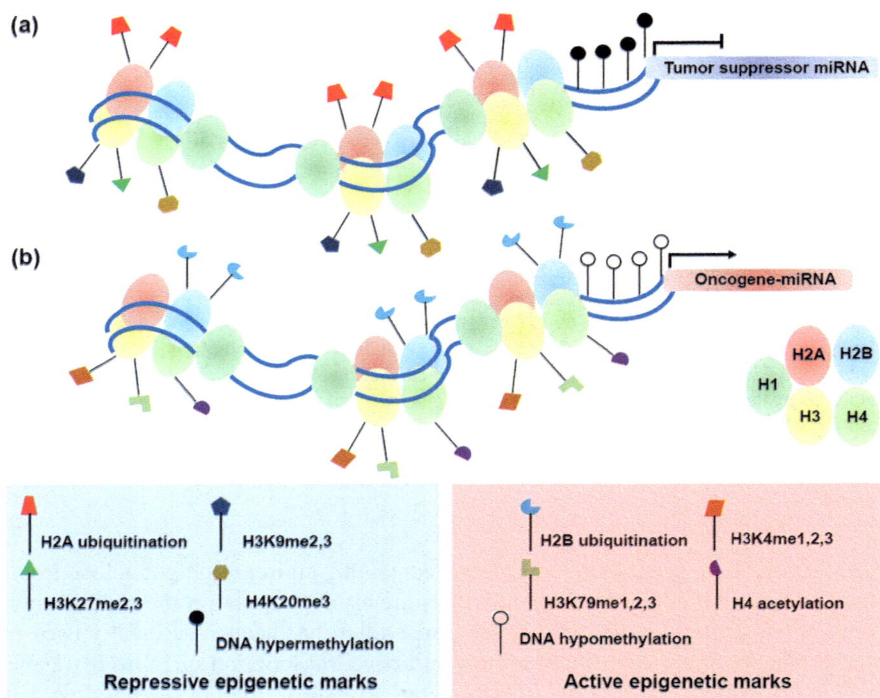


**Fig. 2** Schematic representation of the nucleosome. DNA strands are wrapped around four core histones H2A, H2B, H3, and H4, while histone H1 or the linker protein links the core-histones. Different types of DNA and histone covalent modifications takes place by the catalytic activity of histone modifiers such as histone methyltransferases, demethylases, acetylases, deacetylases etc. Activating DNA and histone marks are summarized in the red box, while the inactivating marks are summarized in the blue box

et al. 2011). Here we tabulate the key histone modifications, frequently reported in the process of oncogenesis (Table 1).

Transcriptionally active euchromatin is marked with high levels of acetylation and trimethylation of H3K4, H3K36 or H3K79, while, the transcriptionally inert heterochromatin is marked with high levels of H3K9, H3K27 and H4K20

**Table 1** A list of key histone modifications and the associated enzymes

| Modifications | Histone | Residue | Enzymes |
|---|---|---|---|
| Methylation | H1 | (Lysine)K26 | EZH2 (enhancer of zeste homolog 2 |
| | H3 | (Arginine)R2, R17, R26 | CARM1 (coactivator-associated arginine methyltransferase-1 |
| | | K4 | MLL (myeloid/lymphoid)-4, SET (lysine methyltransferase)-1, MLL1, SET7/9 |
| | | R8 | PMRT5 (protein arginine N-methyltransferase 5) |
| | | K9 | SUV39h1, SUV39h2, ESET, G9A, EZH2 |
| | | K27 | EZH2, G9A |
| | | K36 | HYPB, NSD1/Set2 |
| | | K79 | DOT1L (disruptor of telomeric silencing 1-like) |
| | H4 | R3 | PRMT1, PRMT5 |
| | | K20 | PR-SET7, SUV4–20/SET9 |
| Acetylation | H2A | K5 | Tip60, Hat1, P300/CBP |
| | H2B | K5 | ATF2 |
| | | K12, K15 | ATF2 (activating transcription factor 2), P300/CBP |
| | | K20 | P300 |
| | H3 | K9 | Gcn5, SRC-1 |
| | | K14 | Gcn5, PCAF, Tip60, SRC-1, hTFIIIC90, TAF1 |
| | | K18, K23 | P300, CBP/Gcn5 |
| | | K27 | gCN5 |
| | H4 | K5 | Hat1, Tip60, ATF2, p300/Hat1, Esa1, Hpa2 |
| | | K8 | Gcn5, PCAF, Tip60, ATF2, p300/Esa1, Elp3 |
| | | K12 | Hat1, Tip60/Hat1, Esa1, Hpa2 |
| | | K16 | MOF, Gcn5, Tip60, ATF2/Gcn5, Esa1, Sas2 |
| Phosphorylation | H2AX | (Serine) S139 | ATM (Ataxia telangiectasia mutated), ATR (ATM and RAD3-related), DNA-PK |
| | H2A | (Threonine) T119 | NHK-1 (Nucleosomal histone kinase-1) |
| | H2B | S14 | Mst1 (macrophage stimulating-1) |
| | H3 | S10 | TG2 (tissue transglutaminase 2), Aurora B, MSK1, MSK2/Snf1 |
| | | T11 | Dlk/ZIP |
| | | S28 | MSK1 (mitogen- and stress-activated kinase1), MSK2 |
| Ubiquitination | H2A | K119 | Ring 1b |
| | H2B | K120 | RNF (ring finger protein) 20/40 |

trimethylation (Li et al. 2007). Transcriptional repression due to loss in histone acetylation in p21, loss of trimethylation of H3K4 in *HOX*, or H4K20 in Sat2 or D4Z4 genes are a few examples, as reported in different types of cancers (Esteller 2007; Fraga et al. 2005). In contrast, gain of H3K9 methylation or H3K27 trimethylation were also demonstrated to be involved in transcriptional repression in key housekeeping genes such as *CDKN2A* or *RASSF1* (Beckedorff et al. 2013; Kondo et al. 2007). In addition to the aforementioned examples in histone modifications, nucleosome positioning in genes such as *BRG1* and *CHD5* have been reported for oncogenic transformation or transcriptional repression (Medina and Sanchez-Cespedes 2008; Mulero-Navarro and Esteller 2008). In summary, aberrant epigenetic changes at the histone tails play a crucial role in the initiation and progression of different types of genetic disorders including cancers. A substantial emphasis hence has been given to develop tools for directed reversal of the anomalous histone marks.

In alignment with the diversity and complexity of histone modifications, a substantial amount of thrust has been made to optimize therapeutic options for correcting the aberrant histone marks. Based on the mode of action, histone modifying epigenetic drugs are broadly classified into (i) histone acetylase (HAT) inhibitors, (ii) histone deacetylase (HDAC) inhibitors, and (iii) histone methyltransferase (HMT) inhibitors.

### 3.2.1 HAT Inhibitors

These are typically used to reactivate gene expression or function by inhibiting the activity of HATs in different cancers. For instance, bisubstrate inhibitors such as curcumin were demonstrated to inhibit histone H3 and H4 acetylation by p300 and PCAF (Balasubramanyam et al. 2004). Another good candidate of a HAT inhibitor is C646, another p300 inhibitor, which is capable of mimicking the propaptotic effect of RNA-mediated p300 knockdown, that has been demonstrated in apoptotic pathways in prostate cancer cells (Santer et al. 2011).

### 3.2.2 HDAC Inhibitors (HDACi)

These are used to reduce chromatin compaction and reverse transcriptional silencing of tumor suppressor genes. HDACi's include vorinostat (suberanilohydroxamic acid), an FDA approved HDACi that is routinely used in clinics to treat persistent and recurring cutaneous T-cell lymphoma. Vorinostat is known to cause hyperacetylation of histones and non-histone proteins, such as p53 to induce apoptosis (Sun et al. 2017). Additionally, ricolinostat (ACY1215) is a selective HDAC6 inhbitor being used in combination with proteasome inhibitors in myeloma treatment. Ricolinostat suppresses proliferation and promotes apoptosis via the PI3K/AKT and ERK pathways.

### 3.2.3 HMT Inhibitors

HMTs are known to methylate the lysine or arginine residues of the histone proteins, which play a crucial role during gene transcription. Experimental studies using HMT inhibitors (HMTi) are showing exciting results in different disease models and are the focus for further optimization. For example, BIX-01294 has been tried as an HMTi against cancer cells. Additionally, several novel small drug molecules are being tested to bring the desired changes at the target site. UNC0646 is one such agent, which has shown HMTi efficacy in the breast cancer cell line MCF7, the prostate cancer cell line 22RV1, and the lung cancer cell line IMR-90 (Olcinia et al. 2015).

## 3.3 *miRNA Epigenetic Mechanisms*

miRNAs are relatively short (typically 22 bp) RNA fragments, which were found to control gene expression. In order to exert their function, miRNAs are first transcribed as primary miRNAs, aided by RNA polymerase II, followed by processing into a precursor miRNA in the nucleus by the RNase II Drosha and a microprocessor complex (DGCR8) (Chuang and Jones 2007). Precursor miRNA are then exported into the cytoplasm by Exportin-5, followed by processing into the mature miRNA by the RNAse III, Dicer, which generates the functionally active form of the miRNA (Kim and Nam 2006; Lee et al. 2002).

miRNAs are mainly explored for their efficacy to inhibit or down-regulate the stability/translation of an active mRNA. They can bind to an mRNA with a complete complementarity or through partial binding to the 3′ UTR regions, which may lead to the translational suppression of the target gene. The impact of miRNAs in different cancers have been well documented both on oncogenes and tumor-suppressor genes (Esquela-Kerscher and Slack 2006; Meltzer 2005). For instance, the anti-apoptotic *BCL2* gene is often downregulated in chronic lymphocytic leukemia from the suppressing effect from miR-15a and miR-16-1 (Cimmino et al. 2005). Upregulation of miR-21 and its antiapoptotic effect is also observed in glioblastoma and aggressive breast cancers (Iorio et al. 2005; Si et al. 2007).

Recent exploration into the biogenesis and mechanism of action of miRNAs suggest that miRNA expression and function can be regulated by both the DNA methylation and histone modifications. For instance, miR-127 expression were found to shoot up remarkably followed by the treatment with a DNA methylation inhibitor (Aza) and a histone deacetylase inhibitor (4-phenylbutyric acid). It was also shown that the HDACi LAQ824 can readily alter the miRNA expression profile of the SKBR3 breast cancer cell line (Saito et al. 2006; Scott et al. 2006). Further research in this direction also suggest that DNA methylation at the intronic CpG islands may promote the transcription of miRNAs (Wutz et al. 1997).

The importance of miRNAs in the development and cancers is now duly acknowledged, which also serve as potential therapeutic targets. For example, knockdown approach of oncogenic miR-21 in glioblastoma, miR-372 and 373 in testicular germ cell tumors, or miR-155 in the lymphomas and breast cancers shown promising result in the clinical setup (Eis et al. 2005; Tam and Dahlberg 2006; Voorhoeve et al. 2006). Cumulatively, miRNA may act as potential regulator of active gene translation and are found to be regulated by both the DNA methylation and histone modifications. Therefore, it's important to investigate the existence, level of expression and functional effect of specific set of miRNAs in a cancer model to determine the effective therapeutic options.

# 4 Epigenomic Landscape in Multiple Myeloma

Hematopoietic stem cells within the bone marrow differentiate into multipotent progenitor and lymphoid progenitor cells, and eventually give rise to the precursor B cells. It has been documented that non-CpG DNA methylation reduces or even completely disappears upon B cell commitment, whereas CpG methylation changes in an accumulative pattern during the B cell maturation (Kulis et al. 2015). It was observed that gene enhancers showed an overall demethylation pattern, which correlates to the upregulation of the B-cell transcription factors and downstream genes. Moreover, an extensive demethylation of heterochromatin and an increase in methylation level at the PRC complexes were observed during later differentiation stages. The multitude of methylation changes in B-cell type specific malignancies such as leukemia, lymphoma or myeloma has been explored further to understand the clonal variation of the B-cell neoplasms, compared to the normal counterparts.

## *4.1 DNA Methylation Aberrations in MM*

Alterations in the DNA methylation pattern have been documented to play an important role in the onset and disease progression in MM (Sharma et al. 2010). Similar to most of cancers, MM is also characterized by acquired global DNA hypomethylation and gene-specific hypermethylation at the promoter of major tumor suppressor genes. Interestingly, the global repetitive elements such as LINE-1 (long interspersed nuclear element-1), Alu and SAT-α (Satellite- α) sequences, which belong to the heterochromatic regions and typically contain high levels of methylation in normal cells, become hypomethylated in MM, which is thought to play a part in genomic instability and chromosomal rearrangements (Aoki et al. 2012; Bollati et al. 2009). In contrast, acquired hypermethylation at the promoter of tumor suppressors and hypomethylation of the growth promoting genes have been identified in different disease stages in MM. Table 2 summarizes a panel of key genes, where changes in the DNA methylation profile have been documented to play a crucial role in their function.

Simultaneous research in this direction has reported that DNA methylation could also serve as a predictor of disease progression in MM. For instance, it was observed

**Table 2** A panel of key genes, which display aberrant DNA methylation pattern at their promoters

| Functional subgroup | Gene symbol | Gene name | References |
|---|---|---|---|
| Tumor suppressor | *CDH1* | Epithelial cadherin or E-cadherin | Braggio et al. (2010) |
| | *p16* $^{INK4a}$ | Cyclin-dependent kinase 4 inhibitor A | |
| | *p15* $^{INK4b}$ | Cyclin-dependent kinase 4 inhibitor B | |
| | *SHP1* | Protein tyrosine phosphatase, non-receptor type 6 (*PTPN6*) | |
| | *ER* | Estrogen receptor | |
| | *BNIP* | BCL2 interacting protein 2 | |
| | *RARβ* | Retinoic acid receptor beta | |
| | *DAPK* | Death-associated protein kinase 1 | |
| | *PTEN* | Phosphatase and tensin homolog | Piras et al. (2014) |
| | *RASSF1* | Ras association (RalGDS/AF-6) domain family member 1 | Galm et al. (2004) |
| | *VHL* | von Hippel-Lindau tumor suppressor, E3 ubiquitin protein ligase | Hatzimichael et al. (2009) |
| | *DLC1* | DLC1 rho GTPase activating protein | de Carvalho et al. (2009) |
| Apoptosis regulation | *GADD45G* | Growth arrest and DNA damage inducible gamma | Heller et al. (2008) |
| | *TP73 (P73)* | Tumor protein p73 | Hatzimichael et al. (2009) |
| | *TP53* | Tumor protein p53 | Hodge et al. (2005) |
| | *XAF1* | XIAP associated factor 1 | Chen et al. (2009) |
| | *RASD1* | RAS, dexamethasone-induced 1 | Kaiser et al. (2013) |
| | *TGFBI* | Transforming growth factor, beta-induced | |
| | *SPARC* | Secreted protein, acidic, osteonectin | |
| | *GPX3* | Glutathione peroxidase 3 | |
| Transcription factors | *PAX1* | Paired box gene 1 | Walker et al. (2010b), Chim et al. (2007) |
| | *JUNB* | Primary growth factor response | |
| Cytokine signaling | *PF4* | Platelet factor 4 | Cheng et al. (2007) |
| | *IL17RB* | Interleukin 17 receptor B | Yuregir et al. (2010) |
| | *CDKN2A* | Cyclin dependent kinase inhibitor 2A | Walker et al. (2010b) |
| | *Socs2* | Suppressor of cytokine signaling 2 | |
| Cell proliferation and adhesion | *IGF1R* | Insulin-like growth factor 1 receptor | Jung et al. (2012) |
| | *DCC* | DCC netrin 1 receptor | de Carvalho et al. (2009) |
| Cell cycle related | *CDKN2A* | Cyclin-dependent kinase inhibitor 2A | Yuregir et al. (2010) |
| | *TGFBR2* | Transforming growth factor, beta receptor II | Kaiser et al. (2013) |

(continued)

**Table 2** (continued)

| Functional subgroup | Gene symbol | Gene name | References |
|---|---|---|---|
| | *CCND2* | Cyclin D2 | Zhan et al. (2006) |
| | *CDKN2B* | Cyclin-dependent kinase inhibitor 2B | Chen et al. (2002) |
| DNA repair | *MGMT* | O-6-methylguanine-DNA methyltransferase | Chim et al. (2004b) |
| | *TDG* | Thymine DNA glycosylase | Peng et al. (2006) |
| JAK/STAT pathway | *PTPN6* | Protein tyrosine phosphatase, non-receptor type 6 | Chim et al. (2004a) |
| | *SOCS1* | Suppressor of cytokine signaling 1 | |
| Wnt pathway | *WIF1* | WNT inhibitory factor 1 | Chim et al. (2007), Walker et al. (2010b) |
| | *SFRP1* | Secreted frizzled-related protein 1 | |
| | *SFRP2* | Secreted frizzled-related protein 2 | |
| | *SFRP4* | Secreted frizzled-related protein 4 | |
| | *SFRP5* | Secreted frizzled-related protein 5 | |
| | *DKK1* | Dickkopf WNT signaling pathway inhibitor 1 | |
| | *DKK3* | Dickkopf WNT signaling pathway inhibitor 3 | |
| | *APC* | Adenomatous polyposis coli | |
| Signal transduction | *ESR1* | Estrogen receptor 1 | Sonaglio et al. (2013) |
| | *CPEB1* | Cytoplasmic polyadenylation element binding protein-1 | de Carvalho et al. (2009) |
| | *IRF8* | Interferon regulatory factor 8 | |

that CD138 positive plasma cells (PC) showed a reducing gradient of DNA methylation from normal PCs to MGUS, SMM, MM and to PCL (Fig. 3). The transition from MGUS to the symptomatic stage is broadly defined with a reduction in global DNA methylation and an increase in promoter hypermethylation in key oncogenes including *CDKN2B*, *GAT4*, *ARID3A*, and *BRCA2* (Walker et al. 2010b) (Fig. 4). A similar trend in the change in DNA methylation was simultaneously reported, based on their observation on 1500 differentially methylated CpG sites (Salhia et al. 2010). It was also realized that changes in DNA methylation are not merely an epigenetic event in MM disease progression, but have a significant impact on gene expression patterns. For instance, expression of a panel of key oncogenes such as *CD38, GPX3, NCAM1/CD56, PDK4, RASD1, RBP1, SPARC and TGFBI* were found to be downregulated due to hypermethylation, mainly at the promoter (Kaiser et al. 2013). In contrast, a simultaneous study reported that gene expression in MM may not always correlate to the promoter methylation of a gene (Jung et al. 2012). However, with the continuous advancement in this field it is now realized that DNA methylation may affect the enhancer regions and thus limit the binding of transcription factors (TFs) to the enhancer or promoter region that regulates gene expression. Gene bodies may also contain multiple enhancer like elements, and

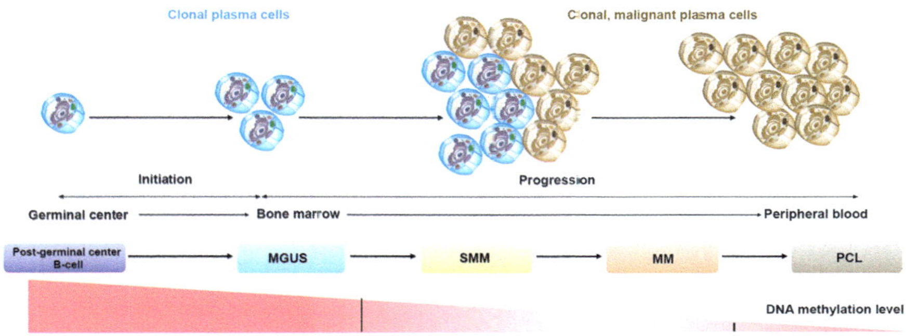


**Fig. 3** DNA methylation level increases along the disease progression in MM. Genome-wide DNA methylation level in monoclonal gammopathy of undetermined significance (MGUS) stage are similar to normal plasma B cells. A reduction in global methylation levels is seen between MGUS samples and smoldering multiple myeloma (SMM) and MM samples, with gene-specific hypermethylation. An increase in the amount of hypermethylation is seen at the PCL stage

changes in the methylation pattern in those regions may potentially influence the expression of those genes in cancer cells (Yang et al. 2014). While the impact of enhancers and superenhancers have been reported to deregulate key oncogenes in MM (Lovén et al. 2013), it is yet to be determined what the impact of DNA methylation level at those regulatory regions has on the expression of key genes in MM.

Another important parameter in the maintenance of DNA methylation is through the catalytic activities from the DNMTs. Previous studies reported that MM cells usually have relatively higher level of both DNMT1, the proofreading member, and DNMT3A/3B, the de-novo methyltransferase members of the DNMT family, compared to normal PC (Bollati et al. 2009; Raimondi et al. 2016). In the scope of prospective therapeutics, it was observed that siRNA based silencing of DNMTs results in the inhibition of cell proliferation. This also signifies the importance of aberrant proactive activity of DNMTs in maintaining the methylation landscape in MM (Amodio et al. 2012). Finally, aberrant mutations in *TET2* or *DNMT3A* have been reported in MM patients, which may also play a crucial role in the function of these enzymes and is yet to be explored with bigger datasets (Heuck et al. 2014).

## 4.2 *Aberrant Histone Modifications in MM*

Histone modifications and several activating or inactivating mutations in the histone modifiers have been widely reported in MM (Pawlyn et al. 2016). For instance, mutations in *HIST1H1E* and other histone H1 family genes were found to impact their interaction with DNMT1 and DNMT3B or altering H3K9 methylation by affecting the binding capacity of SET7/9 (Yang et al. 2013). It was also found that

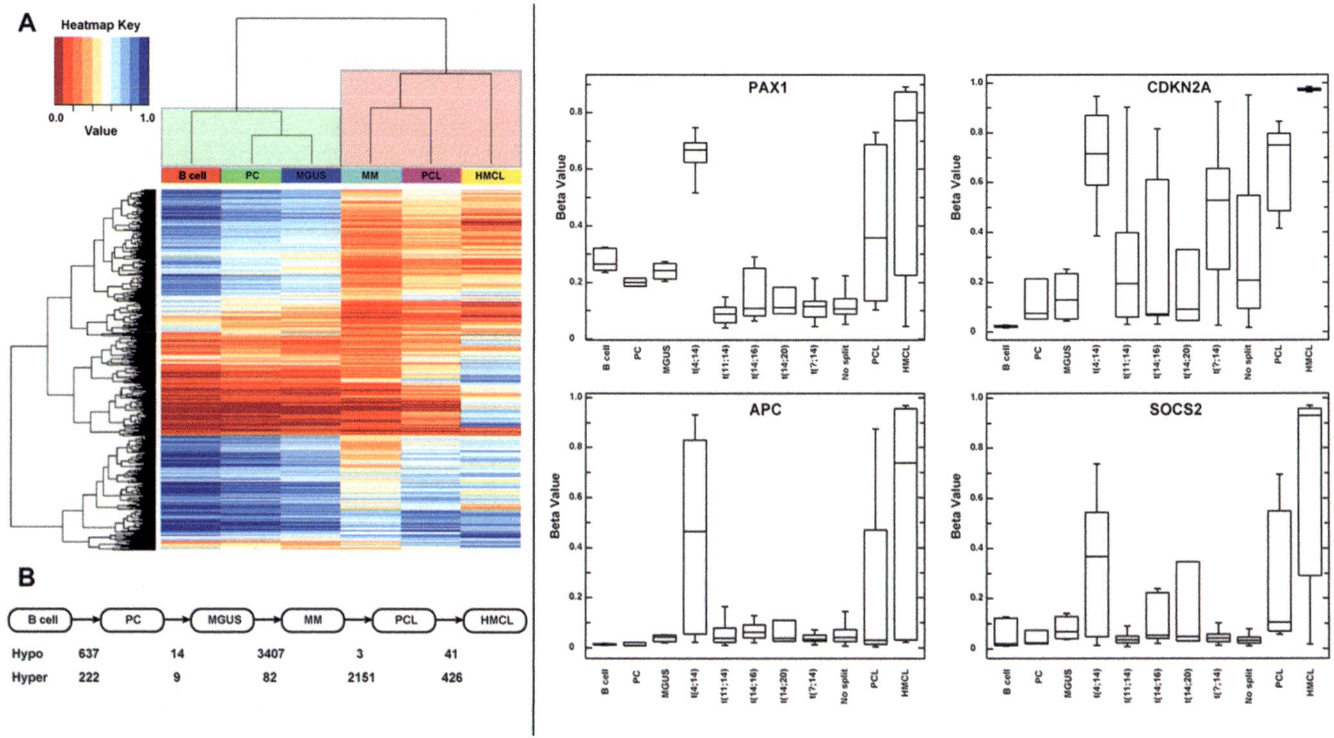


**Fig. 4** Genome-wide and locus specific methylation changes in MM. (**a**) A hierarchical clustering analysis based on a methylation profiling study in patients containing different stages of MM. A trend of hypomethylation was observed along the development of the disease. (**b**) The difference in methylation was determined, based on probesets. Gene-specific increases in DNA methylation were observed at the promoter region of *PAX1*, *CDKN2A*, *APC* and *SOCS2* genes (Walker et al. Blood. 2011 Jan 13; 117(2):553–62). *The figure has been reproduced with permission/ copyright clearance*

*HIST1H1E* and *HIST1H1C* mutations (along with *KDM6A* and *ARID2*) are relatively common in MM. Frequent occurrence of an inactivating lesion in *KDM6A/UTX* is also reported, which is concurrent with the mutations in the DNA methylation modifiers and suggests a tight correlation and histone-DNA epigenetic crosstalk in MM (Pawlyn et al. 2016) (Fig. 5).

### 4.2.1 Histone Modifiers in MM

MMSET

One of the best known histone modifications in MM is characterized by the overexpression of one of the HMTs containing the SET domain, and is popularly referred to as Multiple Myeloma SET (MMSET). The protein is functionally coded by the *NSD2/WHSC1* gene and is known to catalyze dimethylation (me2) of H3K36 of active chromatin (Chesi et al. 1998; Zhan et al. 2006). Overexpression of MMSET in MM is most common in patients (~15% of all MM cases) containing the t(4;14) translocation (Chesi et al. 1998; Martinez-Garcia et al. 2010). This translocation results in the juxtaposition of the *IgH* super-enhancer on chromosome 14 with *MMSET* on chromosome 4, and also results in over-expression of *FGFR3* on the other derivative chromosome. However, MMSET expression is ubiquitous in this translocation group, whereas *FGFR3* may be deleted in 25% of t(4;14) patients (Keats et al. 2003; Santra et al. 2003). MMSET overexpression is also reported to impact gene expression, possibly through H3K36 di- and trimethylation of chromatin and may create a transcriptionally active state for the target genes (Martinez-Garcia et al. 2010; Zhan et al. 2006). To further evaluate the functional effects of *NSD2*, the translocated allele of the gene was knocked out in a lineage of KMS11 cell line containing the t(4;14) translocation to generate TKO (translocated allele knock out) cells. Furthermore, it was quantitatively estimated that H3K36me2 levels were almost two to three-fold higher, while the unmethylated H3K36 and H3K36me1 were one to two-fold lower in the parental KMS11, compared to the TKO-KMS11 cells (Kuo et al. 2011; Mirabella et al. 2013). However, no change in H3K4 or H3K9 methylation was observed as a result of this *NSD2* translocation knockout (Jaffe et al. 2013). These experimental validations strongly corroborates the fact that MMSET in MM cells harboring t(4;14) translocations are actively involved in the catalytic activity of H3K36 and thereby impact gene expression.

MMSET is also reported to regulate the binding and distribution of another HMT, EZH2, resulting in the decrease of inactivating H2K27me marks across the genome (Popovic et al. 2014). The overexpression of MMSET results in a shift of EZH2 function with a reduction of global H3K27me3 levels, but gene-specific increases of H3K27me3, which also facilitate disruption and deregulation of target genes in MM cells (Hernando et al. 2015). Furthermore, MMSET in t(4;14) MM cells was also demonstrated to increase *MYC* and *IRF4* (Interferon regulatory factor-4) expression, which are considered to be important in terms of cell survival and cell proliferation in MM (Min et al. 2013; Xie et al. 2015). In summary, MMSET overexpression has

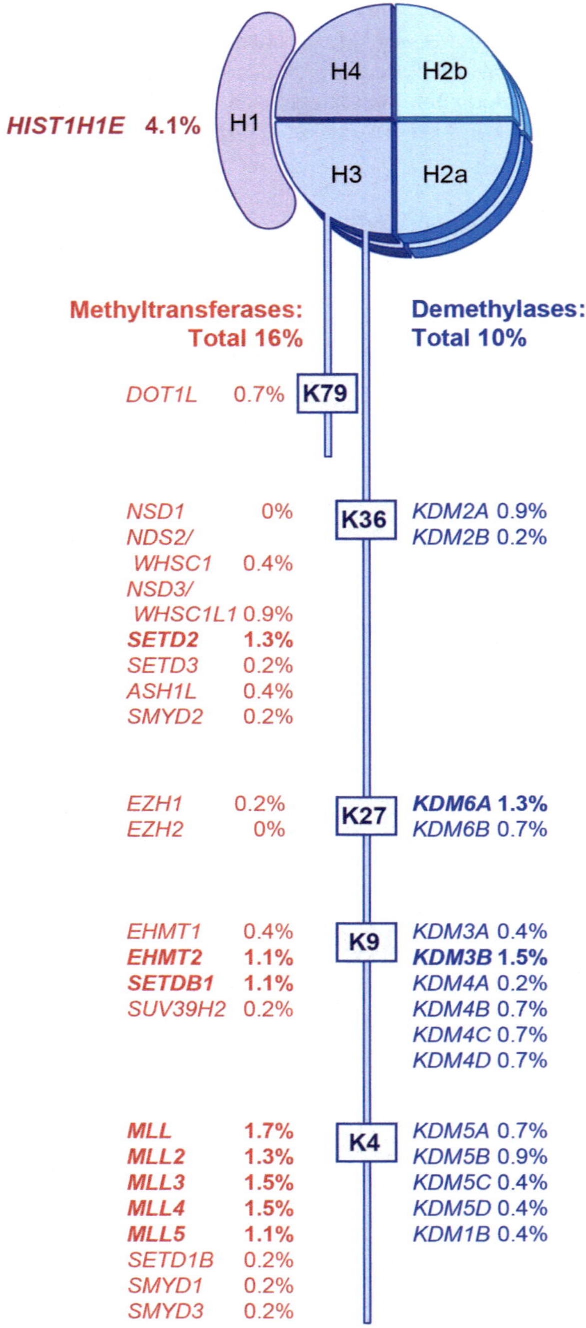


**Fig. 5** Occurrence of mutations in histone methyltransferases and demethylases in MM. The percentage of histone H3 mutations in MM patients (Pawlyn et al. Clin Cancer Res. 2016 Dec 1; 22(23): 5783–5794). *The figure has been reproduced with permission/copyright clearance*

been correlated to increased cell proliferation, chemotherapeutic resistance, poor prognosis and poor survival in MM patients, who harbor the t(4;14) translocation.

### EZH2

Enhancer of zeste homolog 2 is one of the core members of the polycomb repressive complex 2 (PRC2), which maintain a silent transcriptional state to its downstream *HOX* genes through H3K27me3 that contributes to the development and differentiation of tumors (Chase and Cross 2011; Margueron and Reinberg 2011). EZH2 overexpression in MM has been reported in conjunction with disease progression and poor prognosis (Pawlyn et al. 2017; Rastgoo et al. 2018). At the molecular level, EZH2 is thought to interact with the long non-coding RNA *MALAT1*, which can downregulate *miR-29b* and promotes cell growth and survival (Stamato et al. 2017). Recent advancement in the development of EZH2 inhibitors showed promise in the restriction of cell growth and survival, also suggesting its impact as a potential therapeutic target (Agarwal et al. 2016; Kikuchi et al. 2015a).

### PRMTs

PRMT5 or type-II arginine specific methyltransferase promotes mono- or di- methylation of the R8 residue on histone H3 (Table 1) and repress transcription of downstream genes or inhibit activity of other substrate proteins such as p53 and E2F1 in different cancers (Stopa et al. 2015). Overexpression of PRMT5 has been reported in MM and is associated with poor survival in patients. PRMT5 interacts with the E3 ubiquitin ligase, TRIM21, and degrades IKKβ, an activator of the NF-κB signaling pathway (Gullà et al. 2018). Experimental inhibition of PRMT5 resulted in the decrease of MM cell growth, which also suggest its importance in the MM pathogenesis.

### KDMs

Lysine-specific demethylases aid in the demethylation of histone tails at lysine residues and consist of three major enzymes, such as KDM3A, KDM6A and KDM6B. KDM3A is also a member of jumonji C (JMJC) demethylases, which is overexpressed in MM and possibly aids in cell survival (Ikeda et al. 2018). Mechanistically, KDM3A positively regulates *KLF2* and *IRF4* through demethylation of H3K9 at their promoters, and leads to MM cell survival (Ohguchi et al. 2016). KDM3A promotes cell adhesion within the bone marrow microenvironment and plays an important role in cell adhesion and survival. Furthermore, MM cells in hypoxic conditions induce upregulation of *KDM3A* and *MALAT1*, resulting in anti-apoptotic properties in MM cells (Ikeda et al. 2018).

KDM6A or UTX (Ubiquitously transcribed tetratricopeptide repeat, X chromosome) is another member of the JMJC demethylases that preferentially removes di- and tri- methylated H3K27 marks from gene loci, controlling transcriptional activities. KDM6A operates through H3K4 methylation and leads to open and actively transcribed chromatin (Issaeva et al. 2007). KDM6A/UTX inactivating mutations in MM have been associated with increased cell proliferation and shorter survival (Chapman et al. 2011; Ezponda et al. 2017; van Haaften et al. 2009).

KDM6B is another H3K27 demethylase and is closely related to KDM6A. It is observed that KDM6B is mainly involved in inflammatory responses, cellular development and differentiation, and in stress induced senescence (Ramadoss et al. 2012). Previous studies reported that *KDM6B* expression is induced in stromal cells inside the bone marrow, while its activity can be inhibited by targeting IKKβ inhibitors. This also suggests the activation and regulation of this demethylase via NF-κB signaling, a crucial pathway in MM (Ohguchi et al. 2017). In summary, members of JMJC histone demethylases actively control their target gene expression in MM and also take part in cellular differentiation, adhesion, proliferation and survival mechanisms.

### 4.2.2 Modifications in Histone Acetylation and Modifiers in MM

Histone and non-histone deacetylases have emerged as potential drug targets in cancer including MM. Several candidates of HDACs are prime targets in MM, both at the preclinical and clinical settings. Here we summarize key candidates of histone acetylation-deacetylation dynamics in MM and their regulatory roles in disease development.

#### CBP/p300 Family

KAT3 (CBP and its homolog p300) is the most investigated lysine acetyltransferase in cancer, which acetylates and thereby activates H3K18 and H3K27 (Jin et al. 2011). The bromodomain of CBP/p300 is documented to interact with the lymphocyte-specific transcription factor IRF4 in MM cells, impacting its expression through the above mentioned histone modifications at its enhancer and transcription start site (TSS). The same study has tested the sensitivity of MM cells against the bromodomain inhibitor of CBP/p300, which results in the arrest of cell proliferation (Conery et al. 2016).

#### Histone Deacetylases

Histone deacetylases (HDACs) remove the acetylation marks of the histone tails, cause chromatin condensation, and thereby silence the target genes. Based on the clinical relevance, HDACs are the prime target of inhibitory drugs in cancers (Kim and Bae 2011). HDAC inhibitors also reported to impair MM cells, in particular the class-I or class-IIb HDAC inhibitors are being used in the treatment of MM.

HDAC1 and HDAC3 knockdown, or inhibition with small molecules such as MS-275 or BG45, are reported to cause increased death in MM cells. Experimental knockdown of these proteins also showed elevated c-MYC and DNMT1 levels (Harada et al. 2017; Minami et al. 2014). Additionally, HDAC3 inhibition caused a reduction in the phosphorylation level of signal transducer and activator of transcription 3 (STAT3), which may be a possible mechanism of induced cell death. In summary, HDAC1/3 overexpression in MM patients is significantly associated with shorter progression-free survival and remains a prime target for therapeutic interventions.

HDAC4 is another prime target for the class-IIa HDAC inhibitors, which is demonstrated to lead to autophagy and apoptosis in MM (Kikuchi et al. 2015b). HDAC4 is reported to interact with multiple transcription factors, such as Runx2 and MEF2, or may form a complex with the alternative NF-κB factor RelB/p52, that represses the pro-apoptotic genes *Bim* or *BMF* via deacetylation of H3 at their promoter. These multi-tier interactions result in MM cell survival (Vallabhapurapu et al. 2015). HDAC4 has also been shown to be involved in downregulation of the tumor-suppressor miRNA miR-29b by maintaining a condensed inactive chromatin state at its promoter. In addition, HDAC4 is also reported to counteract the ER stress response in MM cells. Knockdown of HDAC4 under ER stress leads to the elevation of ATF4 or CHOP proteins to facilitate apoptosis in MM cells (Kikuchi et al. 2015b).

HDAC6, a key player in protein folding and processing, interacts with polyubiquitinated proteins and motor complexes which are required for aggresome formation. It has been found that HDAC6 aided aggresome formation in MM cells leads to cell death upon treatment with HDAC6 inhibitors (Kawaguchi et al. 2003). Another study has shown that HDAC6 deacetylates HSP90 and maintains the chaperone function of the catalytic subunit (*PPP3CA*) of calcineurin, which aids in the survival of MM cells. Furthermore, treatment with the small molecule ACY-1215 was shown to reduce PPP3CA levels and promote cell death (Richardson et al. 2011).

### SIRT6

SIRT6 deacetylated H3K9 and H3K56 histone or CtIP non-histone marks and contributes to genomic stability through double-stranded DNA break repair, and delays cellular senescence via telomere maintenance (Kaidi et al. 2010; Michishita et al. 2008). Recent studies identified the relevance of SIRT6 in MM. It was reported that depletion or inhibition of SIRT6 causes increased sensitivity of MM cells to melphalan and doxorubicin. It was also reported that SIRT6 knockdown in MM cells stimulates cell growth by activation of the MAPK pathway (Cea et al. 2015; Sebastián et al. 2012).

Collectively, histone acetylation and deacetylation play a pivotal role in operating the transcriptional state in MM, which also aids in tumor progression and cell survival. Based on their functional relevance, drugs affecting histone modifications, in particular HDAC inhibitors, are emerging as a promising therapeutic option in MM (Ohguchi et al. 2018).

## 4.3 Aberrant miRNA Mechanism in MM

Non-coding miRNA mechanisms and their implications in the regulation of cell cycle, proliferation and apoptosis have been widely investigated in MM in conjunction with the development and progression of the disease (Rossi et al. 2014; Tagliaferri et al. 2012). A panel of miRNA such as miR-18, miR-21, miR-125a-5p, miR32, and mirR92 have been identified while profiling MGUS and MM patients, compared to normal PCs, and defined as an 'oncogenic miRNA' signature in MM (Amodio et al. 2012; Zhou et al. 2010). Besides alterations in copy number at the miRNA loci and chromosomal rearrangements, DNA methylation and histone modifications are also reported to impact miRNA expression levels. For instance, miR-34, a tumor suppressor in MM is frequently inactivated by DNA hypermethylation at promoter CpG islands (Di Martino et al. 2012; Wong et al. 2011). Similar epigenetic control of miRNA inactivation were reported at the miR-194-2-192 cluster, miR-203, miR-152, miR124–1, miR-10b, miR-9-3, miR-9-1, miR-155 or miR-23b (Bueno et al. 2008; Wong et al. 2011; Zhang et al. 2015). In all of the above mentioned examples, experimental targeting with demethylating agents resulted in the upregulation of miRNA levels that resulted in a decrease in cell proliferation or apoptosis.

Simultaneous studies reported DNA methylation can also be regulated by miRNAs. For instance, miR-148a, miR-152, miR-29b or miR-222 have been described to target the 3′ UTR of DNMTs and reduces the global methylation level (Amodio et al. 2013; Amodio et al. 2016). In another example, miR-29b was also found to interact with class-II HDAC4 (Amodio et al. 2016). In summary, non-coding miRNAs dynamically interact with DNA and histone marks and influence or influenced by these marks to regulate gene expression, cell growth, cell fate, etc. in MM.

## 4.4 IDH Gene Mutations in MM

Activating mutations in isocitrate dehydrogenase (*IDH1*) and *IDH2* genes were documented to produce high levels of 2-hydroxyglutarate (2HG), and occur in several human cancers (Dang et al. 2009). It was also observed that IDH mutations lead to global DNA hypermethylation, at least in leukemias and gliomas (Kernytsky et al. 2015) through inhibition of *TET2* methylcytosine dehydrogenase (Figueroa et al. 2010). Mutant IDH products also increase histone methylation through competitive inhibition of α-ketoglutarate (αKG)-dependent jumonji C (JmjC) demethylases, thereby activating or inhibiting expression of associated genes (Kernytsky et al. 2015; Tsukada et al. 2006). We recently reported mutations in *IDH1* and *IDH2* at a low frequency (0.6% and 0.3%, respectively) in MM, but the mutations were highly clonal (Walker et al. 2018). *IDH1* or *IDH2* mutations are classified as low frequency gain of function (Whitehall et al. 2014; Zhu et al. 2017) in cancers and are prime epigenetic drug targets (Stein et al. 2017). For instance, ivosidenib is an FDA approved drug, which is being administered to relapsed and refractory patients in acute myeloid leukemia (AML). Mutational activation of IDH1 or IDH2 in MM hence foster hope for the development and application of IDH inhibitory drugs in the disease.

## 5 Epigenetic Therapeutic Options in MM

Epigenetic alterations by drug or novel small-chemical conjugates are achieved at the level of DNA mainly by the demethylating agents or at the level of histones by adding (writers) or removing (erasers) the aberrant histone marks. Additional histone proteins (readers) are also set as the target for epigenetic decoding in cancer. As a result of these targeted approaches, it is now possible to selectively recruit activators or repressors to the altered chromatin structure or a target gene to regulate and/or modify their transcriptional states.

### *5.1 Epigenetic Drug Targeting in MM*

Several drug and small chemical molecules have been in use clinically or experimentally to reverse epigenetic defects in MM. While, most of these drugs are under clinical trial phases, some of them are already approved and being prescribed as the first line of treatment for MM patients. Here we tabulate (Table 3) to summarize the reported therapeutic interventions in MM.

The experimental data however suggests a series of side-effects in MM patients, which includes but is not limited to acute diarrhea, thrombocytopenia, skin and liver complexities, or psychological disorders. However, a recent report from the clinical trial using ACY-1215 with lenalidomide and dexamethasone was shown to be well tolerated and safe in patients with an overall response rate of 55% (Ohguchi et al. 2018; Yee et al. 2016).

### *5.2 Novel Approaches to Treat Epigenetic Aberrations in MM*

Recent development in synthetic biology and protein engineering offer several strategies to precisely target and reverse the aberrant DNA or histone marks in a cell by lowering the adverse consequence of side-effects.

**Table 3** Overview of commonly applied drugs in MM from the published clinical studies

| Drug<br>Chemical name | Target specificity | Clinical trial in MM (phase) | Reference |
|---|---|---|---|
| DNA methylation aberration | | | |
| 5-Azacitidine | Reverse DNA hypermethylation | III (FDA approved) | Kiziltepe et al. (2007) |
| Decitabine (5-azadeoxycitidine) | Reverse DNA hypermethylation | III (FDA approved) | |
| Thalidomide/ lenalidomide/ pomalidomide | Reverse DNA hypomethylation | FDA approved | Zhu et al. (2013) |
| Histone methylation aberration | | | |
| LEM-06 | Inhibit *NSD2* (gene encodes MMSET) | I | di Luccio (2015) |
| GSK126 (GSK2816126) | Inhibit EZH2 | I | Ohguchi et al. (2018) |
| AZD6244 | Inhibit MMSET expression | II | Xie and Chng (2014) |
| Histone acetylation/deacetylation aberration | | | |
| OTX015/MK-8628 | Inhibit BET (Bromodomain and extraterminal domain) | II | Ghoshal et al. (2016) |
| GSK525762 | | I/II | Fratta et al. (2016) |
| CPI-0610 | | I | |
| Curcumin | HAT inhibitor | Preclinical trial | Golombick et al. (2016) |
| Vorinostat (SAHA) | Inhibit class I, II, IV HDACs | III | Fratta et al. (2016) |
| Panobinostat (LBH589) | | III (FDA approved) | |
| Quisinostat (JNJ26481585) | | I | |
| Belinostat (PXD101) | | II | |
| Abexinostat (PCI-24781) | Inhibit class I, II HDACs | I | |
| CKD-581 | Inhibit class I HDACs | I | Lee et al. (2014) |
| Givinostat (ITF2357) | Inhibit class I, II HDACs | II | Ohguchi et al. (2018) |
| Fimepinostat (CUDC-907) | Inhibit class I, II HDACs+PI3K | I | |
| Tinostamustine (EDO-S101) | Bendamustine–vorinostat fusion | I | |
| Romidepsin (FK228) | Inhibit class I HDACs | I/II | |
| Entinostat (MS-275) | Inhibit class I HDACs I | I | Fratta et al. (2016) |
| Tacedinaline (CI-994) | Inhibit class I HDACs | II | |
| Ricolinostat (ACY-1215) | Inhibit HDAC6 | I/II | Ohguchi et al. (2018) |
| Citarinostat (ACY-241) | Inhibit HDAC6 | I | |

The first line of products of these novel therapeutics are some pre-designed miRNAs or 'epi-miRNAs', which can be specifically targeted to suppress DNMT enzymatic activities in cells. For instance, synthetic miR-29b treatment has been demonstrated to downregulate the catalytic activity of DNMT3A and DNMT3B. This inhibition lowers the hypermethylation induced repression of tumor suppressor genes in MM and has also shown to promote the activity of 5-azacitidine and cause increased cell death in MM (Amodio et al. 2013).

In accordance with the epigenetic targeting with CRISPR (Clustered Regularly Interspaced Short Palindromic Repeats) in other cancer models, DNMT or TET proteins may be fused to the Cas9 endonuclease and may be co-treated with the target specific sgRNAs to correct the aberrations at the level of DNA. In contrast, histone reader, writers or erasers may be fused to reverse the aberrant modifications in cells.

## 6 Conclusions and Perspectives

Aberrant epigenetic changes in DNA methylation, histone marks and non-coding miRNA mechanisms have been recognized as a secondary yet crucial driver in the development and progression of MM (Dimopoulos et al. 2014). It has been observed that the prevalence in genome-wide DNA hypomethylation and clustered hypermethylation at the promoter and gene body of tumor suppressor, anti-apoptotic or cell-signaling related genes promote genomic instability with concomitant dysfunction of the genes. Moreover, DNA methylation does not operate on its own but in a tight association with miRNAs and an array of histone modifications. Several drugs and small chemicals are now in use to treat the epigenetic defects in patients with refractory/relapsed MM. However, to achieve more target specific and personalized interventions, we need to understand the genome-wide location and functional network of the DNA and histone modifiers. It is also very important to design and optimize CRISPR or other synthetic proteins (such as TAL effectors) for loci specific editing of the aberrant epigenetic marks. More emphasis on genome-wide epigenome-sequencing on an individual basis for adopting a therapeutic strategy will be the key for future lines of precision medicine in MM.

## References

Agarwal P, Alzrigat M, Parraga AA et al (2016) Genome-wide profiling of histone H3 lysine 27 and lysine 4 trimethylation in multiple myeloma reveals the importance of Polycomb gene targeting and highlights EZH2 as a potential therapeutic target. Oncotarget 7:6809–6823

Allfrey VG, Faulkner R, Mirsky AE (1964) Acetylation and methylation of histones and their possible role in the regulation of RNA synthesis. Proc Natl Acad Sci USA 51:786–794

Amodio N, Leotta M, Bellizzi D et al (2012) DNA-demethylating and anti-tumor activity of synthetic miR-29b mimics in multiple myeloma. Oncotarget 3:1246–1258

Amodio N, Bellizzi D, Leotta M et al (2013) miR-29b induces SOCS-1 expression by promoter demethylation and negatively regulates migration of multiple myeloma and endothelial cells. Cell Cycle 12:3650–3662
Amodio SMA, Gullà N et al (2016) Therapeutic targeting of miR-29b/HDAC4 epigenetic loop in multiple myeloma. Mol Cancer Ther 15:1364–1375
Aoki Y, Nojima M, Suzuki H et al (2012) Genomic vulnerability to LINE-1 hypomethylation is a potential determinant of the clinicogenetic features of multiple myeloma. Genome Med 4:101
Avet-Loiseau H, Li C, Magrangeas F et al (2009) Prognostic significance of copy-number alterations in multiple myeloma. J Clin Oncol 27:4585–4590
Balasubramanyam K, Varier RA, Altaf M et al (2004) Curcumin, a novel p300/CREB-binding protein-specific inhibitor of acetyltransferase, represses the acetylation of histone/nonhistone proteins and histone acetyltransferase-dependent chromatin transcription. J Biol Chem 279: 51163–51171
Bannister AJ, Kouzarides T (2011) Regulation of chromatin by histone modifications. Cell Res 21: 381–395
Beckedorff FC, Ayupe AC, Crocci-Souza R et al (2013) The intronic long noncoding RNA ANRASSF1 recruits PRC2 to the RASSF1A promoter, reducing the expression of RASSF1A and increasing cell proliferation. PLoS Genet 9:e1003705
Benjamin M, Reddy S, Brawley OW (2003) Myeloma and race: a review of the literature. Cancer Metastasis Rev 22:87–93
Bergsagel PL, Kuehl WM (2001) Chromosome translocations in multiple myeloma. Oncogene 20: 5611–5622
Bhutani N, Burns DM, Blau HM (2011) DNA demethylation dynamics. Cell 146:866–872
Bollati V, Fabris S, Pegoraro V et al (2009) Differential repetitive DNA methylation in multiple myeloma molecular subgroups. Carcinogenesis 30:1330–1335
Boyd KD, Ross FM, Chiecchio L et al (2011) Gender disparities in the tumor genetics and clinical outcome of multiple myeloma. Cancer Epidemiol Biomarker Prev 20:1703–1707
Braggio E, Maiolino A, Gouveia ME et al (2010) Methylation status of nine tumor suppressor genes in multiple myeloma. Int J Hematol 91:87–96
Branco MR, Ficz G, Reik W (2012) Uncovering the role of 5-hydroxymethylcytosine in the epigenome. Nat Revs Genet 13:7–13
Brueckner B, Boy RG, Siedlecki P et al (2005) Epigenetic reactivation of tumor suppressor genes by a novel small-molecule inhibitor of human DNA methyltransferases. Cancer Res 65: 6305–6311
Bueno MJ, de Castro IP, Cedrón d et al (2008) Genetic and epigenetic silencing of microRNA-203 enhances ABL1 and BCR-ABL1 oncogene expression. Cancer Cell 13:496–506
Cea M, Cagnetta A, Adamia S et al (2015) Evidence for a role of the histone deacetylase SIRT6 in DNA damage response of multiple myeloma cells. Blood 127:1138–1150
Cedar H, Bergman Y (2009) Linking DNA methylation and histone modification: patterns and paradigms. Nat Rev Genet 10:295–304
Chang H, Qi C, Yi Q-L et al (2005) p53 gene deletion detected by fluorescence in situ hybridization is an adverse prognostic factor for patients with multiple myeloma following autologous stem cell transplantation. Blood 105:358–360
Chapman MA, Lawrence MS, Keats JJ et al (2011) Initial genome sequencing and analysis of multiple myeloma. Nature 471:467–472
Chase A, Cross NC (2011) Aberrations of EZH2 in cancer. Clin Cancer Res 17:2613–2618
Chen W, Wu Y, Zhu J et al (2002) Methylation of p16 and p15 genes in multiple myeloma. Chin Med Sci J 17:101–105
Chen G, Wang Y, Huang H et al (2009) Combination of DNA methylation inhibitor 5-azacytidine and arsenic trioxide has synergistic activity in myeloma. Eur J Haematol 82:176–183
Cheng SH, Ng MH, Lau KM et al (2007) 4q loss is potentially an important genetic event in MM tumorigenesis: identification of a tumor suppressor gene regulated by promoter methylation at 4q13.3, platelet factor 4. Blood 109:2089–2099

Chesi M, Nardini E, Lim RS et al (1998) The t (4; 14) translocation in myeloma dysregulates both FGFR3and a novel gene, MMSET, resulting in IgH/MMSET hybrid transcripts. Blood 92: 3025–3034
Chim CS, Fung TK, Cheung WC et al (2004a) SOCS1 and SHP1 hypermethylation in multiple myeloma: implications for epigenetic activation of the Jak/STAT pathway. Blood 103:4630–4635
Chim CS, Kwong YL, Fung TK et al (2004b) Methylation profiling in multiple myeloma. Leuk Res 28:379–385
Chim CS, Pang R, Fung TK et al (2007) Epigenetic dysregulation of Wnt signaling pathway in multiple myeloma. Leukemia 21:2527–2536
Chuang JC, Jones PA (2007) Epigenetics and microRNAs. Pediatr Res 61:24r–29r
Cimmino A, Calin GA, Fabbri M et al (2005) miR-15 and miR-16 induce apoptosis by targeting BCL2. Proc Natl Acad Sci USA 102:13944–13949
Conery AR, Centor RC, Neiss A et al (2016) Bromodomain inhibition of the transcriptional coactivators CBP/EP300 as a therapeutic strategy to target the IRF4 network in multiple myeloma. elife 5:e10483
Dang L, White DW, Gross S et al (2009) Cancer-associated IDH1 mutations produce 2-hydroxyglutarate. Nature 462:739–744
Davis AJ, Gelmon KA, Siu LL et al (2003) Phase I and pharmacologic study of the human DNA methyltransferase antisense oligodeoxynucleotide MG98 given as a 21-day continuous infusion every 4 weeks. Investig New Drugs 21:85–97
de Carvalho F, Colleoni GW, Almeida MS et al (2009) TGFbetaR2 aberrant methylation is a potential prognostic marker and therapeutic target in multiple myeloma. Intl J Cancer 125: 1985–1991
di Luccio E (2015) Inhibition of nuclear receptor binding SET domain 2/multiple myeloma SET domain by LEM-06 implication for epigenetic cancer therapies. J Cancer Prevent 20:113–120
Di Martino MT, Leone E, Amodio N et al (2012) Synthetic miR-34a mimics as a novel therapeutic agent for multiple myeloma: in vitro and in vivo evidence. Clin Cancer Res 18:6260–6270
Dib A, Gabrea A, Glebov OK et al (2008) Characterization of MYC translocations in multiple myeloma cell lines. J Natl Cancer Inst Monogr 2008:25–31
Dimopoulos K, Gimsing P, Grønbæk K (2014) The role of epigenetics in the biology of multiple myeloma. Blood Cancer J 4:e207
Durie BG, Salmon SE (1975) A clinical staging system for multiple myeloma. Correlation of measured myeloma cell mass with presenting clinical features, response to treatment, and survival. Cancer 36:842–854
Eis PS, Tam W, Sun L et al (2005) Accumulation of miR-155 and BIC RNA in human B cell lymphomas. Proc Natl Acad Sci U S A 102:3627–3632
Esquela-Kerscher A, Slack FJ (2006) Oncomirs-microRNAs with a role in cancer. Nat Rev Cancer 6:259–269
Esteller M (2007) Cancer epigenomics: DNA methylomes and histone-modification maps. Nat Rev Genet 8:286–298
Ezponda T, Dupéré-Richer D, Will CM et al (2017) UTX/KDM6A loss enhances the malignant phenotype of multiple myeloma and sensitizes cells to EZH2 inhibition. Cell Rep 21:628–640
Felsenfeld G (2014) A brief history of epigenetics. Cold Spring Harb Perspect Biol 6:a018200
Figueroa ME, Abdel-Wahab O, Lu C et al (2010) Leukemic IDH1 and IDH2 mutations result in a hypermethylation phenotype, disrupt TET2 function, and impair hematopoietic differentiation. Cancer Cell 18:553–567
Fonseca R, Bergsagel PL, Drach J et al (2009) International myeloma working group molecular classification of multiple myeloma: spotlight review. Leukemia 23:2210–2221
Fraga MF, Ballestar E, Villar-Garea A et al (2005) Loss of acetylation at Lys16 and trimethylation at Lys20 of histone H4 is a common hallmark of human cancer. Nat Genet 37:391–400
Fratta E, Montico B, Rizzo A et al (2016) Epimutational profile of hematologic malignancies as attractive target for new epigenetic therapies. Oncotarget 7:57327–57350

Fuks F (2005) DNA methylation and histone modifications: teaming up to silence genes. Curr Opin Genet Dev 15:490–495
Galm O, Wilop S, Reichelt J et al (2004) DNA methylation changes in multiple myeloma. Leukemia 18:1687–1692
Gardiner-Garden M, Frommer M (1987) CpG islands in vertebrate genomes. J Mol Biol 196: 261–282
Ghoshal A, Yugandhar D, Srivastava AK (2016) BET inhibitors in cancer therapeutics: a patent review. Expert Opin Ther Pat 26:505–522
Goldberg AD, Allis CD, Bernstein E (2007) Epigenetics: a landscape takes shape. Cell 128: 635–638
Golombick T, Diamond TH, Manoharan A et al (2016) Addition of rice bran arabinoxylan to curcumin therapy may be of benefit to patients with early-stage B-cell lymphoid malignancies (monoclonal gammopathy of undetermined significance, smoldering multiple myeloma, or stage 0/1 chronic lymphocytic leukemia): a preliminary clinical study. Integr Cancer Ther 15: 183–189
Gullà A, Hideshima T, Bianchi G et al (2018) Protein arginine methyltransferase 5 has prognostic relevance and is a druggable target in multiple myeloma. Leukemia 32:996–1002
Harada T, Ohguchi H, Grondin Y et al (2017) HDAC3 regulates DNMT1 expression in multiple myeloma: therapeutic implications. Leukemia 31:2670–2677
Hatzimichael E, Dranitsaris G, Dasoula A et al (2009) Von Hippel-Lindau methylation status in patients with multiple myeloma: a potential predictive factor for the development of bone disease. Clin Lymphoma Myeloma 9:239–242
He Y-F, Li B-Z, Li Z et al (2011) Tet-mediated formation of 5-carboxylcytosine and its excision by TDG in mammalian DNA. Science 333:1303–1307
Heller G, Schmidt WM, Ziegler B et al (2008) Genome-wide transcriptional response to 5-aza-2-′-deoxycytidine and trichostatin a in multiple myeloma cells. Cancer Res 68:44–54
Hernando H, Glato KA, Lesche R et al (2015) EZH2 inhibition blocks multiple myeloma cell growth through upregulation of epithelial tumor suppressor genes. Mol Cancer Ther 15: 287–298
Heuck C, Johann D, Walker BA et al (2014) Characterization of the mutational landscape of multiple myeloma using comprehensive genomic profiling. Blood 124:3418–3418
Hodge DR, Peng B, Cherry JC et al (2005) Interleukin 6 supports the maintenance of p53 tumor suppressor gene promoter methylation. Cancer Res 65:4673–4682
Ikeda S, Kitadate A, Abe F et al (2018) Hypoxia-inducible KDM3A addiction in multiple myeloma. Blood Adv 2:323–334
Iorio MV, Ferracin M, Liu CG et al (2005) MicroRNA gene expression deregulation in human breast cancer. Cancer Res 65:7065–7070
Issaeva I, Zonis Y, Rozovskaia T et al (2007) Knockdown of ALR (MLL2) reveals ALR target genes and leads to alterations in cell adhesion and growth. Mol Cel Biol 27:1889–1903
Ito S, Shen L, Dai Q et al (2011) Tet proteins can convert 5-methylcytosine to 5-formylcytosine and 5-carboxylcytosine. Science 333:1300–1303
Jaffe JD, Wang Y, Chan HM et al (2013) Global chromatin profiling reveals NSD2 mutations in pediatric acute lymphoblastic leukemia. Nat Genet 45:1386–1391
Jin B, Robertson KD (2013) DNA methyltransferases, DNA damage repair, and cancer. In: Karpf AR (ed) Epigenetic alterations in oncogenesis. Springer, New York, pp 3–29
Jin Q, Yu L-R, Wang L et al (2011) Distinct roles of GCN5/PCAF-mediated H3K9ac and CBP/p300-mediated H3K18/27ac in nuclear receptor transactivation. EMBO J 30:249–262
Jones PA (2012) Functions of DNA methylation: islands, start sites, gene bodies and beyond. Nat Rev Genet 13:484–492
Jung S, Kim S, Gale M et al (2012) DNA methylation in multiple myeloma is weakly associated with gene transcription. PLoS One 7:e52626
Kaas GA, Zhong C, Eason DE et al (2013) TET1 controls CNS 5-methylcytosine hydroxylation, active DNA demethylation, gene transcription, and memory formation. Neuron 79:1086–1093
Kaidi A, Weinert BT, Choudhary C et al (2010) Human SIRT6 promotes DNA end resection through CtIP deacetylation. Science 329:1348–1353

Kaiser MF, Johnson DC, Wu P et al (2013) Global methylation analysis identifies prognostically important epigenetically inactivated tumor suppressor genes in multiple myeloma. Blood 122: 219–226

Kawaguchi Y, Kovacs JJ, McLaurin A et al (2003) The deacetylase HDAC6 regulates aggresome formation and cell viability in response to misfolded protein stress. Cell 115:727–738

Keats JJ, Reiman T, Maxwell CA et al (2003) In multiple myeloma, t (4; 14)(p16; q32) is an adverse prognostic factor irrespective of FGFR3 expression. Blood 101:1520–1529

Kernytsky A, Wang F, Hansen E et al (2015) IDH2 mutation-induced histone and DNA hypermethylation is progressively reversed by small-molecule inhibition. Blood 125:296–303

Kikuchi J, Koyama D, Wada T et al (2015a) Phosphorylation-mediated EZH2 inactivation promotes drug resistance in multiple myeloma. J Clin Invest 125:4375–4390

Kikuchi S, Suzuki R, Ohguchi H et al (2015b) Class IIa HDAC inhibition enhances ER stress-mediated cell death in multiple myeloma. Leukemia 29:1918–1927

Kim H-J, Bae S-C (2011) Histone deacetylase inhibitors: molecular mechanisms of action and clinical trials as anti-cancer drugs. Am J Transl Res 3:166–179

Kim VN, Nam J-W (2006) Genomics of microRNA. Trends Genet 22:165–173

Kiziltepe T, Hideshima T, Catley L et al (2007) 5-Azacytidine, a DNA methyltransferase inhibitor, induces ATR-mediated DNA double-strand break responses, apoptosis, and synergistic cytotoxicity with doxorubicin and bortezomib against multiple myeloma cells. Mol Cancer Ther 6: 1718–1727

Kondo Y, Shen L, Suzuki S et al (2007) Alterations of DNA methylation and histone modifications contribute to gene silencing in hepatocellular carcinomas. Hepatol Res 37:974–983

Korde N, Kristinsson SY, Landgren O (2011) Monoclonal gammopathy of undetermined significance (MGUS) and smoldering multiple myeloma (SMM): novel biological insights and development of early treatment strategies. Blood 117:5573–5581

Kryukov F, Dementyeva E, Kubiczkova L et al (2013) Cell cycle genes co-expression in multiple myeloma and plasma cell leukemia. Genomics 102:243–249

Kulis M, Esteller M (2010) DNA methylation and cancer. Advs Genet 70:27–56

Kulis M, Merkel A, Heath S et al (2015) Whole-genome fingerprint of the DNA methylome during human B cell differentiation. Nat Genet 47:746–756

Kuo Alex J, Cheung P, Chen K et al (2011) NSD2 links dimethylation of histone H3 at lysine 36 to oncogenic programming. Mol Cell 44:609–620

Lee Y, Jeon K, Lee JT et al (2002) MicroRNA maturation: stepwise processing and subcellular localization. EMBO J 21:4663–4670

Lee C, Ahn K-S, Jung WJ et al (2014) CKD-581, a novel histone deacetylase inhibitor, synergistically enhances bortezomib cytotoxicity in multiple myeloma cells. Cancer Res 74:1695

Li B, Carey M, Workman JL (2007) The role of chromatin during transcription. Cell 128:707–719

Lovén J, Hoke HA, Lin CY et al (2013) Selective inhibition of tumor oncogenes by disruption of super-enhancers. Cell 153:320–334

Margueron R, Reinberg D (2011) The Polycomb complex PRC2 and its mark in life. Nature 469: 343–349

Martinez-Garcia E, Popovic R, Min D-J et al (2010) The MMSET histone methyl transferase switches global histone methylation and alters gene expression in t (4; 14) multiple myeloma cells. Blood 117:211–220

Medina PP, Sanchez-Cespedes M (2008) Involvement of the chromatin-remodeling factor BRG1/ SMARCA4 in human cancer. Epigenetics 3:64–68

Meltzer PS (2005) Cancer genomics: small RNAs with big impacts. Nature 435:745–746

Michishita E, McCord RA, Berber E et al (2008) SIRT6 is a histone H3 lysine 9 deacetylase that modulates telomeric chromatin. Nature 452:492–496

Min D, Ezponda T, Kim MK et al (2013) MMSET stimulates myeloma cell growth through microRNA-mediated modulation of c-MYC. Leukemia 27:686–694

Minami J, Suzuki R, Mazitschek R et al (2014) Histone deacetylase 3 as a novel therapeutic target in multiple myeloma. Leukemia 28:680–689

Mirabella F, Wu P, Wardell CP et al (2013) MMSET is the key molecular target in t(4;14) myeloma. Blood Cancer J 3:e114–e114
Momparler RL (2005) Pharmacology of 5-Aza-2′-deoxycytidine (decitabine). Semin Hemtol 2005: S9–S16
Morgan HD, Santos F, Green K et al (2005) Epigenetic reprogramming in mammals. Hum Mol Genet 14:R47–R58
Mulero-Navarro S, Esteller M (2008) Chromatin remodeling factor CHD5 is silenced by promoter CpG island hypermethylation in human cancer. Epigenetics 3:210–215
Ohguchi H, Hideshima T, Bhasin MK et al (2016) The KDM3A–KLF2–IRF4 axis maintains myeloma cell survival. Nat Commun 7:10258
Ohguchi H, Harada T, Sagawa M et al (2017) KDM6B modulates MAPK pathway mediating multiple myeloma cell growth and survival. Leukemia 31:2661–2669
Ohguchi H, Hideshima T, Anderson KC (2018) The biological significance of histone modifiers in multiple myeloma: clinical applications. Blood Cancer J 8:83
Okano M, Bell DW, Haber DA et al (1999) DNA methyltransferases Dnmt3a and Dnmt3b are essential for de novo methylation and mammalian development. Cell 99:247–257
Olcinia MM, O'Dell S, Hammond EM (2015) Targeting chromatin to improve radiation response. Br J Radiol 88:20140649
Oswald J, Engemann S, Lane N et al (2000) Active demethylation of the paternal genome in the mouse zygote. Curr Biol 10:475–478
Palumbo A, Anderson K (2011) Multiple myeloma. New Eng J Med 364:1046–1060
Parry L, Clarke AR (2011) The roles of the methyl-CpG binding proteins in cancer. Genes Cancer 2:618–630
Pawlyn C, Kaiser M, Heuck C et al (2016) The spectrum and clinical impact of epigenetic modifier mutations in myeloma. Clin Cancer Res 22:5783–5794
Pawlyn C, Bright MD, Buros AF et al (2017) Overexpression of EZH2 in multiple myeloma is associated with poor prognosis and dysregulation of cell cycle control. Blood Cancer J 7:e549
Peng B, Hurt EM, Hodge DR et al (2006) DNA hypermethylation and partial gene silencing of human thymine-DNA glycosylase in multiple myeloma cell lines. Epigenetics 1:138–145
Piras G, Monne M, Palmas AD et al (2014) Methylation analysis of the phosphates and tensin homologue on chromosome 10 gene (PTEN) in multiple myeloma. Clin Epigenetics 6:16
Popovic R, Martyinez-Garcia E, Giannopoulou EG et al (2014) Histone methyltransferase MMSET/NSD2 alters EZH2 binding and reprograms the myeloma epigenome through global and focal changes in H3K36 and H3K27 methylation. PLoS Genet 10:e1004566
Raimondi L, De Luca A, Morelli E et al (2016) MicroRNAs: novel crossroads between myeloma cells and the bone marrow microenvironment. Biomed Res Int 2016:6504593
Rajkumar SV, Dimopoulos MA, Palumbo A et al (2014) International myeloma working group updated criteria for the diagnosis of multiple myeloma. Lancet Oncol 15:e538–e548
Ramadoss S, Chen X, Wang C-Y (2012) Histone demethylase KDM6B promotes epithelial-mesenchymal transition. J Biol Chem 287:44508–44517
Rasmussen T, Kuehl M, Lodahl M et al (2005) Possible roles for activating RAS mutations in the MGUS to MM transition and in the intramedullary to extramedullary transition in some plasma cell tumors. Blood 105:317–323
Rastgoo N, Pourabdollah M, Abdi J et al (2018) Dysregulation of EZH2/miR-138 axis contributes to drug resistance in multiple myeloma by downregulating RBPMS. Leukemia 32:2471–2482. https://doi.org/10.1038/s41375-018-0140-y. [Epub ahead of print]
Richardson PG, Mitsiades CS, Laubach JP et al (2011) Inhibition of heat shock protein 90 (HSP90) as a therapeutic strategy for the treatment of myeloma and other cancers. Br J Haematol 152: 367–379
Robertson KD (2005) DNA methylation and human disease. Nat Rev Genet 6:597–610
Rossi M, Amodio N, Teresa Di Martino M et al (2014) MicroRNA and multiple myeloma: from laboratory findings to translational therapeutic approaches. Curr Pharml Biotechnol 15:459–467
Saito Y, Liang G, Egger G et al (2006) Specific activation of microRNA-127 with downregulation of the proto-oncogene BCL6 by chromatin-modifying drugs in human cancer cells. Cancer Cell 9:435–443

Salhia B, Baker A, Ahmann G et al (2010) DNA methylation analysis determines the high frequency of genic hypomethylation and low frequency of hypermethylation events in plasma cell tumors. Cancer Res 70:6934–6944

Santer FR, Höschele PPS, Oh SJ et al (2011) Inhibition of the acetyltransferases p300 and CBP reveals a targetable function for p300 in the survival and invasion pathways of prostate cancer cell lines. Mol Cancer Thera 10:1644–1655

Santi DV, Norment A, Garrett CE (1984) Covalent bond formation between a DNA-cytosine methyltransferase and DNA containing 5-azacytosine. Proc Natl Acad Sci U S A 81:6993–6997

Santra M, Zhan F, Tian E et al (2003) A subset of multiple myeloma harboring the t (4; 14)(p16; q32) translocation lacks FGFR3 expression but maintains anIGH/MMSET fusion transcript. Blood 101:2374–2376

Sasaki H, Matsui Y (2008) Epigenetic events in mammalian germ-cell development: reprogramming and beyond. Nat Rev Genet 9:129–140

Sawan C, Herceg Z (2010) Histone modifications and cancer. Adv Genet 70:57–85

Scott GK, Mattie MD, Berger CE et al (2006) Rapid alteration of microRNA levels by histone deacetylase inhibition. Cancer Res 66:1277–1281

Sebastián C, Zwaans BMM, Silberman DM et al (2012) The histone deacetylase SIRT6 is a tumor suppressor that controls cancer metabolism. Cell 151:1185–1199

Sharma A, Heuck CJ, Fazzari MJ et al (2010) DNA methylation alterations in multiple myeloma as a model for epigenetic changes in cancer. Wiley Interdiscip Rev Syst Biol Med 2:654–669

Si M, Zhu S, Wu H et al (2007) miR-21-mediated tumor growth. Oncogene 26:2799–2803

Sonaglio V, de Carvalho AC, Toledo SRC et al (2013) Aberrant DNA methylation of ESR1 and p14ARF genes could be useful as prognostic indicators in osteosarcoma. Oncotarget Ther 6: 713–723

Stamato MA, Juli G, Romeo E et al (2017) Inhibition of EZH2 triggers the tumor suppressive miR-29b network in multiple myeloma. Oncotarget 8:106527–106537

Stein EM, DiNardo CD, Pollyea DA et al (2017) Enasidenib in mutant IDH2 relapsed or refractory acute myeloid leukemia. Blood 130:722–731

Stopa N, Krebs JE, Shechter D (2015) The PRMT5 arginine methyltransferase: many roles in development, cancer and beyond. Cell Mol Life Sci 72:2041–2059

Sun J, Cui KQ, Li ZP et al (2017) Suberoylanilide hydroxamic acid, a novel histone deacetylase inhibitor, improves the development and acetylation level of miniature porcine handmade cloning embryos. Reprod Domest Anim 52:763–774

Tagliaferri P, Rossi M, DiMartino M et al (2012) Promises and challenges of MicroRNA-based treatment of multiple myeloma. Curr Cancer Drug Targets 12:838–846

Tam W, Dahlberg JE (2006) miR-155/BIC as an oncogenic microRNA. Genes Chromosomes Cancer 45:211–212

Tan M, Luo H, Lee S et al (2011) Identification of 67 histone marks and histone lysine crotonylation as a new type of histone modification. Cell 146:1016–1028

Tian E, Zhan F, Walker R et al (2003) The role of the Wnt-signaling antagonist DKK1 in the development of osteolytic lesions in multiple myeloma. New Eng J Med 349:2483–2494

Tsukada Y, Fang J, Erdjument-Bromage H et al (2006) Histone demethylation by a family of JmjC domain-containing proteins. Nature 439:811–816

Turner BM (2007) Defining an epigenetic code. Nat Cell Biology 9:2–8

Vallabhapurapu SD, Noothi SK, Pullum DA et al (2015) Transcriptional repression by the HDAC4–RelB–p52 complex regulates multiple myeloma survival and growth. Nat Commun 6:8428

van Haaften G, Dalgliesh GL, Davies H et al (2009) Somatic mutations of the histone H3K27 demethylase gene UTX in human cancer. Nat Genet 41:521–523

Voorhoeve PM, Le Sage C, Schrier M et al (2006) A genetic screen implicates miRNA-372 and miRNA-373 as oncogenes in testicular germ cell tumors. Cell 124:1169–1181

Walker BA, Leone PE, Jenner MW et al (2006) Integration of global SNP-based mapping and expression arrays reveals key regions, mechanisms, and genes important in the pathogenesis of multiple myeloma. Blood 108:1733–1743

Walker BA, Leone PE, Chiecchio L et al (2010a) A compendium of myeloma associated chromosomal copy number abnormalities and their prognostic value. Blood 116:e56–e65
Walker BA, Wardell CP, Chiecchio L et al (2010b) Aberrant global methylation patterns affect the molecular pathogenesis and prognosis of multiple myeloma. Blood 117:553–562
Walker BA, Mavrommatis K, Wardell CP et al (2018) Identification of novel mutational drivers reveals oncogene dependencies in multiple myeloma. Blood 132:587–597
Whitehall VL, Dumenil TD, McKeone DM et al (2014) Isocitrate dehydrogenase 1 R132C mutation occurs exclusively in microsatellite stable colorectal cancers with the CpG island methylator phenotype. Epigenetics 9:1454–1460
Winquist E, Knox J, Ayoub J-P et al (2006) Phase II trial of DNA methyltransferase 1 inhibition with the antisense oligonucleotide MG98 in patients with metastatic renal carcinoma: a National Cancer Institute of Canada clinical trials group investigational new drug study. Investig New Drugs 24:159–167
Wong KY, Yim RLH, So CC et al (2011) Epigenetic inactivation of the MIR34B/C in multiple myeloma. Blood 118:5901–5904
Wutz A, Smrzka OW, Schweifer N et al (1997) Imprinted expression of the Igf2r gene depends on an intronic CpG island. Nature 389:745–749
Xie Z, Chng WJ (2014) MMSET: role and therapeutic opportunities in multiple myeloma. Biomed Res Int 2014:636514
Xie Z, Bi C, Chooi J et al (2015) MMSET regulates expression of IRF4 in t (4; 14) myeloma and its silencing potentiates the effect of bortezomib. Leukemia 29:2347–2354
Yang SM, Kim BJ, Norwood Toro L et al (2013) H1 linker histone promotes epigenetic silencing by regulating both DNA methylation and histone H3 methylation. Proc Natl Acad Sci 110: 1708–1713
Yang X, Han H, De Carvalho DD et al (2014) Gene body methylation can alter gene expression and is a therapeutic target in cancer. Cancer Cell 26:577–590
Yee AJ, Bensinger WI, Supko JG et al (2016) Ricolinostat plus lenalidomide, and dexamethasone in relapsed or refractory multiple myeloma: a multicentre phase 1b trial. Lancet Oncol 17: 1569–1578
Yuregir OO, Yurtcu E, Kizilkilic E et al (2010) Detecting methylation patterns of p16, MGMT, DAPK and E-cadherin genes in multiple myeloma patients. Int J Laboratory Hematol 32: 142–149
Zhan F, Huang Y, Colla S et al (2006) The molecular classification of multiple myeloma. Blood 108:2020–2028
Zhang Q, Wang LQ, Wong KY et al (2015) Infrequent DNA methylation of miR-9-1 and miR-9-3 in multiple myeloma. J Clin Pathol 68:557–561
Zhou Y, Chen L, Barlogie B et al (2010) High-risk myeloma is associated with global elevation of miRNAs and overexpression of EIF2C2/AGO2. Proc Natl Acad Sci 107:7904–7909
Zhu YX, Kortuem KM, Stewart AK (2013) Molecular mechanism of action of immune-modulatory drugs thalidomide, lenalidomide and pomalidomide in multiple myeloma. Leuk Lymphoma 54:683–687
Zhu H, Zhang Y, Chen J et al (2017) IDH1 R132H mutation enhances cell migration by activating AKT-mTOR signaling pathway, but sensitizes cells to 5-FU treatment as NADPH and GSH are reduced. PLoS One 12:e0169038

# Metabolic Deregulations Affecting Chromatin Architecture: One-Carbon Metabolism and Krebs Cycle Impact Histone Methylation

**Francisco Saavedra, Ekaterina Boyarchuk, Francisca Alvarez, Geneviève Almouzni, and Alejandra Loyola**

## Contents



**Abstract** The methylation status of a particular amino acid results from the interplay of two enzymes: "Writers" (methyltransferases) and "Erasers" (demethylases). Methylation of histones in chromatin can be recognized by "Readers" which induce

Authors Francisco Saavedra, Ekaterina Boyarchuk and Francisca Alvarez contributed equally to this work.

F. Saavedra · F. Alvarez · A. Loyola (✉)
Fundación Ciencia & Vida, Santiago, Chile
e-mail: aloyola@cienciavida.org

E. Boyarchuk · G. Almouzni (✉)
Institut Curie, PSL Research University, CNRS, UMR3664, Paris, France

Sorbonne Universités, Paris, France
e-mail: genevieve.almouzni@curie.fr



S. Jurga, J. Barciszewski (eds.), *The DNA, RNA, and Histone Methylomes*, RNA Technologies, https://doi.org/10.1007/978-3-030-14792-1_23

changes in chromatin organization and gene expression, directed by the methylation status. Importantly, the reactions of methylation and demethylation involve several metabolites. Some, such as folate and *S*-adenosyl-L-methionine, act as cofactors for methyltransferases while flavin adenine dinucleotide and α-ketoglutarate act as cofactors for demethylases. Other metabolites, such as succinate and fumarate, function as enzyme inhibitors. Factors that modulate the levels of these metabolites in the cell therefore affect the dynamics of protein methylation. These factors can include diet, as well as altered expression of enzymes involved in cofactor synthesis through mutations and/or post-translational modifications. For example, methionine is a substrate for *S*-adenosyl-L-methionine formation, and reduction in its abundance ultimately induces a global reduction in histone methylation in vitro, affecting gene expression. Changes in the metabolic states of cells in diseases such as cancer, and regulation of metabolites required for histone methylation and demethylation, have thus been highlighted as avenues for therapeutic development. In this review, we evaluate the current knowledge concerning methylation of histones, and also of other protein substrates. We document how this is linked to metabolites such as *S*-adenosyl-L-methionine and other intermediates in the Krebs cycle. Finally, we discuss the implications of deregulation at this level in cancer.

**Keywords** Histone methylation · One carbon metabolism · *S*-adenosyl-L-methionine · Methyltransferases · Demethylases

# 1 Chromatin Methylation: Function and Regulation

## *1.1 Basic Concepts of Chromatin Structure*

The genetic instructions of cells are carried on DNA molecules which encode information relating to the basic processes required for normal cellular function, such as replication, transcription, and DNA repair. In every human cell, a nucleus of around 2 μm diameter contains around 2 m of DNA packaged in a structure called chromatin—a nucleoprotein complex comprising DNA, RNA, and proteins, organized in several hierarchical levels. Correct and dynamic organization of chromatin is necessary for accurate genome functioning. The basic unit of chromatin is the nucleosome, which comprises an octamer with two copies each of the core histones H2A, H2B, H3, and H4, around which is wrapped 147 bp of DNA and a variable linker DNA segment associated with the H1 linker histone (Luger et al. 1997). Additional chromatin-binding proteins including transcription factors and structural RNAs cause the chromatin filaments to fold further, resulting in highly compact DNA. Modulation at each level of chromatin organization ensures that adaptation to environmental cues can occur (Sitbon et al. 2017; Hug and Vaquerizas 2018; Luo et al. 2018; Yadav et al. 2018). Chemical modifications of the histones, termed post-translational modifications (PTM), or onto DNA are major mechanisms of chromatin alteration (Gurard-Levin and Almouzni 2014; Jones 2012). In addition, the properties of nucleosomes can be further modulated by the inclusion of histone variants, which can confer particular properties to chromatin (Sitbon et al. 2017). The expression of these variants differs depending on the cell cycle phase, tissue in which they are expressed, and the mode of their incorporation into the chromatin (Mendiratta et al. 2018). Together, these features are critical for proper chromatin functioning in various processes such as development, aging, or tumorigenesis.

### 1.1.1 Post-Translational Modifications, the Histone Code, and the "Writer-Eraser-Reader" Model

Post-translational modifications can occur throughout the entire lifespan of a protein, from synthesis to degradation (Loyola and Almouzni 2007; Alvarez et al. 2011; Rivera et al. 2015). Methylation of lysine residues in calf thymus was the first histone PTM to be identified (Murray 1964), reported before the discovery of histone acetylation (Allfrey et al. 1964) or phosphorylation (Gutierrez and Hnilica 1967). Since then, over 15 different types of PTMs have been identified on histones (Zhao and Garcia 2015). These modifications provide a stable but reversible system with which the cell can react to external stimuli (Gurard-Levin and Almouzni 2014). Particular PTMs, such as phosphorylation or acetylation, can alter the physical properties of the nucleosome including charge, thereby affecting histone-DNA interactions (Bowman and Poirier 2015).

The most common mechanism of action of histone PTMs is the modulation of protein binding through the recruitment of non-histone proteins, which can consequently modify the chromatin state. The density of a particular PTM at a given chromatin locus can be critical, because a single mark on one histone is unlikely to have significant effects. Rather, it is likely that a certain level of modified histones exists, above which significant effects will be observed with regards to chromatin.

Multiple types of modification can occur at particular residues. For instance, lysines can be methylated, sumoylated, ubiquitinated, or acetylated in an exclusive manner. The large number of possible combinations gave rise to the hypothesis of the "histone code", whereby histone modifications work sequentially or in combination to affect gene regulation (Jenuwein and Allis 2001). Many enzymes have been identified to be involved in catalyzing the chemical modification of histones ("Writers") or removing such modifications ("Erasers") (Kouzarides 2007). Effector proteins ("Readers") recognize and bind to histones or DNA that carry certain chemical modifications, in order to achieve a specific chromatin state at a given locus (Nicholson et al. 2015). Whilst PTMs are generally considered important for the recruitment of proteins, they can also inhibit histone-protein interactions (Wen et al. 2014).

### 1.1.2 Histone Methylation: Effects, Localization, and “Readers”

Of all histone PTMs, methylation has been identified as one of the key modifications in the regulation of gene expression. Methylation predominantly occurs on lysine and arginine residues, but has also been detected on histidine, aspartic, and glutamic acid residues (Zhao and Garcia 2015).

Specific methylation of histone lysine residues has enabled the correlation of methylation at a given locus with its transcriptional activity to be analyzed. Depending on the particular lysine residue, its degree of methylation (mono- [Kme1], di- [Kme2], or trimethylation [Kme3]), and the position of the methylated nucleosome within the gene and genome, this modification can be associated with transcriptionally active or inactive chromatin (Table 1). In general, methylation of the histone H3 lysine 4 (H3K4), H3K36, and H3K79 have been linked to activation of gene expression; whereas di- and trimethylation of H3K9, H3K27, and H4K20 have been associated with gene silencing and/or heterochromatin formation (Mozzetta et al. 2015). In addition, methylation of histone lysine residues has been associated with the regulation of splicing (Luco et al. 2010). For instance, H3K36me3 is present on highly transcribed exons, and is more enriched on constitutive exons compared with alternatively spliced ones (Kolasinska-Zwierz et al. 2009). Moreover, local increases in H3K9me2 and H3K9me3 enhance exon inclusion, whereas H3K9 demethylation is associated with exon skipping (Bieberstein et al. 2016).

These diverse effects require a series of “Readers” that possess methyl-lysine recognition domains. Methylation does not significantly affect the charge of the histone; instead it frequently functions to provide a docking site for Reader proteins. The Reader can then serve as a platform to recruit other effector proteins and form multiprotein complexes to direct either transcriptional activation or repression. Methyl-lysine recognition domains can be divided into four classes: ankyrin repeats, tryptophan-aspartic acid (WD40) repeat domains, plant homeodomain (PHD) fingers, and Royal family proteins. Royal family proteins are classified based on the presence of the conserved barrel-like protein fold called the “Tudor barrel”. This superfamily includes the Tudor domain, chromodomain, malignant brain tumor (MBT) domain, chromo barrel domain, and proline-tryptophan-tryptophan-proline (PWWP) domain families (Teske and Hadden 2017). Each of them exhibit specific binding features which are related to the methylation status of the residue, and whether the modification occurs in *cis* or in *trans* (Teske and Hadden 2017).

The principle of the mechanism behind the modulation of chromatin structure and regulation of transcription by PTMs and associated Readers can be illustrated by H3K9 methylation in heterochromatin. Heterochromatin-meditated gene silencing is thought to result from changes in the packing of nucleosomes to create a dense, compact structure, which prevents transcriptional machinery from accessing the DNA or establishing the modifications that recruit transcriptional activators. At heterochromatin sites that are enriched for H3K9me2/3, direct binding of HP1 via its chromodomain can promote chromatin compaction or phase transition by dimerization or oligomerization of HP1 (Canzio et al. 2011; Machida et al. 2018). This

**Table 1** Histone lysine methylation in mammals

| Histone | Lysine and degree of methylation | Writer | Eraser | Function |
|---|---|---|---|---|
| H3 | K4me1 | KMT7 (SET7)<br>KMT2A (MLL2)<br>KMT2B (MLL3)<br>KMT2C (MLL4)<br>KMT2D (MLL5)<br>KMT2F (SET1A)<br>KMT2G (SET1B) | KDM1A (LSD1)<br>KDM1B (LSD2)<br>KDM5B (JARID1B) | Transcription activation<br>Enhancer function |
| | K4me2 | KMT2A (MLL2)<br>KMT2B (MLL3)<br>KMT2C (MLL4)<br>KMT2D (MLL5)<br>KMT2F (SET1A)<br>KMT2G (SET1B)<br>KMT3E (SMYD3) | KDM1A (LSD1)<br>KDM1B (LSD2)<br>KDM5A (JARID1A)<br>KDM5B (JARID1B)<br>KDM5C (JARID1C)<br>KDM5D (JARID1D)<br>ROIX1 (NO66) | Transcription activation<br>Enhancer function |
| | K4me3 | KMT2A (MLL2)<br>KMT2B (MLL3)<br>KMT2C (MLL4)<br>KMT2D (MLL5)<br>KMT2F (SET1A)<br>KMT2G (SET1B)<br>KMT3E (SMYD3) | KDM2B (JHDM1B)<br>KDM5B (JARID1B)<br>KDM5C (JARID1C)<br>KDM5D (JARID1D)<br>ROIX1 (NO66) | Transcription activation<br>Enhancer function |
| | K9me1 | KMT1E (SetDB1)<br>KMT1C (G9A)<br>KMT1D (GLP)<br>KMT8E (PRDM3)<br>KMT8D (PRDM8)<br>KMT8F (PRDM16)<br>KMT2H (ASH1)<br>KMT3F (NSD3) | KDM1A (LSD1)<br>KMT3A (JLHDM2A)<br>KMT3B (JLHDM2B)<br>KDM3C (JMJD1C)<br>KDM7A (JHDM1D)<br>KDM7B (JHDM1F)<br>KDM7C (JHDM1E)<br>HR (HAIR, hairless) | Transcription repression |

(continued)

**Table 1** (continued)

| Histone | Lysine and degree of methylation | Writer | Eraser | Function |
|---|---|---|---|---|
| | K9me2 | KMT1A/B (SUV39H1/2)<br>KMT1E (SetDB1)<br>KMT1C (G9A)<br>KMT1D (GLP)<br>KMT8A (PRDM2)<br>KMT8D (PRDM8)<br>KMT2H (ASH1L)<br>KMT3F (NSD3) | KDM1A (LSD1)<br>KMT3A (JLHDM2a)<br>KMD4A (JMJD2A)<br>KMD4B (JMJD2B)<br>KMD4C (JMJD2C)<br>KMD4D (JMJD2D)<br>KDM7A (JHDM1D)<br>KDM7B (JHDM1F)<br>KDM7C (JHDM1E)<br>HR (HAIR, hairless) | Transcriptional repression<br>Heterochromatin formation |
| | K9me3 | KMT1A/B (SUV39H1/2)<br>KMT1E (SetDB1)<br>KMT1F (SETDB2)<br>KMT8A (PRDM2)<br>KMT2H (ASH1L)<br>KMT3F (NSD3) | KMD4A (JMJD2A)<br>KMD4B (JMJD2B)<br>KMD4C (JMJD2C)<br>KMD4D (JMJD2D)<br>RIOX2 (MINA) | Transcriptional repression<br>Constitutive heterochromatin formation<br>X-chromosome inactivation |
| | K27me1 | KMT6B (EZH1)<br>KMT1C (G9A)<br>KMT1D (GLP)<br>KMT2H (ASH1L)<br>KMT3G (NSD2)<br>KMT3F (NSD3) | KDM7A (JHDM1D)<br>KDM7C (JHDM1E) | Transcriptional repression |
| | K27me2 | KMT6A (EZH2)<br>KMT6B (EZH1)<br>KMT2H (ASH1L)<br>KMT3G (NSD2)<br>KMT3F (NSD3) | KDM6A (UTX)<br>KDM6B (JMJD3)<br>KDM7A (JHDM1D)<br>KDM7C (JHDM1E) | Transcriptional repression<br>Facultative heterochromatin formation<br>X-chromosome inactivation |
| | K27me3 | KMT6A (EZH2)<br>KMT6B (EZH1)<br>KMT2H (ASH1L) | KDM6A (UTX)<br>KDM6B (JMJD3) | Transcriptional repression<br>Facultative heterochromatin formation |

(continued)

**Table 1** (continued)

| Histone | Lysine and degree of methylation | Writer | Eraser | Function |
|---|---|---|---|---|
| | | KMT3G (NSD2)<br>KMT3F (NSD3) | | X-chromosome inactivation |
| | K36me1 | KMT3A (SET2)<br>KMT3B (NSD1)<br>KMT3G (NSD2)<br>KMT3F (NSD3)<br>KMT2H (ASH1) | KDM2A (JHDM1a)<br>KDM2B (JHDM1b) | Transcription activation |
| | K36me2 | KMT3A (SETD2)<br>KMT3B (NSD1)<br>KMT3G (NSD2)<br>KMT3F (NSD3)<br>KMT2H (ASH1L) | KDM2A (JHDM1a)<br>KDM2B (JHDM1b)<br>KMD4A (JMJD2A)<br>KMD4B (JMJD2B)<br>KMD4C (JMJD2C)<br>KMD4E (JMJD2E)<br>KMD8 (JMJD5)<br>RIOX1 (NO66) | Transcriptional activation<br>Transcription elongation |
| | K36me3 | KMT3A (SETD2)<br>KMT2H (ASH1L)<br>KMT3C (SMYD2) | KMD4A (JMJD2A)<br>KMD4B (JMJD2B)<br>KMD4C (JMJD2C)<br>KMD4E (JMJD2E)<br>RIOX1 (NO66) | Transcriptional activation<br>Transcription elongation |
| | K56me1 | KMT1C (G9A) | KMD4B (JMJD2B)<br>KMD4E (JMJD2E) | DNA replication<br>Heterochromatin formation |
| | K56me3 | KMT1A/B (SUV39H1/2) | KMD4B (JMJD2B)<br>KMD4E (JMJD2E) | DNA replication<br>Heterochromatin formation |
| | K64me | Unknown | Unknown | Heterochromatin formation |
| | K79me1 | KMT4 (DOT1L) | KDM2B (JHDM1b) | Transcriptional activation<br>Telomeric silencing<br>DNA damage response |

(continued)

**Table 1** (continued)

| Histone | Lysine and degree of methylation | Writer | Eraser | Function |
|---|---|---|---|---|
| | K79me2 | KMT4 (DOT1L) | KDM2B (JHDM1b) | Transcriptional activation<br>Telomeric silencing<br>DNA damage response |
| | K79me3 | KMT4 (DOT1L) | KDM2B (JHDM1b) | Transcriptional activation<br>Telomeric silencing<br>DNA damage response |
| H4 | K5me1 | KMT3E (SMYD3) | Unknown | Contributes to cancer phenotype |
| | K20me1 | KMT5A (PR-SET7)<br>KMT3B (NSD1)<br>KMT3G (NSD2) | KDM7A (JHDM1D)<br>KDM7B (JHDM1F) | Transcriptional silencing<br>Mitotic condensation |
| | K20me2 | KMT5B/C (SUV4-20H1/2)<br>KMT3B (NSD1)<br>KMT3G (NSD2) | KDM7C (JHDM1E) | Transcription repression<br>Heterochromatin formation/silencing<br>DNA damage responses |
| | K20me3 | KMT5B/C (SUV4-20H1/2)<br>SMYD5 | KDM7C (JHDM1E) | Transcription repression<br>Heterochromatin formation/silencing<br>DNA damage response |
| H1 | K26me*2/3* | KMT1C (G9A)<br>KMT1D (GLP)<br>KMT6A (EZH2) | KMD4A (JMJD2A)<br>KMD4B (JMJD2B)<br>KMD4C (JMJD2C) | Heterochromatin formation/silencing |
| H2A.Z | K7me1 | SETD6 | Unknown | Transcription repression |

In the table, only methylations with a documented biological outcome and/or modifier are listed. Modified from Allis et al. (2007), Greer and Shi (2012), Wagner and Carpenter (2012), Mozzetta et al. (2015), Zhao and Garcia (2015), Park et al. (2016)

dimerization/oligomerization bridges neighboring nucleosomes that carry H3K9me2/3. In regions of constitutive heterochromatin, HP1 recruits diverse sets of regulators including chromatin modifiers, DNA replication and repair factors, and nuclear structural proteins as well as RNA (Kwon and Workman 2008). These regulators act in combination to mediate the establishment and maintenance of heterochromatin (Probst and Almouzni 2011; Rivera et al. 2014).

Arginine methylation can occur in three different forms: modification of one of the ω-nitrogens to produce monomethyl arginine (MMA, Rme), addition of two methyl groups onto the same ω-nitrogen to produce asymmetric dimethyl arginine (ADMA, Rme2a); or addition of one methyl group to each ω-nitrogens to produce symmetric dimethyl arginine (SDMA, Rme2s). Such modifications do not change the positive charge of arginine, but can affect its involvement in protein-protein

**Table 2** Histone arginine methylation in mammals

| Histone | Arginine and degree of methylation | Writer | Function |
|---|---|---|---|
| H3 | R2me2a | PRMT6 | Transcription repression |
| | R2me2s | PRMT5<br>PRMT7 | Transcription activation |
| | R8me2a | PRMT2 | Transcription repression |
| | R8me2s | PRMT5 | Transcription repression |
| | R17me2a | PRMT4 (CARM1) | Transcription activation |
| | R26me2a | PRMT4 (CARM1) | Transcription activation |
| | R43me2a | PRMT4 (CARM1)<br>PRMT6 | Transcription activation |
| H4 | R3me2a | PRMT1<br>PRMT6 | Transcription activation |
| | R3me2s | PRMT5<br>PRMT7 | Transcription repression |
| | R17me1 | PRMT7 | In vitro substrate |
| | R19me1 | PRMT7 | In vitro substrate |
| H2A | R3me2a | PRMT1<br>PRMT6 | Transcription activation |
| | R3me2s | PRMT5<br>PRMT7 | Transcription repression |
| | R29me2a | PRMT6 | Transcription repression |
| H2B | R29me1 | PRMT7 | In vitro substrate |
| | R31me1 | PRMT7 | In vitro substrate |
| | R33me1 | PRMT7 | In vitro substrate |

Modified from Di Lorenzo and Bedford (2011), Greer and Shi (2012), Alam et al. (2015), Jahan and Davie (2015), Zhao and Garcia (2015)

interactions. As is the case for lysine methylation, the outcome of arginine methylation depends on the particular residue, the degree of methylation, and the symmetry of the modification. The most well-characterized methylated arginine residues include R2, R8, R17, and R26 of histone H3; and R3 of histones H4 and H2A. Key transcriptional activation marks involving arginine methylation include H4R3me2a, H3R2me2s, H3R17me2a, and H3R26me2a; while H3R2me2a, H3R8me2a, H3R8me2s, and H4R3me2s are associated with transcriptional repression (Blanc and Richard 2017) (Table 2).

Arginine methylation affects protein function via at least two different mechanisms. First, methylation can directly alter the ability of arginine to form bonds with hydrogen-bond acceptors by introducing steric constraints. It is noteworthy that unmodified arginine has five potential hydrogen-bond donors. The modification H4R3me2a, for example, prevents recruitment of lysine methyltransferase MLL4, and therefore impairs H3K4 methylation and transcriptional activation (Dhar et al. 2012). This is the mechanism behind H3R2me2a-dependent transcriptional repression, which counteracts H3K4 methylation by inhibiting the binding of the H3K4 methyltransferase MLL1 and several other H3K4me3 effectors (Hyllus et al. 2007).

Interestingly, the opposite is true for symmetrically methylated H3R2, which enhances the binding of H3K4me3 Readers. For example, the RAG2 PHD domain preferentially binds to the H3R2me2sK4me3 modifications, with a 20-fold increased affinity compared to H3K4me3 (Yuan et al. 2012). The second mechanism of action relies on the ability of Tudor domain family proteins to "read" methylated arginine residues and subsequently recruit chromatin modifiers to these residues. Individual PHD and WD40 domains are also able to bind methylated arginines (Gayatri and Bedford 2014).

Interestingly, the majority of methylarginine Readers that have been characterized to date recognize the methylation of non-histone proteins (see below). One of the factors that is recruited by methylated H4R3 is Staphylococcal nuclease domain-containing protein 1 (SND1), also known as Tudor domain-containing protein 11 (TDRD11), which acts as a transcriptional coactivator by recruiting histone acetyltransferases, thereby promoting histone acetylation (Gayatri and Bedford 2014).

### 1.1.3 Histone Methyltransferases: Classification and Recruitment of "Writers"

The human genome encodes around 60 methyltransferases, comprising both SET-domain lysine methyltransferases (KMTs) and seven-beta-strand enzymes that methylate different residues (Clarke 2013).

In mammals, all of the KMTs identified to date are highly specific toward a particular lysine residue within a histone, but also toward a number of non-histone substrates (see below). All KMTs apart from DOT1L belong to a large protein family characterized by the presence of the conserved SET domain, whose name was coined based on the three *Drosophila melanogaster* proteins that were first identified: *S*uppressor of variegation 3–9 (Su(var)3–9), *E*nhancer of zeste (E(z)), and the homeobox gene regulator *T*rithorax (Trx) (Jenuwein et al. 1998). The SET domain catalyzes the transfer of a methyl group to the ε-amino groups of lysine residues using *S*-adenosyl-L-methionine (SAM) as the methyl group donor. Based on the sequence homology within and around the catalytic SET domain, SET-containing KMTs can be divided into six sub-families: SET1, SET2, SUV39, EZH, SMYD, and PRDM (Volkel and Angrand 2007). The features of the SET domain of a protein often reflect its substrate specificity (Herz et al. 2013). The majority of SET-containing KMTs have at least one additional module, which confers the ability to recognize various PTMs, usually including the modification that they catalyze. The coupling of "writing" and "reading" properties provides a mechanism for the nucleation and spreading of lysine methylation along the chromatin. In contrast to SET domain-containing methyltransferases, disruptor of telomeric silencing-like protein (DOT1L, also known as KMT4) contains a domain similar to that of glycine *N*-methylase (Nguyen and Zhang 2011), and mono-, di-, or trimethylates H3K79 in a non-processive manner (Frederiks et al. 2008).

One of the most critical—and debatable—aspects of KMTs functions as regulators is their capacity to target a particular genomic locus. No KMT aside from PRDM family members possess DNA-binding properties, and so they rely on protein-binding partners and other mechanisms to target chromatin. Generally, methyltransferases are recruited to their genomic target *loci* through interaction with sequence-specific transcription factors, other chromatin-binding proteins, and non-coding RNAs, and thus methylate nucleosomal histones (Mozzetta et al. 2015). However, some methyltransferases are involved in the methylation of non-nucleosomal histones, such as SetDB1 (a member of SUV39 family), which binds ribosomes and mono-methylates H3K9 co-translationally (Rivera et al. 2015).

Arginine methylation is catalyzed by a family of enzymes called protein arginine methyltransferases (PRMTs) that belong to the seven-beta-strand group of methyltransferases. PRMTs are generally classified by activity as type I, II, or III. Types I and II catalyze the formation of a mono-methylarginine intermediate, which then gives rise to an asymmetric dimethylarginine in the case of type I PRMTs (PRMT1, 2, 3, 4, 6, and 8), or to a symmetric dimethylarginine in the case of type II PRMTs (PRMT5 and 9). The only known type III PRMT is PRMT7, which exclusively generates mono-methylarginine residues (Morales et al. 2016). A fourth group of arginine methyltransferases, type IV, catalyze the monomethylation of the internal guanidino nitrogen (δ-MMA) of arginine residues. These enzymes have been identified in yeast, but no mammalian homologs have been identified. Nevertheless, such modifications have been recently described in humans (Martens-Lobenhoffer et al. 2016).

### 1.1.4 Histone Demethylases: Classification and Activities of "Erasers"

For about 40 years, histone lysine methylation was considered to be a modification that could not be actively removed, until the discovery of the first histone lysine demethylase (KDM), denoted lysine-specific demethylase 1 (LSD1) (Shi et al. 2004). Other lysine demethylases have since been identified, and there are only a few lysine residues that are not associated with a demethylase (Black et al. 2012). Demethylases can be grouped in two families: LSDs and Jumonji C (JmjC) domain-containing families.

The LSD family consists of two members, LSD1/KDM1A and LSD2/KDM1B, each characterized by the presence of a C-terminal amine oxidase domain (AOD). This domain confers demethylase activity through a flavin adenine dinucleotide (FAD)-dependent amine oxidation mechanism, and a substrate specificity that is limited to mono- and dimethylated lysines (Shi et al. 2007). Via this domain, LSD1 can demethylate mono- and dimethylated H3K4 and H3K9 residues and is thus considered a corepressor or coactivator, respectively (Shi et al. 2004; Metzger et al. 2005). On the other hand, LSD2 can only demethylate mono- and dimethyl marks on H3K4, and is therefore considered a transcriptional corepressor (Fang et al. 2010).

The JmjC domain-containing family includes more than 30 proteins with different substrate specificities and distinct catalytic mechanisms, which are further divided into several subfamilies (KDM2, KDM3, KDM4, KDM5, KDM6, KDM7, and KDM8) (Allis et al. 2007). The JmjC KDMs are dioxygenases that use iron (Fe(II)) and α-ketoglutarate (2-oxoglutarate or 2-OG) as cofactors (Klose et al.

2006). These enzymes can demethylate all three methylation states of lysine on a range of substrates (Table 1). Currently, the KDM for H3K79me remains enigmatic, but a recent report suggests that KDM2B is capable of catalyzing H3K79me2/me3 demethylation (Kang et al. 2018). As for KMTs, the targeting of JmjC KDMs to their *loci* relies on two features of the enzymes. First, they are associated with large multimeric complexes, which may guide them to the histones surrounding specific target genes. Second, other conserved domains such as PHD, Tudor, zinc finger (zf-C2HC4), F-box, and AT-rich interactive (ARID) domains, as well as leucine-rich regions (LRR), participate in the targeting of JmjC KDMs to specific regions (Klose et al. 2006).

The reversibility of arginine methylation is unclear. Several studies have reported the modulation of methylation of particular arginine residues in a window of minutes following induction of transcription, or within one cell cycle, which strongly supports the existence of an active mechanism for arginine demethylation (Metivier et al. 2003; Le Romancer et al. 2008). To date, only a few proteins with potential arginine-demethylating activity have been identified. These include the JmjC protein 6 (JMJD6) (Chang et al. 2007), peptidylarginine deiminase 4 (PAD4) (Wang et al., 2004) and the JmjC protein 1B (JMJD1B) (Li et al. 2018). Notably, JMJD6 also possesses lysine hydroxylase activity, and PAD4 cannot be considered a classical demethylase because it cannot demethylate dimethylated arginines. Notably, although a subset of JmjC KDMs (KDM3A, KDM4E, and KDM5C) are able to demethylate arginine residues in vitro, their in vivo activity is yet to be proved (Walport et al. 2016).

## *1.2 Non-Histone Substrates for HMTs*

Methylation is not restricted to histones. Methylated lysine and arginine residues are found in many cellular proteins including those involved in transcription, RNA processing, DNA repair, cell signaling, and translation. The processes involved in the regulation of methylation should therefore be considered beyond the histones. In recent years, advances in liquid chromatography coupled to tandem mass spectrometry (LC-MS/MS) and the generation of a set of specific antibodies have enabled comprehensive large-scale proteomic analyses of arginine methylation in different organisms (Wesche et al. 2017). Although a lack of good lysine methylation-specific antibodies has limited the proteome-wide analysis of this PTM, promising strategies have been developed for the identification of this modification which involve enrichment of methylated peptides using native methyl-lysine recognition domains (Moore et al. 2013).

Histone methylation modifiers control the methylation state of non-histone substrates, in order to regulate their activities or stabilities. Key components of several signaling pathways are classified as methylated non-histone substrates, including nuclear factor-kappa B (NFκB), estrogen receptor (ERα), β-catenin, and p53 (Alam et al. 2015; Biggar and Li 2015; Mozzetta et al. 2015) (Table 3). Importantly, the

**Table 3** Non-histone substrates of Lys- and Arg- methyltransferases categorized by their biological functions

| Function | Substrate | KMT | Biological outcome |
|---|---|---|---|
| Transcription factors | C/EBPb (K39) | KMT1C (G9a) | Inhibits transactivation activity |
| | MyoD (K104me1/2) | KMT1C (G9a) | Inhibits transcriptional activity |
| | MEF2D (K267 me1/2) | KMT1C (G9a) | Inhibits transcriptional activity |
| | p53 (K373me2) | KMT1D (GLP)<br>KMT1C (G9a) | Inhibits transcriptional activity and p53-dependent apoptosis |
| | p53 (K382) | KMT5A (PR-SET7) | Represses transcriptional activity |
| | p53 (K370) | KMT3C (SMYD2) | Reduces DNA-binding ability and apoptosis |
| | p53 (K372) | KMT7 (SET7/9) | Increases p53 stability and p53-dependent apoptosis |
| | p53 (R333/335/337) | PRMT5 | Alters recruitment to target genes; Inhibits p53 oligomerization |
| | GATA4 (K299me1) | KMT6 (EZH2) | Inhibits transcriptional activity |
| | RORα (K38me1) | KMT6 (EZH2) | Enhances proteasomal degradation |
| | UBF (K232/K254me3) | KMT1E (SetDB1) | Increases nucleolar chromatin condensation; decreases rDNA transcription |
| | TAF10 (K189me1) | KMT7 (SET7/9) | Enhances binding to pol II |
| | ERα (K302me1) | KMT7 (SET7/9) | Stabilizes ERα;<br>Promotes ERα recruitment and ER-dependent gene activation |
| | ERα (K266) | KMT3C (SMYD2) | Inhibits ERα activity |
| | ERα (R260) | PRMT1 | Promotes interactions with PI3K and Src |
| | FOXO3 (K270/271me1) | KMT7 (SET7/9) | Decreases protein stability;<br>Inhibits DNA-binding activity and FOXO3-dependent transcription |
| | RUNX1 (R206/210) | PRMT1 | Abrogates association with co-repressor SIN3A. |
| | RB (K810me1, K873me1) | KMT7 (SET7/9) | Promotes interaction with HP1;<br>Promotes Rb-dependent cell cycle arrest and transcriptional repression |
| | RB (K810me1, K860me1) | KMT3C (SMYD2) | Promotes interaction with transcriptional repressor L3MBTL1 |

(continued)

**Table 3** (continued)

| Function | Substrate | KMT | Biological outcome |
|---|---|---|---|
| | E2F1 (K185me1) | KMT7 (SET7/9) | Stimulates ubiquitination and protein degradation |
| | E2F1 (R111/R113 m2s) | PRMT5 | Promotes protein degradation; Favours cell proliferation |
| | E2F1 (R109m2a) | PRMT1 | Promotes E2F1-dependen expression of genes connected with apoptosis |
| Chromatin-modifiers and chromatin-binding proteins | P300 (R2142) | PRMT4 | Inhibits interaction with glucocorticoid receptor interacting protein (GRIP1) |
| | DNMT1 (K142me1) | KMT7 (SET7/9) | Promotes proteasome-mediated degradation |
| | DNMT3 (K44me2) | KMT1D (GLP) | Stimulates binding of MPP8 |
| | | KMT1C (G9a) | |
| | KMT1D (GLP) (K174) | KMT1D (GLP) | Stimulates binding of MPP8 |
| | KMT1C (G9a) (K165me2/3, K239me3) | KMT1C (G9a) | Stimulates binding of HP1 and CDYL |
| | KMT1A (SUV39H1) (K105/ K123 me1) | KMT7 (SET7/9) | Inhibits methyltransferase activity |
| | CBX4/PC2 | KMT1A (SUV39H1) | Promotes TUG1 ncRNA-dependent recruitment to Polycomb bodies |
| | SMARCC1 (R1064m2a) | PRMT4 | Modulates targeting to subset of genes of c-Myc pathways. |
| | CDYL (K135me3) | KMT1C (G9a) | Decreases interaction with H3K9me3 |
| | RUVBL2 (K67me1) | KMT1C (G9a) | Negative regulates hypoxia-inducible genes |
| | PCNA (K248me1) | KMT5A (PR-SET7) | Stabilizes PCNA |
| | PARP1 (K508me1) | KMT7 (SET7/9) | Stimulates PARP activity and its recruitment to sites of DNA damage. |
| Signaling pathway | STAT3 (K180) | KMT6 (EZH2) | Increases STAT3 phosphorylation;<br>Enhances STAT3 activity |
| | STAT3 (K140me2) | KMT7 (SET7/9) | Inhibits STAT3 activity |
| | p65 (K218, K221) | KMT3B (NSD1) | Activates NF-kB signaling pathway |
| | p65 (K37) | KMT7 (SET7/9) | Activates NF-kB signaling pathway |

(continued)

**Table 3** (continued)

| Function | Substrate | KMT | Biological outcome |
|---|---|---|---|
| | p65 (K314, K315) | KMT7 (SET7/9) | Reduces of p65 stability |
| | p65 (K310) | SETD6 | Inhibits p65-driven transcription |
| | p65 (R30) | PRMT5 | Activates NF-kB signaling pathway |
| | MAP3K2 (K260) | KMT3E (SMYD3) | Activates MAP3K2 |
| | EGFR (R1175) | PRMT5 | Negative regulates EGFR signaling |
| | Axin (R378) | PRMT1 | Negative regulates Wnt signaling |
| RNA binding and processing | SPT5 (R681/696/698) | PRMT1 | Inhibits basal transcription; |
| | | PRMT5 | Decreases interaction with RNA polymerase II |
| | LSM4 (80–139, me2s) | PRMT5 | Stimulates binding to SMN; promotes formation of spliceosome |
| | SNRPD1 (90–119, me2s) | PRMT5 | Stimulates binding to SMN; promotes formation of spliceosome |
| | SNRPD3 (110–126, me2s) | PRMT5 | Stimulates binding to SMN; promotes formation of spliceosome |
| | SNRPB (107–210, me2s) | PRMT5 | Stimulates binding to SMN; promotes formation of spliceosome |
| | SNRPB (PGM motifs, me2a) | PRMT4 (CARM1) | Stimulates binding to SMN; splicing regulation |
| | SNRPC (PGM motifs, me2a) | PRMT4 (CARM1) | Stimulates binding to SMN; splicing regulation |
| | SF3B4 (190–424, PGM motifs, me2a) | PRMT4 (CARM1) | Stimulates binding to SMN; splicing regulation |
| | TAF2S (CA150), (1–136, me2a) | PRMT4 (CARM1) | Stimulates binding to SMN; splicing regulation |
| Other | HSP90 (K615 me1) | KMT3C (SMYD2) | Promotes interaction with titin and its stabilization in myofibers |
| | HSP70 (K561me2) | KMT2F (SETD1A) | Promotes interaction with Aurora kinase B,<br>Stimulates kinase activity |
| | Tat (K50/51) | KMT1E (SetDB1) | Inhibits HIV transcription |
| | Tat (R52/53 me2a) | PRMT6 | Inhibits transactivation activity<br>Inhibits HIV replication |

Modified from Alam et al. (2015), Biggar and Li (2015), Mozzetta et al. (2015)

impact of non-histone protein methylation depends on the exact residue that is modified and its degree of methylation, similarly to histone methylation. A striking example is the case of p53. For instance, the monomethylation of K372 on p53 (p53K372me1) by SET7/SET9 results in increased stability of the protein, enhanced expression of the p53 target gene *p21*, and increased p53-induced apoptosis (Chuikov et al. 2004); while the SMYD2-mediated monomethylation of p53 at K370 (p53K370me1) functions as an inactivating modification, repressing its activity as a transcriptional regulator (Huang et al. 2006). On the other hand, p53K370me2 enhances the transcriptional activity of p53 by promoting its interaction with p53-binding protein 1 (53BP1), which is a p53 coactivator and a regulator of the DNA-damage response (Tong et al. 2015). The p53K382me1 modification (mediated by SET8) and p53K373me2 (mediated by G9a/GLP) both inhibit p53 function (Shi and Whetstine 2007; Huang et al. 2010). Arginine methylation also has a role in the regulation of p53; PRMT5 methylates R333, R335, and R337 in a DNA-damage dependent manner. These residues are located within the oligomerization domain and affect p53 function by interfering with the promoter-binding specificity (Jansson et al. 2008).

The wide variety of cellular processes that are regulated by methyltransferases and demethylases have made these enzymes attractive targets for medical research and therapeutic development. Many of them are altered in several tumor types; for example, KMT2C/MLL3, KMT2D/MLL2, Ezh2, and SETD2 (Lawrence et al. 2014). Therefore, targeting these epigenetic factors presents an opportunity for the development of therapeutics. Although clinical evaluation of drugs that target histone methylation is still in its infancy, promising targets among the KMTs, PRMTs, and KDMs have already been identified (Song et al. 2016; McCabe et al. 2017). However, the modulation of enzymatic activity of methyltransferases and demethylases might be achieved via different mechanisms, and this must be considered for successful drug development.

In the following sections, we review how modulation of the availability of enzymatic cofactors of methyltransferases and demethylases can impact the methylation landscape of chromatin. We discuss examples of pathologies in which such cofactors are deregulated, and demonstrate how this knowledge has been exploited to generate potential therapies.

## 2 Metabolites Involved in the Regulation of Methyltransferases: *S*-Adenosylmethionine and *S*-Adenosylhomocysteine

### *2.1 Regulation of* S*-Adenosylmethionine and* S*-Adenosylhomocysteine Levels: One-Carbon Metabolism*

The production of SAM—the primary methyl group donor for reactions catalyzed by methyltransferases—relies on the use of methionine as a substrate. While plants and bacteria synthesize methionine from aspartate, animals cannot synthesize this amino acid and must acquire it from their diet. Despite this, mammals can regenerate methionine via the one-carbon metabolic pathway which takes carbon groups from nutrient and mediate its incorporation into different outputs, such as nucleotides, glutathione, SAM, and others, occurring mainly in the liver (Suganuma and Workman 2018). This pathway includes two different cycles, the methionine and folate cycle, as illustrated in Fig. 1.

The methionine cycle produces SAH as a byproduct, which is a potent pan-inhibitor of methyltransferases. Thus, the SAM/SAH ratio is an indicator of the "methylation potential" of a cell, and determines the activity of methyltransferases (Caudil et al. 2001). Hydrolysis of SAH to homocysteine (HCY) is important in maintaining the SAM/SAH ratio. Although the reaction is reversible, the equilibrium is shifted toward SAH hydrolysis by the constant removal of HCY via three different mechanisms: (1) methylation of HCY, mediated by methionine synthase (MS) or betaine-homocysteine *S*-methyltransferase (BHMT); (2) the use of HCY in the trans-sulfuration pathway for glutathione synthesis; or (3) release of HCY to the extracellular space (Grillo and Colombatto 2008). Deregulation of the pathways involved in regulating the SAM/SAH ratio—either by increasing or decreasing the ratio—affects the chromatin methylation landscape and may therefore contribute to the development of diseases, especially cancer (Shlomi and Rabinowitz 2013). The deregulation of enzymes involved in one-carbon metabolism also affects histone methylation. This mechanism and its important role in carcinogenesis are discussed below.

## 2.2 Deregulation of One-Carbon Metabolism, Its Impact on Histone Methylation, and Its Association with Diseases

As Fig. 1 illustrates, the synthesis of SAM from methionine is catalyzed by methionine adenosyl transferases (MATs). In humans, three MAT isoforms exist: MATI, MATII, and MATIII. The isoforms MATI and MATIII are liver-specific isoforms, while MATII is expressed in various tissues (Murin et al. 2017). Due to its structure and composition, MATII is the only isoform that is susceptible to inhibition by SAM (Halim et al. 1999). Deregulation of MAT expression has been reported in different types of cancers. For example, an isozyme switch from MATI/III to MATII occurs in hepatocellular carcinomas and bile duct cancer (cholangiocarcinoma), and contributes to the depletion of SAM which results in genome-wide histone hypomethylation, with subsequent activation of oncogenic pathways (Murin et al. 2017). It is hypothesized that this isozyme switch is induced by a reduction in SAM levels. Because of this reduction, the normally hypermethylated *mat2a* promoter, which encodes the catalytic subunit of MATII, becomes hypomethylated during the development of hepatocellular carcinoma, occasioning its upregulation and a further decrease in the SAM levels (Yang et al. 2001).

Under normal conditions, MATII participates in the methylation of specific genes through its SAM-integrating transcription (SAMIT) regulatory module. Thus, MATII physically interacts with methyltransferases and transcription factors at specific chromatin *loci*, providing a direct supply of SAM for histone methylation

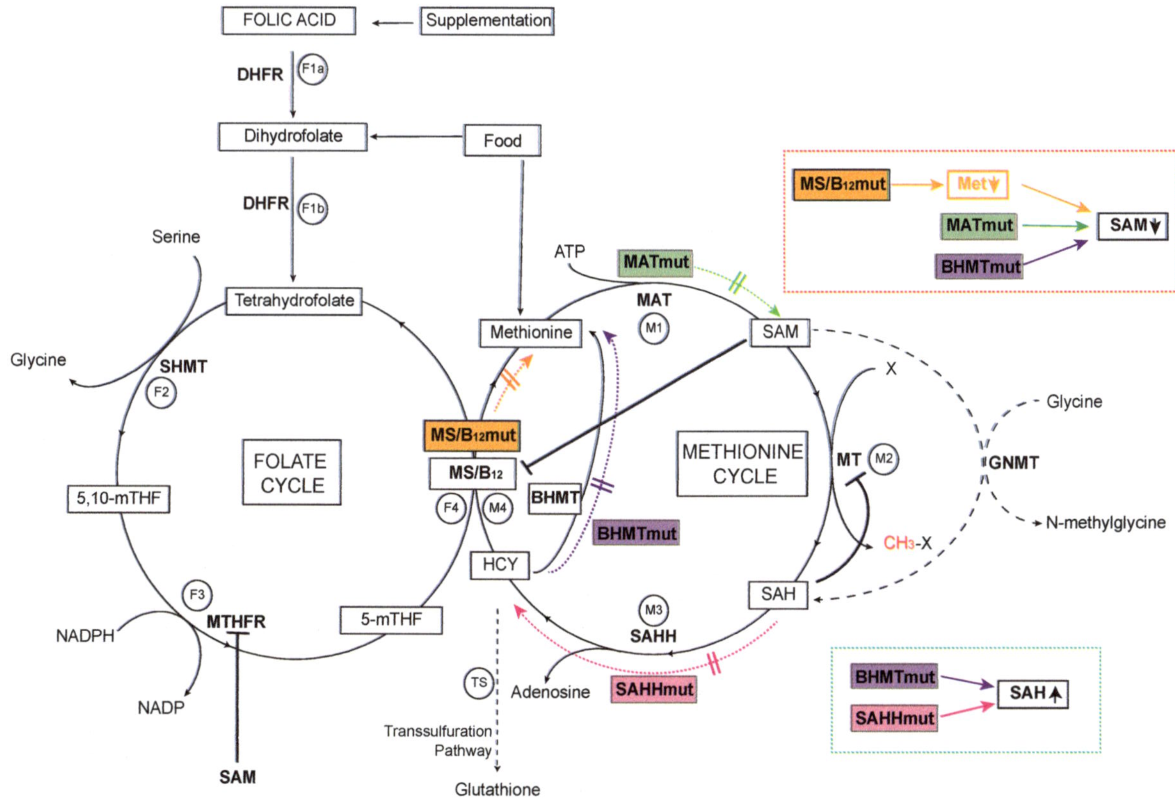


**Fig. 1** One-carbon metabolic pathway. Representative scheme showing the association of the methionine and the folate cycles to synthesize SAM. We numbered the steps of each cycle respectively M1 to M4 for methionine and F1a to F4 for folate. The common point between the two cycles is the re-methylation of homocysteine (Step F4/M4). Abbreviations: *SAM S*-adenosylmethionine, *SAH S*-adenosylhomocysteine, *HCY* homocysteine, *MAT* methionine adenosyltransferase, *SAHH* SAH-hydrolase, *MS* methionine synthase, *B12* vitamin B12, *5,10-mTHF* 5,10-methylene tetrahydrofolate, *5-mTHF* 5-methylene tetrahydrofolate, *DHRF* dihydrofolate reductase, *MTHFR* methylenetetrahydrofolate reductase, *SHMT* serine hydroxymethyltransferase, *ATP* adenosine triphosphate, *NADPH* nicotinamide adenine dinucleotide phosphate. The suffix "mut" inside boxes indicates a mutated version of the corresponding enzyme

(Igarashi and Katoh 2013). For example, the repressive mark H3K9me3 at the COX-2 locus is mediated by SetDB1, and requires expression of the catalytic subunit of MATII. When MATII is silenced, the repressive methylation on COX-2 is absent, the oncogene is upregulated, and carcinogenesis is promoted (Kera et al. 2013).

Another important enzyme of one-carbon metabolism is SAH-hydrolase (SAHH), which catalyzes the hydrolysis of SAH to give HCY and adenosine. In nonalcoholic steatohepatitis (NASH) and nonalcoholic fatty liver disease (NAFLD), the SAHH gene is silenced by DNA hypermethylation (H3K27me3) and/or deacetylation of H4K16. This leads to an accumulation of SAH that not only induces global chromatin hypomethylation, but also deregulates processes such as the trans-sulfuration and transmethylation pathways, thus affecting the redox state of the cell and favoring the development of disease (Pogribny et al. 2018). Despite this, the accumulation of SAH due to the action of SAHH inhibitor, such as neplanocin and 3-deazaneplanocin in mammary adenocarcinoma had resulted in a global decrease in levels of H3K79me2 that are established by the SAM-dependent methyltransferase DOT1; and, ultimately, in a reduction in cancer cell proliferation (Zhang et al. 2014).

Other types of cancer such as adenocarcinoma and squamous cell carcinoma present increased SAM availability through increased one-carbon metabolism. This phenomenon occurs via upregulation of the methionine transporters LAT1 and LAT4, and by redirection of some of the glycolytic intermediates to the serine-glycine biosynthesis pathway. This pathway supports the folate cycle, which in turn leads to aberrant histone methylation (Wong et al. 2017).

Importantly, as well as deregulation of enzymes involved in one-carbon metabolism, environmental factors can also affect the levels of available SAM and, therefore, histone methylation. For example, mice who were fed a diet deficient in choline-methyl showed reduced hepatic H3K9me3 and H4K20me3 due to impairment of the folate and methionine cycles which decreases the SAM/SAH ratio (Pogribny et al. 2012). Similarly, chronic alcohol consumption also leads to SAM depletion, mainly because the metabolism of ethanol induces high oxidative stress in the cells due to increased levels of ROS (Albano 2006). Constant consumption of ethanol results depletion of glutathione (GSH) which is one of the main systems of ROS detoxification, especially in the brain (Mytilineou et al. 2002). Since GSH is synthesized through the trans-sulfuration pathway using HCY as substrate, depletion leads to concomitant depletion of HCY, methionine, and SAM (Fig. 2). Indeed, the amount of SAM is reduced in alcoholic liver disease models, while the amount of SAH is increased and the GSH/GSH disulfide ratio is reduced (Halsted et al. 2002). This ROS-mediated depletion of SAM ultimately leads to global DNA and histone hypomethylation, as well as deregulation of other histone PTMs, including histone acetylation and ubiquitination (Jangra et al. 2016). Importantly, chronic alcohol consumption also affects folate metabolism, reducing uptake and favoring excretion (Medici and Halsted 2013). Through these mechanisms, alcohol induces epigenetic changes that are important for the progression of various cancers including esophageal, hepatic, and colorectal cancers (Dumitrescu 2018).

### 2.3 Deregulation of Enzymes Outside of One-Carbon Metabolism that Affect the SAM/SAH Ratio and Histone Methylation

Other deregulations that affect the SAM/SAH ratio and are observed in cancers include those involving nicotinamide *N*-methyltransferase (NNMT). This enzyme catalyzes the methylation of nicotinamide, consuming the cellular pool of SAM, and has been seen to cause a decrease of up to 50% in the SAM/SAH ratio in certain types of cancers such as liver, kidney, colon, lung, and bladder cancer. This is associated with a significant, genome-wide decrease in histone methylation at H3K4, H3K9, H3K27, and H4K20, resulting in a phenotype which is considered more pluripotent and can, therefore, increase cancer aggressiveness (Ulanovskaya et al. 2013).

As mentioned previously, an abnormal increase in the SAM/SAH ratio can promote carcinogenesis. Glycine *N*-methyltransferase (GNMT) catalyzes glycine methylation using SAM, and it has been suggested that the only purpose of this enzyme is to maintain SAM levels in normal conditions (Martínez-Chantar et al. 2008). Inactivating mutations of this enzyme have been demonstrated to induce a 40-fold increase in SAM levels, leading to enrichment of H3K27me3 on the promoters of tumor suppressor genes such as RASSF1 and SOCS2, causing transcriptional silencing and subsequent activation of oncogenic pathways. This mechanism is particularly common in cells of steatosis and hepatocellular carcinoma (Martínez-Chantar et al. 2008; Luka et al. 2009).

## 3 Metabolites Involved in the Activity of Demethylases: Flavin Adenine Dinucleotide, α-Ketoglutarate, Succinate, and Fumarate

Sugars, fatty acids, and most amino acids are oxidized to $CO_2$ and $H_2O$ via the respiratory chain and the Krebs cycle, also known as acid citric cycle or tricarboxylic acid (TCA) cycle (Fig. 3). Interestingly, except for Fe(II), all the cofactors required by demethylases have a role in the Krebs cycle as intermediaries or products, linking energy metabolism to gene regulation (Nieborak and Schneider 2018). For example, the histone demethylase LSD1 contains a flavin adenine dinucleotide (FAD)-dependent amine oxidase domain (Black et al. 2012), whose activity is dependent on FAD levels. One family of enzymes that are particularly sensitive to the products of the Krebs cycle are the α-ketoglutarate-dependent dioxygenases, especially the JmjC histone demethylases (Black et al. 2012). These enzymes require α-ketoglutarate, $O_2$, and Fe(II) to function; and are inhibited by succinate, fumarate, and 2-hydroxyglutarate (Fig. 3, red dotted box). In this section, we will review how mutations of enzymes involved in the Krebs cycle affect α-ketoglutarate-dependent histone demethylases.

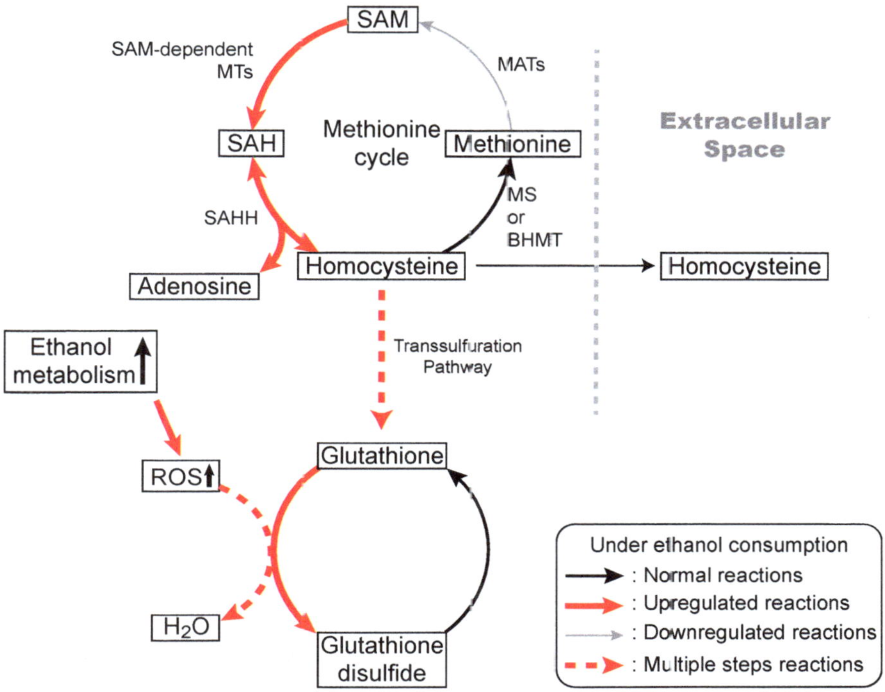


**Fig. 2** Constant alcohol consumption leads to an imbalance in the methionine cycle. ROS generation, caused by ethanol metabolism, leads to an increase in the cellular demand in Glutathione to react to the oxidative stress. Depletion of Glutathione leads to an engagement of homocysteine into the transsulfuration pathway, which in turn, diminishes homocysteine re-methylation and promotes depletion of SAM. Red arrows indicate reactions that are stimulated under alcohol consumption

## *3.1 Oncometabolites Arising from Deregulations in the Krebs Cycle*

### 3.1.1 Accumulation of 2-Hydroxyglutarate: The Prominent Case of Isocitrate Dehydrogenase Mutations in Glioblastoma Multiforme

Intracellular accumulation of 2-hydroxyglutarate is a concern for several reasons. Among them, its activity as a competitive inhibitor of α-ketoglutarate-dependent dioxygenases impedes normal histone demethylation and, therefore, induces chromatin hypermethylation (Xu et al. 2011; Yang et al. 2012).

Glioblastoma multiforme (GBM) is one of the most common adult malignant gliomas, accounting for more than 50% of glioma cases (Alifieris and Trafalis 2015), and representing the most aggressive type of primary brain tumor in humans (Stafford et al. 2016). Due to its aggressiveness and its rapid recurrence following treatment (Stafford et al. 2016), patients have a median survival time of 15 months after diagnosis (Lacroix et al. 2001; Martinez et al. 2010). Among the mutations that have been identified in GBM patients, those occurring in isocitrate dehydrogenase (IDH) genes have caught the attention of researchers (Chesnelong 2015). In 2008, a genome wide analysis identified mutations in the active site of IDH1—specifically, at arginine 132 (R132)—in about 12% of the analyzed samples (Parsons et al. 2008). One year later, the same group identified mutations in the IDH2 gene at codon 172, which encodes an arginine residue analogous to R132 (Yan et al. 2009). Mutations in the IDH1 and IDH2 genes are mutually exclusive and heterozygous, with mutations of IDH1 being more commonly observed (Parsons et al. 2008; Yan et al. 2009). Notably, IDH mutations occur not only in GBM, but also in acute myeloid leukemia, in which they are associated with a worse prognosis (Abbas et al. 2010; Paschka et al. 2010).

The amino acids R132 and R172 in IDH1 and IDH2, respectively, form hydrogen bonds with the isocitrate substrate, suggesting that these mutations affect the catalytic activity of the enzymes (Xu et al. 2004). Indeed, mutation of R132 of IDH1 or R172 of IDH2 result in a loss of the canonical function (Guerra et al. 2009), but confer the ability to reduce α-ketoglutarate to 2-hydroxyglutarate (Dang et al. 2009; Ward et al. 2010). Today, 2-hydroxyglutarate is considered an "oncometabolite", and has been reported to be accumulated in glioma samples that harbor IDH mutations (Xu et al. 2011).

The pathway for intracellular accumulation of 2-hydroxyglutarate is illustrated in Fig. 3 (red dotted box). In normal conditions, isocitrate is converted to α-ketoglutarate by IDHs. This metabolite can then either continue into the Krebs cycle or function as a cofactor for the α-ketoglutarate-dependent dioxygenases of the JmjC family. Glioblastoma cells that have mutations in only one allele of IDH1/2 contain a functional copy of IDH, which acts to maintain the supply of α-ketoglutarate, while the mutated allele converts this continuous supply into 2-hydroxyglutarate (Chesnelong 2015). In this way, IDH mutations lead to major epigenetic deregulations, changing the transcriptional program at a genome-wide scale, with notable effects on tumor suppressors, oncogenes, pro-differentiation genes, DNA repair, and metabolic genes (Chesnelong 2015).

The "hypermethylator" phenotype of IDH-mutant gliomas is associated with genome-wide hypermethylation of CCCTC-binding factor (CTCF)-binding sites, which inhibits the binding of this insulator protein and disrupts the proper establishment of boundary elements that partition topological domains of chromatin. This additional deregulation leads to aberrant upregulation of the canonical glioma oncogene, platelet-derived growth factor receptor A (PDGFRA) (Flavahan et al. 2016).

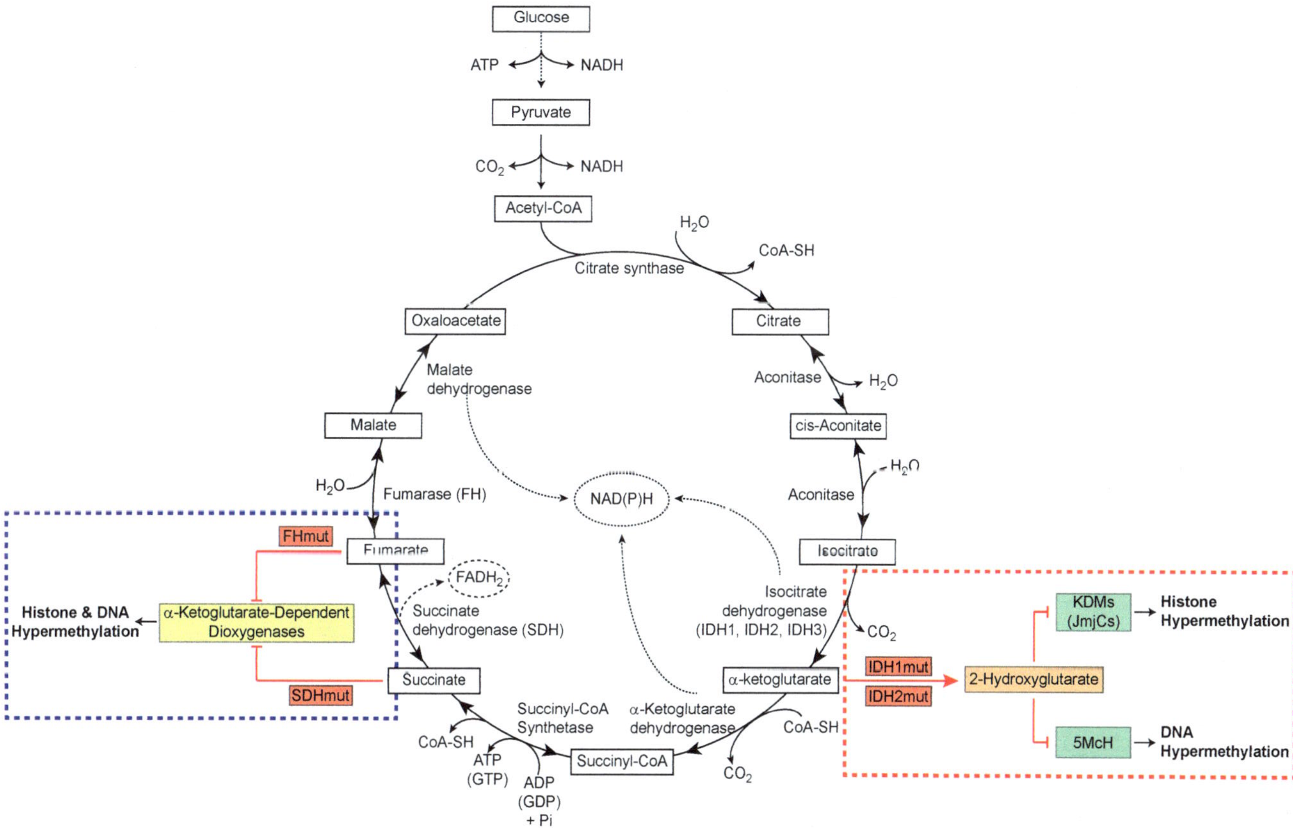


**Fig. 3** Krebs Cycle. Representative scheme of the reactions involved including the steps with two arrowheads for reversible reactions, while those with a single arrowhead indicate irreversibility. Mutations in the enzymes isocitrate dehydrogenase (IDH), fumarase (FH), and succinate dehydrogenase (SDH) affect

Knowing that mutated IDHs could be potential targets for the treatment of glioblastomas and other cancers, multiple clinical trials have focused on different IDH inhibitors such as AG-120, AG-221, or AG-881 to inhibit IDH1, IDH2, or both, respectively. Moreover, other clinical trials have been carried out to study molecules that exploit the metabolic sensitivity of IDH mutated gliomas, such as metformin, or molecules that can revert the hypermethylation of transformed cells (Han and Batchelor 2017). Table 4 includes a summary of 20 ongoing clinical trials of different IDHs inhibitors, demethylating agents, and/or metabolic modulators in different types of cancer.

### 3.1.2 Succinate and Fumarate, Oncometabolites that Promote Histone Hypermethylation

In addition to IDH1 and IDH2, mutations of the *fh, sdha, sdhb, sdhc, sdhd,* and *sdhaf2* genes, which encode subunits of fumarase (FH) and succinate dehydrogenase (SDH) complexes, are also found in some cancers such as paragangliomas, renal cell carcinomas, pheochromocytoma, and gastrointestinal stromal tumors (Toro et al. 2003; Bayley et al. 2008; Hao et al. 2009; Kaelin 2009; Bardella et al. 2011; Gill 2018; Matsumoto et al. 2018). Such mutations correspond to a loss of function, and therefore cause accumulation of fumarate and succinate (Pollard et al. 2005), which disrupts the histone and DNA methylation patterns through inhibition of α-ketoglutarate dependent dioxygenases, in a similar way to 2-hydroxyglutarate (Xiao et al. 2012) (Fig. 3, blue dotted box). Indeed, they are also considered oncometabolites. Mutations of SDHs in samples of paragangliomas and mouse models have been observed to produce a "hypermethylator" phenotype, with global increases in the histone methylation marks H3K9me3, H3K27me2, and H3K27me3. These modifications induce transcriptional changes and cell migration (Letouzé et al. 2013).

In summary, several mutations in various enzymes of the Krebs cycle are involved in cancer development through the accumulation of certain metabolites and intermediaries of the cycle, which in turn promote changes in the epigenetic landscape. This knowledge has informed the design of new strategies to combat these diseases and opened new opportunities which are already being explored.

---

**Fig. 3** (continued) processes highlighted by dotted boxes. Abbreviations: *CoA* co-enzyme A, *ATP* adenosine triphosphate, *ADP* adenosine diphosphate, *GTP* guanosine triphosphate, *GDP* guanosine diphosphate, *NADPH* nicotinamide adenine dinucleotide phosphate, *NADH* nicotinamide adenine dinucleotide, *FADH2* flavin adenine dinucleotide, *IDH* isocitrate dehydrogenase, *FH* fumarase, *SDH* succinate dehydrogenase, *KDMs* histone lysine demethylases, *5McH* 5-hydroxymethylcytosine. Adapted from Nelson et al. (2017)

**Table 4** Clinical trials focused on IDHs inhibitors as therapy for different types of cancer

| Title of the trial | Targeted conditions | Evaluated drugs and therapies | NTC identifier at ClinicalTrails.gov |
|---|---|---|---|
| A Study of FT 2102 in Participants with Advanced Solid Tumors and Gliomas with an IDH1 Mutation | Cohort 1a and 1b: Glioma, cohort 1a and 1b: Glioblastoma Multiforme, cohort 2a and 2b: Hepatobiliary tumors (hepatocellular carcinoma, bile duct carcinoma, intrahepatic cholangiocarcinoma, other hepatobiliary carcinomas), cohort 3a and 3b: Chondrosarcoma, cohort 4a and 4b: Intrahepatic cholangiocarcinoma, cohort 5a: Other solid tumors with IDH1 mutations | FT-2102 (IDH1 inhibitor), Azacytidine (DNA demethylating agent), Nivolumab (monoclonal antibody against PD-1), gemcitabine and cisplatin (standard chemotherapy drugs) | NCT03684811 |
| Treatment with Azacytidine of recurrent gliomas with IDH1/2 mutation | Recurrent IDH1/2 mutated glioma | Azacytidine (DNA demethylating agent) | NCT03666559 |
| IDH1 inhibition using Iopidine as maintenance therapy for IDH1-mutant myeloid neoplasms following allogeneic stem cell transplantation | IDH1 mutation myeloid neoplasms | AG-120 (also known as Ivosidenib, IDH1 inhibitor) | NCT03564821 |
| CB-839 with radiation therapy and Temozolomide in treating participants with IDH-mutated diffuse astrocytoma or anaplastic astrocytoma | Anaplastic astrocytoma with mutant IDH, diffuse astrocytoma with mutant IDH | CB-839 hydrochloride (Glutaminase inhibitor), radiation, Temozolomide (alkylating agent, standard chemotherapy drug) | NCT03528642 |
| IDH1 (AG 120) inhibitor in patients with IDH1 mutated myelodysplastic syndrome | Myelodysplastic syndromes, acute myeloid leukemia | AG-120 (also known as Ivosidenib, IDH1 inhibitor) | NCT03503409 |
| Study of Venetoclax with the mIDH1 inhibitor Ivosidenib (AG120) in IDH1-mutated hematologic malignancies | Other diseases of blood and blood-forming organs, advanced hematologic malignancies, acute myeloid leukemia | AG-120 (also known as Ivosidenib, inhibitor of IDH1), Venetoclax (inhibitor of Bcl-2) | NCT03471260 |

(continued)

**Table 4** (continued)

| Title of the trial | Targeted conditions | Evaluated drugs and therapies | NTC identifier at ClinicalTrails.gov |
|---|---|---|---|
| Study of AG-120 and AG-881 in subjects with low grade glioma | Glioma | AG-120 (also known as Ivosidenib, IDH1 inhibitor), AG881 (pan-mutant IDH inhibitor) | NCT03343197 |
| Study of AG-120 (Ivosidenib) vs. placebo in combination with Azacytidine in patients with previously untreated acute myeloid leukemia with an IDH1 mutation | Newly diagnosed acute myeloid leukemia, untreated acute myeloid leukemia, acute myeloid leukemia arising from myelodysplastic syndrome | AG-120 (also known as Ivosidenib, IDH1 inhibitor), Azacytidine (DNA demethylating agent) | NCT03173248 |
| BAY1436032 in patients with mutant IDH1(mIDH1) advanced acute myeloid leukemia (AML) | Acute myeloid leukemia | BAY1436032 (pan-mutant IDH1 inhibitor) | NCT03127735 |
| Study of DS-1001b in patients with gene IDH1-mutated gliomas | Glioma | DS-1001b (inhibitor of certain mutant forms of IDH1) | NCT03030066 |
| Study of AG-120 in previously treated advanced cholangiocarcinoma with IDH1 mutations (ClarIDHy) | Advanced cholangiocarcinoma, metastatic cholangiocarcinoma | AG-120 (also known as Ivosidenib, IDH1 inhibitor) | NCT02989857 |
| Phase I Study of BAY 1436032 in Patients with IDH1-mutant Solid Tumors | Neoplasms | BAY1436032 (pan-mutant IDH1 inhibitor) | NCT02746081 |
| A safety and efficacy study of Oral AG-120 plus subcutaneous Azacytidine and Oral AG-221 plus subcutaneous Azacytidine in subjects with newly diagnosed acute myeloid leukemia (AML) | Acute myeloid leukemia | AG-120 (also known as Ivosidenib, IDH1 inhibitor), Azacytidine (DNA demethylating agent), AG-221 (mutant IDH2 inhibitor) | NCT02677922 |
| Safety study of AG-120 or AG-221 in combination with induction and consolidation therapy in patients with newly diagnosed acute | Newly diagnosed acute myeloid leukemia, untreated acute myeloid leukemia, acute myeloid leukemia arising from myelodysplastic | AG-120 (also known as Ivosidenib, IDH1 inhibitor), AG-221 (mutant IDH2 inhibitor), Cytarabine, Daunorubicin, Idarubicin, | NCT02632708 |

(continued)

**Table 4** (continued)

| Title of the trial | Targeted conditions | Evaluated drugs and therapies | NTC identifier at ClinicalTrails.gov |
|---|---|---|---|
| myeloid leukemia with an IDH1 and/or IDH2 mutation | syndrome, acute myeloid leukemia arising from antecedent hematologic disorder, acute myeloid leukemia arising after exposure to genotoxic injury | Mitoxantrone, etoposide (standard chemotherapy drugs) | |
| Metformin and chloroquine in IDH1/2-mutated solid tumors | Glioma, cholangiocarcinoma, chondrosarcoma | Metformin (regulator of glucose production in liver and sensitivity to insulin), chloroquine (autophagy inhibitor) | NCT02496741 |
| Study of orally administered AG-881 in patients with advanced hematologic malignancies with an IDH1 and/or IDH2 mutation | Acute myeloid leukemia, myelodysplastic syndrome, hematologic malignancies | AG881 (pan-mutant IDH inhibitor) | NCT02492737 |
| Study of orally administered AG-881 in patients with advanced solid tumors, including gliomas, with an IDH1 and/or IDH2 mutation | Glioma | AG881 (pan-mutant IDH inhibitor) | NCT02481154 |
| Study of orally administered AG-120 in subjects with advanced hematologic malignancies with an IDH1 mutation | Relapsed or refractory acute myeloid leukemia, untreated acute myeloid leukemia, other IDH1-mutated positive hematologic malignancies | AG-120 (also known as Ivosidenib, IDH1 inhibitor) | NCT02074839 |
| Study of orally administered AG-120 in subjects with advanced solid tumors, including glioma, with an IDH1 mutation | Cholangiocarcinoma, chondrosarcoma, glioma, other advanced solid tumors | AG-120 (also known as Ivosidenib, IDH1 inhibitor) | NCT02073994 |
| Study of the Glutaminase inhibitor CB-839 in solid tumors | Solid tumors, triple-negative breast Cancer, non-small cell lung Cancer, renal cell carcinoma, mesothelioma, fumarate hydratase deficient tumors, succinate dehydrogenase deficient gastrointestinal stromal tumors, succinate | CB-839 (Glutaminase inhibitor), paclitaxel, Everolimus, Erlotinib, docetaxel, Cabozantinib (standard chemotherapy drugs) | NCT02071862 |

(continued)

**Table 4** (continued)

| Title of the trial | Targeted conditions | Evaluated drugs and therapies | NTC identifier at ClinicalTrails. gov |
|---|---|---|---|
| | dehydrogenase deficient non-gastrointestinal stromal tumors, tumors harboring IDH1 and IDH2 mutations, tumors harboring amplifications in the c-Myc gene | | |

In the table, 20 trials retrieved from ClinicalTrials.gov at the moment of writing the manuscript. Each trial may be evaluating the IDH inhibitor in addition to DNA demethylating agents and/or metabolic modulators, plus chemotherapy drugs, as indicated

# 4 Final Thoughts on the Topic: Modulation of Metabolism as a Tool to Fight Disease

Methionine and folate, unlike α-ketoglutarate, fumarate, and succinate; cannot be synthesized by humans, meaning that appropriate supplementation is important. As we have discussed, the epigenetic information mediated by histone methylation is highly dependent on an appropriate SAM/SAH ratio. This is, in turn, completely dependent on the availability of methionine and folate, as well as the appropriate functioning of the cycles in which these molecules participate. This is important because epigenetic deregulation can lead to carcinogenesis, but also because epigenetic information is a major influence on embryonic development. Although we did not discuss this aspect here, it is well known that folate is essential for neural tube development and for pregnancy health in general (Greenberg et al. 2011; Viswanathan et al. 2017). Since 2007, the World Health Organization has recommended that pregnant women should take a folic acid supplement of 400 μg daily, from conception until at least 12 weeks of gestation.

Considering the metabolites generated by the Krebs cycle, it is important to study the generation of excess fumarate and succinate due to mutations in the genes *fh*, *sdha*, *sdhb*, *sdhc*, *sdhd,* and *sdhaf2*, and to develop strategies to normalize the levels of these metabolites. Furthermore, the development of drugs that inhibit mutants of IDHs with increased 2-hydroxyglutarate synthesis activity is crucial, as these mutants have key roles in the development of certain types of cancer, particularly those associated with the brain.

Finally, it must be noted that this review focuses on the modulation of histone methylation by specific metabolites, emphasizing the deregulations observed in cancer. However, the contribution of metabolic processes to epigenetic mechanisms and the role of this in human health is beyond the scope of this paper. For example, acetyl-coA metabolism influences histone modifications beyond acetylation. Eight additional types of "acylations" have been recently described, each one with a different effect on gene expression (Sabari et al. 2017). Thus, it is essential that research into diseases in which gene regulation plays a role must also consider the influence of nutrition, gene mutations, and changes in the affinity of metabolic enzymes as well as other potentially related factors such as epigenetic silencing or derepression of genes.

**Funding** This research was funded by grants from CONICYT: Basal Project AFB 170004 (A.L.), FONDECYT 1160480 (A.L.), PCHA/Doctorado Nacional/2014-21140324 (F.A.), PCHA/Doctorado Nacional/2014-21140346 (F.S.). E.B. and G.A. were funded by La Ligue Nationale contre le cancer (Equipe labellisée Ligue), ANR-11-LABX-0044_DEEP and ANR-10-IDEX-0001-02 PSL, ANR-12-BSV5-0022-02 "CHAPINHIP", ANR-14-CE16-0009 "Epicure", ANR-14-CE10-0013 "CELLECTCHIP", ERC-2015-ADG-694694 "ChromADICT", ANR-16-CE15-0018 "CHRODYT", ANR-16-CE12-0024 "CHIFT", ANR-16-CE11-0028 "REPLICAF".

## References

Abbas S, Lugthart S, Kavelaars FG et al (2010) Acquired mutations in the genes encoding IDH1 and IDH2 both are recurrent aberrations in acute myeloid leukemia: prevalence and prognostic value. Blood 116:2122–2126

Alam H, Gu B, Lee MG (2015) Histone methylation modifiers in cellular signaling pathways. Cell Mol Life Sci 72:4577–4592

Albano E (2006) Alcohol, oxidative stress and free radical damage. Proc Nutr Soc 65:278–290

Alifieris C, Trafalis DT (2015) *Glioblastoma multiforme*: pathogenesis and treatment. Pharmacol Therap 152:63–82

Allfrey VG, Faulkner R, Mirsky AE (1964) Acetylation and methylation of histones and their possible role in the regulation of RNA synthesis. Proc Natl Acad Sci USA 51:786–794

Allis CD, Berger SL, Cote J, Dent S, Jenuwien T, Kouzarides T, Pillus L, Reinberg D, Shi Y, Shiekhattar R, Shilatifard A, Workman J, Zhang Y (2007) New nomenclature for chromatin-modifying enzymes. Cell 131(4):633–636

Alvarez F, Munoz F, Schilcher P et al (2011) Sequential establishment of marks on soluble histones H3 and H4. J Biol Chem 286:17714–17721

Bardella C, Pollard PJ, Tomlinson I (2011) SDH mutations in cancer. Biochim Biophys Acta 1807:1432–1443

Bayley JP, Launonen V, Tomlinson IPM (2008) The FH mutation database: an online database of fumarate hydratase mutations involved in the MCUL (HLRCC) tumor syndrome and congenital fumarase deficiency. BMC Med Genet 9(1):–9

Bieberstein NI, Kozakova E, Huranova M et al (2016) TALE-directed local modulation of H3K9 methylation shapes exon recognition. Sci Rep 6:29961

Biggar KK, Li SS (2015) Non-histone protein methylation as a regulator of cellular signalling and function. Nat Rev Mol Cell Biol 16:5–17

Black JC, Van Rechem C, Whetstine JR (2012) Histone lysine methylation dynamics: establishment, regulation, and biological impact. Mol Cell 48:491–507

Blanc RS, Richard S (2017) Arginine methylation: the coming of age. Mol Cell 65:8–24

Bowman GD, Poirier MG (2015) Post-translational modifications of histones that influence nucleosome dynamics. Chem Rev 115:2274–2295

Canzio D, Chang E, Shankar S et al (2011) Chromodomain-mediated oligomerization of HP1 suggests a nucleosome-bridging mechanism for heterochromatin assembly. Mol Cell 41:67–81

Caudil MA, Wang JC, Melnyk S et al (2001) Intracellular S-adenosylhomocysteine concentrations predict global DNA hypomethylation in tissues of methyl-deficient cystathionine beta-synthase heterozygous mice. J Nutr 131:2811–2818

Chang B, Chen Y, Zhao Y et al (2007) JMJD6 is a histone arginine demethylase. Science 318:444–447

Chesnelong C (2015) Isocitrate dehydrogenase (IDH). Mut Gliomas 2:441–458

Chuikov S, Kurash JK, Wilson JR et al (2004) Regulation of p53 activity through lysine methylation. Nature 432:353–360

Clarke SG (2013) Protein methylation at the surface and buried deep: thinking outside the histone box. Trends Biochem Sci 38:243–252

Dang L, White DW, Gross S et al (2009) Cancer-associated IDH1 mutations produce 2-hydroxyglutarate. Nature 462:739–744

Dha SS, Lee SH, Kan PY et al (2012) Trans-tail regulation of MLL4-catalyzed H3K4 methylation by H4R3 symmetric dimethylation is mediated by a tandem PHD of MLL4. Genes Dev 26:2749–2762

Di Lorenzo A, Bedford MT (2011) Histone arginine methylation. FEBS Lett 585:2024–2031

Dumitrescu RG (2018) Alcohol-induced epigenetic changes in cancer. Methods Mol Biol (Clifton, N.J.) 1856:157–172

Fang R, Barbera AJ, Xu Y et al (2010) Human LSD2/KDM1b/AOF1 regulates gene transcription by modulating intragenic H3K4me2 methylation. Mol Cell 39:222–233

Flavahan WA, Drier Y, Liau BB et al (2016) Insulator dysfunction and oncogene activation in IDH mutant gliomas. Nature 529:110–114

Frederiks F, Tzouros M, Oudgenoeg G et al (2008) Nonprocessive methylation by Dot1 leads to functional redundancy of histone H3K79 methylation states. Nat Struct Mol Biol 15:550–557

Gayatri S, Bedford MT (2014) Readers of histone methylarginine marks. Biochim Biophys Acta 1839:702–710

Gill AJ (2018) Succinate dehydrogenase (SDH)-deficient neoplasia. Histopathology 72:106–116

Greenberg JA, Bell SJ, Guan Y et al (2011) Folic acid supplementation and pregnancy: more than just neural tube defect prevention. Rev Obstet Gynecol 4:52–59

Greer EL, Shi Y (2012) Histone methylation: a dynamic mark in health, disease and inheritance. Nat Rev Genet 13:343–357

Grillo MA, Colombatto S (2008) S-adenosylmethionine and its products. Amino Acids 34:187–193

Guerra CA, Noor AM, Myint HY et al (2009) Glioma-derived mutations in IDH1 dominantly inhibit IDH1 catalytic activity and induce HIF-1alpha. Science 324:261–265

Gurard-Levin ZA, Almouzni G (2014) Histone modifications and a choice of variant: a language that helps the genome express itself. F1000Prime Rep 6:76

Gutierrez RM, Hnilica LS (1967) Tissue specificity of histone phosphorylation. Science 157:1324–1325

Halim AB, LeGros L, Geller A et al (1999) Expression and functional interaction of the catalytic and regulatory subunits of human methionine adenosyltransferase in mammalian cells. J Biol Chem 274:29720–29725

Halsted CH, Villanueva JA, Devlin AM et al (2002) Metabolic interactions of alcohol and folate. J Nutr 132:2382–2387

Han CH, Batchelor TT (2017) Isocitrate dehydrogenase mutation as a therapeutic target in gliomas. Chin Clin Oncol 6:33–33

Hao HX, Khalimonchuk O, Schraders M et al (2009) SDH5, a gene required for flavination of succinate dehydrogenase, is mutated in paraganglioma. Science 325:1139–1142

Herz HM, Garruss A, Shilatifard A (2013) SET for life: biochemical activities and biological functions of SET domain-containing proteins. Trends Biochem Sci 38:621–639

Huang J, Perez-Burgos L, Placek BJ et al (2006) Repression of p53 activity by Smyd2-mediated methylation. Nature 444:629–632

Huang J, Dorsey J, Chuikov S et al (2010) G9a and Glp methylate lysine 373 in the tumor suppressor p53. J Biol Chem 285:9636–9641

Hug CB, Vaquerizas JM (2018) The birth of the 3D genome during early embryonic development. Trends Genet 34:903–914

Hyllus D, Stein C, Schnabel K et al (2007) PRMT6-mediated methylation of R2 in histone H3 antagonizes H3 K4 trimethylation. Genes Dev 21:3369–3380
Igarashi K, Katoh Y (2013) Metabolic aspects of epigenome: coupling of S-adenosylmethionine synthesis and gene regulation on chromatin by SAMIT module. Subcell Biochem 61:105–118
Jahan S, Davie JR (2015) Protein arginine methyltransferases (PRMTs): role in chromatin organization. Adv Biol Regul 57:173–184
Jangra A, Sriram CS, Pandey S et al (2016) Epigenetic modifications, alcoholic brain and potential drug targets. Ann Neurosci 23:246–260
Jansson M, Durant ST, Cho EC et al (2008) Arginine methylation regulates the p53 response. Nat Cell Biol 10:1431–1439
Jenuwein T, Allis CD (2001) Translating the histone code. Science 293:1074–1080
Jenuwein T, Laible G, Dorn R et al (1998) SET domain proteins modulate chromatin domains in eu- and heterochromatin. Cell Mol Life Sci 54:80–93
Jones PA (2012) Functions of DNA methylation: islands, start sites, gene bodies and beyond. Nat Rev Genet 13:484–492
Kaelin WG (2009) SDH5 mutations and familial paraganglioma: somewhere warburg is smiling. Cancer Cell 16:180–182
Kang JY, Kim JY, Kim KB et al (2018) KDM2B is a histone H3K79 demethylase and induces transcriptional repression via sirtuin-1-mediated chromatin silencing. FASEB J 32:5737–5750
Kera Y, Katoh Y, Ohta M et al (2013) Methionine adenosyltransferase II-dependent histone H3K9 methylation at the COX-2 gene locus. J Biol Chem 288:13592–13601
Klose RJ, Kallin EM, Zhang Y (2006) JmjC-domain-containing proteins and histone demethylation. Nat Rev Genet 7:715–727
Kolasinska-Zwierz P, Down T, Latorre I et al (2009) Differential chromatin marking of introns and expressed exons by H3K36me3. Nat Genet 41:376–381
Kouzarides T (2007) Chromatin modifications and their function. Cell 128:693–705
Kwon SH, Workman JL (2008) The heterochromatin protein 1 (HP1) family: put away a bias toward HP1. Mol Cell 26:217–227
Lacroix M, Abi-Said D, Fourney DR et al (2001) A multivariate analysis of 416 patients with glioblastoma multiforme: prognosis, extent of resection, and survival. J Neurosurg 95:190–198
Lawrence MS, Stojanov P, Mermel CH et al (2014) Discovery and saturation analysis of cancer genes across 21 tumour types. Nature 505:495–501
Le Romancer M, Treilleux I, Leconte N et al (2008) Regulation of estrogen rapid signaling through arginine methylation by PRMT1. Mol Cell 31:212–221
Letouzé E, Martinelli C, Loriot C et al (2013) SDH mutations establish a hypermethylator phenotype in paraganglioma. Cancer Cell 23:739–752
Li S, Ali S, Duan X et al (2018) JMJD1B demethylates H4R3me2s and H3K9me2 to facilitate gene expression for development of hematopoietic stem and progenitor cells. Cell Rep 23:389–403
Loyola A, Almouzni G (2007) Marking histone H3 variants: how, when and why? Trends Biochem Sci 32:425–433
Luco RF, Pan Q, Tominaga K et al (2010) Regulation of alternative splicing by histone modifications. Science 327:996–1000
Luger K, Mader AW, Richmond RK et al (1997) Crystal structure of the nucleosome core particle at 2.8 A resolution. Nature 389 251–260
Luka Z, Mudd SH, Wagner C (2009) Glycine N-methyltransferase and regulation of S-adenosylmethionine levels. J Biol Chem 284:22507–22511
Luo C, Hajkova P, Ecker JR (2018) Dynamic DNA methylation: in the right place at the right time. Science 361:1336–1340
Machida S, Takizawa Y, Ishimaru M et al (2018) Structural basis of heterochromatin formation by human HP1. Mol Cell 69:385–397, e388
Martens-Lobenhoffer J, Bode-Boger SM, Clement B (2016) First detection and quantification of N (delta)-monomethylarginine, a structural isomer of N(G)-monomethylarginine, in humans using MS(3). Anal Biochem 493:14–20

Martinez R, Rohde V, Schackert G (2010) Different molecular patterns in glioblastoma multiforme subtypes upon recurrence. J Neuro-Oncol 96:321–329
Martínez-Chantar ML, Vázquez-Chantada M, Ariz U et al (2008) Loss of the glycine N-methyltransferase gene leads to steatosis and hepatocellular carcinoma in mice. Hepatology 47:1191–1199
Matsumoto K, Udaka N, Hasumi H et al (2018) Histopathological analysis of aggressive renal cell carcinoma harboring a unique germline mutation in fumarate hydratase. Pathol Int 68:473–478
McCabe MT, Mohammad HP, Barbash O et al (2017) Targeting histone methylation in cancer. Cancer J 23:292–301
Medici V, Halsted CH (2013) Folate, alcohol, and liver disease. Mol Nutr Food Res 57:596–606
Mendiratta S, Gatto A, Almouzni G (2018) Histone supply: multitiered regulation ensures chromatin dynamics throughout the cell cycle. J Cell Biol. https://doi.org/10.1083/jcb.201807179
Metivier R, Penot G, Hubner MR et al (2003) Estrogen receptor-alpha directs ordered, cyclical, and combinatorial recruitment of cofactors on a natural target promoter. Cell 115:751–763
Metzger E, Wissmann M, Yin N et al (2005) LSD1 demethylates repressive histone marks to promote androgen-receptor-dependent transcription. Nature 437:436–439
Moore KE, Carlson SM, Camp ND et al (2013) A general molecular affinity strategy for global detection and proteomic analysis of lysine methylation. Mol Cell 50:444–456
Morales Y, Caceres T, May K et al (2016) Biochemistry and regulation of the protein arginine methyltransferases (PRMTs). Arch Biochem Biophys 590:138–152
Mozzetta C, Boyarchuk E, Pontis J et al (2015) Sound of silence: the properties and functions of repressive Lys methyltransferases. Nat Rev Mol Cell Biol 16:499–513
Murin R, Vidomanova E, Kowtharapu BS et al (2017) Role of S-adenosylmethionine cycle in carcinogenesis. Gen Physiol Biophys 36:513–520
Murray K (1964) The occurrence of epsilon-N-methyl lysine in histones. Biochemistry 3:10–15
Mytilineou C, Kramer BC, Yabut JA (2002) Glutathione depletion and oxidative stress. Parkinsonism Relat Disord 8:385–387
Nelson DL, Cox MM, Lehninger AL (2017) The citric acid cycle (Chapter 16). In: Lehninger principles of biochemistry. W. H. Freeman, New York
Nguyen AT, Zhang Y (2011) The diverse functions of Dot1 and H3K79 methylation. Genes Dev 25:1345–1358
Nicholson TB, Veland N, Chen T (2015) Writers, readers, and erasers of epigenetic marks. Epigenetic cancer therapy. Elsevier, Amsterdam, pp 31–66
Nieborak A, Schneider R (2018) Metabolic intermediates - Cellular messengers talking to chromatin modifiers. Mol Metab. https://doi.org/10.1016/j.molmet.2018.01.007
Park SY, Park JW, Chun YS (2016) Jumonji histone demethylases as emerging therapeutic targets. Pharmacol Res 105:146–151
Parsons DW, Jones S, Zhang X et al (2008) An integrated genomic analysis of human glioblastoma multiforme. Science 321:1807–1812
Paschka P, Schlenk RF, Gaidzik VI et al (2010) IDH1 and IDH2 mutations are frequent genetic alterations in acute myeloid leukemia and confer adverse prognosis in cytogenetically normal acute myeloid leukemia with NPM1 mutation without FLT3 internal tandem duplication. J Clin Oncol 28:3636–3643
Pogribny IP, James SJ, Beland FA (2012) Molecular alterations in hepatocarcinogenesis induced by dietary methyl deficiency. Mol Nutr Food Res 56:116–125
Pogribny IP, Dreval K, Kindrat I et al (2018) Epigenetically mediated inhibition of S-adenosylhomocysteine hydrolase and the associated dysregulation of 1-carbon metabolism in nonalcoholic steatohepatitis and hepatocellular carcinoma. FASEB J 32:1591–1601
Pollard PJ, Brière JJ, Alam NA et al (2005) Accumulation of Krebs cycle intermediates and over-expression of HIF1$\alpha$ in tumours which result from germline FH and SDH mutations. Hum Mol Genet 14:2231–2239
Probst AV, Almouzni G (2011) Heterochromatin establishment in the context of genome-wide epigenetic reprogramming. Trends Genet 27:177–185

Rivera C, Gurard-Levin ZA, Almouzni G et al (2014) Histone lysine methylation and chromatin replication. Biochim Biophys Acta 1839:1433–1439
Rivera C, Saavedra F, Alvarez F et al (2015) Methylation of histone H3 lysine 9 occurs during translation. Nucleic Acids Res 43:9097–9106
Sabari BR, Zhang D, Allis CD et al (2017) Metabolic regulation of gene expression through histone acylations. Nat Rev Mol Cell Biol 18:90–101
Shi Y, Whetstine JR (2007) Dynamic regulation of histone lysine methylation by demethylases. Mol Cell 25:1–14
Shi Y, Lan F, Matson C et al (2004) Histone demethylation mediated by the nuclear amine oxidase homolog LSD1. Cell 119:941–953
Shi X, Kachirskaia I, Yamaguchi H et al (2007) Modulation of p53 function by SET8-mediated methylation at lysine 382. Mol Cell 27:636–646
Shlomi T, Rabinowitz JD (2013) Metabolism: cancer mistunes methylation. Nat Chem Biol 9:293–294
Sitbon D, Podsypanina K, Yadav T et al (2017) Shaping chromatin in the nucleus: the bricks and the architects. Cold Spring Harb Symp Quant Biol 82:1–14
Song Y, Wu F, Wu J (2016) Targeting histone methylation for cancer therapy: enzymes, inhibitors, biological activity and perspectives. J Hematol Oncol 9:49
Stafford JH, Hirai T, Deng L et al (2016) Colony stimulating factor 1 receptor inhibition delays recurrence of glioblastoma after radiation by altering myeloid cell recruitment and polarization. Neuro-Oncol 18:797–806
Suganuma T, Workman JL (2018) Chromatin and metabolism. Annu Rev Biochem 87:27–49
Teske KA, Hadden MK (2017) Methyllysine binding domains: structural insight and small molecule probe development. Eur J Med Chem 136:14–35
Tong Q, Cui G, Botuyan MV et al (2015) Structural plasticity of methyllysine recognition by the tandem tudor domain of 53BP1. Structure 23:312–321
Toro JR, Nickerson ML, Wei M-H et al (2003) Mutations in the fumarate hydratase gene cause hereditary leiomyomatosis and renal cell cancer in families in North America. Am J Hum Genet 73:95–106
Ulanovskaya OA, Zuhl AM, Cravatt BF (2013) NNMT promotes epigenetic remodeling in cancer by creating a metabolic methylation sink. Nat Chem Biol 9:300–306
Viswanathan M, Treiman KA, Kish-Doto J et al (2017) Folic acid supplementation for the prevention of neural tube defects: an updated evidence report and systematic review for the US preventive services task force. JAMA 317:190–203
Volkel P, Angrand PO (2007) The control of histone lysine methylation in epigenetic regulation. Biochimie 89:1–20
Wagner EJ, Carpenter PB (2012) Understanding the language of Lys36 methylation at histone H3. Nat Rev Mol Cell Biol 13:115–126
Walport LJ, Hopkinson RJ, Chowdhury R et al (2016) Arginine demethylation is catalysed by a subset of JmjC histone lysine demethylases. Nat Commun 7:11974
Ward PS, Patel J, Wise DR et al (2010) The common feature of leukemia-associated IDH1 and IDH2 mutations is a neomorphic enzyme activity converting α-Ketoglutarate to 2-hydroxyglutarate. Cancer Cell 17:225–234
Wen H, Li Y, Xi Y et al (2014) ZMYND11 links histone H3.3K36me3 to transcription elongation and tumour suppression. Nature 508:263–268
Wesche J, Kuhn S, Kessler BM et al (2017) Protein arginine methylation: a prominent modification and its demethylation. Cell Mol Life Sci 74:3305–3315
Wong CC, Qian Y, Yu J (2017) Interplay between epigenetics and metabolism in oncogenesis: mechanisms and therapeutic approaches. Oncogene 36:3359–3374
Xiao M, Yang H, Xu W et al (2012) Inhibition of α-KG-dependent histone and DNA demethylases by fumarate and succinate that are accumulated in mutations of FH and SDH tumor suppressors. Genes Dev 26:1326–1338

Xu X, Zhao J, Xu Z et al (2004) Structures of human cytosolic NADP-dependent isocitrate dehydrogenase reveal a novel self-regulatory mechanism of activity. J Biol Chem 279:33946–33957

Xu W, Yang H, Liu Y et al (2011) Oncometabolite 2-hydroxyglutarate is a competitive inhibitor of α-ketoglutarate-dependent dioxygenases. Cancer Cell 19:17–30

Yadav T, Quivy JP, Almouzni G (2018) Chromatin plasticity: a versatile landscape that underlies cell fate and identity. Science 361:1332–1336

Yan H, Parsons W, Jin G et al (2009) Mutations in gliomas. New Eng J Med 360:765–773

Yang H, Huang ZZ, Zeng Z et al (2001) Role of promoter methylation in increased methionine adenosyltransferase 2A expression in human liver cancer. Am J Physiol Gastrointest Liver Physiol 280:G184–G190

Yang H, Ye D, Guan KL et al (2012) IDH1 and IDH2 mutations in tumorigenesis: mechanistic insights and clinical perspectives. Clin Cancer Res 18:5562–5571

Yuan CC, Matthews AG, Jin Y et al (2012) Histone H3R2 symmetric dimethylation and histone H3K4 trimethylation are tightly correlated in eukaryotic genomes. Cell Rep 1:83–90

Zhang L, Deng L, Chen F et al (2014) Inhibition of histone H3K79 methylation selectively inhibits proliferation, self-renewal and metastatic potential of breast cancer. Oncotarget 5:10665–10677

Zhao Y, Garcia BA (2015) Comprehensive catalog of currently documented histone modifications. Cold Spring Harb Perspect Biol 7:a025064

# Histone Methylome of the Human Parasite *Schistosoma Mansoni*

**Ronaldo de Carvalho Augusto, Céline Cosseau, and Christoph Grunau**

## Contents



**Abstract** The trematode *Schistosoma mansoni* belongs to the group of digenetic parasites which need obligatory multiple hosts to develop. They transit between hosts as free-swimming stages in fresh water ecosystems. They generate phenotypically different developmental stages throughout their lifecycle and receive hugly heterogenous environmental cues. Each developmental stage is characterized by specific posttranslational histone modifications, in particular methylations. The combination of the different marks result in stage specific chromatin structure that is essential for development, sexual biology and pathogenesis. Histone methylation also responds to environmental changes and seems to be involved in an adaptive reponse or adjustment to the environment. Histone methylation thus represent promising source of therapeutic targets. In this chapter we will present the state-of-the-art of how the dynamics of histone methylation are involved in multiple factors of the schistosome's development, as well as what is still lacking for better understanding it.

---

R. de C. Augusto (✉) · C. Cosseau · C. Grunau
University of Montpellier, Montpellier, France

CNRS, Perpignan, France

IFremer, Montpellier, France

University of Perpignan Via Domitia, Perpignan, France



S. Jurga, J. Barciszewski (eds.), *The DNA, RNA, and Histone Methylomes*, RNA Technologies, https://doi.org/10.1007/978-3-030-14792-1_24

**Keyword** Histone methylation · Schistosomiasis · Trematode · Development

## 1 Introduction

Schistosomiasis is a chronic parasitic disease caused by trematodes of *Schistosoma* genus, endemic in several tropical and subtropical countries. Schistosomiasis has been reported in 78 endemic countries affecting more than 240 million people, mainly in developing countries, but also in Europe, as an outbreak has been reported recently in Corsica, France (Boissier et al. 2016; Steinmann et al. 2006). *Schistosoma mansoni* is one of the three major human species, besides *S. haematobium* and *S. japonicum*, that exhibit dioecy and have a complex life cycle involving two consecutive obligate hosts (a freshwater snail as intermediate host and a mammal as definitive one) and two transitions between these hosts as free-swimming larvae. Sex is determined genetically with males having two Z chromosomes and females ZW. The life cycle is summarized in Fig. 1.

Epigenetic mechanisms, that we will tentativey define here as any chromatin modification that is potentially affecting gene expression, whether it is heritable of not (Nicoglou and Merlin 2017) play a central role in programmed gene regulation

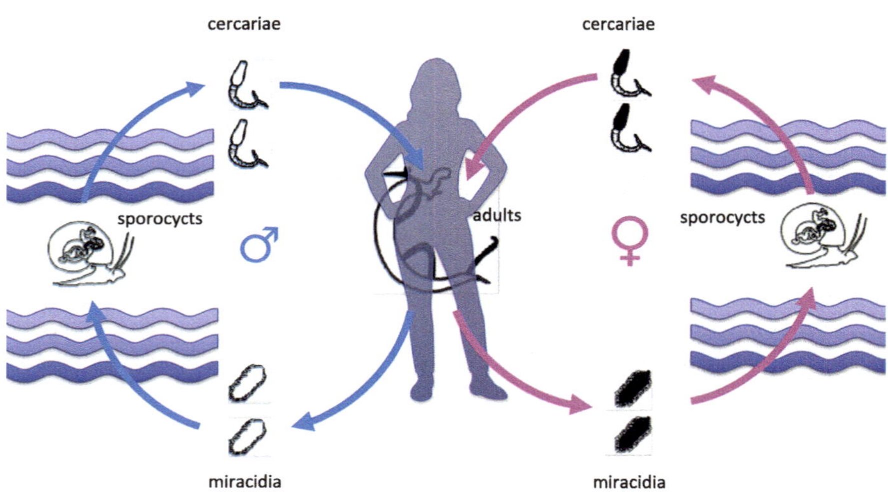


**Fig. 1** Schematic representation of the life cycle of *Schistosoma mansoni*. Male (ZZ) and female (ZW) adults mate in the mesenterial veins of human or rodents. Females produce up to 200 embryonated eggs per day that are excreted with the feces. When eggs touch water, miracidia hatch and seek the freshwater snail of the *Biomphalaria* genus. Miracidia that manage to infect the snail develop into primary and secondary sporocysts and shed human dwelling cercariae into the water. When cercariae penetrate the skin, they develop into somula, migrate in the vertebrate host and develop into adults

and are therefore expected to be key players in organisms with multiple developmental stages. Throughout the development, different epigenetic marks (histone post-translational modifications, small interfering RNAs, genomic RNA/DNA methylation, and topography of the nucleus) could be involved simultaneously and synergeticly in major biological processes. In eukaryotes, the fundamental unit of chromatin is the nucleosome, defined as two of each of the core histones proteins H2A, H2B, H3, H4 wrapped twice by approximately 146 base pairs of DNA (Taube and Barton 2006). The nucleosome establishes the first level of chromatin organization which comprise over 60 modifications in the amino-terminal tail of histones, including methylation, acetylation, phosphorylation, ubiquitination, and others (Nightingale et al. 2006). In this present chapter we will focuss our discussion on the post-translational methylation of histones associated with different parasite phenotypes.

*S. mansoni* shows highly conserved sequences of histones and histone modifying enzymes in comparison to others eukaryotes (Berriman et al. 2009; Protasio et al. 2012). Histone modifications are covalent post-translational modifications which act by two main mechanisms: (i) by decreasing nucleosome contact (between histone-histone or histone-DNA) and therefore altering structure and stability, and/or (ii) by recruting proteins with chromatin modifying enzyme activity (Kouzarides 2007). Both mechanisms involve several chromatin-based processes which can significantly impact gene expression profiles and phenotypes. In the cell nucleus, when histone methylation occurs, one, two, or three methyl groups are transfered from *S*-adenosyl-L-methionine to lysine or arginine residues of histone proteins by histone methyltransferases. It is possibe to study histone methylation by two main approaches: first the use of inhibitors of histone modifying enzymes and by using chromatin immunoprecipitation (ChIP) using antibodies against modified histones.

In this sense, determining the function of histone post-translational modifications often involves investigating the modification's position, abundance, and their characterization. On one hand, position and abundance of histone modifications are mostly studied by techniques based on chromatin immunoprecipitation followed by high-throughput sequencing (ChIP-seq). Briefly, ChIP-seq is based on DNA-bound protein which is immunoprecipitated by antibodies, purification, and then sequencing. Different methods have been described for this propose such as native chromatin immunoprecipitation procedure (N-ChIP) and cross-linked ChIP (X-NChIP) with specific limitations and strengths each (see Sect. 3). On the other hand, the role of histone modifications on the development has been characterized by inhibition of histone modifying enzymes. Since histones and histone modifications are extremely conserved through any eukaryotes, many different histone methyltransferases enzymes inhibitors described to treat human tumor and cancer have been used to understand specific function for the lysine or arginine residue which they modify in schistosomes (Cabezas-Cruz et al. 2014; Ballante et al. 2017; Roquis et al. 2018; Pereira et al. 2018; Padalino et al. 2018).

In many species, specific combinations of histone modifications are associated with transcriptionally permissive or repressive chromatin structures, thus controlling gene expression at the transcriptional level (Strahl and Allis 2000). Methylation of histone H3 at lysine 27 (H3K27me3) accompanied by H3K9me3 are landmarks of heterochromatin region associated with transcriptional repression and silencing of chromatin (Bannister and Kouzarides 2011; Roquis et al. 2015). In contrast, tri-methylation of histone H3 at lysine 4 (H3K4me3) or methylation of histone H4 at lysine 20 (H4K20me1) by histone methyltransferase enzymes (HMT) is a landmark associated with euchromatin regions and activation of transcription. They are important in transcriptional elongation, localized mainly in the 5′ and 3′ portion of active genes respectively (Kouzarides 2007; Roquis et al. 2015). Furthermore, discovery of methylation at both sites H3K4me3/H3K27me3 (bivalent methylation) in schistosomes has attracted wide attention. Bivalent histone methylation are found in several thousands of genes of in embryonic stem cells (ESCs) which can give rise to all lineages of tissues in the developing organism (Mikkelsen et al. 2007; Harikumar and Meshorer 2015). In schistosomes, pluripotent stem cells, called neoblasts, can be found over the life cycle driving cell development and, in adult stage, are responsible for tissue maintenance and surface self-renewing (Collins et al. 2013; Wang et al. 2013). The current view of bivalent methylation is that H3K27me3 represses transcription of lineage control genes during cell development, while H3K4me3 maintains them poised, i.e. transcription machinery is in position at the transcription start site (TSS), ready for activation upon reception of a signal that triggers differentiation (Roquis et al. 2015).

Since histones and histone modifications are extremely conserved through all kingdoms, and histone methyltransferases inhibitors that were described to treat human cancer have been used to understand specific function for the lysine or arginine residue which they modify in schistosomes (Padalino et al. 2018).

## 2 Biological Functions of Histone Methylation on Schistosomes

In *S. mansoni*, histone methylation was detected in all parasite stages by ChIP followed by DNA sequencing (ChIP-Seq). However, only two marks have been described over the entire life cycle: H3K4me3 and H3K27me3 (Roquis et al. 2018). Frequency of the both histone methylation marks starts at low levels in miracidia and increases progressively until the adult stage with two waves of H3K27 methylation/demethylation around the TSS of genes; one wave with a maximum of H3K27 trimethylation in Sp1 and another wave in adult parasites where sexual reproduction occurs (Fig. 2). Additionally, bivalent H3K4me3 and H3K27me3 methylation (at the same locus) starts in sporocysts and continues until the adult stage, with the highest frequency observed at transcriptional start sites in cercariae (Fig. 2). The distinct chromatin profile over the life cycle indicate that histone methylation plays an important role during development. Others marks have been described in some parasite stages and/or involved in a specific biological event, such as H3K9me3 and H4K20me1 (Roquis et al. 2015, 2018; Picard et al. 2016; Cosseau et al. 2017).

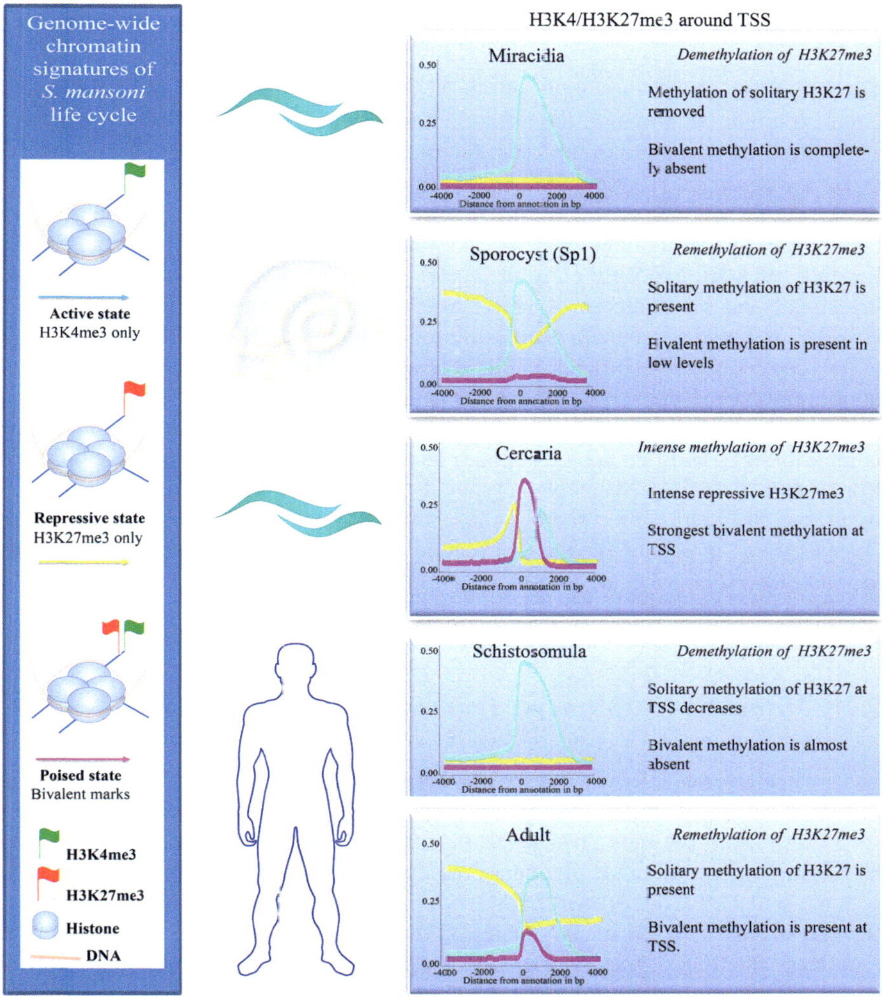


**Fig. 2** Genome-wide frequency of combinatorial states over five developmental stages of *Schistosoma mansoni*. The frequency of histone methylation mark H3K27me3 (yellow line) starts at low levels in miracidia and is characterized by two waves of methylation/demethylation around the transcription start site (TSS) of genes; one wave with a maximum of H3K27 trimethylation in sporocyst and another wave in adults. The histone mark H3K4me3 (blue line) is almost stable over the life cycle. Furthermore, bivalent H3K4me3 and H3K27me3 methylation (pink line) starts in sporocysts and continues until the adult stage, with the highest frequency observed at transcription start sites in cercariae. The distinct chromatin profile over the life cycle indicate that histone methylation plays an important role during development

Thus far, these studies provide enough evidence for the developmental role of specific histone methyltransferases in the flatworm *S. mansoni* biology and also open new perspectives to identify new drug targets based on histones methyl transferases (Cabezas-Cruz et al. 2014; Roquis et al. 2018).

## 2.1 Egg Laying

The life cycle starts when eggs produced by adult females pass from the host's circulation into the lumen of the intestine however as many as half of the parasite's eggs return by vasculare to the liver, the most important tissue for schistosomiasis disease. Reproduction system of female schistosomes are enriched in neoblast-like stem cells which support an intense egg production with an egg laid every one-minute to five-minutes over several years (Basch 1991; Collins et al. 2013). Recently, in vitro exposure of adult female parasites to GSK343 has been used to investigate the involvement of H3K27me3 on egg laying (Pereira et al. 2018). GSK343 is a widely characterized inhibitor of human EZH2 which is the subunit of polycomb repressor complex 2 (PRC2) responsible for trimethylation H3K27 and inhibition of gene transcription. A significant inhibition of H3K27me3 production by 24–48 h incubation in 20 μM GSK343 led to decrease of about 40–60% in the number of eggs laid by exposed females compared to the control. In addition, eggs from exposed females were phenotypically different from control ones showing fissures in the shell. Unfortunately, the authors did not attempt to observe if those changes were significant enough to affect miracidia hatching and compatibility with the snail host (Pereira et al. 2018).

## 2.2 Miracidia-to-Sporocyst Transition

Freshwater contact releases the free-swimming larva miracidia who seek out an intermediate host, a snail of the *Biomphalaria* genus. In the first moment out of egg shell, miracidia presents low level of histone methylation in H3K4me3 and H3K27me3. Genome-wide frequency analyses found H3K4me3 in 3.1% of the genome and H3K27me3 over 2.7%, covering 11.8 kb and 10.6 kb, respectively (Roquis et al. 2018). In many species, H3K4me3 is known to be associated with promoters and TSS of transcriptionally competent genes of vertebrates and invertebrates (Barski et al. 2007; Berger 2007; Gu and Fire 2010; Kharchenko et al. 2011; Zentner and Henikoff 2013). This mark is generally considered a transcriptional activator, despite some recent evidence is hinting that it could be a by-product of transcription instead (Howe et al. 2017). In miracidia, methylation level at TSS varies from over 36.8% in H3K4me3 and 2.2% in H3K27me3 showing a consistent relationship between H3K4me3 to the TSS as observed in other vertebrates and invertebrates (Roquis et al. 2018).

After penetration, miracidia loses its ciliated surface and transform into primary sporocysts (Sp1) which multiply asexually for approximately 10 days and then mature into secondary sporocysts (Sp2). During transition an increase of both marks is observed by a enrichment of H3K4me3 followed by a significant methylation of H3K27me3. In Sp1, H3K4me3 increases to be present at 3.6% of whole genome and H3K27me3 increases to cover 31.3% (14.4 kb and 155.2 kb, respectively). Furthermore, bivalent H3K4me3 and H3K27me3 methylation (i.e. methylation at the same locus) starts in sporocysts and probably promotes cell growth without differentiation (Bao et al. 2017; Roquis et al. 2018). Several genes identified in proliferating sporocyst cells share molecular signatures with neoblast-stem cells genes, mainly planarians neoblasts such as Vasa-like (Smp_068440, Smp_154320, Smp_033710) that are required for proliferation and expansion of neoblasts, putative polo kinase (Smp_009600) that is probably activating mitosis, and fibroblast growth factor receptor-encoding genes (Smp_157300, Smp_175590) that are potentially required for cell cycle and DNA repair machinery (Wang et al. 2013). At the TSS of this subset of genes, an increase of H3K27me3 can be observed starting with a very simple chromatin structure in and around TSS in miracidia becoming more complex in the sporocyst stage. Since trimethylation of H3K27 is a histone modification that undergoes major modifications during miracidia-sporocyst transition, Roquis et al. (2018) used two histone methyltransferase inhibitors to characterize the role of repressive H3K27me3 in this stage. Two H3K27 histone methyltransferase inhibitors targeting G9a/GLP and EZH2 were tested to assess the ability of these epidrugs to block transition efficiency (A366 and GSK343, respectively). Both inhibitors were originally developed to treat human cancers. In the presence of both methyltransferases, the miracidium to sporocyst transformation efficiency was significantly reduced when compared to the controls, even at lower concentration (0.4 μM). The pharmacological inhibition of trimethylation of H3K27 by both A366 and GSK343 blocked life cycle progression suggesting that the activity of histone methyltransferases is also essential for miracidia-sporocyst transition, adding this class of enzyme as a suitable drug target.

## 2.3 *Cercaria-Schistosomula-Adult Parasites Development*

After shed by infected snails, cercariae actively seek a definitive mammalian host (rodent, primate or human) to penetrate into the dermis, reaching the vascular system. The next parasite stage, schistosomula, follow a complex maturation process, ultimately leading to dimorphic adult worms genetically determined by the presence of sex chromosomes ZZ in males or ZW in females. This is the most studied schistosome's transition which is characterized by strong developmental plasticity starting from cercariae to adult stage. In cercariae stage, a characteristic chromatin signature emerged exemplified by strongest bivalency of both, H3K4me3 and H3K27me3 histone marks, occurring in 1/3 at all TSS, starting around 500 bp upstream going to 1000 bp downstream (Roquis et al. 2018). In cercariae H3K4me3 covers 9990 kb of the genome (5.22% of TSS), also starting around 500 bp upstream of at the 5′ region of a subset of genes while the broad H3K27me3 covers 32,756 kb of the genome (5.54% of TSS). Interestingly, bivalent profile is likely to be associated with a poised transcriptional state, and indeed in the cercariae stage, transcription is undetectable (Roquis et al. 2015). Intriguingly, gene transcription is resumed in the schistosomula stage, where the repressive H3K27me3 mark is removed

(Roquis et al. 2015). After schistosomula transformation, an intense demethylation of H3K27me3 is observed with only 2.93% of TSS covered while H3K4me3 reaches coverage levels of 41.81% at TSS. Interestingly, no bivalent state is observed around TSS in this stage.

In adults, 60 days post-infection, a prominent intensification of H3K27me3 is observed upstream of the TSS of a subset of genes with methylation peaks reaching wider than 100 kb in some regions. Adult worms, H3K27me3 covers around 133 kb and the coverage at TSS increases fourfold inside the definitive host in comparison to the schistosomula (from 2.93% to 14.38% of repressive H3K27me3 covering TSS of all genes). However an intense demethylation of H3K27me3 is observed over the gene body followed by an enrichment of sharp peaks of H3K4me3 with a maximum peak located 250 bp (or 1–2 nucleosomes) downstream of the TSS (Roquis et al. 2015, 2018). Intriguingly, the euchromatin histone mark H3K4me3 is almost stable from cercariae to adult maturation (also throughout the life cycle) and only slightly increases at schistosomula stages can be observed (Roquis et al. 2018). In adult's parasites, the active H3K4me3 mark covers around 10 kb (24.1% of TSS) and starts around 500 bp upstream while the repressive HK27me3 mark decreases. The bivalent methylation in adult worm (−500 to +1000 bp) reaches 10% of coverage at TSS. Knowing that histone methylation has an instrumental role in all developmental stages of *S. mansoni*, histone methyltransferases (HMTs) and the histone demethylases (HDMs) families emerge as suitable targets to control the parasite and afterward disease.

Recently, both families of druggable epigenetic targets in *S. mansoni*, the histone methyltransferases (HMTs) and the histone demethylases (HDMs) were re-categorise. Using different bioinformatics approaches, the authors identified 27 HMTs (20 Protein Lysine Methyl Transferases-PKMTs, 1 PR domain containing methyltransferase-PRDM, 1 DOT1 Like Histone Lysine Methyltransferase-DOT1L and 5 Protein Arginine Methyl Transferases-PRMTs) and 14 HDMs (3 Lysine Specific Demethylases-LSDs and 11 Jumonji domain-containing proteins-JMJDs) in *S. mansoni* (Padalino et al. 2018). Among them, Lysine Specific Demethylase 1 (SmLSD1, Smp_150560) homolog was selected showing significant impact against adult worms motility, reproduction and phenotype.

## 2.4 Influence of Histone Methylation on Sexual Biology

The sex of *S. mansoni* gender is genetically determined by the presence of a sex-specific chromosomes ZZ in males and ZW in females which reflects a strongly sex-biased proteomic and transcriptomic profile (Picard et al. 2016; Lu et al. 2016). There is, however no phenotypic dimorphism between males and females in the larval stages, and sexual dimorphism appears only in the vertebrate host, during schistosomula development. Accordingly, sexual differentiation does not only rely on genetic determinant present on the sex chromosome but has also been shown to rely on environmental factors perceived in the host microenvironment. Sex specific epigenetic events have been described during the transition from cercariae to schistosomula and adults and histone methylation seems to be major players suporting these events. In males cercariae, the trimethylation of H3K27 is substantial higher around TSS than in female. Whereas an intense demethylation of H3K27me3 is observed in male's adult development, in females the dynamic of this histone demethylation is less significant (Picard et al. 2016). The euchromatin histone mark H3K4me3 is almost stable from cercariae to adult maturation (throughout the life cycle also) and only slightly increases at schistosomula stages can be observed (Roquis et al. 2018). Furthermore, methylation of H4K20 (monomethylation on lysine 20 of histone H4) was also used to investigate the role of histone methylation on schistosome sexual biology (Roquis et al. 2018). Different context dependent role have been associated to methylation of H4K20: (i) it is associated with transcriptional activation state (Lv et al. 2016), linked to development (Oda et al. 2009), genome integrity and maintenance (Jørgensen et al. 2013), quiescence (Evertts et al. 2013), (ii) it is involved in bivalency with repressive H3K27me3 in early vertebrate embryos (Wang et al. 2008) and (iii) it is associated to repression of X-linked gene expression in XX *Caenorhabditis elegans* hermaphrodites chromosomes during dosage compensation process (Vielle et al. 2012). During female maturation, vast genomic regions (10–100 kb) differentially enriched in H4K20me1 highlighted the importance of this type of histone methylation on Notch signaling pathway. Interestingly, comparative transcriptomics also emphasized genes related to Notch signaling pathway which had emerged as potentially important factor for sex-specific development in schistosomes (Picard et al. 2016; Roquis et al. 2018). These both results support the idea that chromatin structure differences in developmental genes could be the origin of such sex-specific differences in developmental trajectories and the matter shall be analyzed in more details in the future. Furthermore, H4K20me1 in combination with H3K27me3 (heterochromatic signature) have been shown to be depleted in the Z-specific region of the female sexual chromosome, whereas H3K4me3 (euchromatic signature) is enriched in this region and these chromatin signature supports a role for enhanced expression in the Z-specific region of female as it is expected for dosage compensation to occur in *S. mansoni* (Picard M.A.L., personnal communication). Finally, accumulation of heterochromatised repeatitive elements has been described on female schistosome's W chromosome and conforms to the known paradigm on sexual chromosome evolution; Such an accumulation of heterochromatised repeats in sex-determining regions is expected to result in suppression of recombination between the heterochromosome and its homologue (Charlesworth et al. 2005). What makes schistosoma sex chromosomes unique in comparison to other metazoan model species is that some of these W-specific repetitive DNA sequence are transcribed in the larval stages but not in the adults, and this change in transcription level is accompanied by changes in the chromatin structure at these loci (Lepesant et al. 2012). Different histone modifications, including H3K4me3 and H3K27me3 occur concomitantly. In conclusion, these different results support the importance of histone methylation for sexual biology of *S. mansoni* at different level (i) histone methylation are major players of overall structural chromatin changes observed during schistosomula development during which the sexual dimorphism appears, (ii) histone methylation support sex chromosome specific changes which are necessary for evolutionnary mechanisms linked to sexual biology such as dosage compensation and degeneration of sexual chromosomes.

## 2.5 Influence of Histone Methylation on Host-Parasite Interplay

The interaction of hosts and parasites is one of the best-studied examples of evolution in a changing environment. Their reciprocal antagonistic co-evolution can be illustrated by an arms race which occurs over short evolutionary time scales in which epigenetic modifications is expected to provide a source of fast-acting, reversible phenotypic variation (Gómez-Díaz et al. 2012). In this context, epigenetic mechanisms have been studied that are important for the compatibility polymorphism phenotype which occurs between *S. mansoni* and its intermediate snail host *Biomphalaria glabrata*. Compatibility polymorphism means that some parasite strains are compatible with certain hosts (sucessful infection process and parasite transmission) but not with others (and vice versa). *Sm*PoMuc glycoproteins are phenotypic variants which have been shown to be key markers for the compatibility polymorphisms which occurs between *S. mansoni* and *B. glabrata*. Histone modifications, including H3K9me3, clearly display different enrichment profiles over some of the *Sm*PoMuc promoters between the compatible and incompatible strains (Perrin et al. 2013). Changing this histone modification profile was further shown to change the expression of this *Sm*PoMuc phenotypic variants in *S. mansoni* increasing parasite compatibility toward the intermediate mollusk host (Fneich et al. 2016). Another major work has adressed the influence of the snail host environment for epimutations which could happen in the parasite during the interaction with *B. glabrata*. The impact of two host environments (One allopatric versus one sympatric snail host) on different histone marks including H3K4me3, H3K27me3, H3K27ac and H4K20me1 has been observed on cercariae emerging from the two host environments and on the subsequent adult stage as well.

This study has allowed to measure the epimutation rate at the genome wide level (Roquis et al. 2016). An epimutation was considered as a difference in the peak height observed between the two conditions. Three types of epimutations were discovered: (i) environment induced epimutations, (ii) random epimutations and (iii) genotype dependent epimutations. This study has also clearly shown that (i) histone methylations are sensitive to the host environment, (ii) the epimutation rates rely on the histone mark which is targetted, the most sensitive being H3K4me3 and (iii) that environment induced epimutations on histone methylation are not inherited to the adult stages (Roquis et al. 2016). Random epimutations revert with high frequency and are only likely to provide a selective advantage when selection pressure is strong. Probably, developmental plasticity of the *S.mansoni* epigenome is so high that all other epimutations are wiped out rapidly. Taken together, these data suggest that different histone methylation marks in *S. mansoni* are modified in response to the host environment. A functional role for these modifications has not been demonstrated as no functional pathways are associated to genes affected by those changes (Roquis et al. 2016). However, *Sm*PoMuc gene expression are regulated by histone methylation which suggests a functional role regarding the host parasite interplay (Perrin et al. 2013; Fneich et al. 2016). Further support for this idea comes from experiences in which histone deacetylation was inhibited by TSA in miracidia, producing higher phenotypic diversity and higher infection success (Fneich et al. 2016).

While most environmentally induced epimutations appear to be rather ephemeral in *S. mansoni* epimutation that are passed through the germline can arise through paramutations. Paramutations are interactions between the two alleles of a locus, where one allele induces heritable changes in the other allele (reviewed in Chandler 2007). Hybridisation of a compatible and an incompatible (vis-à-vis a reference snail) *S. mansoni* strain led to heritable histone H3K9 acethylation and methylation changes in the abovementioned *Sm*PoMuc loci, and was associated with increased infection success (compatibility) (Fneich et al. 2016). A major threat from the parasites point of view are anti-helminth drugs. Currently, Praziquantel is mainly used to treat the disease, but in the past other drug such as Oxamniquine (OXA) and Hycanthone (HYC) were used (Rosi et al. 1965). The genetic basis of OXA/HYC cross-resistance is known and resides in mutations in the *Sm*SULT-OR gene (Smp_089320), encoding a sulfotransferase that is required for drug activation, (p. E142del and p.C35R) (Valentim et al. 2013; Chevalier et al. 2016). In natural populations the frequencies of these mutations are very low (0.27–0.8%) (Chevalier et al. 2016). However, cure rate of HYC was at maximum 90% in the field (Pellegrino et al. 1969) and resistant strains can readily be produced in the laboratory (Jansma et al. 1977; Brindley and Sher 1987). We have shown that HYC resitance (or tolerance) can be based on chromatin structure changes in H3K4me3, H3K9me3/ac and H3K27me3 in clonal male populations of genetically sensitive *S. mansoni* indicating that histone methylation can be induced by the HYC environmental cue (Roquis et al. 2014). Even though the resitance phenotype might not be inheritable (we did not test this), transient survival to the treatment will be sufficient to ensure higher reproduction success of the epigenetically modified individuals (Fig. 3).

## 3 Technical Remarks

As stated above, posttranslational modifications of histones are widely studied and have been shown to play a key role in the correlation between chromatin structure and gene expression (Lindsay 2007; Dong and Weng 2013). Chromatin immunoprecipitation techniques followed by high-throughput DNA sequencing has been widely used for study transcription factors, histone modifications, chromatin modifying complexes, and other chromatin-associated proteins in a wide variety of

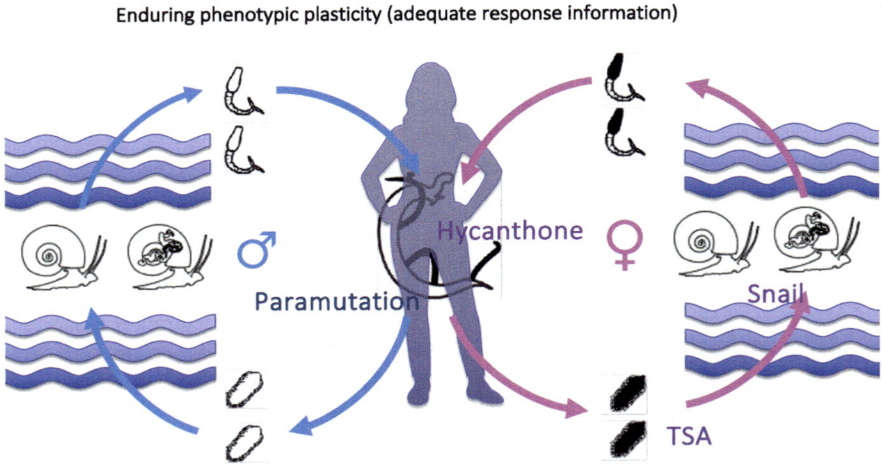


**Fig. 3** Enduring epigenetic plasticity based on genetic and environmental cues. Environmental cues (red) can lead to histone modifications that persist over life cycle stages and/or to the subsequent meiotic generation (violet)

organisms. However, when experiments are designed, executed, and reported in different ways some issues to compare data from multiple studies or to perform integrative analyses across multiple data-types might emerge (Landt et al. 2012). There are many examples of guidelines, practices, and quality metrics in the literature for wide variety of organisms as pionner described by ENCODE and modENCODE consortia (Landt et al. 2012). For *S. mansoni,* comparing the method of chromatin accessibility during fragmentation (enzymatic or physical), a preference for the enzymatic method done by native ChiP-seq shown rather results then the physical delivered by cross-linked ChIP-seq. Taking into consideration: (i) the proteins remain in their native form and there is no danger that crosslinking fixes interactions that do not occur systematically in the cell; (ii) N-ChIP is 10–100 times more sensitive than X-ChIP and less starting material is required; (iii) a better signal-to-noise ratio under the same conditions; and (iv) since enzymatic fragmentation of chromatin is used, no expensive equipment such as a sonicator is necessary. As a result, we strongly recommend the method described by (Cosseau and Grunau 2011) (also available online at http://ihpe.univ-perp.fr/methods/methods/native_chip_sm_3.htm).

Briefly, native ChIP-seq is a technique to extracted chromatin by micrococcal nuclease digestion (MNase), a bacterial enzyme that slices DNA specifically between nucleosomes. An antibody is then used, targeting a histone or a histone chemical modification. The antibody/histone/DNA complex is immunoprecipitated by protein A-coated sepharose microbeads. The microbeads has a very high affinity for the antibodies, and therefore allows the complex to be isolated. The immunoprecipitated DNA fragments are then eluted and followed by high-throughput DNA sequencing (e.g. Illumina sequencing). Their distribution on the reference genome indicates the

**Table 1** Details of the antibodies used for N-ChIP

| Antibody | Supplier | Amount used (μL) | Parasite stage | Biosample |
|---|---|---|---|---|
| *H3K4ME3* | Diagenode | 4 | Primary sporocysts | SAMN08039006 |
| *H3K27me3* | Diagenode | 8 | Primary sporocysts | SAMN08039006 |

location of the histones/targeted histone modification. Below, we report our experience with native ChIP-seq experimental design, execution, and quality assessment and also offer specific recommendations. Here, we will present data from 100 primary sporocysts (in vitro transformed) in two biological replicates. Immunoprecipitation was performed using antibodies against H3K4me3 and H3K27me3. We used a control without antibody to assess nonspecific background (bound fraction) and input (unbound fraction). Inputs should be used for normalization, but are not obligatory. Details for each antibody are in Table 1.

*Native ChIP Procedure* First of all, all buffers that will be cited here should be freshly made and cold down in the ice. Each tube containing 100 primary sporocysts were centrifuged at 4000 rpm for 10 min at 4 °C. The pellets were suspended in 1 ml of buffer 1 (0.3 M sucrose, 30 mM KCl, 7.5 mM NaCl, 2.5 mM $MgCl_2$, 0.05 mM EDTA, 0.1 mM PMSF, 0.5 mM DTT, 7.5 mM Tris–HCl, pH 7.5) containing protease inhibitor cocktail tablets (Roche Applied Science) and 5 mM sodium butyrate as histone deacetylase inhibitor (Sigma). Samples were lysed by adding 1 ml buffer 1 with 0.8% NP40 and homogenized in a SZ22 tissue grinder tube (Kontes Glass Company) using an SC tissue grinder pestle (Kontes Glass Company) on ice for 3 min and later overlaid on 8 ml of buffer 3 (1.2 M sucrose, 30 mM KCl, 7.5 mM NaCl, 2.5 mM $MgCl_2$, 0.05 mM EDTA, 5 mM sodium butyrate, 0.1 mM PMSF, 0.5 mM DTT, 7.5 mM Tris–HCl, pH 7.5) in Falcon tubes and centrifuged for 20 min (8500 rpm at 4 °C). Buffer was removed and the pelleted nuclei were suspended in 1 ml chromatin digestion buffer (0.12 M sucrose, 0.2 mM PMSF, 4 mM $MgCl_2$, 5 mM sodium butyrate, 1 mM $CaCl_2$, 0.05 M Tris–HCl, pH 7.5) and divided into aliquots of 500 μl in 1.5 ml Eppendorf tubes.

Chromatin digestion was performed for 4 min at 37 °C with 1 μl (15U) of micrococcal nuclease (MNase). To stop the reaction, 20 μl of 0.5 M EDTA was added and the tubes were immediately placed on ice. Samples were centrifuged for 10 min (13,000×*g* at 4 °C) and the supernatant (Fraction S1) was transferred into fresh tubes. The pellets (P1) were suspended in 100 μl dialysis buffer (200 mM EDTA, 200 mM PMSF, 5 mM sodium butyrate, 1 mM Tris–HCl, pH 7.5) and dialyzed overnight at 4 °C in a Slide-A-Lyser MINI Dialysis Unit (cut-off at 3500 Da) (Pierce). The fraction that remained in the dialysis tubes was centrifuged for 10 min (13,000×g at 4 °C). The supernatant (Fraction S2) was transferred into fresh tubes. Fractions S1 and S2 were then centrifuged three times for 10 min (13,000×g at 4 °C) and each time the supernatants were transferred into fresh tubes. Chromatin from two combined fractions S1 and S2 were pooled and quantified by measuring the OD at 260 nm. Thirty micro-grams of chromatin were used and antibodies were added in excess (see Table 1).

An appropriate amount of stock solution was added to generate immunoprecipitation incubation buffer (150 mM NaCl, 20 mM sodium butyrate, 5 mM EDTA, 100 mM PMSF, 20 mM Tris–HCl, pH 7.5). Samples were incubated overnight at 4 °C on a rotating wheel. Fifty microliters of protein A-sepharose (Sigma) were added and incubated with the chromatin–antibody complexes for 4 h at 4 °C on a rotating wheel. The chromatin–antibody–protein A bead mixture was centrifuged for 10 min (11,660×g at 4 °C). The supernatant was transferred to a fresh tube, yielding the unbound fraction.

Pellets (chromatin–antibody–protein A bead complex) were suspended in 10 ml washing buffer (50 mM Tris–HCl, pH 7.5, 10 mM EDTA, 5 mM sodium butyrate, 75 mM NaCl) and mixed gently for 10 min on a rotating wheel at 4 °C. The mixture was centrifuged for 10 min at 4000 rpm at 4 °C. The same procedure was repeated twice with increasing stringency conditions by using 125 mM and 175 mM NaCl as wash buffers, respectively. Finally, the pellets were suspended in 500 µl elution buffer (1% SDS, 20 mM Tris–HCl, pH 7.5, 50 mM NaCl, 5 mM EDTA, 20 mM sodium butyrate and 100 mM PMSF) and incubated for 15 min at room temperature on a rotating wheel. The mixture was centrifuged for 10 min at 11,600×g at 18 °C, and the supernatants with the bound fraction was transferred into fresh tubes. DNA from the bound and unbound fractions was extracted with commercial kits.

*Sequencing Parameters* Then ChIP products were sequenced as paired- end 75-bp reads on an Illumina HiSeq 2500. The samples were quantified on a high sensitivity bioanalyser, before being cleaned with Agencourt AMPure XP beads. End repair, A-tailing and adapter ligation were performed using the NEB library prep kit, with Agencourt AMPure XP bead cleaning steps. The amount of template for PCR and the number of PCR cycles required were assessed from a high sensitivity bioanalyser trace post-ligation. Libraries were amplified for 14 cycles. After cleaning with Agencourt AMPure beads, libraries were quantified using a KAPA SYBR FAST ABI Prism qPCR Kit with Illumina GA Primer Premix (10x) and 7x Illumina GA DNA Stan- dards (Kit code: KK4834) on an ABI StepOnePlus qPCR machine. Libraries were diluted into an equimolar pool and run on a HiSeq 2500, generating 75 base pair, paired end reads.

*Quality Control, Alignment and Peak Calling* All data processing was performed on a local GALAXY instance (http://bioinfo.univ-perp.fr). Read quality was verified using the FastQC toolbox (https://www.bioinformatics.babraham.ac.uk/projects/fastqc/). All samples had a minimum average read quality score of 30 over 95% of their length, and no further cleaning steps were performed.

Sequences were aligned to the *S. mansoni* reference genome v5 (Protasio et al. 2012) with Bowtie v2.1 using parameters–end-to-end,–sensitive,–gbar 4. BAM files generated by Bowtie2 were sorted and then filtered for unique matches with samtools v1.3.1(samtools view -Sh -q quality value 40–42—F 0x0004 –| grep -v XS:i). PCR duplicates were also removed using samtools (samtools rmdup). Although not mandatory, we found that performing random sampling to use the same amount of uniquely mapped reads for each sample and each histone mark

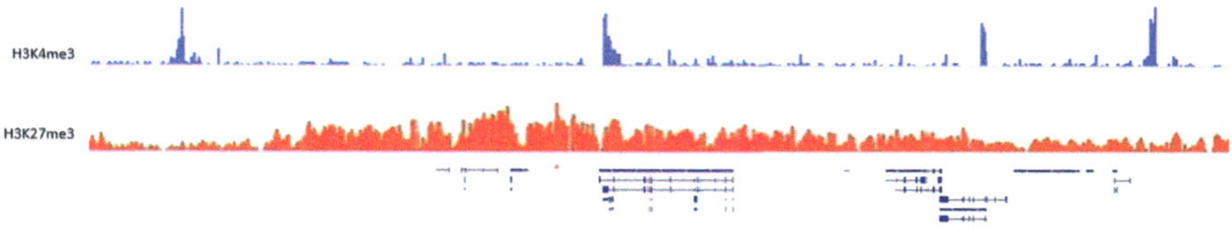


**Fig. 4** Chromatin profile for H3K4me3 and H3K27me3 in *S. mansoni* primary sporocyst. Typical example of the narrow histone mark H3K4me3 (blue) and the broad histone mark H3K27me3

**Table 2** Description of frequency and covered chromatin state (kb) in primary sporocyst for H3K4me3 and H3K27me3, genome-wide and at TSS

| | Covered histone marks (kb) | Frequency of chromatin state genome wide (%) | Frequency of chromatin state at TSS (%) |
|---|---|---|---|
| H3K4ME3 | 14 | 3.6 | 38.7 |
| H3K27me3 | 155 | 31.3 | 15.5 |

improved sensitivity and specificity when looking for chromatin structure differences. We took 1.5 million reads for H3K4me3 and H3K27me3.

*Chromatin Landscape* We conducted our analyses by characterizing the covered chromatin state and their frequency for narrow H3K4me3 and broad H3K27me3, genome-wide and at transcription start site (TSS), using epialleles that chromstaR detected in all replicates for each histone mark (Taudt et al. 2016). In addition, wig files were produced to visualize the distribution of studied histone marks (Fig. 4; Table 2).

## 4 Conclusions

*S. mansoni* needs multiple obligatory hosts plus free-swimming stages on fresh water ecosystem have to address environmental cues and generate strikingly different developmental stages to complete their lifecycle. One aspect of epigenetics is the relative high sensitivity of the epigenotype to external stimulus. Here we summarized the histone methylation over each developmental stage which are crucial for host-parasite compatibility, development, sexual differentiation and pathogenesis and suggest that the enzymes responsible for maintaining these chromatin modifications are suitable targets for anti-schistosomal drugs. In addition, histone methylation is sensitive to environmental cues and could be the bearer of adequate response information. What is important that it is the previous environment that changes developmental trajectories of the subsequent environment, i.e. water quality influences on life stages in the snail and vertebrate host. If this were true then enduring epigenetic reponse must be considered when using control measures. Our recent results that a brief contact of cercariae with a natural molluscicite provokes enduring morphological and physiological changes in the adult worms (de Augusto et al. 2017) lend support to this hypothesis.

## References

Ballante F, Reddy DR, Zhou NJ et al (2017) Structural insights of SmKDAC8 inhibitors: targeting Schistosoma epigenetics through a combined structure-based 3D QSAR, in vitro and synthesis strategy. Bioorg Med Chem 25:2105–2132

Bannister AJ, Kouzarides T (2011) Regulation of chromatin by histone modifications. Cell Res 21:381–395

Bao B, He Y, Tang D et al (2017) Inhibition of H3K27me3 histone demethylase activity prevents the proliferative regeneration of Zebrafish lateral line Neuromasts. Front Mol Neurosci 10:51

Barski A, Cuddapah S, Cui K et al (2007) High-resolution profiling of histone methylations in the human genome. Cell 129:823–837

Basch PF (1991) Schistosomes: development, reproduction, and host relations. Oxford University Press, New York

Berger SL (2007) The complex language of chromatin regulation during transcription. Nature 447:407–412

Berriman M, Haas BJ, LoVerde PT et al (2009) The genome of the blood fluke *Schistosoma mansoni*. Nature 460:352–358

Boissier J, Grech-Angelini S, Webster BL et al (2016) Outbreak of urogenital schistosomiasis in Corsica (France): an epidemiological case study. Lancet Infect Dis 16:971–979

Brindley PJ, Sher A (1987) The chemotherapeutic effect of praziquantel against *Schistosoma mansoni* is dependent on host antibody response. J Immunol 139:215–220

Cabezas-Cruz A, Lancelot J, Caby S et al (2014) Epigenetic control of gene function in schistosomes: a source of therapeutic targets? Front Genet 5:317

Chandler VL (2007) Paramutation: from maize to mice. Cell 128:641–645

Charlesworth D, Charlesworth B, Marais G (2005) Steps in the evolution of heteromorphic sex chromosomes. Heredity 95:118–128

Chevalier FD, Le Clec'h W, Eng N et al (2016) Independent origins of loss-of-function mutations conferring oxamniquine resistance in a Brazilian schistosome population. Int J Parasitol 46:417–424

Collins JJ, Wang B, Lambrus BG et al (2013) Adult somatic stem cells in the human parasite *Schistosoma mansoni*. Nature 494:476–479

Cosseau C, Grunau C (2011) Native chromatin immunoprecipitation. Methods Mol Biol Clifton NJ 791:195–212

Cosseau C, Wolkenhauer O, Padalino G et al (2017) (Epi)genetic inheritance in *Schistosoma mansoni*: a systems approach. Trends Parasitol 33:285–294

de Augusto RC, Tetreau G, Chan P et al (2017) Double impact: natural molluscicide for schistosomiasis vector control also impedes development of *Schistosoma mansoni* cercariae into adult parasites. PLoS Negl Trop Dis 11:e0005789

Dong X, Weng Z (2013) The correlation between histone modifications and gene expression. Epigenomics 5:113–116

Evertts AG, Manning AL, Wang X et al (2013) H4K20 methylation regulates quiescence and chromatin compaction. Mol Biol Cell 24:3025–3037

Fneich S, Théron A, Cosseau C et al (2016) Epigenetic origin of adaptive phenotypic variants in the human blood fluke *Schistosoma mansoni*. Epigenetics Chromatin 9:27

Gómez-Díaz E, Jordà M, Peinado MA et al (2012) Epigenetics of host–pathogen interactions: the road ahead and the road behind. PLoS Pathog 8:e1003007

Gu SG, Fire A (2010) Partitioning the C. elegans genome by nucleosome modification, occupancy, and positioning. Chromosoma 119:73–87

Harikumar A, Meshorer E (2015) Chromatin remodeling and bivalent histone modifications in embryonic stem cells. EMBO Rep 16:1609–1619

Howe FS, Fischl H, Murray SC et al (2017) Is H3K4me3 instructive for transcription activation? BioEssays News Rev Mol Cell Dev Biol 39:1–12

Jansma WB, Rogers SH, Liu CL et al (1977) Experimentally produced resistance of *Schistosoma mansoni* to hycanthone. Am J Trop Med Hyg 26:926–936
Jørgensen S, Schotta G, Sørensen CS (2013) Histone H4 lysine 20 methylation: key player in epigenetic regulation of genomic integrity. Nucleic Acids Res 41:2797–2806
Kharchenko PV, Alekseyenko AA, Schwartz YB et al (2011) Comprehensive analysis of the chromatin landscape in *Drosophila melanogaster*. Nature 471:480–485
Kouzarides T (2007) Chromatin modifications and their function. Cell 128:693–705
Landt SG, Marinov GK, Kundaje A et al (2012) ChIP-seq guidelines and practices of the ENCODE and modENCODE consortia. Genome Res 22:1813–1831
Lepesant JMJ, Cosseau C, Boissier J et al (2012) Chromatin structural changes around satellite repeats on the female sex chromosome in *Schistosoma mansoni* and their possible role in sex chromosome emergence. Genome Biol 13:R14
Lindsay S (2007) Chromatin control of gene expression: the simplest model. Biophys J 92:1113
Lu Z, Sessler F, Holroyd N et al (2016) Schistosome sex matters: a deep view into gonad-specific and pairing-dependent transcriptomes reveals a complex gender interplay. Sci Rep 6:31150
Lv X, Han Z, Chen H et al (2016) A positive role for polycomb in transcriptional regulation via H4K20me1. Cell Res 26:529–542
Mikkelsen TS, Ku M, Jaffe DB et al (2007) Genome-wide maps of chromatin state in pluripotent and lineage-committed cells. Nature 448:553–560
Nicoglou A, Merlin F (2017) Epigenetics: a way to bridge the gap between biological fields. Stud Hist Philos Sci Part C Stud Hist Philos Biol Biomed Sci 66:73–82
Nightingale KP, O'Neill LP, Turner BM (2006) Histone modifications: signalling receptors and potential elements of a heritable epigenetic code. Curr Opin Genet Dev 16:125–136
Oda H, Okamoto I, Murphy N et al (2009) Monomethylation of histone H4-lysine 20 is involved in chromosome structure and stability and is essential for mouse development. Mol Cell Biol 29:2278–2295
Padalino G, Ferla S, Brancale A et al (2018) Combining bioinformatics, cheminformatics, functional genomics and whole organism approaches for identifying epigenetic drug targets in *Schistosoma mansoni*. Int J Parasitol Drugs Drug Resist 8:559–570
Pellegrino J, Katz N, Oliveira CA (1969) Further clinical trials with hycanthone, a new antischistosomal agent. Am J Trop Med Hyg 18:924–929
Pereira ASA, Amaral MS, Vasconcelos EJR et al (2018) Inhibition of histone methyltransferase EZH2 in *Schistosoma mansoni* in vitro by GSK343 reduces egg laying and decreases the expression of genes implicated in DNA replication and noncoding RNA metabolism. PLoS Negl Trop Dis 12:e0006873
Perrin C, Lepesant JMJ, Roger E et al (2013) *Schistosoma mansoni* Mucin gene (SmPoMuc) expression: epigenetic control to shape adaptation to a new host. PLoS Pathog 9:e1003571
Picard MAL, Boissier J, Roquis D et al (2016) Sex-biased transcriptome of *Schistosoma mansoni*: host-parasite interaction, genetic determinants and epigenetic regulators are associated with sexual differentiation. PLoS Negl Trop Dis 10:e0004930
Protasio AV, Tsai IJ, Babbage A et al (2012) A systematically improved high quality genome and transcriptome of the human blood fluke *Schistosoma mansoni*. PLoS Negl Trop Dis 6:e1455
Roquis D, Lepesant JMJ, Villafan E et al (2014) Exposure to hycanthone alters chromatin structure around specific gene functions and specific repeats in *Schistosoma mansoni*. Front Genet 5:207
Roquis D, Lepesant JMJ, Picard MAL et al (2015) The epigenome of *Schistosoma mansoni* provides insight about how Cercariae poise transcription until infection. PLoS Negl Trop Dis 9:e0003853
Roquis D, Rognon A, Chaparro C et al (2016) Frequency and mitotic heritability of epimutations in *Schistosoma mansoni*. Mol Ecol 25:1741–1758
Roquis D, Taudt A, Padalino G et al (2018) Histone methylation changes are required for life cycle progression in the human parasite *Schistosoma mansoni*. PLoS Pathog 14:e1007066
Rosi D, Peruzzotti G, Dennis EW et al (1965) A new, active metabolite of 'Miracil D'. Nature 208:1005–1006

Steinmann P, Keiser J, Bos R et al (2006) Schistosomiasis and water resources development: systematic review, meta-analysis, and estimates of people at risk. Lancet Infect Dis 6:411–425

Strahl BD, Allis CD (2000) The language of covalent histone modifications. Nature 403:41–45

Taube JH, Barton MC (2006) Chromatin and regulation of gene expression. In: Ma J (ed) Gene expression and regulation. Springer, New York, NY, pp 95–109

Taudt A, Nguyen MA, Heinig M et al (2016) chromstaR: tracking combinatorial chromatin state dynamics in space and time. https://doi.org/10.1101/038612

Valentim CLL, Cioli D, Chevalier FD et al (2013) Genetic and molecular basis of drug resistance and species-specific drug action in Schistosome parasites. Science 342:1385–1389

Vielle A, Lang J, Dong Y et al (2012) H4K20me1 contributes to downregulation of X-linked genes for C. elegans dosage compensation. PLoS Genet 8:e1002933

Wang Z, Zang C, Rosenfeld JA et al (2008) Combinatorial patterns of histone acetylations and methylations in the human genome. Nat Genet 40:897–903

Wang B, Collins JJ, Newmark PA (2013) Functional genomic characterization of neoblast-like stem cells in larval *Schistosoma mansoni*. elife 2:e00768

Zentner GE, Henikoff S (2013) Regulation of nucleosome dynamics by histone modifications. Nat Struct Mol Biol 20:259–266

Printed by Printforce, the Netherlands